NORTHERN HORIZON

CASSIOPEIA

CEPHEUS

POLARIS "NORTH STAR"

URSA MINOR "LITTLE DIPPER"

DRACO

CASTOR
POLLUX
GEMINI

DENEB
CYGNUS "NORTHERN CROSS"

URSA MAJOR "BIG DIPPER"

DELPHINUS

VEGA
LYRA
HERCULES
SAGITTA

CANCER

CORONA BOREALIS

LEO

EASTERN HORIZON

AQUILA
ALTAIR

BOOTES

REGULUS

WESTERN HORIZON

ARCTURUS

SERPENS
OPHIUCHUS
SERPENS

VIRGO

SAGITTARIUS

SPICA
CORVUS

HYDRA

LIBRA

ANTARES

SCORPIUS

SOUTHERN HORIZON

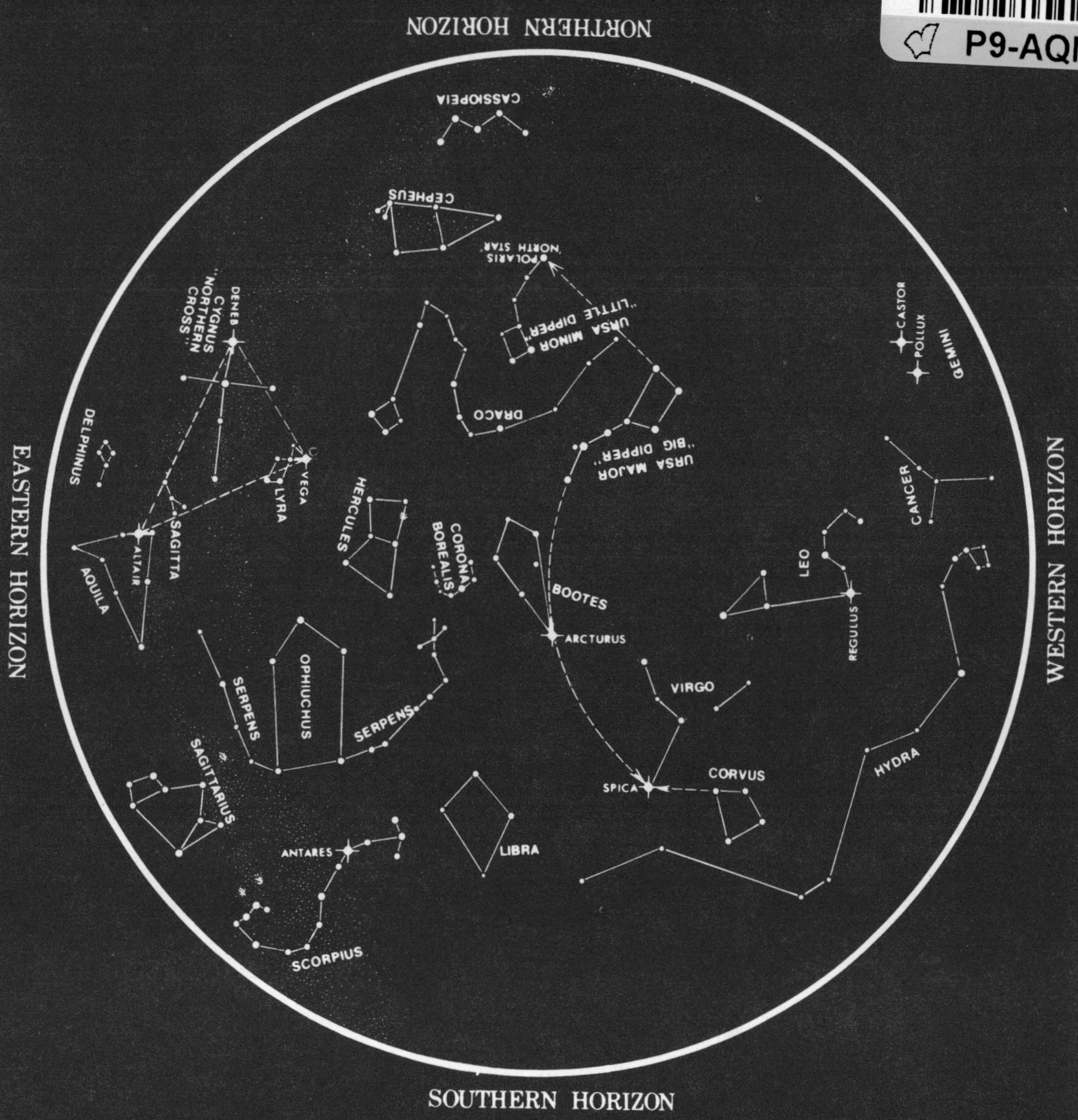

THE NIGHT SKY IN JUNE

Latitude of chart is 34°N, but it is
practical throughout the continental
United States.

To use: Hold chart vertically and turn
it so the direction you are facing
shows at the bottom.

Chart time (Local Standard):

10 p.m. First of month

9 p.m. Middle of month

8 p.m. Last of month

The Dynamic Universe
AN INTRODUCTION TO ASTRONOMY

The Dynamic Universe

AN INTRODUCTION TO ASTRONOMY

THEODORE P. SNOW
University of Colorado at Boulder

WEST PUBLISHING COMPANY
St. Paul   New York   San Francisco   Los Angeles

COPYRIGHT © 1983 By WEST PUBLISHING CO.
50 West Kellogg Boulevard
P.O. Box 3526
St. Paul, Minnesota 55165

Printed in the United States of America

Library of Congress Cataloging in Publication Data

Snow, Theodore Peck.
  The dynamic universe.

  Includes bibliographies and index.
  1. Astronomy.   I. Title.
QB43.2.S66   1983   520   82-21908
ISBN 0-314-69681-4
1st Reprint—1983

Composition: Clarinda Company
Copy editing: Deborah Annan
Text Design: Janet Bollow
Text illustrations: Henry Taly Design

Cover Photo

Front upper: JPL, photograph by H. Arp, processing by Jean L. Lorre, Image Processing Laboratory. *Front bottom:* NASA Photograph. *Back cover:* Courtesy of J. G. Timothy.

Chapter Opening Photos

Chapter 1: NASA photograph. **Chapter 2:** E. C. Krupp. **Chapter 3:** Yerkes Observatory photograph. **Chapter 5:** The Granger Collection. **Chapter 6:** Yerkes Observatory photograph. **Chapter 7:** The Bettmann Archives. **Chapter 8:** NASA photograph. **Chapter 9:** NASA photograph. **Chapter 10:** NASA photograph. **Chapter 11:** NASA photograph. **Chapter 12:** NASA photograph. **Chapter 13:** NASA photograph. **Chapter 14:** Lowell Observatory photograph. **Chapter 15:** The Granger Collection. **Chapter 16:** High Altitude Observatory, National Center for Atmospheric Research, sponsored by the National Science Foundation. **Chapter 18:** Historical Pictures Services, Inc. **Chapter 21:** Lick Observatory photograph. **Chapter 22:** U. S. Naval Observatory photograph. **Chapter 23:** Palomar Observatory, California Institute of Technology. **Chapter 24:** © 1980 Anglo-Australian Telescope Board. **Chapter 25:** *X-ray Image:* Harvard-Smithsonian Center for Astrophysics. **Chapter 26:** © 1980 Anglo-Australian Telescope Board. **Chapter 27:** Computer simulation supplied by A. Toomre and J. Toomre. **Chapter 28:** Harvard-Smithsonian Center for Astrophysics. **Chapter 29:** NASA photograph. **Chapter 30:** JPL, photograph by H. Arp, processing by Jean L. Lorre, Image Processing Laboratory. **Chapter 31:** The Granger Collection. **Chapter 32:** NASA photograph.

Color Photo Credits

ix The Granger Collection. **x** Courtesy of J. G. Timothy. **xi** NASA photograph. **xii** NASA photograph. **xiii** © 1979 Anglo-Australian Telescope Board. **xiv** Photo by F. Espenak. **xv** R. J. Dufour. **xvi** NASA photograph. **Color Plate 1.** *upper:* Photograph by Patrick Wiggins/Hansen Planetarium; *lower:* Sommers-Baush Observatory, University of Colorado, Photograph by J. Kloeppel. **Color Plate 2.** *upper:* © West Publishing Company; *lower:* © Association of Universities for Research in Astronomy, Inc., the Kitt Peak National Observatory. **Color Plate 3.** *upper:* The National Radio Astronomy Observatory, operated by Associated Universities, Inc. under contract with the National Science Foundation; *lower left:* © Association of Universities for Research in Astronomy, Inc., the Kitt Peak National Observatory.; *lower center:* © Association of Universities for Research in Astronomy, Inc., the Cerro Tololo Inter-American Observatory; *lower right:* MMTO: University of Arizona and Smithsonian Institution. **Color Plate 4.** *upper left, upper right, and lower:* NASA photographs. **Color Plate 5.** *upper left, upper right, and lower:* NASA photographs. **Color Plate 6.** *upper, center, and lower:* M. Kobrick, JPL. **Color Plate 7.** *upper:* TASS from SOVFOTO; *lower left:* NASA photograph; *lower right:* Laboratory for Atmospheric and Space Physics, University of Colorado, sponsored by NASA. **Color Plate 8.** *upper left, upper right, and lower:* NASA photographs. **Color Plate 9.** *upper, lower left, and lower right:* NASA photographs. **Color Plate 10.** *upper, lower left, and lower right:* NASA photographs. **Color Plate 11.** *upper, lower left, and lower right:* NASA photographs. **Color Plate 12.** *upper and lower left:* NASA photographs; *lower right:* Laboratory for Atmospheric and Space Physics, University of Colorado, sponsored by NASA. **Color Plate 13.** *upper left:* NASA photograph; *lower right:* © F. Espanek. **Color Plate 14.** *upper left: Solar Maximum Mission,* High Altitude Observatory, National Center for Atmospheric Research, sponsored by NASA; *upper right:* © F. Espanek; *lower:* NASA photograph. **Color Plate 15.** *lower left:* ©Association of Universities for Research in Astronomy, the Kitt Peak National Observatory; *lower right:* Harvard-Smithsonian Center for Astrophysics; *upper:* © 1961, California Institute of Technology. **Color Plate 16.** *upper left:* © 1981 Anglo-Australian Telescope Board; *upper right:* Lick Observatory photograph; *lower left:* N. R. Walborn; *lower right:* Harvard-Smithsonian Center for Astrophysics. **Color Plate 17.** *upper:* © 1981 Anglo-Australian Telescope Board; *lower:* R. D. Gehrz, J. Hackwell, and G. Grasdalen, University of Wyoming. **Color Plate 18.** *upper left and upper right:* © 1980 and 1981 Anglo-Australian Telescope Board; *lower:* © 1959, California Institute of Technology. **Color Plate 19.** *lower left and lower right:* © 1979 and 1981, Anglo-Australian Telescope Board; *upper:* Lick Observatory photograph. **Color Plate 20.** *upper:* J. R. Dickel; *lower:* Fr. R. E. Royer. **Color Plate 21.** *upper left and lower:* R. J. Dufour, Rice University; *upper right:* © California Institute of Technology. **Color Plate 22.** *upper left and upper right:* © 1980, Anglo-Australian Telescope Board; *lower right:* J. D. Wray, University of Texas; *lower left:* JPL, Photo by H. Arp, processing, Jean J. Lorre, Imaging Processing Laboratory. **Color Plate 23.** *upper:* © 1980 Anglo-Australian Telescope Board; *lower left:* The National Radio Astronomy Observatory, operated by Associated Universities, Inc. under contract with the National Science Foundation; *lower right:* Harvard-Smithsonian Center for Astrophysics. **Color Plate 24.** *upper:* Photograph obtained by P. A. Wehringer and S. Wyckoff with the University of Hawaii 2.2-m telescope on Mauna Kea; *lower left:* J. A. Tyson, obtained with the Bell Laboratories CCD camera; *lower right:* Institute for Astronomy and Planetary-Geosciences Data Processing Facility, University of Hawaii.

Figure Credits

Chapter 1

Fig. 1.1. E. E. Barnard Observatory photograph by G. Emerson. **Fig. 1.2.** Lick Observatory photograph. **Fig. 1.3.** NASA photograph. **Fig. 1.5** U.S. Naval Observatory photograph. **Fig. 1.6.** © 1980 Anglo-Australian Telescope Board. **Fig. 1.15.** Courtesy High Altitude Observatory, National Center for Atmospheric Research (sponsored by the National Science Foundation). **Fig. 1.16.** NASA photograph. **Fig. 1.17.** NASA photograph. **Fig. 1.18.** Fr. R. E. Royer. **Fig. 1.22.** E. E. Barnard Observatory photograph by G. Emerson. **Fig. 1.24.** E. E. Barnard Observatory photograph by G. Emerson.

Chapter 2

Fig. 2.1. Reprinted with permission. From C. Ronan, *The Astronomers,* 1964 (New York, Hilland Wang), Fig 2., P. 34. **Fig. 2.2.** The Granger Collection. **Fig. 2.3.** The Granger Collection. **Fig. 2.4.** The Granger Collection. **Astronomical Insight (2.2).** Historical Pictures Services, Inc. **Fig. 2.10.** The Bettman Archive. **Fig. 2.12.** The Granger Collection. **Fig. 2.14.** The Granger Collection. **Fig. 2.15.** Griffith Observatory. **Fig. 2.16.** Robin Rector Krupp. **Fig. 2.17.** From Von Del Chamberlain, 1982, *When Stars Came Down to Earth: Cosmology of the Skidi Pawnee Indians of North America* (Ballena Press: Los Altos, Cal.); Skidi Pawnee

(continued following index)

•CONTENTS IN BRIEF•

•CONTENTS•

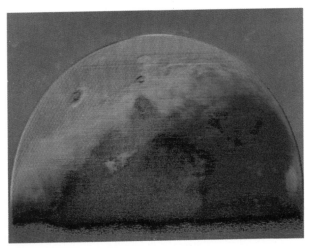

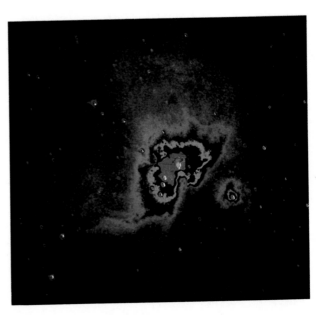

To study astronomy is, in a sense, the quintessential human endeavor. What distinguishes us from lower creatures, if not our curiosity, our compulsion to explore and discover? And what exemplifies this compulsion more than the study of the universe?

We probe the heavens (and the earth) with all possible means, and we do it for no other reason than to learn whatever there is to be known. Astronomy has produced many important and useful by-products, of course, and could be (and often is) justified solely on that basis. That is not the real motivation for astronomy, however.

This textbook represents an attempt by an astronomer to share both the knowledge and the intellectual gratification of our science. There is considerable beauty in the universe, for the eye and the mind to behold. Just as it is visually stimulating to gaze at a great glowing nebula or a colorful moon, it is pleasing to the intellect to grasp a new understanding of one of the grand themes of the cosmos. It is hoped that the reader of this book will gain by doing both.

The text is arranged in a traditional manner, beginning with discussions of the nighttime sky, historical developments, and then the basic tools of astronomy. A thorough discussion of the solar system follows, and then there are sections on the stars, our galaxy, extragalactic astronomy, and life in the universe. There are alternative logical sequences through the material: one may follow the introductory six chapters immediately with the stellar, galactic, and extragalactic sections, for example; or take the solar system and then the closing chapter on life. The content, level, and sequence of topics are intended to be suitable for broad introductory courses, covering all areas of astronomy, or for separate courses on solar system astronomy, and on stellar, galactic, and extragalactic astronomy.

There is a brief introduction for each of the seven major sections, and each is followed by a guest essay, with one exception, these were written especially for this book. The guest essays contain the personal viewpoints of prominent scientists on important current mysteries or controversies in astronomy, and lend to the reader additional excitement and understanding of how research in our science proceeds.

Each chapter begins with an outline of topics to be covered, and ends with a brief "Perspective." These place the chapter in context with those to come. Within the chapters are "Astronomical Insights" which are brief essays on topics related to the main text; they are set apart because they contain anecdotal material, somewhat more technical information than appropriate for the general text, historical background, or additional insights that may apply to other areas of astronomy as well as to the current chapter. At the end of each chapter is a summary of the principal points, an assortment of review questions, and a list of supplementary readings.

Extensive appendixes at the back of the book provide basic data on a variety of topics, such as the elements, telescopes, the planets and satellites, bright stars, the constellations, interstellar molecules, or galaxies of the Local Group. Other appendixes contain detailed explanations of certain concepts that are necessarily treated only qualitatively in the text. These include temperature scales, a logarithmic treatment of the magnitude system, and comprehensive discussion of radiation laws.

Ancillary materials for this text include an Instructor's Manual, written by Stephen J. Shawl of the University of Kansas, and a Study Guide, authored by Catharine D. Garmany and the undersigned, of the University of Colorado. The Instructor's Manual contains helpful discussions of strategies in teaching, provides a large number of possible exam questions (with answers), and gives complete answers to all the problems from the main text. The Study Guide, intended to help the student make the most of the text, contains brief chapter summaries, lists of key words and phrases, self-tests, answers to selected (usually the more difficult) problems from the text, and complete bibliographies of articles on relevent topics, taken from a wide assortment of popular magazines and journals. In addition to the Instructor's Manual and the Study Guide, another aide to teaching from the *Dynamic Universe* is available to large adopters: a set of transparencies for use with overhead projectors, showing a number of useful diagrams and charts from the text.

The art work in the *Dynamic Universe* is both extensive and of great value in illustrating the text material. The line drawings are intended to show, often schematically, the relationships of major concepts, and the photographic reproductions demonstrate the true appearance of the skies and the tools with which they are probed. The color plate sections add beauty, helping the reader gain an esthetic appreciation of astronomy, to complement the intellectual one to be derived from reading the text.

Very little of this book could have been created without the benefit of interaction with other people. It is a pleasure to acknowledge this assistance (with apologies to those inadvertantly omitted).

Many factors motivated the original decision to undertake this project. The most important was the support, enthusiasm, and patience of my wife, Connie, who also did heroic work as manuscript typist. The editor who first rekindled an older interest on my part, and then guided and prodded me throughout, was Denise Simon. When it got to the hard part, where a manuscript with nothing but text is somehow translated into a book, with illustrations, appendixes, captions, and a cover, production editor John Orr made it all come together.

A large number of scientists in our field helped in various indispensable ways, from allowing me to pick their brains, to reviewing portions of the manuscript, to providing illustrations and photographs. I am especially grateful to several of my colleagues at the University of Colorado, who, unable to escape me, were unfailingly generous with their time and energy whenever I sought their assistance. These include J. M. Shull, L. W. Esposito, A. I. Stewart, C. D. Garmany, P. S. Conti, W. C. Cash, T. R. Ayres, R. A. West, R. A. McCray, C. J. Hansen, E. R. Benton, G. E. Thomas, M. LeCompte, J.

Toomre, J. G. Timothy, B. Bohannan, C. McKay, and P. Boston.

Others around the country also assisted with advice, illustrations, or both. While individual photo and illustration credits are listed elsewhere, it is a pleasure to make specific mention here of those who were especially helpful. These include, more or less in the same order as their areas of expertise appear in the book, E. C. Krupp, Griffith Observatory; J. Eddy, High Altitude Observatory; O. Gingerich, Harvard College Observatory; G. Emerson, Ball Aerospace Corp.; A. N. Witt, University of Toledo; R. D. Gehrz, University of Wyoming; H. Masursky, U.S. Geological Survey; G. Newkirk, L. House, and J. K. Watson, High Altitude Observatory; T. Lee, Carnegie Institute of Washington; M. H. Liller, Harvard College Observatory; S. J. Shawl, University of Kansas; R. L. Kurucz, Harvard-Smithsonian Center for Astrophysics; J. C. Wheeler, University of Texas; P. J. Flower, Clemson University; G. A. Wegner, Dartmouth College; J. Heckathorn, Computer Sciences Corporation; N. Walborn, NASA: Goddard Space Flight Center; C. Heiles, University of California, Berkely; P. Thaddeus, Goddard Institute for Space Studies; B. J. Bok, University of Arizona; R. J. Dufour, Rice University; P. J. E. Peebles, Princeton University; and F. D. Drake, Cornell University.

Reviewers of the manuscript, at various stages, were:

Jay S. Boleman
University of Central Florida

Michael Breger
University of Texas

Peter S. Conti
University of Colorado

Richard M. Crutcher
University of Illinois

Kris Davidson
University of Minnesota

Phillip P. Edwards
Illinois State University

Kenneth Fox
University of Tennessee

Stephen A. Gregory
Bowling Green State University

Thomas G. Harrison
North Texas State University

Ronald N. Hartmann
Mt. San Antonio College

William W. Hunt
American River College

Karen R. Johnson
North Carolina State University

Thomas W. Jones
University of Minnesota

Robert D. Kirschner
University of Michigan

Claud H. Lacy
Texas A & M University

Steve Lattanzio
Orange Coast College

Harold A. McAlister
Georgia State University

William H. Murray
Broome Community College

David A. Pierce
El Camino College

Larry Schecter
Oregon State University

Robert D. Schmidt
University of Nebraska

John D. Schopp
San Diego State University

Leon W. Schroeder
Oklahoma State University

Richard D. Schwartz
University of Missouri, St. Louis

Stephen J. Shawl
University of Kansas

J. Michael Shull
University of Colorado

Ron Smith
Santa Monica Community College

David Thieson
University of Maryland

Charles R. Tolbert
University of Virginia

Raymond E. White
University of Arizona

For all of these people, and to the students whose responses to my teaching philosophies have also helped shape this book, I am grateful. With their continued input, I trust that this book will continue to evolve, as does the dynamic universe.

Theodore P. Snow

November, 1982

For Connie

The Nighttime Sky and Historical Astronomy

INTRODUCTION TO SECTION I.

We begin our study of astronomy with an introduction to the nighttime sky. This provides us with an immediate understanding of many of the phenomena that can be seen and appreciated with only our eyes as observing equipment, and arms us with all the knowledge our ancestors had as they attempted to develop a successful picture of the cosmos and their place in it. In chapter 1 we go well beyond simply describing the sky; we also uncover the explanations for the observed phenomena. Thus we start with knowledge that took mankind millennia to develop.

We then trace the historical development of astronomy, beginning with the earliest recorded astronomical observations and hypotheses, and following the ofttimes slow growth of science through the ages. The basis of our modern science lies in the Mediterranean region of Europe; therefore we emphasize developments there. We also explore the genesis of astronomy in other parts of the world, where sophisticated knowledge ultimately either leads to dead ends or merges with the mainstream of development.

In the third chapter we discuss the groundwork for modern science that was laid during the Renaissance, when fresh ideas arose in astronomy, as in all forms of human endeavor. We learn to appreciate the awesome breakthroughs made by giants such as Copernicus, Brahe, Kepler, and Galileo, who led the way toward a correct understanding of the universe and the place of man and his planet in it.

CHAPTER 1

The Naked-Eye View: Practical Astronomy

We view the heavens from a moving platform. The earth
spins while it travels around the sun in a nearly circular
path, and these motions create both daily and yearly
cycles of celestial events as seen from the earth's surface.
The other eight planets all behave in a similar fashion,
and most, including the earth, have one or more satellites orbiting them. Because we are viewing the skies
from a moving vantage point, and because many of the
prominent celestial luminaries have motions of their
own, our impression is that the heavens are very complex. It took mankind many centuries of thought, observation, and technological development to arrive at
the simple explanation outlined in this paragraph, an
explanation that most of us, in these modern times,
understand from early childhood.

In this chapter we will learn about the nighttime sky
as it is seen with the unaided eye. We will develop a
modern understanding of the objects that can be seen
without a telescope, and how their simple motions create complex paths through the sky, as seen from the

earth. In the process we will become familiar with the nighttime sky, and we will develop an appreciation for the task of the ancient philosophers who strove to comprehend the workings of the universe, for the view of the heavens described here is the sum total of all the evidence available to pretechnological cultures.

The Nighttime Sky

On a clear night the heavens are a splendid sight, overflowing with enough visual effects to rival any manmade light show. If we sit outdoors on a fine evening, our gaze is inevitably drawn upward and our thoughts outward, and we feel as one with the ancient scientists and philosophers. We cannot help but think of larger things than our everyday affairs, and we understand the human drive to seek out our origins and contemplate our future. Let us imagine that we are outdoors on a clear night.

When we look at the sky, we do not see it in three dimensions, because there are no obvious clues to tell us the distances to the objects we see. This fact led, long ago, to the concept of the **celestial sphere,** in which the stars and other objects in the sky are said to lie on the surface of a sphere that is centered on the earth. Although we no longer think of this as literal truth, it is still a convenient device for describing the heavens. Positions of objects on the celestial sphere are measured in angular units, because without knowledge of how far away things are, we have no means of determining their actual separations in any true distance units such as kilometers.

The most obvious objects on the celestial sphere, as long as the moon is not in one of its brightest phases, are the stars (Fig. 1.1). They appear in profusion, scattered across the sky, displaying a wide range of brightnesses and subtle variations in color. They twinkle, giving the appearance of vitality. Here and there we may see concentrations of stars in clusters (Fig. 1.2), and possibly a few dimly glowing gas clouds.

If it is a moonless night, we see a broad, diffuse band of light stretching across the celestial sphere; this is the Milky Way, our own galaxy of stars, seen edgewise from an interior position.

We may also see a few bright, steady objects that do not appear to twinkle. These are the planets, and we may find as many as five to be visible on a given night, distributed along a great arc through the sky. Careful

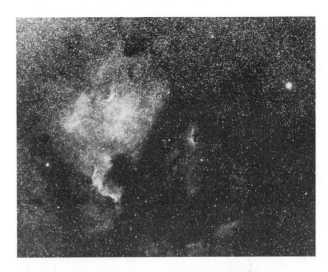

FIG. 1.1. A STAR CLOUD AND NEBULA IN THE MILKY WAY. This photograph shows a typical region of star clouds and interstellar gas and dust, along with glowing hot gas in the plane of the Milky Way.

FIG. 1.2. A GLOBULAR STAR CLUSTER. This cluster, called M13, is one of the most prominent of its type in the sky.

observation over several nights will reveal that the planets are all moving gradually along this arc, changing their positions relative to the fixed stars.

The most prominent object in the nighttime sky is usually the moon, which shines so brightly when near full (Fig. 1.3) that it drowns out all but the brightest stars and planets. The moon, about one–fourth the diameter of the earth but some 250,000 miles away, has various appearances, depending on how much of its sunlit portion we see. The moon, too, is always to be found somewhere along the east-west strip where the planets travel.

Occasionally we may see brief flashes or trails of light known as meteors. These "shooting stars" can be rather spectacular events, particularly when they arrive with great frequency, as they do during a meteor shower.

Comets are occasional visitors to our sky; perhaps once or twice a year we find one that is bright enough to be seen with the naked eye. These largely gaseous bodies orbit the sun, as do the earth and other planets, but in very elongated paths that bring them close enough to the sun to heat up and glow visibly only for brief periods of days or weeks. Some of the more spectacular, bright comets were interpreted in ancient times as harbingers of catastrophe.

The Motions of the Heavenly Bodies

The observed motions of celestial bodies result from a combination of rotational and orbital motions, including the spin of the earth and its annual movement around the sun as well as the individual motions of the moon and the planets. We can best understand the overall machinery of the solar system by examining the individual parts.

DAILY MOTIONS

The most obvious of the many motions that affect our view of the universe are the daily cycles of all celestial objects, resulting from the rotation of the earth. The earth spins on its axis in twenty-four hours, so we, on its surface, see a continuously changing view of the heavens. We see the sun rise and set, along with the moon, the planets, and most of the stars. Even though we understand that these daily, or **diurnal,** motions are the result of the earth's spin, we still refer to them as though the objects themselves were moving. The rotation of the earth means that a person standing on the equator covers the entire circumference (about 25,000 miles) in twenty-four hours, yet our senses give us no feeling of motion.

The rotation of the earth forms the basis of our timekeeping system, since the length of a day is a natural unit of time on which to base our lives; one to which, indeed, nearly all earthly species have adapted. The day is divided into twenty-four hours, each containing sixty minutes, and each minute consisting of sixty seconds. These divisions are based on the numbering system developed several thousand years ago, largely by the Babylonians.

Careful observation shows that the sun and the moon take longer to complete their daily cycles than do the stars. Each has its own motion that carries it in the same direction as the earth's rotation, opposite to the daily rising and setting of the stars. The sun and the moon therefore "get ahead" of the stars a little each day, so that it takes a little longer for each to return to the same position, from our point of view (see Fig.

FIG. 1.3. THE FULL MOON. This view, taken from space, shows our satellite from the same side that faces the earth.

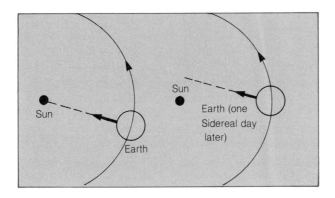

FIG. 1.4. THE CONTRAST BETWEEN SOLAR AND SIDEREAL DAYS. The arrow indicates the overhead direction from a fixed point on the earth. From noon one day (left), it takes one sidereal day for the arrow to point again in the same direction, as seen by a distant observer. Because the earth has moved, however, it will be about four minutes later when the arrow points directly at the sun again; hence the solar day is nearly four minutes longer than the sidereal day.

navigation. Today's official time standards are based on a combination of these celestial clocks and on the vibration frequencies of certain kinds of atoms, which are very precise over long periods.

Let us return now to the diurnal motions of celestial objects. We have noted that not all the stars rise and set with the daily rotation of the earth. Why not? The answer is most easily visualized by imagining we are at the North Pole—that we are located exactly at one end of the earth's rotation axis. Directly overhead is the **north celestial pole,** the point where the earth's pole is projected onto the celestial sphere. From here we see the stars circling overhead as the earth spins, but they do not dip below the horizon. Now if we return to a location at some intermediate latitude and look toward the north, we find that stars lying near the celestial pole can be seen all night, and that they circle a fixed point. On a time-exposure photograph of the north polar region, the paths of the stars can clearly be seen

1.4). Thus, the sun rises about four minutes later each day, in comparison to the stars, and the moon rises almost an hour later each day. The planets also move, but so slowly that their diurnal motions are not easily distinguished from those of the stars.

Our "official" day, the one that forms the basis of our timekeeping system, is based on the **solar day** rather than the **sidereal day,** which is the true rotation period of the earth (the term *sidereal* means "with respect to the stars"). Because the earth's orbital speed is not precisely constant, the length of the solar day varies a little throughout the year. It would be inconvenient to allow our hour, minute, and second to vary along with it, so the average length of the solar day, called the **mean solar day,** has been adopted as our timekeeping standard. The mean solar day is 3^m56^s longer than the sidereal day.

Time was once kept exclusively by use of special telescopes, called **transit telescopes,** that could measure the precise time at which certain bright stars passed overhead. The method was used to measure directly the sidereal day, and from this measurement, allowances had to be made to determine the mean solar time. Transit telescopes are located at many sites throughout the world; in the United States they are operated by the timekeeping division of the U.S. Naval Observatory (Fig. 1.5), which got into the business because of the close connection between timekeeping and celestial

FIG. 1.5. A TRANSIT TELESCOPE. Such a device points straight up. It is used to record the times when certain reference stars pass over the meridian and is therefore helpful in measuring the sidereal day.

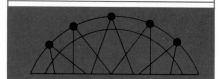

What Is an Astronomer Anyway?

Throughout this text we will be referring to astronomers, scientists, astrophysicists, and physicists. In a book that endeavors to summarize all that we know about the universe and its contents, we can hardly omit a description of the people who devote their time to developing this knowledge.

Thousands of people in the United States study astronomy. The amateur astronomers, who outnumber the professionals, enjoy a variety of activities (either on their own, or through local or even nationwide organizations), including telescope making, astrophotography, long-term monitoring of variable stars, public programming, and just plain stargazing. If you wish to join such a group, get advice on buying a telescope, or learn techniques such as photography of celestial objects, your best bet is to get in touch with an amateur astronomy group. Local clubs exist in most major cities, and regional associations are everywhere. You might have difficulty finding such groups in the telephone book; you would have better luck if you contacted workers at a telescope shop or planetarium.

There are some 3,500 members

of the American Astronomical Society, the principal professional astronomy organization in the United States. These people, as a rule, make up a limited number of categories: those teaching and doing research at colleges and universities; those doing research at federal observatories and laboratories, such as the National Aeronautics and Space Administration (NASA); and those doing research and related engineering activities in private industry, most often with companies involved in aerospace projects.

About 8 percent of the members of the American Astronomical Society are women. A recent study indicated there is little discrimination in the hiring of professional astronomers, so this low percentage reflects the small number of women who enter the field of astronomy at the graduate level.

Most of the funding for people doing research in astronomy, even for those not working directly for federal laboratories and observatories, comes from the government. A significant function of an astronomer on a university faculty is to write proposals, usually to the National Science Foundation or to NASA, for support of the research programs he wishes to carry out.

The terms *astronomer* and *astrophysicist* have come to mean pretty much the same thing in modern usage, although historically, there was a difference. An astronomer was one who studied the skies and gathered data, but did relatively little interpretation; an astrophysicist primarily was interested in under-

standing the physical nature of the universe, and therefore he carried out comprehensive analyses of astronomical data or did theoretical work, in both cases applying the laws of physics to phenomena in the heavens. Nearly all modern astronomers are to some degree astrophysicists; they range from the observational astronomer at one end of the spectrum to the pure theorist at the other. The two terms are therefore used interchangeably today. Many modern astronomers call upon the fields of engineering (for instrument development), chemistry (in studying planetary and stellar atmospheres and the interstellar medium), geophysics (in probing interior conditions in planets and other solid bodies), and possibly even biology, but always with an underlying foundation in physics.

If you want to become a professional astronomer, you should be aware from the outset that the field is small and job opportunities are both limited and often subject to the vagaries of federal funding. If you persist, the best course is to study physics, at least through the undergraduate level, and then plan to attend graduate school in an astronomy department or in physics (the latter option may make you a bit more versatile, and would not be a handicap when you enter astronomy later). Demographic studies have shown that there may be a shortage of astronomers for a period beginning in the late 1980s and extending through the rest of this century. Perhaps the timing will be right for you.

FIG. 1.6. STAR TRAILS ILLUSTRATING THE EARTH'S
ROTATION. The circular trails are created by stars near the south
celestial pole, which completed about half of a full circle during
this all-night exposure.

to describe complete circles about either pole (Fig. 1.6).
It happens that a bright star, called Polaris, lies very
close to the position of the north pole, and is therefore
almost stationary throughout the night. If we are located
at 40° north latitude, for example, we can see all the
sky within 40° of the north pole throughout the entire
day and night; therefore all these stars stay up all night.
At locations farther than 40° from the pole, they will
rise and set as portions of their circular paths are
blocked from our view by the solid earth. If we were
at the earth's equator, all the stars visible to us would
rise and set each day, because we would not be able
to see beyond either pole.

The portion of the sky that we can see depends on
our latitude; our position north or south of the equator.
For those of us living in the northern hemisphere, a

large region of the southern sky is forever beyond our
view. The constellations we see vary as we travel north
or south, a fact not lost on early astronomers, who concluded that the earth is round.

Our discussion of diurnal motions, latitude, and the
changing appearance of the sky as we move over the
surface of the earth has led us to a point where it is
convenient to introduce the notion of coordinate systems for the skies. Just as it is useful to be able to
describe the position of a place on the surface of the
earth, it is also important to be able to record the position of a particular object in the sky, so that the astronomer may return to it later, or tell another where
it is.

The standard coordinate system used in astronomy
is based on the equator of the earth, along with a fixed
direction that establishes the east-west coordinate. This
system, analogous to latitude and longitude, is called
the **equatorial coordinate system.** It is imagined that
the earth's equator is projected onto the celestial sphere
(Fig. 1.7), forming a *celestial equator.* More precisely,
the celestial equator is the intersection of the earth's

FIG. 1.7. THE CELESTIAL SPHERE. For practical reasons, the sky
can be imagined as a sphere centered on the earth. The earth's
poles and equator, projected onto the celestial sphere, provide the
basis for measuring star positions.

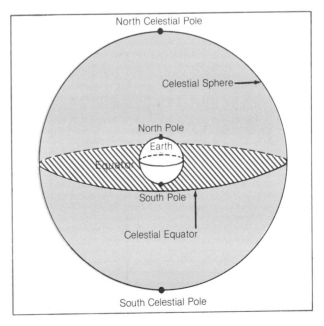

equatorial plane with the celestial sphere. A star's north-south position, called its **declination,** is the angular distance north or south of the celestial equator. Thus a star that passes directly overhead at New York City, whose latitude is 40°40′ N, has a declination of +40°40′, and one that passes overhead at Rio de Janeiro has a declination of −22°50′ (plus and minus signs are used to indicate north and south).

A star's position in the east-west direction, called its **right ascension,** cannot be so directly based on the longitude system used on the earth's surface, since the earth is rotating. The standard of reference for longitude measurements is a fixed meridian called the **prime meridian,** passing through Greenwich, England. If we imagined this to be projected onto the sky, we would find that it sweeps across the heavens as the earth rotates, making it very awkward indeed for us to try to keep track of star positions with respect to this moving reference point. Instead, astronomers have established a reference direction that remains fixed in the sky. This direction is defined by the line of intersection of the earth's equatorial plane and the plane of its orbit. Because it takes twenty-four sidereal hours for a given star to make a complete circuit of the sky as the earth rotates, its position in the east-west direction is measured in units of sidereal hours, minutes, and seconds, rather than in angular units. Thus a star located in the direction of our reference point has a right ascension

of $0^h0^m0^s$, whereas a star one-quarter of the way around the sky to the east of that point has a right ascension of $6^h0^m0^s$. The concept of a **meridian** is still used; it refers to the north-south line on the celestial sphere that passes from the celestial poles through the **zenith,** the point directly overhead. When a celestial object is on this line; that is, when it is due north or south of the point directly overhead, it is said to be on the meridian. This means, for example, that at local noon the sun is on the meridian.

There is one complication with using the equatorial coordinate system. The earth slowly wobbles on its axis (Fig. 1.8), in a motion called **precession,** and this causes the celestial equator to move slowly through the sky. It takes some 26,000 years for the earth to complete one cycle of this motion, which is exactly like the wobbling of a spinning play top or gyroscope. The motion of our coordinate system is therefore gradual, causing star positions to shift very slowly. Despite the small magnitude of the effect, it was noticed more than two thousand years ago, and must be considered by modern astronomers in planning observations. One of the typical preparations for observing is to "precess" the coordinates of all the target stars; that is, to calculate their current locations. Catalogues of star positions always specify the date for which they are valid, so that astronomers know how much precession to allow for.

FIG. 1.8. PRECESSION. The earth's axis is tilted 23½° with respect to its orbital plane, and it wobbles on this axis, so that an extension of it describes a conical pattern (left) in a time of about 26,000 years. The north celestial pole therefore follows a circular path on the sky (right).

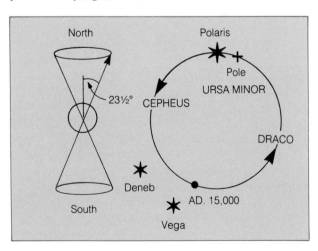

ANNUAL   MOTIONS: THE   SEASONS

We turn our attention now to celestial phenomena that are caused by the earth's motion as it orbits the sun. One aspect of this, the daily eastward motion of the sun with respect to the stars, has already been mentioned, in connection with the difference between the solar and sidereal day. It must be emphasized that this is only an apparent motion, caused by our changing angle of view as we move with the earth in its orbit. As the earth moves about the sun, it travels in a fixed plane, so that the apparent path of the sun through the constellations is the same each year. The apparent path of the sun is called the **ecliptic,** and the sequence of constellations through which it passes is called the **zodiac** (Fig. 1.9). There are twelve principal constellations of the zodiac, identified since antiquity, and once thought by astronomers to have significance in our daily lives. To this day, astrologers will tell you that your fate is influenced by the constellation that the sun hap-

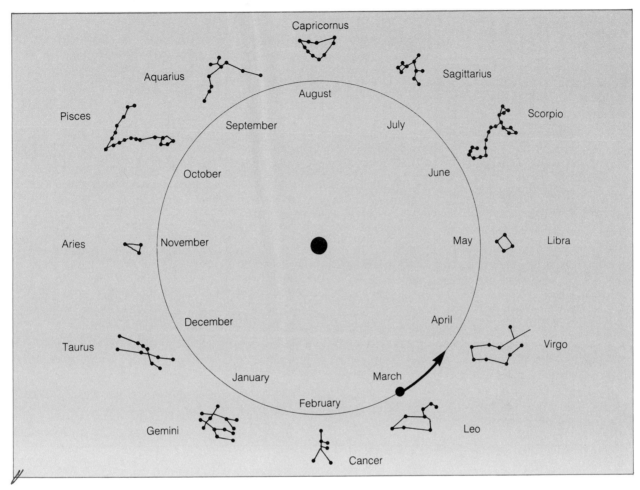

FIG. 1.9. THE PATH OF THE SUN THROUGH THE CONSTELLATIONS OF THE ZODIAC.

pened to be in on the day you were born. This is a very old idea; astronomers have refined their understanding of the heavens in the last two thousand years, even if astrologers have not.

Because we can only see stars in the nighttime sky, the constellations most easily visible to us are the ones at least a few hours of right ascension to the east or west of the sun. This means that most stars, except for those near the poles, cannot be observed year-round.

It happens that the orbital planes of the other planets and the moon are closely aligned with that of the earth, so that the planets and the moon are always seen near the ecliptic. Hence all the major objects in the solar system that can be seen by the naked eye pass through the same sequence of constellations, the zo-

diac, so it is not surprising that ancient astronomers attached great significance to this sequence.

Besides causing the apparent annual motions of the sun and planets, the earth's orbital motion has a second, far more significant, effect on us; it creates our seasons. The earth's spin axis is tilted with respect to its orbital plane, so that during the course of a year, portions of the earth away from the equator are exposed to varying amounts of daylight. Summer in the northern hemisphere occurs when the north pole is tipped toward the sun; winter occurs during the opposite part of the earth's orbit, as the pole, which remains fixed in orientation, is tilted away from the sun. The tremendous seasonal variations in climate at intermediate latitudes are caused by a combination of two

ASTRONOMICAL INSIGHT 1.2

Astronomy and Astrology

There is an unfortunate tendency in modern society to confuse astrology and astronomy, or, worse yet, to consider one a legitimate alternative to the other.

Astrology, the pseudoscience based on the belief that human lives are influenced by the configurations of heavenly bodies, arose at a primitive stage in development of mankind, at a time when the earth was thought to be a flat disk under the dome of the heavens. Although ancient Greek astronomers did much to raise the study of the heavens to a scientific level, during the subsequent Dark Ages astrology and the governing of human lives by the stars became once again the primary basis for study of the heavens. This trend changed radically with the Renaissance, when the true nature of the heavens and the motions of astronomical objects were untangled. It is unfortunately true, however, that a segment of the human populace has continued to profess a faith in astrology.

One of the basic lessons learned about the universe is that it is easy to make mistakes unless we are careful to be objective, accept only conclusions that can be verified by repeated observations or experiments, and make predictions that can be tested. Astrology, utterly and abysmally, fails to meet these criteria. Serious attempts have been made to test astrological lore by statistical analysis of people born under different signs, with no trace of a correlation ever being seen.

There certainly are many phenomena that defy the understanding of modern science, and many of them deserve more attention than they are getting. Astrology is not one of them. As a phenomenon, it is worthy of study only by the fields of sociology and psychology, for its effects do not exist in the physical universe. It may be interesting at a party to compare astrological signs, but it would be good also to keep in mind the difference between objective reality and subjective impressions. Failure to do so would be failure to understand what science is.

---

effects: (1) the length of the day varies, so that in summer, for example, the sun has more time to heat the earth's surface; and (2) the angle at which the sun's rays strike the ground is more nearly perpendicular in the summer, so that the sun's intensity is much greater, heating the surface more efficiently.

The earth's axis is tilted 23½° from the perpendicular to the orbital plane. Therefore during the year the sun, as seen from the earth's surface, can appear directly overhead as far north and south of the equator as 23½° (see Figs. 1.10 and 1.11), defining a region called the **tropical zone** (Fig. 1.12). For those of us who live outside the tropics, the sun can never be directly overhead. When the north pole is tilted toward the sun, an occasion occurring around June 21 and called the **summer solstice,** the sun passes directly overhead at 23½° north latitude at local noon, but passes to the south of the zenith for anyone at more northerly latitudes. In Boulder, Colorado, for example, which is right on the fortieth parallel, the sun on this day reaches a point 40° − 23½° = 16½° south of the zenith, or 90° − 16½° = 73½° above the southern horizon.

Traveling farther north, we see that at the time of the summer solstice, sunlight covers the entire north polar region to a latitude as far as 23½° south of the pole. This defines the **arctic circle,** and at the time of the solstice the entire circle has daylight for all twenty-four hours of the earth's rotation. At the pole itself there is constant daylight for six months.

During winter in the northern hemisphere the sun does not rise as high above the southern horizon at local noon as it does during summer, because the south pole is now tilted toward the sun. At the **winter solstice,** when the south pole is pointed most closely in the direction of the sun, the sun's midday height above the southern horizon, as viewed from the northern hemisphere, is the lowest of the year. Returning to Boulder, we find that on this day, which usually is near

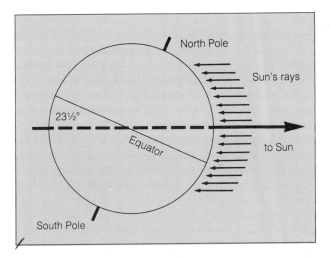

FIG. 1.10. THE EFFECT OF THE EARTH'S TILTED AXIS. In general, the sun is not directly overhead at the equator, but at some latitude between 23½° S and 23½° N. The sun's rays reach the ground at varying angles, depending on the latitude, and therefore the solar heating of the earth's surface varies with latitude.

December 21, the sun is $40° + 23½° = 63½°$ south of the zenith, or $90° - 63½° = 26½°$ above the horizon.

If we follow the sun's motion north and south of the equator throughout the year, we find that it follows a graceful curve as it traverses its range from $+23½°$ (north) declination to $-23½°$ (south) declination. The sun crosses the equator twice in its yearly excursion, at the times when the earth's north pole is pointed in a direction 90° from the earth-sun line. At these times the lengths of day and night in both hemispheres are equal, and these occasions are referred to as the **vernal** (spring) and **autumnal** (fall) **equinoxes,** taking place

FIG. 1.11. THE PATH OF THE SUN THROUGH THE SKY. Because of the earth's orbital motion and the tilt of its axis, the sun's annual path through the sky has the shape illustrated here.

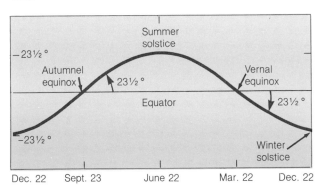

on about March 21 and September 21, respectively. The direction to the sun at the time of the vernal equinox coincides exactly with the direction of 0^h right ascension (remember, the definition of this direction is that it lies along the line of intersection of the earth's equatorial plane and the orbital plane).

THE MOON AND ITS PHASES: ECLIPSES

Except for the sun, the moon is the brightest object in the sky, and it, too, has its own complex motions as seen from the earth. By contrast with the sun, whose apparent motions are entirely a result of our own changing viewpoint from the moving earth, the moon's observed path through the sky is created almost wholly by its own movements. The moon appears to be going around the earth, and indeed it is; the ancient astronomers solved this part of the celestial mystery long ago.

Of course, as in the case of the sun, the first thing

FIG. 1.12. THE DEFINITION OF LATITUDE ZONES ON THE EARTH. As shown, at summer solstice, the sun is overhead at 23½° north latitude, its northernmost point. This defines the Tropic of Cancer, the northern limit of the tropical zone. At the same time, the entire area within 23½° of the North Pole is in daylight throughout the earth's rotation, and the boundary of this region is the Arctic Circle. Similarly, the Antarctic Circle receives no sunlight at all during a complete rotation of the earth. Six months later, the sun is overhead of the Tropic of Capricorn.

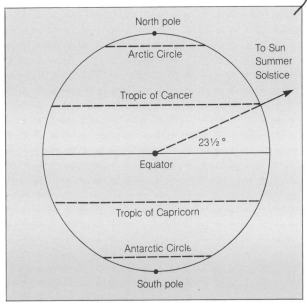

we notice about the moon's motions is that it rises and sets every day. If we pay a bit more attention, however, we find that from day to day the moon's position with respect to the stars is changing. It is not difficult, in fact, to notice the lunar motion with respect to the stars during the course of an evening. The moon moves about 13° across the sky every day, or more than 1° every two hours. Its angular diameter is about ½°, so this means that the moon moves a distance about equal to its own diameter every hour.

The plane of the moon's orbit is closely aligned with that of the earth's orbit, so the moon stays near the ecliptic, as seen from the earth. It takes the moon a little over twenty-seven days ($27^d7^h43^m11.^s5$) to make

one trip around the earth. Because the earth moves at the same time, however it appears to us that the moon takes longer than twenty-seven days to make one complete circuit; from our point of view, the moon takes $29^d12^h44^m2.^s8$ to complete its full cycle of phases (Fig. 1.13). The true orbital period is called the **sidereal period,** because it is the time required for the moon to go around the earth, as seen in a fixed reference frame with respect to the stars. The observed cycle of phases, which is the time required for the moon to return to a given alignment with respect to the sun, is called the **synodic period** or the **lunar month.** The difference is akin to the distinction between the solar and sidereal days discussed earlier. In both cases it is

FIG. 1.13. SIDEREAL AND SYNODIC PERIODS OF THE MOON. Because the earth moves in its orbit while the moon orbits it, the moon must move more than one full circle (as seen by an outside observer) to go from one full moon to the next. Hence the lunar (or synodic) month is about two days longer than the moon's sidereal period.

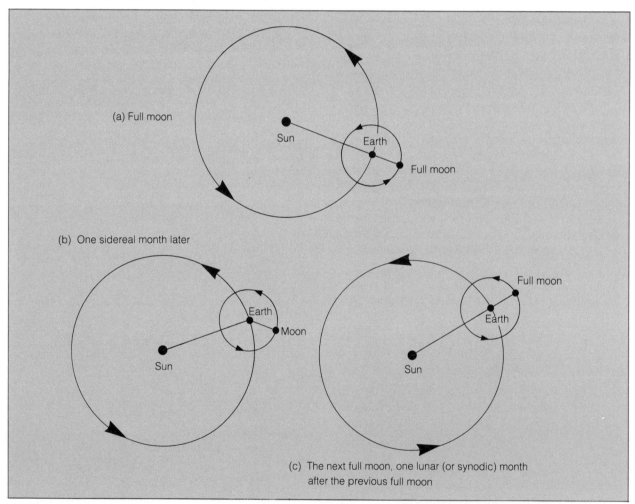

the earth's motion about the sun that lengthens the time it takes to complete a full cycle as we see it.

The moon always keeps the same side facing the earth, because its rotation period is equal to its orbital period. It is a common misconception that the moon is not rotating; actually, if it did not rotate, we would see all its sides as it circled the earth. The fact that the orbital and spin periods are equal is not a coincidence, but is a result of the tidal forces exerted on the moon and the earth by their mutual gravitational pull. The phenomenon of matching orbital and spin periods is called **synchronous rotation,** and it is common in the universe, both within the solar system and in double stars. Synchronous rotation and tidal forces are discussed more fully in chapter 4.

Since the moon does not emit light of its own, but instead shines by reflected sunlight, we easily see only those portions of its surface that are sunlit. As the moon orbits the earth and our viewing angle changes, we see varying fractions of the daylit half of the moon. This causes the moon's apparent shape to change drastically during the month, the sequence of shapes being referred to as the **phases** of the moon (Fig. 1.14). The full cycle of phases is completed during the one synodic period, or **lunar month,** of about twenty-nine and a half days.

The extremes of the cycle are represented by the **full moon,** occurring when it is directly opposite the sun, so that we see its entire sunlit hemisphere; and the **new moon,** when it is between the earth and the

FIG. 1.14. LUNAR PHASES AND CONFIGURATIONS. As the moon orbits the earth, we see varying portions of its sunlit side. The phases sketched here (outside the circle representing the moon's orbit) show the moon as it appears to an observer in the Northern Hemisphere.

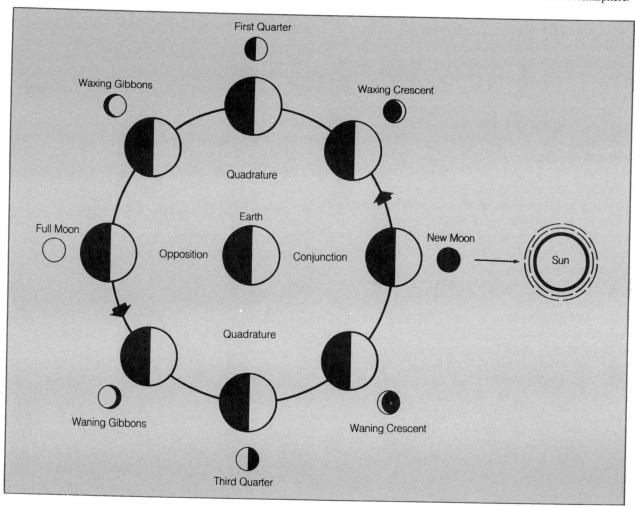

sun, with its dark side facing us. The full moon is on the meridian at local midnight, whereas the new moon is on the meridian at local noon. The new moon cannot be observed, because it is dark and is so close to the sun.

Just as we speak of the moon's phases, which really refer to its apparent shape as seen from the earth, we can also speak of its **configurations,** which describe its position with respect to the earth-sun direction (Fig. 1.14). For example, a full moon occurs at **opposition** when the moon lies in the direction opposite that of the sun; and a new moon occurs at **conjunction,** when the moon lies in the same direction as the sun. We can follow the moon through its phases as we trace its configurations, beginning with the new moon, which takes place at conjunction. During the first week following conjunction, as the moon moves toward **quadrature** (the position 90° from the earth-sun line), it appears to have a crescent shape that grows in thickness each night. This is called the **waxing crescent** phase, and when the moon reaches quadrature, so that we see exactly one-half of the sunlit hemisphere, it has reached the phase called **first quarter.** For the next week, as the moon goes from quadrature toward opposition, and we see more and more of the daylit side, its phase is said to be **waxing gibbous.** After the full moon, as it again approaches quadrature, the phase is **waning gibbous,** and this time at quadrature, the phase is called **third quarter.** We can easily distinguish first quarter from third quarter by noting the time of night: if the moon is already up when the sun sets, it is first quarter, but if it does not rise until later, it is third quarter. After third quarter, as the moon moves closer to the sun, we see a diminishing slice of its sunlit side, and it is in the **waning crescent** phase.

From what we have stated so far, it may seem that the moon should pass directly in front of the sun on each trip around the earth, and through the earth's shadow at each opposition, producing alternating solar and lunar eclipses at two-week intervals, but this is obviously not the case. The reason is that the moon's orbital plane does not lie exactly in the ecliptic, but is tilted by about 5°. Therefore the moon usually passes just above or below the sun as it goes through conjunction, and similarly misses the earth's shadow at opposition. The moon passes through the ecliptic at only two points on each trip around the earth, where the planes of the earth's and the moon's orbits intersect.

Because the moon's orbital plane wobbles slowly, in a precessional motion similar to that of the earth's spin axis, the line of intersection with the earth's orbital plane slowly moves around. The combination of this motion, the moon's orbital motion, and the movement of the earth around the sun creates a cycle of eclipses, with the same pattern recurring every eighteen years. This cycle, called the **Saros,** was recognized in antiquity.

It is purely coincidental that the moon and the sun have nearly equal angular sizes, so that the moon neatly blocks out the disk of the sun during a solar eclipse (Figs. 1.15 and 1.16 and Color Plate 1). This coincidence was undoubtedly given great significance by the ancient astronomers, reinforcing the notion that the sun and the moon were closely related.

The angular diameter of an object is inversely proportional to its distance, meaning that the farther away it is, the smaller it looks. The sun is much larger than the moon, but is also much more distant. The two objects have the same angular diameter because the ratio of the sun's diameter to that of the moon just happens to be compensated by the ratio of the sun's distance to that of the moon. Because the coincidence is so precise, a solar eclipse is seen only from a highly localized region as the moon's shadow moves across (Fig. 1.17).

If a total solar eclipse occurs at the time when the moon is farthest in its slightly noncircular orbit from the earth, it does not quite block all of the sun's disk,

FIG. 1.15. A TOTAL SOLAR ECLIPSE, which occurs when the moon entirely blocks our view of the sun's disk.

but rather, leaves an outer ring of the sun visible. This is called an **annular** eclipse. Because a total (or annular) solar eclipse requires a precise alignment of sun and moon, an eclipse will only appear total along a well-defined, narrow path on the earth's surface. There is a wider zone outside of that where the moon appears to block only a portion of the sun's disk; people in this zone see a partial solar eclipse.

During a lunar eclipse, when the moon passes through the earth's shadow, observers everywhere on the earth see the same portion of the moon eclipsed. If the moon passes through the **umbra,** the dark inner portion of the earth's shadow, then the eclipse is total, as no part of the moon's surface is exposed to direct sunlight. If the moon misses being fully immersed in the umbra, it may pass through the **penumbra** and undergo a partial eclipse (Color Plate 1).

FIG. 1.17. THE MOON'S SHADOW DURING A SOLAR ECLIPSE. This sequence shows the path of the moon's shadow across the surface of the earth during a solar eclipse. The eclipse appeared total only to observers on the earth who were located directly in the center of the shadow's path (that is, in the umbra).

PLANETARY MOTIONS

Like the sun and the moon, the planets move with respect to the background stars. It was this fact that lent the planets their generic name, since *planet* is the Greek word for *wanderer.*

The observed motions of the planets are primarily a result of their orbital movement about the sun, although, as we shall see, one important aspect of the motion of certain planets is a reflection of the earth's motion. The planets all orbit the sun in the same direction as the earth and, as mentioned earlier, in nearly the same plane, so that they appear to move nearly in the ecliptic (Fig. 1.18), through the constellations of the zodiac.

The two planets lying within the orbit of the earth, Mercury and Venus, are called **inferior planets,** and can never appear far from the sun in our sky. For Mercury, the greatest angular distance from the sun, called the **greatest elongation,** is about 28°, whereas for Venus, this distance may be as great as 47°. Like the moon, the planets have specific configurations, referring to their positions with respect to the sun-earth line (Fig. 1.19). An inferior planet is said to be at **inferior conjunction** when it lies directly between the earth and sun, and at

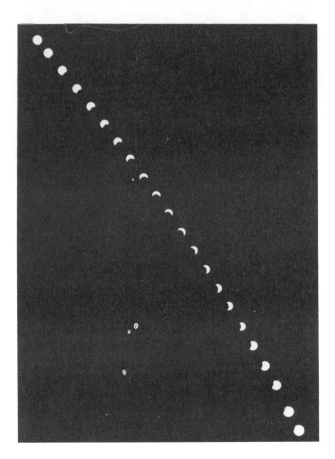

FIG. 1.16. A SOLAR-ECLIPSE SEQUENCE, illustrating the moon's progression across the disk of the sun during a total solar eclipse.

FIG. 1.18. THE ECLIPTIC. This photo illustrates that the planets and the moon follow a narrow path across the sky. Here we see three planets: Venus, Jupiter, and Mars.

superior conjunction when it is aligned with the sun, but lying on the far side.

The outer, or **superior,** planets can be seen in any direction with respect to the sun, including opposition, when they are in the opposite direction from the sun. Conjunction for a superior planet can occur only when the planet is aligned with the sun but on the far side of it (in analogy with a superior conjunction for an inferior planet), and quadrature occurs when a superior planet is 90° from the direction of the sun.

Each planet has a sidereal period and a synodic period, the former being the true orbital period as seen in the fixed framework of the stars, and the latter being the length of time it takes the planet to pass through the full sequence of configurations, as from one conjunction or one opposition to the next (Fig. 1.20). The situation is much like that of two runners on a track. The time it takes the faster runner to lap the slower one is analogous to the synodic period, whereas the time it takes the runners simply to circle the track corresponds to the sidereal period.

The movement of the earth has one very important effect on planetary motions. Going outward from the sun, each successive planet has a slower speed in its orbit (see the discussion of Kepler's law of planetary motion in chapter 3). This means that the earth, moving faster than the superior planets, periodically passes each of them (this occurs once every synodic period). As the earth overtakes one of the superior planets, there

is an interval of time during which our line of sight to that planet sweeps backward with respect to the background stars, making it appear that the planet is moving backward (Fig. 1.21). The same thing happens when, in a rapidly moving automobile, we pass a slowly moving vehicle: for a brief moment, the other vehicle appears to move backward with respect to the fixed background. This apparent backward movement, called **retrograde motion,** was thought in ancient times to be a real motion of the superior planets rather than a reflection of the earth's motion. This idea very much

FIG. 1.19. PLANETARY CONFIGURATIONS.

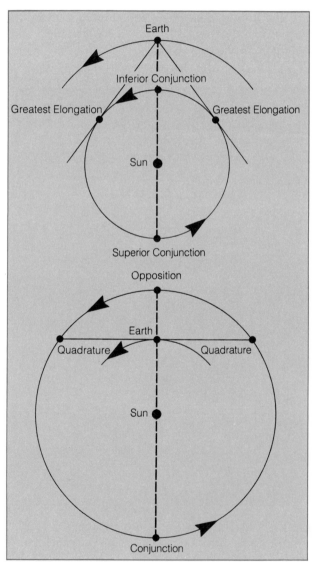

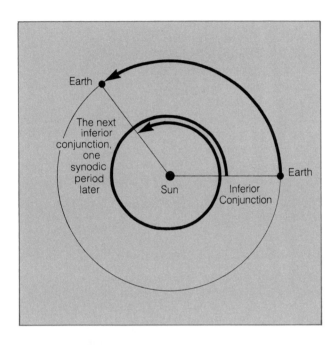

FIG. 1.20. THE SYNODIC PERIOD OF AN INFERIOR PLANET. The inner planets travel faster than the earth in their orbits, and therefore "lap" the earth, much as a fast runner laps a slower one on a track. This illustration shows approximately the situation for Mercury, which has a synodic period of about 116 days, or roughly one-third of a year.

FIG. 1.21. RETROGRADE MOTION. As the faster-moving earth overtakes a superior planet in its orbit, the planet temporarily appears to move backward with respect to the fixed stars. This sketch illustrates the modern explanation of something that took ancient astronomers a long time to correctly understand.

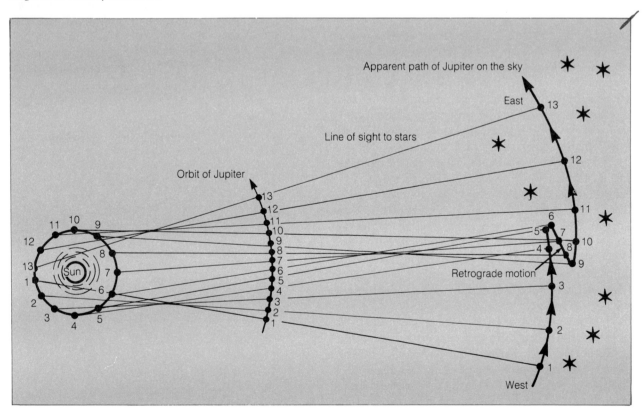

complicated many of the early cosmologies, giving rise, in fact, to a whole class of theories in which the planets were thought to move in small circles called epicycles, which in turn orbited the earth (see the discussions of Hipparchus and Ptolemy in chapter 2).

Even the inferior planets undergo retrograde motion as they overtake the earth, but this is difficult to observe since it occurs near the time of inferior conjunction, when they lie near the direction of the sun. They also move in the retrograde direction when on the far sides of their orbits, when they actually do move in the opposite direction as we see it.

Celestial Visitors: Comets and Meteors

So far we have discussed only regular, cyclical motions of heavenly objects. The universe has a clockwork regularity reminiscent of a precision machine. Occasionally, however, events take place in the skies that contradict this notion; they occur randomly, with no respect for the usual order. **Comets** and the tiny objects that create meteors are among the most obvious of these celestial anarchists.

To say that comets lack any regularity in their motions is incorrect, because they do orbit the sun and, barring disturbances, would do so forever, retracing their orbits with as much precision as the planets. The orbits of comets are very elongated, carrying them to the far reaches of the solar system, where they spend most of their time as dark, frozen chunks of rock and ice. It is only when a comet passes near the sun (so that it is heated and ejects dust and glowing gases) that it takes on the familiar form of a bright object attended by a long tail (Fig. 1.22). The orbit of a comet, when undisturbed, has a period of many hundreds of thousands of years. Gravitational forces created by the planets (especially massive Jupiter) can, however, alter the orbit (Fig.1.23). A comet can be nudged into a smaller orbit around the sun with a period of only a few years or decades, or it can be speeded up so that it escapes the solar system entirely. Thus some comets reappear regularly, with periods ranging from several months to many years, whereas others make just one appearance.

Meteors are the bright, brief flashes of light that can be seen occasionally on almost any clear night. They occur when small chunks of rock, ice, or metallic debris enter the earth's upper atmosphere from space at such

FIG. 1.22.   A COMET. Comet West, one of the more easily observed and spectacular comets of the last decade.

high velocities that the friction causes them to burn up. On rare occasion one of these chunks, called a **meteroid** when in space, reaches the ground, where it may later be found and labeled a **meteorite.** Meteoroids, which become meteors upon entering the earth's atmosphere, are ancient material left over from the formation of the solar system. Some of them originate in the asteroid belt, a swarm of loose, chunky debris orbiting the sun between Mars and Jupiter; others are the

FIG. 1.23.   A TYPICAL COMETARY ORBIT. Comets follow high elongated orbits about the sun, and are only visible as comets when in the inner portion of the solar system, where heating by the sun forces the ejection of the familiar tail.

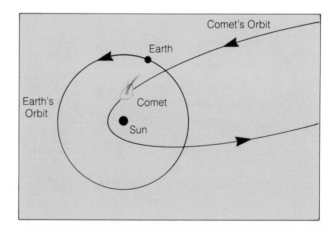

FIG. 1.24. A METEOR SHOWER. This is a time-exposure photograph of a particularly active meteor shower.

remains of comets that have gradually distintegrated. When the earth enters the orbital path of a comet, there is a high frequency of meteors. Such an occurrence is called a **meteor shower** (Fig. 1.24), which can be quite a spectacular sight, with several meteors appearing every minute. Cometary debris rarely survives to reach the ground, however; meteorites almost always originate as chunks of asteroids, rather than comets.

Other random, but much less frequent, celestial events may take the form of novae or supernovae, fantastically powerful explosions that can take place in dying stars. Only a handful of these events bright enough to be obvious to the naked eye have occurred throughout recorded history.

Perspective

We have now developed a modern picture explaining all of the astronomical phenomena visible to the naked eye. There is much left for us to learn about the universe and the laws that govern it, but already we have available to us all of the information upon which the first astronomers based their crude theories of cosmology. In the next two chapters we shall see how, presented with this evidence, the ancients fared.

Summary

1. Many beautiful and significant sights can be seen in the nighttime sky on any clear night.

2. The earth's rotation causes diurnal motions: the daily rising and setting of the sun, the moon, the stars, and the planets.

3. The solar day is a few minutes longer than the sidereal day, because of the earth's orbital motion about the sun.

4. The most commonly used astronomical coordinate system, called the equatorial coordinate system, is based on the projection of the earth's equator onto the celestial sphere, from which declination is measured, and on a fixed direction in the sky, from which right ascension is measured.

5. Precession is a slow drift of the positions of stars with respect to this earth-based coordinate system, and is caused by the wobbling of the earth's rotation axis.

6. The orbital motion of the earth about the sun causes annual motions such as the apparent motion of the sun through the constellations of the ecliptic, and is responsible, along with the tilt of the earth's axis, for our seasons.

7. The moon orbits the earth while the earth orbits the sun, and from the earth we see the various phases of the moon as it passes through different configurations.

8. Solar and lunar eclipses occur when the moon passes directly in front of the sun or through the earth's

shadow, respectively. The occurrence of these alignments is affected by the tilt and precession of the moon's orbit.

9. The planetary orbits lie nearly in the same plane, so they are always seen along the ecliptic, in various configurations with respect to the sun-earth line.

10. The planets go through temporary retrograde motion resulting from the relative speed with which they pass or are passed by the earth.

11. Comets and meteors are transitory phenomena caused by chunks of debris in interplanetary space.

Review Questions

1. Explain why it is necessary to have astronomical observatories in both the northern and southern hemispheres.

2. A transit telescope can be used for timekeeping, by observing the time at which selected stars pass directly overhead. A given star will pass overhead once each night, as the earth's rotation sweeps the telescope's field of view across the sky. Does this define the solar day or the sidereal day? Why are planets never used for timekeeping purposes? Is it possible, at a given location, to use the same star all year round for timekeeping?

3. Imagine that you live at a latitude of 30°N. At the time of the winter solstice (on or about December 21), how far above the southern horizon would the sun rise at midday? How would your answer differ if you lived at 30°S latitude?

4. Imagine that the earth's rotation axis is perpendicular to the ecliptic, instead of being tilted 23½° away from perpendicular. What would be the length of the day at the time of the summer solstice for a person living at 40°N latitude?

5. Suppose a certain star is located 30° east of the line where the right ascension is defined to be $0^h0^m0^s$. What is the right ascension of this star? What is the right ascension of a star that is 30° to the west of this line?

6. What is the declination of the sun at the time of the summer solstice? At the winter solstice? At the vernal equinox?

7. The moon's synchronous rotation means that its spin and orbital periods are equal. Is it the synodic or the sidereal orbital period that is equal to the rotation period?

8. A careful astrologer, one who makes his own observations of the heavens, will notice that the sun is never actually in the constellation that is associated with a particular date. For example, a person born in late January is said to be born under the sign of Aquarius, yet the sun at this time of the year is not in Aquarius. How can you explain this discrepancy?

9. Explain why lunar and solar eclipses often occur in pairs: on many occasions one occurs shortly after the other, but then there may be a long period with neither type of eclipse occurring.

10. Lunar eclipses always occur at the same phase of the moon. Which phase is it? At which lunar phase do solar eclipses always take place?

Additional Readings

There are a number of magazines, for example, *Mercury, Sky and Telescope,* and *Astronomy,* that contain practical information for the sky-watcher, such as planetary positions and the seasonal appearances of the constellations. In addition, there are various handbooks with similar data. One of the most widely used is the *Observer's Handbook* (published yearly) by Roy L. Bishop (Toronto: Royal Astronomical Society of Canada). Some practical exercises in astronomy can be found in books such as *Astronomy: A Self-Teaching Guide,* by Dinah L. Moche (New York: Wiley, 1981).

Early Astronomy

It is likely that mankind was preoccupied with the heavens from the time he first became aware of his environment. The speculative mood that we, in these modern times, can conjure up only by disregarding our daily pressures and escaping into the countryside on a clear night to look at the stars must have dominated the nighttimes of the earliest cultures.

For the most part we can only speculate about the astronomical knowledge of the prehistoric peoples who left no written records. We must assume that our distant ancestors in all parts of the world developed some awareness of astronomy, but we can discuss with real knowledge only those cultures which left records of their achievements. We find that parallel developments occurred in several parts of the world, including Asia, India, the Americas, and of particular importance to us, the Middle East and Mediterranean regions, wherein lie the roots of our modern astronomical science.

To develop the historical sequence leading to the present, we devote this chapter to astronomy in the vicinity of the Mediterranean, the center of the regions where the Babylonian, Egyptian, and Greek civilizations flourished (Fig. 2.1). Later in the chapter, accounts of the early Asian and American astronomies are presented, but largely for historical interest, since these cultures, by accident of geography, did not contribute to the mainstream development.

Babylonian Astronomy

In what is now Iraq, along the valleys of the Tigris and Euphrates rivers, there arose one of the earliest civilizations known to have left written records. The first people in this region were the Sumerians, who occupied the southern portion by about 3000 B.C. and invented a crude written language, *cuneiform script,* which was used to record astronomical data on clay tablets (Fig. 2.2).

The entire region was united in around 1700 B.C. under the leadership of King Hammurabi, who established his capital at the city of Babylon (not far from the present Baghdad), lending the kingdom the name Babylonia. The clay tablets from that era reveal that the movements of the planets were given great significance. Tables were constructed for astrological purposes, showing that the Babylonians knew a great deal about the various cyclical motions of the heavens. They recognized that the sun moves through the sky, and they divided the zodiac into twelve parts, probably because the year contains twelve lunar months (although not precisely twelve). The length of the year was known to within four minutes' accuracy, but the Babylonian calendar contained twelve thirty-day months, creating a discrepancy that was corrected by the addition of an extra month every few years. A numbering system was developed based on powers of 60, much as our own system is based on powers of 10, and it was the Baby-

FIG. 2.1. THE ANCIENT MEDITERRANEAN. This map shows the locations of many of the sites mentioned in the text.

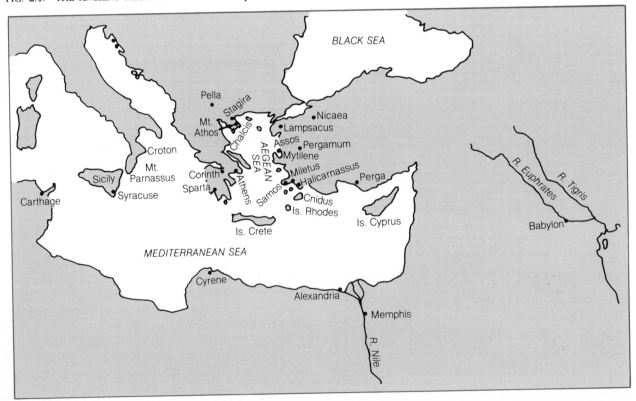

The primary motivation throughout these times was the application of astronomical information to astrology. The Babylonians seem to have speculated very little about the nature of the universe, apparently envisioning the heavens as consisting of a dome supported by the mountains surrounding their region. No evidence has been found for any theories of the motions of the heavenly bodies.

Some astronomical development occurred contemporaneously in Egypt, but little was accomplished there beyond the establishment of timekeeping techniques also known to the Babylonians, who used sundials to record local solar time. In Egypt the main purpose in studying astronomy was neither philosophical nor astrological, but was practical and religious, centered on anticipating the seasons and other events of importance to agriculture and commerce.

As the Babylonian science developed and the region underwent several successive conquests, a separate civilization began to organize itself along the shores of the Mediterranean, a civilization that was to reach the greatest heights of scientific accomplishment in ancient times. The Greek empire that arose there was greatly influenced in its early astronomical development by the Babylonians.

FIG. 2.2. A CUNEIFORM CLAY TABLET. Thousands of tablets such as this one, containing Babylonian records of astronomical data, have been uncovered in what is now Iraq.

The Early Greeks and the Development of Scientific Thought

The foundations of the Greek civilization were established some five thousand years ago in the eastern Mediterranean. There arose a seafaring culture, whose earliest home was the island of Crete, where the legends were born that gave us the names of most of our constellations. This civilization prospered around 1600 B.C. during the reign of King Minos, and it was about this time that the settlement of what is now the Greek mainland began. Much of what we know of this era is gleaned from the writings of Homer in the *Iliad* and the *Odyssey* (Fig. 2.3), written later, around 900–800 B.C. In these epic poems the early Greek view of the universe is described, and the constellations are named. The cosmology of this time consisted of the dome of the heavens, with a disk-like earth floating on water. This was similar to the Babylonian view, developed at about the same time, but the Greeks went further, hypothesizing, for what were essentially theoretical rea-

lonians who bequeathed to us our units of angular measure (360 degrees in a circle corresponding to the 360 days in their year; 60 minutes of arc per degree; and 60 seconds per minute).

The land of the Tigris and the Euphrates, known to the Greeks as Mesopotamia, saw several empires come and go, well into the first centuries A.D. The Assyrians dominated in the era around 600 to 800 B.C., and they preserved the clay tablets containing astronomical data from earlier times. Around 200 B.C., after the center of power had moved west to Greece, a massive amount of data was compiled by the then-occupants of Babylonia, the Chaldeans. The Chaldean tables were sufficiently accurate so that solar, lunar, and planetary motions could be predicted, as could the occurrence of eclipses.

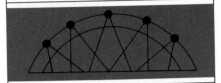

ASTRONOMICAL INSIGHT (2.1)

The Measurement of Angles

We are indebted to the Babylonians for our system of measuring angles, a system that in many ways is very cumbersome, but that historically was developed on the basis of astronomical observations.

The Babylonian calendar consisted of 12 30-day months. Although these people knew that the length of the year is actually 365¼ days, the 360 days of the calendar year led them to divide the celestial sphere into 360 degrees. It is not so clear why the Babylonians chose to subdivide the degree into 60 minutes, and the minute into 60 seconds, but this they did. The number 60 was especially significant to the Babylonians, perhaps because it was evenly divisible into 360; and at the same time 12, the number of months in a year, was divisible into it. Indeed the entire Babylonian counting system was based on the number 60, and it played much the same role that the number 10 does in our modern system.

When we speak of positions or distances in the sky, we always speak in terms of angular measures. It makes no sense to measure one star's position relative to another in miles or inches; what we see and measure is their separation in degrees, minutes, and seconds. Often we refer to angular distances as measures of arc, since we are, after all, speaking of segments of a circle. Thus, we may say that one star is 15 minutes of arc, or 15 arcminutes (15'), from another. In modern times, measurement techniques have become sufficiently accurate to allow us to speak of arcseconds, an arcsecond being only the 1,296,000th part of a circle, and much too small to be measured by the ancients.

To put these measurements into perspective, we will find it helpful to recall that the disk of the full moon, as well as that of the sun, have angular diameters of about ½°, or about 30'. The planet Jupiter, when seen near opposition, has an angular diameter of about 40". By contrast, the disk of a typical star, seen at a distance of a few hundred light years, is less than 0.05 arcseconds (0".05), which is too small to be measured, except by very modern techniques (see chapter 6).

---

sons, that there was an underworld comparable in scope and complexity to the heavens. Apparently these people had developed a sense of aesthetics and unity in the universe, which shows that their thinking went beyond mere description of the visible skies.

The first formal scientific thought evolved after the time of Homer's writing, and its inception is associated with the philosopher **Thales** (c.624–547 B.C.), who lived in Miletus, on the eastern shore of the Aegean Sea. His principal contribution was the idea that rational inquiry can lead to *understanding* the universe, going beyond describing it. This general precept was adopted by followers of Thales, who formed the so-called Ionian school of thought, named for that part of Greece (now Turkey) surrounding Miletus. One of Thales's most singular claims to fame as an astronomer was his supposed prediction of a solar eclipse that took place in 585 B.C., a feat that some historians believe did not ac-

tually occur. Much of the Greek knowledge of the motions of the heavens at that time came from the Babylonians, and it is not certain that even they were yet able to predict eclipses accurately.

Thales did create a primitive cosmology, encompassing the idea that all the basic elements of the universe were formed from water, the primeval substance. An associate of his, **Anaximander** (c.611–546 B.C.), altered this view and adopted the idea of a universal medium, of unknown properties, from which all material substance was made. Anaximander also developed the notion that outside the earth the firmament is filled with flame, contained in rotating tubes with holes that allowed the light within to be seen (Fig. 2.4). Apparently it was his concept that the sun, moon, and stars were openings in the sides of these hoops, so that their rotation created the observed motions of the heavenly bodies.

THE PYTHAGOREANS

Pythagoras (c.570–500 B.C.) was born on the island of Samos, off the Aegean coast not far from Miletus. His early life is obscure, but at around the age of forty, following several years of travel and study, he established a school at Croton, on the southern tip of Italy, in which scientific and religious teachings were intermixed. The lasting contribution of this school was the notion that all natural happenings can be described by numbers, laying the foundation for geometry and trigonometry. Pythagoras himself is thought to have been the first to assert that the earth is round and that all heavenly bodies move in circles, ideas that never lost favor thereafter in ancient times.

Most of what we know about Pythagoras, who himself left no written records, comes from the works of his follower **Philolaus** (c.500–400 B.C.), who added the idea that the earth moves about a central fire (not the sun), always hidden on the far side of the earth. Both he and Pythagoras believed the planetary distances to correspond to the lengths of vibrating strings that produced harmonious musical notes. This led to the concept of a celestial harmony which later had considerable influence in philosophical thinking.

There were numerous followers of the Pythagorean school who in various ways refined this view of the universe. One of particular note was **Anaxagoras** (c.500–428 B.C.), who is credited with the realization that the moon shines by reflected sunlight, rather than by its own power. This awareness allowed him to correctly attribute solar and lunar eclipses to the passage of the moon in front of the sun and through the earth's shadow, respectively.

PLATO AND HIS FOLLOWERS

On the Greek mainland a number of democratically ruled city-states arose, of which Athens became preeminent by about the fifth century B.C., dominating artistic and literary affairs. Even after the military defeat in 404 B.C., at the hands of the rival city of Sparta, Athenian culture continued to flourish. It was in this environment that **Plato** (428–347 B.C.; Fig. 2.5) reached maturity, having studied philosophical thought under the tutelage of Socrates.

After some years of traveling, Plato established an academy outside of Athens, in about the year 387 B.C.,

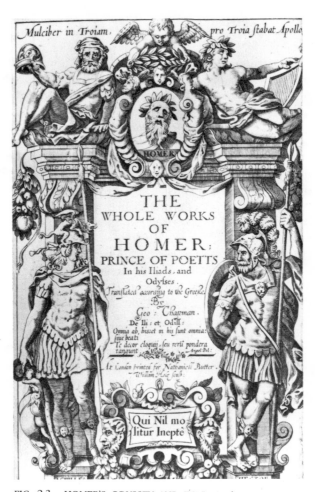

FIG. 2.3. HOMER'S ODYSSEY AND ILIAD. In these two epic poems, Homer described the much more ancient legends of astronomical lore from the civilization that had flourished on the isle of Crete.

FIG. 2.4. HOOPS OF ANAXIMANDER. In this construction, the sun and the planets are envisioned to be holes in fire-filled tubes that rotate about the earth.

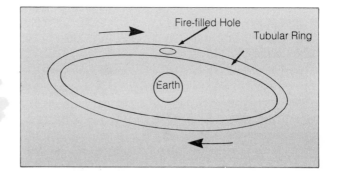

where he taught his ideas of natural philosophy. His fundamental precept was that what we see of the material world is only an imperfect representation of the ideal creation. The corollary of this doctrine was that we can learn more about the universe by reason than by observation, since, after all, observation can give us only an incomplete picture. Hence Plato's ideas of the universe, described in his *Republic,* were based on certain idealized assumptions that he found reasonable. One of the most important was that all motions in the universe are perfectly circular and that all astronomical bodies are spherical. Thus he adopted the Pythagorean view that the sun, moon, and planets move in various combinations of circular motion about the earth. It seems Plato thought of these objects as affixed to clear, ethereal spheres that rotated. The teachings of Plato, particularly the premise that the mysteries of the cosmos should be solved through reason rather than by direct observation, dominated much of western thinking until the Renaissance, nearly two thousand years after his time.

A pupil of Plato's was **Eudoxus** (c.408–356 B.C.), who made the first attempt to put Plato's view of the universe on a firm mathematical footing; he constructed a series of concentric spheres on which the sun, moon, and planets moved in perfect circular motions (Fig. 2.6). Each body required three or four spheres, each with its own rotation rate, to produce the composite motion necessary to duplicate the overall motion. A total of twenty-seven spheres was needed for all the known heavenly bodies. It is not clear whether Eudoxus thought of these spheres as material objects or merely mathematical constructs.

The most renowned student of Plato was **Aristotle** (c.384–322 B.C.) who, after the death of Plato in 347 B.C., established his own small academy across the Aegean Sea at Assos. He then moved to Pella, where he tutored Alexander, the son of Phillip of Macedonia, king of the Greek empire. Aristotle later created a school in Athens, where he taught his own view of the universe, one formed from Plato's mold of reason rather than observation.

Aristotle was the first to adopt physical laws and then to show why, in the context of those laws, the universe works as it does. He taught that circular motions are the only natural motions and that the center of the earth is the center of the universe. He also believed that the world is composed of four elements: earth, air, fire,

FIG. 2.5. PLATO.

and water. He demonstrated, in the context of his adopted physical laws, that both the universe and the earth are spherical. He had three ways of proving that the earth is spherical: (1) only at the surface of a sphere

FIG. 2.6. THE SPHERES OF EUDOXUS.

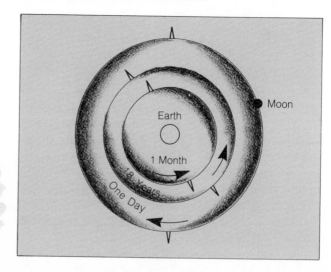

ASTRONOMICAL INSIGHT (2.2)

The Mythology of the Constellations

As noted in the text, the writings of Homer in the *Iliad* and the *Odyssey,* which appeared around 900–800 B.C., contain descriptions of the constellations and their meanings. These descriptions were based on legends from the Minoan civilization that were already ancient in Homer's time.

Little precise information is available as to when and how the mythology of the constellations arose. However, it has been possible to deduce roughly the era that gave birth to them. Careful examination of the ancient constellations shows them to be distributed symmetrically about a point in the sky that was at the north celestial pole around 2600 B.C. (but it has since shifted away from the pole because of precession), so it is probable that the constellations were invented at about that time. There are few ancient constellations near the southern pole, in regions not visible from the latitude of Crete, a fact that gives independent support to the supposition that the legends arose in the Minoan culture.

It is likely that the constellations were not taken as literally as is usually supposed. Rather than being considered faithful depictions of the people and events with which they are associated, the constellations should more properly be viewed as symbolic representations. They were probably not originally designated on the basis of their imagined resemblance to certain characters; instead, areas of the sky were dedicated in honor of prominent figures of mythology, and the familiar pictures of these figures were then fitted to the patterns of bright stars. This helps account for the lack of obvious resemblance between the star patterns and the figures and events they supposedly represent.

The constellation names were translated from the Greek of Homer to Latin when the Roman Empire rose to dominance, and the names with which we are familiar today are the Latin ones. Interestingly, the names of prominent stars went through another transformation, into

Arabic, and today's star names are Arabic designations which usually are literal indications of the place these stars hold in their constellations. Betelgeuse, for example, the name of the prominent red star in the shoulder of Orion, is translated as "the armpit of the giant" or "the armpit of the central one." Thus, in today's standard usage, we adopt Arabic names for stars in constellations with Latin designations, although both translations are based on Greek descriptions of the ancient Minoan legends.

Ancient and nonscientific though they are, the constellations have a significant impact on the nomenclature of modern astronomy. The modern constellations refer to very specific regions of the sky with well-defined boundaries that have been agreed on by the international community of astronomers. The "official" constellations are based on those of legend, containing within them the ancient figures of mythology, but their boundaries have been extended so that every part of the sky falls within one constellation or another. A map of the constellations looks a bit like a map of the western United States, where the boundaries are generally straight lines, but the shapes are irregular. As an alternative to the Arabic names for the brightest stars, modern astronomers often use designations based on a star's rank within its constellation, with letters of the Greek alphabet used to indicate the brightness rank. Thus Betelgeuse, the brightest star in Orion, is also called α Orionis.

do all falling objects seek the center by falling straight down (it was another premise of his that falling objects are following their natural inclination to reach the center of the universe); (2) the view of the constellations changes as one travels north or south; and (3) during lunar eclipses it can be seen that the shadow of the earth is curved (Fig. 2.7). By relating his theories to observation in this manner, Aristotle broke with the tradition of Plato to some extent, although, as we have seen, he still approached the problem in the same manner, letting reason rather than observation guide the way.

Other tenets of Aristotle included the conclusion that the universe is finite in size (this led directly to his belief that the heavenly bodies can follow only circular motions, because otherwise they might encounter the edge of the universe), and that the heavenly bodies are made of a fifth fundamental substance, which he called the aether.

THE GEOMETRIC GENIUS OF ARISTARCHUS AND ERATOSTHENES

After Aristotle's period, the center of Greek scientific thought moved across the Mediterranean Sea to Alexandria (the capital city established in 332 B.C. by Aristotle's former pupil, Alexander the Great), near the site of the present-day city of Cairo.

The most prominent astronomer of this era, however, was born on the Mediterranean island of Samos, and it is not well known where he lived and worked, although it may have been in Alexandria. This was **Aristarchus** (c.310–230 B.C.), the first scientist to adopt the idea that the sun, not the earth, is at the center of the universe. This was some seventeen hundred years before the time of Copernicus, who usually is given credit for this discovery. The heliocentric hypothesis of Aristarchus failed to attract many followers at the time because of a lack of concrete evidence that the earth was in motion, and a general satisfaction with the Aristotelian viewpoint, which had no recognized flaws.

It is interesting to consider how Aristarchus reached his revolutionary idea. By geometrical means he was able to deduce the relative sizes and distances of the moon and the sun; and although his results were not very accurate, he did correctly conclude that the sun is far larger than either the moon or the earth. Accepting this, he found it difficult to imagine that the sun should

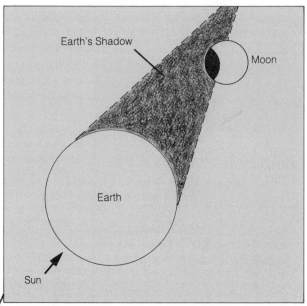

FIG. 2.7. CURVATURE OF THE EARTH'S SHADOW ON THE MOON. Only a spherical body can cast a circular shadow for all alignments of the sun, moon, and earth.

TABLE 2.1. ANCIENT GREEK ACHIEVEMENTS		
DATE	PHILOSOPHER	DISCOVERY OR ACHIEVEMENT
c.900–800 B.C.	Homer	*Iliad* and *Odyssey;* summaries of legends
c.624–547 B.C.	Thales	Rational inquiry leads to knowledge
c.611–546 B.C.	Anaximander	Universal medium; primitive cosmology
c.570–500 B.C.	Pythagoras	Mathematical representation; round earth
c.500–400 B.C.	Philolaus	Earth orbits central fire
c.500–428 B.C.	Anaxagoras	Moon reflects sunlight; eclipses explained
c.428–347 B.C.	Plato	Material world imperfect; learn by reason
c.408–356 B.C.	Eudoxus	First mathematical cosmology
c.384–322 B.C.	Aristotle	Concept of physical laws; round earth
c.310–233 B.C.	Aristarchus	Size of sun; first heliocentric theory
c.273–? B.C.	Eratosthenes	Size of earth
c.265–190 B.C.	Apollonius	Introduction of the epicycle
c.200–100 B.C.	Hipparchus	Many astronomical firsts; epicyclic theory
A.D. c.100–200	Ptolemy	*Almagest;* elaborate epicycle model

orbit the lesser earth, so he adopted the contrary point of view.

A younger contemporary of Aristarchus was **Eratosthenes** (c.273 B.C.–?), who worked in Alexandria during his long career of research in geometry and astronomy. His greatest achievement was the determination of the size of the earth, by a method of great simplicity (Fig. 2.8). It was known that on a certain day in the summer, sunlight penetrated all the way to the bottom of a deep well located at Syene. On the same day in Alexandria, the sun's midday position was some 7° south of the zenith, indicating that the distance between the two cities corresponded to 7° of the 360° circle of the earth's circumference. Hence the circumference was equal to 360°/7° times the distance between Alexandria and Syene, which was a little less than 5,000 stades, a *stade* being a unit of measure equal to about one-tenth of a mile. The circumference of the earth, as measured by Eratosthenes, was 252,000 stades, or just under 25,000 miles—within 2 percent of the correct value!

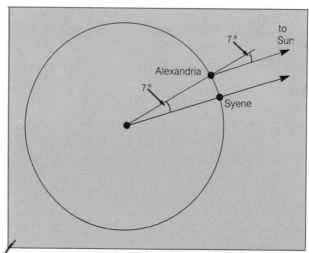

FIG. 2.8.   THE METHOD OF ERATOSTHENES FOR MEASURING THE SIZE OF THE EARTH.

GEOCENTRIC COSMOLOGIES: HIPPARCHUS AND PTOLEMY

By the third century B.C. the need for more precise mathematical models of the universe became apparent, as better observing techniques were developed. A mathematical concept that provided the needed precision while still preserving the precepts of Aristotle was the **epicycle** (Fig. 2.9). It was **Apollonius** (c.265–190 B.C.) who put the epicycle on a firm geometric footing, providing the first mathematically viable alternative to the idea of spheres developed by Eudoxus. An epicycle is a small circle on which a planet moves, whose center in turn orbits the earth following a larger circle called a **deferent.** This construct allows a combination of motions whose speeds and circular radii can be adjusted as needed to provide a close match with the observed motions. The epicycle provided a clear advantage for explaining the retrograde motions of the superior planets, something that confounded Eudoxus's theory that the planets move on rotating spheres.

The epicyclic motions of the celestial bodies were refined further by **Hipparchus** (Fig. 2.10, who was active during the middle of the second century B.C. (very little is known about his life, even the dates of his birth and death). Hipparchus, most of whose work was done at his observatory on the island of Rhodes, was one of

the greatest astronomers of antiquity. Among his major contributions were: (1) the first use of trigonometry in astronomical work (in fact, he is largely credited with its invention, although many of the concepts were developed earlier); (2) the refinement of instruments for measuring star positions, along with the first known use of a celestial coordinate system akin to our modern equatorial coordinates, enabling him to compile a catalogue of some 850 stars; (3) the refinement of the

FIG. 2.9.   THE EPICYCLE. Apollonius realized that planetary motions could be explained by a combination of motions involving an epicycle, which carries a planet with it as it spins while orbiting the earth.

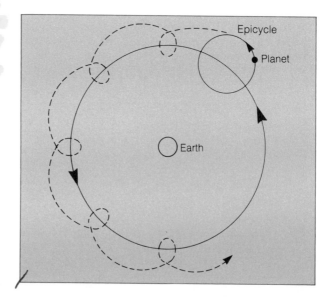

FIG. 2.10. HIPPARCHUS AT HIS OBSERVATORY.

FIG. 2.11. STELLAR PARALLAX. As the earth orbits the sun, our line of sight toward a nearby star varies, causing the star's position (with respect to more distant stars) to change. The change of position is *greatly* exaggerated in this sketch; in reality, even the largest stellar parallax displacements are too small to have been measured by ancient astronomers.

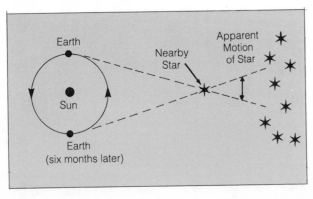

methods of Aristarchus for measuring the relative sizes of the earth, moon, and sun; (4) the invention of the stellar magnitude system for estimating star brightnesses, a system still in use today (with minor modifications); and, perhaps most impressively, (5) the discovery of the precession of star positions, which he accomplished by comparing his observations with some that were made 160 years earlier.

Although Hipparchus did not add much that was new to the then-current models of planetary motions, he did compile an extensive series of observations of the solar motion. This led him to a refinement of the epicycle theory; namely, that the earth was off-center in the large circle on which the sun moved. This idea of an **eccentric,** or off-center, orbit accounted rather accurately for the observed variations of the speed of the sun in its annual motion.

It is worthwhile to consider why Hipparchus, with access to the work of Aristarchus (including the knowledge that the sun is much larger than the earth), did not adopt the heliocentric hypothesis. Apparently he rejected this idea because of one very significant observation: he could not detect any shifting of star positions during the course of the year, shifting that he realized should appear as a result of the changing point of view if the earth moved about the sun (Fig. 2.11). The lack of any detectable **stellar parallax,** as such a motion is called, caused many to resist the heliocentric viewpoint for many centuries to come.

After the great work of Hipparchus, almost three hundred years passed before any significant new astronomical developments occurred. Claudius Ptolemaeus (Fig. 2.12), or simply **Ptolemy,** lived in the middle of the second century A.D... Attempting to summarize all the world's knowledge of astronomy, he published a treatise called the *Almagest,* which was based in part on the work of Hipparchus, but which also contained some new developments of his own. The thirteen books of the *Almagest* range from a summary of the observed motions of the planets to a detailed study of the motions of the sun and moon, and from a description of the workings of all the astronomical instruments of the day to a reproduction of Hipparchus's star catalogue. Most important, in his treatise, however, are the detailed models of the planetary motions. Here Ptolemy made his greatest personal contribution, for these models were so accurate in their ability to predict the planetary positions that they were used for the next thousand years.

To provide this accuracy, Ptolemy had to adopt and extend an idea from Hipparchus, that of the eccentric (Fig. 2.13). To satisfactorily account for all the observed irregularities in a planet's motion, Ptolemy assumed not only that the center of the deferent was off-center from the position of the earth, but also that the rate of motion of the epicycle as it orbited the earth was constant as seen from another off-center point, on the opposite side of the earth from the geometrical center of the deferent. This combination of offsets, with properly chosen dimensions and rotational speeds, allowed good representation of the observed motions while preserving the hypothesis that all the heavenly motions are circular. It is interesting that other doctrines of Aristotle, such as the precise centering of all the planetary orbits on the earth, were not strictly followed.

It seems from some of Ptolemy's writings that he may not have considered his model of the universe to represent physical reality, but instead may have viewed it as simply a mathematical tool for predicting planetary positions. This is something modern scientists do when confronted with data on phenomena for which the underlying physical principles are not clear, and in this sense, Ptolemy may be considered as one with modern astronomers. Whether Ptolemy believed his model to represent physical reality or not, it was taken literally in the centuries following his time. The notion of a geocentric universe with only circular motions dominated for some fourteen centuries afterward, and was expunged only after great turmoil.

The history of Greek astronomy came to an end with the work of Ptolemy, as did most significant astronomical development in Europe and western Asia until the time of the Renaissance. We will return to pick up the thread of astronomical knowledge and growth during the Dark Ages, after we digress and consider what was going on in the rest of the world.

Early Astronomy in Asia

CHINESE ACCOMPLISHMENTS

One of the most ancient recorded astronomies outside the Middle East arose in China. Legends there refer to astronomical activities as far back as 2000 B.C., although the oldest writings are much more recent.

FIG. 2.12. CLAUDIUS PTOLEMY.

FIG. 2.13. THE COSMOLOGY OF PTOLEMY. To account for nonuniformities in the planetary motions, Ptolemy devised a complex epicyclic scheme. The deferent, the large circle on which the epicycle moves, is not centered on the earth, and has a rotation rate that is constant as seen from another displaced point called the equant. This meant that from the earth, the planet appeared to move faster through the sky when on one side of its deferent than when on the other. Each planet had its own deferent and epicycle, with motion and offsets adjusted to reproduce the observations.

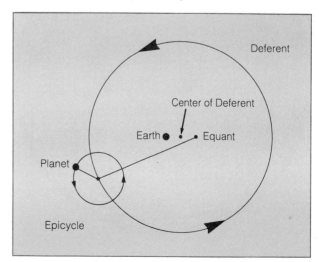

Ancient China had a tumultuous history, with periods of peace, prosperity, and cultural advance occurring throughout numerous centralized dynasties, interspersed with eras of disarray resulting from barbarian invasions and civil unrest. Historical records were sometimes lost during the transitions; thus it is not known just how far in the past the first significant astronomical activity began. Legend refers to a pair of astronomers named Hi and Ho, who, having failed to prevent a solar eclipse that occurred in 2137 B.C., were put to death.

We know that timekeeping was well established in China by the fourth century B.C., since records from that era indicate that the length of the year and the lunar month were known with considerable accuracy. The principal roles of astronomers were to keep the time and to officiate over state functions, ensuring that all was done in accordance with the rules of the gods. Astrology played a major role, and the constellations were identified with the emperor and with government organization. The north polar region of the sky, for example, was thought to represent the seat of the emperor, and within a constellation, stars were given royal rank according to their brightness. A catalogue of

FIG. 2.14. ANCIENT CHINESE DEPICTION OF THE CELESTIAL SPHERE.

some 809 stars was constructed in the fourth century B.C. by the astronomer Shih-shen, predating by about two hundred years Hipparchus's catalogue.

The prevalent concept of the universe, however, was not nearly as advanced as that of the Greeks. The cosmos was thought to consist simply of the sphere of the sky with its daily rotation, and little significance was attached to the motion of the planets. The poles were considered to be the exalted regions of the sky, whereas the celestial equator and the ecliptic seem to have been given no great importance. Heaven and earth were thought to be closely related, with events in one domain affecting those in the other; thus, irregularities in the heavens were taken as omens of calamities on earth.

In later centuries observations became more accurate, with the employment of simple measuring devices. By the first century A.D., accurate tables of lunar motions had been constructed and eclipses could be predicted. Apparently precession was known to the Chinese astronomers of this era as well.

Contact with the western world developed during the first few centuries A.D., as a result of Chinese conquests to the west and the eastward diffusion of Greek culture, which reached China by way of India. From this time on, it is difficult to separate independent Chinese developments from those introduced from the west, but the records indicate that numerous important advances were made, particularly in the improvement of measuring devices.

Records of celestial events were kept continuously until medieval times, providing later generations with important data to be found nowhere else. Among the events mentioned in the Chinese records is the appearance of a ''guest star'' in A.D. 1054 (Fig. 2.14), identified in modern times with the supernova explosion that created the Crab Nebula.

Because of China's remote location as seen from Europe, astronomical developments there had little influence elsewhere, except for the maintenance of observing records during the Dark Ages in Europe, when no one else was recording what was seen in the skies.

HINDU ASTRONOMY

In India, as in China, it is evident that some astronomical lore had developed by very early times, but our only knowledge of the oldest accomplishments is based on ancient legends. There are written records of lunar motions dating back to about 1500 B.C., and by 1100 B.C. a calendar had been established, based on the solar year. Apparently all the prominent celestial bodies were named for gods or goddesses and were thought to represent their character.

Astronomy in India flourished during and immediately after the time of the great prince Rama, who was born in the year 961 B.C. During this period tables of planetary motions were constructed for use in astrology, since the Hindus thought that heavenly occurrences represented earthly affairs. A very complex calendar was adopted that was based on an observed 247-year cycle of solar and lunar motions of great significance. The construction of this calendar, as well as the discovery of precession, show that the Hindu astronomers made use of positional records that must have been maintained over hundreds of years.

Sometime in the first millennium B.C. there was established cultural contact between India and the Middle Eastern civilizations of the Babylonians and the Greeks, and it is difficult to be certain of the direction in which information flowed. Some people claim that the entire basis of the Babylonian and Greek astronomical development was borrowed from India in about the eighth century B.C., whereas others state that the developments in the two regions were quite independent. In any case, it is clear that by A.D. 500, western influence dominated Hindu astronomy, and that few new developments took place in India after that time.

Ancient Astronomy in the Americas

A modern appreciation of ancient astronomical developments in the new world has been slow in coming. The principal reasons for this were the near-complete destruction of the writings of the Mayan civilization at the time of the Spanish conquest, and the lack of a written language of the North American natives.

Archeological studies have revealed that in Central and South America, sophisticated cultures existed before the time of Columbus. The ancient cities of the Mayan and Incan civilizations, now in ruins, show highly complex and organized structures requiring sound knowledge of astronomical phenomena. It has been discovered that in some cases, alignments of principal buildings and avenues were designed to point toward significant astronomical locations, such as the rising point of the sun on the day of its passage through the zenith. A great deal of what we know about astronomy in these ancient cultures is read from the record of their architecture, rather than from their writings.

The center of the Mayan civilization was located in what is now Mexico and several Central American nations to the south, in a region known today as Mesoamerica. Other cultures in this region, which flourished between 500 B.C. and A.D. 1500, were the Olmec, Zapotec, and Aztec.

As in other ancient cultures, the principal motivations behind the Mesoamerican astronomy were timekeeping and astrology. The Mayans developed a very accurate calendar, and measured with great precision the motions of the celestial bodies. A particular significance was associated with the planet Venus, which inspired sacrificial rites and other ceremonies. The motions of Venus were extensively tabulated in a surviving document called the *Dresden Codex* (Fig. 2.15).

In some locations special astronomical observatories were constructed, usually with many significant alignments made in the directions of important celestial events. The most famous is the Caracol tower at Chichén Itzá, in the Yucatán, a building whose strange architecture defied explanation until the numerous astronomical sightlines it entails were discovered in the twentieth century (Fig. 2.16).

The Mayan cosmology apparently consisted of a layered universe, each level of which contained one type of celestial body. Above the earth was the domain of the moon, then the clouds, the stars, the sun, Venus, comets, and so on. Similarly, there was thought to be an underworld consisting of nine levels, providing symmetry with the heavens, an idea parallel to that of the early Greek cosmologies described by Homer.

In South America, in the Andean territory where modern-day Peru lies, the Incan civilizations flourished in the centuries before the Spanish conquest. Here, too, great attention was paid to astrological omens, and much of the astronomical effort was devoted to the construction of accurate timekeeping systems. The Incans had a rather complex calendar that consisted of twelve months, but it was otherwise quite unlike the modern calendar, not even containing the same number of days in a week that we have today. The Incans, like their Measoamerican neighbors to the north, often laid out their temples and cities in accordance with astronomi-

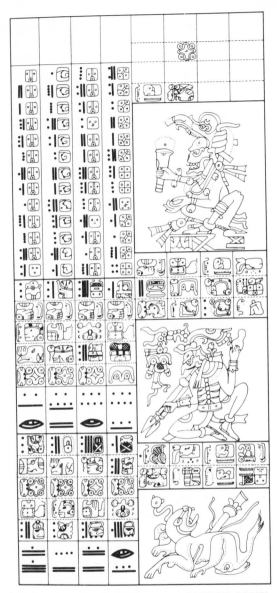

FIG. 2.15. THE VENUS TABLES IN THE DRESDEN CODEX (MAYA). These tables show that the Mayans attached particular significance to the planet Venus.

FIG. 2.16. THE CARACOL TOWER AT CHICHÉN ITZÁ.

artifacts such as mounds, religious symbols and artwork (Fig. 2.17). It seems evident, however, that the Indians of North America were intimately familiar with the skies and significant events such as the solstices that foretold the changing of the seasons. Among the few records that do exist are illustrations either carved or painted on rocks (Fig. 2.18), found in the Southwest, some of which may depict the appearance of the supernova of A.D. 1054. This was the same event observed by the Chinese, but except for the possible native-American sighting, it was not mentioned in the records of any other cultures. That the petroglyphs do represent the supernova of 1054 is by no means established, however.

FIG. 2.17. PAWNEE INDIAN SKY MAP. This chart, embossed on hide, appears to depict constellations of the northern hemisphere.

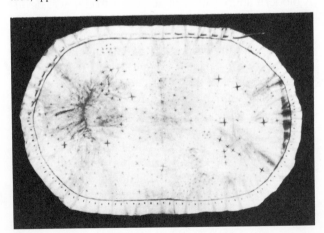

cal reference points, in at least one case using markers set out over an entire valley as a framework for computing dates.

Astronomy practiced by North American natives in ancient times is much more difficult to assess, there being no written language and few records of any kind left, with the exception of scattered rock drawings and

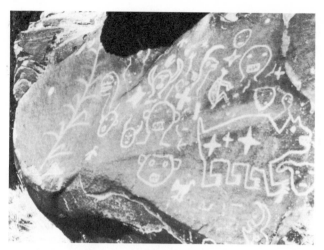

FIG. 2.18.   ROCK DRAWINGS BY NORTH AMERICAN NATIVES.

FIG. 2.19.   THE BIGHORN MEDICINE WHEEL, BIG HORN MOUNTAINS, WYOMING. This is a prominent and well-studied example of a North American Indian medicine wheel.

Some structures survive that appear to have astronomical significance, including the sun temple at Mesa Verde and other ruins in the Southwest that may have been influenced by the Mesoamerican culture to the south. In addition there are earthen mounds of various descriptions located in many parts of the Midwest, and some seem to represent celestial events or follow astronomically significant alignments.

In the American West there are several circular monuments marked out in rock, called medicine wheels. These were probably left behind by the Plains Indians, and they show astronomical influences to varying degrees. Among the best studied is the Bighorn medicine wheel, located high in the Big Horn Mountains of Wyoming (Fig. 2.19). This is a circular structure some 90 feet in diameter, with rock cairns around its perimeter. Lines traced through these piles of rock coincide with a number of astronomically significant directions, including a primary line pointing in the direction of the rising sun on the morning of the summer solstice.

Unfortunately, it is impossible to be certain when these monuments were constructed, so we do not know just how long ago our predecessors on this continent developed their understandings of the heavens.

The Mystery of Stonehenge

By now we are aware that some form of astronomical influence affected nearly every ancient culture. The timing of major developments seems to have been sur-

prisingly similar in most parts of the world. The earliest records of accurate positional measurements and the establishment of calendars often date from the second millennium B.C., with sufficient sophistication for predictions of eclipses and planetary motions not developing until the last few centuries B.C.

It is most impressive, then, to consider the stone circle monuments of the British Isles, for they predate recorded astronomical achievements in the rest of the world, yet they show that their builders had substantial knowledge of astronomical skills, possibly including the prediction of eclipses.

Of the nine hundred or so of these monuments, by far the best known is *Stonehenge,* located on Salisbury Plain, about seventy miles west of London. This is a large and complex monument characterized by immense upright stones arranged in various circular patterns, in some instances with horizontal lintels connecting the vertical members (Fig. 2.20).

Little is known about the builders of Stonehenge. The monument was built in three stages, the first commencing about 2800 B.C., and the most recent being completed by 2075 B.C. The sophisticated astronomical principles that guided the construction were incorporated from the beginning, even though it appears that the people inhabiting the region changed from the original Stone Age civilization to a new group, called the Beaker People, who arrived from what is now the Netherlands in approximately 2500 B.C. Subsequent

FIG. 2.20. STONEHENGE. This view is to the northeast from the center, looking toward the Heelstone.

stages were built by these later arrivals on the scene, who apparently embraced the logic that had led to the original construction.

It is not known how the stones, which weigh up to fifty tons, were moved to Salisbury Plain. Some came from Wales, possibly by sea, and many came from a region some twenty miles distant. The monument is now in a state of partial ruin and disarray, with some stones missing. Besides the large stones, there are a number of holes that may have once contained wooden posts to mark certain directions. Even today new parts of Stonehenge are occasionally found, in the form of postholes or distant mounds that may have been used to mark significant alignments.

There is no question that Stonehenge was laid out according to astronomical principles (Fig. 2.21). There is substantial controversy, however, about the details, and despite a great deal of publicity in the 1960s over interpretations of Stonehenge as an early computer used to predict eclipses, today there is little agreement as to what is the correct interpretation.

One axis of Stonehenge, delineated by the sight line to the Heelstone from the center of the monument, aligns with the position of sunrise at the summer solstice. However, because of precession the alignment is better now than it was in 2800 B.C., when the first stage was built (this has led at least one student of Stonehenge to hypothesize that this axis was meant to signify the midwinter rising of the full moon, with which the alignment was better). A circle of fifty-six holes around the center of Stonehenge has also attracted a great deal

of attention; some people thought it possible that the holes were used to predict eclipses, although it appears unlikely.

Whatever the correct detailed interpretation of Stonehenge, the fact remains that extensive knowledge of astronomical phenomena was available to its builders as early as 2800 B.C., some two thousand years before other cultures such as the Babylonians are known to have developed such a high level of sophistication. It is unfortunate that we do not know more about the remarkable people who built it.

Astronomy in the Dark Ages

Let us now pick up the main historical thread, with a brief mention of events after the time of Ptolemy. The Christian influence began to be felt throughout what had been the Greek empire, turning people away from the belief that astronomical events dominated their fates, and leading to the dissipation of efforts to improve the understanding of the heavens. The prophet Mohammed was born in the mid-seventh century A.D., and within a century, his followers had conquered most of the Mediterranean and some of southern Europe.

The need for accurate calendars caused some revival of interest in astronomy, and the Arabs resurrected a great deal of the Greek knowledge. During this period our modern numbering system, brought by the Arabs from India, was established, and new mathematical techniques, notably algebra, were developed. Although some new astronomical observations were made by Arab scientists, their principal contribution was to keep alive the Greek knowledge of astronomy and to pass it on to the western European civilization in the twelfth century, through Spain.

In the thirteenth century, Spanish King Alfonso X ordered the construction of new tables of planetary motions, the old tables of Ptolemy finally having begun to show inadequacies. The new tables, derived by a group of astronomers using the epicyclic model of Ptolemy, became known as the *Alfonsine Tables,* and were used for the next three hundred years.

Meanwhile, the philosophical teachings of Aristotle, adhered to by the Islamic astronomers, were adopted by the scientists of the Christian world as well. Thus as the sixteenth century dawned, European astronomy was based almost entirely on the teachings and observations of the ancient Greeks.

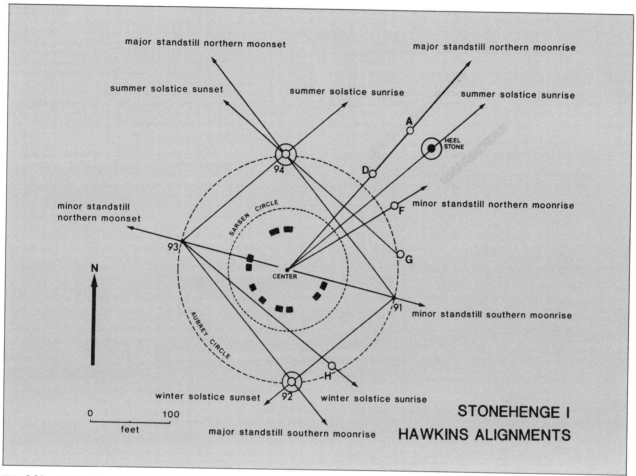

FIG. 2.21. POSSIBLE ASTRONOMICAL ALIGNMENTS OF STONEHENGE. This sketch shows several astronomically significant lines of sight as Stonehenge, identified by G. Hawkins.

Perspective

We have traced the development of astronomy, beginning with mankind's earliest attempts to understand the world around him. These attempts were based on fear of the unknown and the primitive belief that events in the heavens ruled the inhabitants of the earth. This kind of astrological outlook emerged independently in many parts of the world, yet only in the vicinity of the Mediterranean, with the rise of the Greek philosophers, did there exist the intellectual stimulus to understand the universe just for the sake of understanding. Political, economic, and most of all, religious developments then brought about a prolonged intercession, when no real progress was made. Nearly fourteen hundred years after the death of Ptolemy, the first stirrings of intellectual revolution began to be felt.

Summary

1. The earliest recorded astronomical data were found in the region known as Babylonia, now Iraq.

2. The first rational inquiry and the earliest cosmological theories arose in the ancient Greek empire, on the shores of the Mediterranean.

3. From 570 B.C., Pythagoras and his followers developed the notion that the universe can be described by numbers, and adopted the belief that the earth is spherical.

4. Plato, and his pupil Aristotle, stated underlying principles (which could not be tested by observation) that they believed to govern the universe; their principles included the beliefs that the earth is the center of the universe and that all heavenly bodies are spherical.

5. In the third century B.C., Aristarchus used geometrical arguments to conclude that the sun, not the earth, is at the center of the universe.

6. During the same period, Eratosthenes made an accurate measurement of the size of the earth, by comparing the directions toward the sun from two widely separated points on the earth's surface.

7. Hipparchus, in the second century B.C., made precise observations, applied new mathematical techniques to astronomy, compiled a large star catalogue, developed a system for measuring star brightnesses, discovered precession, and developed the epicyclic theory of planetary motion.

8. Ptolemy, in the second century A.D., summarized all astronomical knowledge in the *Almagest,* and used the epicyclic theory to construct tables of planetary motion that were used throughout the Dark Ages.

9. Astronomy in ancient China and India led to significant observational achievements such as the compilation of large star catalogues and the development of accurate calendars, but did not lead to sophisticated cosmological theories.

10. Astronomy in the Americas was highly developed, but few records were left other than astronomical alignments of buildings and monuments.

11. The great megaliths of Stonehenge were constructed along sophisticated astronomical alignments, and may even have been used to predict eclipses millennia before these things were known in the rest of the world.

Review Questions

1. The Babylonians adopted a 360-day year, divided evenly into 12 months. Since the year is actually 5¼ days longer than this, how often did they have to add an extra month, in order to make things come out evenly? How do you suppose they allowed for the remaining error that accumulated over many years?

2. Calculate how many minutes of arc there are in a 90° angle. How many seconds of arc?

3. What was the significance for later scientific development of the ancient Greek concept, described in Homer's writings, of an underworld?

4. If the sun had been more than 7° from the zenith as seen from Alexandria when it was overhead at Syene, would the size of the earth as calculated by the method of Eratosthenes have been larger or smaller than the size he found?

5. Even the best measuring instruments at the time of Hipparchus could only measure angular separations or positions with an accuracy of several arcminutes. If you assume that the error or uncertainty of a single measurement was 40', how many years of observation would be required in order for precession, which occurs at a rate of around 50" per year, to be noticed? How much precession had occurred over the 160-year period covered by the observational data available to Hipparchus?

6. Explain how the epicyclic theories of Hipparchus and Ptolemy satisfied the principles of Plato and Aristotle, yet in certain ways violated some of them as well.

7. Although Aristotle was a follower of Plato, he developed some philosophical views that were different from the teachings of his mentor. Summarize these differences in outlook.

8. In deducing the size of the earth, Eratosthenes made the assumption that the sun's rays are parallel when

they reach the earth. What was the implication of this in terms of the distance of the sun from the earth?

9. Why did Hipparchus, who was aware of the work of Aristarchus, not accept the idea that the earth orbits the sun?

10. Compare the achievements of the ancient Chinese astronomers with those of the Greeks, and explain why we say that the Greeks have had more influence on the development of modern astronomy.

Additional Readings

Here are some readings on ancient astronomical history. Others may be found by looking up references cited in these articles and books. It is also useful to browse through journals, such as *Vistas in Astronomy* and the *Journal for the History of Astronomy,* both of which can be found in most science-oriented libraries.

Heath, T. L. 1969. *Greek astronomy.* New York: AMS.

Krupp, E. C. 1978. *In search of ancient astronomies.* Garden City, N.J.: Doubleday.

———. 1983. *Echoes of the ancient skies.* New York: Harper and Row.

Neugebauer, O. 1969. *The exact sciences in antiquity.* New York: Dover.

·CHAPTER 3·

The Renaissance

The fifteenth century saw the beginnings of a reawakening of intellectual spirit in Europe. Some scientific studies began at the major universities, increased maritime explorations brought demands for better means of celestial navigation, and the art of printing was discovered, opening the way for widespread dissemination of information.

Copernicus:
The Heliocentric View Revisited

Some nineteen years before the epic voyage of Columbus, Niklas Koppernigk (Fig. 3.1) was born in Toruń, in the northern part of Poland. As a young man he attended the university in Cracow, where his entrancement with Latin, the universal language of scholars, led him to change his surname to Copernicus. At Cracow he nourished an abiding interest in astronomy, becoming fully acquainted with the Aristotelian view as well as the Ptolemaic model of the planetary motions.

Copernicus continued his academic pursuits in Italy, studying ecclesiastical law at Bologna and Ferrara, and medicine at Padua, before returning to Poland to assist his uncle, the bishop (and therefore chief government administrator) of a region called Ermland. During all this time he persisted in his studies of astronomy, and it is known that, by 1514, he had developed some doubts about the validity of the accepted system.

His reasons for doing so have been the subject of some uncertainty and misconception. It was long assumed that Copernicus was encouraged to adopt the sun-centered view of the universe because he recognized shortcomings in the geocentric model of Ptolemy. There is, however, no evidence of widespread dissatisfaction with the Ptolemaic system, nor any records indicating that Copernicus himself found serious inaccuracies in it. His reasons for adopting the heliocentric viewpoint were more subtle.

The basis for his conversion was primarily philosophical. The new system, in the mind of Copernicus, presented a pleasing and unifying model of the universe and its motion. Although he was, no doubt, encouraged by the climate of change and cultural revolution that was sweeping Europe with the advent of the Renaissance, he did not adopt the heliocentric hypothesis just to be different, nor did he do it to improve the accuracy or reduce the complexity of the accepted view. The Copernican model was, in fact, no more accurate in predicting planetary motions than was the Ptolemaic system, and it was just as complex. Copernicus adhered to the notion of perfect circular motions, and was obliged to include small epicycles in order to match the observed planetary orbits. Apparently one of the most pleasing aspects of the sun-centered system to Copernicus was the fact that the relative distances of the planets could be deduced (Fig. 3.2), and were found to have a certain regularity, the spacings between planets growing systematically with distance from the sun. Copernicus was also able to determine the relative speeds of the planets in their orbits, finding that each planet moves more slowly than the next one closer to the sun.

Copernicus first circulated his ideas informally sometime before 1514, in a manuscript called *Commentariolis,* and it drew increasing attention during the next several years. The Church voiced no opposition, despite the fact that the ideas expressed in the work strongly contradicted commonly accepted Church doctrine. Copernicus was reluctant to publish his findings in a more formal way, for fear of raising controversy, and was continually rechecking his calculations. Eventually a young Protestant scholar known as Rheticus

FIG. 3.1. NICOLAUS COPERNICUS.

encouraged him to publish his work. Having been persuaded of the validity of the heliocentric theory, Rheticus had published his own summary of the ideas of

FIG. 3.2. COPERNICUS'S METHOD FOR FINDING RELATIVE PLANETARY DISTANCES FROM THE SUN. For an inferior planet such as Mercury or Venus, Copernicus knew the angle of greatest elongation, and was therefore able to reconstruct the triangle shown, providing him with the sun-Mercury (or sun-Venus) distance relative to the sun-earth distance (that is, the astronomical unit). For superior planets, similar but slightly more complicated considerations provided the same information.

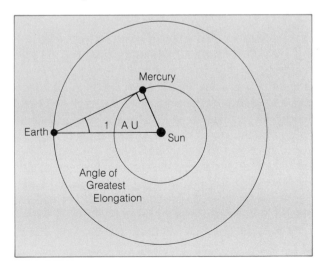

Copernicus, and then gave the manuscript written by Copernicus to a Lutheran priest named Osiander, who assumed the responsibility for publishing it.

The printer was fearful of strong religious opposition, and possibly with Osiander's collusion, added an unsigned preface in which he stated that the ideas within did not represent physical reality, but should be used only as a mathematical tool for predicting planetary motions. There is no doubt that Copernicus himself believed in the literal truth of his heliocentric model, however. Osiander titled the work *De Revolutionibus Orbium Coelestium (On the Revolution of the Celestial Sphere)*. Copernicus died at just about the time his work was published, so he did not live to see its impact.

The impact was profound, although not immediately so. The many important improvements in understanding the solar system that were embodied in the new theory eventually made a strong enough impression on other scientists to win them to this view. Among the persuasive aspects were the natural ordering of the planets (the relationship between orbital period and distance from the sun) and the simple explanation for the differing sizes of the retrograde loops that the planets go through (the extent of the backward motion observed for each planet, if it is a result of the earth's relative motion as Copernicus believed, depends on how far away the planet is). Copernicus also was able to explain the cause of the seasons, and made the daring suggestion that the reason stellar parallax could not be observed was that the stars are simply too far away to have perceptible displacements resulting from the earth's motion around the sun.

Tycho Brahe: Advanced Observations and a Return to a Stationary Earth

Although Copernicus had contributed little observationally, and had not demanded a close match between theory and observation as long as general agreement was found, there was, in fact, a strong need for improved precision in astronomical measurements. The next influential character in the historical sequence filled this need; if he had not, further progress would have been seriously delayed, as would have the eventual acceptance of the Copernican doctrine.

Tycho Brahe (Fig. 3.3) was born in 1546, some three

years after the death of Copernicus, in the extreme southern portion of modern Sweden (the region was at the time part of Denmark). Of noble descent, Tycho spent his youth in comfortable surroundings, and was well educated, first at Copenhagen University, then at Leipzig, where he insisted on studying mathematics and astronomy despite his family's wish that he pursue a law career. Interruptions created by war and plague caused Tycho to change universities several times, with most of his formal training eventually taking place at the university in Wittenberg. He developed a strong reputation as an astronomer (and astrologer; the distinction was still scarcely recognized), and attracted the attention of the Danish king, Frederick II. In 1575 the king ceded to Tycho the island of Hveen, about fourteen miles north of Copenhagen, along with enough servants and financial assistance to allow him to build and maintain his own observatory (Fig. 3.4).

Tycho had, even before this time, shown an acute interest in astronomical instruments, and with the grant to build his observatory, this interest bore fruit. He devised a variety of instruments which, although they really did not encompass any new principles, were capable of more accurate readings than any before his time.

FIG. 3.3. TYCHO BRAHE.

FIG. 3.4.   TYCHO'S OBSERVATORY AT HVEEN.

FIG. 3.5.   THE GREAT MURAL QUANDRANT AT HVEEN.   This instrument was used to measure the angular positions of stars and planets with respect to the horizon.

The largest was a quadrant built into a wall of the observatory (Fig. 3.5), with a 13½-foot diameter. A quadrant is a quarter-circle with graduated markings indicating degrees, minutes, and, if it is large enough, seconds of arc around its circumference. A sight located at the center allows the observer to read the angular heights of celestial objects above the horizon. Another instrument used by Tycho was a sextant, a portable device that works just like the quadrant, except that it only encompasses a 60-degree arc. Because a sextant is portable, it can be used to measure angular distances between objects in the sky, whereas the fixed quadrant could only measure angular heights above the horizon. The principal improvements of Tycho's instruments over those of his contemporaries were that the angular scales were more precisely marked, and that a new sighting design allowed the scales to be more accurately read.

Tycho made two specific observational discoveries of importance, in addition to amassing a large body of data on planetary motions. One, the observation of a supernova, occurred in 1572, and the other, the observation and analysis of a bright comet, took place in 1577, soon after the observatory became active. By comparing his data with those of astronomers else-

where, he was able to determine in both cases that the new object was very distant, well beyond the orbit of the moon (Fig. 3.6). This showed that the heavens were not perfect and immutable. The comet was particularly significant, because Tycho's analysis showed that it passed right through the crystalline spheres of the planets, demonstrating that they could not be the solid material objects supposed to exist by the ancient Greeks. Tycho's analysis of the supernova of 1572 was also quite significant, for it was in recognition of this achievement that King Frederick II granted Tycho the resources to build and operate his observatory. The supernova also provided evidence against the accepted doctrine, which held that the heavens were immutable.

Tycho was unable to accept the heliocentric view, primarily because he could find no evidence that the

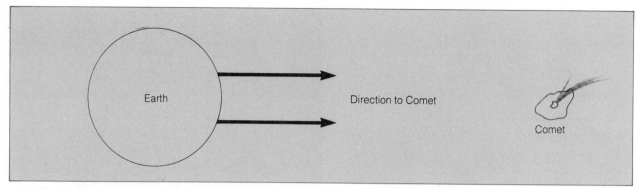

FIG. 3.6. TYCHO'S ANALYSIS OF THE COMET OF 1577. The position of the comet as measured from widely separated points on the earth appeared to be the same, which demonstrated that the comet was very distant. This contradicted Aristotle, who had taught that comets were atmospheric phenomena.

earth was moving. He tried and failed to detect stellar parallax, which he supposed he should see with his accurate observations if the earth really moved. Furthermore, as a strict Protestant, he found it philosophically difficult to accept a moving earth, when the Scriptures stated that the earth is fixed at the center of the universe.

On the other hand, he realized that the Copernican system had advantages of mathematical simplicity over the Ptolemaic model, and in the end he was ingenious enough to devise a model that satisfied all of his criteria. He imagined that the earth was fixed with the sun orbiting it, but that all the other planets orbited the sun (Fig. 3.7). Mathematically, this is equivalent to the Copernican system in terms of accounting for the motions of the planets as seen from the earth. The idea

FIG. 3.7. TYCHO'S MODEL OF THE UNIVERSE. The earth was held fixed, and the sun was thought to orbit the earth. The planets in turn were envisioned to orbit the sun. This system was never worked out in great mathematical detail, but it successfully preserved both the advantages of the Copernican system and the spirit of the ancient teachings. Most significant, it also accounted for the lack of observed stellar parallax, since the earth was fixed.

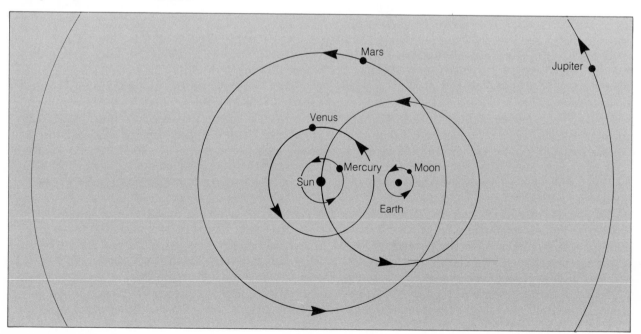

never received much acceptance, however, and Tycho is remembered primarily for his fine observations.

His observational contribution consisted of in part the unprecedented accuracy of his data, and in part the completeness of his records. Until his time, it had been the general practice of astronomers to record the positions of the planets only at notable points in their travels, such as when a superior planet comes to a halt just before beginning retrograde motion. Tycho made much more systematic observations, recording planetary positions at times other than just the significant turning points in their motions. He also made multiple observations in many instances, allowing the results to be averaged together to improve their accuracy. Tycho himself did not attempt any extensive analysis of his data, but the vast collection of measurements that he gathered over the years contained the information needed to reveal the basis of the planetary motions. The task of unraveling the secrets contained in Tycho's data was left to those who followed him, particularly a young astronomer named Johannes Kepler.

FIG. 3.8. JOHANNES KEPLER.

Kepler and the Laws of Planetary Motion

Until now it has been possible to follow developments in a straight sequence, but here we must begin to cover events that occurred at nearly the same time, although in different locations. To complete the thread begun with our discussion of Tycho Brahe, we now turn to Johannes Kepler (Fig. 3.8), who worked briefly at the side of Tycho and then spent many years analyzing the great wealth of observational data that Tycho had accumulated. We must keep in mind, however, that during the same period a scientist and philosopher named Galileo Galilei was at work to the south, in Italy, and that the two were aware of each other's accomplishments.

Kepler was born in 1571 in Weil, Germany, a year before Tycho's supernova was observed. A sickly youngster, he seemed headed for a career in theology, but while attending the university in Tübingen, he encountered a professor of astronomy who inspired in him a strong interest in the Copernican system.

Having completed his education, Kepler accepted a teaching position in the Austrian city of Graz in 1594. While there he published his first scientific work, a book called the *Mysterium Cosmigraphicum*, in which he

outlined several arguments favoring the Copernican theory, and stated his own view that the distances of the planets from the sun were determined by a series of regular geometric solids separating the crystalline spheres of the planets (Fig. 3.9). Kepler's reasons for adopting this notion and the Copernican system as well were based on his love of mathematical simplicity and his intuitive feeling that the universe was constructed on a harmonious, unified plan. He devoted his life to seeking out the underlying principles of this universal harmony, and, as we shall see, this led him to a number of profound discoveries. The hypothesis put forth in the *Mysterium Cosmigraphicum* was, of course, not correct, but the book had major impact anyway, especially because of its eloquent arguments in favor of the sun-centered universe.

In 1598 a decree banning all Protestants forced Kepler out of Graz and, after first visiting extensively, he accepted a post as chief assistant to Tycho, who had by then moved his observatory to Prague following the death of his sponsor, King Frederick II of Denmark. Tycho died the next year, and Kepler succeeded him as court mathematician to Emperor Rudolph.

Kepler's main mission, owing to both Tycho's wishes and his own, was to develop a refined understanding

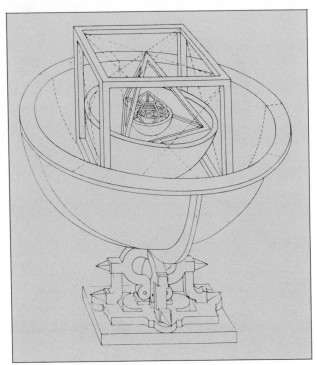

FIG. 3.9. KEPLER'S NESTED GEOMETRIC SOLIDS. Kepler expended considerable effort in exploring the possibility that the planetary distances from the sun were determined by the spacings of nested regular geometric solids (that is, figures whose faces were regular polyhedrons; the number of possible solids of this sort was equal to the number of known planets). Kepler eventually abandoned this particular scheme, but never relented in his search for regularity in the solar system.

of the planetary motions and to upgrade the tables used to predict their positions. He set to work first on the planet Mars, for which the data were particularly extensive, and whose motions were among the most difficult to explain in the established Ptolemaic system (and, in fact, in the Copernican system, with its requirement of only circular motions). By a very complex process, Kepler was able to separate the effects of the earth's motion from those of Mars itself, so that he could map out the path that Mars followed with respect to the sun.

He soon realized that the sun must play a central role in making the planets move, and this inspired him to look for a physical link between the sun and planets. In this sense he was like a modern scientist, and quite unlike other astronomers of his time and before. Whereas they sought only a mathematical device for predicting motions, Kepler believed in a true physical machinery of the solar system, and he sought the underlying principles.

The data on Mars led Kepler quickly to the realization that the sun could not be at the true center of the planet's orbit. He experimented with various geometric shapes, including a brief attempt with epicycles, which he considered to be contrived and not likely to represent physical reality. At one point, he was forced to reject a system he had devised because of a small error in predicting the position of Mars, an error that would have been acceptable to anyone else who had worked with the planetary motions.

By 1604 Kepler had determined that the orbit of Mars was some kind of oval, and further experimentation revealed that it was fitted precisely by a simple geometric figure called an **ellipse.** This is a closed curve defined by a fixed total distance from two points called **foci,** and indeed Kepler found that the sun was at the precise location of one focus of the ellipse (Fig. 3.10).

Further analysis of the motion of Mars revealed a second characteristic, having to do with the fact that the planet moves fastest in its orbit when it is nearest the sun, and slowest when it is on the opposite side of its orbit, farthest from the sun. Mathematically, Kepler's second discovery was that a line connecting Mars to the sun sweeps out equal areas of space in equal intervals of time (Fig. 3.10).

These two properties of the orbit of Mars were later generalized as Kepler's first two laws of planetary motion, applicable to all the planets. Kepler's results on the orbit of Mars were published in 1609 in a book entitled *The New Astronomy: Commentaries on the Motions of Mars.* The book received a great deal of attention.

In 1612 Emperor Rudolph, who had been supporting Kepler's research, died, and Kepler moved to Linz, in northern Austria. There his work on the planetary motions continued, as he still sought the true nature of the relationship among the planets. In 1619 he published a book entitled *The Harmony of the World,* in which is reported his discovery of a simple relationship between the orbital periods of the planets and their average distances from the sun. Now known as Kepler's third law, or simply the Harmonic law, it states that the square of the period of a planet is proportional to the cube of the semimajor axis (which is half of the long axis of an ellipse). In other words, this says

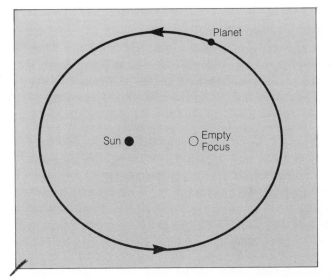 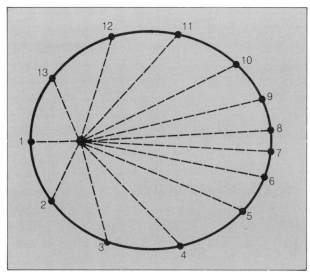

FIG. 3.10. THE ELLIPSE. At left is an exaggerated ellipse representing a planetary orbit with the sun at one focus; at right is a similarly exaggerated ellipse with lines drawn in to illustrate Kepler's second law.

that $P^2 = a^3$, where P is the sidereal period of a planet in years, and a is the semimajor axis in terms of the sun-earth distance (that is, in terms of the **astronomical unit**).

In another major work called the *Epitome of the Copernican Astronomy,* published in parts in 1618, 1620, and 1621, Kepler presented a summary of the state of astronomy at that time, including Galileo's discoveries. In this book Kepler generalized his laws, explicitly stating that all the planets behaved similarly to Mars, something that had clearly been his belief all along.

We can now state Kepler's first two laws of planetary motion in general terms: (1) The orbits of the planets are ellipses, with the sun at one focus; and (2) for each planet, the area swept out in space by a line connecting that planet to the sun is equal in equal intervals of time.

For most of the planets, the orbits are more nearly circular than in the case of Mars, and it is indeed fortunate that Kepler chose to work with the Mars data first, or he might not have been able to discover his first two laws.

By the time the *Epitome* was published, the Roman Catholic church was in a very intolerant frame of mind, in contrast with the situation at the time of Copernicus, nearly a hundred years before. Kepler's treatise soon found itself on the *Index of Prohibited Books,* along with *De Revolutionibus.*

In 1627 Kepler published his last significant astronomical work, a table of planetary positions based on his laws of motion, which could be used to accurately predict the planetary motions. These tables, which he called the *Rudolphine Tables* in honor of his former benefactor, were used for the next several years. The Rudolphine Tables represented an improvement in accuracy over any previous tables by nearly a factor of 100, a resounding and remarkable confirmation of the validity of Kepler's laws. In a very real sense the Rudolphine Tables represented Kepler's life's work, since with their publication he completed the task set before him when he first went to work for Tycho. Kepler died in 1630, at the age of fifty-nine.

TABLE 3.1 TESTING KEPLER'S THIRD LAW			
PLANET	A (AU)	P (YEARS)	A^3	P^2
Mercury	0.387	0.241	0.058	0.058
Venus	0.723	0.615	0.378	0.378
Earth	1.000	1.000	1.000	1.000
Mars	1.523	1.881	3.533	3.538
Jupiter	5.203	11.86	140.85	140.66
Saturn	9.539	29.46	867.98	867.89
Uranus	19.18	84.01	7055.79	7057.68
Neptune	30.06	164.8	27162.32	27159.04
Pluto	39.44	248.4	61349.46	61762.56

ASTRONOMICAL INSIGHT (3.1)

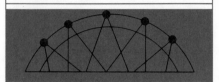

The Geometric Properties of the Ellipse

In the text we did not specify the properties of an ellipse in any great detail. We do so here.

An ellipse is one example of a **conic section,** meaning that it is one of several figures that may be obtained by cutting through a cone; more precisely, a conic section is the figure formed by the intersection of a plane with a cone. Other possible figures formed this way include a circle (if the plane intersects the cone exactly at right an-gles to its axis), a parabola, and a hyperbola (the latter two are open figures, rather than being closed like a circle or an ellipse). Laws of physics developed well after the time of Kepler show that all the possible trajectories of two bodies in space that are subject to each other's gravitational forces follow one of these conic sections. When two objects are in mutual orbit, the general form is an ellipse.

The geometrical definition of an ellipse may be stated as follows: an ellipse is a figure such that the sum of the distances from any point on it to two fixed points called *foci* is constant.

The degree of elongation (called the **eccentricity**) of an ellipse is defined by the separation of the foci relative to the size of the ellipse. The closer together the foci are, the more nearly circular the ellipse; indeed, if we imagine that the foci merge together into a single point, the re-sulting figure is precisely a circle.

The long axis of an ellipse is the **major axis,** and half of that is the **semimajor axis.** In terms of plane-tary orbits, the semimajor axis can be viewed as the average distance of a planet from the sun (although strictly speaking it not an average, because the planet moves more slowly and hence spends more time in the part of its orbit on the oppo-site side of the focus where the sun is located, and spends less time on the near side).

To draw an ellipse is quite sim-ple. All that is needed is a board with two tacks or pins representing the two foci, a pencil, and a piece of string. The string is looped over the two pins, and then the pencil is used to keep the string taut as it is traced around the pins. Because the string is fixed in length, this ensures that the total distance from any point to the two foci is constant, and therefore that the resulting figure is an ellipse.

Galileo, Experimental Physics, and the Telescope

Very strong contrasts can be drawn between Kepler and his great contemporary, Galileo (Fig. 3.11). Whereas Kepler was fascinated with universal harmony and therefore with the underlying principles on which the universe operates, Galileo was primarily concerned with the nature of physical phenomena and was less de-voted to finding fundamental causes. Galileo wanted to know how the laws of nature operated, whereas Kep-ler sought the reason for the existence of the laws. Gal-ileo's approach was level-headed and rational in the extreme. He used simple experiment and deduction in advancing his perception of the universe. He has fre-quently been cited as the first truly modern scientist, although others of his time probably deserve a share of that recognition.

Galileo was born in 1564 at Pisa, in northern central Italy. Although his family had intended for him a ca-reer in some trade, his obvious intellectual abilities were soon recognized, and in 1581 he began formal training at the university in Pisa. There his reluctance to accept dogmatic ideas without question was soon noted, and he earned the nickname "The Wrangler" for his ten-dency toward debate. Poverty forced him to live at home for awhile following his fourth year of study, and for some time he worked at a teaching post in Pisa. In

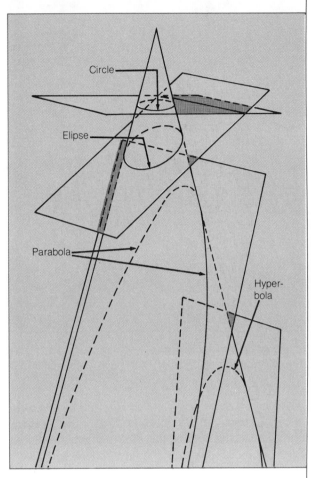

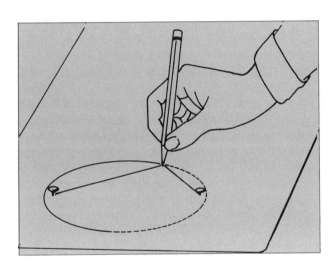

1592 he moved to the University of Padua, where he was to stay until 1610.

During his early academic years he carried out numerous experiments in physics, discovering many simple laws, and in the process completely overturning the accepted method of studying science. Whereas a follower of Plato and Aristotle (whose works still dominated in Galileo's time) would proceed by rational thought from standard unproven assumptions, Galileo found it much more sensible to begin with experiment or observation, and to work from there toward an understanding of the underlying principles. In doing so, Galileo founded an entirely new basis for scientific inquiry, an achievement in many ways more profound than his contributions to astronomy, which were considerable.

Galileo's discoveries in physics, having to do with the motions of objects, were published in his later years, after his astronomical career had been forcibly ended by Church decree. It was, in fact, an early interest in **mechanics,** the science of the laws of motion, that lured Galileo away from a career in medicine, the subject of his first studies.

Among his most important contributions to mechanics was his elucidation of a principle called **inertia,** the tendency of a moving object to remain in motion or of an object at rest to resist being moved. Galileo was the first to clearly recognize this natural

FIG. 3.11. GALILEO GALILEI.

tendency, and to understand that an object sliding along a surface stops only because the force of friction acts upon it. He later applied this idea to the motion of the earth, in explaining how the atmosphere and all the objects on its surface could move along with it.

Galileo also understood the concept of **acceleration,** the rate of change of velocity, and showed that falling objects are subject to a constant rate of acceleration, regardless of their **mass,** the total quantity of matter they contain. Supposedly this point was made in a graphic manner when Galileo is said to have dropped balls of unequal weight from the Leaning Tower of Pisa, demonstrating to a crowd of onlookers that the balls hit the ground at the same time. There is some doubt as to whether, or under what circumstances, this experiment was actually performed, although it is certain that Galileo did carry out comparable experiments, perhaps without attendant ceremony.

Galileo's other contributions to mechanics include the discovery of the nature of the pendulum, whose oscillation period he found to be independent of how far it swings, showing that it could be used as a time-keeping device.

Galileo's astronomical discoveries were quite sufficient to earn him a major place in history, even if he had never done his work in mechanics. It was his astronomical studies, along with his flair for debate and his habit of ridiculing those with whom he disagreed, that made him famous (though not universally loved) in his own time.

His first astronomical contribution was the observation of a supernova in 1604, one also seen by Kepler. Both men showed that its lack of parallax placed it in the distant realm of the stars.

In 1609 Galileo learned of the invention of the telescope, and devised one of his own, which he soon put to use by systematically viewing the heavens with it. Despite the poor quality of the instrument, Galileo made a number of important discoveries almost at once from his observation tower at Padua (Fig. 3.12), and reported them in 1610 in a publication called the *Starry Messenger*. Here Galileo showed that the moon was not a perfect sphere, but instead was covered with craters and mountains (Fig. 3.13). He also reported the existence of many more stars than could be seen with the unaided eye. Most significant of all, he showed that Jupiter was attended by four satellites, whose motions he observed long enough to establish that they orbited the parent planet (Fig. 3.14). All these discoveries, but especially the latter, violated the ancient philosophies of an idealized universe centered on the earth. The satellites of Jupiter showed beyond reasonable question that there were centers of motion other than the earth.

Once Galileo had begun his astronomical observations, he became fascinated with the structure of the universe, and for a time his work in mechanics was put aside. He spent much of the next several years in pursuit of evidence for the Copernican model.

Mostly as a result of the reputation Galileo earned with the publication of the *Starry Messenger,* he was able to negotiate successfully for the position of court mathematician to the Grand Duke of Tuscany, and he moved in 1610 to Florence, where he was to spend the remainder of his long career. Once established there, Galileo continued his observations, and soon added new discoveries to his list. The first of these was the varying aspect of Venus as its position changed with respect to the sun. Galileo quickly realized that this was analo-

FIG. 3.12. GALILEO'S OBSERVATORY AT PADUA. In a sense, this tower is the world's first modern astronomical observatory, since it was the first to be used for telescopic observations.

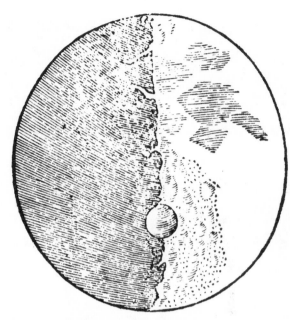

FIG. 3.13.   GALILEO'S SKETCH OF THE MOON.

FIG. 3.14.   GALILEO'S SKETCHES OF THE SATELLITES OF JUPITER.

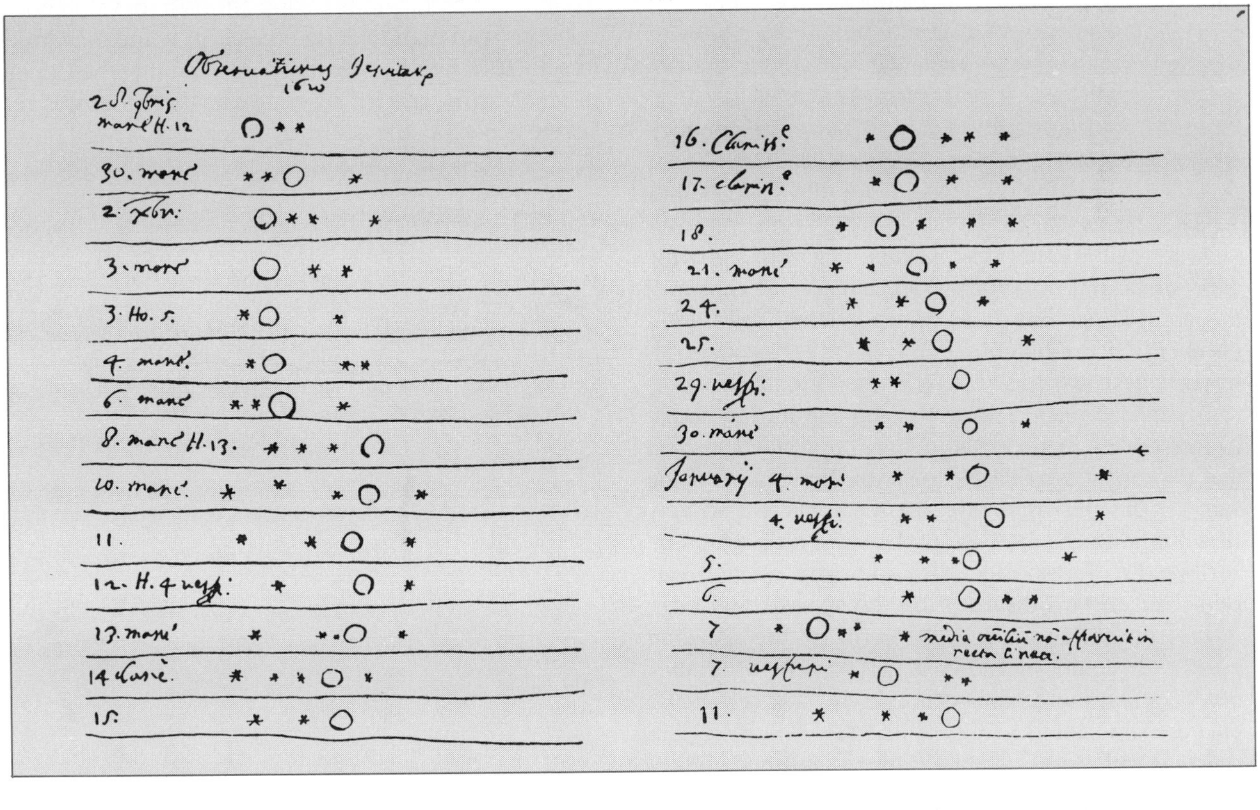

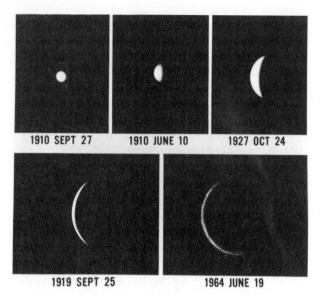

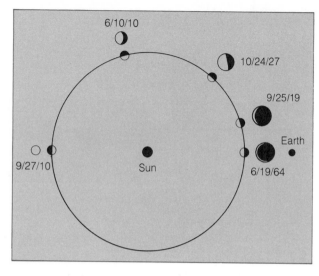

FIG. 3.15.   THE PHASES OF VENUS. As Venus orbits the sun and its position changes relative to the earth-sun line, its phase varies as we see differing portions of its sunlit side. In addition, its apparent diameter varies because of its varying distance from the earth.

gous to the phases of the moon, as we see from the earth varying portions of the sunlit side of Venus (Fig. 3.15). This had two important implications: it showed that Venus shines by reflected sunlight, rather than by its own power; and it demonstrated that Venus is orbiting the sun instead of the earth, since the apparent size of Venus varies (from being smallest when Venus is fully sunlit to being largest when the dark side of Venus is facing the earth).

At about the same time, Galileo began to study the dark spots that occasionally appeared on the face of the sun. These had been independently seen by a number of astronomers, and there was substantial controversy as to what they were. Some thought the spots to be small planets close to the sun that moved across its face during their orbital motion, but Galileo showed that they must actually be spots on the surface of the sun. He did this by noting that the spots moved very slowly when near the edge of the sun's disk, which he attributed to foreshortening as the sun's rotation carried them toward the earth on the approaching edge of the sun and away from the earth on the receding edge, with significant sideways motion only when they moved across the central portion of the disk (Fig. 3.16). That the sun should have blemishes on its surface, and rotate as well, were contrary to the established view.

In the years immediately following these discover-

ies, Galileo began to draw increasingly heavy criticism from the Church, and he made efforts to develop good relations with high-ranking officials in Rome. Nevertheless, in 1616 he was pressured into refuting the Copernican doctrine, and for several years thereafter was relatively quiet on the subject. Except for a well-publi-

FIG. 3.16.   SUNSPOT MOTION. As the sun rotates, a spot on its surface moves at a steady rate, but as seen from the earth, its speed varies. In this sketch, the spot moves from *a* to *b* in the same length of time that it moves from *b* to *c,* but from the earth, it appears to move through angles *A* and *B,* respectively. Because of foreshortening, angle *A* is smaller than angle *B,* and the spot therefore appears to move more slowly from *a* to *b* than from *b* to *c.* Galileo's explanation of this effect convinced him that the sunspots are truly located on the surface of the sun and are not planets orbiting farther out.

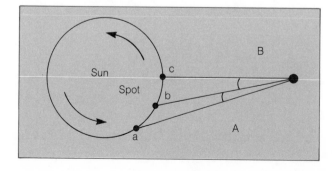

ASTRONOMICAL INSIGHT (3.2)

An Excerpt from Dialogue on the Two Chief World Systems

During the course of three days of discussion, the three characters in Galileo's *Dialogue* thoroughly air all the available evidence and arguments bearing on the understanding of mechanics and the structure and nature of the universe. About midway through, a discussion develops that is central to Galileo's principal point: the sun, not the earth, is at the center of the universe. In the following excerpt,* we see an example of Salviati's persuasive style, Simplicio's dogged reluctance to give up the ideas of Aristotle, and Sagredo's ready comprehension of Salviati's arguments.

> *Salviati:* Now if it is true that the center of the world is the same

about which the circles of the mundane bodies, that is to say, of the Planets, move, it is most certain that it is not the Earth but the Sun, rather, that is fixed in the center of the World. So that as to this first simple and general apprehension, the middle place belongs to the sun, and the Earth is as far remote from the center as it is from that same Sun.

Simplicio: But from whence do you argue that not the Earth but the Sun is in the center of planetary revolutions?

Salviati: I infer the same from the most evident and therefore necessarily conclusive observations, of which the most potent to exclude the Earth from the said center, and to place the Sun therein, are that we see all the planets sometimes nearer and sometimes farther off from the Earth, with so great differences, that, for example, Venus when it is at the farthest is six times more remote than when it is nearest, and Mars rises almost eight times as high at one time as at another. See therefore whether Aristotle was somewhat mistaken in thinking that it was at all times equidistant from us.

Simplicio: What in the next place are the tokens that their motions are about the sun?

Salviati: It is shown in the three superior planets, Mars, Jupiter, and Saturn, in that we find them always nearest to the Earth when they are in opposition to the Sun and farthest off when they are towards the conjunction; and this

approximation and recession imports thus much, that Mars near at hand appears sixty times greater than when it is remote. As to Venus, in the next place, and to Mercury, we are certain that they revolve about the Sun in that they never move far from it, and in that we see them sometimes above and sometimes below it, as the mutations in the figure of Venus necessarily prove. Touching the Moon, it is certain that it cannot in any way separate itself from the Earth, for the reasons that shall be more distinctly alleged hereafter.

Sagredo: I expect that I shall hear more admirable things that depend upon this annual motion of the Earth than were those dependent upon the diurnal revolution.

In this exchange, Galileo, in the guise of Salviati, advances several arguments based on observations, including the well-known phases of Venus, and the less widely quoted arguments having to do with the varying distances of the planets from the earth. In this and other passages, Galileo went out of his way to mock the followers of Aristotle, who, according to Salviati in an earlier paragraph, "would deny all the experiences and all the observations in the world, nay, would refuse to see them, that they might not be forced to acknowledge them, and would say that the world stands as Aristotle writes and not as Nature will have it."

cized debate on the nature of comets, Galileo spent most of his time preparing his greatest astronomical treatise, which was finally published, after some difficulties with Church censors, in 1632. To avoid direct violation of his oath not to support the Copernican heliocentric view, Galileo wrote his book in the form of

a dialogue among three characters, one of whom, named Simplicio, represented the official position of the Church; another, Salviati, that of Galileo (although this was, of course, not stated explicitly); and a third, Sagredo, who was always quick to see and agree with Salviati's arguments. In this treatise, called the *Dia-*

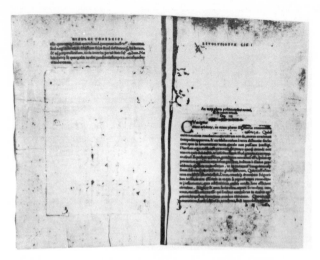

FIG. 3.17. A CENSORED VERSION OF GALILEO'S *DIALOGUE*. This photo shows an early edition, as it was censored by the Church.

logue on the Two Chief World Systems, Galileo, through the character Salviati and at the expense of Simplicio, systematically destroyed many of the traditional astro-nomical teachings of the Church. The book was published in Italian, rather than the scholarly Latin, and its contents were therefore accessible to the general populace.

Despite a lengthy preface in which Galileo disavowed any personal belief in the heliocentric doctrine, the Church reacted strongly, and within a few months Galileo was summoned before the Roman Inquisition, while existing copies of the book were heavily censored (Fig. 3.17) and further publication was banned. It joined the works of Copernicus and Kepler on the *Index* (where it was to remain until the 1830s), and Galileo himself was sentenced to house arrest, eventually serving this punishment at his country home near Florence. This is where he spent the remainder of his days.

Before his life came to a close, however, Galileo published one more major work, this being a summary of his studies of mechanics, based largely on experiments carried out many years before. The book on mechanics appeared in 1638, and Galileo died in 1642, having suffered blindness the last four years of his life.

Perspective

We have seen how, in the space of less than 150 years, the science of astronomy advanced from adherence to the 1,500-year-old doctrines of Plato, Aristotle, and Ptolemy to a nearly modern view. Despite the suppression of Galileo by the Church, enough questions had been raised through his work and that of Copernicus, Tycho Brahe, and Kepler, that serious consideration of the nature of the cosmos thereafter centered on the heliocentric system.

It has been interesting to examine the lives, and especially the frame of mind, of the characters who contributed to this revolution, ranging from Copernicus, whose intuition forced him to reject the established view and adopt an entirely new one; to Tycho, who philosophized little, but contributed much through his observations; Kepler, whose search for universal harmony led him to the laws of planetary motion; and finally Galileo, whose implacable common sense and brilliant deduction put it all on a firm logical footing. The stage was now set for the missing link, the physical cause of the motions in the universe, to be discovered.

Summary

1. Copernicus developed the heliocentric theory because it provided a unifying picture of the universe, not because it was more accurate or simpler than the earth-centered theory.

2. With his new heliocentric theory, Copernicus was able to calculate the relative distances of the planets from the sun, and he arrived at the correct explanation of the seasons and lack of detectable parallax.

3. Tycho Brahe made vast improvements in the quantity and quality of astronomical observations of stars and planets, accomplishing this by making more precise and complete measurements than his predecessors.

4. Tycho was unable to detect stellar parallax, and therefore rejected the heliocentric hypothesis.

5. Kepler sought the underlying harmony among the planets and its physical basis, and in the process he discovered three laws of planetary motion.

6. Kepler's first law states that each planet orbits the sun in an ellipse, with the sun at one focus.

7. Kepler's second law states that a planet moves fastest in its orbit when it is closest to the sun, and slowest when it is farthest away, in such a manner that the area swept out in space by a line connecting the planet to the sun is equal in equal intervals of time.

8. Kepler's third law relates the periods and semi-major axes of the planets; it states that the square of the period (in years) is equal to the cube of the semi-major axis (in AU).

9. Galileo, by carrying out experiments, defined several important concepts of physics, including most notably the ideas of inertia and acceleration.

10. Galileo's use of a telescope to observe the heavens led to numerous discoveries that he used to refute the geocentric hypothesis.

11. Despite Church opposition, Galileo brought his ideas before the public in the form of a published fictional dialogue between believers of opposing points of view.

Review Questions

1. In what way did the heliocentric hypothesis of Copernicus provide a more unified overall concept of the universe than the earth-centered theory?

2. In what ways was the theory of Copernicus more consistent with the teachings of Aristotle than was the model later developed by Kepler?

3. How does the theory of Copernicus account for the retrograde motion of the outer planets?

4. To illustrate the technique of averaging measurements to improve accuracy, we will suppose that the true separation of two stars is $4°13'$ of arc. Using instruments that are accurate to $4'$ of arc, we will make the following measurements of the separation: $4°10'$, $4°14'$, $4°17'$, and $4°10'$. Compute the average of these measurements, and discuss the use of this technique for improving the accuracy of measurements.

5. Kepler's outlook was similar to that of Plato and Aristotle in some ways, and rather different in others. Briefly discuss the similarities and differences.

6. Suppose there were a planet at a distance of 2 AU from the sun. According to Kepler's harmonic law, what would be its sidereal period?

7. Aristotle and Galileo had completely different explanations for the fact that an object that is being pushed along a surface will stop when the force that is pushing it is discontinued. What were their contrasting explanations?

8. How did Galileo's observational discoveries about the moon and the sun violate the teachings of Plato and Aristotle?

9. Why did Tycho adopt the geocentric viewpoint, despite his knowledge of Copernicus and the many advantages of the heliocentric concept?

10. Why did Copernicus retain the concept of epicycles in his heliocentric theory?

Additional Readings

The references listed here are sources of biographical and historical data on the people discussed in this chapter, and in many cases they also contain extensive lists of additional material.

Beer, A., and Strand, K. A., eds. 1975. *Copernicus*. Vistas in Astronomy, vol. 17. New York: Pergamon Press.

Beer, A., and Beer, P., eds. 1975. *Kepler*. Vistas in Astronomy, vol. 18. New York: Pergamon Press.

Galileo, G. 1632. *Dialogue on the two chief world systems.* Translated by G. de Santillana, Chicago: University of Chicago Press.

(*Note:* There are other translations and editions of this work that are readily available.)

Gingerich, O. 1973. Kepler. In *Dictionary of scientific biography,* vol. 7, ed. C. G. Gillispie, p. 289. New York: Scribner's.

Gingerich, O., ed. 1975. *The nature of scientific discovery.* Proc. of a symposium in honor of Copernicus. Washington: Smithsonian Institution Press.

Draper, J. L. E. 1963. *Tycho Brahe: A picture of scientific life and work in the sixteenth century.* New York: Dover.

Kuhn, T. 1957. *The Copernican revolution.* Cambridge: Harvard University Press.

The Curious Case of Claudius Ptolemy

Owen Gingerich

Dr. Gingerich is an astrophysicist at the Smithsonian Astrophysical Observatory, and a professor of astronomy and the history of science at Harvard University. He has made many significant contributions to stellar astrophysics, particularly in the construction of theoretical models of stellar atmospheres, but is probably best known for his studies of the history of astronomy. One of his special interests is the development and transmission of planetary theory from antiquity to the present.

Around A.D. 150, the Alexandrian astronomer Claudius Ptolemy wrote the first comprehensive astronomical textbook. By the Middle Ages, Islamic scientists held it in such esteem that they called it the *Almagest,* meaning "the greatest."

Ptolemy's book first outlined the basic principles of geocentric astronomy: the earth could be considered a mere point within the celestial sphere. On this sphere were the stars, turning with the sky every twenty-four hours, and against this whirling starry backdrop the planets traveled along the ecliptic. Ptolemy not only provided tables showing the relationships of the equator and ecliptic, and of the oblique path of the ecliptic with respect to the horizon for different latitudes, but he even described how to calculate the trigonometric tables that he needed to accomplish this. He described various observational techniques, gave a table of over a thousand star positions, and finally he presented his epicyclic model of planetary motions, including tables from which it was possible to calculate the positions of the sun, moon, and planets for any time past or future. Because he also included elaborate tables of the lunar parallax (the apparent displacement in the sky of the moon that depends on the observer's location), for the first time it became possible to predict the circumstances of solar eclipses.

The relative accuracy of Ptolemy's results was somewhat erratic: the moon, Jupiter, and Saturn were generally predicted within 40′ of their correct places, but for Mars the error could on occasion exceed 1°. The accuracy of these predictions was not substantially improved for well over fourteen hundred years, until the time of Kepler.

Despite Ptolemy's great achievement, his work contains some serious flaws and curious puzzles that have made his reputation today quite controversial. Some of his critics have called him the greatest fraud in the history of science, whereas others, including myself, regard him as the greatest astronomer of antiquity.

Ptolemy used as a starting point the solar theory of Hipparchus, an astronomer who had lived nearly three centuries earlier. Because Hipparchus's theory worked fairly well for predicting lunar eclipses, Ptolemy thought that the length of the year given by Hipparchus was sufficiently accurate, not realizing that it was a little too long and that the moon's period had a similar error that apparently kept the two in step. Thus, although Ptolemy describes making observations to determine the times of the equinox (which could establish how long the year actually is), his numbers were actually calculated from the Hipparchan theory rather than genuine observations. These data were wrong by a little over a day, representing the accumulated error in the 275 years since Hipparchus had made his own calculations. Since the sun moves 360 degrees in about 365 days, it goes about a degree per day. An error of a day in Ptolemy's reported time of the equinoxes meant that everything else in his *Almagest* was thrown off by about 1 degree, and this in turn generated a very poor value for the rate of precession.

Furthermore, the planetary positions recorded in the *Almagest* have varying errors (from a few minutes of arc up to a degree, not counting the systematic error of 1 degree in Ptolemy's coordinate system), but almost all of them agree perfectly well with the predictions of his own tables. This situation has raised a considerable suspicion that none of his reported observations were observations at all, but were merely calculated. One detractor has declared that it would have been better for astronomy if Ptolemy had never written his *Almagest*.

On the other hand, the basic numbers in his planetary theory are surprisingly accurate, seemingly more accurate than the observations on which the theory is purportedly based. How could Ptolemy have achieved such a successful result unless he had a sound observational basis for his work? It is quite possible that Ptolemy had many genuine observations available on which his theory was based, but in compiling his textbook, he wanted to give only "perfect" data to show how the underlying parameters could be derived. Hence, perhaps he used his theory to "correct" his reported observations in an attempt to smooth out any observational errors. Astronomers today might follow much the same procedure, except they would be careful to explain what smoothing had been carried out. By modern standards an undisclosed "laundering" of the data would be considered reprehensible, but in Ptolemy's day there were very few models for scientific (as opposed to geometrical) treatises. Probably it did not occur to Ptolemy that astronomers centuries in the future would want many of the details he failed to provide.

The Tools of Astronomy

INTRODUCTION TO SECTION II.

Before we can plunge into a full description of the universe and an understanding of it, we must first learn about the methods and tools used by astronomers to gain this understanding. One of the principal goals of this text is to describe *how* we know what we know, going well beyond a simple summary of the phenomena that are observed.

The first of the three chapters in this section is devoted to a description of some of the basic physical principles that underlie the motions and interactions of bodies large and small. We start with a discussion of Isaac Newton, whose intellectual powers unearthed many of these principles, and then we move on to the laws of physics that govern such diverse phenomena as planetary orbits, the motions of molecules in a gas, and the tides on the earth and other bodies.

We would know nothing of the external universe were it not for the light that reaches us from far-away objects, and the next chapter describes the nature of light and the way we decipher its messages. We find that an amazing variety of information can be derived from the spectra of objects like planets and stars. Things once considered forever beyond our grasp are now routinely measured, and in chapter 5 we learn how this is done.

The final chapter of this section is devoted to the instruments that are used to collect light from afar. Telescopes on the ground and in space are the material monuments to mankind's quest for knowledge of the universe, and in chapter 6 we will learn the principles of their design and the practical considerations of their construction and use. Having done this, we will, at last, be ready to explore the dynamic universe.

.CHAPTER 4.

Isaac Newton And the Laws of Motion

.CHAPTER PREVIEW.

Isaac Newton: his life and discoveries
Newton's laws of motion
  Mass and inertia
  Force and acceleration
  Action and reaction
The law of universal gravitation
  The evidence for a central force
  The inverse-square dependence on distance
A modern view of orbits
  The concept of energy; kinetic and potential
  Orbits involving two bodies and a center of mass
  Angular momentum and its preservation
  Kepler's third law revised
Gas kinetics and escape velocity
  The speed needed to escape
  Particle motions, temperature, and the atmosphere
Tidal forces
  Differential gravitational forces
  Bulges on the earth and the daily tides

After the publication in 1632 of Galileo's *Dialogue on the Two Chief World Systems,* a fifty-year period elapsed during which no major discoveries were made, but substantial progress nevertheless occurred. Telescopic observations were extended to a greater variety of objects, the surface of the moon was mapped, star catalogues were expanded, and the planets were studied in more detail. Perhaps the greatest astronomer of the period was Christiaan Huygens (1629–1695), who refined the telescope and was able to detect a moon of Saturn. He also discovered the true nature of the misshapen appearance of the planet (Galileo's crude telescope had only revealed a distortion in the shape of Saturn, whereas Huygens was able to determine that the planet is encircled by a great ring system). The earth was accurately measured by an assortment of people, and it was even discovered that it is not precisely spherical, but instead bulges at the equator.

The Life of Newton

In the year 1643, a few months after the death of Galileo, Isaac Newton (Fig. 4.1) was born, in Woolsthorpe, England, into a climate of inquiry and scientific ferment. Newton's childhood was unremarkable, except that he showed a growing interest in mathematics and science, and at age eighteen he entered college at Cambridge, receiving a bachelor's degree in early 1665. He then spent two years at his home in Woolsthorpe, largely because the plague made city living rather dangerous, and it was during this time that he made a remarkable series of discoveries in the fields of physics, astronomy, optics, and mathematics, in what surely must have been one of the most intense and productive periods of individual intellectual effort in human history.

Newton had a tendency to exhaust a subject, get bored with it, and go on to new fields, so that nothing of his work was published for some time, during which some of his discoveries were repeated independently by others. Finally, after persistent urging by his friend (and fellow astronomer) Edmund Halley, in 1687 Newton published a massive work called *Philosophiae Naturalis Principia Mathematica,* now usually referred to as the *Principia* (Fig. 4.2). In this three-volume book, Newton established the science of mechanics (which he viewed as merely background material and relegated it to an introductory section), and applied it to the motions of the moon and the planets, developing the law of gravitation as well. His work in optics was published separately in 1704, although it was probably written much earlier than that.

The *Principia* received great notice, particularly in England. As a result, Newton's later life was a public one, with various government positions which left less and less time for scientific discovery. With the help of younger associates, he did revise the *Principia* on two occasions, in 1713 and in 1726, making some improvements each time. Newton died in 1727, at the age of eighty-four. His impact lives today, in our modern understanding of physics and mathematics. Newton's conclusions on the nature of motions and gravity are still viewed as correct, although it is now realized that there are circumstances in which more complex theories (such as Einstein's relativity) must be used.

Newton's Laws of Motion

INERTIA AND THE FIRST LAW

We have already discussed the concept of inertia, first developed by Galileo. Newton expanded the idea, rec-

FIG. 4.1.   ISAAC NEWTON.

FIG. 4.2.   THE TITLE PAGE FROM AN EARLY EDITION OF THE *PRINCIPIA.*

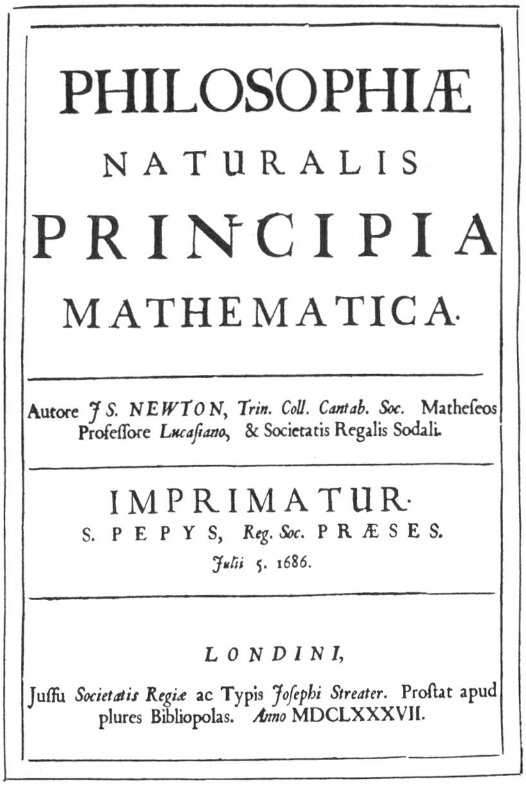

PHILOSOPHIÆ

NATURALIS

PRINCIPIA

MATHEMATICA.

Autore *J* S. *NEWTON,* Trin. Coll. Cantab. Soc. Matheseos Professore *Lucasiano,* & Societatis Regalis Sodali.

IMPRIMATUR·

S. P E P Y S, *Reg. Soc.* P R Æ S E S.

Julii 5. 1686.

L O N D I N I,

Jussu *Societatis Regiæ* ac Typis *Josephi Streater.* Prostat apud plures Bibliopolas. *Anno* MDCLXXXVII.

ASTRONOMICAL INSIGHT 4.1

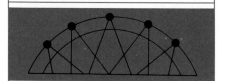

The Controversy over Calculus*

Among Newton's great achievements was the development of **calculus,** which is a mathematical method of great power that is especially useful in solving problems involving quantities that vary with time or distance. The approach is to consider what happens in infinitesimally small steps, and then to

*Based on information from Hall, A.P. 1980, *Philosophers at War: The Quarrel Between Newton and Leibniz* (New York: Cambridge University Press)

sum up the effect of variations over a large number of steps. The tiny steps are called **differentials,** and their sum an **integral.** This general approach to solving physical problems has itself become a major branch of mathematics.

Although Newton apparently invented calculus (which he called the method of fluxions) during the plague years 1665–1666, while residing at his family's estate in the English countryside, he published nothing about it for many years. Even the *Principia,* which appeared in 1687, made almost no mention of it. Some of Newton's acquaintances and scientific colleagues, particularly in England, became aware of some aspects of his mathematical discovery only through correspondence with him, which sometimes included examples of problems and their solutions.

Meanwhile the eminent German

mathematician G. W. Leibniz quite independently discovered the principles of calculus and began publishing articles as early as 1684. Subsequently Leibniz and his followers developed the method to a far greater extent than Newton had. Thus there is no doubt that Leibniz was more responsible than Newton for the widespread adoption of calculus and its continued study throughout learned circles.

Initially the two men made little reference to each other's work, and when they did, both appeared to acknowledge the likelihood that each individually and independently had invented the method (for example, Newton, in his only brief mention of his method of fluxions in the *Principia,* stated so explicitly). Beginning in the 1690s, however, there arose a controversy over the discovery of calculus, and it eventually became extremely bitter and

ognizing inertia as just one in a series of physical principles that govern the motions of objects, and adding the all-important notion of **mass.**

The mass of an object reflects the amount of matter it contains, which in turn determines other properties such as weight and momentum. It is the other properties that are easily observed, not the mass, so mass can be a difficult concept. We can make useful illustrations by considering situations where we alter these other properties but not the mass. For example, the weight of an object varies according to the strength of the gravitational field to which it is subjected, but its mass does not change. An astronaut may be weightless, but in space he contains just as much mass as he does on the ground.

Mass is usually measured in units of **grams** or **kilograms.** One gram is the mass of a cubic centimeter of water, and a kilogram, which weighs about 2.2 pounds

at sea level, is the mass of 1,000 cubic centimeters (or one **liter**) of water.

An observed property that is very closely related to mass is **inertia,** which may be defined as an object's resistance to having its state of motion or rest altered. The more massive an object is, the more inertia it has, and the more difficult it is to start it moving, or to stop or alter its motion once it is moving (Fig. 4.3). For example, imagine two crates of equal size, one containing books, the other, pillows. Each has wheels on the bottom, minimizing friction. You may be able to move the box of pillows rather easily, but not the box of books. The box of books contains more mass, and therefore has more inertia; more resistance to being moved. Similarly, a massive object in motion is more difficult to turn or stop than a less massive one. Imagine, for example, the deftness required to bring a large ship to rest at a pier without destroying the pier.

ugly. In the end both of these great scientists acted very dishonorably.

Apparently the trouble began in 1693 when, an English publication compared the works of Newton and Leibniz for the first time, and, noting the similarities of their mathematical methods, stated that the calculus of Leibniz was really the method of fluxions of Newton, with a different name. This seemed to assert Newton's priority and at the same time to imply indirectly that Leibniz got his idea from Newton's work. There followed an incredible series of letters and articles from both sides, all widely publicized, in which the followers of Newton and of Leibniz gradually escalated the affair into outright charges of plagiarism. The fact that most of the vitriolic statements were made not by the two principal characters, but instead by their supporters, may seem to exonerate Newton and Leibniz from responsibility for the sordid affair, but this is a false impression. Despite their efforts to conceal their roles, both Newton and Leibniz, it later became clear, behaved very badly. Newton was in fact the behind-the-scenes director of all the attacks on Leibniz, though none actually bore his name, and as president of the Royal Society, he oversaw (again, from outside the public view) an inquiry into the matter that was totally biased in his favor from the start. For his part, Leibniz wrote anonymous public statements impugning Newton's character, and, like Newton, had friends and supporters carry the brunt of the attack. The controversy reached its peak in the years between 1710 and 1715.

In the end no satisfactory settlement was reached, and bitterness smoldered for many decades after the deaths of both men, that of Leibniz in 1716 and Newton eleven years later. Modern historians have done a remarkable job of uncovering and analyzing the correspondence and secret writings from the years of turmoil, and discovering the true roles of both Newton and Leibniz. It has also been established beyond reasonable doubt that both of them deserve credit for the discovery of calculus; from references in Newton's correspondence, it is clear that he had indeed developed it to a well-advanced state before 1670, but it is equally clear, on the other hand, that Leibniz could not have known of Newton's work at the time of his own invention of the method.

From the entire incident, which was surely the greatest scientific controversy of all time, we learn that even the gifted among us are subject to common human foibles.

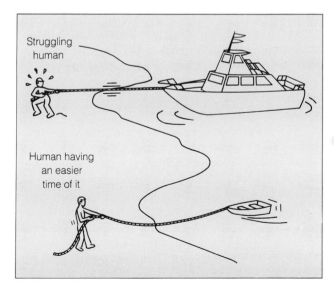

FIG. 4.3.   INERTIA. Here two people are attempting to pull two boats of rather different masses to shore. Both boats are at rest initially, and the more massive one resists being started in motion much more strongly than the less massive one.

Newton summarized the concept of inertia in what has become known as his first law of motion:

A body at rest or in a state of uniform motion tends to stay at rest or in uniform motion unless an outside force acts upon it.

As we have seen, this had been stated by Galileo, although he did not make it quite so general as Newton, who realized that it applies to the earth, planets, and all celestial bodies. One very important consequence of the first law of Newton is the implication that a planet

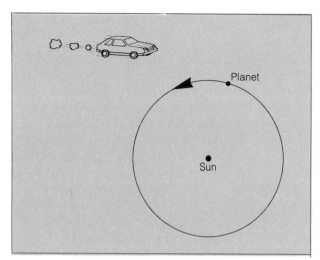

FIG. 4.4. ACCELERATION. A car that is speeding up is accelerating, as is a planet that follows a curved path, even if its speed is constant. *Any* change of speed *or* direction is an acceleration.

FIG. 4.5. ACTION AND REACTION. Some manifestations of Newton's third law are rather subtle, whereas others are not. Here a book exerts a force on a table, and the table exerts an equal and opposite force on the book. This is a static situation; there is no motion. When a cannon is fired, however, the cannonball is accelerated one way and the cannon the other. The force applied to each is the same, but the cannon has more mass than the cannonball and is therefore accelerated less, as Newton's second law states.

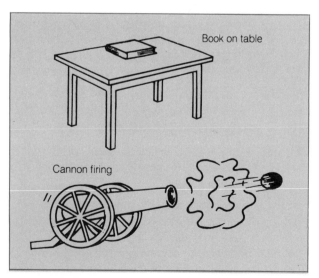

should follow a straight line, rather than a curved path around the sun, unless some force acts upon it. We shall return to this point shortly.

FORCE AND ACCELERATION

Having stated that a force is required to change an object's state of rest or uniform motion, Newton went on to determine the relationship between force and the change in motion that it produces. To understand this, we must discuss the idea of acceleration.

Acceleration is a general word for *any* change in the motion of an object. It is acceleration when a moving object is speeded up or slowed down, or when its direction of motion is altered (Fig. 4.4). It is also acceleration when an object at rest is put into motion. A planet orbiting the sun is undergoing constant acceleration; otherwise it would fly off in a straight line.

Another way of stating Newton's first law is that in order for an object to be accelerated, a force must be applied to it. His second law spells out the relationship among this force, the resultant acceleration, and the mass of the object:

> An object is accelerated when a force is applied to it, and the rate of acceleration is equal to the force divided by the mass of the object.

This may be written mathematically as $a = F/m$, where a is the acceleration, F is the force, and m is the mass. More commonly it is written in the equivalent form $F = ma$.

We can illustrate the second law as follows: If one object has twice the mass of another and equal forces are applied to the two, the more massive one will be accelerated only half as much. Conversely, if unequal forces are applied to objects of equal mass, the one to which the greater force is applied will be accelerated to a greater speed.

ACTION AND REACTION

Newton's third law of motion is probably more subtle than the first two, although in some circumstances it is quite obvious. It states:

> For every action there is an equal and opposite reaction.

In other words, when a force is applied to an object, it pushes back with an equal force (Fig. 4.5). This may

FIG. 4.6. NEWTON'S THIRD LAW APPLIED TO THE LAUNCH OF A ROCKET. Hot gases are forced out of a nozzle (or several, as in this case), and in return they exert a force on the rocket that accelerates it. This is the first launch of the Space Shuttle *Columbia*.

sound confusing, because the acceleration is not necessarily equal, and because in most common situations other forces such as friction complicate the picture.

We can visualize the third law by considering instances where friction is not important. For example, imagine standing in a small boat and throwing overboard a heavy object, such as an anchor. The boat will move in the opposite direction from the anchor, because the anchor exerts a force on you as you throw it. The "kick" of a gun when it is fired is another example of action and reaction. When a person jumps off the ground by pushing against the earth, both he and the earth are accelerated by the mutual force, but of course the immensely greater mass of the earth prevents it from being accelerated noticeably.

The third law of motion states the principle on which a rocket works. In this case hot, expanding gas is allowed to escape through a nozzle, creating a force on the rocket. The gas is accelerated in one direction and the rocket is accelerated in the opposite direction (Fig. 4.6). Anyone who has inflated a balloon and then let

go of it, allowing it to zoom through the air, is familiar with the operating principle of a rocket.

Gravitation and Orbits

We return now to a point raised earlier; namely, that the planets would fly off along straight lines if no force were acting upon them. Newton's first law says that this should happen, yet it obviously does not. Newton realized that the planets must be undergoing constant acceleration toward the center of their orbits; that is, toward the sun (Fig. 4.7). He set out to understand the nature of the force that creates this acceleration.

If you feel confused about the direction of this force, remember what you have just learned about acceleration and inertia: a planet needs no force to keep it

FIG. 4.7 PLANETARY ORBITS AND CENTRAL FORCE. A planet would fly off in a straight line if no force attracted it toward the center of its orbit. From diagrams like the one at right, Newton was able to determine the amount of acceleration required by an orbiting body to keep it in its orbit. This led him to discover the law of gravitation.

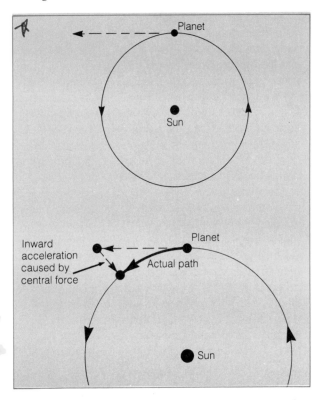

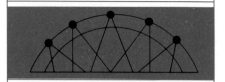

The Forces of Nature

One of the most difficult concepts to grasp is that of a force. We think of a force as a push or a pull, and we can visualize it quite easily in certain circumstances, such as when a gardener pushes a wheelbarrow and it moves. It is perhaps less obvious when no movement results, and even less so when no tangible agent exerts the force. In the case of gravity, the force is exerted invisibly and over great distances.

In general terms, a force exerted by a concrete object is referred to as a **mechanical force,** whereas one exerted without any such agent is a **field force.** Gravity is the most common force created by a field.

There are four kinds of field forces in nature, and they form the basis for *all* forces, mechanical or field. Gravity was the first to be discovered, although in "discovering" gravity, and in mathematically describing its behavior, Newton did not develop a real fundamental understanding of how it works, or why. The reason gravity was discovered first is simple: no special conditions are required for gravitational forces to be exerted (*all* masses attract each other), and it operates over very great distances. After all, Newton deduced its properties by noting how it controls the motions of plan-

ets that are separated from the sun by as much as a billion kilometers.

The second field force to be discovered and described mathematically, is the **electromagnetic force.** There is a close relationship between electric and magnetic fields. The interaction of these fields with charged particles, first described mathematically by the Scottish physicist James Clerk Maxwell in the late 1800s, creates forces. The electromagnetic force is actually much stronger than gravity; when binding an electron to a proton in the nucleus of an atom, this force is 10^{39} times stronger than the gravitational force between the two particles. The electromagnetic force, like gravity, is inversely proportional to the square of the distance between charged particles. Therefore we might expect this force to always dominate over gravity, as it does on a subatomic scale, but it does not. The reason is that most objects in the universe, composed of vast numbers of atoms containing both electrons and protons, have little or no electrical charge, whereas they always have mass and are therefore subject to gravitational force. If the planets and sun had electrical charges in the same proportion to their masses as the electrons and protons do, then electromagnetic forces, rather than gravity, would control their motions. Besides being immensely stronger, the electromagnetic force also differs from gravity in that it can be either repulsive or attractive, depending on whether the electrical charges are the same or opposite.

The other two forces involve interactions at the subatomic level,

and it was not until the science of quantum mechanics was developed in the 1930s that they were discovered. One is the **strong nuclear force,** which is responsible for holding together the protons and neutrons in the nucleus of an atom, and the other is the **weak nuclear force,** some 10^{-5} times weaker than the strong nuclear force. The strong nuclear force in turn is about a hundred times stronger than the electromagnetic force, making it the strongest of all, but it operates only over very small distances. Within an atomic nucleus, where protons with their like electrical charges are held together despite their electromagnetic repulsion for each other, the strong nuclear force acts as the glue that keeps the nucleus from flying apart. The weak nuclear force plays a more subtle role, showing its effects primarily in certain modifications of atomic nuclei during radioactive decay.

From the smallest to the largest scales, these four forces appear to be responsible for all interactions of matter. It is ironic that we do not yet understand the mechanism that makes the forces work; to do so is one of the principal goals of modern physics. One hope is to develop a mathematical framework that encompasses all the forces, a framework first sought more than sixty years ago by Albert Einstein, and referred to as a **unified field theory.** Progress has been made since Einstein's time, particularly toward unifying the electromagnetic and nuclear forces, but the ultimate goal still eludes us.

moving, but it does require a force to keep its path curving as it travels around the sun. What is needed is something to push the planet inward, toward the sun. This force can be compared to the tension in a string tied to a rock that you whirl about your head; if you suddenly cut the string, the rock would fly off in whatever direction it happened to be going at the time.

What, then, is the string that keeps the planets whirling about the sun? Newton realized that the sun itself must be the source of this force, and he made use of Kepler's third law, as well as observations of the moon's orbit and falling objects at the earth's surface, to discover its properties. He was led to formulate his law of universal gravitation:

> Any two bodies in the universe are attracted to each other with a force that is proportional to the masses of the two bodies and inversely proportional to the square of the distance between them.

The law of gravitation is one of the fundamental rules by which the universe operates. As we shall see in later chapters, it explains the motions of stars about each other or about the center of the galaxy, the movements of the moon and planets, and the motions of galaxies about one another. Gravity, in fact, is the dominant factor that will determine the ultimate fate of the universe.

It is useful to consider a few examples illustrating how the law of gravitation is applied. The weight of an object is simply the gravitational force between it and the earth. If, for example, the diameter of the earth were suddenly doubled (while its mass remained constant), our weight would decrease by a factor of four. If the earth were three times smaller, our weight would be $3^2 = 9$ times greater. If we climb to the top of a high mountain our weight decreases, but not very much, because even the highest mountains are small compared to the radius of the earth. The earth exerts a gravitational force on an astronaut in orbit, but there is no acceleration relative to his spacecraft, so he is weightless in his local environment (Fig. 4.8).

The law of gravitation states that the force also depends on the masses of the two objects that are attracting each other. It is easy to imagine that a person's weight would double if his mass doubled; similarly, the force between the earth and the moon depends on the masses of these two bodies. If we triple the mass of the moon, the force is tripled; if we triple the mass

FIG. 4.8. WEIGHTLESSNESS. Although astronauts orbiting the earth are subject to the earth's gravitational force and are "falling around" the earth along with their spacecraft, they are weightless because they experience no gravitational acceleration relative to their surroundings. This photo shows two astronauts aboard *Skylab*, an orbiting scientific research station, in 1976.

of the moon but decrease the mass of the earth by a factor of three at the same time, the force remains unchanged.

When we consider the surface gravity of a planet, we must take into account both its size and its mass. The moon, for example, has 0.273 times the radius of the earth, and 0.0123 times the mass. Thus the acceleration of gravity at the moon's surface is $0.0123/0.273^2 = 0.165$ times the acceleration at the surface of the earth. Therefore an astronaut on the moon weighs approximately one-sixth as much as he does on the earth.

Galileo had shown that the acceleration of gravity at the surface of the earth does not depend on the mass of the object that is falling. This was shown mathematically by Newton, using the law of gravitation and the second law of motion. The second law states that the acceleration of a falling object is equal to the gravitational force acting on it, divided by its mass. The gravitational force, in turn, is proportional to the mass of the object, so when we divide the force by the mass of the object, the mass cancels out and the acceleration does not depend on mass at all. In equation form, we say that the force is

$$F = GmM/R^2, \text{ and that the acceleration is}$$
$$a = F/m = GM/R^2,$$

where m represents the mass of the falling object, M the mass of the earth, R the earth's radius, and G the gravitational constant. The m representing the mass of the object does not appear in the final expression for the acceleration. Normally the gravitational acceleration is designated by the lowercase letter g, which could have been used in place of a in these equations.

Energy, Angular Momentum, and Orbits: Kepler's Laws Revisited

Even though Newton made use of Kepler's third law in deriving the law of gravitation, the latter is in fact more fundamental. It was soon possible for Newton to show that all three of Kepler's laws follow directly from Newton's laws of motion and gravitation. Kepler's studies of planetary motions revealed the *result* of the laws of motion and gravitation, whereas Newton found the *cause* of the motions. To appreciate how this was done, we must further discuss some basic physical ideas.

An important concept in understanding not only orbital motions but also many other aspects of astrophysics is **energy.** In an intuitive sense, we can define as energy as the ability to do work. Energy can take on many possible forms, such as electrical energy, chemical energy, heat, and others. All forms of energy can be classified as either **kinetic energy,** the energy of motion, or **potential energy,** which is stored energy that must be released (converted to kinetic energy) if it is to do work. A speeding car has kinetic energy because of its motion, whereas a tank of gasoline has potential energy in the form of its chemical reactivity, a tendency to release large amounts of kinetic energy if ignited. Thus a car operates when this potential energy is converted to kinetic energy in its cylinders.

The units used for measuring energy can be ex-

pressed in terms of the kinetic energy of specified masses moving at specified speeds. The kinetic energy of a moving object is $\frac{1}{2}mv^2$ where m is the mass of the object and v is its velocity. In astronomy the most commonly used unit of energy is the **erg,** a small amount of energy that is equivalent to the kinetic energy of a mass of two grams moving at a speed of one centimeter per second (the technical definition is that an erg is the amount of energy required to move a mass of one gram a distance of one centimeter). A larger unit often used by physicists is the **joule,** the equivalent of a mass of two kilograms moving at a speed of one meter per second. Thus one joule is equal to 10^7 ergs.

We often speak in terms of **power,** which is simply energy expended per second. The familiar **watt** is one joule per second; astronomers, however, tend to use units of ergs per second, which have no special name as a unit. Thus we will speak of the power, (or, equivalently, of the **luminosity**), of a star in terms of its energy output in ergs per second.

Using this understanding of energy, we can now discuss orbital motions in a much more general way than we have previously done. Two objects subject to each other's gravitational attraction have kinetic energy resulting from their motions, and potential energy because they each feel a gravitational force. Just as a book on a table has potential energy that can be converted to kinetic energy if it is allowed to fall, an orbiting body also has potential energy by virtue of the gravitational force acting on it.

Newton's laws and the concepts of kinetic and potential energy can be used to show that there are many types of possible orbits when two bodies interact gravitationally. Not all are ellipses, because if one of the objects has too much kinetic energy (exceeding the potential energy caused by the gravity of the other), it will not stay in a closed orbit, but will instead follow an arcing path known as a hyperbola, and will escape after one brief encounter. Some comets have so much kinetic energy that after one trip close to the sun, they escape forever into space, following hyperbolic paths. A rocket launched from earth with sufficient velocity will escape into a similar orbit (Fig. 4.9).

If the kinetic energy is less than the potential energy, as it is for all the planets, then the orbit is an ellipse, as Kepler found. It is technically correct to say that a planet and the sun orbit a common **center of mass** (Fig. 4.10), rather than saying that the planet orbits the sun. The center of mass is a point in space between the two bodies where their masses are essentially balanced; more specifically, it is the point where the product of mass times distance from this point is equal for the two objects. Since the sun is so much more massive than any of the planets, the center of mass for any sun-planet pair is always very close to the center of the sun, so the sun moves very little, and we do not easily see its orbital motion. It is true, however, that the sun's position wiggles a little as it orbits the centers of mass established by its interaction with the planets, especially the most massive ones. In a double-star system, where the two masses are more nearly equal, it is easier to see that both stars orbit a point in space between them. Thus Kepler's first law as he stated it requires a slight modification: each planet has an elliptical orbit, with the center of mass between it and the sun at one focus.

FIG. 4.9. ORBITAL AND ESCAPE VELOCITIES. A rocket launched with insufficient speed for circular orbit would tend to orbit the earth's center in an ellipse, but would intersect the earth's surface. Given the correct velocity it will follow a circular orbit. A somewhat larger velocity will place it in an elliptical orbit that does not intersect the earth. If given enough velocity so that its kinetic energy is greater than its gravitational potential energy, however, it will escape entirely.

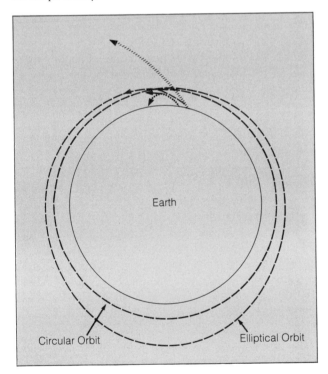

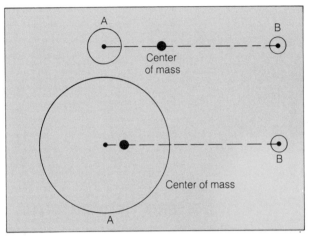

FIG. 4.10. CENTER OF MASS. The upper sketch depicts a double star where star A has twice the mass of star B, so the center of mass, about which both stars orbit, is one-third of the way between the centers of the two stars. The lower sketch shows a case where star A has ten times the mass of star B, so the center of mass is very close to the center of star A. The sun is so much more massive than any of the planets that the center of mass for any sun-planet pair is very near the center of the sun, so the sun's orbital motion is very slight.

The second law also can be restated in terms of Newton's mechanics. Any object that rotates or moves around some center has **angular momentum.** This is dependent on its mass, speed, and distance from the center of motion. In the simple case of an object in circular orbit, the angular momentum is the product mvr, where m is its mass, v its speed, and r its distance from the center of mass.

The total amount of angular momentum in a system is always constant. Because of this, a planet in an elliptical orbit must move faster when it is close to the center than when it is farther away, so that its velocity compensates for the changes in distance (Fig. 4.11). Thus a planet moves faster in its orbit near **perihelion** (its point of closest approach to the sun) than at **aphelion** (its point of greatest distance from the sun). Kepler's second law really stated that angular momentum is constant for an orbiting object.

Kepler's third law was also revised by Newton, and in this case the revision is especially important. Newton discovered that the relationship between period and semimajor axis depends on the masses of the two objects. Kepler had not realized this, primarily because the sun is much more massive than any of the planets, so that the differences among the masses of the planets

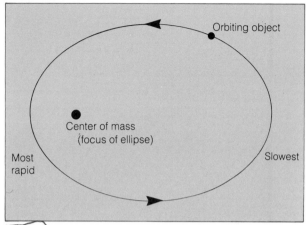

FIG. 4.11. CONSERVATION OF ANGULAR MOMENTUM. An object moves in its elliptical orbit with varying speed, and the product of its mass times its velocity times its distance from the center of mass (that is, its angular momentum) is constant. Kepler's second law of planetary motion is a rough statement of this fact.

have only a very small effect. Kepler's form of the third law can be written $P^2 = a^3$, where P is the planet's period in years and a is the semimajor axis in astronomical units. Newton revised this to:

$$(m_1 + m_2)P^2 = a^3,$$

where m_1 and m_2 represent the masses of the two bodies, for example the sun and one of the planets. The masses must be expressed in terms of the sun's mass in this equation. If we use other units, such as grams for the masses, seconds for the period, and centimeters for the semimajor axis, then the equation is complicated by the addition of a numerical factor and is written $(m_1 + m_2)P^2 = 4\pi^2 a^3/G$, where G is the gravitational constant.

The practical importance of expressing Kepler's third law this way is that it becomes possible to use the law to determine the masses of distant objects. In any case where the period and the semimajor axis of an orbiting object can be observed directly, the equation can be solved for the sum of the masses:

$$m_1 + m_2 = a^3/P^2.$$

For example, consider how Newton could have derived the mass of Saturn. Titan, the largest of Saturn's moons, has an orbital period of 15.945 days = 0.04365 years, and a semimajor axis of 1,222,000 km = 0.00817 AU. Substituting these values for P and a in the expres-

sion just cited leads to $m_1 + m_2 = 0.000286$ solar masses; in other words the sun of the masses of Titan and Saturn is only 0.000286 times the mass of the sun. Since Titan is very much smaller than Saturn, we can assume that its mass is negligible and that we have found the mass of Saturn.

It is possible to determine the masses of orbiting stars or even orbiting galaxies by the same technique. We will find occasions to use Kepler's third law in this manner throughout our study of astronomy.

Escape Velocity and Gas Motions

The consideration of orbital motions in terms of kinetic and potential energy leads to the concept of an **escape velocity.** If an object in a gravitational field has greater kinetic than potential energy, it will entirely escape the gravitational field. To launch a rocket into space (that is, completely free of the earth) therefore requires giving it enough speed at launch so that its kinetic energy is greater than its potential energy caused by the earth's gravitational attraction (Fig. 4.9). It so happens that the speed required to accomplish this is the same for any mass of object. In equation form it is $v_e = \sqrt{2GM/R}$, where v_e is the escape velocity, G is the gravitational constant, and M and R are the earth's mass and radius, respectively. For the earth this velocity is 11.2 kilometers per second, or slightly more than 40,000 kilometers per hour.

The particles in a gas such as the earth's atmosphere move all the time, and the speed with which they do so is related to the temperature of the gas (Fig. 4.12). In a hot gas the atoms and molecules move more rapidly than in a cool gas. This is why a heated gas expands; the individual particles move more rapidly and therefore exert greater pressure on their surroundings. In a strict sense, temperature can be defined in terms of molecular motions (this leads directly to the concept of **absolute zero,** the temperature at which all molecular activity ceases). We can speak of the kinetic energy of individual particles in a gas, or more realistically, the average kinetic energy, which in turn is based on the average particle velocity. For any gas, the temperature is proportional to average kinetic energy.

In physics and astronomy, temperatures are usually

expressed in terms of the absolute scale, in which zero is absolute zero, and the degrees, equal to one one-hundredth of the difference between the freezing and boiling points of water, are called **Kelvins** (K). Water freezes at 273 K, and boils at 373 K. Room temperature is about 295 K. For details, see Appendix 3.

When we discuss the planets, we will be concerned with the likelihood of given gases escaping into space, thus leaving the atmosphere devoid of those gases. We can see from this discussion that the probability of a given gas escaping a planet depends on the temperature of the atmosphere. Recall also that the kinetic energy of a moving particle depends on its mass as well as its velocity; this means that in a gas where the particles have a uniform average kinetic energy, the more massive particles, on average, are moving more slowly than the light particles. Therefore the lightest gases are the most likely to escape, because their higher velocities are more likely to exceed the planet's escape velocity. We will find, for example, that some of the planets (including the earth) have lost their light gases such as hydrogen and helium while retaining the heavier ones. We will speak often of this concept in section III, where we describe the solar system and the properties of the planets.

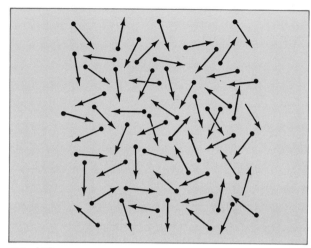

FIG. 4.12.   GAS MOTIONS. The average kinetic energy of particles in a gas is proportional to the temperature. If the gas particles, either because of high temperature or low mass, exceed a planet's escape velocity, they can escape into space. In any gas, there is a range of particle velocities, and if the average velocity for a particular gas is as much as one-sixth of the escape velocity, it will eventually escape altogether.

Tidal Forces

A number of important astronomical phenomena can now be understood in terms of gravitational forces. One is tides, which, as we will see, occur in many situations besides our earthly oceans.

We have seen that the gravitational force resulting from a distant body decreases with distance. This means that an object subjected to the gravitational pull of such a body feels a stronger pull on the side nearest that body, and a weaker pull elsewhere. For example, the side of the earth facing the moon feels the strongest attraction toward the moon, and the side opposite the moon feels the weakest force. The earth is therefore subjected to a **differential gravitational force,** which tends to stretch it along the line toward the moon. Of course the sun exerts a similar stretching force on the earth, but it is too distant to have as strong an effect as the moon (but its *total* gravitational force on the earth is much greater than that of the moon). A **tidal force,**

as such forces are called, depends on how close one body is to the other, because the key is how rapidly the gravitational force drops off over the diameter of the body subject to the tidal foce, and it drops off most rapidly at close distances.

The earth is more or less a rigid body, so it does not stretch very much as a result of the differential gravitational force of the moon. Nevertheless the tidal forces exerted on the earth by the moon create net forces that tend to make the liquid oceans flow in the direction of the points facing directly toward the moon and directly away from it (Fig. 4.13). As the earth rotates, the water in the oceans tends to follow the tidal forces created by the moon, so that in effect the oceans have two huge ridges of water that flow around the earth as it rotates. Since there are two ridges of water on opposite sides, and the earth rotates in twenty-four hours, at any given point on its surface one of these ridges passes roughly every twelve hours. Thus we have the ocean's tides, with high tides at any given location separated by about twelve hours (the separation is actually a bit longer than this, because the moon moves along in its orbit while the earth rotates).

It is interesting to consider what is happening to the moon at the same time. It is subjected to a more in-

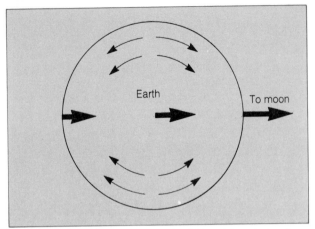

FIG. 4.13. THE EARTH'S TIDES. The differential gravitational force caused by the moon tends to stretch the earth (large arrows). Seawater at any given point on the earth is subjected to a combination of vertical and horizontal forces, causing it to flow toward either the side of the earth facing the moon or the side opposite it (light, curved arrows).

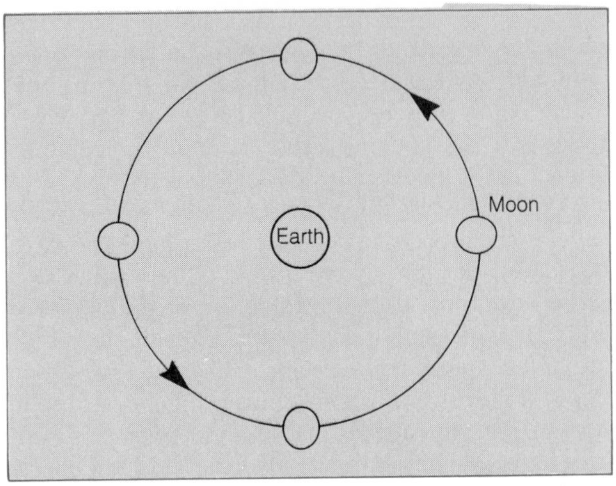

FIG. 4.14. SYNCHRONOUS ROTATION OF THE MOON. The moon is subjected to a tidal force caused by the earth that is strong enough to slightly deform the shape of the moon. This sketch shows how the elongated (here greatly exaggerated) moon keeps one bulge facing the earth at all times.

tense differential gravitational force than the earth, since the earth is more massive than the moon. Even though the moon is a solid body, its shape is deformed by this force, and it has tidal bulges. The moon is slightly elongated along the line toward the earth. As we shall see in chapter 8, this has had drastic effects on the moon's rotation, causing it to keep one side facing the earth (Fig. 4.14).

There are many other examples of tidal forces, both in the solar system and outside it. The satellites of the massive outer planets are subjected to severe tidal forces, and there are double-star systems where the two stars are so close together that they are stretched into elongated shapes. We will discuss these in more detail later, along with galaxies that are affected by tidal forces, sometimes even tearing each other apart.

Perspective

We have traced the development of astronomy to the point where we can now describe, in terms of a few simple laws of physics, the motions of the bodies in the solar system. These laws also provide a basis for understanding the motions of more distant astronomical objects. The law of universal gravitation will be invoked time and time again in our study of the solar system and the rest of the universe, because so many important phenomena are explained by it.

Newton's laws of motion and gravitation have stood the test of time rather well. Einstein's theory of general relativity, as we will learn in chapters 23 and 31, may be viewed as a more complete description of gravity and its interaction with matter. For most situations on the earth's surface and in space, however, Newton's laws are perfectly adequate.

We are ready now to discuss other laws of physics, particularly those which govern the emission and absorption of light.

Summary

1. Isaac Newton, in the late 1600s, developed the laws of motion, the law of gravitation, and calculus, and made many important contributions to our knowledge of the nature of light and telescopes.

2. Newton's first law states that an object at rest or in a state of uniform motion tends to remain in that state; this is the principle of inertia discovered by Galileo.

3. The second law of motion states that an object is accelerated by a force, the amount of acceleration being equal to the force divided by the mass of the object.

4. Newton's third law states that for every action, there is an equal and opposite reaction.

5. The law of universal gravitation states that any two objects in the universe attract each other with a force that is proportional to the product of their masses and inversely proportional to the square of the distance between them.

6. Newton's laws, along with the concepts of kinetic and potential energy and angular momentum, can be used to explain orbital motions.

7. Kepler's third law was modified by Newton to show that the relationship between the period and semimajor axis of an orbit is dependent on the sum of the masses of the two objects; this is an important tool for measuring the masses of distant objects.

8. Every body such as a planet or a star has an escape speed, the speed at which a moving object has more kinetic than potential energy and will therefore escape into space.

9. The average speed of particles in a gas is a function of the temperature of the gas and the mass of the particles; hence the probability that a given gas will escape a planetary atmosphere depends on these two quantities.

10. Differential gravitational forces are responsible for tides on the earth and in the interiors of other planets and satellites.

Review Questions

1. Explain, in your own words, why the law of inertia (Newton's first law) implies that the planets must experience a force attracting them toward the sun, and not in some other direction.

2. A book lying on a table is subject to a gravitational force of the earth, yet it is stationary. Explain why, in terms of Newton's laws of motion.

3. Assume that Saturn is ten times as far from the sun as the earth and that it has a hundred times the mass of the earth. Compare the gravitational force between Saturn and the sun with that between the earth and the sun.

4. If the diameter of the earth were suddenly reduced to one-fourth its present size (while the earth's mass remained constant), how would your weight be altered?

5. The summit of Mt. Everest is 8.84 kilometers above sea level. The radius of the earth is 6,371 kilometers. How much less would you weigh on top of Mt. Everest than at sea level?

6. Imagine a double star in which star A has twice the mass of star B. If the two stars are 3 AU apart, how far from star A is the center of mass between the two stars?

7. Suppose, as is believed to be the case, that the moon was once much closer to the earth than it now is. How would this have affected the tides in the earth's oceans?

8. The hydrogen molecule consists only of two protons and two electrons, whereas a nitrogen molecule contains a total of fourteen of each, as well as fourteen neutrons. Which gas, hydrogen or nitrogen, is more likely to escape from the atmosphere of a planet like the earth, and why?

9. When an ice skater in a spin pulls his arms in close to his body, his spin rate speeds up. Explain why, in terms of the conservation of angular momentum.

10. How would Kepler's laws have been different if we lived on a very massive planet, for example a planet with half as much mass as the sun?

11. How many high tides per day would there be if the earth rotated once every twelve hours instead of once every twenty-four hours?

Additional Readings

Authoritative and thorough information on Newton, as well as many additional references, can be found in the following works.

Cohen, I. B. 1960. Newton in light of recent scholarship. *Isis* 51:489.

———. 1974. Newton. In *Dictionary of scientific biography,* vol. 10, ed. C. G. Gillispie. New York: Scribner's.

Whiteside, D. T. 1962. The expanding world of Newtonian research. *History of Science* 1:15.

There are many books on conceptual understanding of physics in which more detailed explanations of the principles discussed in this chapter may be found. Many are used as textbooks in introductory physics courses, and should be easy to locate in a bookstore or library.

·CHAPTER 5·

The Nature of Light

·CHAPTER PREVIEW·

The electromagnetic spectrum
    Wavelengths, colors, and the spectrum
    Particles, waves, and the concept of the photon
    Continuous radiation and spectral lines
The continuous spectrum
    Temperature, color, and Wien's law
    Luminosity and the Stefan-Boltzmann law
    The effect of distance: the inverse-square law
The atom and spectroscopy
    Spectral lines and Kirchhoff's laws
    Electron energy levels and atomic structure
    The formation of spectral lines
    Ionization and excitation
    The Doppler effect and velocity measurements

Some of the tools for unlocking the secrets of the universe became available with the publication of Newton's *Principia* in the 1680s, but others had to wait two hundred years or more to be discovered. The laws of motion allow astronomers to understand how the heavenly bodies move, and were of fundamental importance in unraveling the clockwork mechanism of the solar system. To understand the essential nature of a distant object, however, to learn what it is made of and what its physical state is, requires an understanding of what light is and how it is emitted and absorbed. The only information we can obtain on the nature of a distant object is conveyed by the light from it. Fortunately, an enormous amount of information is there, if we know how to dig it out.

The Electromagnetic Spectrum

One characteristic of light is that it acts as a wave. It is possible to think of light as passing through space like ripples on a pond (although, as we will discuss shortly, the picture is actually somewhat more complicated than that). The distance from one wavecrest to the next, called the **wavelength** (Fig. 5.1), distinguishes one color from another. Red light, for example, has a longer wavelength than blue light. It is possible to spread out the colors in order of wavelength, using a prism to obtain the traditional rainbow. Newton was the first to dis-

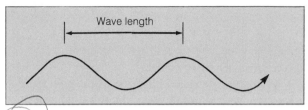

FIG. 5.1 PROPERTIES OF A WAVE. Light can be envisioned as a wave moving through space at a constant speed, usually designated *c*. The distance from one wavecrest to the next is the wavelength, often denoted λ. The frequency ν is the number of wavecrests to pass a fixed point per second; it is related to the wavelength and the speed of light by $\nu = c/\lambda$.

cover that sunlight contains all the colors, and he did so by carrying out experiments with a prism (Fig. 5.2). Whenever light is spread out by wavelength, the result is called a **spectrum;** more technically, a spectrum is the arrangement of light from an object according to wavelength. The science of analyzing spectra is called **spectroscopy,** and will be discussed at some length later in this chapter.

Let us consider for a moment what lies beyond red at one end of the spectrum or beyond violet at the

FIG. 5.2. NEWTON EXPERIMENTING WITH A PRISM.

other end. By the mid-1800s experiments had been carried out to demonstrate that there is invisible radiation from the sun at both ends of the spectrum. At long wavelengths, beyond red, is **infrared** radiation, and at short wavelengths is **ultraviolet** radiation. The spectrum continues in both directions, virtually without limit. Going toward long wavelengths, after infrared light, are microwave and radio waves; and toward short wavelengths, after ultraviolet, come X-rays and then gamma rays (γ-rays). All these kinds of radiation are just different forms of light, distinguished only by their wavelengths, and together they form the **electromagnetic spectrum** (Fig. 5.3). Electromagnetic radiation is a general term for all forms of light, whether it is visible, X-rays, radio, or anything else.

The range of wavelengths from one end of the electromagnetic spectrum to the other is immense. Visible light has wavelengths ranging from 0.00004 to 0.00007 centimeters. A special unit called the **Angstrom (Å)** is used to measure light, defined such that 1 cm = 100,000,000 Å, or 1 Å = 0.00000001 cm = 10^{-8} cm. Thus, visible light lies between 4,000 Å and 7,000 Å in wavelength.

A variety of other special units are used by astronomers in referring to electromagnetic radiation of different types, but to reduce confusion, we will use only Angstoms and centimeters. Even this is perhaps a bit complicated, but the usage is so standard that the student of introductory astronomy should be comfortable with both units.

Infrared light has wavelengths between about 7,000 Å and a few million Angstoms; that is, between 7 × 10^{-5} cm and 2 − 3 × 10^{-2} cm. Microwave radiation (which includes radar wavelengths) lies roughly between 0.1 and 50 cm, with no well-defined boundary separating this region from radio, which simply includes all longer wavelengths, up to many meters or even kilometers.

At the other end of the spectrum, ultraviolet light is usually considered to lie between 100 Å and 4,000 Å, whereas X-rays are in the range of 1 to 100 Å, anything shorter than that being considered γ-rays.

Thus the entire electromagnetic spectrum has a wavelength range from less than 1 Å to many kilometers; that is, from 10^{-10} cm to 10^{6} cm. In principle, it could extend even farther at both ends, but no known physical process is capable of producing such radiation. The portion of the spectrum to which the human

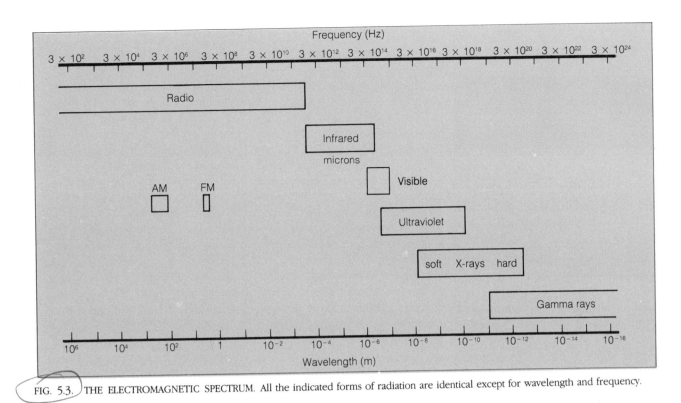

Frequency (Hz)

FIG. 5.3. THE ELECTROMAGNETIC SPECTRUM. All the indicated forms of radiation are identical except for wavelength and frequency.

eye responds is extremely limited, compared with the full range. The eye undoubtedly has evolved to be sensitive to just this region for a combination of reasons: (1) the sun emits most strongly in visible wavelengths; (2) the earth's atmosphere blocks out many wavelengths of radiation, but visible light lies in one of the wavelength regions that can pass through to the ground. Therefore, it was advantageous for creatures dwelling on the earth's surface to develop the ability to see radiation at these particular wavelengths.

The concept of **frequency** is often used as an alternative to wavelength in characterizing electromagnetic waves. The frequency is the number of waves per second that pass a fixed point, and it is determined by the wavelength and the speed with which the waves move. The speed of light, usually designated c, is constant, and the frequency of light (ν) with wavelength λ is $\nu = c/\lambda$. The standard-unit for measuring frequency is the **hertz** (Hz), one hertz being equal to one wave per second. The frequency of a photon of visible light with a wavelength of 5,000 Å is 6×10^{14} Hz. For a radio photon in the AM part of the spectrum, frequen-

cies are typically around 10^6 Hz, and at the other extreme, γ-rays have frequencies around 10^{19} Hz.

PARTICLE OR WAVE?

So far we have discussed light only as a wave, but there are situations in which it acts more like a stream of particles. Newton developed a corpuscular theory of radiation in which he assumed that light consisted of particles, but at the same times others, including most notably Huygens, carried out experiments showing light to have definite wave characteristics.

The fact that light bends around sharp corners, in a process called **diffraction** (which also affects water waves), is evidence for a wave nature, as is the fact that light waves can interfere with each other if their waves and crests match up in such a way as to cancel each other out. On the other hand, it has been found that light can carry energy only in specific amounts, as though it came in individual packets, and that it can travel in a total vacuum, with no medium to transmit it.

ASTRONOMICAL INSIGHT (5.2)

The Perception of Light

Newton's experiments in optics are commonly thought to have marked the beginning of scientific inquiry into the nature of light, but in fact Newton was preceded in this area by many scientists and philosophers. Much of the early research, and even to some extent Newton's work, did not deal with the physical properties of light, but rather was aimed at understanding how the human eye senses light. The ancient Greeks, the earliest philosophers who studied the nature of light, did not recognize the distinction between light and the act of seeing it.

The Greeks studied the principles of vision in much the same way they studied astronomy: they carried out very few experiments or even systematic observations, choosing instead to unravel the secrets of the universe by reason and logic. The prevalent concept, described by such notable figures as Hippocrates and Aristotle, was that the eye somehow emitted rays or beams with which it sensed things. This was analogous to the sense of touch, where a hand is extended to feel objects. It was known that objects appeared distorted or bent when viewed under water, but this was attributed to bending of the rays that came from the eye, not to effects on something coming from the object being viewed. It would have been possible for the Greeks to experiment with optics by using natural crystals or eyes removed from dead animals, but no such experimentation was done.

In a definitive summary of all that was known about vision, the Greek scientist Galen, in the second century A.D., wrote of an "animal spirit" that flowed from the brain along the optic nerve to the eye, where it was converted into a "visual spirit" in the retina, a membrane covering the rear interior portion of the eyeball. The lens, in the front of the eyeball, was thought to be responsible for sending out the beam with which the external world was perceived.

In analogy with the astronomical work of Ptolemy, the concepts of the nature of light and vision that were summarized by Galen were accepted as doctrine for some fifteen centuries to come, until the time of the Renaissance. Some of the ancient Greek ideas, such as the no-

---

Out of all the seemingly contradictory evidence has developed the concept of the **photon.** A photon is thought of as a particle of light that has a wavelength associated with it. The wavelength and the amount of energy contained in the photon are closely linked; in general terms, the longer the wavelength, the lower the energy. The energy can be expressed mathematically as $E = hc/\lambda$, where h is called the Planck constant, c is the speed of light, and λ is the wavelength. It is important to understand that a photon carries a precise amount of energy, not some arbitrary or random quantity, and that when light strikes a surface, this energy arrives in discrete bundles like bullets, rather than in a steady stream. When a photon is absorbed, this energy can be converted into other forms, such as heat.

From this discussion it may be seen that short-wave-length radiation, such as ultraviolet light, X-rays, and γ-rays, has high energy associated with it, whereas longer-wavelength photons have low energies. It is extremely important to remember the relationship between wavelength and energy in the discussions of the emission and absorption of light later in this chapter.

CONTINUOUS RADIATION AND SPECTRAL LINES

Newton, in examining the spectrum of light obtained by passing sunlight through a prism, saw only the smooth spread of light over all the colors. Later observers, however, notably W. H. Wollaston and Joseph Fraunhofer, were able to see dark lines across the sun's spectrum at certain fixed positions. Fraunhofer noted

tion that the optic nerve was hollow so that the animal spirit could flow along it from the brain, were especially persistent. Even Leonardo da Vinci, who in the late fifteenth and early sixteenth centuries carried out pioneering work in anatomy, accepted much of the old thinking. Da Vinci took a more modern view in that he believed that light rays passing from an object to the eye play an important role, but he was unwilling to entirely give up the idea that other rays emanate outward from the eye.

In the early seventeenth century, none other than Johannes Kepler and the great French philosopher René Descartes developed an understanding of refraction and the formation of an image by a lens, and both correctly viewed vision as the process of image formation on the retina by the lens in the front of the eyeball. Thus when Christiaan Huygens and then Newton carried

out their experiments later in the seventeenth century, it was already well established that seeing was accomplished by sensing something that passes from the object to the eye.

Both men were more concerned with the nature of this radiation than with the properties of the eye, although Newton did discuss the perception of color, which he realized to be entirely a function of the eye. This is an important concept: the rays of light are not colored, nor, technically, is the object being viewed; it is the eye that senses wavelength differences and the brain that interprets them as colors. Light rays consist simply of alternating electric and magnetic fields.

Other advances in understanding the physical properties of light, made by scientists who followed Newton, are described in the text.

Research has also been aimed at learning how the human eye perceives light. We now know that the eye has a logarithmic response, meaning that it does not perceive the true range of brightnesses that may be displayed by a series of objects, but instead perceives a lesser range. This point is explained more fully in chapter 18, in the discussion of the stellar magnitude system.

Entire books have been written on the complex interplay between what the brain perceives and physical reality, and of course, this interplay is important in astronomy. To a large extent, modern observational astronomy is the science of recognizing and bypassing the limitations or distortions created by our natural light-gathering system.

some six hundred of these lines, and in 1817 published a catalogue of many of the stronger ones (today these features are still called Fraunhofer lines).

It must be emphasized that the position of each of these lines corresponds to a specific wavelength of light. Remember that the spectrum is simply a spread of light according to wavelength, and that each position in the spectrum corresponds to a certain wavelength of light. The fact that the lines are dark means that somehow the sun emits less light at these particular wavelengths than at other wavelengths on either side.

In analyzing the light from a distant object such as a star, an astronomer may focus his attention on these lines, from which certain types of information can be derived, or he may ignore the lines and concentrate instead on the overall distribution of light with wavelength. Here we will make a distinction between **continuous radiation,** as the overall distribution is called, and spectral lines (Fig. 5.4).

CONTINUOUS RADIATION

A number of consistent properties of glowing objects can be applied to the study of stars. Most of these properties were discovered in the mid–1800s through laboratory experiments by people such as Wien, Stefan, Boltzmann, and Rayleigh, and in the 1920s were placed in proper theoretical framework by Bohr and Planck.

Any object with a temperature above absolute zero emits radiation over a broad range of wavelengths, simply because it has a temperature (Fig. 5.5). Thus, not only stars, but also such commonplace objects as

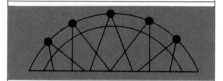

Electromagnetic Waves and Polarization

We have spoken of the wave nature of light, but we have not specified what it is that undulates or oscillates. It is easy to visualize waves in a fluid medium such as water, where we can see the motion of the water as the wave passes by. Electromagnetic waves are quite different, however. No medium is required: light waves can travel through a total vacuum, unlike most waves with which we are familiar.

Rather than having physical motions, electromagnetic waves consist of electric and magnetic fields that oscillate back and forth at right angles to the direction of motion. The electric and magnetic fields lie in planes that are also perpendicular to each other, so that the electric field flips back and forth in one plane, with the magnetic field doing so in the plane that lies perpendicular to it, and both are transverse to the direction of travel.

A photon, therefore, can be viewed as a packet of electromagnetic energy consisting of alternating fields, which, by virtue of their wavelength and frequency, provide the photon with its wave properties.

Light from most sources (such as stars) consist of vast numbers of photons, each with a specific, constant orientation of its electric and magnetic planes. Ordinarily the ori-entation of the various photons is random, but under some circumstances it is not. If light passes through or reflects from a medium that preferentially absorbs photons with a certain orientation, what is left is light in which all the photons are aligned; that is, the planes of the electric and magnetic fields of the photons are parallel. In this case, the light is said to be **polarized.** Many of the sunglasses sold today consist of polarizing filters, and they have the effect of screening out light in a given orientation. In astronomy we will learn that polarization occurs naturally in many situations, sometimes because a natural filtering occurs, as in the interstellar medium (see chapter 25), and in other cases because the source of the light emits it with a preferred orientation.

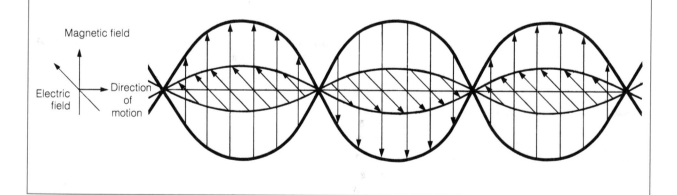

Magnetic field

Electric field

Direction of motion

the walls of a room or a human body, emit radiation. The continuous radiation produced because of an object's temperature is called **thermal radiation.** (We will discuss only this type of continuous radiation here; in other chapters, such as 12 and 30, we will examine nonthermal sources of radiation.)

For glowing objects such as stars, thermal radiation is emitted over a broad range of wavelengths (technically, in fact, at least some radiation is emitted at *all* wavelengths), with a peak in intensity at some particular wavelength. For the sun this peak is near a wavelength of 5,500 Å, in the green portion of the spectrum,

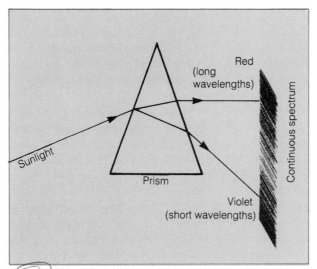

FIG. 5.4. CONTINUOUS SPECTRUM AND SPECTRAL LINES. When sunlight is dispersed by a prism, the light forms a smooth rainbow of continuous radiation, gradually merging from one color into the next, with a maximum intensity in the yellow portion of the spectrum. Superimposed on this continuous spectrum are numerous spectral lines, wavelengths where little or no light is emitted by the sun.

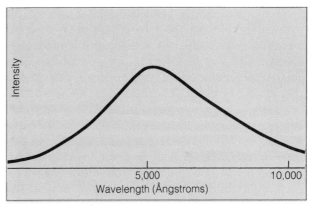

FIG. 5.5. AN INTENSITY PLOT OF A CONTINUOUS SPECTRUM. This kind of diagram shows graphically how the brightness of a glowing object varies with wavelength. The curve roughly represents the sun, whose continuous radiation peaks near 5,500 Å in the yellow-green portion of the spectrum.

but to our eyes the sun appears yellow in color because what we see as green corresponds only to a very narrow range of wavelengths, whereas the sun emits almost as strongly over the much broader range of wavelengths that we see as yellow.

A simple relationship between the wavelength of maximum emission and the temperature of an object was discovered in 1893 by W. Wien, and is now referred to as **Wien's law.** The law states that the wavelength of maximum emission is inversely proportional to the absolute temperature, that is, the hotter an object is, the shorter the wavelength of peak emission (Fig. 5.6). This explains the variety of stellar colors as being a result of a range in stellar temperatures. A hot star emits most of its radiation at relatively short wavelengths, and thus appears bluish in color, whereas a cool star emits most strongly at longer wavelengths and appears red. The sun is intermediate in temperature and color.

When speaking of the colors of stars, we must keep in mind that a star emits light over a broad range of wavelengths, so we do not have pure red or pure blue stars. Our eyes receive light of all colors, and stars therefore are all essentially white. Our impression of color arises from the fact that there is a wavelength

(given by Wien's law) at which a star emits more strongly than at other wavelengths. It is not as though the star emits *only* at that wavelength.

Wien's law applies as well to radiation beyond the visible range. The hottest stars actually emit most strongly in ultraviolet wavelengths, and very cool ones emit mostly in the infrared. Indeed, for every object with a temperature above absolute zero, the wavelength of maximum radiation is dictated by the temperature of the object. The human body, as well as all objects in the normal room-temperature environment,

FIG. 5.6. CONTINUOUS SPECTRA FOR OBJECTS OF DIFFERENT TEMPERATURES. This diagram illustrates Wien's law, which states that the wavelength of maximum emission is inversely proportional to the temperature (on the absolute scale). The arrows indicate the wavelengths of maximum emission for objects of various temperatures.

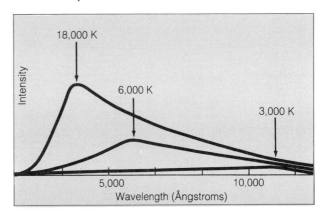

radiates in far infrared wavelengths. Imagine the problems this creates for the infrared astronomer, whose telescope glows at the very wavelengths he is trying to observe! (We will describe the solution to this problem in chapter 6.)

A second property of glowing objects, known as either Stefan's law or the Stefan-Botzmann law, has to do with the total amount of energy emitted over all wavelengths, and how this total energy is related to the temperature of an object. This law says: the total energy radiated per square centimeter of surface area is proportional to the fourth power of the temperature. This shows that the total energy emitted is very sensitive to the temperature; if we change the temperature a little, we change the energy a lot. If, for example, we double the temperature of an object (such as the electric burner on our stove), we increase the total energy it radiates by $2^4 = 2 \times 2 \times 2 \times 2 = 16$. If one star is three times hotter than another, it emits $3^4 = 3 \times 3 \times 3 \times 3 = 81$ times more total energy per square centimeter of surface area.

Notice that we have been careful to express this law in terms of the surface area. The total energy emitted by an object depends also on how much surface area it has. If we talk of stars or other spherical objects, which have a surface area of $4\pi R^2$, where R is the radius, we can say that the total energy emitted is proportional to the fourth power of the temperature and to the square of the radius. This can be illustrated by considering two stars, one that is twice as hot but has only half the radius of the other. The hotter star emits

$2^4 = 16$ times more energy per square centimeter of surface, but has only $\frac{1}{2}^2 = \frac{1}{4}$ as much surface area, hence it is $16 \times \frac{1}{4} = 4$ times brighter overall. If, on the other hand, this star were twice as hot and three times as large in radius as the other, it would be $2^4 \times 3^2 = 144$ times brighter.

Both Wien's law and the Stefan-Boltzmann law were first derived experimentally, much in the same manner as Kepler's discovery of the laws of planetary motion. In the case of planetary motions, it remained for Newton to find the underlying reasons for the laws, and he was able to derive them strictly on a theoretical basis. Analogously, Max Planck, the great German physicist who was active early in the twentieth century, found a theoretical understanding of thermal emission, and was able to derive Wien's law and the Stefan-Boltzmann law from purely theoretical considerations. The basis of Planck's new understanding was the **quantum** nature of light: the fact, already discussed, that light has a particle nature and carries only discrete, fixed amounts of energy.

Finally, in addition to taking into account the temperature and size of a glowing object, we must consider the effect of its distance. So far we have discussed the energy as it is emitted at the surface, but not how bright it looks from afar. What we actually observe, of course, is affected by our distance from the object. For a spherical object that emits in all directions, the brightness decreases as the square of the distance, as illustrated in Fig. 5.7 (this should remind you of the law of gravitation). Thus, if we double our distance from

FIG. 5.7. THE INVERSE-SQUARE LAW OF LIGHT PROPAGATION. This shows how the same total amount of radiant energy must illuminate an ever-increasing area with increasing distance from the light source. The area to be covered increases as the square of the distance; hence the intensity of light per unit of area decreases as the square of the distance.

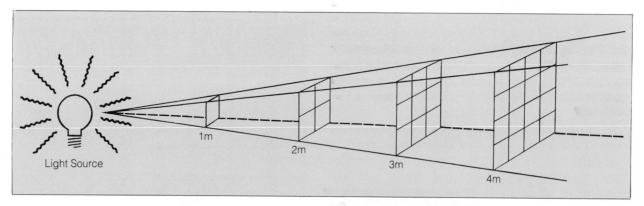

Light Source
1m
2m
3m
4m

a source of radiation, it will appear $\frac{1}{2}^2 = \frac{1}{4}$ as bright. If we approach, reducing the distance by a factor of two, it will appear $2^2 = 4$ times brighter. As shown in later discussions of stellar properties, we must know something about the distances to stars before we can compare other properties having to do with their brightnesses.

The Atom and Spectroscopy

We turn our attention now to the spectral lines first noted by Wollaston and Fraunhafer in the spectrum of the sun. These lines are difficult to see, and are not apparent in the spectrum of light obtained with an ordinary prism. However, with the development of superior techniques for examining the spectrum, their existence became well established.

In the late 1850's the German scientists R. Bunsen and G. Kirchhoff performed experiments and developed theories that made clear the importance of the Fraunhofer lines. Bunsen observed the spectra of flames created by burning various substances, and found that each chemical element produced light only at specific places in the spectrum (Fig. 5.8). The spectrum of such a flame in this case is dark everywhere except at these specific places, as though the flame were emitting light only at certain wavelengths. The bright lines seen in this situation are called **emission lines** for that reason.

It was soon noticed that some of the dark lines seen in the sun's spectrum coincide exactly in position with some of the bright lines seen by Bunsen in his laboratory experiments. Kirchhoff studied this in detail, and was able to show that a number of common elements such as hydrogen, iron, sodium, and magnesium must be present in the sun because of the coincidence in wavelengths of the lines. This was the first hint that the chemical composition of a distant object could be determined, and as such it was a major advance.

Kirchhoff further analyzed the absorption and emission of light and, through a series of experiments, was able to develop a set of rules describing when continuous radiation will be observed, when emission lines will be seen, and when **absorption lines** will form instead (see Fig. 5.9):

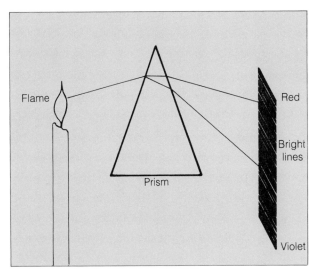

FIG. 5.8. EMISSION LINES. The spectrum of a flame is devoid of light except at specific wavelengths where bright emission lines appear.

1. A hot, dense gas or a hot solid produces a continuous spectrum (one with no lines).
2. A hot, rarefied gas produces emission (bright) lines.
3. A cool gas in front of a continuous source of light produces absorption (dark) lines.

These rules explain why Bunsen saw only emission lines when he observed the spectra of flames, which are rarefied, hot gases, and they tell us something important about the sun. The fact that the sun's spectrum contains absorption lines implies that the outer layers of the sun must be cooler than the interior (Fig. 5.10), so that the continuous radiation is formed inside, where it is hot and dense, and absorption lines are formed as this continuous radiation passes out through the outer layers. Thus a simple understanding of how gases in the laboratory emit and absorb light tells us something about the structure and internal temperature of the sun.

ENERGY LEVELS AND PHOTONS

The work of Bunsen and Kirchhoff demonstrated that each element has its own unique spectrum. As we have seen, this discovery had profound implications, for it meant that astronomers could gain information about distant objects previously thought to be forever be-

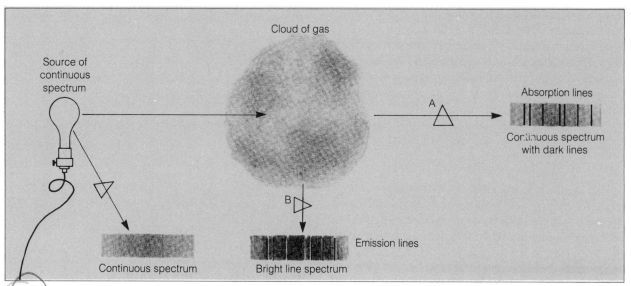

FIG. 5.9. CONTINUOUS, EMISSION-LINE, AND ABSORPTION-LINE SPECTRA. The positions of the emission and absorption lines match because the same element emits or absorbs at the same wavelengths. Whether it emits or absorbs depends on the physical conditions, as described in Kirchhoff's laws.

yond their grasp. Although astronomers in the latter half of the nineteenth century were able to use these astrophysical fingerprints, no real understanding of their nature and cause was developed until the early years of this century.

Spectral lines were studied by a number of scientists, and various regularities in the arrangement of the lines from a given element were noted. Although it was suspected that these regularities reflected some aspect of the structure of atoms, it was not until 1913 that the

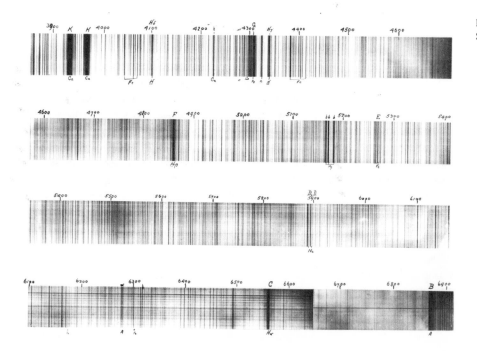

FIG. 5.10.   DARK LINES IN THE SUN'S SPECTRUM.

true relationship was discovered by Niels Bohr. By that time it had been established that an atom consists of a small, dense nucleus surrounded by a cloud of negatively charged particles called electrons. Bohr found that these electrons were responsible for the absorption and emission of light, gaining energy (absorption) or losing it (emission) in the form of photons.

The key to the fixed pattern of spectral lines for each element lies in the fixed pattern of energy levels the electrons can be in. We can visualize an atom as a miniature solar system, with electrons orbiting the nucleus. Each kind of atom (each element) has its own characteristic number of electrons, and in each case the electrons have a certain set of electron orbits. It may be helpful to visualize a ladder, with each rung representing an orbit, or energy level (Fig. 5.11). The spacing between the rungs of the ladder for one element, say hydrogen, are different from those of the ladder for any other element. The energy associated with each level increases, the higher up the ladder, or the farther from the nucleus, the electron goes.

An electron can absorb a photon of light *only* if the photon carries the precise amount of energy needed to move the electron to some higher level than the one it is in. Imagine an atom sitting in space, with light streaming past it: the atom will let pass every photon that comes by, until it finds one with just the right amount of energy to boost one of its electrons to a higher level. Since the energy of a photon is closely related to its wavelength, this means that the atom will only absorb photons that have specific wavelengths, corresponding to the spacings between its energy levels. Because each kind of atom has it own unique set of energy levels, each can absorb light only in its own unique set of wavelengths. Thus Bohr was able to explain the findings of Bunsen and Kirchhoff.

Emission is the reverse process of absorption. Emission will occur whenever an electron in an upper level drops down to a lower one, and the energy of the photon emitted corresponds, as before, to the energy difference between the two levels. Since the energy levels are fixed for a given element, the emission lines and the absorption lines occur at the same wavelengths. Whether an atom produces emission lines or absorption lines depends on whether the electrons are moving downward or upward.

We can now state Kirchhoff's three rules in a difference form:

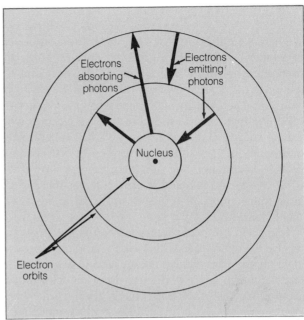

FIG. 5.11. THE FORMATION OF SPECTRAL LINES. An electron must be in one of several possible orbits, each representing a different electron energy. If an electron absorbs exactly the amount of energy needed to jump to a higher (more outlying) orbit, it may do so. This is how absorption lines are formed, because the wavelength of photon absorbed corresponds to the energy difference between the two electron orbits, which is fixed for any given kind of atom. Conversely, when an electron drops from one orbit to a lower one, it emits a photon whose wavelength corresponds to the energy difference between orbits, and this is how emission lines form.

1. In a hot, dense gas or a hot solid, the atoms crowd together so that their enegy levels overlap and all their lines blend together,[1] and we see a continuous spectrum.

2. In a hot, rarefied gas, the electrons tend to be in high energy states, and create emission lines as they drop to lower levels.

3. In a cool gas in front of a hot, continuous source of light, the electrons tend to be in low energy levels, and absorb radiation from the background continuous source.

[1]The production of a continuous spectrum is actually a bit more complex than this. When free electrons combine with ions, they emit at any wavelength, not just in spectral lines. See the discussion of ionization in the following section.

IONIZATION AND EXCITATION

We have seen that electrons can move from one level to another by giving up or gaining energy. Even though each atom has many levels, there is always a limit to the range of levels an electron can have. If the electron gains too much energy, more than what is necessary to reach the highest level the atom has, the electron will escape entirely (Fig. 5.12). This process is called **ionization,** and it leaves the atom with one electron short of its usual number. The atom then has a positive electrical charge and is called an **ion.** It is possible for an atom to lose more than one electron, becoming more highly ionized.

Ionization can occur in two ways. The electron can receive the energy needed for escape by either absorbing a photon with enough energy or colliding with another atom. In any gas, the individual particles move about randomly, occasionally colliding with each other. Energy may be passed from one atom to another in such a collision, and when it is, the electrons of the atom that gained the energy may move up a few rungs in the ladder of energy levels. If the collision takes place at sufficiently high velocity, one or more of the electrongs may gain enough energy to escape. The speed of motion of the atoms depends on the temperature of the gas, so the hotter the gas is, the more ionizations

occur. Hence a hot star has a higher degree of ionization in its outer layers than a cool star does.

The spectrum of an atom changes drastically when it has been ionized, because the arrangement of energy levels is altered, and because different electrons are now available to do the absorbing and emitting of photons. The spectrum of atomic helium, for example, is quite different from that of ionized helium, so the astronomer not only can see that helium is present in the spectrum of a star, but also can tell whether it is ionized. This provides information on the temperature in the outer layers of the star, where the absorption lines are formed. By analyzing the degree of ionization of all the elements seen in the spectrum of a star, the astronomer can determine the gas temperature quite precisely.

Let us now consider an atom that is not ionized, and which collides at relatively low velocity with another atom. Its electrons may gain energy, but if it is not enough to cause ionization, the electrons may land in energy levels that are higher than those they were in before the collision (Fig. 5.12). Ordinarily an electron will tend to stay in the lowest available energy state, and when it gets into a higher state, it is said to be **excited,** or in an **excited state.** When an electron does move to an excited state, it usually will drop back down immediately, and it will emit a photon each time it descends from one level to a lower one. In gas where collisions are rare, at any given time very few electrons will be in excited states. In a dense gas, however, where collisions occur often, at any moment a certain fraction of the atoms will have their electrons in excited states. It should be clear that if an electron is in an excited state, the wavelengths at which it can absorb photons are different from those at which the electron can absorb when it is in the lowest level. The amount of energy required for an electron to move upward in the ladder of energy levels depends on which rung the electron is on to begin with; hence the wavelength of photon that it can absorb depends on which level it is in. Therefore the spectrum of an atom depends on how excited the atom is (Fig. 5.13). Observation of which lines appear in the spectrum of a star can tell the astronomer how many electrons are in excited states. This in turn reveals how frequently collisions are taking place in the star's outer layers, so in the end it is possible to infer the density of the gas.

From careful analysis of a star's spectrum, then, we can determine both the temperature of the outer lay-

FIG. 5.12.   EXCITATION AND IONIZATION. In the energy-level diagram at left, electrons are jumping up from the lowest level to higher ones. This is excitation, and it can be caused by either absorption of photons or collisions between atoms. On the right, electrons are gaining sufficient energy to escape altogether, a process called ionization.

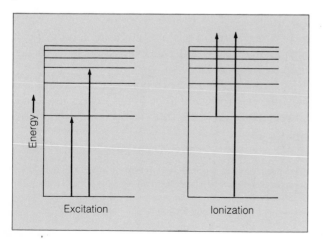

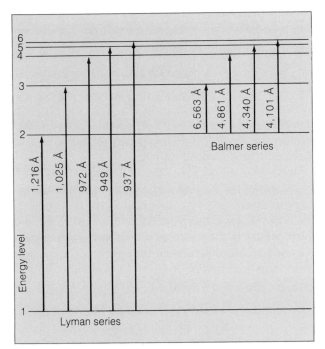

FIG. 5.13. THE EFFECT OF EXCITATION ON THE SPECTRUM OF HYDROGEN. As shown here, the absorption lines that can be formed by an atom depend on its degree of excitation. At left are the wavelengths of absorption lines originating in the lowest energy level of hydrogen, and at right are the wavelengths of absorption lines arising from an electron in the first excited level. Note that the Lyman lines are in ultraviolet wavelengths, whereas the Balmer lines are in the visible portion of the spectrum. This sketch shows only a few of the many energy levels of hydrogen.

ers of the star (from the ionization) and the density (from the excitation). Although most of the examples used in the discussion have referred to the sun or the stars, the same considerations can be applied to the spectra of the planets.

THE DOPPLER EFFECT

There is something else that spectral lines can tell us about a distant object, in addition to all the physical data we have been discussing. We can also learn how rapidly an object such as a star or planet is moving toward or away from us.

To understand this, we must visualize light from a distant object as though it were a series of concentric waves moving away from the source like ripples on a pond where a pebble has been dropped in (Fig. 5.14).

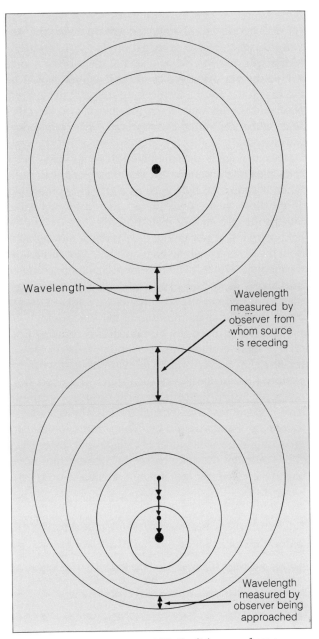

FIG. 5.14. THE DOPPLER EFFECT. The light waves from a stationary source (upper) remain at constant separations (that is, constant wavelength) in all directions, whereas those from a moving source get "bunched up" in the forward direction and "stretched out" in the trailing direction. This causes a blueshift or a redshift for an observer who approaches or recedes from a source of light (note that it does not matter which is moving, source or observer).

If the source of light is moving, the waves will tend to get bunched up in the direction toward which the motion is occurring, and stretched out in the opposite direction. An observer sitting ahead of the moving source will see waves that are closer together than would be the case if the source were sitting still; that person will observe a shorter wavelength. An observer on the other side, with the source moving away, will see longer wavelengths.

This analogy is oversimplified in some ways, but it does illustrate an effect that really happens with light. When a source of light such as a star is approaching, all the lines in its spectrum are shifted toward wavelengths that are shorter than those observed when the light source is at rest, and if the source is moving away from the observer, all the lines are shifted toward longer wavelengths (Fig. 5.15). These two situations are called **blueshifts** (approach) and **redshifts** (recession), respectively, because the spectral lines are shifted toward either the blue or the red end of the spectrum.

A general term for any wavelength shift resulting from relative motion between source and observer is **Doppler shift,** in honor of the German physicist who first explored the properties of such shifts. The effect does not apply just to light waves. Most of us, for example, have noticed the Doppler shift in sound waves when a source of sound passes by. The whistle on an approaching train will suddenly change to a lower pitch at the moment the train passes by, because the wavelength received by the listener suddenly shifts to a longer one.

FIG. 5.15. THE DOPPLER SHIFT IN A STELLAR SPECTRUM. The upper spectrum shows the Balmer lines of hydrogen in a laboratory spectrum, whereas the lower spectrum is that of a star that has prominent hydrogen lines. The stellar lines are shifted uniformly toward longer wavelengths than the laboratory lines, indicating that this star is moving away from the earth.

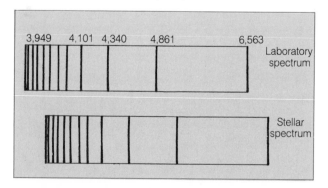

With the Doppler shift of light, we can determine the speed with which the source of light is approaching or receding, from the formula:

$$v = (\Delta\lambda/\lambda)c,$$

where v is the relative velocity between source and observer, $\Delta\lambda$ is the shift in wavelength (the observed wavelength minus the rest, or laboratory, wavelength of the same line), λ is the laboratory wavelength of the line, and c is the speed of light.

Suppose that we find a star with a spectral line at a wavelength of 5,994 Å. Examination of the pattern of lines in this star's spectrum shows that this is a line of iron, which is measured in the laboratory to have a wavelength of 6,000 Å. Then the speed of the star with respect to the earth is:

$$
\begin{aligned}
v &= (5{,}994 \text{ Å} - 6{,}000 \text{ Å}) \times 300{,}000 \text{ km/sec} \\
&= (-6 \text{ Å}/6{,}000 \text{ Å}) \times 300{,}000 \text{ km/sec} \\
&= -0.001 \times 300{,}000 \text{ km/sec} \\
&= -300 \text{ km/sec.}
\end{aligned}
$$

In this example we find a negative velocity, because the observed wavelength is smaller than the rest wavelength. This means that the star is approaching the earth, and we have a blueshift. If, on the other hand, the observed wavelength of this line had been 6,006 Å, we would have found a velocity of +300 km/sec, and the star would have been moving away from the earth.

The Doppler shift tells us only about *relative* motion between the source and the observer: we cannot distinguish whether it is the star or the earth that is moving, or a combination of the two (which is actually most likely the case). It is also important to keep in mind that the Doppler shift tells us only about motion directly toward or away from the earth (Fig. 5.16). There is no Doppler shift resulting from motion perpendicular to our line of sight. If a star is moving at some intermediate angle with respect to the earth, as is usually the case, we can determine the part of its velocity that is directed straight toward or away from us, but we cannot determine its true direction of motion or its speed transverse to our line of sight.

The Doppler shift is another powerful tool for astronomers. From it we learn about the motions not only of stars, but also of planets, galaxies, and any other objects in the universe that have lines in their spectra.

FIG. 5.16. THE LINE-OF-SIGHT VELOCITY. The Doppler shift only reflects motion along the line of sight. Here a star is moving at an angle to the line of sight, so that from the Doppler shift we can measure only a fraction of its true velocity.

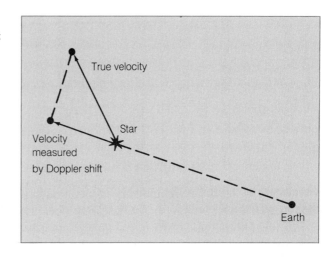

True velocity

Velocity
measured
by Doppler shift

Star

Earth

Perspective

We now have learned how light is absorbed and emitted, and in doing so have begun to see how astronomers can derive great amounts of information about distant objects. We have seen how the continuous spectrum and overall intensity of light from a star depend on its temperature and size, and how the spectral lines provide information on temperature, density, and motion of a star, as well as chemical composition.

We are almost ready to get on with the job of seeing what is out there, but first we must learn how the light is captured by astronomers so that it can be analyzed.

Summary

1. Visible light is just one part of the electromagnetic spectrum, which extends from the γ-rays to radio wavelengths.
2. Light has properties associated with both waves and particles, and these aspects are combined in the concept of the photon.
3. Astronomers gain information about a source of light, such as a star, from both the continuous radiation and the spectral lines.
4. The continuous radiation provides information on the temperature, luminosity, and radius of a star, through the use of Wien's law, the Stefan-Boltzmann law, and the more general Planck's law.
5. The observed brightness of a glowing object is inversely proportional to the square of the distance between the object and observer.
6. Spectral lines are produced by transitions of electrons between energy levels in atoms, molecules, and ions. Absorption occurs when an electron gains energy, and emission occurs when it loses energy, the wavelength in both cases corresponding to the energy gained or lost as the electron changes levels.
7. Each chemical element has its own distinct set of spectral-line wavelengths, and therefore the composition of a distant object can be determined from its spectral lines.
8. The ionization and excitation of a gas can be inferred from its spectrum, the former yielding information on the temperature of the gas, and the latter providing data on its density.
9. Any motion along the line of sight between an observer and a source of light produces shifts in the wavelengths of the observed spectral lines, and measurement of this Doppler effect, as it is called, can be used to determine the relative speed of source and observer.

Review Questions

1. Summarize the evidence that light has both a wave nature and a particle nature.

2. Suppose your favorite radio station has a frequency of 1,200 Khz (that is, 1,200 kilohertz, where one kilohertz equals 1,000 hertz). What is the wavelength at which this station transmits?

3. Compare the energy of an infrared photon, with wavelength $\lambda = 20,000$ Å, and that of an ultraviolet photon with wavelength $\lambda = 2,000$ Å. How does the energy of a radio photon, with $\lambda = 2$ meters, compare with that of the ultraviolet photon with $\lambda = 2,000$ Å?

4. Using Wien's law, calculate the wavelength of maximum emission for the following objects, and comment on whether they would appear to glow in visible light: (1) a star with surface temperature 25,000 K; (2) the walls of a room with temperature 300 K; (3) the surface of a planet whose temperature is 100 K; (4) gas being compressed as it falls into a black hole, so that its temperature is 10^6 K; and (5) liquid helium, with a temperature of 4 K.

5. Suppose a white dwarf star has a surface temperature of 10,000 K, and a red giant has a temperature of 2,000 K. Compare their surface brightnesses (the energy emitted per square centimeter). Now suppose that the white dwarf has a radius of 10^8 cm, and the red giant has a radius of 10^{13} cm. Compare the luminosities (the total energy emitted over the entire surface) of the two stars.

6. Now suppose that the red giant described in the previous question is fifty times farther away than the white dwarf. Compare the brightnesses of the two stars, as seen from the earth.

7. Normal stars have *absorption* lines in their spectra, because the relatively cool gas in the outer layers, lying above the hotter interior, absorbs light at specific wavelengths. What would you conclude about a star that has *emission* lines in this spectrum?

8. Suppose a certain element will lose an electron (that is, will become ionized) if it receives energy in the amount of 2×10^{-11} erg. What wavelength of photon does this energy correspond to? Will this element be ionized by radiation from a star whose temperature is 3,000 K?

9. Suppose the elements iron and beryllium each have a spectral line at the same wavelength. If you found a line at this wavelength in the spectrum of a star, how could you decide which element was responsible?

10. The hydrogen atom has a strong spectral line at 6,563 Å, in the red portion of the spectrum. This line represents a transition between the first excited level of the electron and the next (between levels two and three in the energy-level structure, where level one is the lowest state the electron can be in). Explain why this line is often seen in the spectra of stars, but never in the absorption spectrum of the very rarefied gas that lies in interstellar space.

11. The strong line of hydrogen, whose rest wavelength is 6,563 Å, is observed in a number of stars. Calculate the line-of-sight velocity of a star where the line is observed at (a) 6,561.8 Å; (b) 6,567 Å; and (c) 6,593 Å.

Additional Readings

The best sources for further explanation of the properties of light and radiation are elementary physics books of the sort used as texts in introductory courses. In addition there are a few nontechnical books on the nature of light, such as the following works.

Bragg, W. 1959. *The universe of light*. New York: Dover.
Rublowsky, J. 1964. *Light* New York: Basic Books.
Ruechardt, E. 1958. *Light, visible and invisible*. Ann Arbor: University of Michigan Press.

CHAPTER 6

Telescopes on Earth and in Space

The preceding two chapters illustrate the techniques used by astronomers to determine the properties of faraway objects. These techniques, however, require that certain measurements be made, and we have not yet discussed the instruments and methods used to make these measurements. Hence this chapter will outline the principles of telescopes and their use.

The Need for Telescopes

Although we will discuss a wide variety of telescopes, ranging from those used in space for X-ray and ultraviolet observations, to traditional ground-based telescopes for visible light, to radio antennae, all perform essentially the same task: they gather as much light as possible and bring it to a focus. If we want to observe faint objects, it helps to collect light from as large an area as possible. The human eye has a collecting area only a fraction of a centimeter in diameter, whereas the largest telescopes are several meters in diameter.

Another basic difference between the eye and the telescope is that the telescope can be equipped to record light over a long period of time, through the use of photographic film or a modern electronic detector, whereas the eye has no capability for storing light. A long-exposure photograph taken through a telescope allows us to obtain pictures of objects that we could not see even when looking through the same telescope.

Both their large size and their capability of making long exposures make it possible with telescopes to detect objects some 100,000,000 times fainter than what we can see with the unaided eye, even under the best conditions.

A third major advantage of large telescopes is that they have superior **resolution,** the ability to discern fine detail. For visible-wavelength telescopes, the earth's atmosphere creates practical limitations on how fine the resolution can be, but for radio telescopes and optical telescopes in space, the atmosphere is not a problem.

One problem must be overcome in making long exposures: the earth rotates, and if nothing were done to compensate for this, a star would quickly move out of the field of view. So that this problem is avoided, telescopes are mounted so that they can be moved by a motor in the direction opposite the earth's rotation, keeping a target object centered in the field of view (Fig. 6.1). The telescope mounting usually allows the telescope to move independently in declination (north-south motion) and right ascension (east-west motion), so the drive mechanism that compensates for the earth's rotation needs to operate only in the east-west direction. The need to do this, of course, is avoided for telescopes that are not attached to the earth, such as the various orbiting observatories (discussed later in this chapter).

Refractors and Reflectors

We begin by discussing the principles of telescopes used for visible light, since these have the longest history and because many of the basic ideas can be used for infrared and ultraviolet telescopes as well.

The telescope used by Galileo utilized lenses to bring light to a focus and to magnify the image (Fig. 6.2). This technique was later refined by Huygens and studied by Newton, who realized it has a fundamental shortcoming: because different wavelengths of light are bent to different degrees by passing through glass, it is impossible to bring all the light of a star to a focus at a single point. The blue light focuses closer to the lens than the longer-wavelength red light; the image of a star gets spread out into a rainbow. This problem, called **chromatic aberration,** was solved much later by the use of lenses made of combinations of different kinds of glass, so that the separation of colors caused by a layer of one kind of glass could be reversed by a layer of another kind of glass.

Telescopes that use a lens to bring light to a focus are called **refractors.** A large lens, called the primary lens, at the end of the telescope tube bends the light rays so that they come together and form an image at a point called the **focus.** The size and curvature of the primary lens determine the focal length; that is, the distance from the primary lens to the focus point. At this point (where the other end of the tube is located) is an eyepiece to magnify the image for the human observer.

The power of a telescope has nothing to do with

FIG. 6.1.   COMPENSATING FOR THE EARTH'S ROTATION. In order for the telescope to stay pointed at a target star, the instrument must be rotated at just the right rate to counteract the earth's rotation.

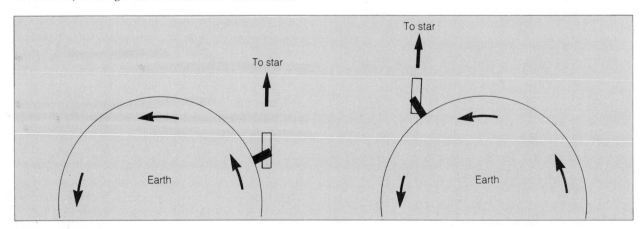

Largest refractor- Yerkes Obser. Williams Bay, Wisconson 40" diameter

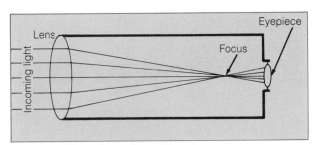

FIG. 6.2. THE REFRACTING TELESCOPE. The path of light is bent when it passes through a surface such as that of glass. A properly shaped lens thus can bring parallel rays of light to a focus, where the image of a distant object may be examined. A second lens is used to magnify the image.

magnification, as is commonly thought. It is the light-gathering power that is crucial, and this is determined by the size of the principal light-collecting element, in this case the primary lens. This is what determines how sensitive the telescope will be for observing faint objects. The eyepiece does nothing but magnify the image to a scale that is easier to discern; it does not enhance the light-gathering power of the telescope.

The largest refractors ever made are less than four feet in diameter. The biggest is the refractor at the Yerkes Observatory at Williams Bay, Wisconsin, with a primary lens of 40-inch diameter (Fig. 6.3). Another large refractor, with a 36-inch diameter, is located at the Lick Observatory, on Mount Hamilton, east of San Jose, California.

Newton, having discovered (but not solving) the problem of chromatic aberration in refracting telescopes, devised an alternative type of telescope that utilizes a concave mirror to bring the light to a focus (Figs. 6.4, 6.5, and 6.6). This arrangement, called a **reflecting telescope,** has the advantage that all wavelengths of light reflect from a mirror at the same angle, so no separation of colors occurs. A disadvantage, however, is that for us to look at the image created by a concave mirror we need to block the incoming light, because the image forms in front of the mirror. One solution to this problem was suggested by J. Gregory, who placed a **secondary mirror** in front of the primary to reflect the image back through a hole in the primary to a point behind it, where the observer could look at it without blocking the incoming light. This design is similar to the modern **cassegrain focus** arrangement (Fig. 6.7), and is used in nearly all large reflecting telescopes. Newton developed a different means of deflecting the image outside the telescope

FIG. 6.3. THE 40-INCH (1-METER) REFRACTOR AT THE YERKES OBSERVATORY IN WILLIAMS BAY, WISCONSIN. This is the largest refractor ever built.

tube; he placed a flat mirror at a 45-degree angle inside (just before the focal point) so that the light passed out through a hole in the side of the tube before coming to a focus. This **Newtonian focus** is used in many small- to moderate-sized reflectors, particularly of the sort often used by amateur astronomers.

Another, more complex, design is sometimes available with large telescopes. Called a **coudé focus,** this

FIG. 6.4. THE RELECTING TELESCOPE. A properly shaped concave mirror can be used instead of a lens to bring light to a focus. Usually an additional mirror is used to reflect the image outside the telescope tube.

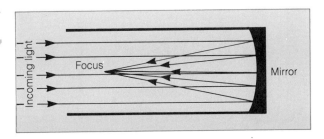

FIG. 6.5.   A REFLECTING TELESCOPE BUILT BY ISAAC NEWTON.

FIG. 6.6.   A LARGE PRIMARY MIRROR. This photo shows that the proper shape for a telescope mirror is not highly concave. In this case only the distorted reflections reveal that the mirror is not flat. This is the 2.3-meter mirror of the University of Wyoming's infrared telescope.

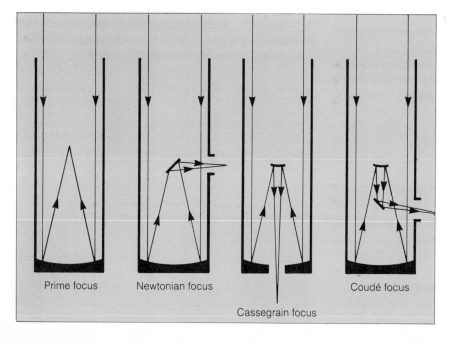

FIG. 6.7.   VARIOUS FOCAL ARRANGEMENTS FOR REFLECTING TELESCOPES.

arrangement utilizes up to five mirrors (including the primary) to reflect the image out of the telescope tube along one of the main axes of motion to a focus in a separate room, some distance from the telescope. The reason for going to all this trouble is to allow the use of heavy instruments in the analysis of light. In the cassegrain or Newtonian designs, any equipment needed to record or measure the light must be attached directly to the telescope and light enough to move with it. That is not practical if the instrument is very massive (Fig. 6.8). In the coudé design, however, the image does not move, no matter where the telescope is pointed, so large instruments can be fixed in place to analyze it.

Yet another arrangement, possible only with the largest telescopes, is called **prime focus.** In this case the light is allowed to focus inside the tube, and the astronomer actually sits there, in a small cage, with his equipment for analyzing the light (usually a simple camera in this instance). The advantage of this design is that it reduces the number of reflections to just one, thereby minimizing the loss of light that inevitably occurs at each reflection. Prime focus is used, therefore, for observations of especially faint objects.

All the focal arrangements used in reflecting telescopes have the disadvantage that some fraction of the incoming light has to be blocked by a secondary mirror or by the observer's cage (if the prime focus is used). This disadvantage is far outweighed by the size advantage offered by the use of mirrors instead of lenses. A lens can be supported only by its edges, since light must be free to pass through. If a lens is large enough to be very heavy, it may sag a little under its own weight, which distorts its light-focusing properties. Even worse, the amount and direction of sag will vary as the telescope is pointed in different directions. A mirror, on the other hand, can be supported from behind and can therefore be kept much more rigid. All the largest telescopes in the world are reflectors for this reason (see, for example, Fig. 6.9). Another advantage of reflectors is that only one surface has to be shaped to precision, rather than two, as in a lens (or four, in the composite lenses made of two layers of glass to prevent chromatic aberration). Also, because light only reflects from the surface and does not pass through the glass, flaws within the material of the mirror can be tolerated. The biggest telescope for many years was the 200-inch (5-meter) reflector at Mt. Palomar, in southern California, but recently a 6-meter telescope was completed in the Soviet Union (Fig. 6.10).

FIG. 6.8. TELESCOPE CONSTRUCTION. Building a large telescope requires precision handling of massive components and usually must be done at remote, difficult-to-reach sites. Here part of the mount for the Wyoming Infrared Telescope is being lifted into the dome.

FIG. 6.9. THE 4-METER MAYALL TELESCOPE, KITT PEAK NATIONAL OBSERVATORY. Kitt Peak is the principal observatory funded by the National Science Foundation, and this is its largest telescope. Kitt Peak is located about fifty miles west of Tucson, Arizona.

FIG. 6.10.   THE 6-METER TELESCOPE. This is the world's largest telescope, located in the Caucasus Mountains of the southern central Soviet Union.

A very promising new kind of telescope was completed in 1979 at Mt. Hopkins, Arizona. This is the *Multiple-Mirror Telescope,* which consists of six separate 72-inch primary mirrors with secondary mirrors arranged so that each focuses at the same point (Figs. 6.11, 6.12). The total collecting area of the six mirrors is equivalent to a single mirror of 176-inch diameter. This sys-

FIG. 6.11.   THE MULTIPLE-MIRROR TELESCOPE, MT. HOPKINS, ARIZONA. The observatory at Mt. Hopkins is operated by the University of Arizona and the Smithsonian Institution.

FIG. 6.12.   A SCHEMATIC OF THE MULTIPLE-MIRROR TELESCOPE DESIGN. This drawing shows how the images from two of the six mirrors are brought to a common focus.

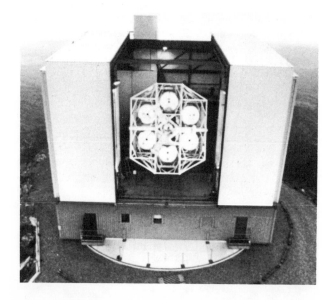

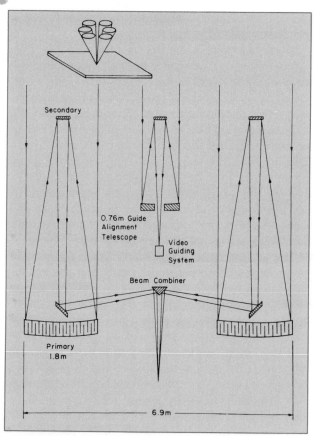

tem is very complicated, and the difficulties of perfectly aligning the six mirror are enormous, but the savings in cost and the relative ease with which the smaller mirrors can be constructed make the effort worthwhile.

AUXILIARY INSTRUMENTS

Once the light is brought to a focus, it must then be recorded in some way. The old-fashioned stereotyped image of the astronomer peering into the eyepiece all night, perhaps taking notes on what he sees, is replaced in reality by the picture of the astonomer sitting before a rather complex and impressive control panel, operating sophisticated electronic equipment.

Generally speaking, two kinds of things are done to the light after it is focused: (1) it is often spread out or sorted according to wavelength; and (2) it is then recorded by film or electronic detector.

The first of these steps may be accomplished in a variety of ways, including the use of a prism or other device to spread out the light, or the use of filters to allow only certain wavelengths to pass through. The device that spreads out, or **disperses,** the light according to wavelength is called a **spectrograph.** Modern spectrographs usually utilize grooved surfaces called **gratings** to disperse the light, rather than prisms. The effect of a grating may be visualized by holding a phonograph record so that light glances off of it, forming a rainbow of colors.

Besides spectrographs, other common kinds of auxiliary instruments include ordinary cameras, to photograph the field of view (sometimes this is done through a series of filters so that star brightnesses at different wavelengths can be measured); and **photometers,** which use simple photocells like those which control some kinds of automatic doors. In this case the photocell produces an electric current proportional to the brightness of the starlight that hits it, so the brightness is determined by measuring the current.

The final stage of any instrument is the device that actually records the intensity of light. Traditionally this job has been done with photographic film, and film is still widely used today, but several alternatives have been developed in recent years. These include a variety of electronic instruments that produce pictures like film does, but do so by storing an image electrically so that it can be analyzed easily by computer. Some of these

detectors, as any device for measuring the intensity of light is called, are very similar to ordinary television cameras, whereas others operate on quite different principles.

The auxiliary instruments used in astronomy range from rather compact, lightweight devices that can be easily mounted directly on the telescope for use at the cassegrain focus, to bulky, large instruments (usually spectrographs) that must be used at the coudé focus.

absorption of light
latitude of sight.

Choosing a Site: Major Observatories

Most major observatories are located in remote regions, for a number of reasons. It is best to minimize absorption of light by the atmosphere, so an observatory should be situated on a high mountain, above as much of the atmosphere as possible. Of course, a site with generally clear weather is imperative, for only radio telescopes can peer through clouds. It is also desirable to build observatories well away from large cities, because the diffuse light from a densely populated area can drown out faint stars. Another consideration is the latitude of the site, which should not be too far north or south, since that would exclude a large portion of the sky.

Other factors that affect the quality of an observing site are the cleanliness of the air, for pollution can reduce the transparency of the atmosphere; and the amount of turbulence in the air above the site, because turbulence creates the twinkling effect known to astronomers as "seeing." This twinkling, which tends to smear out an image, reduces the quality of photographs or spectra. On a good night the diameter of a star image is 1 or 2 arcseconds, whereas when the seeing is bad, images may blow up to diameters of 5 arcseconds or more.

The largest observatories in the world generally are located along the western portions of North and South America, in Hawaii, and in Australia, although the biggest single telescope is in the Caucasus Mountains of the southern central Soviet Union. The U.S. government, through the National Science Foundation, supports large observatories in North America (Kitt Peak National Observatory, about fifty miles west of Tucson, Arizona) and South America (the Cerro Tololo Inter-American Observatory, on the crest of the Andes, northeast of Santiago, Chile).

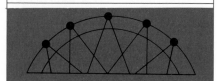

A Night at the Observatory

A typical astronomer's night at the observatory will vary quite a bit from one individual to another, and especially from one observatory to another. There is a big difference between sitting in a heated room viewing an array of electronics panels and standing in a cold dome peering into an eyepiece while recording data on a photographic plate, yet both extremes (and various possibilities in between) are part of modern observational astronomy.

Despite apparent differences in practice, however, the principles are very much the same, regardless of where the observations take place or with what equipment. Here is a brief outline of a typical night at a large observatory, where a modern electronic detector is used to record spectra of stars.

By mid-afternoon the astronomer and the assistants (an engineer who knows the detector and its workings, and a graduate student) are in the dome, making sure the detector is working properly. To do this requires various tests of the electronics, and the recording of test spectra from special calibration lamps. Everything in order, they eat supper in the observatory dining hall a couple of hours before sunset, then return to the dome to complete preparations for the night's work. A night assistant (an observatory employee) arrives shortly before dark and is given a summary of the plans for the night. It is the night assistant's job to control the telescope and the dome, and he is responsible for the safety of the observatory equipment. At large observatories, the telescope is considered so complex and valuable that a specially trained person is always on hand so that the astronomer (who may use a particular telescope only a few nights a year) does not handle the controls, except for fine motions required to maintain accurate pointing during the observations.

Before it is completely dark, a number of calibration exposures are taken, using lamps and even the moon, if it is up. These measurements will help the astonomer analyze the data later, by providing information on the characteristics of the spectrograph and the detector.

Finally, everything is in order, and if things are going smoothly, the telescope will be pointed at the first target star just as the sky gets dark enough to begin work. Observing time on large telescopes is difficult to obtain and is very valuable; not a moment is to be wasted.

A pattern is quickly established and repeated throughout the night. First, after the night assistant has pointed the telescope at the requested coordinates, the astronomer looks at the field of view to confirm which star is the correct one (this is often done by consulting a chart prepared ahead of time). When the star is properly positioned so that its light enters the spectrograph, the observation begins. Depending on

Telescopes in Space: Infrared and Ultraviolet Astronomy

As we saw in our discussion of the emitting properties of glowing objects, the radiation from a star or planet extends beyond the limits detectable by the human eye. It also extends beyond the limits set by the earth's atmosphere, which absorbs all wavelengths shorter than about 3,100 Å, and absorbs in several wavelength regions in the infrared as well. Some portions of the infrared can penetrate to the ground, so infrared astronomy can be done at the major observatories (two of the three large telescopes on Mauna Kea, for example, were built specifically for infrared observations, as was the large telescope at the University of Wyoming). To cover the entire infrared spectrum or any of the ultraviolet, we must, however, place telescopes in orbit above the earth's atmosphere. This is accomplished through the use of high-altitude planes (Fig. 6.13) or balloons (which do not quite get above all of the atmosphere), sounding rockets (providing only five minutes or so of observing time), or satellites.

Simply putting an infrared telescope above the atmosphere does not solve all the problems. As shown

the brightness of the star and the efficiency of the spectrograph and the detector, the exposure time may be anywhere from seconds to hours. During this time, the astronomer or the graduate-student assistant continually checks to see that the star is still properly positioned, making corrections as needed by pushing buttons on a remote-control device.

If the observations are made through the use of an instrument at the coudé focus, the astonomer and his assistants spend the entire night in an interior room in the dome building, emerging occasionally only to view the skies, give instructions to the night assistant, or go to the dining hall for lunch (usually served around midnight).

As dawn approaches and the sky begins to brighten, a last-minute strategy is devised to get the most out of the remaining time. Finally, the last exposure is completed, and the night assistant is given the OK to close the dome and park the telescope in its rest position (usually pointed straight overhead, to minimize stress on the gears and support system). The astronomer and assistants make final calibration exposures before shutting down their equipment and trudging off to get some sleep. The new day for them will begin at noon or shortly thereafter.

The scenario just described depicts a typical night at a major modern observatory, such as Kitt Peak, with extensive resources and the largest telescopes. At smaller facilities, such as a typical university observatory, it is more common to find the astronomer working alone, most often using a cassegrain-focus instrument. In this case the observer must spend the night out in the dome with the telescope (where it can be very cold in the winter), and must refresh himself with a homemade lunch. The work pattern is similar, however, as is the desire not to waste any telescope time.

in the discussion of Wien's law (chapter 5), a telescope at room temperature glows at infrared wavelengths, which can drown out the infrared light from a target object. We solve this problem, on the ground or in space, by cooling the telescope, or at least the associated instruments, so that the emission is shifted to much longer wavelengths and does not confuse the observations. The cooling is usually done by circulation of liquid nitrogen (which has a temperature of 77 K) or liquid helium (4 K) through a series of tubes surrounding the instrument.

Infrared and ultraviolet telescopes work on the same principles as a reflector designed for visible wavelengths, bringing the light to a focus by means of a similar arrangement of mirrors. Radio telescopes operate in essentially the same way, but X-rays and γ-rays require rather different designs.

Ultraviolet astronomy began in the late 1940s with sounding-rocket observations of the sun, and has progressed steadily ever since, with the development of a series of satellite observatories for ultraviolet spectroscopic observations. Currently one such spacecraft, the **International Ultraviolet Explorer** (Fig. 6.14) is in operation, and the **Space Telescope** (described briefly

FIG. 6.13.   THE KUIPER AIRBORNE OBSERVATORY. The aircraft is used for infrared observations at altitudes above much of the water vapor in the atmosphere. Water vapor absorbs infrared light at many wavelengths and interferes with ground-based observations at these wavelengths.

later in this chapter) will have a variety of ultraviolet capabilities.

Infrared astronomy has until recently been carried out from the ground where possible, and also from high altitude balloons and planes. Only in the 1980's

have there been satellite missions for this purpose. The **Infrared Astronomical Satellite,** designed to map the sky in infrared wavelengths, was scheduled for launch in 1982, and later in this decade the **Shuttle Infrared Telescope Facility** will be flown.

THE X-RAY UNIVERSE

Extremely hot gases, according to Wien's law, emit most strongly at very short wavelengths. Temperatures of a million degrees or more produce emission in the X-ray portion of the electromagnetic spectrum, between 1 Å and 100 Å. There are even spectral lines in this wavelength region, produced by electrons jumping between orbits with very large energy differences.

It is not simple to build an X-ray telescope, however. This is because most materials are not shiny enough for X-rays; a mirror used in the usual way simply absorbs X-rays rather than reflecting them. It is possible, though, to reflect X-rays if the radiation strikes the reflecting surface at a very oblique angle, and X-ray telescopes can be designed this way (Fig. 6.15).

Text continues on page 102

FIG. 6.14.   THE *INTERNATIONAL ULTRAVIOLET EXPLORER.* This NASA-operated spacecraft has been in operation since 1978, obtaining ultraviolet spectroscopic data for hundreds of astronomers.

FIG. 6.15.   A TYPICAL X-RAY TELESCOPE DESIGN. The radiation is brought to a focus by a series of grazing-incidence reflections; that is, reflections at oblique angles. Several mirrors may be "nested" concentrically, to increase the light-gathering power.

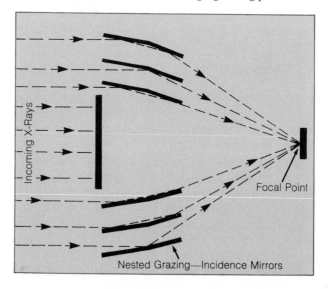

Astronomy in the Tropics: The Mauna Kea Observatory

The observatory atop the 13,796 foot dormant volcano Mauna Kea, on the island of Hawaii, has rapidly reached a position of preeminence in the world of astronomy. Twenty years ago there were no telescopes on the gigantic cinder cone that forms the summit of the mountain, yet today the total light-collecting area of the telescopes there surpasses that of any other observatory in the world, and will continue to grow. Mauna Kea is becoming recognized as the finest site of all, because of its unique combination of extreme altitude and high atmospheric transmissivity, the stability of the air overhead, its high percentage of clear nights, and its low latitude, providing access to much of the southern sky.

The potential of the Hawaiian Islands as a possible observatory site was recognized nearly a hundred years ago. When David Kalakaua, king of the Hawaiian empire, stopped in California in the 1880's on his way to visit England, he was given a tour of the Lick Observatory. He was impressed with what he saw, and, realizing that sites in the Islands would satisfy many of the criteria for a good observatory, he later remarked in correspondence that the possibility of stationing a telescope there should be considered. Although a solar observatory was eventually established on the extinct volcano Haleakala on the island of Maui, nothing happened on Mauna Kea until the early 1960's, when a group of scientists at the University of Hawaii won a grant from the National Science Foundation to establish an observatory there, beginning with the construction of an 88-inch (2.2-meter) telescope that finally went into operation in 1970. Before the University of Hawaii astronomers took their bold action, it had been assumed that the nearly 14,000-foot altitude of Mauna Kea and the rugged terrain leading to the summit would make it impossible to build and operate an observatory there. It had generally been thought that any new observatory development would take place on Haleakala, with its 10,000-foot altitude and easier access.

The high altitude of Mauna Kea certainly has an effect on human efficiency, and for that reason construction goes slowly there and observers must be extra careful not to make careless mistakes. It is also true that the road to the summit is rough and hazardous and that the weather can be harsh (there are many tales of observers stranded at the summit by snowstorms). Nevertheless Mauna Kea has more than lived up to the expectations of King Kalakaua and those of the astronomers at the University of Hawaii who established the observatory.

There are now three very large telescopes on Mauna Kea, in addition to the University's 2.2-meter telescope and two smaller ones. Two of the new large telescopes were constructed specifically for infrared observations. These are the 3.0-meter *Infrared Telescope Facility*, sponsored by the National Aeronautics and Space Administration (NASA), and the 3.8-meter *United Kingdom Infrared Telescope*, sponsored by the British government, which is the largest telescope on Mauna Kea. The newest telescope on the mountain is the 3.6-meter *Canada-France-Hawaii Telescope*, sponsored jointly by the governments of Canada and France and operated in collaboration with the University of Hawaii, which provides dormitory and food services at all the mountain's telescope sites.

The Mauna Kea Observatory is probably not finished growing. The so-called *Next Generation Telescope* (described in the text), which may have an aperture 10 to 15 meters in diameter, will most likely be located on Mauna Kea, and plans are being formulated to place a large radio telescope there as well. We can only wonder what King Kalakaua would have thought had he been able to see these developments.

Sometimes several reflections are necessary to bring the X-rays to a focus, because only a small deflection of the radiation can be accomplished at each reflection. The process of dispersing X-rays by wavelength is also more complicated than for visible light, again because most substances absorb X-rays.

Despite the difficulties, a number of rocket and satellite observatories have successfully probed the heavens in X-ray wavelengths, and a wide range of phenomena have been discovered (Fig. 6.16). Sites of violent explosions and other highly energetic regions are the main sources of X-rays, so these space-borne observations have opened a new universe for astronomers, one whose existence was only hinted at previously.

Currently X-ray astronomy is in a lull between major instruments, but within a few years the **Advanced X-Ray Astrophysics Facility** will be launched, with a variety of X-ray detectors.

The shortest wavelengths of electromagnetic radiation, the γ-rays, also require special techniques for detection. Here even grazing-incidence reflections will not work, and as yet no instrument has been developed that can bring γ-rays to a precise focus. So far only rather primitive γ-ray telescopes have been flown, but plans are being made for a major satellite mission called the **Gamma Ray Observatory,** which promises to provide important new data on the extremely violent regions in space where nuclear reactions are taking place, such as inside of stars and in supernova explosions.

Radio Telescopes

We turn now to a discussion of telescopes for another invisible portion of the spectrum, the radio wavelengths. Radio astronomy has a relatively long history, dating back well before World War II. In 1931 Karl Jansky, a scientist with Bell Laboratories, noticed upon testing a radio antenna that the static he was receiving originated in the Milky Way (Fig. 6.17). He was unable to pursue this discovery much further, but another American, Grote Reber of Wisconsin, became fascinated by Jansky's work and built his own radio antenna. Reber systematically mapped the sky and found many intense sources of radio emission; until the war was over, his work constituted the whole of the science of radio astronomy. British radar technicians during the war found that the sun is a radio emitter. When radio techniques were further refined in the late 1940s, astronomers began to take more interest in the possibilities. One particularly provocative theoretical discov-

FIG. 6.16.   AN X-RAY PICTURE. This image of a region of hot gas and bright stars was obtained with the *Einstein Observatory,* a major X-ray satellite that operated in the late 1970's and early 1980's.

FIG. 6.17.   KARL JANSKY WITH THE RADIO ANTENNA HE USED TO DISCOVER THE FIRST CELESTIAL RADIO SOURCE.

ery, the prediction by the Dutch astronomer H. C. van de Hulst that hydrogen atoms in space should emit at a wavelength of 21 centimeters, spurred a great deal of activity, as several research groups worked to develop antennae capable of measuring this emission line. When the 21-centimeter line was detected in 1951, the immense promise of radio astronomy as a probe of the galaxy was realized. Now there are several major radio observatories around the world, and they provide insights into such diverse objects as the sun, exploding galaxies, and molecules deep inside dark interstellar clouds.

The daytime sky is dark in radio wavelengths, so radio observatories need not shut down when the sun rises, and most are on a 24-hour operating schedule.

A radio telescope works much like a reflecting telescope for visible wavelengths, except that generally everything is on a larger scale. The primary reflecting element need not be a smooth surface; instead it is usually a wire mesh. To radiation with such long wavelengths, a mesh acts like a shiny surface, as long as the wavelength observed is much longer than the spacing between wires in the mesh. At a point centered above the radio antenna, at what would be called the prime focus in a visible-wavelength reflector, is placed a **receiver,** a device that measures the intensity of the radiation. This receiver usually can be tuned so as to screen out unwanted wavelengths, serving in a sense as a filter. The intensity data can then be stored directly in a computer for later analysis.

A radio antenna must be very large in diameter to produce clear pictures of the sky. The resolution of a telescope is usually expressed in terms of how close together two objects can be and still be seen as separate objects, and is determined by the ratio of the wavelength being observed to the diameter of the telescope. Thus if long wavelengths, such as radio emission, are to be observed, the telescope must be very large in order to yield good resolution (that is, sharp images with fine detail). For a visible-wavelength telescope, a diameter of 10 centimeters (about 4 inches) provides a resolution of about 1 arcsecond, but a radio telescope would have to be several kilometers in diameter to achieve comparable resolution. Therefore even the largest radio telescopes (up to around 100 meters in diameter) have resolutions of several minutes of arc, and a radio picture of the sky tends to look fuzzy, with all the fine details smeared out.

A solution to this problem has been found and is being exploited at several radio observatories. By using two or more radio antennae simultaneously, it is possible to simulate the effect of one very large antenna. The farther apart the antennae are, the better the resolution that can be achieved. Whereas a visible-wavelength telescope looking through the earth's atmosphere typically has a resolution of 1 to 2 seconds of arc (that is, two stars about 1 or 2 arcseconds apart can be distinguished as separate objects), radio telescopes used in combination have managed resolutions of a small fraction of an arcsecond. The technique, called **interferometry,** is obviously very powerful, but it also is very complex, requiring precise knowledge of when particular radio waves arrive at the separate antennae. Radio telescopes at various separations ranging from a few hundred feet to many thousands of miles have been employed; imagine the complexity involved in simultaneously using radio telescopes on opposite sides of the Atlantic!

The largest radio observatory in the United States is the National Radio Astronomy Observatory (Fig. 6.18), sponsored by the National Science Foundation, with a 300-foot antenna at Green Bank, West Virginia; a 30-foot dish at Kitt Peak (used mostly for observing interstellar molecules, whose emission lines occur at relatively short radio wavelengths; see chapter 25); and a

FIG. 6.18. AN ASSORTMENT OF RADIO TELESCOPES. These antennae range in diameter from 26 to 91 meters, and are part of the U.S. National Radio Astronomy Observatory in Green Bank, West Virginia.

new observatory, called the **Very Large Array** (VLA), near Socorro, New Mexico (Fig. 6.19). The VLA has 27 movable 82 foot antennae arranged in a Y-shaped pattern spanning some 15 miles of the New Mexico desert, and is designed for interferometry, ultimately achieving a resolution of better than 0.1 arcseconds.

Another large U.S. radio observatory, at Arecibo, Puerto Rico, consists of a 1,000-foot dish built into a natural bowl in the mountains. This large diameter provides fairly high resolution and sensitivity, but since the antenna cannot be pointed (it is immovable), observational coverage is somewhat limited in the north-south direction.

Major radio observatories outside the United States include the Jodrell Bank Observatory, operated by Cambridge University in England; the Westerbork Observatory in the Netherlands; the Parkes Observatory in Australia; and the 100-meter dish at Bonn, West Germany.

FIG. 6.19. THE VERY LARGE ARRAY. This overall view shows the 27 antennae (each about 25 meters in diameter) in their Y-shaped arrangement. The longer arms of the Y are 21 kilometers long, and the antennae can be moved along rails to achieve optimum separation for interferometric observations.

The Next Generation

What kinds of new telescopes will be developed? Astronomers have a flair for dreaming up new and better ways to overcome technical difficulties, so in some cases plans for innovative instruments are already well developed.

There has been widespread discussion of the **Next Generation Telescope,** a visible-wavelength telescope with a diameter of 10 to 15 meters (remember, the largest telescope today is the Russian 6-meter one). It is unlikely that such a large mirror can be produced as a solid surface with the required precision, so two novel alternatives are being considered. One, the multiple-mirror concept, has already been put into successful operation on a smaller scale, as described earlier in this chapter. The other idea, not yet well tested, is to build a single mirror of a relatively thin, flexible metal, and place behind it a number of automatically activated pushrods that will bend the mirror into the proper shape, compensating for sag caused by gravity as the telescope is pointed in different directions. This **rubber-mirror** concept obviously is complicated in practice, but it may be feasible. A merging of the two concepts, creating the **segmented mirror,** may even-

tually prove best. In this design, a number of hexagonal mirrors are constructed and shaped separately, then mounted together in a large honeycomb-like structure, with each segment being adjustable in keeping all the light focused at a single point (Fig. 6.20).

In ultraviolet astronomy, scientists are looking forward to the **Space Telescope,** a 2.4-meter orbiting telescope that will have with it a variety of cameras, photometers, and spectrographs (Fig. 6.21). This facility, scheduled for launch on the Space Shuttle in the mid-1980's, is intended to have a long lifetime, since it will be possible for astronauts to visit it occasionally to make repairs. It is even planned that the Space Telescope will be able to return to earth if necessary to make repairs or refurbish the instruments. In addition to providing opportunities for ultraviolet observations, the Space Telescope will have greater resolution and be more sensitive in visible wavelengths than ground-based telescopes, because it will avoid problems introduced by the earth's atmosphere.

Even farther in the future, we might expect someday to see a very large visible- and ultraviolet-wavelength telescope in space, to capitalize further on the advantages of getting above the atmosphere. And discussions are under way of an unmanned interstellar

FIG. 6.21. THE SPACE TELESCOPE. This is an artist's sketch of the Space Telescope in its operating configuration. The open end of the 2.4-meter telescope is pointed away from us in this view; the near end houses control systems, cameras, and other scientific instruments. The large "wings" are solar panels, which provide electrical power for the spacecraft. A Space Shuttle is seen in the distance.

FIG. 6.20. THE NEXT GENERATION TELESCOPE. This model illustrates the segmented-mirror design, one of several proposed methods for constructing a very large telescope.

mission, in which a probe would be launched toward alpha Centauri (the nearest star) and would radio back information on interstellar gas and dust, as well as data on alpha Centauri itself, as seen from close up. Such a mission would take several years, since alpha Centauri is four light-years away, so the scientists waiting on earth will have to be patient.

Perspective

We have seen, in broad outline, how astronomers gather the light, in all wavelengths, that carries the information they seek. We have learned that, for many wavelengths, the same basic telescope design will work, although it may be necessary to place the instrument in orbit, above the earth's atmosphere. To observe other portions of the electromagnetic spectrum, new technology, such as radio telescopes and X-ray and γ-ray instruments, has had to be developed. Today astronomers can look at the universe in nearly any wavelength. We are ready now to see what has been learned.

Summary

1. Telescopes are better than the human eye for astronomical observations because they have larger collecting areas, have superior resolution, and can collect light over long periods of time.

2. Telescopes for visible light use lenses or mirrors to bring light to a focus, but all big telescopes are reflectors, which can be made larger and are relatively free of aberrations.

3. Reflecting telescopes can bring light to a focus in a variety of ways for different purposes.

4. Auxiliary instruments are used to analyze the light once it is brought to a focus, by photographing the image, measuring its brightness, or dispersing the light for spectral analysis.

5. An observatory site must be chosen for clear weather, high atmospheric transmissivity, low turbulence in the air overhead, minimal pollution, remoteness from city lights, and the proper latitude for viewing the desired part of the sky.

6. Infrared and ultraviolet observations can be made through the use of the same optical designs as visible-light telescopes, but must be carried out from above the earth's atmosphere, which blocks these wavelengths from reaching the ground.

7. Observations in X-ray and γ-ray wavelengths require novel telescope designs and must be made from space.

8. Radio telescopes are similar in design to reflecting telescopes for visible light, but must be built very large in order to provide good resolution.

9. In all wavelengths, but most commonly in the radio portion of the spectrum, several telescopes can be used together to produce high-resolution images by interferometry.

10. Planned future telescopes include the Space Telescope, a giant reflector (the Next Generation Telescope) that will be built on the ground, and a variety of radio, X-ray, ultraviolet, infrared, and γ-ray telescopes, many of them to be launched into space on the Space Shuttle.

Review Questions

1. Summarize the advantages of using a telescope, rather than the unaided eye, in making astronomical observations.

2. Compare the light-gathering power of a 1-meter telescope to that of a 4-meter telescope.

3. Summarize the advantages of reflecting telescopes over refractors.

4. Explain why the coudé-focus arrangement is usually not used for observing very faint objects.

5. What is the advantage of placing a telescope in orbit to make observations in visible wavelengths of light, which also can be observed from the ground?

6. Suppose you want to make observations of an interstellar dust cloud, whose temperature is 500 K. Explain why this would be difficult to do with a telescope at room temperature. How would the situation be improved by cooling the telescope with liquid nitrogen, which has a temperature of 77 K?

7. Explain why ultraviolet telescopes are needed for observations of very hot stars.

8. The resolution of a telescope is proportional to λ/D, where λ is the observed wavelength and D is the diameter of the telescope. If a 5-meter telescope used for observing visible light (for example, $\lambda = 5,000$ Å) has a resolution of 0″01, what would be the resolution of a radio telescope 10 meters in diameter, making observations at a wavelength of 21 centimeters? How big would the radio telescope have to be to have the same resolution as the 5-meter visible-light telescope?

9. Suppose the so-called Next Generation Telescope is to be built following the multiple-mirror concept. How many 4-meter mirrors would be required to build the equivalent of a single 15-meter telescope?

Additional Readings

General references on telescopes and their principles include the following works.

Asimov, I. 1975. *Eyes on the universe*. Boston: Houghton Mifflin.

King, H. C. 1979. *The history of the telescope*. New York: Dover.

Kirby-Smith, H. T. 1976. *U.S. observatories: A directory and travel guide*. New York: Van Nostrand.

Meyer-Arendt, J. R. 1972. *Introduction to classical and modern optics*. Englewood Cliffs, N.J.: Prentice-Hall.

Miczaika, G. R. 1961. *Tools of the astronomer*. Cambridge: Harvard University Press.

The Universe As Home for Man*

John Archibald Wheeler

Following a lengthy tenure as Joseph Henry Professor of Physics at Princeton University, Dr. Wheeler is now professor of physics at the University of Texas. Throughout much of his career, he has been a leader in developing an understanding of the atomic nucleus, and has made many fundamental contributions to advances in nuclear technology. More recently, he has been occupied with gravitational physics and the application of gravitational theory to astrophysics. He has led the way toward the understanding of compact stellar remnants, gravitational radiation, quantum mechanics and relativity, and, among other things, is credited with the invention of the term black hole to describe stars that have completely collapsed under their own gravitation. The following essay has been excerpted, with the permission of the Smithsonian Institution Press, from an article written by Dr. Wheeler for a symposium celebrating the five-hundredth anniversary of the birth of Copernicus.

The discoveries of the future will outnumber those of the past, we can believe, because of the pace in today's science, the tools, and the pressure, but will they be

*By permission of the Smithsonian Institution Press from *The Nature of Scientific Discovery,* "The Universe as Home for Man," by J. A. Wheeler, p. 261. Copyright 1975 Smithsonian Institution, Washington, D.C.

greater? Nothing so much encourages an affirmative answer as the mysteries encountered wherever we turn. Each one of us has his own catalog of the great unknowns: three at the top of my own list are the mind, the universe, and the quantum. I know no area where the mystery is greater than it is in these three fields, or any where the linkage between observer and system observed is stranger, or any more suggestive of things hidden beyond present imagination.

Wilder Penfield has told us of the patient lying on the operating table, with brain case open, following past and present in an amazing stream of double consciousness, a consciousness moreover strangely observing

consciousness. Today no mystery more attracts the minds of distinguished pioneers from the field of molecular biology than the mechanism of brain action. Participating in the exploration are workers from fields as far removed from one another as neurophysiology, anatomy, chemistry, circuit theory, and mathematical logic. Many feel that the decisive step forward is waiting for an idea, an as-yet undiscovered concept, a central theme and thesis. Whatever it will prove to be, we can believe that it will somehow touch the tie between mind and matter, between observer and observed.

The brain is small; the universe is large. In what way, if any, is the universe, the observed, affected by man, the observer? Is the universe deprived of all meaningful existence in the absence of mind? Is it governed in its structure by the requirement that it give birth to life and consciousness? Or is man merely an unimportant speck of dust in a remote corner of space? In brief, are life and mind irrelevant to the structure of the universe—or are they central to it? Lack of conclusive evidence on so cosmic an issue suggests that something is still to be learned about how the universe came into being.

It is agreed in science today that it is not sufficient to answer a question with a number only. We must specify in addition the limits of error of that answer. On no question do the limits of error encompass a wider range of uncertainty than on the importance of life and mind for the constitution of the world: Zero? Or everything?

I see no more hopeful sign that we can and will make our way into this unknown land than the immense progress we have already made into the world of the quantum, where the observer and the observed turn out to have a tight and totally unexpected linkage. The quantum principle has demolished the once-held view that the universe sits safely "out there," that we can observe what goes on in it from behind a foot-thick slab of plate glass without ourselves being involved in what goes on. We have learned that to observe even so miniscule an object as an electron we have to shatter that slab of glass. We have

to reach out and insert a measuring device. We can install a device to measure position or insert a device to measure momentum; but the installation of the one precludes the insertion of the other. We ourselves have to decide which it is that we will do. Whichever it is, it has an unpredictable effect on the future of that electron, and to a degree the future of the universe is changed. We changed it. We have to cross out that old word "observer" and replace it by the new word "participator." In some strange sense the quantum principle tells us we are dealing with a participatory universe.

Three mysteries have passed in review that call out for clarification: the quantum, the universe, and the mind. All three lie at that point where, in the phrase of Fred

Hoyle, "mind and matter meld." All three threaten that clean separation between observer and observed which for so long seemed the essence of science. Consciousness can analyze the world around; but when will consciousness understand consciousness? The universe runs its course from big bang to collapse; but what part do the future requirements for life and mind have in fixing the physics that comes into being at that big bang? The quantum principle says, "No physics without an observer"; but from what comes the necessity of this principle in the construction of the world? Of all challenges ever confronted by the adventurous fellowship of science, I know none more deserving than these three to be called the heart of darkness.

The Solar System

INTRODUCTION TO SECTION III.

In this portion of the book, we will examine the properties of our system of sun and planets, with a view toward understanding how this system came to be. We will also learn about the fascinating phenomena that have been revealed through its exploration.

The solar system is complex, and we will see that it is made up of a variety of individual parts. We will also find, however, that there are a number of underlying, general processes at work, so that as we go through the system, examining individual objects, we will be able to make comparisons, to see how these general processes have influenced different situations.

Among the events that we will encounter time and again are tidal forces and other gravitational effects which are responsible for an astonishing assortment of phenomena, such as the spin rates of many of the planets and satellites, the locations of their orbits, and the failure of the asteroids and the ring particles of the outer planets to form large bodies. We also will allude often to the process by which the solar system formed (namely, from the collapse of a rotating interstellar cloud), because on our tour of the universe we will encounter many recognizable artifacts of this process. Another phenomenon that will be mentioned often is the solar wind, a steady stream of subatomic particles from the sun that pervades the solar system, creating important effects in the outer atmospheres of many of the planets.

As we study the planets, we will make a number of comparisons among them. To establish some of the standards for these comparisons, we will begin with the best-studied and understood of all the planets, the earth. We will then tour the moon and the other terrestrial planets, the innermost four (including the earth), which have similar overall properties. Following this, we will turn our attention to the outer giant planets, starting with Jupiter, the biggest of them all. Again, the first example will serve as the standard for comparison with the others. Finally, we will discuss other bodies in the solar system, which are important leftovers from its formation, and then the sun itself, before wrapping up the section with a description of the formation and evolution of the solar system.

CHAPTER 7

The Earth as a Planet

Of all the bodies in the heavens, the earth, of course, has been the best studied, although it is equally true that many mysteries remain. To understand the planet we live on has practical, as well as philosophical, importance. Furthermore, working at such close quarters with the object of study provides us with numerous advantages, including the ability to observe in great detail and over long periods of time, and the possibility of making direct experiments by probing and sampling the surface of the earth.

It would be beyond the scope of this text to attempt to summarize all that is known about the earth. Here our discussion will emphasize those aspects that can be compared with what we know of the other planets. We should keep in mind, however, that not only is the earth much more complex than it is portrayed here to be, but so, undoubtedly, are the other planets. Each is a unique and varied world, probably as diverse as the earth, except for the influences of life.

From afar, the earth is a brilliant planet, with an overall bluish color created by its oceans and by the scattering of light in its atmosphere (Fig. 7.1). Much of its usually cloud covered, and where it is clear, brownish land masses appear. The earth appears so bright because it has a high **albedo,** the fraction of light that is reflected from it. The earth's albedo is 0.39, meaning that almost 40 percent of the light that strikes it is reflected.

The Atmosphere

We begin by examining the thin layer of gas that surrounds the solid earth. The earth's atmosphere is composed of a variety of gases, as well as a distribution of suspended particles called **aerosols.** The gases, which are evenly mixed together up to an altitude of about 80 kilometers, are primarily nitrogen and oxygen, in the form of the molecules N_2 and O_2. Nearly 80 percent of the gas is nitrogen, about 20 percent is oxygen, and all the other species exist only as traces, although in many cases they are important traces. In addition to this standard mixture of gases, a number of others appear in the atmosphere, but are highly variable in quantity. Among these, water vapor (H_2O) is important, because of its direct role in supporting life, and others, such as ozone and carbon dioxide, are also essential, but for less direct reasons. Ozone (designated O_3) is a form of oxygen, in which three oxygen atoms are bound together instead of two. Unlike ordinary oxygen (O_2), ozone in the upper atmosphere acts as a shield against ultraviolet light from the sun, which could be harmful to life forms if it penetrated to the ground.

The importance of carbon dioxide (CO_2), which is produced by respiration in animals and by many industrial processes, lies in its ability to trap the sun's radiant energy within the atmosphere, creating overall heating by a process called the **greenhouse effect.** This heating process has progressed to a much greater extent on the planet Venus than on the earth (and will be discussed more fully in chapter 9).

The principal atmospheric constituents, nitrogen and oxygen, are both released into the air by life forms on the surface. This process has been important in the evolution of the earth's atmosphere, and leads to the expectation that other planets in the solar system are unlikely to have the same composition (although nitrogen is produced also by processes that do not involve life forms, so we might find it elsewhere). Nitrogen is released into the earth's atmosphere by the decay of biological material and by emission from volcanic

FIG. 7.1. THE EARTH AS SEEN FROM SPACE.

eruptions. Oxygen comes almost exclusively from the photosynthesis process in plants, a process in which carbon dioxide is converted into oxygen with the assistance of radiant energy from the sun.

HEATING AND ATMOSPHERIC MOTIONS

Although there are obvious fluctuations in the temperature of the atmosphere, particularly near the surface, there is a definite, stable temperature structure as a function of height (Fig. 7.2). This structure includes several distinct layers in the atmosphere: the **troposphere,** from the surface to about 10 km altitude, where the temperature decreases slowly with height and where the phenomenon that we call **weather** occurs; the **stratosphere,** between 10 and 50 km, where the temperature increases with height; the **mesosphere,** extending from 50 to 80 km, where the temperature again decreases; and the **thermosphere,** above 80 km, where the temperature gradually rises with height to a constant value above 200 km. In layers where the temperature decreases with height, there can be substantial vertical motion of the air, whereas in layers where it gets warmer with height, the air is stable, with only lateral motions.

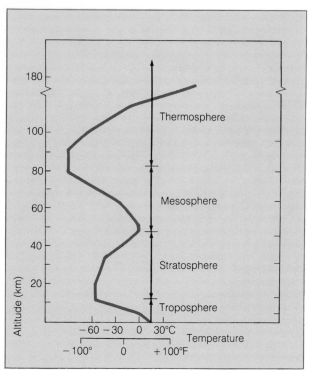

FIG. 7.2. THE VERTICAL STRUCTURE OF THE EARTH'S ATMOSPHERE.

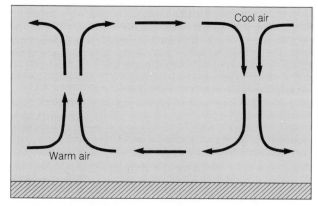

FIG. 7.3. CONVECTION IN THE EARTH'S ATMOSPHERE. This is a simplified illustration of the principle of convection, in which warm air rises, cools, and then descends.

The principal source of heating throughout the atmosphere is the sun's radiation, but the energy is deposited in different ways at different levels. Much of the heat is absorbed by the ground. This heating, which is concentrated at low and middle latitudes on the earth, is responsible for driving the large-scale motions of the atmosphere, which produce the global wind patterns.

There are many scales of motion in the atmosphere, ranging from small gusts a few centimeters in size to the continent-spanning flows thousands of kilometers in scale. Here we will discuss only the large-scale motions, since generally we are able to study only motions of similar scale in the atmospheres of other planets.

The primary influences on the global motions in the earth's atmosphere are heating from the sun, as already noted, and the rotation of the earth. The heating creates regions where the air rises. The air must later cool and fall somewhere else, and a pattern of overturning motions called **convection** is created (Fig. 7.3). Generally, air rises in the tropics and descends at more northerly latitudes, although the situation is more com-

plicated than that. For example, the continents tend to be warmer than the oceans, so that high-pressure regions (characterized by descending air) preferentially lie over water, whereas low-pressure regions (where air rises and cools) tend to be over land.

These tendencies, combined with the rotation of the earth, create horizontal flows at several levels in the atmosphere. Air that is rising or falling is forced into a rotary pattern by the earth's rotation. Where air rises in a low-pressure zone, the resultant swirling motion

TABLE 7.1.   EARTH
Orbital semimajor axis: 1.000 AU (149,600,00 km)
Perihelion distance: 0.983 AU
Aphelion distance: 1.017 AU
Orbital period: 365.256 days (1.000 years)
Orbital inclination: 0°0′0″
Rotation period: 23^{h}56^{m}4.1^s
Tilt of axis: 23°27′
Diameter: 12,734 km (1.000 D$_\oplus$)
Mass: 5.974 × 10^{27} grams (1.000 M$_\oplus$)
Density: 5.517 grams/cm^3
Surface gravity: 980 cm/sec^2 (1.000 earth gravity)
Escape velocity: 11.2 km/sec
Surface temperature: 200–300 K
Albedo: 0.39 (average)
Satellites: 1

is called a **cyclone.** In the northern hemisphere, the motion of a cyclone is counterclockwise; south of the equator, it is clockwise (Fig. 7.4). Descending air flows, in high-pressure regions, are forced into oppositely directed circular patterns, and are called **anticyclones.** Cyclonic flows sometimes intensify into storms of great strength; hurricanes and monsoons are examples.

The net result of the vertical motions caused by the sun's heating and the spiral motions caused by the earth's rotation is the creation of a complex pattern of flows (Fig. 7.5) with both vertical and horizontal components.

Seasonal shifts in the distribution of the sun's energy input cause changes in the flow patterns, creating our well-known seasonal weather variations. Apparently other factors can influence the location of the high- and low-pressure regions, because there are longer-term fluctuations in the climate whose cause is not well understood. Some of these may be related to cyclic behavior in the sun (see chapter 16).

In later chapters, we will examine the flow patterns in the atmospheres of the other planets, and we will see that in many ways they behave like the earth, except for differences in the amount of solar-energy input and the rate of rotation.

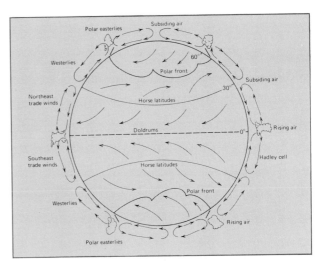

FIG. 7.5. THE GENERAL CIRCULATION OF THE EARTH'S ATMOSPHERE.

FIG. 7.4. A STORM SYSTEM ON EARTH. This is the Southern Hemisphere, so the circulation around this low-pressure region is clockwise.

The Earth's Interior

At first it may seem surprising that much is known about the deep interior of the earth. Substantial quantities of data have been collected, however, mostly by indirect means. It is interesting that the same indirect techniques could be applied (through the use of unmanned space probes) to other planets, and in principle it is possible for us to know as much about their interiors as that of our own planet (this would not be true of the outer, gaseous planets, but does apply to the other terrestrial planets and the moon).

The primary probes of the earth's interior are **seismic waves** created in the earth as a result of major shocks, most commonly earthquakes. These waves take three possible forms, called the *P, S,* and *L* waves. Studies of wave transmission in solids and liquids have shown that the P waves, which are the first to arrive at a site remote from the earthquake location, are **compressional** waves, which means that the oscillating motions occur parallel to the direction of motion of the wave, creating alternating regions of high and low density without any sideways motions (Fig. 7.6). Sound waves are examples of compressional waves. The S waves are **transverse,** or **shear,** waves, the vibrations

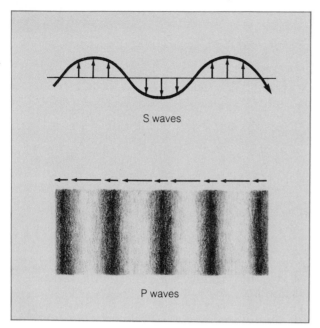

FIG. 7.6. S AND P WAVES. An S, or shear, wave (left) consists of alternating motions transverse (that is, perpendicular) to the direction of propagation. Waves in a tight string or on the surface of water are S waves. Compressional, or P, waves have no transverse motion, but instead consist of alternating dense and rarefied regions created by motions along the direction of propagation. The arrows at the top represent the relative speeds in and between the dense zones. Sound waves are P waves.

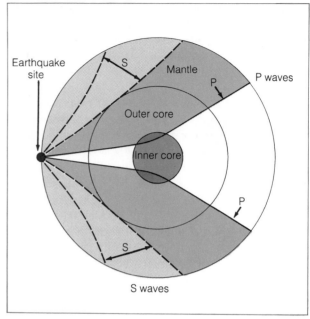

FIG. 7.7. SEISMIC WAVES IN THE EARTH. This simplified sketch shows that P waves can pass through the core regions (although their paths may be bent), whereas S waves cannot. This demonstrates that the outer core is liquid, because S waves require a rigid medium.

FIG. 7.8. THE INTERNAL STRUCTURE OF THE EARTH.

occurring at right angles to the direction of motion (Fig. 7.6). These waves require that the material they pass through have some rigidity and, unlike the P waves, they cannot be transmitted through a liquid. The L waves travel only along the surface of the earth and thus do not provide much information on the deep interior.

By measuring both the timing and the intensity of these seismic waves at various locations away from the site of an earthquake, scientists can determine what the earth's interior is like. The speed of the P waves depends on the density of the material they pass through, and the distribution of the P and S waves reaching remote sites provides data on the location of liquid zones in the interior (Fig. 7.7).

The general picture that has developed from studies is that of a layered earth (Fig. 7.8). At the surface is a crust whose thickness varies from a few kilometers beneath the oceans to perhaps 60 kilometers under the continents (Fig. 7.9). The crust is made of solid rock of two general types: **sial,** or silicon-aluminum rock,

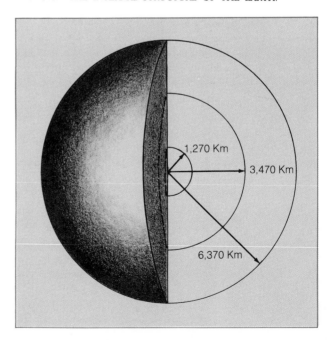

which is predominant in the upper continental crust; and **sima,** or silicon-magnesium rock, in the deeper portions of the crust and the ocean floors.

There is a sharp break between the crust and the underlying material, which is called the **mantle.** The mantle transmits S waves, so it must be solid, but on the other hand, it undergoes slow, steady flowing motions in its uppermost regions. Perhaps it is best viewed as a plastic material, one that has some rigidity, but which can be deformed, given sufficient time.

The uppermost part of the mantle and the crust together from a rigid zone called the **lithosphere.** The part of the mantle itself where the fluid motions occur, just below the lithosphere, is called the **asthenosphere.** Below the asthenosphere, there is a more rigid portion of the mantle that extends nearly halfway to the center of the earth. The lower mantle is called the **mesosphere** (not to be confused with the level in the earth's atmosphere bearing the same name).

Beneath the mantle is the **core,** consisting of an **outer core** (between 2,900 and 5,100 km in depth), and an inner core. The S waves are not transmitted through the outer core, which is therefore thought to be liquid. As a result, the inner core can be probed only with the P waves. Since these waves travel more rapidly through the inner core, its density is thought to be greater than that of the outer core, and the innermost region is thought to be solid.

The overall density of the earth is 5.5 times that of water (that is, 5.5 grams per cubic centimeter, or 5.5 gm/cm^3). The rocks that make up the crust have typical densities of less than 3 gm/cm^3, and the mantle material is thought to have relatively low density also, perhaps 3.5 gm/cm^3. From these facts it is deduced that the core must have a density roughly 15 gm/cm^3. The most likely substance that could give the core this high density is iron, probably with some nickel and perhaps cobalt.

The high density and probable metallic composition of the core are quite significant, for they show that the earth has undergone **differentiation,** a sorting out of elements according to their weight (Fig. 7.10). This can only happen when the planet is in a molten state, so it shows that the earth once was largely liquid, probably early in its history, soon after it formed. The principal cause of the earth's molten state most likely was radioactive heating in the interior. Several naturally occurring elements, such as uranium, thorium, and potassium, are radioactive, which means that their nuclei

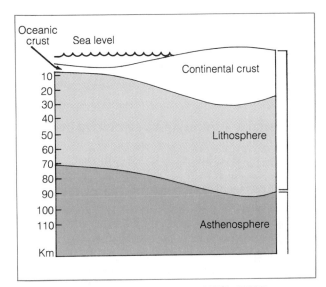

FIG. 7.9. THE STRUCTURE OF THE EARTH'S OUTER LAYERS. This illustrates the relative thickness of the crust in continental areas as compared to the seafloor.

spontaneously emit subatomic particles and, over a long period of time, produce substantial quantities of heat. Another source of heat early in the earth's history was created by the energy of impact as the smaller bodies that merged to form the earth came together. Additional heating was caused by the process of differentiation itself, as the heavy elements sank to the center of

FIG. 7.10. DIFFERENTIATION. In a fluid, some separation of heavy and light elements will occur as the heavy ones sink. The earth's interior is molten in places today, and much more of it was molten early in the planet's history. Much of the supply of heavy elements such as nickel, iron, and cobalt has therefore become concentrated in the core.

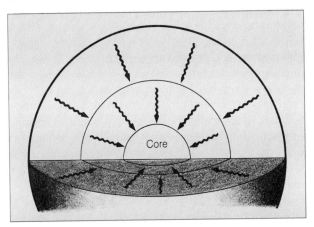

the earth, releasing gravitational potential energy. We shall see that differentiation has not occurred in all the bodies in the solar system, and this tells us something significant about their histories.

Surface Terrain and Tectonic Activity

As in previous sections, here we will deal primarily with large-scale features. One reason for this is that many of the small-scale details of the surface structure of the earth have been modified by processes, such as flowing water, that do not occur on the other planets (with the exception of Mars; see chapter 10), and our principal goal here is to describe features of the earth that can be compared with structures on the other planets.

About three-quarters of the earth's surface is covered by water. The remainder is covered by a few large continents, whose relative positions have changed steadily over past millennia and are still changing today.

The evidence favoring the idea that continental drift occurs includes the obvious fit between land masses on opposite sides of the Atlantic, the similarities of mineral types and fossils, and the alignment of vestigial magnetic fields in rocks that once were together in the same place. In modern times, more direct evidence, such as the detection of seafloor spreading away from undersea ridges and the actual measurement (through the use of sophisticated laser-ranging techniques) of continental motions, has removed any trace of doubt that pieces of the earth's crust are in motion, and that the continents were once arranged differently (Fig. 7.11).

From all this evidence has arisen a theory of **plate tectonics,** which postulates that the earth's crust (that is, the lithosphere) is made of a few large, thin, pieces that float on top of the asthenosphere (Fig. 7.12). The flowing motions in the asthenosphere cause these plates to constantly move about and occasionally to crash into each other. The rate of motion is only a few centimeters per year at the most, and the major rearrangements of the continents have taken many millions of years to occur (the Americas and the European-African system became separated from each other 150 to 200 million years ago).

The driving force in the shifting of the plates is not well understood. The most widely suspected cause is

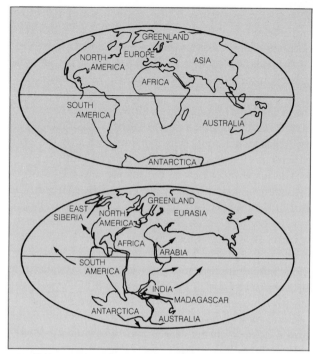

FIG. 7.11. CONTINENTAL DRIFT. These maps show the distribution of the continents today and as it was some 200 million years ago, before continental drift rearranged things.

convection currents in the asthenosphere (Fig. 7.13). This is the same process that occurs in the earth's atmosphere: temperature differences between levels cause an overturning motion. If a fluid is hot at the bottom and much cooler at the top, warm material will rise, cool, and descend again, creating a constant churning. The speed of the overturning motions depends largely on the **viscosity** of the fluid; that is, the degree to which it resists flowing freely. The earth's mantle, as we have already seen, is sufficiently rigid to transmit S waves, and must therefore have a high viscosity as well. Hence if convection is occurring in the mantle, it is reasonable that the motions should be very slow.

Whether convection is the cause or not, the effects of tectonic activity are becoming well known. When plates collide, one may submerge below the other in a process called **subduction,** which creates an undersea trench, or, particularly if both plates carry continents, the collision may force the uplifting of mountain ranges. The lofty Himalayas are thought to have been created when the Indian plate collided with the Eurasian plate some fifty million years ago. The bounda-

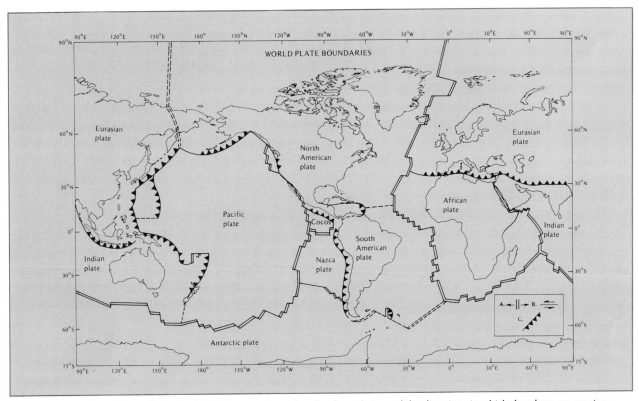

FIG. 7.12.   THE EARTH'S CRUSTAL PLATES. This map shows the plate boundaries and the directions in which the plates are moving.

FIG. 7.13.   SCHEMATIC OF THE MECHANISMS OF CONTINENTAL DRIFT. This sketch shows how continental drift is responsible for uplifted mountain ranges and parallel undersea trenches, where one crustal plate sinks below another (subduction zone), and how a mid-ocean ridge is built up where two plates move away from each other.

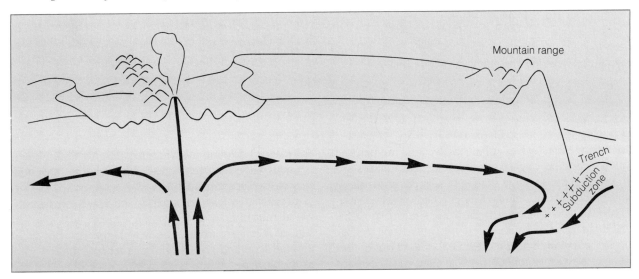

ASTRONOMICAL INSIGHT 7.1

The Difficulty in Seeing the Obvious

Given the rather obvious evidence for continental drift that can be seen any time we look at a world map, it may seem odd that the idea was not accepted by the scientific community for a long period of time. The reasons for this are complex and are related as much to human nature as to the geological evidence.

Although scientists are generally open-minded and willing to let the evidence guide their thinking, they are sometimes influenced by other factors as well. In the case of continental drift, an idea that radically contradicted conventional scientific thought, most were too cautious to embrace it at first. The notion that the earth was fluid and dynamic enough for the continental masses to float around like ice floes was just too much of a departure from the accepted view, that the earth was a

rigid body. Things might have been different if there had been any hint of a possible *cause* for continental drift, but for a long time there was no reasonable theory about this. The lack of a driving mechanism, along with a natural human reluctance to go out on a limb, slowed the acceptance of continental drift.

The man now recognized as having stimulated development of the modern theory of plate tectonics was not trained as a geologist, but was instead a meteorologist, schooled in scientific methodology but not heavily immersed in conventional geological wisdom. Alfred Wegener, a German, working in the early decades of the twentieth century, collected and wrote about the evidence favoring continental drift. He was by no means the first to suggest the idea (there were others who predated him by centuries), but his was the first extensive effort to unearth sound scientific data. Despite his success in showing that rocks, geological structures, and fossils on both sides of the Atlantic matched, and despite the fact that his work was widely discussed in scientific circles, the notion that the continents had drifted was still universally rejected for nearly half a century.

The acceptance of Wegener's idea, when it did come, occurred only gradually, as additional ideas and evidence were accumulated. As early as 1928 the suggestion was made that convection currents in the earth's mantle might be responsible, but even the author of that idea, a British geologist named Arthur Holmes, disclaimed it as being without sound basis. Only after World War II did the tide of opinion begin to turn, when the first maps of the seafloors revealed the mid-Atlantic ridge and the deep trenches along some continental boundaries. In the early 1960's the American geologist Harry Hess developed the concept of seafloor spreading, and this was soon confirmed by studies of the magnetic alignments of rocks near the undersea mid-Atlantic ridge. Finally there was direct evidence that a dynamic process is at work; that the earth really is undergoing internal upheavals that make the continents move.

Since the 1960's the theory has been refined, with the specific identification of tectonic plates and their boundaries, and the development of a much more detailed picture. Today continental drift is part of the established scientific wisdom.

---

ries where plates either collide or separate are marked by a wide variety of geological activity. Earthquakes are attributed to the sporadic shifting of adjacent plates along fault lines, and volcanic activity is common where material from the mantle can reach the surface, most often along plate boundaries. It has been long recognized, for example, that a great deal of earthquake and volcanic activity is concentrated around the shores of

the Pacific, defining the so-called Ring of Fire. Now it is understood that this region represents the boundaries of the Pacific plate.

Chains of volcanoes, such as the Hawaiian Islands, are also attributed to the action of plate tectonics. In several locations around the world, volcanic hot spots that lie deep and are fixed in place bring molten rock to the surface. As the crust passes over one of these

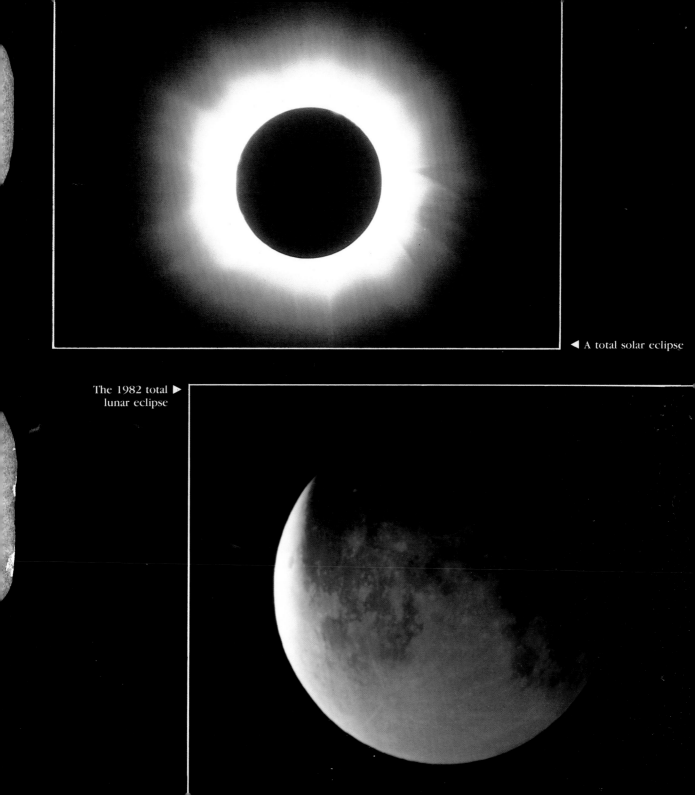

◀ A total solar eclipse

The 1982 total ▶
lunar eclipse

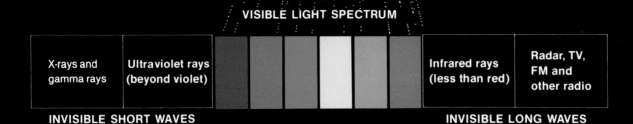

VISIBLE LIGHT SPECTRUM

| X-rays and gamma rays | Ultraviolet rays (beyond violet) | | | | | | | Infrared rays (less than red) | Radar, TV, FM and other radio |

INVISIBLE SHORT WAVES   **INVISIBLE LONG WAVES**

When a white light is directed through a prism,
the visible light spectrum results.

The electromagnetic spectrum ▲

◄ Northern lights over Kitt Peak

The 4-m telescope
at Kitt Peak ▼

The Cerro Tololo
Inter-American Observatory ▼

The Multiple-Mirror Telescope ▼

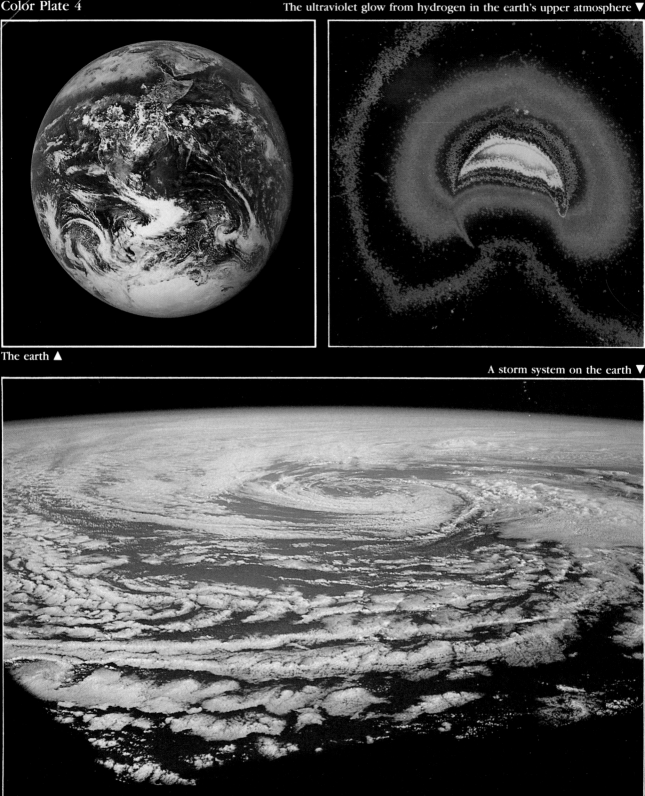

Color Plate 4

The ultraviolet glow from hydrogen in the earth's upper atmosphere ▼

The earth ▲

A storm system on the earth ▼

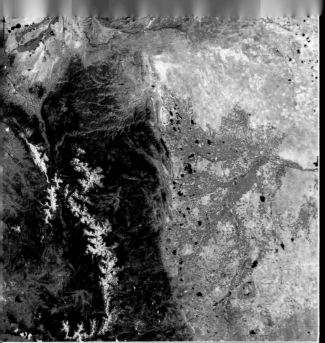

The first launch of the Space Shuttle *Columbia* ▲

Earthrise ▼

Color Plate 6

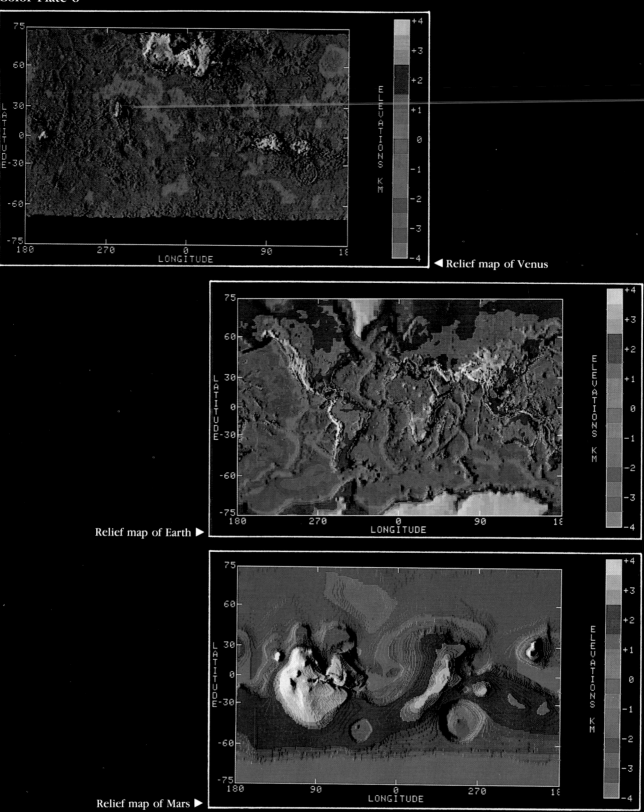

◄ Relief map of Venus

Relief map of Earth ▶

Relief map of Mars ▶

The Colorado Rocky Mountains seen from space ▼

The first launch of the Space Shuttle *Columbia* ▲

Earthrise ▼

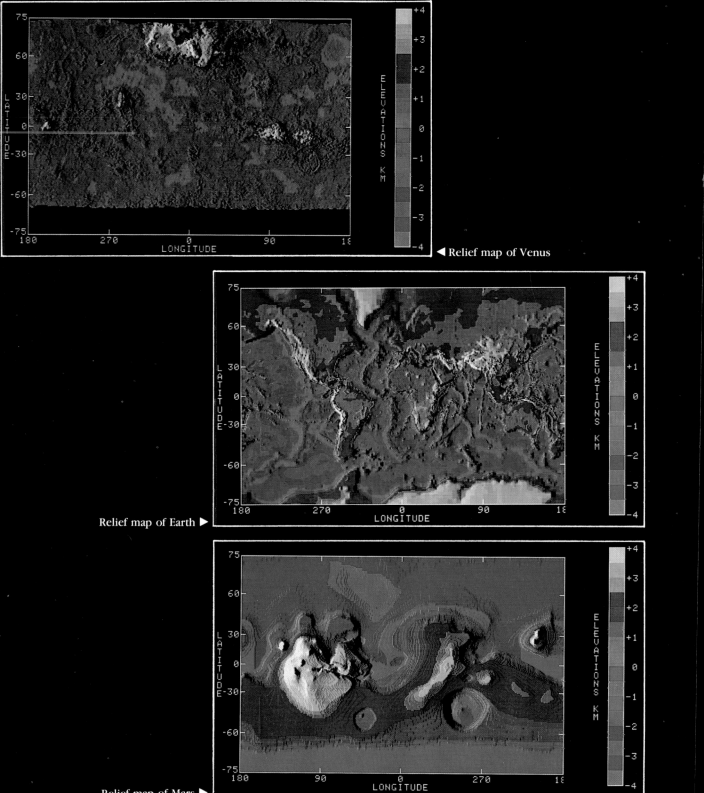

◄ Relief map of Venus

Relief map of Earth ▶

Relief map of Mars ▶

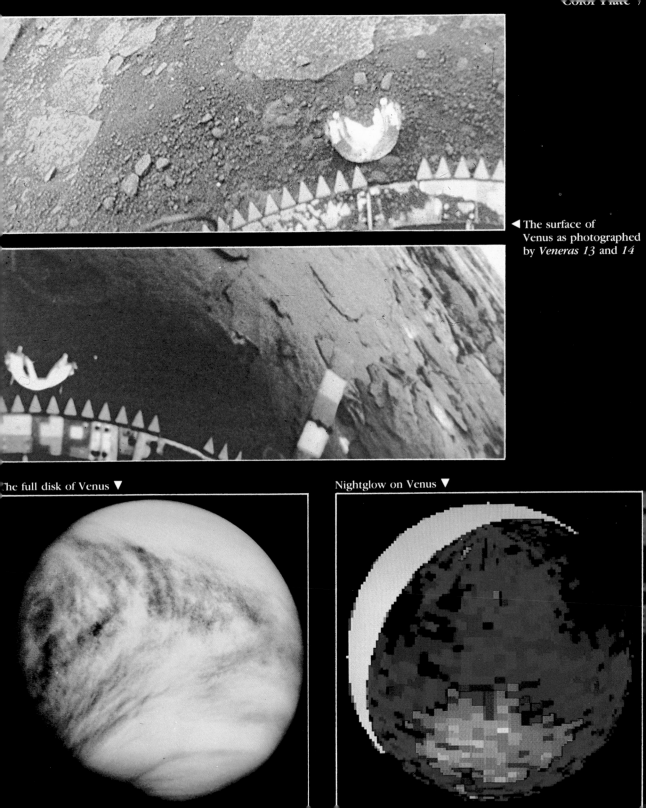

◄ The surface of Venus as photographed by *Veneras 13* and *14*

The full disk of Venus ▼

Nightglow on Venus ▼

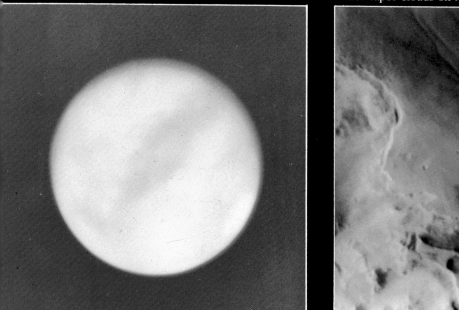

Mars as seen from earth ▼

Water vapor clouds on Mars ▼

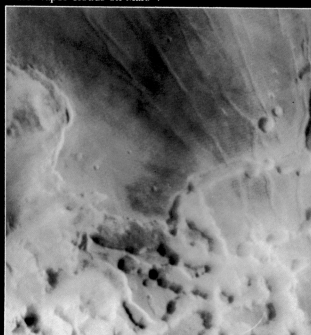

hot spots, volcanoes are formed and then carried away, and a chain of mountains is created as new material keeps coming to the surface over the hot spot. This process is still active, and the Hawaiian Islands, for example, are still growing. There is volcanic activity on the southeast part of the largest (and youngest) island, Hawaii. Furthermore, a new island, already given the name Loihi, is rising from the seafloor about 20 kilometers south of Hawaii. Its peak has already risen 80 percent of the way to the surface and only has about 1 kilometer to go, but it will take an estimated 50,000 years to break through.

Plate tectonics, then, accounts for many of the most prominent features of the earth's surface. Of course, other processes, such as running water, wind erosion, and glaciation, are also important in modifying the face of the earth.

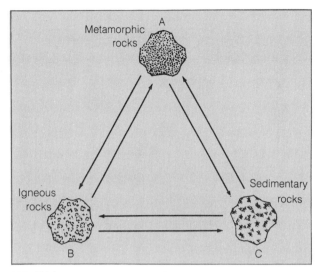

FIG. 7.14. THE ROCK CYCLE. This shows schematically how the three basic types of rocks can be converted from one to another.

Rocks, Minerals and the Age of the Surface

When we examine the surface of the earth, we see little outward evidence of the large-scale processes. We do not see the continents moving, and we cannot directly perceive the interior. What we do see are rocks. The basic substance of the earth's crust is rock, although in many areas soil covers the bedrock (even in those places, however, gravel, boulders, and rocky stream and lake beds are evident).

Rocks fall into three basic groups according to their origin, called **igneous, sedimentary,** and **metamorphic.** Igneous rocks, formed from volcanic activity, consist of cooled and solidified **magma,** the molten material that flows to the surface during volcanic eruptions. Sedimentary rocks are formed from deposits of gravel and soil that have hardened, usually in layers where old seabeds or coastlines lay. Metamorphic rocks are those which have been altered in structure by heat and pressure created by movements in the earth's crust.

All three forms can be changed from one to another in a continual recycling process (Fig. 7.14). Igneous and metamorphic rocks can be eroded by wind and water into gravel and soil, later to become sedimentary rocks; igneous and sedementary rocks can be transformed into metamorphic rocks; and sedimentary and metamorphic rocks can be melted by pressures in the earth's crust to become magma and reappear later as igneous rocks.

Because of these evolutionary processes, the surface of the earth is continually being renewed. Whereas the age of the earth itself is thought to be some 4.5 billion years, the ages of most surface rocks can be measured in the millions or hundreds of millions of years. The distinction between the age of a planet and the age of its surface is an important one, and it will reappear as we discuss the surfaces of other bodies.

The surface rocks and soil are composed of **minerals,** which are chemical compounds, each consisting of a particular arrangement of atoms. The various kinds of atoms in a mineral are regularly spaced in a set crystal pattern (Fig. 7.15). In most cases these crystal patterns are confined to very small, microscopic regions that are randomly stuck together in a rock, giving it an overall amorphous appearance.

Some two thousand minerals have been identified in rocks, but the bulk of the earth's surface is composed of a very limited number, about a dozen. The most common, accounting for about 90 percent of the earth's rocks, are the **silicates,** minerals containing silicon, oxygen, and some metallic elements. Other classes of common minerals include the **oxides,** simple combinations of oxygen and other elements; the **sulfides,** similar combinations of sulfur and other elements; and the **carbonate** and **sulfate** minerals, which contain special combinations of carbon and oxygen or sulfur and oxygen, along with metallic elements. Finally, there are the **halides,** minerals formed by combinations of

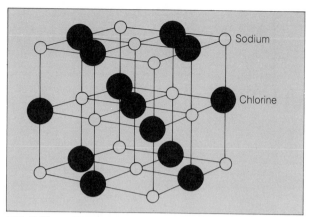

FIG. 7.15.   A TYPICAL CRYSTAL. This shows the regular arrangement of atoms in the crystalline structure of sodium chloride, common table salt.

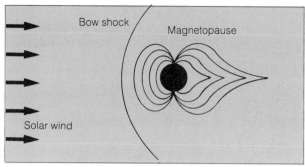

FIG. 7.16.   THE EARTH'S MAGNETIC FIELD. This cross-section shows the earth's field and how its shape is affected by the stream of charged particles from the sun known as the solar wind.

the haline elements chlorine, iodine, bromine, and fluorine with other elements. Common table salt, sodium chloride, is a halide. Any rock can be classified according to both its type of origin and its mineral content. Two very common examples are basalt and granite, which are both igneous rocks with silicate compositions.

The Magnetic Field

The magnetic field of the earth has guided travelers since antiquity. Certain metallic substances, usually containing iron, are sensitive to the presence of a magnetic field and are used as pointers in compasses; thus we take advantage of the near alignment of the earth's magnetic field with the poles (the magnetic axis is tilted 11½ degrees with respect to the rotation axis). It is convenient to imagine magnetic lines along which iron filings lie when placed near a small bar magnet.

OUTER STRUCTURE
AND THE MAGNETOSPHERE

In cross-sectional view, the earth's magnetic field is reminiscent of a cut apple, but one that is lopsided, because of the flow of charged particles from the sun that constantly sweeps past the earth (see the discussion of the solar wind in chapter 16). The region en-

closed by the field lines is called the **magnetosphere** (Fig.7.16), and it acts as a shield, preventing the charged particles from reaching the surface.

The past alignment of the magnetic field can be ascertained from studies of certain rocks that contain iron-bearing minerals whose crystalline structure is aligned with the direction of the magnetic field at the time the rocks solidify from a molten state. Traces of the earth's ancient magnetic activity, called **paleomagnetism,** thus can be deduced from the analysis of magnetic alignments of rocks. Such studies reveal not only that the continents have moved, but also that the magnetic poles of the earth have moved. Furthermore, the north and south magnetic poles have completely and rather suddenly reversed from time to time, so that the North Pole has moved to the south and vice versa. This flip-flop of the magnetic poles seems to have happened at irregular intervals, typically thousands or hundreds of thousands of years apart.

While a reversal of the poles is taking place, there is, for a short period of time, a much-weakened magnetic field. The magnetosphere is greatly diminished, and charged particles from the solar wind and from **cosmic rays,** highly energetic particles from interstellar space (see chapter 25), are more likely to penetrate to the ground. These particles can bring about important effects on life forms, including genetic mutations. The sporadic reversals of the earth's magnetic field may have played a major role in shaping the evolution of life on the surface of our planet.

The source of the earth's magnetic field is a mystery, although it is nearly certain that it has to do with the nickle-iron core. A magnetic field is produced by flowing electrical charges, as in the current flowing

The Ages of Rock

We have spoken of the ages of rocks and of the earth itself but we have said little about how these ages are measured. The most direct technique, *radioactive dating,* involves the measurement of relative abundances of closely related atomic elements present in rocks.

Recall from our earlier discussions of atomic structure (chapter 5) that the nucleus of an atom consists of protons and neutrons, and that the number of protons (that is, the **atomic number**) determines the identity of the element. Different **isotopes** of an element have the same number of protons but differing numbers of neutrons. In most elements the number of protons is not very different from the number of neutrons, but there are exceptions. If the imbalance between protons and neutrons is large enough, the element is unstable, which means that it has a natural tendency to correct the imbalance by undergoing a spontaneous nuclear reaction. The reactions that

occur in this way usually involve the emission of a subatomic particle, and the element is said to be **radioactive.** The emitted particle may be an **alpha particle** (that is, a helium nucleus, consisting of two protons and two neutrons), or, as is more often the case, an electron or a **positron,** a tiny particle with the mass of an electron but with a positive electrical charge. When a positron or an electron is emitted, the reaction is called a **beta decay,** and at the same time a proton in the nucleus is converted into a neutron (if a positron is emitted) or a neutron is changed into a proton (if an electron is released). If the number of protons in the nucleus is altered, the identity of the element is changed. A typical example is the conversion of potassium 40 (^{40}K, with 19 protons and 21 neutrons in its nucleus) into argon 40 (^{40}Ar, with 18 protons and 22 neutrons).

Several elements are known to be radioactive, naturally changing their identity by emitting particles. The rate of change, expressed in terms of the **half-life,** is known in most cases. The half-life is the time it takes for half of the original element to be converted, and it may be as short as fractions of a second or as long as billions of years. The very slow reactions are the one that are useful in measuring the ages of rocks.

If we know the relative abun-

dances of the elements that existed in a rock when it formed, then measurement of the ratio of the elements in that rock at the present time can tell us how long the radioactive decay has been at work; that is, the age of the rock. In a very old rock, for example, there might be almost no ^{40}K, but a lot of ^{40}Ar. The decay of ^{40}K to ^{40}Ar has a half-life of 1.3 billion years, so from the exact ratio of these two species in a rock, we can infer how many periods of 1.3 billion years have passed since the rock formed.

Other decay processes that are useful in dating rocks include the decay of rubidium 87 (^{87}Rb) into strontium 87 (^{87}Sr), which has a half-life of 47 billion years; and the decays of two different isotopes of uranium to isotopes of lead, ^{235}U to ^{207}Pb and ^{238}U to ^{206}Pb, with half-lives of 700 million years and 4.5 billion years, respectively (note that these decays each involve several intermediate steps, and that the half-lives given represent the total time for all those steps).

These dating techniques have shown that the oldest rocks on the earth's surface are about 3.5 billion years old, and that ages of a few hundred million years are more common. The age of the solar system, and therefore of the earth itself, is estimated, from isotope ratios in meteorites, to be about 4.5 billion years.

through a wire wound around a metal rod, which is the basis of an electromagnet. Convection and the earth's rotation may combine to create systematic flows in the liquid outer core, giving rise to the magnetic field, if the core material carries an electrical charge. This general type of mechanism is called a **magnetic dynamo.** The cause of the reversals of the poles is not well understood, but presumably could be related to occasional changes in the direction of the flow of material in the core.

RADIATION BELTS

We have referred to the magnetosphere and its ability to control the motions of charged particles. When the first U.S. satellite was launched in 1958, zones high above the earth's surface were discovered to contain intense concentrations of charged particles, primarily protons and electrons. There are several distinct zones containing these particles, and they are now called the **Van Allen belts,** after the physicist who in the late 1950's first recognized their existence and deduced their properties.

The charged particles, or ions, in these radiation belts are captured from space (primarily from the solar wind), and are forced by the magnetic field to spiral around the lines of force. The portion of the outer atmosphere where the belts lie is called the **ionosphere,** and it extends upward from a height of about 60 kilometers.

The ionosphere has important effects for us on the surface, particularly in terms of our radio communications. Short-wave radio signals are able to travel around the world because they are reflected back to the ground from the underside of the ionosphere. When the ionosphere is disrupted, as it can be after a particularly violent solar flare, its reflective properties may be destroyed temporarily, interrupting communications (this problem has been largely overcome by communications satellites, which do not rely on the ionosphere at all). At the same time, emissions caused by the charged particles may be enhanced, creating displays of the **aurora borealis** (Fig. 7.17), or northern lights (in the Southern Hemisphere, they are called the **aurora australis,** or southern lights). The aurorae are always most prominent close to the poles, because that is where the earth's magnetic field lines, and the charged particles that are forced to travel along them, come closest to the surface. In middle latitudes, these particles are kept well outside of the atmosphere, and aurorae are rarely seen.

The Evolution of the Earth

A great deal has been learned about the general evolution of the earth. We know it was once largely a molten body with an atmosphere rather different from that of the present-day earth; along the way the atmosphere has undergone major changes, some of them triggered by the development of life on the surface.

The age of the earth, estimated from geological evidence, is about 4.5 billion years. The earth probably formed from the coalescence of a number of **planetesimals,** small bodies that were the first to condense out of the debris that swarmed about the infant sun in the earliest days of the solar system. As a result of the heat of impact as these bodies merged, and in time, the heating caused by radioactivity, most or all of the earth was molten at some point during the first billion years. During this early period differentiation occurred, with the densest materials (such as iron) sinking to the core. At the same time, **volatile** gases were emitted by the newly formed rocks at the surface. Volatile gases are those which vaporize most easily when heated; they formed the earliest atmosphere of the earth. The most common of these gases were hydrogen (H_2), ammonia (NH_3), methane (CH_4), and water vapor (H_2O). It is noteworthy that the present-day atmospheres of the cold, outer planets are very similar to that of the primitive earth. We will explore this point later (for example, in chapter 17).

FIG. 7.17.   THE AURORA BOREALIS. These colorful displays are triggered by the impact of electrically charged particles from space on the earth's upper atmosphere.

The dominant element in the earth's early atmosphere was hydrogen, but nearly all of it escaped into space by the time the earth was about a billion years old (this is about when the first simple life forms apparently arose). Recall, from chapter 4, that individual molecules in a gas whiz around randomly at speeds that depend on both the temperature of the gas and the mass of the particles. At a given temperature, the lightweight molecules move fastest and are therefore most likely to escape the planet's gravitational attraction. This is what happened to the hydrogen in the early atmosphere of the earth (whereas the heavier gases that now dominate the atmosphere are too massive to escape). Another gas that escaped early in the earth's history is helium, the second most abundant element in the universe, but one so rare on earth that is was not discovered until spectroscopic measurements revealed its presence in the sun.

The critical reactions that led to the development of life must have taken place before all the hydrogen escaped, because the types of reactions that were probably responsible involve this element. The earliest fossil evidence for primitive life dates back at least three billion years, when some hydrogen was still left.

Essentially no free oxygen was present in the atmosphere until life forms had developed that release this element as a by-product of their metabolic activities. Most plants release oxygen into the atmosphere, and as a supply of this element built up, the opportunity arose for complex animal forms to evolve. Besides being a substance that was necessary for the metabolisms of living animals, the oxygen buildup created a reservoir of ozone (O_3) in the upper atmosphere, which in turn began to screen out the harmful ultraviolet rays from the sun. Once life forms gained a toehold on the continental land masses, the process of converting the atmosphere to its present state began to accelerate. Soon nitrogen from the decomposition of organic matter began to be released into the air in large quantities, and by the time the earth was perhaps two billion years old, the atmosphere had reached approximately the composition it has today.

By this time also the mantle had solidified and the crust had hardened (the oldest known surface rocks are nearly 3.5 billion years old). The interior has remained warmer than the surface, because heat can only escape slowly through the crust, and because radioactive heating of the interior took place over a long period of time and probably is still occurring today.

Perspective

We have itemized the earth's overall properties, from its outermost atmosphere to its core, always keeping in mind the facets of its character that will provide a basis for comparison with what we know about the other planets. We have seen that the earth is an active world, seemingly placid only because of the long time scales over which significant changes occur. We are ready now to examine other bodies in the solar system, to see whether they are similarly dynamic. We begin with the moon, our nearest neighbor.

Summary

1. The earth, seen from space, is brilliant because of its high albedo.
2. The atmosphere consists of 80 percent nitrogen, and about 20 percent oxygen, traces of H_2O and CO_2, and fine suspended particles.
3. The atmosphere is divided vertically into four temperature zones: the troposphere, stratosphere, mesosphere, and thermosphere.
4. The sun's heating and the earth's rotation create the global wind patterns.
5. The earth's interior, explored with seismic waves, consists of a solid inner core, a liquid outer core, a mantle, and a crust.
6. The crust is broken into tectonic plates that shift around, which accounts for continental drift and most of the major surface features of the earth.

7. Rocks can be igneous, metamorphic, or sedimentary by origin, and most are silicate minerals by composition.

8. The earth has a magnetic field, probably created by currents in the molten core, which traps charged particles in zones, called radiation belts, above the atmosphere.

9. The earth's evolution from a largely molten planet with a hydrogen-dominated atmosphere to its present state was caused by the loss of lightweight gases into space and by the development of life forms on the surface of the earth, which helped convert the atmospheric composition to nitrogen and oxygen.

Review Questions

1. In the first photographs from deep space that showed the earth and the moon together, the earth was very bright but the moon was dark and difficult to see. Both receive about the same intensity of sunlight. Explain why the moon looked so much darker.

2. What is the major contrast between the origin of the gases that make up the earth's atmosphere and the origins of those which are dominant on other planets?

3. Explain why the sun heats the earth's surface more effectively near the equator than near the poles.

4. The mesosphere, the level of the atmosphere between 50 and 80 kilometers in altitude, is warmer at the bottom than at the top. This is also the zone where the atmospheric ozone is most heavily concentrated. How are these two facts related?

5. It was noted in the text that the continents tend to be warmer than the oceans. What does this tell us about the albedo of land, compared with that of the ocean surface?

6. Suppose the earth's interior had a liquid layer just below the crust but above the mantle, so that no S waves could penetrate into the deeper zones. Would it be possible in that case to learn anything at all about the interior from observations of seismic waves?

7. Why is the east coast of North America relatively free of earthquakes and volcanoes, whereas the west coast has a great deal of activity?

8. Why is the earth not still molten throughout its interior?

9. If exploration of a planet revealed no sedimentary or metamorphic rocks, what would you conclude about the geological history of that planet?

10. Would you expect a planet with no molten core to have radiation belts?

Additional Readings

Here are a few texts and articles in which more thorough discussion of the earth may be found.

Battan, L.J. 1979. *Fundamentals of meteorology*. Englewood Cliffs, N.J.: Prentice-Hall.

Carrigan, C. R., and Gubbins, D. 1979. The source of the earth's magnetic field. *Scientific American* 240(1):118.

Leet, L. D., Judson, S., and Kauffman, M. E. 1978. *Physical geology*. 5th ed. Englewood Cliffs, N.J.: Prentice-Hall.

Siever, R. 1974. The steady state of the earth's crust, atmosphere, and oceans. *Scientific American* 230(6):72.

Wilson, J. T., ed. 1972. *Continents adrift*. San Francisco: W. H. Freeman. (A collection of readings from *Scientific American*)

CHAPTER 8

The Moon

Our closest neighbor in the solar system is the earth's satellite, the moon. The fact that the moon orbits the earth was correctly recognized in the earliest days of civilization, when scientists first began to speculate on the nature of the heavenly bodies. In more modern times, the moon and its motions have helped astronomers understand the workings of the laws of physics, particularly gravity, that govern orbital motions; and concepts derived from studies of the lunar orbit have been applied to many kinds of faraway objects. Thus the moon served as a sort of laboratory for earthly scientists even before experiments could be performed on its surface.

We have already discussed the moon's phases and other celestial phenomena associated with its motions, such as eclipses and the Saros cycle. In this chapter we will concentrate instead on the nature of the moon itself.

General Properties

Seen from afar, the moon is a very impressive sight (Figs. 8.1 and 8.2), except in comparison with the earth, which far outshines it. The moon's albedo is only 0.07, compared with the earth's albedo of 0.39. In the first

127

FIG. 8.1. THE FULL MOON. This is a high-quality photograph taken through a telescope on the earth.

photographs that showed both objects together, taken by the *Voyager* spacecraft, the image of the moon had to be artificially enhanced to make it show up. Our normal view of the moon from the earth's surface shows

FIG. 8.2. THE MOON AT TWO PHASES.

it to be a brightly lit object, especially impressive when full. Its ½° diameter is pockmarked with a variety of surface features, the most prominent of which are the light and dark areas. The latter, thought by Renaissance astronomers to be bodies of water, are called **maria** (singular: **mare**), after the Latin word for seas. Closer examination of the moon with primitive telescopes also revealed numerous circular features called **craters,** named after similar-looking structures on volcanoes.

The pattern of surface markings on the face of the moon is quite distinctive, and because it never changes, it was recognized long ago that the moon always keeps one face toward the earth. The far side cannot be observed from the earth but it has been thoroughly mapped by spacecraft. The synchronous rotation of the moon (described in chapter 4), is attributed to tidal forces exerted on the moon by the earth. (Later in the chapter, the history of this situation will be discussed, as we learn of the moon's formation and evolution.)

With no trace of an atmosphere (because the escape velocity is so low that all the volatile gases were able to escape long ago), erosion caused by weathering does not occur on the moon, so craters, mountain ranges, and other features retain a jagged appearance indefinitely. Other types of features seen on the moon include **rays,** light-colored streaks emanating from some of the large craters, and **rilles,** winding valleys that resemble earthly canyons.

The moon's mass is only about 1.2 percent of the mass of the earth, and its radius is a little more than one-fourth the earth's radius. The moon's density, 3.34 grams/cm^3, is below that of the earth; it more closely resembles the density of ordinary rock. This indicates that the moon probably does not have a dense core, and that it therefore most likely did not undergo strong differentiation. Infrared measurements showed, even before the space program allowed for direct measurements to be taken, that the lunar surface is subject to hostile temperatures, ranging from 100 K (-274°F) during the two-week night to 400 K (273° F) during the day.

Of course, all the long-range studies of the moon were almost instantly antiquated by the development of the space program, which has featured extensive exploration of the moon, first by unmanned robots, then by men. The moon was the first celestial body to be probed by spacecraft sent from earth, and it is still the

TABLE 8.1.   MOON
Mean distance from earth: 384,401 km (60.4 $R_\oplus$)
Closet approach: 363,297 km
Greatest distance: 405,505 km
Orbital sidereal period: 27^d 7^h 43^m 12^s
Synodic period (lunar month): 29^d 12^h 44^m 3^s
Orbital inclination: 5°8′43″
Rotation period: 27.32 days
Tilt of axis: 6°41′ (with respect to orbital plane)
Diameter: 3476 km (0.273 $D_\oplus$)
Mass: 7.35 × 10^{25} grams (0.0123 $M_\oplus$)
Density: 3.34 grams/cm^3
Surface gravity: 0.165 earth gravity
Escape velocity: 2.4 km/sec
Surface temperture: 400 K (day side); 100 K (dark side)
Albedo: 0.07

FIG. 8.3.   THE MOON AS SEEN FROM SPACE. Here is a view obtained by one of the Apollo missions, showing portions of the near (left) and far (right) sides. On the left horizon is Mare Crisium, and in left center are Mare Marginis (upper) and Mare Smythii (lower). No maria are seen in the right half of this image; there are almost none on the entire lunar farside.

only one to be visited by manned missions. The Soviet Union had the first successes, obtaining photographs of the far side in 1959 (Fig. 8.3), and the United States launched spacecraft that began to carry out many complex and sophisticated observations and experiments in the 1960s. The *Ranger* spacecraft were designed to simply crash-land on the moon, taking photographs on the way. Following the first success of this program in 1964, the United States launched several *Surveyor* and *Lunar Orbiter* missions, all of which were spectacularly successful. Five *Surveyors* landed on the moon between 1966 and 1968, with the primary goal of analyzing surface conditions for future manned landings, and five *Lunar Orbiter* missions circled the moon during the same time period, mapping it photographically (Fig. 8.3), again in preparation for the manned missions to come.

The historic first manned landing occurred on July 20, 1969. This was the *Apollo 11* mission, and was followed by five more manned landings (Figs. 8.4 and 8.5), the last being *Apollo 17,* which took place in late 1972. Each mission incorporated a number of scientific experiments; some involved observations of the sun and other celestial bodies from the airless moon, but most were devoted to the study of the moon itself.

The following sections describe our current understanding of the moon, based to a very large extent on the information gleaned from the Apollo program.

FIG. 8.4.   MAN ON THE MOON. The Apollo missions, six of which included successful manned landings on the moon, represent mankind's only attempt so far to visit another world.

FIG. 8.5.   THE LUNAR ROVER. The later Apollo missions used these vehicles to travel over the moons's surface, allowing the astronauts to explore widely in the vicinity of the landing sites.

FIG. 8.6.   THE LUNAR "SEAS." Here is a broad vista encompassing portions of three maria: Mare Crisium (foreground); Mare Tranquilitatis (beyond Mare Crisium); and Mare Serenitatis (on the horizon at upper right). These relatively smooth areas are younger than most of the lunar surface, having been formed by lava flows after much of the cratering had already occurred.

FIG. 8.7.   LUNAR CRATERS. This is a view of the far side of the moon, where there is little to interrupt the nearly total coverage by impact craters. The prominent crater in the upper center is Kohlschutter, named by Soviet scientists who first charted the far side of the moon.

A BATTLE-WORN SURFACE

Viewed on any size scale, from the largest to the smallest, the moon's surface is irregular, marked throughout by a variety of features. We have already mentioned the maria, the large, relatively smooth, dark areas (Fig. 8.6). The maria appear darker than their surroundings because they have a relatively low albedo (that is, they reflect less sunlight). Despite their smooth appearance relative to the more chaotic terrain seen elsewhere on the moon, the maria are marked here and there by craters.

Much of the rest of the lunar surface is covered by rough, mountainous terrain. Even though the maria dominate the near side of the moon, the highland regions actually cover most of its surface. The Lunar Orbiter missions found that, oddly enough, there are no maria on the far side.

Craters are seen literally everywhere (Fig. 8.7). They range in diameter from hundreds of kilometers (Fig. 8.9) to microscopic pits that can be seen only under intense magnification. In some regions the craters are

FIG. 8.8. THE CRATER ERATOSTHENES. This is a prominent feature on the near side of the moon. Here the raised rim and central peak of the crater are clearly visible.

so densely packed together that they overlap. As noted earlier in this chapter, the lack of erosion on the moon means that craters can survive for billions of years, and there is plenty of time for younger craters to form within the older ones (Fig. 8.9). The fact that relatively few craters are seen in the maria indicates that the surface in these regions has been transformed in recent times, after most of the cratering had already occurred.

All the craters on the moon are *impact craters,* formed by collisions of interplanetary rocks and debris with the lunar surface, rather than by volcanic eruptions. The particular shapes of the craters, the central peaks in some of them (Fig. 8.8), and the trails left by **ejecta,** or cast-off material created by the impacts (Fig. 8.10), all suggest that the craters were formed in this

FIG. 8.9. CRATERS WITHIN CRATERS. A small crater is seen here near the center of a larger one. What is not so obvious in this photograph is that both are embedded in the floor of a much larger crater, which extends beyond the borders of this image. The central peak of this giant crater, which is 175 kilometers in diameter, is seen at upper left.

FIG. 8.10. A CRATER WITH EJECTA. This crater on the lunar far side is a good example of a case where material ejected by the impact has created rays of light-colored ejecta. Close examination of such features often reveals secondary craters, formed by the impacts of debris blasted out of the lunar surface by the primary impact.

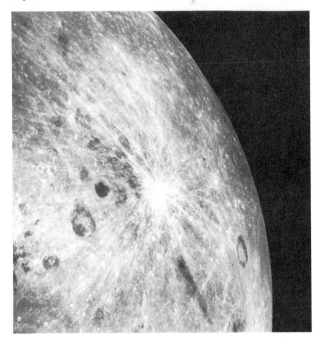

manner. The rays seen stretching away from some craters are strings of smaller craters formed by the ejecta from the larger, central crater.

Some of the lunar mountain ranges reach heights greater than any on earth. They are more jagged than earthly mountains, again because there is no erosion, and they lack the prominent drainage features usually found in terrestrial ranges.

The **rilles** are rather interesting features, resembling dry riverbeds (Fig. 8.11). Apparently they were formed by flowing liquid, but rather than water, lava was responsible. In some cases there are even lava tubes that have partially collapsed, leaving trails of sinkholes. The rilles and, as we shall see, the maria as well, indicate that the moon has undergone stages when large portions of its surface were molten.

FIG. 8.11. A RILLE. The sinuous feature meandering through this image is Hadley Rille, one of many such features seen in the maria. One of the Apollo missions landed at the edge of this rille, so that it could be examined at close range. The adjacent mountains are the Apennines.

Rocks and Minerals on the Moon

The Apollo astronauts found a surface that in many areas is strewn with loose rock, ranging in size from pebbles to boulders as big as a house (Figs. 8.12 and 8.13). The rocks generally show sharp edges (owing to the lack of erosion), and occasional cracks and fractures. In most cases the large boulders appear to have been ejected from nearby craters, and are therefore thought to represent material originally from beneath the surface.

More than 4,500 pounds of smaller lunar rocks were returned to earth by the Apollo astronauts. This gave scientists the opportunity to study the lunar surface characteristics in detail and to compare them with earth rocks and soils (Fig. 8.14). Thus extensive chemical analyses, as well as close-up observation of rocks in place on the moon, were possible. At present, the rock samples from the moon are housed in numerous scientific laboratories around the world, where analysis continues. Specimens have also found their way into museums, and in at least one case (the Smithsonian Institution's Air and Space Museum in Washington, D.C.), a lunar rock can be touched by visitors.

The lunar soil, called the **regolith,** consists of loosely packed rock fragments and small glassy minerals probably created by the heat of meteor impacts (Fig. 8.15).

FIG. 8.12. A FIELD OF BOULDERS. This jumbled region is inside a relatively young crater, and the rocks lying around were disrupted by the impact.

FIG. 8.15. THE LUNAR SOIL. The fine soil of the lunar surface was found to be firm and cohesive; no doubt the astronauts' footprints scattered here and there on the moon will last for millennia.

FIG. 8.13. A LARGE BOULDER. Rocks on the lunar surface range in size from tiny pebbles to massive objects such as this.

In addition to the loose soil, there are three morphological types of surface specimens. The most common are the *breccias*, which consist of small rock fragments cemented together and resembling chunks of concrete. There are similar kinds of rock on earth, except that those on earth are formed in stream beds where

Fig. 8.14. A MOON ROCK. This is one of thousands of lunar samples brought back to earth by the Apollo astronauts. This is an example of a breccia.

water plays a role in shaping them. The lunar breccias contain jagged, sharp rock fragments, and they are probably fused together by pressure created in meteor impacts. The other common morphological types of lunar rocks are two basalts with small openings permeating their interiors (recall that basalt is a silicate rock common on earth).

All lunar rocks are pitted, on the side that is exposed to space, with tiny craters called **micrometeorite craters.** These are formed by the impact of tiny bits of interplanetary material no bigger than grains of dust.

The simple classification of lunar rocks according to shape does not reveal much of significance in terms of the composition and history of the moon itself. A more meaningful analysis of samples brought to earth was carried out, in which their chemical makeup and mineral structure were examined. Several important discoveries were made. First, the relative abundances of some atomic isotopes were found to be very similar to those of earthly rocks. Of particular importance are the oxygen isotopes, which are known to vary in relative abundance throughout the solar system. The fact that these ratios are the same on the earth and the moon indicates that the two bodies must have formed out of very similar material, probably in the same part of the

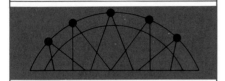

Lunar Geography*

All the major topographic features on the moon, and many minor ones, have names. These names are used in maps of the moon and are recognized by the international community of astronomers. It is interesting to examine how they were assigned.

Galileo was the first to have a chance to name lunar features, because he was the first to look at the moon through a telescope that enabled him to see them. He is responsible for the terms **maria** (the dark areas that he thought were

*Based on information from El-Baz, F. 1979, Naming moon's features created "Ocean of Storms", *Smithsonian*, 9(10):96.

seas) and **terrae** (the highlands). Following Galileo's lead, as telescope technology improved and more and more people examined the moon, a number of early lunar cartographers chose names for prominent craters, mountain ranges, and other regions. Some of the names that were assigned, such as those devised by the court astronomer to the king of Spain (who named features after Spanish nobility), were doomed to be forgotten. Others, such as the suggestion by a German astronomer that lunar mountain ranges be named after terrestial features, are still used today.

The most influential of the lunar cartographers of medieval times was the Italian priest Joannes Riccioli, who, with his pupil Francesco Grimaldi, assigned names to the maria after human moods or experiences, and named craters after famous scientists. Thus we have names for maria such as Mare Imbrium (Sea of Showers), Mare Tranquilitatis (Sea

of Tranquility, where the first manned moon landing occurred), and Mare Serenitatis (Sea of Serenity); and craters named Tycho, Hipparchus, and Archimedes. There were political overtones to Riccioli and Grimaldi's nomenclature; they published their map in 1651, at a time when the heliocentric hypothesis was still not something that people publicly embraced. With this in mind, Riccioli and Grimaldi gave geocentrists like Ptolemy and Tycho very large, prominent craters, but assigned Galileo only a very small one, and placed the name of Copernicus on a crater in the Sea of Storms (Mare Nubium, which can also be translated as Sea of Clouds).

Other names were added during the eighteenth and nineteenth centuries, but until 1921, no formal international agreement was made to honor a specific, uniform naming system. In that year, the newly formed International Astronomical Union designated a committee to oversee and standardize lunar no-

solar system. This has some impact on theories of the origin of the moon, as discussed later.

A second discovery made from the chemical analysis of moon rocks is that they contain lower abundances of the volatile elements than do earthly rocks. This probably implies that the moon's volatile elements escaped sometime before the moon formed.

The ages of lunar rocks could also be estimated. This was done by using the radioactive-dating techniques described in chapter 7. The lunar rock and soil samples from highland regions were found to be very old by earthly standards; as old as 3.5 to 4 billion years. Rocks from the maria are not quite so old, but still date back several billion years. In addition, a correlation was found between the ages of the rocks and the types of minerals they contain. This provides detailed informa-

tion on the history of the moon, because the conditions needed to form the different minerals are known, and the correlation with age reveals when the various required conditions existed.

Clues to Interior Conditions

Some of the experiments carried out on the moon by the Apollo astronauts were aimed at revealing the interior conditions. As in the case of the earth, the best way to do this is to monitor seismic waves caused by earthquakes on the moon. Therefore the astronauts carried with them devices for sensing vibrations in the lunar crust and, because it was not known whether nat-

menclature. Things were uneventful thereafter, until 1959, when the Soviets obtained the first photographs of the far side of the moon, and promptly named numerous features after prominent Russians. Many of these names (but not all) were later approved by the International Astronomical Union. Perhaps the most controversial was the Sea of Moscow, which reportedly was finally approved (amid laughter) when the Soviet representative declared, in response to criticism that it was inconsistent with tradition to name a mare after a city, that Moscow is a state of mind, just as tranquility and serenity are.

More recently, numerous additional features have been mapped and named, as the sophistication of lunar probes has improved, and especially since manned exploration of the moon has occurred. The Apollo astronauts assigned rather colloquial names to features at each landing site, and because of all the attendant publicity, these names became well known before there was time for the International Astronomical Union to even consider them for approval. Most of them were based on characteristics of the features themselves, such as a terraced crater that was called Bench, and one with a bright rim that was dubbed Halo. Groups of craters were named for their pattern, such as a pair called Doublet and a group named Snowman, which included individual craters with names such as Head. Some of the terrain mapped by the Apollo astronauts received names of prominent scientists, in keeping with the tradition established by Riccioli and Grimaldi more than three hundred years earlier. Many of the names invented by the Apollo astronauts were eventually approved by the International Astronomical Union, as were a dozen craters named for American and Soviet astronauts and cosmonauts.

Just about the time that lunar cartography was getting resolved (after the years of confusion brought on by the lunar exploration program), unmanned probes began to obtain images of Mercury, Venus, and Mars, and the whole problem of naming extraterrestrial landmarks arose again. Except for Mars, whose large-scale light and dark regions had been named on the basis of telescopic observations from earth, no tradition existed for naming features on other planets. After considerable discussion, the International Astronomical Union adopted rules for doing so. New features on Mars will be named after villages on earth; Venus will bear the names of prominent women and radio and radar scientists; Mercury will have features named after people famous for pursuits other than science; and features of the outer planets (or, as we shall see, the satellites thereof) will have names assigned from mythology. Perhaps it will thus be possible to map the rest of the solar system with a minimum of controversy.

---

ural moonquakes occurred frequently, they also brought along devices for thumping the surface to make it vibrate. They were even able to take advantage of quakes caused by the impact of man-made objects on the surface, such as a discarded fuel tank that crashed into the moon. It turned out that natural moonquakes do occur, although not with great violence. The seismic measurements continued after the Apollo landings, with data radioed to earth by instruments left in place on the lunar surface.

The measurements showed that the regolith, or surface soil, is typically about 10 meters thick, and is supported underneath by a thicker layer of loose rubble (Fig. 8.16). The crust is 50 to 100 kilometers thick at the Apollo sites, and may be somewhat thicker on the far side, where there are no maria. Beneath the crust is a mantle, consisting of a well-defined lithosphere, which is rigid, and beneath it an asthenosphere, which is semiliquid. The innermost 500 kilometers consist of a relatively dense core, but not as dense as that of the earth.

The seismic measurements indicate no truly molten zones in the moon at the present time, although temperature sensors on the surface discovered a substantial heat flow from the interior. This is probably caused by radioactive minerals below the surface.

There is no detectable overall magnetic field, further evidence against a molten core. The chief distinction among the internal zones in the lunar interior is density, with the densest material closest to the center. Thus moderate differentiation has occurred in the moon, which implies that it was once molten.

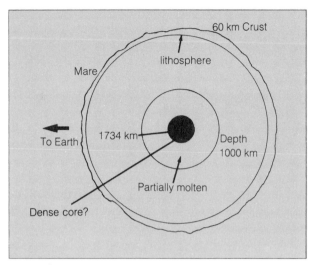

FIG. 8.16. THE INTERNAL STRUCTURE OF THE MOON. This cross-section illustrates the interior zones inferred from seismological studies. The existence of a dense core is not certain. Note that the maria lie almost exclusively on the side facing the earth, where the lunar crust is relatively thin.

There is no trace of present-day lunar tectonic activity, perhaps the greatest single departure from the geology of the earth. The force that has had the greatest influence in shaping the face of the earth has no role today on the moon. Instead, the lunar surface is entirely the result of the way the moon formed (which may have involved some tectonic activity in the early stages) and the manner in which it has been altered by lava flows and by the incessant bombardment of rocks and debris from space.

Formation and History

The wealth of detailed information about the moon has allowed scientists to piece together a comprehensive life story. There have been three competing theories of the moon's origin: (1) that the moon was once part of the earth, but fissioned, or split off, from it; (2) that the moon formed elsewhere and was later captured by the earth's gravitational attraction; and (3) that the earth and the moon formed together, and have remained in mutual orbit ever since. The first two are thought unlikely, on the basis of the Apollo data and other arguments.

The fission theory would be reasonable only if it

were assumed that the moon split off very early in the history of the earth, when the earth was still pliable enough to reestablish a spherical shape. The chief arguments against this idea are the lack of any known force that could cause the fission, and the chemical differences, particularly in the quantities of volatile elements, between rocks on the moon and on earth. If the two bodies had had identical histories until the split occurred, the rocks should be more similar in content than they are.

The second theory, that the moon was captured by the earth after having formed elsewhere in the solar system, has similar problems. Studies of gravitational interaction between two bodies moving through space show that it is not possible for one to capture the other unless something intervenes to slow their speed of approach to each other. It is easy to imagine how this could happen for very small bodies approaching a large one, where collisions between the small objects could slow some of them down, or an extended atmosphere of the large body could brake the incoming small objects. However, for a body the size of the moon, no reasonable mechanism for allowing capture has been found. A second argument against this theory has to do with the oxygen isotopes in lunar rocks: they have the same ratios as in earth rocks, which implies that both were formed in the same region of the solar system. It is possible, however unlikely it may seem, for three bodies to encounter each other simultaneously, resulting in two of them becoming trapped in mutual orbit. Thus the moon could have been captured, but most scientists do not believe that it was.

This leaves us with the third theory of lunar formation, that the earth and moon formed together (Fig. 8.17). The modern view of this idea actually includes a bit of the second theory. It is thought that when the earth formed out of rocky debris circling the sun, from the beginning it may have been able to capture small chunks of material that were slowed sufficiently by the interplanetary gas that filled the solar system at that time. A disk of orbiting material, vaguely resembling the rings of Saturn, formed around the earth. Dynamical studies have shown that if a few relatively large chunks were orbiting among the smaller pieces, in rather short order the smaller bits and pieces would be swept up by these large ones, which in turn would quickly collide and coalesce into a single massive satellite. The entire process would take only a few thousand years, which is such a short time geologically that

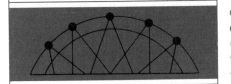

A Contrary Notion

The summary of the formation and evolution of the moon that is given in the text represents the most widely accepted viewpoint, but it is not without problems and conflicts. Some scientists find shortcomings in it so severe that they say the origin of the moon is unknown.

As noted in the text, studies of the moon's internal structure show that it is somewhat differentiated, but that it has not developed an extremely dense core; this illustrates that the process by which dense elements sink to the center did not occur to any large degree. Thus the moon's dense elements such as iron have not become highly concentrated at its center. On the other hand, samples from the lunar surface reveal a relatively low iron content there. If the moon and the earth formed from similar materials (as was the case if the two formed together, as described in the text), the moon should contain the same proportion of iron as the earth. It does not. The composition of moon rocks is nearly identical to that of the earth's mantle (as represented by rocks on the seafloor), where the iron content is low because differentiation caused much of it to sink to the earth's core.

This evidence has provoked a reconsideration of the fission hypothesis, which holds that the moon split off from the earth at some point in the past. If fission occurred some time after the earth had differentiated, this would account for the low iron content of the moon. The moon rocks are also low in volatile elements, compared with the earth, but these elements could have been evaporated from the cast-off material before it recondensed to form the moon.

The fission theory still has the problem of explaining the mechanism that could cause a substantial portion of the primitive, molten earth to be cast off, and primarily for this reason, it is not widely accepted. Studies of the effects of rotation offer some hope, however; if the newborn earth rotated very rapidly (with a period of only about two and a half hours, as is assumed by some), instabilities could have developed which forced the ejection of matter.

Perhaps it is wise to recall the case of continental drift, which scientists rejected for decades because they did not understand how it could occur. The lunar fission theory has problems similar to those of the continental drift theory after the work of Wegener: there is strong cicumstantial evidence supporting it, but it is not accepted because it lacks a known cause. It may be that our ideas of the moon's origin will be radically altered, as was our picture of the earth, when such a mechanism is found.

---

it is fair to say that the earth and moon formed together. Radioactive dating indicates that this took place about 4.5 billion years ago.

It is very likely that the moon was originally much closer to the earth. Evidence for this is found in fossil records, which indicate that the lunar month used to be much shorter than it now is. (This will be discussed more fully a little later in the story.)

After the moon coalesced from debris orbiting the newborn earth, its crust was molten for a time as a result of heating caused by the heavy bombardment of the surface by objects from space. After a time, the crust cooled and hardened, but gradually the interior became molten because of radioactive heating. Some dif-

ferentiation occurred, so to some extent the crust became depleted of heavy elements, which sank into the core.

The next major stage in the moon's life involved extensive volcanic activity. Molten material from the interior flowed across the surface over large areas, particularly in the lowlands. Thus the maria were formed. Apparently some of the huge lava flows were triggered by meteorite impacts, which cracked the crust in places where it was thin, providing an outlet for the molten material beneath.

Throughout the early history of the moon, it was continually bombarded by the rocky material still floating around the young solar system. Once the lunar crust

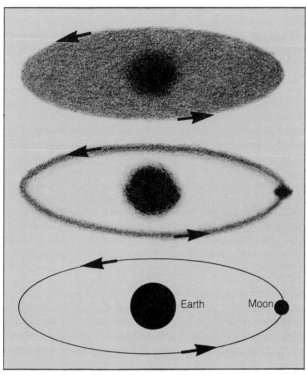

FIG. 8.17. COEVAL FORMATION OF THE EARTH AND
MOON. In this theory, the moon condensed from a disk of debris
that orbited the newly formed earth.

the moon had formed. When the lava flows that created the maria occurred, they obliterated craters in the
lowlands, so today there are few craters in the maria.
Those which are seen there were formed by meteorites that hit the moon after the lava flows had cooled.

The creation of the maria accentuated a tendency
that the moon may have had already, by making it more
lopsided. The maria exist only on the near side, which
is more dense than the far side. Because of this asymmetry, tidal forces exerted on the moon by the earth's
gravitational pull (see chapter 4) worked to keep the
heavy side of the moon pointed toward the earth. As
we learned in chapter 4, in time the internal friction
created in the lunar interior by these tidal forces slowed
the moon's rotation to the point where it now keeps
one side (the heavy side) always facing the earth.

The angular momentum of rotation that was lost by
the moon in this process was transferred to the moon's
orbital motion, forcing it to move gradually away from
the earth as its spin slowed (recall, from our discussion in chapter 4, that angular momentum is always
constant). Thus the moon is now much farther from
the earth than it once was, when it was newly formed
and rotating more rapidly.

After the creation of the maria, some three billion
years ago, little else happened to modify the moon's
structure. The rate of cratering decreased as the interplanetary debris either escaped the solar system or got
swept up by the planets and other large bodies, so only
a moderate amount of cratering has occurred since the
maria formed. The moon's interior gradually cooled,
eventually reaching its present nonmolten state.

had cooled, craters formed by the impacts of material
hitting the surface were able to survive, and the heavily
cratered regions of the moon that we see today were
shaped at that time, within the first billion years after

Perspective

The moon is the best-studied object in the solar system, other than the earth. The fact that its fate is so
closely intertwined with that of the earth enhances our
interest in it, and it is important to note the similarities
and differences between the two bodies. They are alike
in age and composition, except for the lower quantities of volatile elements on the moon. The lunar interior is geologically very quiet, whereas the earth is still
a dynamic, evolving body.

We are now ready to take a larger step out into the
solar system, to visit and explore earth's near-twin, the
planet Venus.

Summary

1. The general nature of the moon (its surface topography and its physical conditions), was ascertained from earth-based observations.

2. Much more detailed information has come from the Soviet and U.S. space programs, which have sent unmanned and manned probes to the moon.

3. The lunar surface consists of relatively smooth areas called maria, as well as highland, mountainous regions, and is marked everywhere by impact craters.

4. The lunar soil is called the regolith, and all the rocks on the surface are igneous (most are silicates), with low abundances of volatile gases.

5. Seismic data show that the moon has a crust 50 to 100 kilometers thick, a mantle, and a core extending about 500 kilometers from the center.

6. The moon has no present-day tectonic activity and no magnetic field, indicating that there is probably no liquid core.

7. The moon probably formed together with the earth, rather than splitting off from it or being gravitationally captured by it.

8. The moon's evolution consisted of a molten state, followed by hardening of the crust and subsequent large-scale lava flows, which created the maria. Since that time (about one billion years after the moon's formation), the moon has been geologically quiet.

Review Questions

1. Suppose you examine a portion of the moon where craters are so closely packed together that they actually overlap. How could we determine which were formed most recently?

2. The moon has much more severe temperature fluctuations between night and day than does the earth. Explain why.

3. The moon has no sedimentary or metamorphic rocks on its surface. What does this tell us about its geological history?

4. The fact that the moon is heavily cratered shows that it was bombarded with debris from space during some period in its past. Why is the earth not similary cratered?

5. Why is the moon's surface constantly being hit by charged particles from space, whereas the earth's surface is not?

6. Summarize the differences in the way the surfaces of the moon and the earth have been shaped.

7. Compare the mineral composition of moon rocks and earth rocks.

8. Why are there relatively few craters in the maria?

9. The earth's tidal force acting on the moon has slowed the moon's rotation to the point where it is synchronous; the moon always has one face toward the earth. The moon similarly exerts a tidal force on the earth. Explain why the earth is not also in synchronous rotation.

Additional Readings

Burgess, E, and Oberg, J. E. 1976. Science on the moon. *Astronomy* 4:684.

French, B. M. 1977. What's new on the moon? *Sky and Telescope* 53:164.

Hammond, A. L. 1973. Lunar science: analyzing the Apollo legacy. *Science* 179:1313.

Goldreich, P. 1972 Tides and the earth-moon system. *Scientific American* 226(4):42.

Goles G. G. 1971. A review of the Apollo project. *American Scientist,* 59:326.

Taylor, S. R. 1975. *Lunar science: a post-Apollo view.* Elmsford, N.Y.: Pergamon Press.

Wood, J. A. 1975. The moon. *Scientific American* 233(3)92.

CHAPTER 9

Venus: A Cloud-Covered Inferno

Of all the planets, Venus is the most prominent in our nighttime sky, and it has played a role in the astronomical lore since antiquity. Because Venus is an inferior planet (Fig. 9.1), it never appears far from the sun, as seen from the earth (its greatest elongation is 47 degrees). Thus it is visible to us either early in the evening, shortly after sunset, or in the wee hours just before dawn. In ancient times, the morning and evening stars were thought to be two separate planets, named Phosphorous and Hesperus, but by the sixth century B.C. the two were identified as a single planet alternately appearing on either side of the sun.

Bright as it is, Venus has become a natural subject for careful scrutiny by astronomers throughout history, and it has played an important role in some major developments. For example, the fact that it never wanders far from the sun prompted early speculation that perhaps it orbits the sun rather than the earth, and Galileo's observation of the phases of Venus was an important bit of evidence favoring the heliocentric theory.

In modern times a great deal of scientific effort has gone into studies of Venus, because its proximity to earth, both in location and in general characteristics, means that Venus may teach us valuable lessons about our own planet.

FIG. 9.1. **THE CRESCENT VENUS.** This photograph, taken from earth, shows Venus when it is near inferior conjunction, so that we see only a sliver of its sunlit side.

TABLE 9.1.   VENUS

Orbital semimajor axis: 0.723 AU (108,200,000 km)
    Perihelion distance: 0.718 AU
    Aphelion distance: 0.728 AU
Orbital period: 224.7 days (0.615 years)
Orbital inclination: 3°23'40″

Rotation period: 243 days (retrograde) *Tilt of axis:* 3°
Diameter: 12,104 km (0.951 $D_\oplus$) *Mass:* 4.87 × 10^{27} grams (0.851 $M_\oplus$) *Density:* 5.3 grams/cm^3 *Surface gravity:* 0.90 earth gravity *Escape velocity:* 10.3 km/sec
Surface temperature: 750 K *Albedo:* 0.76
Satellites: none

Observations from Near and Far

A substantial body of knowledge about Venus was developed, particularly within the last two decades, from remote observation. Its size and orbital parameters, as well as some data on the composition of its upper atmosphere and atmosphere and an estimate of its mass, all were derived from earth-based observations, as were some surprising discoveries about its rotation and its surface temperature.

Venus is so brilliant because it is shrouded in clouds that efficiently reflect sunlight. Its albedo is 0.76. The same clouds that create the brilliance also hide the surface from telescopic view, and for this reason special techniques have been required to penetrate the clouds.

Although it was once popular to imagine that Venus is a pleasantly tropical planet, with lots of water and, no doubt, plant and animal life, what was discovered from early radio observations was vastly different. Radio waves penetrated the cloud cover, and the message they brought to earth is that the surface of Venus is very hot, about 750 K (477°C, or about 900°F). This discovery made it obvious that, despite outward appearances of similarity, Venus and the earth had followed different paths. No longer could Venus be viewed as a likely abode for life.

The reason for the high temperature was quickly ascertained (Fig. 9.2). Spectroscopic measurements dating back to the 1930's showed that carbon dioxide (CO_2) is a major constituent of the atmosphere. This molecule has many spectral lines in the infrared part of the spectrum, so that a CO_2 atmosphere is effectively opaque to infrared light.

Visible light from the sun can penetrate the clouds of Venus from above, and is absorbed at the surface, heating it. When the ground responds by emitting infrared radiation, however, this radiant energy cannot rapidly escape because of absorption by the CO_2, and heat is trapped near the surface. This heating mechanism is called the **greenhouse effect,** because it is similar to what happens in a greenhouse when light enters through the glass and heats the interior.

A second surprising discovery was made from earth, again through the use of radio (specifically radar) wavelengths. The first radar measurements of the surface were made in 1961, and the Doppler effect was used to determine the rate of rotation. Contrary to expectations, the planet was found to rotate very slowly, and even more astonishingly, to do so in the backward direction, the direction opposite that of the other orbital and rotational motions in the solar system. This is called **retrograde rotation.**

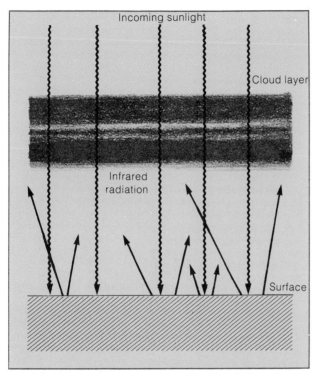

FIG. 9.2. THE GREENHOUSE EFFECT. Visible light from the sun reaches the surface of Venus and heats it, causing the ground to emit infrared radiation. The CO_2 in the atmosphere of Venus efficiently absorbs infrared radiation, however, so heat is trapped near the surface.

The spin of Venus takes 243 earth days, so slow that its day is actually longer than its year (the orbital period is 225 days). Here we must distinguish between the sidereal day, the 243 days just mentioned, and the solar day, the apparent time it takes for the sun to go through one daily cycle, as seen from a point on the surface of Venus. The length of the solar day on Venus is 116.8 days.

The reason for the unusual rotation of Venus is not understood. Because it is so strongly at odds with the spins of the other planets, it seems likely that something unusual happned either during the formation of Venus or at some later time.

Other data on the atmosphere of Venus were obtained from earth-based observations. Ultraviolet photographs, which reveal features in the cloud structure, showed that the cloud tops are moving rapidly, circling the planet every four days at speeds of about 100 meters per second (360 kilometers per hour). This atmospheric circulation occurs in the retrograde direction.

The temperature of the cloud tops was derived from infrared measurements, which led to an estimate of 240 K (− 27°F).

Both the Soviet Union and the United States focused on Venus as a logical and interesting target for exploration by spacecraft, and a number of missions have been flown there by both nations. Spacecraft can be launched toward Venus only when the relative positions of the earth and Venus are favorable. The United States launched *Mariner 2* toward Venus on such an occasion in 1962, and this spacecraft flew within 35,000 kilometers of the cloud tops, making various measurements. The next opportunity came in 1965–1966, when both the United States and the Soviet Union took advantage by launching Venus probes. The U.S. *Mariner 5* mission was another flyby, but the Soviets chose instead to have their *Venera 4* probe descend into the atmosphere of Venus, and it relayed back to earth important information on conditions there, before it failed on its way down to the surface. *Veneras 5* and *6,* launched in 1969, also entered the atmosphere of Venus, as did *Veneras 7* and *8* in 1970 and 1972.

The U.S. *Mariner 10* mission in 1973–1974 flew by Venus (and later Mercury), obtaining detailed data on the structure of the clouds and their motions. This and earlier *Mariner* spacecraft showed that Venus has no magnetic field.

The most recent probes sent to Venus have provided spectacular new information on conditions beneath the clouds. Whereas the earlier *Venera* instruments had failed (as a result of the intense heat and pressure) before reaching the surface, the *Venera 9* and *10* missions (in the mid–1970s), *Veneras 11* and *12* (1978), and most recently, *Veneras 13* and *14* (1982), all succeeded in landing and operating on the surface. *Veneras 9, 10, 13,* and *14* carried cameras and sent back photographs of the ground surrounding the landing sites. These images showed that the surface is brightly lit, with enough sunlight penetrating the clouds to actually cast shadows, and that the ground is strewn with rocks, many of them with surprisingly sharp edges. This indicates that despite the hostile conditions, there is very little erosion acting on the surface of Venus.

The most recent U.S. probe to Venus was called *Pioneer Venus,* and consisted of two separate spacecraft, an orbiter (Fig. 9.3, and 9.4) and a package of probes designed to penetrate the atmosphere. The orbiter is still in operation (in late 1982), gathering data on the cloud structure, compositon, and motions, as well as

Velikovsky and Venus

The planet Venus played a major role in a scientific controversy that began in the 1950's. The controversy was not so much over the proper interpretation of scientific data, although that played a role, but was centered instead on questions of the proper role of science in responding to pseudoscientific theories, which occasionally catch the public imagination. Science must be open to new ideas, but at the same time it must be on watch against those which are not supported by evidence or rational analysis. In the case of Immanuel Velikovsky and the origin of Venus, scientists perhaps stepped beyond the boundaries of proper response.

In 1950, following years of study of ancient records, Velikovsky, trained as a medical doctor and psychologist, published a book called *Worlds in Collision*. In this book, extrapolating from records of catastrophic upheavals that apparently affected many different ancient cultures at about the same time (between the fifteenth and eighth centuries B.C.), Velikovsky suggested that these catastrophes had extraterrestrial origins. He described a sequence of events involving a "comet," released from the atmosphere of Jupiter, that twice nearly collided with the earth, causing its rotation to be stopped temporarily, and which then lost its tail in a collision with Mars before entering a circular orbit about the sun and becoming the planet Venus. Velikovsky hypothesized that the interactions between the sun and planets were dominated by electromagnetic forces, that the motions of both Mars and the earth were altered to their present state by the near-collisions with the comet, and that these near-collisions (there was also supposed to have been a collision between the earth and Mars in 687 B.C.) accounted for the upheavals in human society whose occurrence had originally inspired Velikovsky.

Despite its absolute and total disregard for astronomical observation and for the laws of physics, *Worlds in Collision* attracted a great deal of public attention and interest. Fearing that this attention would be followed by general acceptance, the scientific community reacted quickly. Numerous well-known astronomers, historians, and physicists made public statements criticizing the book, its publisher was effectively boycotted, and in general the scientists gave the appearance of feeling strongly threatened. Unfortunately, critics of Velikovsky often used questionable tactics. Also, in many cases they were not as well-versed in his writings as they should have been, and they frequently left themsleves open for countercriticisms. Matters became more heated when new evidence, in the form of data from planetary space probes launched in the early 1960's, supported some of Velikovsky's predictions. These included the fact that Venus has a high surface temperature, that the earth's magnetic field extends very far into space, and that Jupiter is an emitter of radio waves.

The Velikovsky affair has by now faded from public view, but it has not been forgotten. There are probably people who still believe the ideas expressed in *Worlds in Collision* and in Velikovsky's second book, *Earth in Upheaval*. The continued development of the space program and the successful probes of other planets (including, of course, Venus) have helped a great deal, since other predictions of Velikovsky, such as his notion that electromagnetic forces control planetary motions, have now been refuted by direct measurement. There is much less room now for Velikovsky's supporters to avoid conflict with observation.

Responding to works such as *Worlds in Collision* and similar unfounded pseudoscientific treatises (popular books on ancient astronauts and astrology, for example) is a difficult problem for scientists. There is clearly an obligation to keep the public as well-informed as possible, so that people can see for themselves the folly of unsupportable and unprovable ideas, but this must be done in a dignified manner, without creating the impression of having something to hide. Thus, Velikovsky could not be ignored, but he and his ideas probably should have been treated with less emotion and more logic.

FIG. 9.3. A CLOSE-UP PORTRAIT OF VENUS. This is the first full-disk image of Venus obtained by the *Pioneer Venus* spacecraft, and it shows structure in the clouds.

FIG. 9.4. THE *PIONEER VENUS* ORBITER. This spacecraft, with its wide assortment of scientific instruments, is still in operation, orbiting Venus every twenty-four hours.

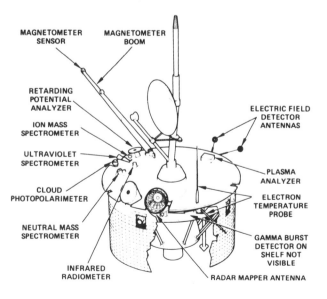

on conditions in the upper atmosphere, above the clouds. The five small probes that descended into the atmosphere when the spacecraft arrived at Venus in late 1978 gathered data on physical conditions such as temperature and pressure, and data on chemical composition of the atmosphere as a function of altitude. The *Pioneer Venus* probes reached the ground safely, and continued to operate there for awhile (up to an hour).

The Atmosphere of Venus

We can summarize what has been learned about the atmosphere of Venus by discussing three distinct aspects: its composition, vertical structure, and motions.

The most important ingredient in the atmosphere of Venus, already identified from earth-based measurements, is carbon dioxide. The *Venera* and *Pioneer Venus* probes showed that CO_2 dominates all through the atmosphere to the ground, and makes up about 96 percent of the atmosphere. It is interesting that the overall quantity of CO_2 on the earth is about the same as that on Venus, except that on earth the CO_2 is mostly in carbonate rocks or dissolved in seawater, and not in the atmosphere.

The next most abundant gas in the atmosphere of Venus is nitrogen (in the form of N_2), which accounts for most of the remaining 4 percent or so. In addition, traces of water vapor and other species such as oxygen, argon, neon, and sulfur (in the form of compounds such as sulfur dioxide and sulfuric acid) exist there. The sulfuric acid (H_2SO_4) apparently condenses out of the lower clouds, forming droplets like rain. These circumstances, along with the high temperature and pressure at the surface, make it difficult for spacecraft to operate on Venus.

The Structure of the Atmosphere

We have already seen that the atmospheric conditions on Venus vary quite a lot from the top of the clouds to the surface. The temperature increases from 240 K at cloud tops to 750 K at the ground, and the pressure at the bottom of the atmosphere reaches the rather incre-

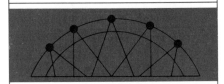

ASTRONOMICAL INSIGHT (9.2)

Argon, Tracer of Planetary Genealogy

One of the measurements of atmospheric composition led to a surprise for the scientists studying *Pioneer Venus* data. Although most gases in the atmosphere of Venus have been added or modified since the time the planet formed, there are some species that have probably been present in the same form and amount from the beginning. One such gas is a particular isotope of argon, called argon 36 or ^{36}Ar (argon has 18 protons in its nucleus, and this isotope has 18 neutrons, unlike the most common form of argon, ^{40}Ar, which has 22 neutrons). The ^{36}Ar atoms that were present when Venus formed are still in the atmosphere, because they are unable to combine chemically with other species, they are unaffected by any radioactive-decay pro-

cesses (unlike ^{40}Ar, which can be formed by the decay of potassium 40), and they are too heavy to escape from Venus.

The rocky debris from which the planets formed was itself created by condensation of the gas that floated around the infant sun. The higher the temperature at the time of condensation, the fewer volatile elements end up in solid form. Since it was thought that the inner portions of the solar system were hotter than regions farther out, it was expected that the abundance of ^{36}Ar, a volatile gas, would be lowest for the planets close to the sun, and would increase outward. Contrary to this expectation, however, it had already been found from the *Viking* missions in 1975–1976 that Mars has a lower, rather than a higher, abundance of ^{36}Ar than the earth. This may not have been considered too great a puzzle, since both the earth and Mars are relatively far from the sun, but it was thought that surely Venus would show the expected effect.

It was a great surprise, therefore, when the *Pioneer Venus* data showed that ^{36}Ar is even more abundant on Venus than on the earth. It appears that the quantity of

this volatile element decreases steadily with distance from the sun, just the opposite of the expected trend.

A probable explanation has been found. Although temperature is the most important factor determining how much of the volatile element will condense into solid form, pressure is another condition that has some influence. If the temperature in two regions was the same, the greater volatile abundance would occur where the pressure was higher. Thus if the temperature throughout the inner portion of the primordial solar system was constant (rather than decreasing with distance from the sun), and the pressure was highest in the center, then the greatest abundances of volatiles would have condensed near the center, as observed. Thus a seemingly innocuous measurement of a single isotope of argon has rather substantially altered our theories of how the solar system formed, by showing that the condensation of the first solid materials occurred before there was strong heating at the center caused by the presence of a hotly glowing sun.

dible value of 90 earth atmospheres, or more than 1,300 pounds per square inch, equivalent to the pressure about 3,000 feet below the surface of the ocean. The temperature decreases smoothly from the surface to the top of the clouds.

The clouds in the atmosphere of Venus are composed of small droplets of sulfuric acid and some other, as yet unidentified, kinds of particles. The H_2SO_4 drop-

lets are about 2 millionths of a meter (about 0.0002 centimeters) in diameter.

There are three distinct zones of clouds, at separate altitudes (Fig. 9.5). The uppermost layer, the one visible from afar, lies some 60 km above the surface of Venus and averages about 10 km in thickness. The middle cloud layer, about 6 km thick, is suspended at an altitude of roughly 53 km, whereas the lowest layer

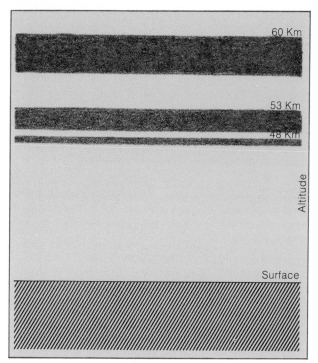

FIG. 9.5. THE CLOUDS OF VENUS. The sulfuric-acid clouds are separated into three distinct layers, as shown here.

lies at 48 km and is relatively thin, extending only 2 km or so vertically. Below the lower cloud is clear atmosphere, all the way to the ground. Above the uppermost clouds is a haze layer, consisting of much smaller particles than the droplets that form the clouds, but of unknown composition.

The clouds also contain quantities of the gas sulfur dioxide (SO_2), which is an efficient absorber of ultraviolet light. We see from ultraviolet photographs that there is considerable structure in the clouds, because the SO_2 abundance varies with depth, and therefore in places where the winds stir up regions with quantities of SO_2, or where the cloud structure allows us to see to greater depths, the clouds look dark (Fig. 9.3). Thus ultraviolet photographs show some contrast in the clouds, and are useful for mapping the motions of the upper regions.

Clouds form in the atmosphere wherever the combination of temperature, pressure, and relative abundance of H_2SO_4 corresponds to the conditions required for this gas to condense. This is exactly analogous to the formation of clouds in the earth's atmosphere, except that on the earth, the gas that condenses

is water vapor rather than sulfuric acid. It happens that the proper conditions for condensation occur at three different levels in the atmosphere of Venus.

The atmosphere seems to be less changeable than that of the earth, where cloud structures come and go with changes in the weather. The relative stability of Venus is probably a result of a combination of factors, including the greater density and pressure of the atmosphere, and the slow rotation of the planet, which minimizes circulation patterns driven by rotation.

Another important distinction between earth and Venus is that there is no magnetic field around Venus. There is no magnetosphere and no protective shield of magnetic field lines to prevent charged particles from entering the atmosphere. This creates a complex electrical environment at the top of the atmosphere, including lightning discharges, which probably affects the chemical processes there as well.

ATMOSPHERIC MOTIONS

We have seen that the upper atmosphere of Venus has a general circulation pattern at the cloud tops, moving in the same direction as the planet's rotation (Fig. 9.6). The velocity is high at the cloud tops, but diminishes to nearly zero at the surface.

There are complex vertical motions in addition to these horizontal ones. Because of solar heating near the equator, there is a rising current there that spreads out in all directions, descending when it reaches cooler regions near the poles and on the dark side of the planet. This pattern is similar to the circulation cells in the earth's atmosphere, except that on Venus there is insufficient rotation of the planet to break up the flow into numerous small cells. On Venus it is more of a global circulation pattern, with warm gas rising in the location called the **subsolar point,** where the sun is directly overhead, and descending in a very broad region away from that point (Fig. 9.7). Besides minimizing rotational forces that might otherwise break up this simple flow pattern into smaller cells, the slow spin of Venus also means that the subsolar point moves very slowly as the planet rotates. Thus the heating from the sun lingers, enhancing the flow pattern.

Some of the gas that rises at the subsolar point flows all the way around to the dark side of Venus before

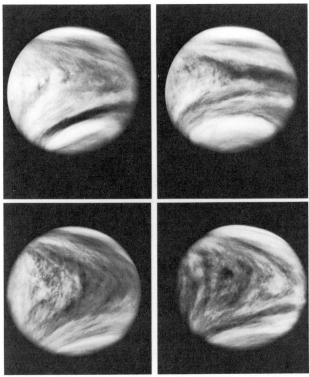

FIG. 9.6. CIRCULATION OF THE ATMOSPHERE. These four ultraviolet views cover two rotations of the planetary cloud cover, the upper two being separated by one day, and the lower two, obtained about a week after the first pair, also being separated by a day. The atmospheric motion is from right to left in these images.

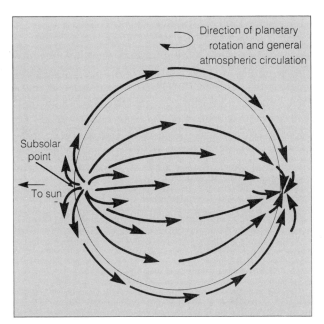

FIG. 9.7. HIGH-ALTITUDE CIRCULATION. Well above the clouds (which are rotating from right to left in this sketch), gases heated at the subsolar point circulate around the planet toward the dark side, where they cool and descend.

FIG. 9.8. AIRGLOW ON THE NIGHT SIDE OF VENUS. Atoms produced by the disruption of molecules on the sunlit side circulate around Venus to the dark side, where they cool and combine once again into molecules. The newly formed molecules have excess energy which is released in the form of light, creating a glow in the upper atmosphere on the night side.

cooling and descending. As it does so, free atoms combine to re-form molecules, which then emit light as they cool further. This creates aurorae on the dark side of Venus, a phenomenon that was suspected from earth-based observations, but which was not confirmed until the flight of *Pioneer Venus* (Fig. 9.8). (These aurorae are quite distinct from those in the earth's upper atmosphere, which are created by charged particles from space.)

In and between clouds, additional vertical circulation patterns exist, depending on where the sun's heat energy is deposited. This in turn depends on the cloud composition; we have already seen that SO_2 is particularly effective as an absorber of solar energy, so there is extra heating in levels were SO_2 is abundant. There appear to be at least four flow patterns at different levels involving circulation between the equatorial and polar regions.

The Surface of Venus

Radar experiments have provided most of the information available on the surface of Venus. Observations made from earth had insufficient resolution to reveal much in the way of surface features, although there were some indications of highlands and lowlands. The radar instrument on the *Pioneer Venus* orbiter, however, is capable of making fairly detailed measurements and can detect features as small as 30 kilometers. As a result, global relief maps of Venus have been constructed, showing many interesting features (Fig. 9.9).

The surface has been classified into three general types of terrain: (1) rolling plains, covering about 65 percent of the planet; (2) highlands, covering about 8 percent; and (3) lowlands, occupying the remaining 27 percent.

The rolling plains are characterized by numerous craters and circular basins, many of them apparently lava-filled, resembling small-scale lunar maria. There is uncertainty about whether the craters are all caused by meteor impacts, or whether some may be volcanic in origin.

The highlands are concentrated in three major regions, two of which, called Ishtar Terra (Fig. 9.10) and Aphrodite Terra, are comparable in size to the continents of Africa and Australia on the earth, whereas the third, Beta Regio, is much smaller. The maximum elevation above the mean surface level is greater than that of Mt. Everest above sea level, but the numerous can-

FIG. 9.9. A RELIEF MAP OF VENUS. This image, constructed on the basis of *Pioneer Venus* radar maps of the surface, shows the three major highland areas: Ishtar Terra at the top, Aphrodite Terra at right center, and Beta Regio at left center.

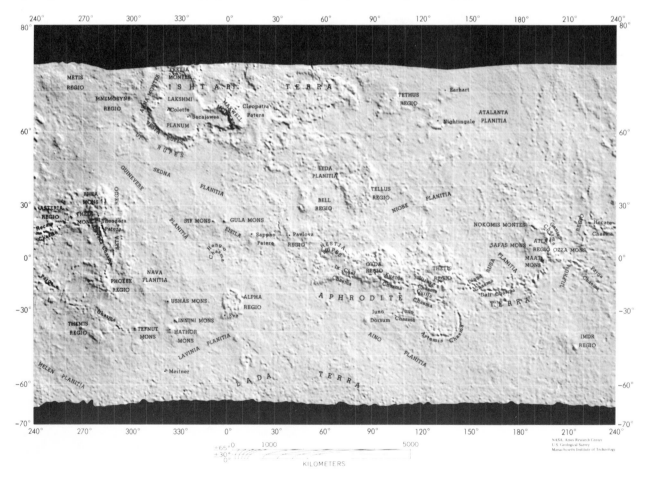

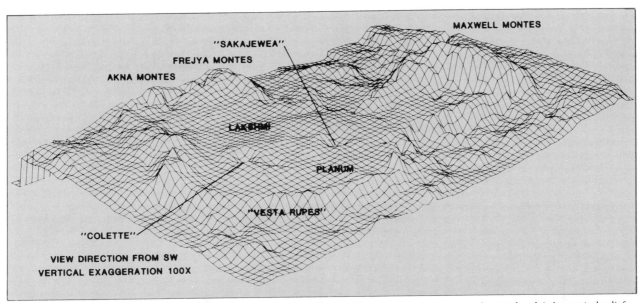

FIG. 9.10. A PORTION OF ISHTAR TERRA. This computer reconstruction illustrates (on an exaggerated vertical scale) the vertical relief of this highland region, and includes the largest mountain on Venus, Maxwell Montes, more than 11 kilometers above the mean surface level.

yons are much shallower than the deepest undersea trenches on earth. There are many adjacent parallel valley and trench systems in the highlands of Venus, forming systems up to 9,000 kilometers in length.

Given the similarities in size and average density of Venus and the earth, it is interesting to consider whether the internal structure and processes are also similar. One point of comparison is tectonic activity. Venus apparently does not have such active continental drift as the earth; there is not a planetwide system of plate boundaries and ridges and trenches, comparable to that of the earth. The reasons for this difference between the earth and Venus are not clear.

There is evidence, however, in the form of the valley and trench systems in the highlands, that tectonic activity has played some role in the formation of these regions. The highlands consist in part of large volcanic mountains that built up as a result of prolonged lava-flow episodes. The smallest of the three highland areas, Beta Regio, appears to consist entirely of a few large volcanoes. The combination of the valley and ridge systems and the large volcanoes indicates that there are (or were in the past) convection currents in the upper mantle, and that there are (or were) volcanic hot spots where lava flowed upward for long periods. Thus Venus's internal activity and structure may be much like the earth, but its crust has not broken up into plates that move about as has happened on the earth.

The reason for the lack of continental drift on Venus is not known, although there are some interesting speculations. One suggestion is that because Venus has such a high surface temperature, its crust is warmer and therefore more buoyant than that of the earth. On both planets the highlands (or continents) consist of old material that essentially floats, in equilibrium with the denser material of the lowlands (or seafloors). On earth the seafloor material is as dense as the underlying asthenosphere, and therefore is easily forced below continental margins in the subduction zones that produce the great undersea trenches. On Venus the lowland material is less dense than the underlying mantle, and therefore cannot easily be forced downward. The result is that convection currents inside Venus have not succeeded in creating crustal plates that can move about by slipping over one another, because subduction cannot take place.

Some information about surface rocks was derived from the photographs and other measurements obtained by the *Venera* landers (Fig. 9.11, 9.12). As we saw earlier, the rocks have sharp edges, indicating a lack of erosion, something that was a surprise until scientists realized that the winds at the surface of Venus are almost nonexistent.

Measurements of surface rocks show a basaltic composition, similar to the predominant crustal rocks on the earth and the moon. Basalt, as noted in chapter 7,

FIG. 9.11.   SURFACE ROCKS ON VENUS. These views are from the first Soviet *Venera* landers. They show that sunlight penetrates strongly to the surface, and that the surface rocks have surprisingly sharp edges, indicating a lack of erosion.

FIG. 9.12.   *VENERA 13* and *14* PHOTOS OF SURFACE ROCKS. These two missions, which took place in the spring of 1982, obtained more extensive and better-quality images of surface rocks than had their predecessors, *Veneras 9* and *10*.

is an igneous silicate rock with a moderate density. This, combined with the fact that the average density of Venus is comparable to that of the earth (5.3 grams/cm^3 compared with 5.5 for the earth), implies that Venus is differentiated, with a large, dense core.

Other than this very indirect evidence, very little is known about the interior of Venus. The fact that its internal structure is similar to that of earth might make us suspect that Venus has a magnetic field, but as we have seen, it does not. This may be explained by the slow rotation of the planet, which could be insufficient to create a magnetic dynamo in its core.

The Earth and Venus: So Near and yet So Different

We are now equipped with enough information to discuss the history of Venus and to develop some understanding of what might be responsible for the extreme differences between conditions there and on the earth. This is important, for we need to know how precarious our situation is, so that we know how to avoid turning the earth onto the path that Venus has followed.

Venus is thought to have formed from rocky debris orbiting the infant sun in a great disk, just as the earth did. The early evolution also must have been similar, with Venus undergoing a molten period when its dense elements sank to the center, leaving a lighter crust composed largely of carbonates and silicates. Before the crust cooled, volatile gases escaped from the surface, forming a primitive atmosphere of hydrogen compounds much like the earliest atmosphere on earth.

At this point a major contrast developed between the earth and Venus. As oxygen escaped from the rocks on earth and combined with hydrogen to form H_2O, much of it could persist in the liquid state, and the

earth had oceans from that time onward. On Venus, however, which is closer to the sun, liquid water could not exist, but instead stayed in the atmosphere as water vapor. Venus is 0.72 AU from the sun, so the intensity of sunlight at its surface is $1/.72^2 = 1/.52 = 1.92$ times greater than at the surface of the earth (recall the inverse-square law, discussed in chapter 5). This is not a very great difference, but it was extremely significant.

Once the atmosphere of Venus became contaminated with water vapor, the greenhouse effect went into operation, further heating the surface. The temperature got sufficiently high to bake the carbon dioxide out of the carbonate rocks, and with no oceans to dissolve it and return it to rock, soon this gas dominated the atmosphere. On the earth the CO_2 that was outgassed was largely dissolved in seawater and then deposited in rocks, where it still exists. The development of life on earth also had a major effect that was absent on Venus, gradually altering the earth's atmospheric composition to one dominated by oxygen and nitrogen, which do not create a strong greenhouse effect.

The increasing CO_2 content in the atmosphere of Venus enhanced the heating caused by the greenhouse effect, leading to the extreme surface conditions known today. Eventually the water vapor in the atmosphere was dissociated by the high temperatures, releasing the hydrogen atoms so that they escaped into space.

Meanwhile the surface of Venus started to follow an evolution much like that of the earth. Convection apparently developed soon after the interior differentiated, giving rise to the tectonic activity that has shaped the highland masses, although on Venus, unlike on the earth, these continental areas did not begin to move about. Volcanic activity has been a continuing feature of the environment, however, just as it has on earth. In a strictly geological sense, Venus may almost be a twin of the earth.

Perspective

The study of Venus has been very instructive, not only for what it tells us of the planet itself, but also for the lessons learned about the earth. We must not forget that, except for a rather small difference in the intensity of sunlight at an early stage in its history, the earth might have ended up with a crushing, hot CO_2 atmosphere like that on Venus.

We turn our attention next to another of the terrestrial planets, the next closest to earth. In examining Mars we will be able to draw other parallels and contrasts with earth.

Summary

1. Observations of Venus from earth, particularly radio data, revealed that it is extremely hot at the surface, and rotates slowly in the retrograde direction.

2. Soviet and U.S. spacecraft have flown by Venus and probed its atmosphere. Several Soviet probes have landed on the surface and sent back information on conditions there.

3. The atmosphere of Venus consists mostly of carbon dioxide, with a bit of nitrogen and traces of sulfur dioxide and sulfuric acid.

4. The atmospheric temperature increases from 240 K at the cloud tops to 750 K at the surface, where the pressure is 90 times that at sea level on the earth. The heating is caused by the greenhouse effect.

5. The clouds, composed of droplets of sulfuric acid, form three distinct layers between 48 and 60 km in altitude. The atmosphere is clear below the clouds.

6. At the cloud tops, high winds circle the planet in four days, whereas it is calm at the surface. Above the clouds there are global convection currents.

7. The surface of Venus, consisting of rolling plains, highlands, and lowlands, shows evidence of limited tectonic activity but no continental drift.

8. Venus evolved in a manner very differently from the earth, primarily because it is so close to the sun that liquid water could not exist on its surface. The lack of water allowed CO_2 to stay in the atmosphere and discouraged the development of life.

Review Questions

1. Why could the mass of Venus not be deduced precisely from earth-based observations in the same manner as the masses of most of the other planets?

2. What is the wavelength of maximum emission for the surface of Venus, where the temperature is 750 K?

3. On the earth a solar day is a little longer than a sidereal day, but on Venus it is shorter. Explain why.

4. Suppose a radar echo from one edge of Venus has its wavelength shifted by $\Delta\lambda/\lambda = 6 \times 10^{-14}$. What is the rotational velocity (that is, the speed of rotation at a point on the surface) of Venus? Given that the radius is about 6,000 kilometers, use the velocity you just calculated to determine what the rotation period is.

5. Compare the role of SO_2 in the atmosphere of Venus to that of ozone in the atmosphere of earth.

6. Why does the fact that Venus has a large, dense core lead to the expectation that the surface rocks probably have a low iron content?

7. Why does Venus not have oxygen and nitrogen in its atmosphere in large quantities?

8. Why do conditions on Venus cause scientists on earth to be concerned about increasing levels of CO_2 in the earth's atmosphere?

9. Aurorae in the earth's atmosphere are concentrated in the polar regions. Would you expect this to be the same on Venus?

Additional Readings

Beatty, J. K., O'Leary, B., and Chaikin, A., eds. 1981. *The new solar system*. Cambridge, England: Cambridge University Press.

Pettingill, G. H., Campbell, D. B., and Masursky, H. 1980. The surface of Venus. *Scientific American* 243(2):54.

Schubert, G., and Covey, C. 1981. The atmosphere of Venus. *Scientific American* 245(1):66.

Young, A., and Young, L. 1975. Venus. *Scientific American* 233 (3):70.

• CHAPTER 10 •

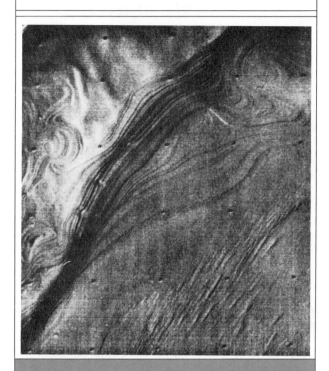

Mars: The Red Planet

The planet Mars, although not as brilliant as Venus, is nevertheless a prominent object in the heavens (Fig. 10.1). The fourth planet from the sun, and our next nearest neighbor after Venus, Mars has played an important role in the development of astronomy, first lending itself to mythological interpretations, and later providing the basis of Kepler's discoveries on the motions of the planets. In rather recent times, Mars again became a center of considerable human speculation, as people mused over the possibility that life might exist there. The most recent chapter in the story, the exploration of Mars by unmanned spacecraft, has dispelled such notions but has not diminished our fascination.

Percival Lowell and the Martian Myth

From antiquity, Mars has attracted special attention because of its distinct and unusual reddish color, apparent to even the unaided eye. The Greeks named this

TABLE 10.1   MARS
Orbital semimajor axis: 1.523 AU (227,900,000 km)
Perihelion distance: 1.381 AU
Aphelion distance: 1.665 AU
Orbital period: 1.881 years (687.0 days)
Orbital inclination: 1°51′0″
Rotation period: 24^h 37^m 22.^{s}6
Tilt of axis: 23°59′
Diameter: 6,774 km (0.532 D$_\oplus$)
Mass: 6.42 × 10^{26} grams (0.107 M$_\oplus$)
Density: 3.96 grams/cm^3
Surface gravity: 0.38 earth gravity
Escape velocity: 5.0 km/sec
Surface temperature: 130–290 K
Albedo: 0.15 (average)
Satellites: 2

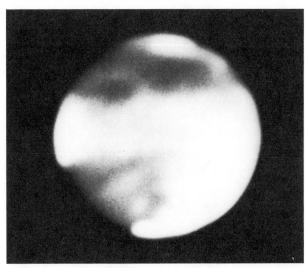

FIG. 10.1.   THE EARTH-BASED VIEW OF MARS. A red filter was used to make this photograph in order to enhance the clarity with which surface markings are seen.

planet Ares, after their god of war, and the Roman name Mars has the same connotations.

The first telescopic observations of the red planet were made by Galileo, who noted that its disk varies in angular diameter as its position with respect to the sun varies. About fifty years later, Huygens became the first to sketch surface features as he saw them on the planet's disk, something which later led to one of the more fascinating scientific debates since medieval times. Successive generations of astronomers attempted to study the surface markings, a very subjective business, especially since there was no photographic film with which to record the image. Photographs are not necessarily superior to the human eye in this case anyway, because variations in the turbulence of the earth's atmosphere may allow momentary clear glimpses, but these few instances of high clarity will be smeared out in a photograph obtained over a long exposure time.

Huygens noted a dark patch on the surface, which was named Syrtis Major. Additional details were seen later, and by the 1870s observers had named a number of features, using Latin words related to bodies of water, which they thought the features to be. At about this time, Father Secchi, the Roman Catholic priest who later was to play a pioneering role in the development of stellar spectrum analysis, saw what he believed to be linear features on the surface of Mars, and he referred to them as *canali,* an Italian word referring generally to natural channels of water. Giovanni Schiaparelli, who was then the director of an observatory in Milan, imagined that he saw a network of these features on the surface of Mars, and in 1877, when Mars was at opposition, he produced a detailed sketch mapping them (Fig. 10.2). Soon other observers around the world were seeing the canali, and in due course the name began to be interpreted as meaning canals, artificially constructed channels for transporting water.

At this point a wealthy Boston aristocrat named Percival Lowell entered the story, fantasizing an extensive and well-developed Martian civilization that had built the canals for irrigation, in response to the growing dryness of the planetary climate. Lowell (Fig. 10.3) published a book full of these speculations in 1895, which created a groundswell of public interest in the notion. Despite skepticism on the part of some astronomers who either could not see the canals or realized how unreasonable it was that Martian canals should be visible from earth even if they really existed, the idea of a Martian civilization became fixed in the public mind. To this day a wide variety of science-fiction novels and movies (starting with H. G. Welles's *The War of the Worlds*) have been inspired by these beliefs.

Lowell probably invested more personal energy and effort in the Martian civilization than anyone else. He

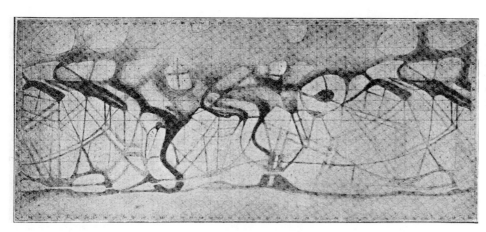

FIG. 10.2. THE MARTIAN "CANALS." These sketches, made by various early observers of Mars, show the linear features eventually interpreted to be artificial channels in which water flowed.

almost literally dedicated the rest of his life to the study of Mars, establishing an observatory near Flagstaff, Arizona, for this purpose. The Lowell Observatory has since grown to become a major research facility, still supported largely by bequests from the Lowell family. Although his contributions to modern knowledge of Mars were ultimately minimal, Lowell did carry on observations of other planets as well, with a bit more scientific detachment. Some valuable discoveries were made, most notably his development of evidence for the existence of a ninth planet (see chapter 14).

FIG. 10.3. PERCIVAL LOWELL.

Observations of Mars from Far and Near

As long ago as 1666, the rotation period of Mars (slightly more than twenty-four hours) was determined from observations of motions of the surface features. In the late 1700's the British astronomer Sir William Herschel deduced the angle of inclination of the Martian axis of rotation, finding it very similar to the tilt of the earth's axis; this fact therefore explained the seasonal variations already seen in the polar ice caps (Fig. 10.4). In 1877 the two moons of Mars, Phobos and Deimos, were discovered, which allowed, through the use of Kepler's third law, accurate determinations of the mass of the planet to be made.

Spectral analysis of the light reflected to earth from Mars started in the first decades of this century. Photographs taken through red and blue filters revealed a haze in the Martian atmosphere that scatters blue light, and the first spectroscopic measurements in the 1930s showed an absence of water vapor and oxygen, of particular interest in view of the prevailing idea that life might exist on Mars. More recently, absorption lines of carbon dioxide were identified in the spectrum of Mars.

Seasonal changes in the coloration of certain regions on Mars, well away from the polar caps, were once interpreted as being caused by foliation of green plants, which were thought to undergo the same sort of annual cycle as earthly plants. By the 1960's good photographs had revealed evidence of large-scale dust storms in the Martian atmosphere, and this inspired the American astronomer Carl Sagan to explain the seasonal color variations in terms of seasonal shifts in the Martian wind patterns, covering and uncovering

FIG. 10.4. VARIATIONS IN THE MARTIAN SURFACE MARKINGS. This series of photos shows the changes in the polar ice caps and surface markings that occur with the seasons on Mars.

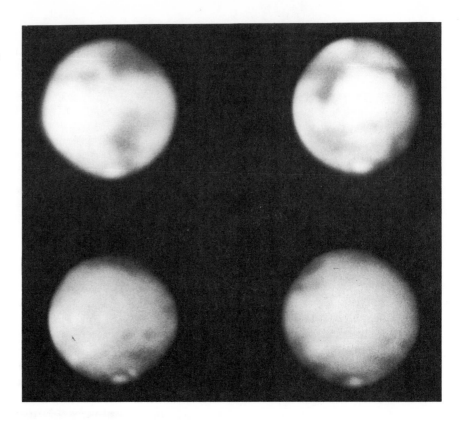

some areas of the planetary surface. This explanation was later verified by the space probes sent to observe Mars at close range.

Infrared observations of Mars yielded data on its surface temperature, found to vary between 130 K (about −225°F) and 290 K (63°F). At high noon in the summer near the equator of Mars the temperature is comfortable, by earthly standards.

Given our profound interest in Mars and the history of speculation over the possibility that life exists there, it is not surprising that the red planet became an early target for unmanned interplanetary probes. The first successful mission to Mars was *Mariner 4,* launched in 1964. The images sent back to earth by this spacecraft forever dispelled any notions that Mars might be the home of an advanced civilization, for the planet was revealed to be totally barren, its surface covered with craters and deserts. More detailed photographs were later obtained by *Mariners 6* and *7,* which flew by the red planet in 1967.

Mariner 9 was launched toward Mars in 1971. Designed to go into orbit around Mars so that it could map the entire surface, the spacecraft arrived at Mars

just as a major dust storm fully engulfed the planet (Fig. 10.5) and obscured it totally for the next three months. Eventually the storm subsided, and *Mariner 9* succeeded in its mission. Among its most striking discoveries were a number of tremendously large volcanoes and an enormous valley that dwarfs the Grand Canyon in size.

The crowning achievement of the exploration program came some four years later, with the launching of two *Viking* spacecraft in the late summer of 1975. Each consisted of an orbiter, destined to circle Mars as *Mariner 9* had done; and a lander, designed to descend to the surface and land intact (Fig. 10.6). The orbiters and landers each carried a complex assortment of instruments designed for a variety of measurements, including tests for the presence of life forms. Upon arrival at Mars the craft, seeking appropriate landing sites, went into orbit and began to take photographs with very high-resolution cameras. These were the first close-up color photographs of Mars and they showed that even at close range, the surface looks red everywhere.

Viking 1 landed on the seventh anniversay of the

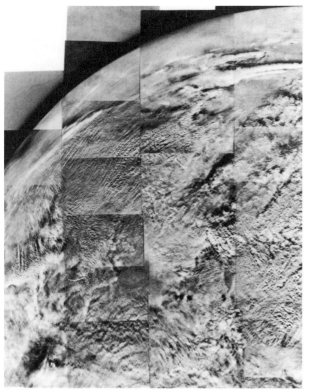

FIG. 10.5. A GLOBAL DUST STORM. This is the view of Mars that confronted the *Mariner 9* spacecraft for the first several weeks after it went into orbit about the planet.

first manned landing on the moon, and *Viking 2* followed forty-five days later. Both landed in plains regions, one in an area called Chryse and the other in a region known as Utopia. The first images sent back to earth from the landers were panoramic views of the landscape, showing again the red color and revealing a dusty surface with rocks lying everywhere.

Once the surroundings had been photographed, other experiments were undertaken, including the tests for life forms. Each lander was equipped with a scoop for digging up soil samples to be analyzed by the instruments on board the craft. By the time the *Viking* landers had completed their missions, our knowledge of Mars was nearly as complete as our data on the moon, which had been explored by humans. Every facet of the Martian environment, including its climate, weather patterns, atmospheric makeup, surface conditions, interior state, and geologic history, was understood in some detail.

The Martian Atmosphere and Seasonal Variations

Mars has a very thin atmosphere compared to the earth, with a surface pressure only 0.006 times the sea-level pressure on earth. The Martian air is about 95 percent

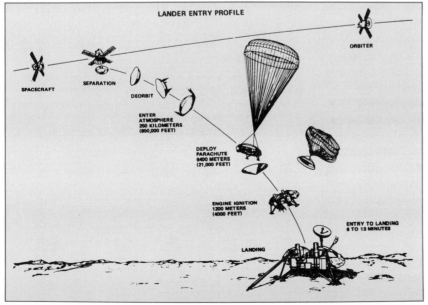

FIG. 10.6. THE *VIKING* LANDERS. This schematic illustration shows the sequence of events as one of the *Viking* spacecraft descended toward its soft landing on the Martian surface.

Mars.

Thin atmosphere compared to earth

air - 95% carbon dioxide

carbon dioxide (CO_2), and is similar to Venus in this regard. Besides CO_2, other species include nitrogen, which accounts for 2.7 percent; argon, which constitutes another 1.6 percent; and other trace gases, which together total less than 1 percent. Water vapor is present in variable quantities, reaching levels of a few percent in the Martian summer. This is occasionally sufficient to form clouds.

The polar caps of Mars have been somewhat controversial, because of our difficulty in determining whether they are made of water ice or frozen CO_2. Temperature measurements have shown, however, that dry ice (that is, frozen CO_2) accounts for most of the cap material, but some water ice is also present. Apparently the seasonal variations in the polar caps are caused by the accumulation of CO_2 ice during the Martian winter, followed by evaporation in the summer. The residual caps that remain throughout the year are probably water ice.

The general question of water on Mars is complex and, for a long time, was the basis of a serious scientific mystery, quite separate from the earlier speculations about canals. Enough water should have been released from surface rocks early in the history of the planet to cover its entire surface to a depth of several hundred meters, so the lack of water on the planet was difficult to explain. Temperatures on Mars probably were never hot enough to dissociate H_2O, as was the case on Venus, but on the other hand, the atmospheric pressure is too low to allow liquid water to exist on the surface. The atmosphere, which gets rather cold at night, cannot support much water in the form of vapor, and the amount in the ice caps is much too small to explain the discrepancy. The answer had to be sought elsewhere. (We will discuss current ideas about the fate of the Martian water supply later in the chapter.)

The existence of clouds in the atmosphere of Mars was recognized long ago from earth-based observations. Although some are caused by blowing dust, many are white clouds formed of ice crystals, primarily H_2O, although some CO_2 clouds also exist. The clouds are formed in lowlands in the early Martian morning, and in regions where winds push the air up mountain slopes, forcing the H_2O vapor to cool and condense. In late summer in each hemisphere, extensive cloud systems form over the polar caps.

Winds on Mars are always present, often with substantial velocities (although the thinness of the atmosphere reduces the winds' impact on surface features). The global flow patterns are created by the daily variations in solar heating at the surface, and are strongly modified by the terrain, such as the large mountains that exist on Mars. The prevailing winds tend to move from west to east, but with a veer either toward or away from the poles, depending on the season. Typical wind velocities are 35 to 50 kilometers per hour.

The fact that Mars has seasons was recognized long ago, when variations in the sizes of the polar ice caps were noticed. The cause of the seasons is a bit more complex than on earth, however, because of the elliptical shape of the orbit of Mars (see Fig. 10.7). Earth's orbit is so nearly circular that the intensity of sunlight hitting its surface is almost constant year-round. (In fact, the earth is closest to the sun in early January, and the slight extra solar input surely does not moderate the winter weather in the northern hemisphere.) The orbit of Mars is noticeably flattened, as we have already learned, with the result that at closest approach to the sun Mars is about 17 percent nearer than when it is farthest away, on the other side of its orbit. Therefore the sun's intensity on the Martian surface is $1/.83^2 = 1.45$ times greater at closest approach, which is enough to have a pronounced effect on the climate. Mars approaches the sun most closely during winter in the northern hemisphere, and is farthest away during the northern summer; hence the shape of the orbit tends to moderate both seasons in the north. By the same token, the southern seasons are enhanced by the orbital shape, so that this hemisphere has much more extreme variations from summer to winter.

FIG. 10.7. THE MARTIAN SEASONS. This exaggerated view shows that in the northern hemisphere summer and winter are moderated by the varying distance of the planet from the sun, whereas in the southern hemisphere both seasons are enhanced. The extreme temperature fluctuations in the southern hemisphere give rise to the winds that cause the seasonal dust storms.

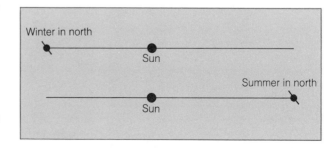

polar caps made of dry ice

Water: atmosph. pressure to low to allow water to exist on surface

Clouds Mostly H_2O

Winds high speeds

The southern polar cap grows larger than the northern one in winter, but diminishes to a smaller remnant in summer.

The most striking seasonal variations on Mars are the widespread dust storms, which have been observed for several decades to occur at the time when Mars is closest to the sun, during the southern summer. The extra solar heating in the south causes very rapid evaporation of the polar cap, and this in turn creates strong winds between the cap and the surrounding territory, winds driven by the extreme temperature differences of the two adjacent regions. Before long, the winds are rapid enough to begin to raise fine particles off the ground, and quickly the air is filled with dust, which reaches high altitudes. The storm grows, usually covering the southern hemisphere; it often expands to smother the entire planet and lasts for weeks. As we noted earlier, one result of these annual storms is seasonal variations in the dust cover on the ground in certain regions, creating the well-known seasonal color variations seen from earth.

Water, Tectonics, and the Martian Surface

The surface of Mars can be classified into two types of areas: plains, which dominate much of the northern hemisphere; and rough, cratered terrain, which covers much of the south. The plains are low-elevation regions covered with lava flows, and altogether they spread over some 40 percent of the planet (Fig. 10.8). The rest of Mars is covered by the rougher cratered area, which is older than the plains, and mostly consists of highlands. One of the highland regions, called Tharsis, encompasses about a quarter of the entire Martian surface, and averages some 6 kilometers' elevation above the mean surface level. This massive plateau, which is very ancient, may be a Martian analogue to earthly continents, although there are some differences. The formation of this gigantic bulge on Mars is not well understood, but most theories suggest it is a result of tectonic activity driven by convection in the mantle that occurred very early in the planet's history.

A major valley system, extending some 4,500 kilometers in length, was named Valles Marineris in honor of *Mariner 9,* whose photographs revealed its gargantuan proportions (Fig. 10.9). This valley would dwarf

FIG. 10.8. A BARREN SURFACE. The first close-up photos of Mars revealed it to be arid and cratered, more closely resembling the moon than the site of teeming civilization it was once imagined to be. This photo was taken by one of the *Viking* orbiters.

the Grand Canyon, which is five to ten times smaller in every dimension. Valles Marineris is 700 kilometers across at its widest, and 7 kilometers deep at its deepest. It probably was formed when the Tharsis highland was uplifted, creating extensive cracks in the crust.

Another major feature of the surface is created by the astounding volcanoes that exist on Mars, primarily in three locations. These are the most conspicuous features on the planet, except for the polar caps. The grand champion is Olympus Mons, a monster with a height of 27 kilometers (nearly 90,000 feet) above the mean surface level, and a base some 700 kilometers across (Fig. 10.10). Olympus Mons is in the northern hemisphere, not far from the Tharsis plateau. The other great volcanic structures are formed by chains of peaks stretching across vast areas of the plains.

The giant volcanoes apparently formed because of a

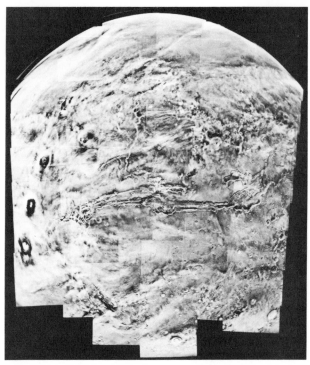

FIG. 10.9.   VALLES MARINERIS. This tremendous valley, as long as the breadth of North America, is put into perspective in this global mosaic of *Viking* photos. The three dark spots at left are the giant volcanoes Ascraeus Mons, Pavonis Mons, and Arsia Mons (top to bottom).

FIG. 10.10.   OLYMPUS MONS. This is the largest mountain known to exist in the entire solar system. Its base is comparable in size to the state of Colorado, and its height is about three times that of Mt Everest.

lack of continental drift on Mars. If they had formed on moving plates in the crust, they should have moved away from the subsurface hot spots that created them, before they got so big. This and other evidence indicate that Mars probably has a thicker and more rigid crust than the earth, such that there is not a planetwide system of moving tectonic plates. The Tharsis highland region was uplifted some 4 billions years ago and shows some effects of tectonic activity, with rift valleys and ridges, but certainly when the great volcanoes formed, between 2.5 and 3.5 billion years ago, there was no widespread crustal motion.

The present-day lack of tectonic activity is also consistent with the absence of any detectable magnetic field. Most theories of planetary magnetism invoke a fluid core, something else that Mars may not have.

Close examination of the surface of Mars reveals some interesting details. There are craters in most regions, created by impacts, as on the moon. Some regions appear completely disorganized, resembling loosely piled rocks and rubble, as though landslides had occurred there. In many locations, often adjacent to this **chaotic terrain** (Fig. 10.11), there are winding valleys that appear to have been formed by flowing water (Figs. 10.12, 10.13). These features have caused a great deal of excitement because, as we learned earlier, the lack of water on Mars has been a major puzzle.

It now appears that a large amount of water has been there all along, but hidden underground. The exact form

FIG. 10.11.   CHAOTIC TERRAIN. This *Viking* orbiter photo shows a region that apparently subsided when permafrost below the surface suddenly melted and flowed away toward the left.

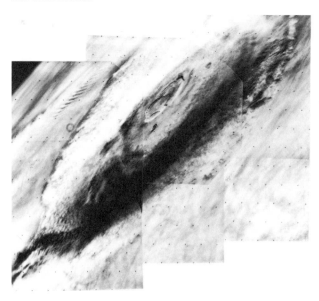

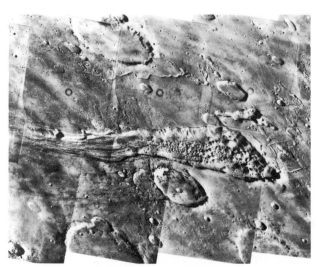

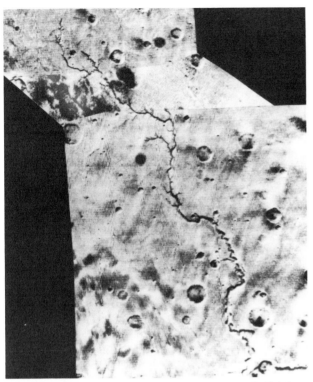

FIG. 10.12. AN ANCIENT RIVER CHANNEL ON MARS. This is one of the most striking examples of evidence for flowing water in the Martian past.

it may take is still uncertain, but the prevailing suggestion is that Mars has a thick layer of **permafrost** beneath the surface, comparable to the permafrost found in the frozen tundra of the far north on the earth. The presence of underground water on Mars, whether it is in the form of permafrost or not, was deduced from the close association of the dry river valleys and the chaotic terrain.

The idea is that a large reservoir of underground water is suddenly released in liquid form, most likely as a result of volcanic heating. It quickly runs off, carving a channel in the surface and leaving behind empty caverns or porous soil in the subsurface volume that it evacuated. The soil there collapses, creating the kind of jumbled, disorganized appearance associated with the chaotic terrain. This theory explains the fact that the dry riverbeds most often are found leading away from the borders of chaotic regions.

Another class of surface features on Mars are those created by winds. We have already learned about the seasonal dust storms and the variations in the surface coloration that they cause, but there are other features created by the winds, including light and dark streaks and the **laminated terrain,** interesting areas where the surface has apparently been deposited in layers and then eroded at the edges (Fig. 10.14). The stair-step structures that result are especially prominent around

FIG. 10.13. A FLOOD PLAIN. This is evidence that water once inundated this region of the Martian surface. The surface patterns indicate that the prominent craters already existed when the water flowed.

FIG. 10.14. LAMINATED TERRAIN. This photo shows some of the interesting layered surface features that exist near the edges of the polar ice caps.

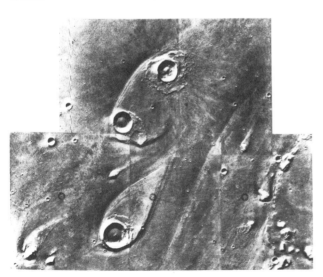

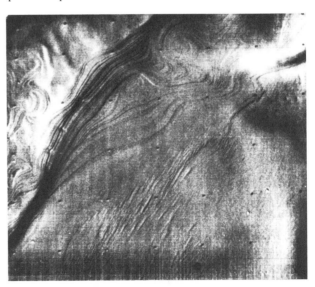

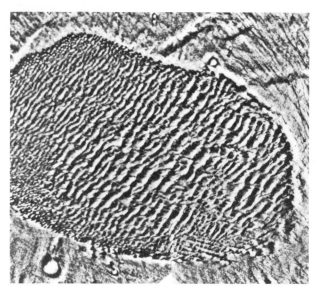

FIG. 10.15. MARTIAN SAND DUNES. These dunes are testimonials to the incessant action of winds on Mars.

FIG. 10.16. A GROUND-LEVEL PANORAMA. This is the first photograph made by the *Viking 1* lander, and it shows a rock-strewn plain extending in all directions.

the edges of the polar caps, and may be formed of thin layers of dust and ice that are precipitated by the annual storms.

There are also sand-covered regions, where the surface has been shaped into dunes just like those seen on earthly deserts and beaches (Fig. 10.15). The sand in the dune regions is distinct from the much finer dust that is suspended in the air during the major storms. A vast field of dunes girdles the planet just outside the north polar cap, creating both a dark belt visible from afar and a mystery for scientists, who can offer no plausible explanation of its origin.

Surface Rocks and the Interior of Mars

The *Viking* landers were able to carry out some close-up analyses of Martian soil and surface rocks. Both landing sites were in the plains regions of the northern hemisphere, and at both places the landscape consisted of dusty soil, with low ridges, a dense scattering of rocks and boulders, and craters here and there (Figs. 10.16 and 10.17). The rocks lying around were probably ejected by the impacts that formed the nearby craters.

The chemical composition of the surface minerals was measured, and striking differences were found between Mars and the earth. The soil of Mars appears to

have formed from the breakdown of igneous rocks such as basalt, but it contains unusually high concentrations of silicon and iron. The iron, in the form of iron oxide (rust), is responsible for the reddish color.

The high abundance of iron is quite significant, for it indicates that Mars is not highly differentiated. Apparently the crust was not sufficiently molten for long enough to allow the heavy elements to sink to the center of the planet, and Mars does not have a dense nic-

FIG. 10.17. DRIFTING DUST ON MARS. The large boulder in this view is about two meters across. The fine dust seen here shows evidence of nearly continuous change, as winds rearrange the drifts.

kle-iron core, in contrast with the earth and Venus. The lack of a large, dense core is also indicated by the relatively low average density of Mars (3.9 grams/cm^3, whereas the earth's density is 5.5 grams /cm^3 and that of Venus is 5.3 grams/cm^3), and the absence of a detectable magnetic field.

These data lead to a picture of an early Mars which, because of its low mass or its distance from the sun, was not fully molten for long, and which formed a very thick crust early in its history. Some tectonic activity has occurred, but there have not been extensive motions of the crust. Thus massive volcanoes formed, building up to tremendous heights because there was no continental drift to carry them away from the hot spots that created them. Geologically speaking, Mars is apparently very much like Venus.

Prospecting for Life: The *Viking* Experiments

The most widely publicized aspect of the dual *Viking* missions was the attempt made by the robot landers to detect evidence of life forms on Mars. This was done in several ways, the first and most straightforward being simply to look with the television cameras for any large planets or animals. None were seen, and more sophisticated tests were tried.

The tests for life were carried out by three different experiments on board the landers, each of which used soil scooped up by a mechanical arm (Fig. 10.18) and deposited in containers for the analysis. The basic idea of all three was similar: to look for signs of metabolic activity in the sample. All living organisms on earth, even microscopic ones, alter their environment in some way just by existing. Usually the effects involve chemical changes as the organism derives sustenance from its surroundings and ejects waste material.

In the **labeled release experiment,** a sample of Martian soil was sealed in a container to which a nutrient solution was added. The nutrient contained a trace of radioactive carbon (the isotope ^{14}C, consisting of 6 protons and 8 neutrons in the nucleus instead of the usual 6 of each). After the sample was incubated at a warm temperature for several days, detectors were activated to look for traces of ^{14}C in the gas in the container, on the theory that if life forms were present, they would have ingested some of the nutrient and excreted waste products containing ^{14}C. As a control on the experiment, separate samples were sterilized by high

FIG. 10.18. SAMPLING THE MARTIAN SOIL. Here the scoop on *Viking 2* digs up a sample of Martian soil for one of the life-detection experiments.

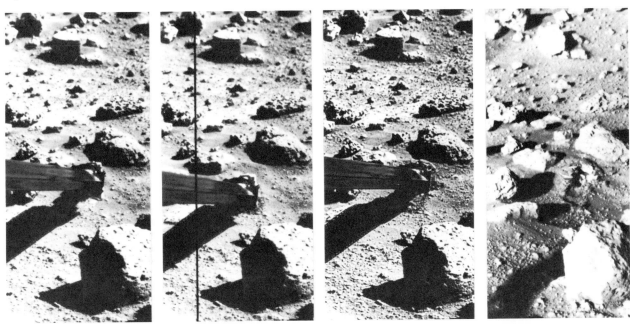

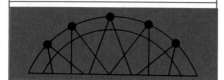

Re-creating Mars in Earth's Image?

Although speculations that an advanced civilization might exist on Mars have been laid to rest, the possibility that one might prosper there in the distant future is very much alive. We speak here not of a native culture that might arise to populate the planet, but of a future time when the human race might settle there in large numbers.

Mars has a relatively hospitable climate, by extraterrestrial standards, but it is hostile enough that human settlers there would have to rely on sealed buildings, pressure suits (or at least helmets), and imported foods and other supplies. This may prove feasible, but an idea is developing that might, in the distant future, make it possible to live on Mars without dependence on artificial means.

The idea, familiar to science-fiction readers, is that it might be possible to deliberately alter the atmosphere and climate of Mars to simulate that of the earth. A new word,

terraforming, has been added to the English language to describe this process. In science-fiction novels this is usually represented as a nearly instantaneous transformation achieved with powerful machinery and unlimited supplies of energy, but in the more serious discussions being undertaken by astronomers and engineers, natural forces would instead be used to remodel Mars in earth's image.

Recall that the earth's atmosphere was modified by the presence of life forms that released oxygen and nitrogen into the air. It has been hypothesized that the introduction of life on Mars might have the same effect there, given enough time. To that end, experiments were performed in which earth plants and microorganisms were sealed in containers that simulated the atmospheric conditions of Mars, and the life forms grew successfully. It is possible to simulate the low pressure, the cold temperatures, and the less intense sunlight of Mars; the only essential ingredient missing is genuine Martian soil. This may not be a serious problem, however, in view of the vigor with which plants have been found to grow in samples of lunar soil. Perhaps if widespread crops of plants could be made to grow on Mars, eventually enough oxygen would be produced to make the Martian air breathable without artificial aid. Calculations

show that this is possible, despite the lower gravity (and consequently low escape velocity) of Mars. The increased atmospheric pressure, along with the addition of water vapor and other gases to the Martian atmosphere, would produce a greenhouse effect that could warm up the surface to comfortable temperatures.

Other necessities for life could be derived from the ground. Water, as we have seen, is probably present in plentiful quantities beneath the surface, in the form of permafrost. Food could be produced in sufficient supply through farming, perhaps initially indoors but eventually out in the open.

The scenario just described is, of course, very speculative, and would take centuries to run its course. For the time being, there are less ambitious steps that could be taken, starting with the first manned missions to Mars. Although it would take a very long time to terraform the entire planet, and doing so would require of the human race great patience and vision, it is perhaps not unrealistic to start by producing localized habitable climates, within sealed buildings, by use of the techniques described here. To do so would greatly enhance the practicality of maintaining at least an outpost on Mars, even if we did not yet colonize it wholesale.

temperatures and then treated with the nutrient and checked for ^{14}C. This was done to be sure that the soil did not undergo nonorganic chemical reactions that mimicked life forms by releasing ^{14}C.

Much to everyone's surprise, large quantities of ^{14}C were released in the sample that was not sterilized, and very little ^{14}C was released in the sample that was sterilized. This is just what had been expected from life forms, except that the amount of ^{14}C that was released was far more than what had been thought possible if microscopic organisms in the soil had been responsible. There were other inconsistencies as well, particularly in the fact that a given sample would only produce the released ^{14}C once, even if nutrient were added a second time. It began to look as though some kind of rapid chemical reaction was responsible after all, one that was hindered by the sterilization process. If there are highly oxidized materials in the Martian soil, they might have reacted by violently bubbling and fizzing when the liquid nutrient was first added, and in the process have released large quantities of gas. If the reaction used up all the oxidized material the first time, it would not occur again when more nutrient was added to the sample.

The labeled release experiment might have been viewed more optimistically if the other two had also given positive signs of life, but they did not. The **pyrolytic release experiment** was designed to look for a reverse of the reaction sought in the labeled release experiment. In this case a soil sample was placed in a container filled with gases (mostly carbon monoxide and carbon dioxide that contained ^{14}C), with the hope that any organisms in the soil would ingest these gases, just as earthly plants do. A lamp was used to illuminate the sample, on the theory that it might stimulate photosynthesis. After several days of exposure to the soil, the tracer gases were flushed out of the container so that there would be no ^{14}C left inside it, unless some had been absorbed or ingested into the soil sample. The sample was then heated, so that any ^{14}C in it would be baked out and detected.

A little ^{14}C was found to have been absorbed into the soil, but in quantities thought too small to be caused by living organisms. Furthermore, the addition of some water to the samples did not increase the amount of ^{14}C that was absorbed, whereas it should have had some effect if organisms in the soil were responsible.

The third attempt to detect life was carried out by the **gas exchange experiment,** which simply looked for signs of respiration. The soil sample was placed in a container of inert gases; these are gases that do not easily combine with others to form molecules and compounds. A nutrient was then added, and detectors in the container looked for any new gases that might appear as a result of biological activity. Some carbon dioxide, nitrogen, and oxygen were released from the sample, but only in quantities expected from nonorganic chemical reactions. Again, no traces of life forms were found.

A further test for evidence of life was carried out by a device called a **mass spectrometer,** which is capable of analyzing a sample to determine the types of molecules it contains. The mass spectrometers aboard the *Viking* landers found no evidence in Martian soil samples of **organic molecules;** that is, those containing certain combinations of carbon atoms that are always found in plant or animal matter.

The lack of evidence for life on Mars is not the same as evidence for lack of life. After all, these experiments only tested samples at two localities on the entire planet, and further more, the tests were predicated on the assumption that Martian life forms, if they exist, would be in some way similar to those on earth. Now that a great deal is known abut the chemical properties of the Martian soil, new experiments probably could be devised that would be relatively free of the confusion created by the nonbiological activity detected in the *Viking* experiments. Perhaps one day soon further attempts to find living organisms on Mars will be made.

The Martian Moons

The two satellites of Mars, discovered telescopically in 1877, are rather insignificant compared with the earth's moon or the major satellites of the outer planets. Both are very near to the parent planet, and therefore have relatively short periods: Phobos (Fig. 10.19), the inner moon, circles Mars in just 7 hours, 39 minutes; and Deimos (Fig. 10.20) has a period of 30 hours, 18 minutes. Because Phobos orbits the planet in a time less than a Martian day, to an observer on the planet's surface, this satellite would cross the sky twice daily, in the west-to-east direction.

Close-up photographs of Phobos and Deimos reveal that each is a small, irregularly shaped chunk of rock, pitted with craters, and, in the case of Phobos, covered by linear grooves. Both satellites are somewhat elongated; Phobos has a length of some 28 km and a width of 20 km, and Deimos measures roughly 16 by 10 km. Both are in synchronous rotation, keeping one end permanently toward Mars.

A striking and unusual characteristic of the surface of Phobos is that most of it is covered with parallel grooves some 100 to 200 meters wide and 10 to 20 *Text continues on page 167.*

ASTRONOMICAL INSIGHT (10.2).

Predicting the Martian Moons*

There are numerous examples in science of discoveries that were made by accident, as the by-product of the search for something else, or as the result of research based on improper foundations. The general term for such discoveries is **serendipity,** which technically means the ability to find valuable things not sought. The discovery of the two Martian moons, following centuries-old and largely mythological predictions of their existence, is commonly thought of as serendipitous, but was in fact based on sound scientific reasoning.

The most well-known prediction that Mars should have moons was a reference in Jonathan Swift's *Gulliver's Travels,* which was published in 1726, some 150 years before the discovery of the moons was made by the American astronomer Asaph Hall. In Swift's satirical work there is a passage that describes the two moons of Mars, providing such details as the orbital periods (given as 10 hours and 21½ hours,

*Based on information from Gingerich, O. 1970, The Satellites of Mars: Prediction and Discovery, *J. Hist. Astro.,* 1(2):109.

not terribly far from the correct values). Contrary to popular impression, historical research shows that Swift was not entirely guessing when he chose the orbital characteristics for these fictional moons. Their relative distances from Mars were very similar to those of two of the prominent moons of Jupiter, and the periods followed from the distances according to Kepler's third law, which was well known in Swift's time.

The prediction that there should be two moons did not originate with Swift. A number of people had made this suggestion, and in fact some took it to be inevitable. The source of this idea may have been Kepler himself, who, in the course of his numerological theorizing, thought that two moons for Mars would be appropriate, since Mercury and Venus have none, the earth has one, Jupiter has four (many more are now known), and Saturn has five (again, many more have been discovered since). It just made sense to Kepler that Mars should have two moons, and this idea eventually became prevalent.

The actual discovery of the two moons was perhaps inspired in part by these unscientific musings, but was in fact based on sound reasoning. The discovery was made at the U.S. Naval Observatory in 1877, when Mars was at opposition. The satellites were actually quite easy to see, and it may seem surprising, particularly in view of all the earlier speculation, that they were not found much sooner. Ironically, the principal reason for this was that no one had taken Swift's prediction seri-

ously enough: no one before 1877 took the pains to look for satellites very close to Mars, despite the fact that their proximity was described clearly in *Gulliver's Travels.* It is interesting that people were willing to let this fictional work inspire them to consider the possibility of Martian moons, but were not willing to take it seriously enough to think of moons with exactly the predicted properties.

Asaph Hall deliberately looked for moons very close to Mars, not because of the prediction in *Gulliver's Travels,* but because he had made calculations showing that the gravitational attraction of Mars for a satellite would be sufficient only if the satellite were close to the planet (this was based on knowledge of the small size of Mars and the assumption that its mass was in similar ratio to its size as was the case with earth). At moderate and larger distances from Mars, Hall calculated that the gravitational force caused by the sun would exceed that caused by Mars, and no satellite could be held in orbit. Hall therefore used a special disk to block out the light of the planet itself, enabling him to see dim objects very close to it. This was the simple step that no one had previously taken, yet they might have, if they had followed the mythology faithfully enough. Hall found the first of the moons on his first attempt. Thus, in the end, the discovery of Phobos and Deimos was not serendipitous, but instead was an example of sound scientific judgment followed to its logical conclusion.

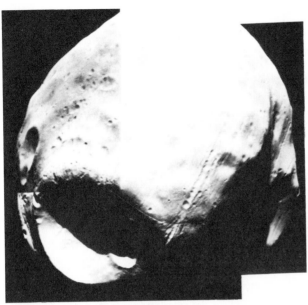

FIG. 10.19. THE MARTIAN MOON PHOBOS. This *Viking* mosaic shows two interesting features: the parallel grooves that mark the surface of Phobos, and the major impact crater whose formation may have caused the stresses that created the grooves.

FIG. 10.20. DEIMOS. The more distant of the two moons from Mars, Deimos has a relatively featureless surface.

meters deep. These features are most prominent in the vicinity of a very large crater on Phobos, and least evident on the opposite side. There are several proposed explanations of their origin, but the one that appears to fit the data best is that they were created by the same impact that carved out the major crater. It has been suggested that this impact fractured Phobos along planes where its material was weakest, and that heat from the impact caused some melting of subsurface rock, which then partially filled the fractures. The reason Deimos does not have similar grooves, (according to this view), is that it has never suffered such a major impact, as shown by its lack of large craters.

Clearly Phobos and Deimos were not formed in the same way as the earth's moon. The moon has a much larger mass relative to its parent planet, and it evidently went through some geological evolution, whereas Phobos and Deimos are tiny and show no signs of having been acted on by the kinds of forces, such as volcanism, that have forged the surface of the moon. The origin of the two Martian moons is not well understood, but many have pointed out the general similarities between them and typical asteroids.

Perspective

The mysterious red planet is still mysterious, but in new ways. Most of the age-old questions related to its changing appearance, possible civilizations thriving on its surface, and its distinctive color have been answered through close-up examination. What we have found is a planet like the earth, except that it did not develop as fully, in the geological sense. Tectonic activity was arrested in the planet's youth, and the low temperatures and low escape velocity combined to leave Mars with only a thin atmosphere of CO_2, its water hidden underground.

We are almost done with our tour of the terrestrial planets, except for Mercury, the closest to the sun but the farthest of the inner four from earth.

Summary

1. Mars has held a fascination for humans, particularly during the time when its surface markings were widely interpreted to be canals built by a Martian civilization.

2. Telescopic observations showed a thin carbon-dioxide atmosphere, polar caps that vary in size with the Martian seasons, and a very low abundance of water vapor.

3. U.S. space probes have flown by Mars, orbited it, and landed and operated on its surface.

4. The atmosphere on Mars has almost the same composition as that of Venus, but has only 0.6 percent of earth's sea-level pressure.

5. Seasonal variations on Mars are strongly influenced by the elliptical shape of its orbit.

6. The surface on Mars consists of plains and rough highlands, with evidence for ancient tectonic activity but no continental drift, similar to Venus.

7. Water on Mars is probably stored beneath the surface, in the form of permafrost.

8. Surface rocks have a high iron content, and Mars has only a moderate average density, both facts indicating that the planet has undergone little differentiation.

9. The *Viking* experiments found no evidence for life forms in the Martian soil.

10. Mars has two tiny satellites, resembling asteroids in size and shape.

Review Questions

1. Before the *Mariner* and *Viking* missions brought back close-up photographs of the Martian surface, what were the best arguments against the idea that the dark markings seen on Mars were canals filled with water?

2. Summarize the similarities and contrasts between the atmospheres of Mars and Venus.

3. To show why the earth's seasons are not strongly affected by the shape of its orbit, calculate how much more intense sunlight is at the earth's surface when the earth is closest to the sun than when it is farthest away. At closest approach, the earth is 0.983 AU from the sun, and at its farthest is 1.017 AU from it.

4. Summarize the similarities and differences between the winds on Mars and on the earth.

5. Compare Mars, Venus, and the earth in terms of tectonic activity.

6. Mars has no detectable magnetic field. What effect would this have on any life forms that might exist there, and on humans who might settle there?

7. Summarize the evidence supporting the suggestion that the Martian water supply exists just beneath the surface, in the form of permafrost.

8. The lack of a magnetic field on Mars was explained in a different manner than was the lack of a magnetic field on Venus. What were these contrasting explanations?

9. In testing for the possible presence of life on Mars, the designers of the *Viking* experiments made certain assumptions about the nature of any life forms that might be there. Summarize these assumptions.

10. The Martian moons are both somewhat elongated. Explain why they each keep one end pointed toward Mars at all times, rather than being in synchronous rotation with some other orientation.

Additional Readings

Arvidson, R. E., and Binder, A. B. 1978. The surface of Mars. *Scientific American* 238(3):76.

Horowitz, N. H. 1977. The search for life on Mars. *Scientific American* 237(5):52.

Leovy, C. B. 1977. The atmosphere of Mars. *Scientific American* 237(1):34.

Murray, B. C. 1973. Mars from Mariner 9. *Scientific American* 228(1):48.

Pollack, J. B. 1975. Mars. *Scientific American* 233(3):106.

Veverka, J. 1977. Phobos and Deimos. *Scientific American* 236(2):30.

Young, R. S. 1976. Viking on Mars: A preliminary survey. *American Scientist* 64:620.

•CHAPTER 11•

Mercury

Mercury is the closest of the planets to the sun, and the most difficult of the four terrestrial planets to observe from earth (Fig. 11.1). Its greatest elongation is only 28 degrees (Fig. 11.2), so it is always so near the sun in our sky that it can be viewed only at dawn or dusk, rising or setting at most only two hours before or after the sun.

As in the case of Venus, Mercury for a long time was thought of as two distinct planets, given the names Mercury and Apollo.

The Earth-Based View

Many people never see Mercury, despite the fact that it is rather bright (a little brighter than Sirius, the brightest star in the sky). The difficulty of observing it has made Mercury the most enigmatic of the terrestrial planets, with little of its nature observable by earth-based telescopes. The best photographs, taken during daylight so that Mercury can be viewed well above the distortion caused by the earth's atmosphere near the horizon, are degraded by the high degree of atmospheric turbulence or poor "seeing" that prevails during the day, and show little except a fuzzy disk with a hint of surface markings possibly resembling the lunar maria.

The orbit of Mercury is unusually eccentric; that is, its elliptical shape is more highly elongated than that

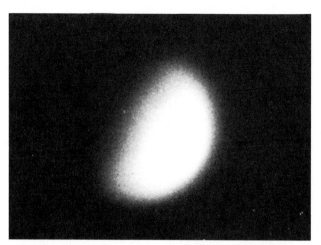

FIG. 11.1. MERCURY, AS SEEN FROM EARTH. The planet is never far from the sun, and is therefore difficult to observe.

of all the other planets except distant Pluto. Mercury is almost 50 percent farther away from the sun at its greatest distance than when it is closest. As we will see, this variation in distance from the sun has had some important consequences. The orbit is also tilted by the unusually large angle of 7 degrees with respect to the ecliptic. At an average distance from the sun of 0.387 AU, Mercury has an orbital period of 88 days, the shortest of all the planets.

Mercury is small, as planets go, with a diameter only slightly larger than the moon. At least three satellites in the solar system, Ganymede of Jupiter, Titan of Saturn, and Triton of Neptune, are bigger. The mass of

FIG. 11.2. MERCURY'S GREATEST ELONGATION. The farthest the planet ever gets from the sun is 28 degrees, which means that it always rises and sets within two hours of the sun.

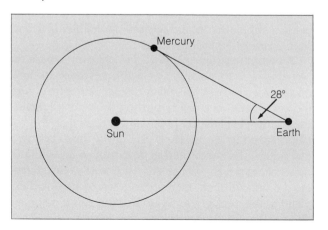

Mercury has been determined from its gravitational influence on nearby objects such as comets, and especially the asteroid Icarus, which passed nearby in 1968. More accurate measurements were later possible, when the U.S. spacecraft *Mariner 10* flew by Mercury. Mercury's mass is about 5.5 percent of the earth's mass, and its density is 5.4 grams /cm^3, very similar to that of the earth, implying that Mercury probably has a large, dense core. No trace of an atmosphere was detected from earth-based observations.

ORBITAL ECCENTRICITY AND ROTATIONAL RESONANCE

As we have seen, observations of Mercury are difficult to make, and no surface features on the planet can be seen distinctly. Careful observations of the fuzzy markings that do show up led to the conclusion that Mercury was in synchronous rotation, always keeping the same side facing the sun. As close to the sun as Mercury is, this was not a surprising idea, for the tidal forces acting on it are immense.

Radio observations of Mercury carried out in the early 1960's allowed its surface temperature to be measured, through the use of Wien's law (see chapter 5). The daylit side was hot, as expected, with a temperature of about 700 K. Because observers thought that the other

TABLE 11.1. MERCURY
Orbital semimajor axis: 0.387 AU (57,900,000 km)
Perihelion distance: 0.467 AU
Aphelion distance: 0.307 AU
Orbital period: 87.97 days (0.241 years)
Orbital inclination: 7°0′15″
Rotation period: 58.65 days
Tilt of axis: ? (less than 28°)
Diameter: 4,864 km (0.382 $D_\oplus$)
Mass: 3.32 × 10^{26} grams (0.0555 $M_\oplus$)
Density: 5.42 grams/cm^3
Surface gravity: 0.38 earth gravity
Escape velocity: 4.2 km/sec
Surface temperature: 700 K (day side); 100 K (dark side)
Albedo: 0.06
Satellites: None

side never saw the light of day, they expected to find very low temperatures there. It was a surprise, therefore, when a temperature of about 100 K was deduced for the dark side. This is certainly cold, but not as cold as it would have been if this side of the planet never faced the sun.

Radar observations soon provided the solution to this mystery. The Doppler shift of radar signals reflected off of Mercury showed that the planet's rotation period is shorter than its 88-day orbital period. Mercury is not in synchronous rotation, but instead rotates once every 59 days, so that all portions of its surface are exposed to sunlight some of the time.

But why 59 days? This was a bit of a puzzle in itself, because most planets spin much more rapidly. Again, the answer was forthcoming, when it was noted that 59, or, more precisely, 58.65 days is precisely two-thirds of the orbital period of 88 (actually 87.97) days. Thus Mercury rotates exactly three times for every two trips around the sun (Fig. 11.3).

This is no mere coincidence; clearly the gravitational tidal forces exerted by the sun are at work. The

FIG. 11.3. SPIN-ORBIT COUPLING. This sketch shows how Mercury spins 1½ times while completing one orbit around the sun. At perihelion, it always has the same axis aligned with the sun.

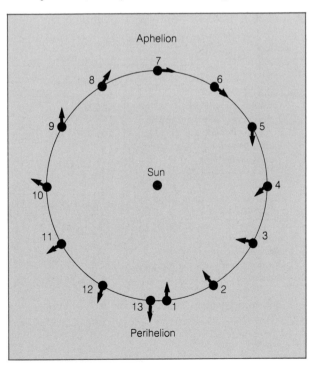

key is what happens at the point where Mercury is closest to the sun, called **perihelion,** for it is here that the tidal forces are the strongest. If Mercury has a heavy side, like our moon, then the tidal forces exerted by the sun will act to ensure that this side is aligned with the sun's direction when Mercury is at perihelion. One way to accomplish this, of course, would be synchronous rotation, so that this side would always point toward the sun, but because Mercury's orbit is so elongated, the urge to keep its heavy side facing the sun is most significant at perihelion.

The planet spins one-and-a-half times during each orbit; thus the heavy side either faces the sun or exactly the opposite direction at each perihelion passage. In either case, the tidal force on Mercury is balanced, so that there is no tendency to alter the rotation further. Apparently the planet once rotated much faster, but whenever it passed through perihelion with the heavy side pointed in some random direction, tidal forces exerted a tug that tended to change its spin. This went on until the rotation slowed to the present rate, so that the tidal forces were always balanced when Mercury was at perihelion. If the orbit had not been so elongated, the tidal forces would have been more uniform throughout and Mercury would no doubt be in synchronous rotation.

The three-to-two relationship between Mercury's orbital and spin periods is an example of **spin-orbit coupling,** a general term applied to any situation where the spin of a body has been modified by gravitational forces so that a special relationship between the orbital and spin periods is maintained. Synchronous rotation is the most common example, and we will find several cases of it as we explore the rest of the solar system.

One remaining question regarding Mercury and its spin is why it should have a heavy side, rather than being a symmetric sphere. We will find a hint of a possible answer in a later section.

The combination of orbital and rotation speeds on Mercury produces a very unusual occurrence for anyone who might visit this planet. At closest approach to the sun, when Mercury is moving most rapidly in its orbit, the orbital speed is actually greater than the rotation speed at its surface. For a short time (lasting a few hours) the sun would appear to turn around and move backward across the sky, from west to east. Imagine the difficult time our ancestors would have had explaining this retrograde motion of the sun if a similar phenomenon had occurred on the earth!

Rendezvous with *Mariner*

Mercury has not drawn the concentrated attention of the U.S. space program that Venus and Mars have, but nevertheless it was observed at close range by one probe, *Mariner 10*. This spacecraft, launched in late 1973, flew by Venus in February of 1974 and then set its course for Mercury. The gravitational pull of Venus was used to alter *Mariner 10's* path so that this could be done. The spacecraft arrived at Merury in late March, 1974.

An extraordinary opportunity presented itself when *Mariner 10* reached Mercury and continued into solar orbit. It proved possible to adjust its orbit so that the period was 176 days, exactly twice that of Mercury. This meant that the spacecraft made one orbit about the sun for every two of Mercury, and the two therefore encountered each other repeatedly (Fig. 11.4). There was sufficient power to operate the instruments aboard *Mariner 10* on each of the next two meetings with Mercury, so there were actually three close encounters during which observations were made. The approaches to the planet were very close on the first and third enounters, when *Mariner 10* passed only a few hundred kilometers above the surface, and much further away on the second flyby, when the spacecraft passed by some 50,000 kilometers away.

There was one drawback to the *Mariner 10* orbit. Although it did allow ample opportunity for close-up observation of Mercury, only one side of the planet could be viewed. Because the spacecraft visited Mercury every other orbit, the same side of the planet was facing the sun each time, since Mercury rotates three times in every two orbits around the sun. If *Mariner 10* could have been placed in an orbit with an 88-day or a 264-day period, it could have viewed opposite sides of Mercury each time the two met. As it was, though, only one side of the planet was charted photographically (Fig. 11.5); close-up examination of the other side will have to wait for some future mission.

Magnetic Field and Internal Conditions

The relatively high density of Mercury, known before the *Mariner 10* encounters, indicated the likelihood of a dense core inside this planet, similar to the earth and Venus. This might have led us to believe

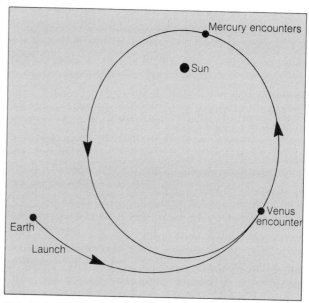

FIG. 11.4. THE TRAJECTORY OF *MARINER 10*. The spacecraft flew by Venus and then encountered Mercury repeatedly, as its orbital period was adjusted to coincide with twice the orbital period of Mercury.

Mercury had a magnetic field, except for the fact that the slow rotation was thought insufficient to produce the internal electrical currents needed to activate a magnetic dynamo.

FIG. 11.5. MERCURY AS SEEN FROM SPACE. This *Mariner 10* view is a mosaic of several images. The planet bears a strong resemblance to the moon, although there are significant differences, particularly in internal structure.

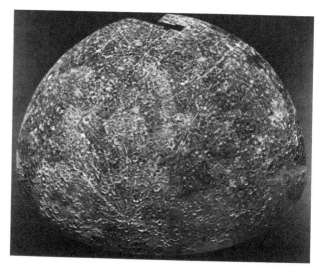

Thus it was a bit of a surprise when *Mariner 10* detected a magnetic field around Mercury. Its strength is not great, about 1 percent as strong as that of the earth, but nevertheless it is there. Mercury has a magnetosphere, just as the earth does, whose shape is modified by the charged particles flowing past from the sun.

Scientists concluded that Mercury's core must be relatively large (Fig. 11.6), in order to produce a magnetic field despite the slow rotation of the planet. The core apparently extends about three-fourths of the way out from the center to the surface, whereas the earth's core is contained within the inner half of its radius.

The implication of such a large core is not only that Mercury is differentiated, but that it contains a relatively high overall abundance of heavy elements to begin with. This says something about its formation, namely, that a greater portion of its volatile, lightweight gases escaped than was the case for the other terrestrial planets. Thus the temperature in the vicinity of Mercury must have been relatively high early in the history of the solar system, higher than at the places where Venus, the earth, and Mars formed. As in the other cases where well-developed cores exist, Mercury

must have been fully molten, staying in that state long enough for differentiation to occur.

Not much known about the mantle of Mercury, which occupies the outer 25 percent or so if its radius. The density there is lower than in the core, but its other properties are unknown. The photographs taken by *Mariner 10* show no obvious evidence of tectonic or volcanic activity, although we must keep in mind that half of the surface has yet to be examined.

Surface Features and Geological History

The surface of Mercury, at first glance, strongly resembles the lunar landscape. It even looks quite similar on second glance, the differences being revealed only under rather detailed scrutiny.

Perhaps the most obvious distinction is found in the prominent and extensive cliffs, referred to as scarps (Fig. 11.7), which exist in many locations and form a network that girdles the planet. These are quite unlike anything seen on the other terrestrial planets or the moon. Their appearance suggests that Mercury shrank a little after its crust hardened, causing it to shrivel and crack.

Another less obvious difference between Mercury and the moon is that on Mercury the craters tend to be more widely separated, with more smooth space be-

FIG. 11.6. MERCURY's INTERNAL STRUCTURE. The presence of a magnetic field, along with the planet's relatively high density, implies that Mercury has a large core.

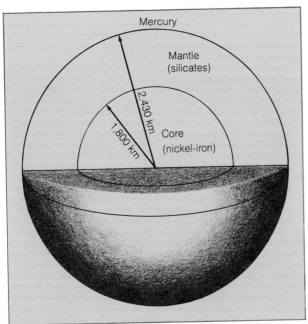

FIG. 11.7. A SCARP SYSTEM ON MERCURY. This lengthy series of cliffs (seen here extending from upper left to lower right) is about 300 km long, and is one of many on the planet's surface.

tween them (Fig. 11.8). Apparently when impacts occur on Mercury and form craters, the ejecta do not travel as far as on the moon, so there are less extensive rays or other ejected debris around the large craters (Fig. 11.9). The reason for this is clear: Mercury has a higher surface gravity than the moon, by about a factor of three, so that the rubble blasted out of the surface by an impact does not travel as far before falling back down. The craters on Mercury also tend to have a flatter appearance than those on the moon, for the same reason. There are extensive smooth areas on Mercury, much like the lunar maria, which were formed by lava flows.

One very large impact crater was seen by *Mariner 10*, just at the dividing line between daylight and darkness (Fig. 11.10). The basin, formed by the tremendous crash of a massive body hitting the surface, resembles a giant bull's-eye, with concentric rings around it, and it extends over a diameter of some 1,400 kilometers. The gigantic crater, called Caloris Planitia, happens to lie on the side of Mercury that is facing the sun at perihelion every other orbit (the name Caloris was given to the crater for this reason, because this position is the hottest place on Mercury every other time it passes close to the sun). We noted earlier that Mercury is apparently lopsided, with an asymmetric distribution of *Text continues on page 177.*

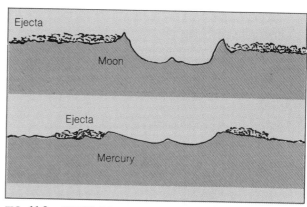

FIG. 11.9.   CRATER FORMATION OF THE MOON AND ON MERCURY. The moon's lower surface gravity allows higher rim walls to remain, and allows ejecta to travel farther from the point of impact than on Mercury.

FIG. 11.10.   CALORIS PLANITIA. This photo shows half of the immense impact basin known as Caloris Planitia. This region is directly facing the sun at perihelion on every other orbit.

FIG. 11.8.   CRATERS ON MERCURY. Because Mercury has a greater surface gravity than the moon, impact craters have lower rims and are shallower, and ejecta do not travel as far. A scarp is also visible in this photo, in the upper left.

Visiting the Inferior Planets

In chapter 4 there was some discussion of the mechanics of placing a satellite in the earth's orbit, but nothing on how to aim a spacecraft that is designed to visit another planet. The principles involved are particularly simple when an inferior planet is to be the target.

The concept of energy was discussed in chapter 4, where we pointed out that an orbiting body has both kinetic energy due to its motion, and potential energy due to the gravitational field of the object that it orbits. For a planet orbiting the sun, the total energy (the sum of kinetic plus potential) is greater, the larger the orbit. Thus for an object to go from the orbit of the earth to that of an inner planet, it must lose energy.

A spacecraft sitting on the launchpad is moving with the earth in its orbit, at a speed of 29 km/sec. To make the rocket fall into the path that intercepts the orbit of an inner planet, we must diminish this speed. Therefore we launch the rocket backward with respect to the earth's orbital motion, thereby lowering its velocity with respect to the sun, and decreasing its orbital energy. If the launch speed is properly chosen,

the spacecraft will fall into an elliptical orbit that just meets the orbit of the target planet. In order to reach Mercury from the earth, the rocket must be launched backward with a speed relative to earth of 7.3 km/sec.

There are, of course, a few additional considerations. For one thing, the rocket has to be launched with enough speed to escape the earth's gravity. The escape velocity for the earth is 11.2 km/sec, so in fact we have to launch our Mercury probe with a speed in excess of this value, which is more than the speed needed to attain our trajectory to Mercury. The launch speed must therefore be calculated to take the earth's gravity into account, so that the rocket escapes the earth, but in the processed is slowed just the right amount to give it the proper course. The gravitational pull of the target planet must also be taken into account, for this speeds the spacecraft up as it approaches.

Another important consideration is timing; the target planet must be in the right spot in its orbit at just the moment when the spacecraft arrives. It is for this reason that the term *launch window* is used. The earth and the target planet must be in specific relative positions at the time of the launch. The interval between launch windows for a given planet is simply its synodic period. For a probe to Mercury, the travel time is about 106 days, so Mercury actually makes a little more than one complete orbit while the probe is on its way.

The gravitational pull of the target planet can be used to good ad-

vantage in modifying the orbit of a spacecraft that flies by. If the probe is aimed properly, its trajectory may be altered in just the right way to send it on to some other target. This technique was used with *Mariner 10,* first to send it from Venus to Mercury, and then to modify its orbit again so that it returned to Mercury repeatedly thereafter.

Sending spacecraft to the outer planets is a bit more difficult, because the spacecraft has to have more energy than what it gets from the earth's motion. The launch is therefore made in the forward direction, so that the rocket has the speed of the earth in its orbit plus its own launch speed with respect to the earth.

Ironically, the seemingly simple task of launching a probe to the sun is one of the most complex. The most straightforward solution would be to launch the spacecraft backward from the earth with a speed of 29 km/sec, entirely canceling out the earth's orbital speed, so that the probe would then fall straight in toward the sun. This is a prohibitively high launch speed, however, so in practice we will probe the sun by first sending the spacecraft out around Jupiter. The spacecraft will fly by the massive planet in such a way (backward with respect to Jupiter's orbital motion) that its own orbital energy is reduced; the craft can then fall into the center of the solar system. A mission to explore the outer layers of the sun in this manner is currently in the planning stages.

mass; the position of this huge impact basin suggests that perhaps the object that crashed into Mercury was responsible for making the planet heavier on one side.

Directly opposite Caloris Planitia, on the far side of Mercury, is a curiously wavy region, so strange and unprecedented in appearance that is called the *weird terrain*. The rippled appearance there was probably created by seismic waves that raced around the planet when the impact occurred that formed Caloris Planitia. When the waves met on the far side of the planet, they distorted the surface there, resulting in the odd terrain seen today. Similar, but less prominent, features are seen on the moon, at positions opposite some of the largest impact craters. On alternate orbits of Mercury about the sun, when Caloris Panitia is on the dark side, the weird terrain is at the point directly beneath the sun, and temporarily holds the dubious distinction of being the hottest place on the planet.

Infrared measurements made by *Mariner 10* indicate that there is a fine layer of dust on Mercury, much like that on the moon. The surface rocks themsleves are probably also similar to earth and moon rocks, except perhaps for a lower abundance of volatile elements.

Despite the combination of relatively low surface gravity and high temperature, which allowed nearly all the gases present on Mercury to escape long ago, a trace of an atmosphere does appear to exist. *Mariner 10* detected very small quantities of a number of gases, which are thought to originate in the solar wind, the steady stream of particles that sweeps past Mercury.

All these data indicate that Mercury's formation and evolution must have been quite similar to that of the moon. Mercury and the moon went through comparable stages of accretion and melting at the beginning, followed by hardening of the crust along with major lava flows (perhaps triggered by large meteorite impacts), and succeeded finally by a long period of cooling and decreased cratering.

We are not sure just when the tremendous impact that created Caloris Panitia took place, nor how long ago tidal forces succeeded in locking the planet's rotation into its 3:2 coupling with the orbital period. We can surmise that there must have been substantial internal heating before then, as tidal forces exerted by the sun created internal friction. Perhaps there is still some excess internal heating caused by tidal forces, which helps to maintain the large molten core needed to create the planet's magnetic field.

Perspective

In Mercury we have found a world that is more like the moon than a planet in some ways, yet in others is quite similar to the earth. It is Mercury's surface features that are similar to the moon's (and in this sense the resemblance is superficial), whereas it is Mercury's interior, with its large dense core and magnetic field, that is apparently more like the interior of the earth. In discussing Mercury, we have seen a rather unique example of spin-orbit coupling, which emphasizes the importance of tidal forces.

Having completed our examination of the terrestrial planets, we turn our gaze outward, bypassing the asteroid belt for now, to focus on the outer planets. We will begin our exploration of this part of the solar system with Jupiter, the king, or at least the prince (after the sun) of the solar system.

Summary

1. Mercury is so close to the sun that it is difficult to observe from earth, and only very general properties have been derived from telescopic observations.
2. Radar measurements showed that Mercury rotates exactly one-and-a-half times for each orbit around the sun, in a form of spin-orbit coupling that was created by a combination of the sun's tidal force on Mercury and the highly elliptical orbit of the planet.
3. The *Mariner 10* space probe, which made three close flybys of Mercury, provided the most detailed information on the planetary properties.
4. Mercury has a magenetic field and a high density,

indicating that it has a large, dense core, probably partially molten.

5. The surface of Mercury is nearly identical to that of the moon, except for a planetwide system of scarps, and minor differences in the heights of impact craters and the distances traveled by ejecta.

6. The great impact basin Caloris Planitia lies at the subsolar point at perihelion on alternate orbits of the sun, and may be the site of the mass imbalance that caused Mercury's spin to fall into resonance with its orbital period.

Review Questions

1. How much more intense is sunlight on the surface of Mercury at perihelion than at aphelion? Assume that the planet is 50 percent farther away at aphelion.

2. Given what you know of the surface albedos and the diameters of the moon and Mercury, would the moon be visible from earth if it orbited Mercury?

3. What do you think the rotation period of Mercury would be if the planet had a perfectly circular orbit?

4. Compare Mercury and the moon in terms of surface terrain and probable internal structure.

5. Mercury has about the same average density as the earth, yet its core occupies a greater fraction of its volume than does that of the earth. What does this tell you about the density of Mercury's core compared with the density of the earth's core?

6. Why does Mercury not have a significant atmosphere?

Additional Readings

Beatty, J. K., O'Leary, B., and Chaikin, A., eds. 1981. *The new solar system.* Cambridge, England: Cambridge University Press.

Murray, B. C. 1975. Mercury. *Scientific American* 233(3):58.

Weaver, K. F. 1975. *Mariner* unveils Venus and Mercury. *National Geographic* 147:848.

•CHAPTER 12•

Jupiter: Giant among Giants

The largest planet in our solar system, Jupiter is completely unlike any of the inner four. In discussing the terrestrial planets we could draw comparisons with the earth, but our modest home is not in the same league with Jupiter, and we will find that almost no aspect of this gaseous giant can be compared with the earth. Entirely new standards must be adopted here, and Jupiter, the closest of the giants to earth and the best studied, will serve as the model for our later discussions of the others.

Jupiter is very complex; it has many fascinating satellites, a ring surrounding it, extensive and intense radiation belts, and a fantastic banded appearance. Like Venus and Mars, Jupiter has played a role in the historical development of astronomy, particularly with Galileo's discovery of the planet's four large moons, whose existence he cited as an argument against the earth being the center of all heavenly motions.

Jupiter truly is a giant planet, containing more mass than all the other planets together. It orbits at an average distance of 5.2 AU from the sun, so that it is never

TABLE 12.1. JUPITER
Orbital semimajor axis: 5.203 AU (778,300,000 km)
Perhelion distance: 4.951 AU
Aphelion distance: 5.455 AU
Orbital period: 11.86 years (4,333 days)
Orbital inclination: 1°18′17″
Rotation period: 9^h 50^m
Tilt of axis: 3°5′
Mean diameter: 137,400 km (10.79 D$_\oplus$)
Polar diameter: 133,100 km
Equatorial diameter: 141,700 km
Mass: 1.901 × 10^{30} grams (318.1 M$_\oplus$)
Density: 1.33 grams cm^3
Surface gravity: 2.64 earth gravities
Escape velocity: 60 km sec
Surface temperature: 130 K (cloud tops)
Albedo: 0.51
Satellites: 14

FIG. 12.1.   THE VIEW OF JUPITER FROM EARTH. This is a good photograph from an earth-based telescope.

closer to the earth than 4.2 AU. Its mass (318 times that of the earth) is spread over such a large volume that the average density is only 1.3 grams/cm^3, barely greater than that of water.

Observation and Exploration

Even a small telescope reveals colorful structure on the Jovian surface (Fig. 12.1); reddish-brown **belts** and light-colored **zones** encircle the planet. The Great Red Spot, a giant reddish oval in the southern hemisphere that dwarfs the earth in size, has apparently been a constant feature for at least three hundred years. The thick atmosphere makes it impossible for us to see a solid surface underneath; as we will learn, modern theories hold that there is no real surface anyway. Spectroscopy of Jupiter's atmosphere reveals compounds of the general type that are thought to have been present in the primitive atmospheres of the terrestrial planets: hydrogen-bearing molecules such as methane (CH$_4$), ammonia (NH$_3$), and hydrogen itself (H$_2$), as well as helium. This composition is rather significant, for it means that the evolution of the Jovian atmosphere was ar-

rested at a very early stage. Jupiter is certainly a cold planet by our earthly standards, with a temperature at the cloud tops of about 130 K.

Jupiter's rotation is very rapid, despite its large size. The rotation period is difficult to pinpoint, because the planet spins differentially, meaning that it goes around more quickly at the equator than near the poles. This can only happen in a nonrigid object. The Jovian day is just under ten hours long, which means that the cloud tops near the equator must travel at a speed close to 45,000 kilometers per hour. This rapid spin has flattened the planet a bit, giving it an oblate shape, so that the diameter through the equator is slightly larger than the diameter through the poles (Fig. 12.2).

Quite accidentally, Jupiter was discovered to be a source of radio emission, and a variety of activity in the radio portion of the spectrum has been revealed, some of it understood, and some of it not. A general radio glow from the planet, at a wavelength of a few centimeters, is simply the thermal radiation expected according to Wien's Law (although the peak emission occurs at shorter wavelengths, in the infrared).

In addition to the thermal radiation, Jupiter emits bursts of radio static much like the emission associated with lightning in the earth's atmosphere, and it is now known that there is lightning on Jupiter. A form of ra-

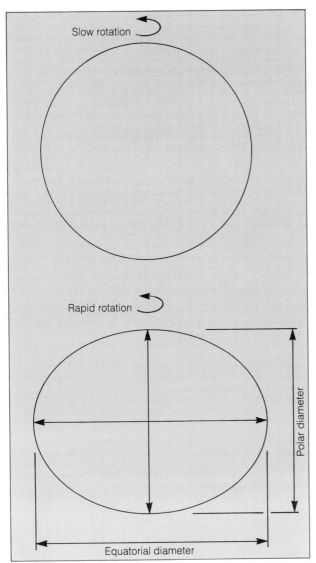

Slow rotation

Rapid rotation

Polar diameter

Equatorial diameter

FIG. 12.2. THE EFFECT OF RAPID ROTATION. An elastic or fluid body is distorted out of a spherical shape by rotation. Jupiter is "flattened" by its rotation so much that its equatorial diameter is 6 percent larger than its polar diameter.

dio emission was also detected that is caused by charged particles moving through a magnetic field. This **synchrotron radiation** showed that Jupiter has both a strong magnetic field and intense radiation belts surrounding the planet, analogous to the Van Allen belts of earth.

The intensity and spectrum of the infrared radiation from Jupiter imply that the giant planet is producing some excess energy in its interior. The total amount of energy being emitted by the planet is about two-and-a-half times greater than the amount it receives from the sun. (The probable explanation for this is discussed later.)

Observations from earth also revealed that Io, the innermost of the four Galilean satellites, has strange properties. This satellite modulates the Jovian radio emission as it passes in front of the planet; also, Io is accompanied by a cloud of atomic and ionized gas that streams along behind it in its orbital path around Jupiter.

Like the terrestrial planets, Jupiter has been a target of the U.S. space program. Sending probes to the outer planets requires some modifications to the technology that has been so successful in reaching the inner planets. The spacecraft must endure a passage through the asteroid belt, have their own internal power source (small nuclear reactors have been used) because they travel too far from the sun to use solar panels, and be able to communicate with the earth over great distances.

These challenges were successfully met by the *Pioneer 10* and *11* missions, and more recently by the *Voyager 1* and *2* spacecraft. *Pioneers 10* and *11* reached Jupiter in late 1973 and late 1974, and carried out an assortment of studies of the planet's atmosphere, magnetic field, radiation belts, and satellites (Fig. 12.3). The *Voyager* spacecraft flew past Jupiter in March and July of 1979, and carried out similar types of research and sent back hundreds of sensational images of Jupiter and its moons, obtained with their fine high-resolution cameras. Perhaps more than any other facet of the planetary exploration program, these pictures of Jupiter (and later of Saturn) have caught the public imagination, and today copies are seen everywhere (Fig. 12.4 and Color Plates 9 and 10).

Pioneer 11, as well as both *Voyager* spacecraft, went on to visit Saturn, using the gravitational pull of Jupiter to correct their courses and provide a boost along the way. Throughout much of the 1980's, all the outer planets will be on the same side of the sun, and this has led to the idea of a "grand tour." It was realized that a space probe could visit all the outer planets, using each one along the way to provide a gravitational boost toward the next. Although the full tour is not being tried, most of it is being traveled by *Voyager 2.* This spacecraft, having flown by Saturn in August of

ASTRONOMICAL INSIGHT (12.1)

Jovian Life Forms?

Any hope we might have had of finding living organisms on the terrestrial planets has been dashed by close-up examination. Venus is far too hot to support life on its surface, and Mars, apparently not so hostile, has revealed not a trace of biological activity. At first thought, it might seem exceedingly unlikely that far-away Jupiter, so much colder at its cloud tops than any of the terrestrial planets, could be an abode for life. On second thought, however, there is some basis for considering the possibility.

The clouds on Jupiter contain the same chemical compounds thought to have fostered the creation of life on earth; these are mainly hydrogen-bearing molecules such as methane and ammonia. Lightning discharges are common in the Jovian clouds, so the energy necessary to spark the chemical reactions could also be readily available. Temperatures within the clouds may be hospitable because, as we have seen, the atmosphere gets increasingly warmer with depth.

Of course, any life forms that might exist on Jupiter would be radically different from anything we are familiar with on earth. Their chemical makeup and respiration would be different, and they probably would have physical shapes adapted for floating about in the currents of the Jovian atmosphere. Perhaps they would subsist on molecular species abundant in the atmosphere, or on smaller life forms. Maybe the Jovian atmosphere hosts a biological zoo as complex as that on earth. It will be interesting to see whether future probes sent to Jupiter, such as the *Galileo* mission planned for the late 1980's, will reveal any evidence of the existence of life.

FIG. 12.3. A CLOSE-UP VIEW. This image of Jupiter was obtained by the *Pioneer 11* spacecraft from a distance of slightly more than a million kilometers. Although far superior to any earth-based photos, this image was soon to be surpassed by those from the *Voyager* missions.

FIG. 12.4. A *VOYAGER* IMAGE. This photo was obtained by *Voyager 1* when it was some 35 million kilometers from Jupiter. A vast amount of detail was already evident in the atmospheric structure.

1981, will reach Uranus in 1986, and, if all goes well, will then visit Neptune in 1989.

Atmosphere in Motion

Let us now return to Jupiter itself, and consider the structure and especially the complex motions in its atmosphere. The upper portion of the atmosphere is made of hydrogen and hydrogen compounds, as already mentioned. There are also large quantities of helium. The presence of these gases represents a major contrast with the terrestrial planets, which long ago lost to space all of their free hydrogen and helium. The overall chemical composition of Jupiter is essentially identical to that of the sun, except that Jupiter is cold enough for most of the material to be in molecular form, whereas the sun is so hot that few molecules can exist, and most atoms are ionized. It is noteworthy that the principal species in the Jovian atmosphere, CH_4, NH_3, H_2, and helium, are the ones thought to have been present in the earth's atmosphere at the time life forms first appeared.

Heavy elements, such as metals, have not been observed in the Jovian atmosphere, but are no doubt present, having sunk to the core of the planet. Jupiter, being fluid nearly all the way to the center, is highly differentiated.

The bright and dark bands (Fig. 12.5) in the upper atmosphere are probably caused by slight differences in molecular composition, which in turn are created by subtle differences in temperature and pressure. The dark belts are regions of low pressure at the cloud tops,

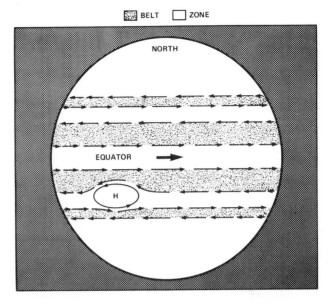

FIG. 12.6. CIRCULATION OF THE JOVIAN ATMOSPHERE. This sketch illustrates the direction of flow in the belts and zones.

where the atmospheric gas is descending, and the light-colored zones are high-pressure regions where the gas is rising (Figs. 12.6, 12.7). The banded appearance of the planet is created by the rapid rotation, which stretches what would otherwise be circular flow patterns into elongated strips that circle the planet. Just as on the earth, air that is descending forms a cyclonic flow, and it rotates in a clockwise direction in the southern hemisphere, and counterclockwise in the north. When cool gas is ascending, the anticyclonic flow, as it is called, occurs in the opposite direction.

FIG. 12.5. BELTS AND ZONES ON JUPITER. This schematic drawing names the major features that give Jupiter its banded appearance.

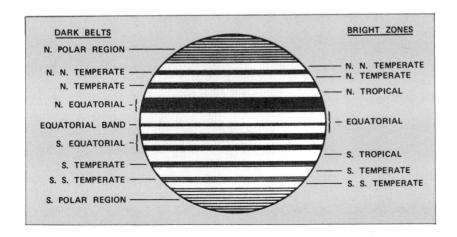

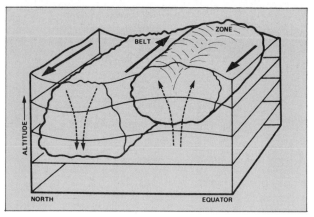

FIG. 12.7. VERTICAL MOTIONS IN THE JOVIAN ATMOSPHERE. This cross-section indicates how the horizontal circulation pattern is related to vertical convective motions in the outer layers of the atmosphere. This is analogous to circulation in the earth's atmosphere, except that the rapid rotation of Jupiter stretches rotary patterns into belts and zones.

Besides the regular pattern of belts and zones in Jupiter's atmosphere, there is a wide variety of spots and other features, the Great Red Spot being the largest and most prominent. There are also a number of smaller spots, both dark and light in color. Just like the belts and zones, these are cyclonic (if light-colored) or anticyclonic (if dark). Thus the Great Red Spot (Fig. 12.8) may be characterized as a storm system, although it is a very long-lived one, having been present for at least three hundred years. The Great Spot is not well understood; it has properties that are unusual even for Jupiter. One surprising fact is that the central portion of the spot consists of gas that is rising, and this gas reaches greater heights than the gases in the belts and zones, yet it is dark-colored, in contrast with the rising gas in the light-colored zones.

Very detailed photographs of Jupiter's cloud tops reveal a fantastic picture of turbulent, chaotic flows where the belts, zones, and spots interact (Fig. 12.9 and Color Plate 9). An incredible complexity of motions all happen at once, as in some gigantic machinery of gears and wheels.

Internal Structure and Excess Radiation

No direct probes of the interior of Jupiter have been made, so what we know has largely been surmised from theoretical calculations. This may seem like an exceed-

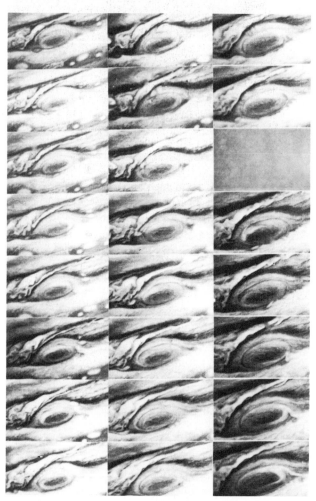

FIG. 12.8. ROTATION OF THE GREAT RED SPOT. The sequence starts at the upper left and goes down each column. Each frame is separated from the ones above and below it by about twenty hours. Note the two white spots in the upper left, and trace how they enter the spot, move around it counterclockwise, and then move out of view at the lower left.

ingly uncertain basis for any detailed theories about interior conditions, and of course there is room for error, but in fact the laws of physics dictate rather explicitly what the internal structure must be like. The general procedure for calculating a theoretical model of Jupiter is very much like that which will be described in chapter 20 for determining the internal properties of a star; a set of equations describing the relationships among various properties such as pressure and temperature, as well as processes such as heat flow, is solved, starting with the known size, mass, composition, and surface characteristics of the planet.

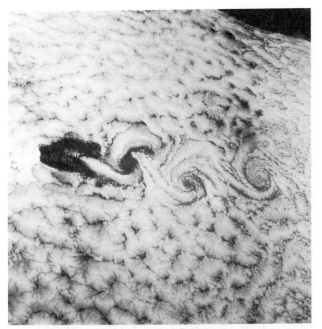

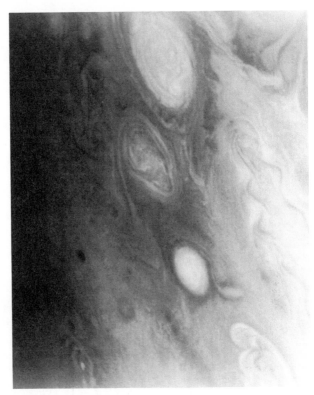

FIG. 12.9. CIRCULATION ON THE EARTH AND ON
JUPITER. Here we see similar weather patterns in two different
atmospheres. At left is a system of vortices photographed by *Skylab*
astronauts, and at right is a similar sequence of vortices in the
Jovian atmosphere. Convection and planetary rotation are
responsible for both.

Although different assumptions made by people
working in this field lead to somewhat different results, the major features turn out to be the same (Fig.
12.10). Below a relatively thin layer of clouds (perhaps
1,000 kilometers thick) is a zone consisting of hydrogen in liquid form, which extends more than one-third
of the way toward the center. The pressure in this region is as great as three million times the sea-level
pressure on earth, and the temperature is as high as
10,000 K or more. An even thicker zone below the liquid hydrogen has such extreme pressure that hydrogen is forced into a state called **liquid metallic hydrogen,** which has a somewhat rigid structure like
certain metals. Liquid metallic hydrogen cannot be
produced on earth, because there is no known way to
create the necessary pressure, but the simplicity of hydrogen lends credence to the theoretical calculations
that predict its behavior under these extreme conditions. The central core of Jupiter, below the liquid metallic hydrogen, is probably solid rocky material of very
high density, containing all the heavy elements of the
entire planet, which sank to the center as the planet
differentiated. In the core the temperature is expected

to be as great as 30,000 K. The mass of the core is
probably in the range of ten to twenty times the mass
of the earth, indicating that as much as 6 percent of the
planet's mass is contained in only 0.5 percent of its
volume. Except possibly for the boundary of this rocky
central zone, Jupiter has no solid surface. Rather, there
is (inward from the top), gas, then liquid, and finally,
metallic liquid.

As noted earlier, the planetary radio emission shows
the presence of an internal heat source, and this was
included in the theoretical calculations just described.
The source of this internal heat is difficult to understand, however, because Jupiter is not sufficiently massive to have nuclear reactions taking place in its core,
as in a star. For a while it was thought that Jupiter might
be shrinking gradually, converting gravitational energy
into heat in the process, but this idea now appears to
be ruled out. The most likely explanation, based on
calculations of the heat capacity and compressibility of
the interior, is that the interior is still hot from the
time of the planet's formation, some 4.5 billion years
ago. Such a prolonged retention of heat is made possible by the characteristics of the outer layers, which

FIG. 12.10. POSSIBLE INTERNAL STRUCTURE OF
JUPITER. This cutaway illustrates the rough model of
the Jovian interior described in the text.

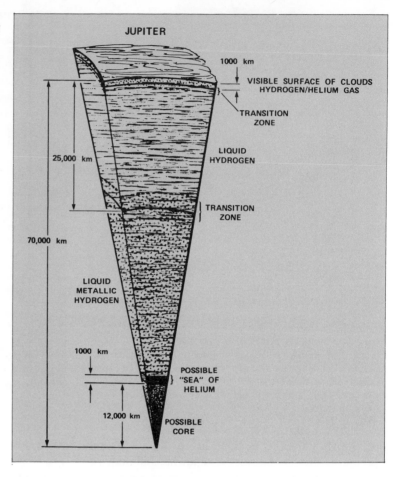

do not efficiently transmit heat from the interior to be
lost to space. The atmospheric circulation, already described, is caused in part by heating of the lower atmosphere from below.

The Magnetic Field and Radiation Belts

It has been known for some time that Jupiter has a
magnetic field, as mentioned earlier. This was originally deduced from observations of radio emission, and
was later verified by the instruments on board the *Pioneer* and *Voyager* probes.

The Jovian magnetic field at the planet's cloud tops
is about ten times the strength of the earth's magnetic
field, and is oriented in the opposite direction, with

the north magnetic pole pointing south, and vice versa.
This is not too surprising, since we know that the earth's
magnetic poles have reversed themselves from time to
time. Perhaps the Jovian field does the same, and has
its poles aligned with those of the earth some of the
time.

The magnetic axis of Jupiter does not exactly parallel the rotation axis, but is instead tilted by 11 degrees.
This situation is also similar to the earth's, but on Jupiter it has some important effects on the distribution
of charged particles in the radiation belts.

The origin of the magnetic field is thought to be
roughly similar to that of the earth's field; that is, circular motions of electrical currents inside the planet
create a magnetic dynamo. Because of certain complexities in the field structure, it is inferred that the
dynamo operates not far below the surface of the planet,
probably in the liquid hydrogen zone.

Within the magnetosphere of Jupiter are very intense radiation belts, with much higher densities of charged particles than in the Van Allen belts of earth (Fig. 12.11). The passage of the *Pioneer* and *Voyager* spacecraft through these zones was rather perilous, since electronic circuits can be damaged by an environment full of speeding electrons and ions. Some data transmissions were garbled or lost, but on the whole the craft survived intact, and were able to continue functioning normally.

One mystery concerning the Jovian radiation belts was the origin of the great quantities of charged particles. A number of possibilities were considered, but most were not feasible, and gradually it was realized that the answer had to do with the inner Jovian satellites. The five closest to the planet all orbit within the magnetosphere, where they sweep up charged particles as they move along in their paths. The discovery (from earth-based observations), that Io is accompanied by a cloud of glowing gas led to the speculation that some of the satellites also emit particles that enrich the radiation belts. As we shall see in the next section, the *Voyager* missions were soon to provide spectacular confirmation that just such a process is occurring.

The rapid rotation of Jupiter forces the radiation belts to be confined to a thin layer lying in the equatorial plane of the planet. This layer is called the **current sheet,** because it is sheetlike and the electrons and ions move about within it, creating electrical currents.

The Satellites of Jupiter

Jupiter has fourteen satellites, ranging from the most recently discovered J–14, the innermost, to faraway Ananke, the outermost. Most of the satellites are small, rocky objects, with diameters of less than 100 kilometers in many cases. The outermost four have retrograde orbits that are substantially elongated in shape, and inclined by several degrees to the orbital plane of the more regular inner satellites. The fact that the outer four vary from the norm may suggest their origin is different from the others, which probably formed from interplanetary debris swirling around the infant Jupiter at the time the solar system coalesced.

Four of the Jovian satellites, the third through the sixth in order from Jupiter, are much larger than the rest, all being comparable to or larger than the earth's moon (Fig. 12.12). These are the four spotted by Galileo with his primitive telescope, and they are easily bright enough to be seen with the unaided eye, except that the brilliance of nearby Jupiter drowns them out.

The *Pioneer* and *Voyager* spacecraft were sent along trajectories aimed at providing close-up views of these Galilean satellites. Of particular interest was Io, the innermost of the four, because of its mysterious influence on radio emissions from Jupiter and because of the glowing cloud of gas that surrounds it.

The sizes of the Galilean satellites, as well as their masses, have been most accurately determined by the *Pioneer* and *Voyager* probes. Although each of the four

FIG. 12.11.   THE JOVIAN MAGNETOSPHERE AND RADIATION BELTS. The shape of the magnetosphere is influenced by the solar wind and by the rapid rotation of Jupiter. The result is a sheet of ionized gas that is closely confined to the equatorial plane, but which wobbles as the planet's off-axis magnetic field rotates.

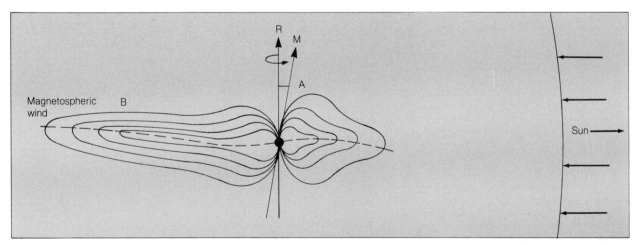

FIG. 12.12. THE MAJOR
JOVIAN SATELLITES. These
profiles indicate the relative
distances of the satellites from
Jupiter and their relative sizes.
The earth's moon and the planet
Mercury are included for
comparison.

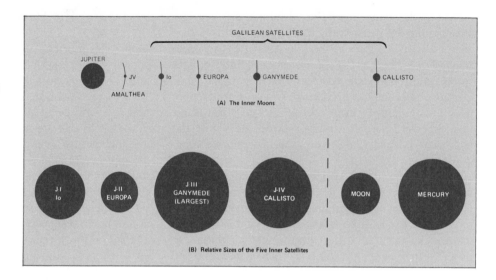

has its own distinctive character, the Galilean satellites fit a pattern, with systematic trends from one to the next. Their densities decrease steadily with distance from Jupiter, Io having the greatest density and Callisto, the outermost of the four, the least. Their albedos also decrease in the same manner. These two facts, along with measurements of the surface temperatures, indicate that the outer satellites contain large quantities of water ice, which can be very dark when seen from afar, especially if the ice contains some gravel and grit. The inner Galilean satellites contain less ice and greater fractions of rock. The surfaces of these bodies also show

marked contrasts from one to another. The fine color photographs taken by the *Voyager* probes showed tremendous complexity, revealing the Galilean satellites to be wondrous worlds, each with its own special effects (Color Plate 10).

Callisto, the farthest from Jupiter, is covered entirely with craters, and has the appearance of a celestial target range that for ages has been bombarded with rocks and missiles from space (Fig. 12.13). The impact points look white, in contrast to the overall dark coloration of the surface. This contrast occurs because the impacts have removed the weathered, old surface layer of the

NO.	NAME	DISTANCE	PERIOD	DIAMETER	MASS	DENSITY
			TABLE 12.2. SATELLITES OF JUPITER			
14		1.80 R_J*	$0^d.297$	40 km		
5	Amalthea	2.54	0.489	240		
1	Io	5.90	1.769	3,640	8.89×10^{22} gm	3.53 gm/cm^3
2	Europa	9.40	3.551	3,130	4.85×10^{22}	3.03
3	Ganymede	14.99	7.155	5,280	1.49×10^{23}	1.93
4	Callisto	26.33	16.689	4,840	1.07×10^{23}	1.70
13	Leda	156	240	2–14		
6	Himalia	161	150.6	170		
10	Lysithia	164	260	6–32		
7	Elara	165	260.1	80		
12	Ananke	291	617	6–28		
11	Carme	314	692	8–40		
8	Pasiphae	327	735	8–46		
9	Sinope	333	758	6–36		

*The symbol R_J stands for the equatorial radius of Jupiter, which is equal to 70,850 km.

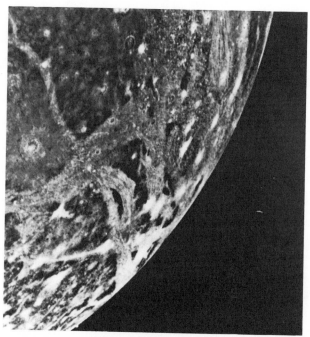

FIG. 12.13. CALLISTO. The outermost of the Galilean satellites, Callisto has the oldest surface. Here numerous white impact craters are seen, as is most of a huge impact basin (at right).

FIG. 12.14. GANYMEDE. The surface of this moon shows evidence of less geological activity than the inner two Galilean satellites, but it does have banded segments indicative of surface motions and stresses.

ice, and because the fractured ice has created multifaceted cracks and fissures, which reflect and scatter sunlight. At one point on the surface of Callisto is a tremendous circular feature resembling a giant bull's-eye, where a very large chunk of material from space left its mark.

Ganymede, next in from Callisto, is the second largest satellite in the solar system (Fig. 12.14). Until recently Titan, the largest of the satellites orbiting Saturn, was thought to be bigger, but *Voyager* measurements showed that the solid body hidden within the dense atmosphere of Titan is actually smaller than Ganymede (Triton, a satellite of Neptune, is the largest of all). The surface of Ganymede is dark, but is marked with linear, light-colored streaks that show evidence of tectonic activity in the satellite's past, with discontinuities and fractures where the crustal plates have shifted (Fig. 12.15). Numerous impact craters are seen, though they are not as dense as on Callisto, indicating that some geological process has eradicated the oldest craters on Ganymede.

The next satellite in sequence is Europa, whose surface is reminiscent of the old sketches of canals on Mars (Fig. 12.16). Europa is covered by random linear features, and resembles nothing so much as a cloudy crystal ball that has been cracked throughout. The dark lines appear to be rifts in the rocky crust where water has oozed to the surface and frozen, leaving a smooth

FIG. 12.15. GANYMEDE. This close-up of the surface shows that a complex pattern of grooves and ridges probably was created by deformations of an icy crust. Some impact craters are seen.

FIG. 12.16. EUROPA. This is the second of the Galilean satellites (in progression outward from Jupiter). The linear features and relative lack of craters indicate that this is a tectonically active moon with a young surface.

surface, but one with strong color contrasts. Like Ganymede, Europa shows evidence of tectonic activity, particularly in this pattern of linear fractures. The surface of Europa has very few craters, which implies that some process has removed the traces of most of the impacts that occurred in the past.

FIG. 12.17. AMALTHEA. Orbiting very close to Jupiter, this tiny, potato-shaped moon always keeps the same end pointed toward Jupiter.

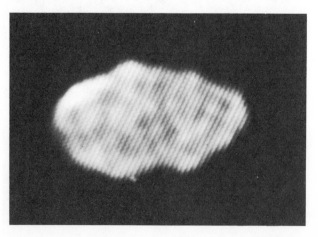

FIG. 12.18. IO. The innermost of the four Galilean satellites, Io has long been a source of mystery. Here we see a number of volcano vents (dark markings) amid a surface of sulfur compounds. No impact craters are visible, indicating that this is a very young surface.

We finally reach Io, the seat of great mystery. Except for the oblong rocky body called Amalthea (Fig. 12.17) and the tiny J–14, Io is the closest satellite to Jupiter, orbiting at a distance of about five times the radius of the planet. Seen close-up, Io is a brilliantly colorful object with a wonderfully variegated red-orange-yellow surface, marked here and there by black smudges and circular features resembling volcanic craters (Fig. 12.18). White frosty patches are seen in places. No impact craters have been found, indicating that Io has a very young surface, constantly being reprocessed.

The explanation for all these fantastic features was discovered by *Voyager 1,* in a photograph taken as the spacecraft left Io behind and looked back toward it with the sunlight passing the edge of the satellite. There, illuminated by the sun's rays, was a plume of gas streaming out of Io's interior, in a volcanic eruption (Fig. 12.19 and Color Plate 10). This was the first detection of active volcanism on any body other than the earth, and soon nearly ten active regions were found on Io's surface (Fig. 12.20).

Apparently Io is undergoing incessant volcanic activity, and the surface is continually being coated with deposits ejected from its interior. The dominant colors are caused by sulfur compounds, which can range from black through red to yellow, depending largely on temperature. Perhaps even more than Venus, Io resem-

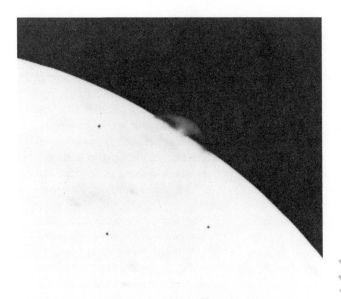

FIG. 12.19. AN ERUPTION ON IO. This image dramatically illustrates Io's present state of dynamic activity, as it shows an eruption from a volcanic vent, with a plume of ejected gas above.

FIG. 12.20. MULTIPLE ERUPTIONS. Here three volcanoes are seen simultaneously erupting on Io; up to ten were found to be active by *Voyager* scientists. Individual vents appear to stay active continuously over periods of at least several months.

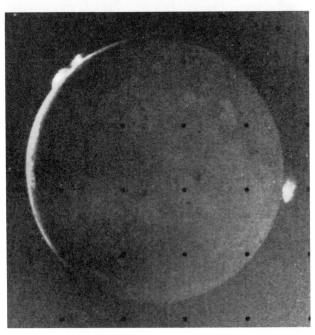

bles the classical picture of hell, with its tortured landscape and sulfurous fumes.

TIDAL FORCES AND
THE VOLCANOES ON IO

We have seen that the four Galilean satellites show some clear-cut trends in properties as we progress from one to the next: densities and surface albedos decrease outward while the ice content increases, and at the same time the ages of the surfaces increase steadily.

When all of this was discovered by the *Voyager* probes, scientists did not have to wait long for an explanation, for it had already been published. The presence of volcanoes on Io had actually been predictd, from a consideration of the effects on the satellite of the tremendous gravitational force of Jupiter. As close to the massive planet as Io is, it is subject to very strong tidal forces. One result is that Io, as well as the other Galilean satellites and Amalthea, are in synchronous rotation, always keeping the same side facing the parent planet, just as the earth's moon does. In addition to this, because of the force exerted by Jupiter and the smaller tugs of the other satellites, Io is being constantly squeezed and stretched in different directions, and this has created internal friction and heating. Europa, which has a period almost exactly twice that of Io and therefore is aligned with it frequently, is particularly effective in helping keep Io in a vise of tidal forces. As a result, the interior of Io is probably fully molten, accounting for all the volcanic activity. The heating and continual eruptions also explain why Io has such a high density compared with the other Galilean satellites, since any water or other volatiles that initially might have been present were long ago vaporized and expelled fom the satellite.

Similar processes have affected the other three large satellites, but to a decreasing degree for each one in sequence going outward from Jupiter. Thus Europa, Ganymede, and Callisto (in order) show fewer signs of volcanic and tectonic activity; have older, less processed surfaces; and contain higher abundances of ice and volatiles.

The eruptions on Io help explain another puzzle that we have already mentioned. Io is attended by a cloud of gas that fills its orbital path in a gigantic doughnut-shape called a torus (Fig. 12.21). The discovery of volcanoes on Io offered an immediate explana-

ASTRONOMICAL INSIGHT (12.2)

Weather Seen from the Outside Looking In

When we discussed weather patterns and atmospheric motions on the earth, in chapter 7, we put everything in the perspective of someone living at the bottom of the atmosphere, looking up. When we discuss a distant planet, we see things from the top, looking down.

There are some problems with terminology that arise because of these different perspectives. The most potentially confusing has to do with high- and low-pressure regions. When we spoke of the earth, we described a high-pressure area as one where cool air descends, creating an anticyclonic flow. When discussing Jupiter, on the other hand, we said that cool air descends in *low*-pressure regions, which seems to contradict what we said about the earth. The difference is that in the case of Jupiter, we are considering the top of the atmosphere, whereas on earth we are talking about conditions at the bottom. Everything becomes clear when we realize that when a column of cool air descends, it creates high pressure at the bottom and low pressure at the top. Conversely, where warm air rises, low pressure is created at the bottom and high pressure is created at the top, and we have an anticyclonic-flow pattern.

The terms *clockwise* and *counterclockwise* are used in the same way for both the earth and Jupiter, referring to the direction of flow as seen from a distance, as though the hypothetical clock lies face-up on the planetary surface.

tion of the source of this cloud. Recall that Jupiter rotates very rapidly, taking only about ten hours to spin once. This is much shorter than the orbital period of Io, so that as Jupiter rotates, its magnetic field sweeps past Io at high speed. Charged particles from the volcanoes are captured by the magnetic field and carried along with it, quickly being spread out along the full length of Io's orbital path. The particles become more highly ionized in the acceleration process and eventually become part of the Jovian radiation belts. Since Io lies in Jupiter's equatorial plane, the torus is thickened by the wobbling of the magnetic equator of Jupiter, in turn caused by the 11-degree misalignment of the magnetic and rotation axes. The surface and the immediate surroundings of Io are filled with charged particles, accounting for the electrical properties that modify radio emission from Jupiter as Io plows through the radiation belts.

The *Voyager* missions were not yet finished making major discoveries about the complex and fascinating Jovian system. A bright streak seen in some of the photographs turned out to be a ring around the planet, very close in, well within the orbit of Io and even inside of Amalthea's distance from Jupiter (Figs. 12.22, 12.23 and Color Plate 10). The ring is very thin, and is not bright enough to be seen from earth (although its presence was suspected from *Pioneer* observations). Its discovery made Jupiter the third of the four giant planets found to have at least one ring. Whatever the origin of planetary rings, they could no longer be viewed as special, singular features, but rather as a natural consequence of the formation of the giant planets. Presumably the ring around Jupiter, like those encircling Saturn, consists of small icy particles.

FIG. 12.21.   THE IO TORUS. This sketch illustrates the geometry of the ring of gas that fills the orbital path of Io.

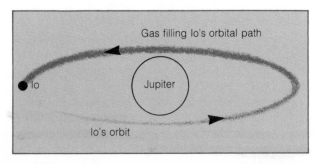

FIG. 12.19. AN ERUPTION ON IO. This image dramatically illustrates Io's present state of dynamic activity, as it shows an eruption from a volcanic vent, with a plume of ejected gas above.

FIG. 12.20. MULTIPLE ERUPTIONS. Here three volcanoes are seen simultaneously erupting on Io; up to ten were found to be active by *Voyager* scientists. Individual vents appear to stay active continuously over periods of at least several months.

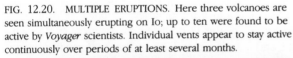

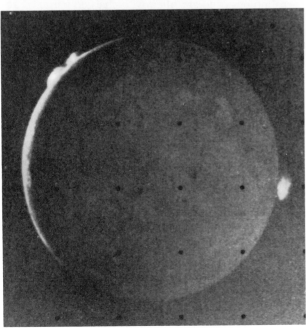

bles the classical picture of hell, with its tortured landscape and sulfurous fumes.

<h2>TIDAL FORCES AND
THE VOLCANOES ON IO</h2>

We have seen that the four Galilean satellites show some clear-cut trends in properties as we progress from one to the next: densities and surface albedos decrease outward while the ice content increases, and at the same time the ages of the surfaces increase steadily.

When all of this was discovered by the *Voyager* probes, scientists did not have to wait long for an explanation, for it had already been published. The presence of volcanoes on Io had actually been predictd, from a consideration of the effects on the satellite of the tremendous gravitational force of Jupiter. As close to the massive planet as Io is, it is subject to very strong tidal forces. One result is that Io, as well as the other Galilean satellites and Amalthea, are in synchronous rotation, always keeping the same side facing the parent planet, just as the earth's moon does. In addition to this, because of the force exerted by Jupiter and the smaller tugs of the other satellites, Io is being constantly squeezed and stretched in different directions, and this has created internal friction and heating. Europa, which has a period almost exactly twice that of Io and therefore is aligned with it frequently, is particularly effective in helping keep Io in a vise of tidal forces. As a result, the interior of Io is probably fully molten, accounting for all the volcanic activity. The heating and continual eruptions also explain why Io has such a high density compared with the other Galilean satellites, since any water or other volatiles that initially might have been present were long ago vaporized and expelled fom the satellite.

Similar processes have affected the other three large satellites, but to a decreasing degree for each one in sequence going outward from Jupiter. Thus Europa, Ganymede, and Callisto (in order) show fewer signs of volcanic and tectonic activity; have older, less processed surfaces; and contain higher abundances of ice and volatiles.

The eruptions on Io help explain another puzzle that we have already mentioned. Io is attended by a cloud of gas that fills its orbital path in a gigantic doughnut-shape called a torus (Fig. 12.21). The discovery of volcanoes on Io offered an immediate explana-

ASTRONOMICAL INSIGHT (12.2)

Weather Seen from the Outside Looking In

When we discussed weather patterns and atmospheric motions on the earth, in chapter 7, we put everything in the perspective of someone living at the bottom of the atmosphere, looking up. When we discuss a distant planet, we see things from the top, looking down.

There are some problems with terminology that arise because of these different perspectives. The most potentially confusing has to do with high- and low-pressure regions. When we spoke of the earth, we described a high-pressure area as one where cool air descends, creating an anticyclonic flow. When discussing Jupiter, on the other hand, we said that cool air descends in *low*-pressure regions, which seems to contradict what we said about the earth. The difference is that in the case of Jupiter, we are considering the top of the atmosphere, whereas on earth we are talking about conditions at the bottom. Everything becomes clear when we realize that when a column of cool air descends, it creates high pressure at the bottom and low pressure at the top. Conversely, where warm air rises, low pressure is created at the bottom and high pressure is created at the top, and we have an anticyclonic-flow pattern.

The terms *clockwise* and *counterclockwise* are used in the same way for both the earth and Jupiter, referring to the direction of flow as seen from a distance, as though the hypothetical clock lies face-up on the planetary surface.

tion of the source of this cloud. Recall that Jupiter rotates very rapidly, taking only about ten hours to spin once. This is much shorter than the orbital period of Io, so that as Jupiter rotates, its magnetic field sweeps past Io at high speed. Charged particles from the volcanoes are captured by the magnetic field and carried along with it, quickly being spread out along the full length of Io's orbital path. The particles become more highly ionized in the acceleration process and eventually become part of the Jovian radiation belts. Since Io lies in Jupiter's equatorial plane, the torus is thickened by the wobbling of the magnetic equator of Jupiter, in turn caused by the 11-degree misalignment of the magnetic and rotation axes. The surface and the immediate surroundings of Io are filled with charged particles, accounting for the electrical properties that modify radio emission from Jupiter as Io plows through the radiation belts.

The *Voyager* missions were not yet finished making major discoveries about the complex and fascinating Jovian system. A bright streak seen in some of the photographs turned out to be a ring around the planet, very close in, well within the orbit of Io and even inside of Amalthea's distance from Jupiter (Figs. 12.22, 12.23 and Color Plate 10). The ring is very thin, and is not bright enough to be seen from earth (although its presence was suspected from *Pioneer* observations). Its discovery made Jupiter the third of the four giant planets found to have at least one ring. Whatever the origin of planetary rings, they could no longer be viewed as special, singular features, but rather as a natural consequence of the formation of the giant planets. Presumably the ring around Jupiter, like those encircling Saturn, consists of small icy particles.

FIG. 12.21. THE IO TORUS. This sketch illustrates the geometry of the ring of gas that fills the orbital path of Io.

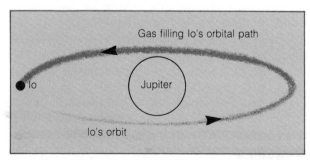

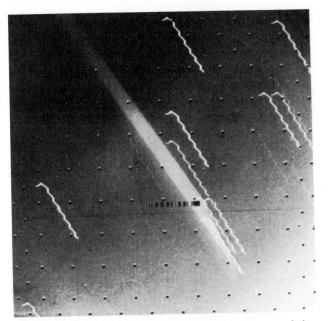

FIG. 12.23.   THE GEOMETRY OF THE JOVIAN RING. Here the ring has been drawn in, showing its size relative to Jupiter. The ring radius is about 1.8 times the radius of the planet, and it is estimated to be no more than 30 km thick.

FIG. 12.22.   THE JOVIAN RING. This *Voyager 1* image revealed the first evidence of a ring girdling Jupiter. The ring is the broad light-colored band extending from upper left to lower right; the bright, wavy lines are star images, distorted by spacecraft motion during the exposure.

Perspective

The planet Jupiter, with its satellites, ring, and radiation belts, is the most complex and varied member of the solar system. It is like a universe all its own, with a distinctive and unique family of celestial bodies. Although it is modified by the sun, which provides some heat energy as well as solar wind particles to deform and feed the Jovian magnetosphere, Jupiter would probably exist in very much the same condition with no help from the sun. It is interesting to speculate on whether the universe contains lots of objects like Jupiter, just short of being stars in their own right.

Our journey through the realm of the giant planets now continues, with our sights set next on Saturn, another fascinating world that has recently been studied by space probes.

Summary

1. Jupiter is gigantic, with 318 times the earth's mass, but so large that its average density is only one-fourth that of the earth.
2. Jupiter is totally unlike the terrestrial planets, having no solid surface, a thick atmosphere of hydrogen compounds, and rapid rotation.
3. The *Pioneer* and *Voyager* spacecraft have flown by Jupiter on four separate occasions, providing detailed data on the planet and its satellites.
4. Jupiter's atmosphere has a pattern of belts and zones, which are planet-girdling wind systems created by a combination of convection and rapid rotation of the planet.
5. Internally, Jupiter has a small, solid core formed

by differentiation, a thick zone of liquid metallic hydrogen outside the core, an even thicker layer of liquid hydrogen above that, and a thin layer of clouds at the top.

6. Jupiter's strong magnetic field and its rapid rotation combine to create and maintain an intense radiation belt that is confined to the planet's equatorial plane.

7. Jupiter has fourteen satellites, of which the four Galilean moons are by far the largest.

8. The Galilean satellites show a range of properties indicating a decreasing degree of tectonic and geological activity from the innermost (Io) to the outermost (Callisto).

9. Io, the innermost Galilean moon, has volcanic activity and is accompanied in its orbit by a cloud of gas that forms a torus around Jupiter. Io's size and orientation are controlled by Jupiter's magnetic field.

10. Jupiter has a thin, dim ring.

Review Questions

1. Summarize the differences in the compositions of Jupiter and Mercury, and comment on the cause of the differences.

2. Calculate the wavelength of maximum emission for the Jovian cloud tops, whose temperature is 130 K. Why would it be difficult to observe this radiation from the earth?

3. Since Jupiter and the sun have the same composition, and the light from Jupiter is reflected sunlight, you might expect a great deal of confusion in attempting to distinguish lines formed in the Jovian atmosphere from those formed in the sun. Explain why this is not a problem.

4. Summarize the differences between the atmospheric circulation pattern on the earth and on Jupiter, and give the principal reasons for these differences.

5. Suppose a spot seen on Jupiter is found to be an anticyclone. If it is in the southern hemisphere of the planet, which way does it rotate, and is the material within the spot rising or falling? How does the Great Red Spot depart from the expected behavior of an anticyclone?

6. Calculate the average density of the core of Jupiter, if its mass is twenty times that of the earth, and its radius is 20,000 kilometers.

7. What would you conclude about the formation of the solar system if Jupiter had the same composition as the earth?

8. Why are the particle belts of Jupiter confined largely to the equatorial plane of the planet, whereas those of earth are distributed all around it?

9. Why does the density of surface craters decrease steadily from Callisto, the outermost Galilean satellite, to Io, the innermost?

10. Contrast the properties and probable geological history of the earth's moon and Europa, which is very nearly the same size.

Additional Readings

Beatty, J. K. 1979. The far-out worlds of Voyager 1. *Sky and Telescope* 57:423 (Part 1); 57:516 (Part 2).

Beatty, J. K., O'Leary, B., and Chaikin, A., eds. 1981. *The new solar system*. Cambridge, England: Cambridge University Press.

Morrison, D., and Samz, J. 1980. *Voyage to Jupiter*. NASA Special Publication SP–439. Washington, D.C.: NASA.

Soderblom, L. A. 1980. The Galilean moons of Jupiter. *Scientific American* 242(1):68.

•CHAPTER 13•

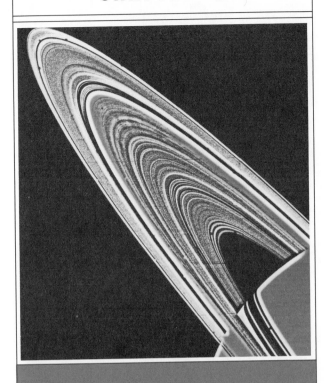

Saturn: The Ringed Planet

Perhaps the most awe-inspiring sight in the solar system, one that can be appreciated with even a modest telescope, is the planet Saturn. With its banded atmosphere and delicate ring system (Figs. 13.1 and 13.2), this planet, more than any other celestial object, has come to represent the realm of outer space in our imagination. Saturn is unusual in almost every respect. Besides the fantastic ring structure, it has a less dense constitution than any other planet, it has an assortment of mysterious moons, and one of its satellites, Titan, is one of only two nonplanetary objects in the solar system known to have an atmosphere.

For a long time, Saturn, at its distance of nearly 10 AU from the sun, was the outermost planet known. Its long orbital period, close to thirty earth years, means that its synodic period is not much longer than a year, and we get annual opportunities to view it in our nighttime skies.

In this chapter we will discover Saturn's nature, both in the distant view and as it is seen from nearby, for space probes have visited it, too. Many comparisons will be made with the other giant planet we have stud-

FIG. 13.1. SATURN AS SEEN FROM EARTH. This photo was taken under good conditions, and shows some structure in the ring system.

TABLE 13.1   SATURN
Orbital semimajor axis: 9.539 AU (1,427,000,000 km)
Perihelion distance: 9.008 AU
Aphelion distance: 10.070 AU
Orbital period: 29.46 years (10,759 days)
Orbital inclination: 2°29′33″
Rotation period: 10^h39^m
Tilt of axis: 26°44′
Mean diameter: 113,450 km (8.91 $D_\oplus$)
Polar diameter: 106,900 km
Equatorial diameter: 120,660 km
Mass: 5.69 × 10^{29} grams (95.2 $M_\oplus$)
Density: 0.68 grams/cm³
Surface gravity: 1.13 earth gravities
Escape velocity: 36 km/sec
Surface temperature: 95 K (at cloud tops)
Albedo: 0.50
Satellites: 17 (one additional satellite is suspected)

ied, and some similarities with Jupiter will be noted, along with numerous differences.

General Properties

As in many previous cases, it was Galileo who made the first important discoveries about Saturn, using his telescope to observe it in 1610. He saw a brilliant yellowish object that appeared greatly elongated. He later referred to Saturn as the planet with "ears."

Some fifty years after Galileo's observations, Huygens, with better data, was able to deduce the ringlike

FIG. 13.2. A sketch of Saturn as seen from earth, with the rings tilted to the greatest possible extent. This was drawn by E. E. Barnard in 1898.

structure of the extensions on the planet, and thus explained the shape noted by Galileo. Extended observations showed the rings to be so thin that when viewed edge-on, they are nearly impossible to see. Saturn's rotation axis is tilted to its orbital plane by 27 degrees, and the rings, which lie in its equatorial plane, are therefore tilted alternately 27 degrees above and 27 degrees below the orbital plane as Saturn goes around the sun.

Huygens also discovered Titan, the largest of Saturn's satellites, and shortly thereafter the French astronomer G. D. Cassini found four more satellites and a prominent gap in the rings (to this day called the Cassini division).

From the satellite orbits, observers deduced the mass of Saturn, using Kepler's third law, and this led to the realization that the planet's density is very low, only about 0.7 grams/cm³. This is about half the density of the sun and a little more than half that of Jupiter. Saturn is the only body in the solar system that is actually less dense than water.

Careful study of photographs revealed that the rings, as seen from the earth, appear to be three in number, with two inner ones (one of which is brighter than the

density very low
only planet in s.s. that's
less denser than water

Text continues on page 198.

ASTRONOMICAL INSIGHT (13.1)

The Elusive Rings of Saturn*

In the text we said little about the discovery of the ring system encircling Saturn, except that Galileo was perplexed by the appearance of the planet, and that, some years after his death, the problem was solved by Christiaan Huygens. The story is actually much more complex and interesting. The mystery of Saturn's appearance, and the extensive efforts devoted to solving it, were the leading concerns in astronomy from the 1630's to the end of the seventeenth century.

Galileo first observed Saturn through a telescope in 1610, and he immediately saw that its appearance was strange. He believed that he saw a spherical planet attended on either side by smaller spherical bodies, and he concluded that these were two satellites of Saturn. He made repeated observations thereafter, but to his surprise, he saw no changes in the positions of the suspected satellites, at least not for

*Based on information from Van Helden, A. 1974, Saturn and his anses, *Journal Hist. Astronomy*, 5(2):105; and Annulo Cingitur: the solution to the problem of Saturn, *Journal Hist. Astronomy*, 5(3):155.

awhile. Frustrated by the lack of change, Galileo stopped observing Saturn regularly, and was therefore shocked when, in 1612, he found it to be a simple sphere, with no trace of the peculiar extensions or satellites seen earlier (the rings were edge-on to the earth at that time). Galileo resumed frequent observations, and saw the shape of Saturn change steadily. In 1616, he saw an elliptical overall shape, with dark spots near each end, and this is what gave him the impression of ears (the dark spots were the small gaps between the rings and the planet, where the blackness of space can be seen in the background).

Galileo exchanged correspondence with other observers, and soon Saturn was being watched by a number of astronomers, all of whom were puzzled by its unorthodox behavior. When the planet regained a simple spherical appearance in 1642, interest was further heightened, and by this time Saturn was the central problem in astronomical research and discussion. After 1642 numerous sketches of the planet were published by a number of observers, and soon a variety of suggestions were made as to the cause of its variable and strange appearance.

It is interesting that the resolution of most of the telescopes used was actually good enough to provide some detail, and the explanation of the phenomenon in terms of a ring circling Saturn quite possibly could have been developed years before it actually was put forth by

Huygens. The problem was that the idea of a ring was completely alien in terms of the usual astronomical phenomena, and nobody made the mental leap to this concept. Meanwhile, the data on the question continued to accumulate.

In early 1655, soon before the ring was to disappear again as it presented itself edge-on to the earth-based observer, Huygens had his first look at Saturn. His telescope, contrary to popular belief, was really little better than those of many of his contemporaries, and he was able to see no more detail than they. Furthermore, he made his first observation at a time when the rings were so nearly edge-on that he had no chance of seeing the true nature of Saturn's peculiar shape. Nevertheless, on the basis of his own limited data and especially his knowledge of Saturn's appearance as seen by others, Huygens deduced that all the observations could be explained by a flat ring that circled the planet in its equatorial plane (the inclination of the equator of Saturn had been deduced from the orbital motions of the satellite, Titan, which had been discovered by Huygens during his observations of 1655).

A number of alternative views were proposed at about the same time, and it took years for Huygens's solution to become generally accepted. It is interesting that it was essentially a theoretical solution, rather than something that he saw simply by looking through a good telescope, as is commonly assumed.

other), and a third that lies outside of Cassini's division. Observations made when the rings are viewed edge-on showed that they can be no more than about 10 kilometers thick. It was also discovered that when they are seen nearly face-on, background stars and the disk of Saturn itself are visible through them, as though the rings were made of sheer fabric. This supported a suggestion made in the late 1800's that the rings actually consist of innumerable small chunks of debris, each orbiting Saturn in accordance with Kepler's laws. Spectroscopic measurements, and later radar observations, used the Doppler effect to determine the speeds of the particles, and confirmed this idea by directly showing that they do travel at orbital speeds consistent with Kepler's laws. The radar data also revealed that at least some of the particles are chunks of material some centimeters or meters in size.

The planet itself rotates very rapidly, with a period of about ten hours, similar to that of Jupiter, and resulting in a similarly oblate shape. The rotation is also differential, with a shorter period at the equator than at the poles.

When spectroscopic measurements of the atmosphere were first carried out, familiar chemical compounds such as methane (CH_4), molecular hydrogen (H_2), and some ammonia (NH_3) were found. The lower temperature of Saturn compared to Jupiter causes some differences, however, in that much of the ammonia on Saturn has apparently crystallized and precipitated out of the atmosphere in the form of snow, and a few more complex molecular species are present on Saturn. No doubt the overall composition of Saturn is the same as that of Jupiter, since both formed out of the same material originally, and neither has lost any significant fraction of its gases to space. With their large masses and low temperatures (about 130 K at the Jovian cloud tops, and 95 K in the upper atmosphere of Saturn), the combination of high escape velocity and low particle speed has prevented any large quantity of particles from escaping.

Like Jupiter, Saturn is a source of radio emission, with a combination of different origins. Synchrotron emission signifies the presence of radiation belts, where the planetary magnetic field has captured an extensive cloud of electrons and ions, and thermal radiation shows that Saturn also emits excess energy from its interior.

Another similarity between Saturn and Jupiter is the attention each has received from the U.S. space program. Three of the four probes sent to Jupiter so far have also visited Saturn, using the gravitational pull of Jupiter to assist and guide them on their way. Only *Pioneer 10* did not do this, having taken a path through the Jovian system that was unsuited for such a maneuver.

Pioneer 11, with its charged-particle monitors, radio antennae, and camera, uncovered a number of new details about Saturn and its attendant rings and satellites. At least one new moon was discovered and an additional ring was found, lying outside of those previously known or suspected. The magnetic field of Saturn was directly measured for the first time, and was found to be somewhat stronger than that of the earth, with the north and south magentic poles upside down compared to earth (as in the case of Jupiter), and the magnetic axis well aligned with the rotation axis, unlike both Jupiter and the earth. The radiation belts are largely confined to the equatorial plane by the rapid rotation of Saturn, again mimicking Jupiter.

Hardly more than a year after *Pioneer 11* encountered Saturn in September, 1979, *Voyager 1* arrived (Fig. 13.3), and a great wealth of additional information quickly followed *Pioneer's* discoveries. The high-resolution color cameras on *Voyager* provided new revelations as sweeping in impact as their previous portraits of Jupiter. *Voyager 2* reached Saturn in August, 1981, following a different trajectory and viewing it from a different perspective (Fig. 13.4).

As in the case of Jupiter, the *Voyager* missions have added breadth, color, and a certain amount of fantasy

FIG. 13.3. *VOYAGER* ENCOUNTERS SATURN. This artist's rendition shows one of the *Voyager* spacecraft approaching the ring system, in preparation for passing through the ring plane and behind the planet.

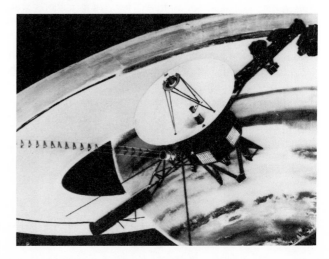

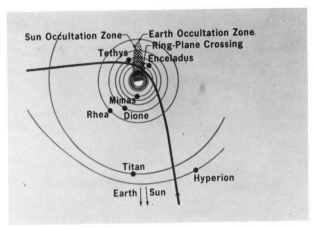

FIG. 13.4. THE PATH OF *VOYAGER 2* through the rings and satellites of Saturn.

to our picture of a newly explored planetary system (Color Plates 11 and 12). Now we can discuss in some detail the weather patterns on Saturn, the composition and structure of its atmosphere and interior, and the nature of its dazzling array of rings and satellites (Fig. 13.5).

Atmosphere and Interior: Jupiter's Little brother?

To the faraway observer, the atmospheres of Jupiter and Saturn look similar, in terms of their overall color and banded appearance. There are also many contrasts, however.

FIG. 13.5. A *VOYAGER* PORTRAIT. This image of Saturn was obtained from 5.3 million kilometers away, and it shows much more detail than can be seen in the best earth-based photos.

As noted earlier, the chemical compositions of the two atmospheres are much alike, except that the colder temperature on Saturn has caused much of the ammonia to precipitate out, and has allowed the formation of somewhat more complex molecules. An example that was detected from earth-based spectroscopic data is ethane (C_2H_6), and other derivatives of methane are probably present.

Whereas Jupiter has a colorful and distinct contrast between belts and zones, and turbulent, swirling winds that show up with great clarity, Saturn has a muted overall appearance (Color Plate 12). The belts and zones are not so distinct, and the dark and light spots that can be seen do not contrast strongly with their surroundings.

Observations show that Saturn's cloud layer is much thicker than that of Jupiter (Fig. 13.6), although it is lacking the high haze layer that characterizes the larger planet. The lower temperature of Saturn allows the cloud layer to persist to much greater heights, with the result that the structure of the lower atmosphere is much more heavily obscured than on Jupiter. Hence Saturn does not have the striking banded appearance of Jupiter, and

FIG. 13.6. THE CONTRASTING CLOUD THICKNESSES ON JUPITER AND SATURN. The deeper cloud zone on Saturn is largely responsible for the relative lack of contrast in its visible surface features.

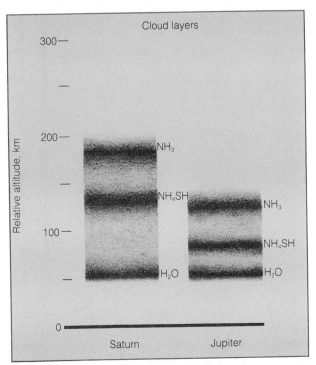

the atmospheric flow patterns are correspondingly more difficult to discern (Fig. 13.7).

Enough can be seen through the haze on Saturn for us to determine that the atmospheric flow patterns, although similar in some ways to those on Jupiter, also display significant differences. As on Jupiter, rising gas forms cyclonic flows, and descending cool gas creates anticyclones, in each case stretched by the rapid planetary rotation into elongated strips circling the planet (Fig. 13.8). The speed of the flow patterns on Saturn is much greater than on Jupiter, however. At the equator, for example, where Jupiter has wind velocities of about 100 meters/sec (360 km/hr), Saturn has winds moving as rapidly as 400 to 500 meters/sec (up to 1,800 km/hr). The belt and zone structure on Saturn extends closer to the poles than on Jupiter, while the light and dark strips are broader in the north, and narrower in the south. These contrasts between Saturn and Jupiter may be caused by the fact that Saturn, with its axial tilt of nearly 27 degrees, has seasons and therefore variations in the sun's intensity, whereas Jupiter, whose axis is nearly perpendicular to the ecliptic (its tilt is only a little more than 1 degree), does not. To directly observe the effects of Saturn's seasons will take a long time, in view of its nearly thirty-year orbital period.

FIG. 13.7. BANDS IN THE ATMOSPHERE OF SATURN. This view of the planet's northern polar region shows the muted appearance of the belts and zones.

There are differences in the internal structure as well. First and foremost, the density inside Saturn is much lower, because its much smaller masss is spread out over nearly the same volume as Jupiter. The solid, rocky core inside Saturn, on the other hand, is probably as dense as the core of Jupiter, and is correspondingly smaller. Outside of this relatively tiny core, the internal pressure, although not as great as inside Jupiter, is sufficient to form liquid metallic hydrogen, which is thought to inhabit a zone extending about halfway out

FIG. 13.8. CIRCULATION IN THE ATMOSPHERE OF SATURN. Even though the contrast is not as vivid on Saturn as it is on Jupiter, the ringed planet does have complex atmospheric motions, driven by a combination of convection and rapid planetary rotation. Bright spots in the upper left of the upper frame are seen here to move in a counterclockwise (anticyclonic) direction.

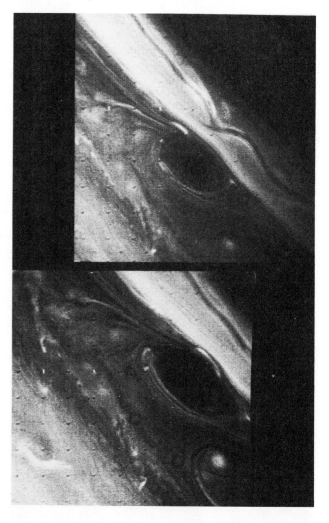

to the surface. Above this is an equally thick layer of liquid hydrogen and helium, topped off by the layer of clouds that is visible from afar.

Like Jupiter, Saturn also radiates substantially more energy than it receives from the sun, by a factor of 2.2 in this case. The explanation offered for Jupiter does not work as well here; Saturn, with its lower mass, could not have retained enough primordial heat from its formation to still be as warm internally as the excess radio emission implies. Some additional source of heat must be present, and its nature is not clear. One suggestion is that differentiation is still going on in the metallic and liquid hydrogen zones, as the heavier helium atoms gradually sink toward the center of the planet, releasing gravitational energy as they do so. The likelihood that this process is really at work inside Saturn is enhanced by the observation that Saturn has relatively less helium in its outer layers than does Jupiter, consistent with the idea that Saturn's helium has sunk, whereas Jupiter's has not. The probable reason that this has not occurred inside Jupiter is that the interior of the larger planet is warmer and more turbulent, preventing the helium from quietly sinking toward the center.

The magnetic field of Saturn is, no doubt, formed by a dynamo in the interior, perhaps at a deeper level than in Jupiter, since the field of Saturn is not so extensive. The zone of trapped ions and electrons around Saturn is still quite large, and most of the satellites and all of the rings are within this zone, sweeping up charged particles as they move along their orbits. Because of variations in the flow of ions from the sun, the magnetosphere of Saturn fluctuates in size, with the result that Titan, the largest satellite, is sometimes within the magnetosphere and sometimes outside of it.

Filling the orbit of Titan, and extending quite a bit farther inward and outward, is a cloud of hydrogen atoms resembling in shape the Io torus that girdles Jupiter. In this case, however, the gas is not ionized, so its structure is not controlled by the planetary magnetic field.

The Satellites of Saturn

Before the *Pioneer* and *Voyager* encounters, Saturn was thought to have at least ten satellites, some of them small and rather difficult to see from earth (an eleventh satellite was suspected, having been reported in 1978, but was not considered confirmed until *Pioneer 11* also detected it). The first of Saturn's moons to be discovered was Titan, noted by Huygens in 1655. Four more were found by Cassini in the latter part of the

TABLE 13.2. SATELLITES OF SATURN							
NO.	NAME	DISTANCE	PERIOD	DIAMETER	MASS	DENSITY	ALBEDO
15		$2.282R_S$*	$0^d.602$	40×20 km			0.4
14		2.310	0.613	$140 \times 100 \times 80$			0.6
13		2.349	0.629	$110 \times 90 \times 70$			0.6
10		2.510	0.694	$140 \times 120 \times 100$			0.5
11		2.511	0.695	$220 \times 200 \times 160$			0.5
1	Mimas	3.075	0.942	392	4.5×10^{22} gm	1.44 gm/cm^3	0.7
2	Enceladus	3.946	1.370	500	8.4×10^{22}	1.16	1.0
16		4.884	1.888	$34 \times 28 \times 26$			0.6
17		4.884	1.888	$34 \times 22 \times 22$			0.8
3	Tethys	4.884	1.888	1,060	7.6×10^{23}	1.21	0.8
4	Dione	6.256	2.737	1,120	1.05×10^{24}	1.43	0.5
12		6.267	2.739	$36 \times 32 \times 30$			0.5
5	Rhea	8.737	4.518	1,530	2.44×10^{24}	1.33	0.6
6	Titan	20.253	15.945	5,150	1.35×10^{26}	1.88	0.2
7	Hyperion	24.55	21.277	$410 \times 260 \times 220$			0.3
8	Iapetus	59.022	79.331	1,460	1.88×10^{24}	1.16	0.5/0.05
9	Phoebe	214.7	550.45	220			0.06

*The symbol R_S stands for the equatorial radius of Saturn, taken here to be equal to 60,330 km.

seventeenth century, and in the next three hundred years, an additional five were seen. The innermost was thought for a time to be the satellite Janus, first reported in 1966, but the suspected eleventh satellite, simply called S–11, was actually even closer in, orbiting very near to the outer portion of the rings visible from earth.

The *Pioneer 11* and *Voyager 1* and *2* encounters have clarified the situation quite a bit. *Pioneer 11* confirmed the presence of S–11 (Fig. 13.9), but failed to detect Janus, which therefore was officially ruled not to exist. Thus only ten moons remained of those detected from the earth. In addition, five new satellites were discovered by the *Voyager* missions and two from the earth-based observations, so the current total is seventeen. The confounding former eleventh satellite turned out to be sharing the same orbit with another satellite, something previously never seen. Both of these moons, now designated S–10 and S–11, are elongated in shape, and orbit with their long axes pointed toward Saturn, in synchronous rotation. Another of the new satellites, probably equally as small as the 200-kilometer diameter of these two, was found orbiting in the same path as Dione, one of the larger satellites. This moon, S–12, is held in position about 60 degrees ahead of Dione by the combined gravitational forces of Saturn and Dione. (See the discussion of the Trojan asteroids in chapter 15 for an explanation of how this happens.) Two small satellites, S–16 and S–17, similarly share an orbit with Tethys, another of the relatively large moons. There is suspicion that an eighteenth satellite, a companion to the larger Mimas, may also exist.

Another pair of small satellites, designated S–13 and S–14, were found to be orbiting at nearly the same distance from Saturn, with a thin ring, itself discovered by *Pioneer 11,* between them. As we shall see in the next section, these satellites probably formed and now maintain this ring by their gravitational forces. The last of the new satellites discovered by *Voyager 1,* designated S–15, is the closest of them all to Saturn, actually orbiting well inside the three rings that lie farthest out, although it is still just outside the broad, bright ring that is the outermost one visible from earth.

The major satellites, those easily observed from earth, turned out to be as varied in their individual qualities as the Galilean satellites of Jupiter. The biggest of them all, Titan (Fig. 13.10), was especially important for *Voyager 1* to examine closely, since it was already known to have an atmosphere, making it nearly unique in that respect among all the satellites of the solar system. Recall that until *Voyager 1* revealed the great depth of Titan's atmosphere, this satellite was thought to be larger than Ganymede, whereas in fact Ganymede has the honor of being the second largest satellite in the solar system, after Neptune's satellite Triton.

FIG. 13.9. SOME OF THE SMALL SATELLITES OF SATURN. Taken at various ranges, these images do not reflect the true relative sizes. The upper middle and lower left moons are the co-orbital satellites (S–10 and S–11), and the two at lower right (S–13 and S–14) are the pair that "shepherd" the F ring (discussed later in this chapter). The upper left satellite (S–12) is the one that shares the orbit of Dione, and at upper right and lower left center are the pair (S–16 and S–17) that share the orbit of Tethys.

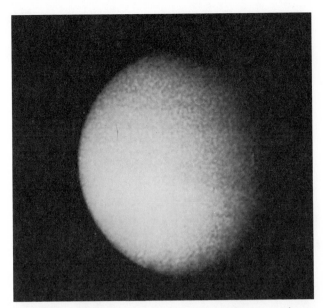

FIG. 13.10.   TITAN. The largest of Saturn's moons, Titan is almost unique among all the satellites in the solar system in that it has an atmosphere.

It was hoped that the vertical structure of Titan's atmosphere could be probed, and the surface below could be examined, but it turned out that the atmosphere is so thick that it is impossible to see to any great depth. In addition, there is an obscuring haze layer at the top of the atmosphere, making direct observations more difficult. Nevertheless, radio and infrared sensors could penetrate deeply, providing data which, along with spectroscopy of the upper layers, were sufficient to allow some rather detailed estimates of the surface conditions to be made.

First, the density of the solid body of Titan itself was ascertained to be just under 2 grams/cm^3, corresponding to an internal composition of about half rock and half ice. This is very similar to Ganymede, the largest Jovian satellite.

We can only speculate as to what the surface of Titan might be like. There is methane in the atmosphere, and radio data showed that the surface temperature is about 90 K, very near the value at which methane can exist in any of three forms: liquid, solid, or vapor. Thus there may be liquid methane on the surface, along with methane ice formations such as glaciers, as well as gaseous methane in the atmosphere. In short, methane on Titan may play very much the same role as water on the earth. If it does, there is probably substantial erosion, so that old impact craters and tectonic formations, if they exist, are continually being smoothed over.

There is also the possibility of liquid nitrogen lakes on Titan, since it was discovered that nitrogen is by far the most abundant gas in the atmosphere, and that the surface temperature is not far above the point where nitrogen condenses into a liquid form. Besides the earth, only Mars, Venus, and now Titan are known to have any significant abundance of nitrogen in their atmospheres. The origin of the nitrogen on Titan may be volcanic, since it is a volatile gas that can be emitted from certain kinds of rocks when they are heated. Therefore, there may also be volcanic features on the surface of this satellite.

The atmospheric pressure is about 50 percent greater at the surface of Titan than at sea level on earth, and Titan's atmosphere extends to five times the height of earth's atmosphere above the surface (Fig. 13.11). It is not well understood why Titan should have been able to retain such a thick atmosphere, when no other satellite (except Triton) has any atmosphere at all. Certainly the low temperature and large mass have helped, but in these regards Titan does not appear to have any great advantage over Ganymede, for example. Perhaps the atmosphere of Titan is being continually replenished by some ongoing process, such as volcanic activity.

The satellites of Saturn are generally classified into four groups: the tiny inner moons that orbit near to and between the rings (Fig. 13.9); Titan, which is in a class by itself; distant Phoebe, a weird misfit of a satellite that also is without peer; and the seven remaining satellites, which are intermediate in size and similar in many respects. We have already discussed the first two of these categories, and we turn our attention now to the intermediate-sized moons.

These seven, named Mimas, Enceladus, Tethys, Dione, Rhea, Iapetus, and Hyperion, are actually quite varied in some respects. They have diameters ranging from about 400 to 1,500 km (Titan's diameter is 5,150 km, by comparison). Masses for most of them were determined from their gravitational influences on the trajectory of the *Pioneer* and *Voyager* spacecraft, and from these measurements the densities were estimated. In all cases the densities are rather low, indicating a high ice content, and a few of these moons may be nearly pure ice, with densities near 1 gram/cm^3. In this regard they are quite unlike any other bodies in the solar system.

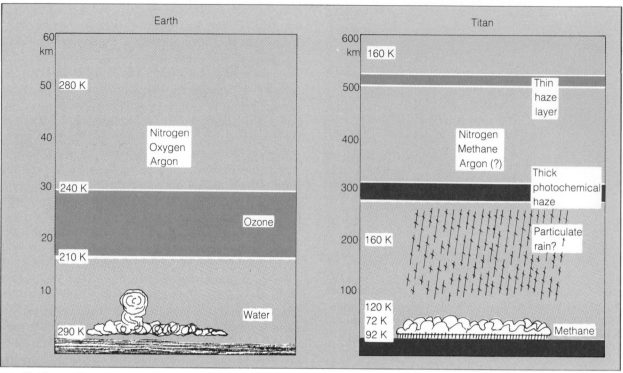

FIG. 13.11.   TITAN'S ATMOSPHERE. This diagram compares conditions in the atmospheres of Titan and the earth. Note that the vertical scales are not the same; Titan's atmosphere is much deeper.

FIG. 13.12.   DIONE. This intermediate moon of Saturn has wispy, light-colored features on its surface that may be icy deposits of material that escaped from the interior and crystallized. This side of Dione, which is in synchronous rotation, always trails as the satellite orbits Saturn; the leading side is more heavily cratered.

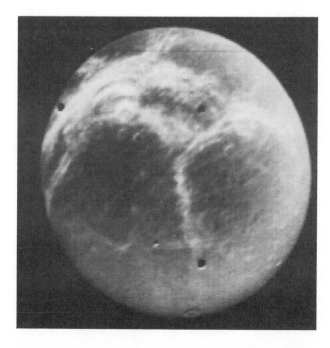

Among the intermediate-sized satellites, a variety of surface features were seen in the *Voyager* photographs, including heavy cratering in many cases, and also some peculiar structures. On Dione (Fig. 13.12) were found sinuous, branching valley systems and white, wispy patterns that look deceptively like the high-altitude cirrus clouds of earth. These may be frozen gases that crystallized immediately upon being emitted from surface rocks or volcanoes, falling to the ground in cloudlike arrangements. Linear trenches were found on a few of the satellites, including Dione, Tethys (Fig. 13.13), and Rhea (Fig. 13.14), possibly the result of fractures caused by massive impacts. The innermost of this group of satellites, Mimas (Fig. 13.15), is marred

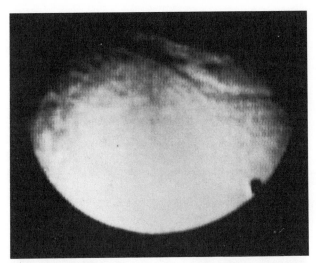

FIG. 13.13. TETHYS. Another of the intermediate moons, Tethys is characterized by large linear trenches, an example of which is seen here.

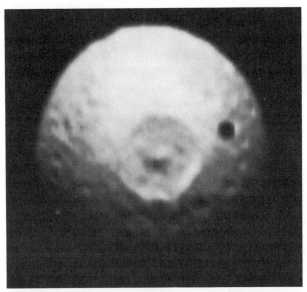

FIG. 13.15. MIMAS. This satellite shows many impact craters, including one very large one that has about one-fourth the diameter of the satellite itself. This moon is probably composed chiefly of ice, as are several of the other satellites of Saturn.

by a tremendous impact crater that is fully one-fourth as large as the diameter of the satellite itself.

Mysteries abound. Iapetus (Fig. 13.16), one of the larger members of this group, has a bright and a dark

FIG. 13.14. RHEA. One of the seven intermediate-sized moons, Rhea shows evidence of cratering, as well as light-colored areas that probably represent ice.

side, the one some ten times more reflective than the other. *Voyager 2* images indicate that the dark side of Iapetus may be covered by some sort of deposit, perhaps resembling soot. Enceladus (Fig. 13.17 and Color Plate 11) is the shiniest object in the solar system, with

FIG. 13.16. IAPETUS. This satellite has very unusual bright and dark areas. It appears that the dark material may be some sort of deposit.

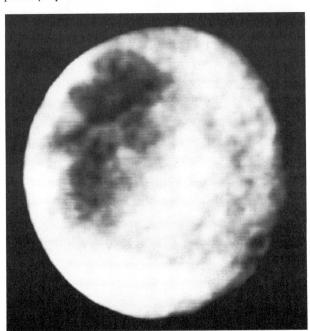

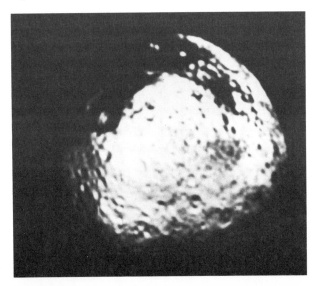

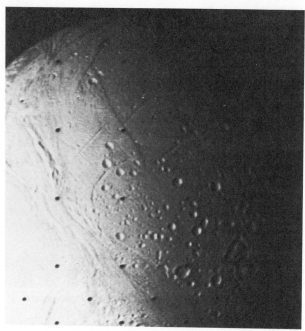

FIG. 13.17. ENCELADUS. This is the shiniest object in the solar system, with an albedo of approximately 1.0, just like a mirror. The surface may have been coated with deposits resulting from internal melting and outgassing.

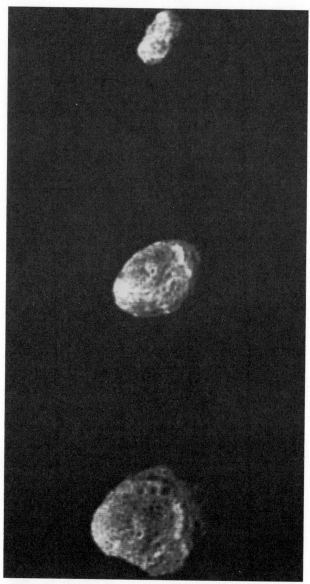

FIG. 13.18. THREE VIEWS OF HYPERION. This satellite has a very unusual shape, and is marked by several impact craters.

an albedo near 1, meaning that it reflects just about all the light that hits it. The orbit of this satellite is coupled to that of Dione, with Enceladus orbiting Saturn once for every two orbits of Dione; thus the two are frequently lined up. This may create sufficient tidal friction inside Enceladus to make it volcanically active, much like Io, so its surface may be covered by volcanic ejecta. Another hypothesis is that Enceladus is being continuously coated by ice particles from the tenuous E ring, within which its orbit lies. Another satellite, Hyperion (Fig. 13.18), was found to be strangely asymmetric in shape despite its moderately large size, indicating that it possibly was never molten.

Perhaps the strangest of all the satellites of Saturn is Phoebe, nearly four times farther out than all the rest, and orbiting in the retrograde direction, with the plane of its orbit tilted nearly perpendicular to the others. This oddball among oddballs, which has a dark surface, may be a stray asteroid that was captured by Saturn's gravitational field and forced into orbit. It has also been suggested that Phoebe is a cometary nucleus that was captured.

The Rings

We turn our attention now to the stunning rings of Saturn, probably the most spectacular sight of all (Fig. 13.19 and Color Plate 11). The *Voyager* data proved these structures to be even more complex and fascinating than had been imagined.

The rings consist of countless individual particles, ranging in diameter from a fraction of a centimeter in some portions of the system to a few meters in others. Each particle has its own orbit, with its period determined by its distance from Saturn according to Kepler's third law. Thus the rings do not rotate like rigid hoops around the planet, in which case all the particles would have the same period (meaning that the outer ones would have to travel fastest). Instead they are fluid structures, moving slowest in the outermost portions.

The three distinct rings visible from the earth had been labeled the A, B, and C rings (in order, from the outer one inward). The possibility of a faint ring inside the C ring was raised some time ago, and this D ring was actually found, as were E, F, and G rings outside the A ring (the letters are assigned in the order of discovery).

Even before the full complexity of the ring system was known, some idea of how the rings formed had been developed. The rings are all very close to Saturn, and this led to the suspicion that tidal forces have had something to do with their formation. There is a certain distance from any massive body, called the **Roche limit,** inside of which the tidal forces pulling apart an object of a given size are greater than the gravitational force holding it together. The exact location of the Roche limit depends on the mass and size of the orbiting object. We have discussed the fact that the moon has a tidal bulge exerted by the earth's gravitational pull; to envision what the Roche limit means, imagine that the bulge becomes so severe that the moon is actually pulled apart and broken into pieces.

The rings of Saturn are all within the Roche limit for a satellite of any significant size. Therefore it may be that there once was a satellite that wandered too close to Saturn and was broken apart, leaving behind the debris that now forms the rings. More likely, the rings are made of particles that have been orbiting Saturn since its formation, but are so close to it that tidal forces have prevented their ever merging to form a satellite.

The major gap in the ring system of Saturn was also explained by gravitational forces. The Cassini division, between the A and B rings, lies at just the right distance from Saturn that any particle orbiting there would be in orbital resonance with Mimas, the innermost of

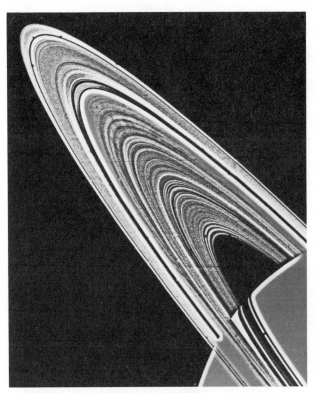

FIG. 13.19. THE RINGS OF SATURN. This *Voyager* image reveals some of the fantastically complex structures of the ring system. A small satellite (S–14) is visible just inside the F ring at upper left.

TABLE 13.3 THE RINGS OF SATURN	
FEATURE	DISTANCE
D ring, inner edge	1.11 R_S*
C ring, inner edge	1.233
B ring, inner edge	1.524
B ring, outer edge	1.946
A ring, inner edge	2.021
A ring, gap center	2.212
A ring, outer edge	2.265
F ring, center	2.326
G ring, center	2.8
E ring, inner edge	3
E ring, outer edge	8

*The symbol R_S stands for the equatorial radius of Saturn, taken here to be equal to 60,330 km.

the larger satellites. By this we mean that the orbital period of such a particle would be precisely half the period of Mimas, so that every two trips around, the particle and Mimas would line up on the same side of Saturn. On each of these occasions, the particle would feel a little extra tug exerted by Mimas. Over a long period of time, the cumulative effect of these tugs would be sufficient to change the shape of the particle's orbit such that the particle would stray out of its former circular path and eventually collide with other particles and be deflected out of that part of the ring system entirely. In this way Mimas has maintained a clear gap at the position of Cassini's division. Other satellites, or other fractions of the orbital period of Mimas (such as the location where an orbiting particle would have exactly one-third the period of Mimas) could create other gaps in the rings, and indeed such gaps have been found. Orbital resonances with satellites of Saturn led to the expectation that the rings have complex structure, even before *Pioneer* and *Voyager* data were available.

Despite this expectation, when the first *Voyager* photographs of the rings were obtained, scientists were amazed by what they saw. None of the rings had a smooth appearance when viewed from nearby; the prominent ones visible to us from earth are each composed of a number of thin rings that came to be called "ringlets." The photopolarimeter experiment on *Voyager 2* revealed much more detail in the rings, showing structures as small as 100 meters in size (Color Plate 12). There are thousands of ringlets in the entire system, some appearing dark and others light. Even Cassini's division is not completely empty, but rather is inhabited by several dark ringlets. There apparently is also a narrow gap that truly is empty, probably created by Mimas in the manner just described.

Whether a ring appears dark or light depends on how the particles in it reflect sunlight, and the angle from which it is viewed. Some of the rings tend to reflect light straight back, whereas others allow it to go through in a forward direction. A ring that reflects sunlight back will appear bright when viewed from the earth, but will look dark when viewed from the far side of Saturn, looking towards the sun. Conversely, a ring that allows light to pass through will appear dark from the earth (as do the narrow ringlets in Cassini's division) and bright if viewed from behind. Whether a given ring will scatter light forward or backward depends on the size, shape, and composition of the particles it is made of.

From considerations of their light-scattering properties, observers have been able to deduce the average sizes of the particles in the various rings of Saturn. The outermost ones tend to be made of very tiny particles, a few ten-thousandths of a centimeter in diameter, whereas the inner rings, in contrast, seem to consist mostly of chunks of material a few meters across. All the particles are probably mostly composed of ice, and their shapes are irregular. The total mass of all the ring material together is very small, only about a millionth of the mass of our moon, so if the ring particles could have formed into a satellite of Saturn, it would have been a rather small one.

The complex pattern of rings and ringlets seen by the *Voyager* cameras required a more sophisticate explanation than the orbital-resonance hypothesis, although that process is undoubtedly at work. Apparently at least three other mechanisms exist for shaping the rings of Saturn.

One of these mechanisms is also caused by gravitational effects of Saturn's moons. A clue to this was presented by the two small satellites, discovered by *Voyager 1,* which orbit just inside and just outside the F ring (Fig. 13.20). The gravitational interaction between

FIG. 13.20. THE SHEPHERD SATELLITES AND THE F RING. The two small satellites (S–13 and S–14) that maintain the F ring through gravitational interaction are both visible in this image.

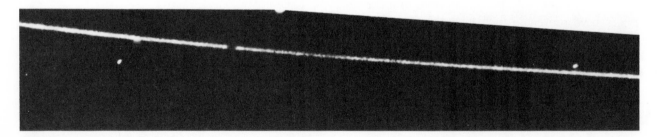

the ring particles and these satellites keeps the particles confined to orbits between them. The moon just outside the ring, according to Kepler's laws, moves a little more slowly than the ring particles. Whenever a ring particle overtakes this satellite at a close distance, the particle loses energy because of the gravitational tug of the satellite, and drops to a slightly lower orbit. On the other hand, the satellite on the inside of the ring slowly overtakes the ring particles, and when it passes close by one, the particle gains energy and moves out to a higher orbit. In this way the two satellites ensure that the ring particles stay in orbit between them, because any particle that strays too close to either one will have its orbit altered in just such a way that it moves back into the ring. These two satellites are called "shepherd" satellites, because they behave like sheep dogs, keeping their flock in line by nipping at their heels.

Voyager 2 data revealed that another mechanism (which was suspected after the *Voyager 1* flyby) was also influencing the structure of the rings. Mathematical calculations show that in any flattened disk system such as Saturn's rings, small gravitational forces can create a spiral wave pattern, something like the grooves on a phonograph record. The waves take the form of alternating dense and rarefied regions, and the entire spiral pattern rotates about the center of the system. Similar spiral density waves have long been thought responsible for the spiral arm structure of galaxies like our own; the *Voyager* data have shown that the rings of Saturn are a small-scale model of a spiral galaxy.

It was predicted from calculations that the spacing between spiral density waves should increase steadily with distance from the center. The photopolarimeter data from *Voyager 2* revealed that the expected increase in spacing between ringlets occurs in several portions of the ring system (Fig. 13.21). This confirms that at least some of the ring structure is created by spiral density waves. The ringlets that are formed by this mechanism are not separate, circular structures, but instead are part of a continuous, tightly-wound spiral pattern that slowly rotates. Individual ring particles, by contrast, follow elliptical orbits, and alternately pass through the ringlets and the spaces between. (It is important here to distinguish between the motion of the wave and the motions of the particles, which do not move with it; a reasonable analogy would be a cork bobbing on water waves, but not moving along with them.) The reason there are more particles in the ring-

FIG. 13.21. VERY FINE STRUCTURE IN THE RINGS. This is a synthetic image of a small section of the ring system, reconstructed from stellar occultation data obtained by the photopolarimeter experiment on *Voyager 2*. The finest details visible here are only about 500 meters in size, whereas the best *Voyager* camera images have a resolution of about 10 kilometers. Data like these have established that spiral density waves play a role in shaping the rings.

lets than between them at any given moment is that they move more slowly as they pass through the density waves, and tend to congregate there, just like cars in a traffic jam that builds up at a point where the flow is slowed down.

The gravitational disturbances that create the density waves in portions of Saturn's rings system are caused by satellites. The strongest disturbances occur at resonance points where the particles have orbital periods that are a simple fraction of the periods of satellites orbiting at greater distances. In some cases, these are also the locations of gaps in the rings (as explained earlier), so we sometimes find that a gap lies just at the inner edge of a series of ringlets that are created and maintained by a spiral density wave.

A fourth type of force that may affect the shape of the rings is not gravitational at all. Some of the rings have nonsymmetric shapes that cannot be formed by any simple gravitational forces exerted by Saturn or its

Text continues on page 211.

Resolving the Rings

The *Voyager* missions to Saturn have produced a wealth of new information on the rings, primarily because of the high resolution of the images that they have sent back to earth. Recall, from our discussions in chapter 6, that resolution is the ability to make out details. In our past discussion, we emphasized that resolution can be enhanced by the use of large telescopes, by placing observatories above the earth's turbulent atmosphere, or by the use of special interferometry techniques. There is another way to achieve high resolution, however: get close to the subject of the observations. This is what the *Voyager* spacecraft did.

The cameras aboard *Voyagers 1* and *2* yielded images of Saturn's rings with a maximum resolution of about 10 km; that is, details as small as 10 km in size could be seen. This was quite an improvement over the best photographs taken from earth, which have a resolution of about 4,000 km. The *Space Telescope,* when it is placed in earth orbit in the late 1980s, will resolve details in the ring system about ten times smaller than this, roughly 400 km, still far inferior to *Voyager* images.

There was an instrument on the *Voyager* spacecraft that provided much better resolution than even the cameras. This was the photopolarimeter, a device for measuring the intensity of light, as well as its polarization. The photopolarimeter on *Voyager 1* failed before that spacecraft reached Saturn, but the one on *Voyager 2* was able to carry out an observation that has provided by far the most detailed information yet on the ring system.

The observation was rather simple: as *Voyager 2* approached Saturn, the photopolarimeter measured the brightness of a star (Delta Scorpii) that happened to lie behind the rings, from the point of view of the spacecraft. As *Voyager 2* swept by the planet with the photopolarimeter trained on the star, the spacecraft motion caused the ring system to move across in front of the star. As it did so, the ringlets and gaps caused the brightness of the star to fluctuate. When a dense ringlet passed in front of the star, its brightness temporarily dimmed; when a gap passed in front, the star appeared brighter. This kind of observation is known as a stellar occulation.

The photopolarimeter was able to record the changes in the star's brightness very rapidly. This, combined with the rate at which the rings appeared to move across in front of the star, determined how fine the details were that could be detected in the ring structure. As it turned out, the resolution of this experiment was about 100 meters, a factor of 100

better than the best images obtained by the *Voyager* cameras! This meant that ringlets or gaps as small as the length of a football field could be detected.

Of course, the photopolarimeter could only measure the ring structure along a single cross-section; it could not provide a full picture of the entire system. Its image of the rings was one-dimensional, but it nevertheless led to profound new understanding, particularly of the spiral density waves described in the text.

The data produced by the photopolarimeter were originally in the form of plots showing the variations in the star's intensity as a function of time (see Fig. 13.21). Dips in the intensity corresponded to the passage of ringlets in front of the star. This information was then translated into a plot of ring density as a function of distance from Saturn. The data were processed a step further so that things were easier to visualize; synthetic images of the rings, simulating those taken by a camera, were produced (it was assumed that the rings were circular, and different colors were used to represent varying degrees of ring density; see Color Plate 12). In this way, false-color images of the rings were produced, somewhat similar to the actual photographs obtained by the cameras, but showing details a factor of 100 smaller.

moons. For example, one of the inner ringlets is not circular (Fig. 13.22), but is instead flattened into a slightly oval shape (although it does not seem to be part of a spiral density wave); and the F ring, the one that is held in place by the pair of shepherd satellites, has a very peculiar braided structure, as though three rings had been interwoven (Fig. 13.23). No good explanation has been found for either kind of behavior. Comparison of *Voyager 1* and *Voyager 2* photographs showed that the irregular structure of the F ring varies with time.

One possibility has been raised, however, but not yet worked out in detail, which takes note of the fact that the ring particles all orbit Saturn within the magnetosphere. The ring particles sweep up charged subatomic particles such as electrons as they travel along, becoming electrically charged themselves in the process. If a ring particle is sufficiently small, once it has an electrical charge its motion can be controlled by magnetic forces, rather than gravitational ones. It is therefore possible that Saturn's magnetic field exerts some influence over the smaller particles in the ring system, distorting the shapes of some of the rings. The magnetic field rotates with Saturn much faster than the particles move in their orbits, and the magnetosphere itself varies in size and intensity with fluctuations in the stream of charged particles coming from the sun. Thus it is possible to imagine nonsymmetric forces acting to shape the rings.

It is almost certain that magnetic forces are responsible for the distribution of particles in another phenomenon associated with the rings. A haze layer of very fine particles is suspended above and below the plane of the rings, apparently held there by the magnetic field. These suspended particles were first noticed because they create spoke-like structures that point away from Saturn and move around it (Fig. 13.24). The origin of these dark, asymmetric features was quite a mystery until it was realized that the particles there are so tiny that the small electrical charge they would acquire due to ions striking them would be enough to allow electromagnetic forces to control their motions.

A great deal of work lies ahead before astronomers fully understand all the influences acting on the myriad particles and rings orbiting Saturn. Meanwhile we can be content to marvel at the wondrous appearance of this planet and its satellites, and to speculate about new mysteries yet to be revealed.

FIG. 13.22. ASYMMETRIES IN THE RINGS. This composite shows that the rings are not perfectly circular; we see here that the thin ring within the dark gap is thinner on one side of the planet than on the other, and slightly displaced.

FIG. 13.23. THE F RING. Here we see the unusual braided appearance of this thin, outlying ring. The cause of the asymmetric shape is not well understood.

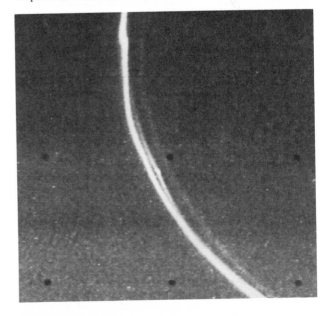

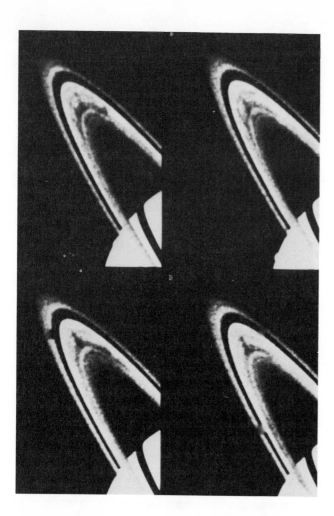

FIG. 13.24. THE DARK SPOKES. This sequence shows the dark, hazy radial features referred to as "spokes." These features, probably a result of very fine particles suspended above and below the ring plane by electromagnetic forces, are seen here to vary in appearance with time.

Perspective

Our visit to the ringed planet has taught us much about how complex nature can be. Every planet, even every satellite, that we have discussed has been revealed as a distinct individual, and Saturn itself is perhaps the most unique of the lot.

Although there are many similarities between Saturn and Jupiter, in almost every respect, from the internal structure to the atmospheric motions, from the satellites to the rings, there are also many differences. Saturn may be a cousin of Jupiter, but it is certainly not an identical twin.

We have almost completed our tour of the planets, for the remaining three are not yet well known to us as individuals, and we can only discuss them as seen from afar.

Summary

1. The general properties of Saturn and its ring system, including the planet's size, mass, and very low density, as well as the particulate nature of the rings, were all determined from earth-based observations.

2. One of the *Pioneer* and both *Voyager* spacecraft made close-up observations of Saturn, its satellites, and its rings, revolutioninzing our knowledge of each.

3. Saturn's atmosphere and internal structure are different from Jupiter: Saturn lacks contrast in its global wind patterns, and it has seasonal variations, greater wind velocities, and a different source of excess internal heat.

4. Saturn has at least seventeen satellites, including the giant Titan, seven intermediate moons, and a large number of tiny irregular satellites.

5. Titan's thick atmosphere is dominated by nitrogen, but it also includes methane, which may exist in gaseous, liquid, and frozen states on the surface.

6. The intermediate-sized moons all have very low densities, indicating that they have icy compositions; and some show evidence that they were resurfaced, either by frozen internal gases that were released or by debris from space.

7. The complex structure of Saturn's ring system is created by a combination of effects: orbital resonances with the satellites, spiral density waves, shepherding by pairs of satellites, and electromagnetic forces acting on charged particles.

Review Questions

1. How often are Saturn's rings seen edge-on, as viewed from the earth?

2. As a check on the mass of Saturn, given in table 13.1, calculate it by using satellite orbital data from table 13.2.

3. Rank, in order of decreasing orbital speed, the following: the satellite Mimas, a particle at the inner edge of the A ring, Tethys, a particle in the F ring, and the satellite S–15. Explain how you decided on the ranking. Hint: you will need to use data from tables 13.2 and 13.3.

4. Compare the atmosphere of Saturn with that of the primitive earth.

5. Summarize the differences between the atmospheres of Jupiter and Saturn, in terms of structure and motions.

6. Compare the internal structures and heating mechanisms of Jupiter and Saturn.

7. Why do none of the satellites of Saturn have an ionized torus like that of Io?

8. Why does the large abundance of nitrogen in the atmosphere of Titan imply that there might be present-day geological activity on this satellite?

9. Calculate the location of a gap in the ring system that might be created by the 3:1 orbital resonance with the satellite Mimas (that is, calculate the orbital radius where a ring particle would have exactly one-third the orbital period of Mimas).

10. Summarize the processes thought to be responsible for the complex structure in the rings of Saturn.

Additional Readings

Beatty, J. K., O'Leary, B., and Chaikin, A., eds. 1981. *The new solar system.* (Cambridge, England: Cambridge University Press.

Morrison, D. 1981. The new Saturn system. *Mercury* 10(6):162.
———. 1982. *Voyages to Saturn.* NASA Special Publication SP–451. Washington, D.C.: NASA.

•CHAPTER 14•

The Outer Planets

•CHAPTER PREVIEW•

Uranus
  Discovery by Herschel
  Planetary properties
  Anomalous inclination
  Atmospheric properties
  Satellites and rings
Neptune
  Discovery: victory for Newton
  Planetary properties
  Satellites in chaos
Pluto
  Discovery: an inspired accident
  Unusual properties
  The double planet

The three planets that we have not yet discussed, orbiting in the distant reaches of the solar system far from the warmth of the sun, are not well understood. Uranus, Neptune, and Pluto all lie well beyond the orbit of Saturn, and were not known to the ancient astronomers. They are grouped together in this chapter, not because of intrinsic similarities (although, as we will see, two of the three are virtually identical in many respects), but because of our similarly limited knowledge of them.

In discussing the outer planets, we will gain an appreciation of the precision and power of Newton's laws of motion on the one hand, and an understanding of the perversity of human intuition on the other.

Uranus

The seventh planet from the sun, Uranus is marginally bright enough to be seen with the naked eye, and was actually included in a number of star charts compiled from telescopic observations in the seventeenth and eighteenth centuries. At its distance of nearly 20 AU, its period is so long (about eighty-four years) and its motion so slow that for quite awhile it was not noticed that its position constantly changes with respect to the background stars.

The planet was finally discovered when, in 1781, the English astronomer William Herschel found an object

214

during a routine sky survey that seemed unusual for a star. It had a disk-like appearance, rather than being a single, twinkling point of light, and it appeared blue-green in color. Herschel at first suspected that he had found a comet, but after observing it for an extended period of time and then calculating its orbit, he found it to be moving about the sun in a nearly circular path, about twice as far out as Saturn. This established that Herschel had found a planet, for no comet is visible at such a great distance from the sun, nor would a comet move in a circular orbit. Herschel initially named the new planet after King George III of England, but this was never widely accepted, and the recognized name became Uranus, after the Greek god representing the heavens.

The size of Uranus was not easily estimated, because its image is very indistinct in even the best telescopic views. Photographs taken from a high-altitude balloon in 1972 (Fig. 14.1) allowed a fairly accurate measurement to be made of its angular diameter, which turned out to be about three seconds of arc, corresponding to a true diameter of 51,800 kilometers. A more accurate size determination was made in 1977, when Uranus moved in front of a background star, so that its diameter could be deduced directly from knowledge of its speed and how long it took to pass in front of the star, occulting its light. Uranus is a little less than half the size of Saturn, and only about one third as big as Jupiter.

Uranus has five satellites, and these allowed observers to determine, using Kepler's third law, the mass of the planet. It was found that Uranus is about fifteen times more massive than the earth; it has an estimated density of 1.2 grams/cm^3, very similar to that of Jupiter.

In 1986, the *Voyager 2* spacecraft will reach Uranus on its long trek through the solar system. If all goes well, and the spacecraft systems are still operating, we will get close-up portraits of yet another member of the sun's family.

ANOMALOUS INCLINATION

The rotation period of Uranus, determined from spectroscopic measurements using the Doppler effect, is 16.16 hours. This is similar to the other giant planets, which was no surprise. However, the data revealed something else about the planet's rotation that is very peculiar indeed: Uranus is tipped over so far on its axis

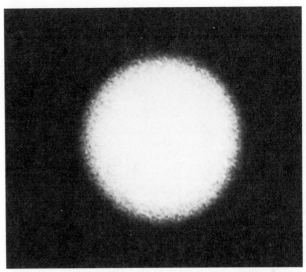

FIG. 14.1. URANUS. This exceptionally fine photograph was obtained from a high-altitude balloon, which was able to avoid much of the blurring effect of the earth's atmosphere. No surface markings are evident, except for some darkening at the edges of the disk.

that its north pole actually points a little below the plane of the ecliptic. The inclination of its rotation axis is 98 degrees, whereas for most of the other planets it is less than 30 degrees. The north pole of Uranus points al-

TABLE 14.1.   URANUS
Orbital semimajor axis: 19.18 AU (2,869,000,000 km)
Perihelion distance: 18.27 AU
Aphelion distance: 20.09 AU
Orbital period: 84.01 years (30,685 days)
Orbital inclination: 0°46'23"
Rotation period: 16^{h}10^m
Tilt of axis: 97°55'
Mean diameter: 52,300 km (4.10 D$_\oplus$)
Mass: 8.72 × 10^{28} grams (14.6 M$_\oplus$)
Density: 1.2 grams/cm^3
Surface gravity: 1.07 earth gravities
Escape velocity: 21 km/sec
Surface temperature: 95 K
Albedo: 0.66
Satellites: 5

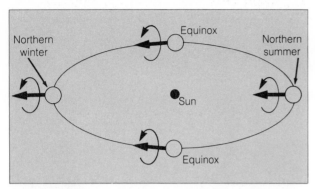

FIG. 14.2. SEASONS ON URANUS. The unusual tilt of Uranus creates bizarre seasonal effects. The solid arrow in each case indicates the direction of the planet's north pole, and the curved arrow, the direction of its rotation.

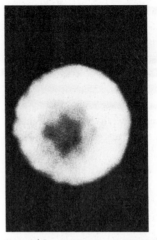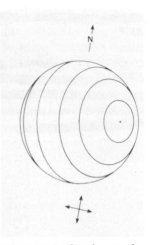

FIG. 14.3. SURFACE MARKINGS ON URANUS. This photograph, obtained at the infrared wavelength of an absorption band of gaseous methane, shows some contrast between different portions of the planetary surface. The sketch at right shows the orientation of the planet at the time.

most directly at the sun at one point in its 84-year orbit, and directly away 42 years later. In between, the sun is overhead at middle latitudes on the planet. Needless to say, this must create rather strange seasonal variations (Fig. 14.2).

The extreme tilt of Uranus stands as an anomaly in the solar system, and is not easily explained by theories of planetary formation (see chapter 17). The fact that all five of the known satellites, as well as the recently discovered rings of Uranus, orbit in the equatorial plane of the planet (that is, the plane of the satellite and ring orbits is nearly perpendicular to the ecliptic) suggests that whatever made the planet deviate from the usual orientation must have happened at some time before the satellites had formed, perhaps when they were still part of a disk of orbiting debris.

ATMOSPHERIC PROPERTIES

Even the best photographs of Uranus show a featureless disk, with no obvious bands or other markings of the sort that appear on Jupiter and Saturn. Because of the similarly rapid rotation, however, we might expect that some similarities are present in the atmospheric flow patterns, and indeed some observers claim to have seen banded surface features during moments of unusual clarity. Infrared observations have shown variations that can be attributed to motions in the atmosphere (Fig. 14.3).

The temperature was found from infrared observations (through the application of Wien's law) to be rather low, as we might expect at such a great distance from the sun. The best estimate is 95 K at the top of the atmosphere, which means Uranus is comparable in temperature to Saturn. Spectroscopic observations show that hydrogen (H_2) and methane (CH_4) are present, but no ammonia. This molecule (NH_3) probably is there, but as in the case of Saturn, has precipitated out of the atmosphere in the form of crystalline snow.

Although little is known about the atmosphere, even less is known about the internal structure of Uranus. It probably resembles that of Jupiter and Saturn, with the possible exception that inside Uranus the pressure may not be sufficient to form liquid metallic hydrogen. There is probably a rocky core, since differentiation has most likely occurred, allowing the heavy elements to sink to the center. Outside of this there may be a zone of ice, and above that liquid and then gaseous hydrogen, mixed throughout with other light elements such as helium.

We have no information on the question of whether Uranus has a magnetic field. With its rapid rotation and possible liquid zone inside, it seems likely that it does, but close-up measurements will have to be made before we can answer this question.

SATELLITES AND RINGS

Uranus has five known satellites (Fig. 14.4). The two largest, named Oberon and Titania, were found by Herschel in 1787, and the most recently discovered,

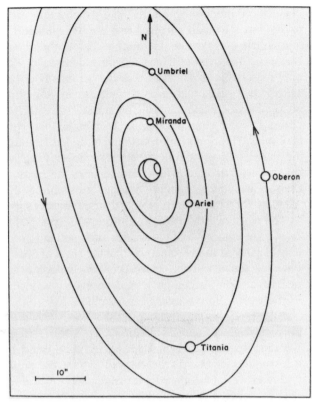

FIG. 14.4. URANUS AND ITS SATELLITES. All five moons are visible in the photo (left), and are identified in the sketch (right). The direction indicated as north is in the earth's reference frame; the planet and its entire satellite (and ring) system are tilted nearly perpendicular.

Miranda, was spotted in 1948. If Saturn and Jupiter can be used as examples, it may be that a number of small moons orbit Uranus, in addition to these five satellites, whose diameters range between about 600 and 1,600 kilometers. Perhaps *Voyager 2* will reveal the presence of others.

Little is known about the satellites except their distances from Uranus and their approximate sizes. No mass or density measurements will be possible until their gravitational effects on a passing object can be determined, and this will not be feasible until *Voyager 2* arrives in 1986. It is not known whether the satellites are in synchronous rotation, but in view of their distances from Uranus and the large mass of the planet, it seems very likely that at least the innermost satellites do keep one face pointing perpetually inward.

Uranus was the second planet discovered to have a ring system (Fig. 14.5). In 1977, observers measured the diameter of the planet by timing how long it took Uranus to pass in front of a background star (the stellar

FIG. 14.5. THE RINGS OF URANUS. This artist's sketch illustrates the diameters but exaggerates the thicknesses of the rings of Uranus. Also shown is the *Voyager 2* spacecraft, as it will approach the planet in 1986.

occultation mentioned earlier in this chapter), and found, much to their surprise, that the star blinked off and on five times during the minutes before the planet was to pass in front of it, and five times afterward. The only reasonable explanation was that Uranus is encircled by five rings, with gaps between them. After this accidental discovery, careful observations directly revealed the five rings, as well as four additional, very faint ones. The ring system was difficult to see because the particles do not reflect light as efficiently as the ring particles around Saturn, and because the rings of Uranus are much thinner, perhaps resembling the "ringlets" of Saturn more closely than the main rings.

The rings of Uranus lie well inside the orbit of Miranda, the innermost of the known satellites, and probably originated in the same way that the rings of Saturn did (see the discussion in chapter 13), where it appears that tidal forces prevented particles inside the Roche limit from coalescing into satellites.

It is clear that the ring system of Uranus has complex structure, in some ways similar to Saturn's rings. We can speculate that gravitational effects created by the satellites of Uranus are responsible for the structure, and we look forward with anticipation to the arrival of *Voyager 2,* which may reveal direct evidence of this, as it did in the case of Saturn.

Neptune

Perhaps more than any other two planets, Uranus and Neptune can be regarded as twins. Neptune, the eighth planet, orbits at a distance of some 30 AU from the sun, taking 165 years to make one circuit. Its size is very similar to that of Uranus, but its mass is somewhat greater, giving it a density of 1.7 grams/cm^3, the greatest of the gaseous giants.

Since Neptune is even farther from the sun than Uranus, less is known about it. To find out more about this planet, we will have to wait until 1989, when the well-traveled *Voyager 2* spacecraft is to fly by Neptune.

THE DISCOVERY: A VICTORY FOR NEWTON

Neptune is a rather dim object, too faint to be seen with the unaided eye, and even too inconspicuous to have been included on star charts made before the mid-

1800's. Rather than being found by accident, as Uranus was, the existence of Neptune was first deduced indirectly.

When Uranus was found by Herschel in 1781, and its orbit subsequently computed, astronomers naturally were eager to see whether its path conformed perfectly to the predictions of Newton's laws of motion and gravitation. Since Uranus had been included on star charts dating back some ninety years before Herschel's discovery, it was possible to compare its observed and computed positions over a long period. Much to the dismay of the scientists who made the comparisons, a discrepancy was found. Although it was not very large, amounting to two minutes of arc by 1840, it nonetheless was considered very serious, for this was too large an error to be allowed by the laws of motion. Something was amiss, and people began to question the validity of Newton's work.

It was soon suggested, however, that the discrepancies could be resolved if another body were exerting a gravitational influence on the motion of Uranus. Calculations carried out independently in England and France (see page 219) showed that an eighth planet could be responsible, if it were located at a particular position in the constellation Aries. In 1846 a planet was found at the predicted position, and what had begun as a major problem for Newton's laws turned into a resounding victory. This experience showed that the

TABLE 14.2.   NEPTUNE
Orbital semimajor axis: 30.07 AU (4,498,000,000 km)
Perihelion distance: 29.80 AU
Aphelion distance: 30.34 AU
Orbital period: 164.8 years (60,188 days)
Orbital inclination: 1°46'22"
Rotation period: 18^h 12^m
Tilt of axis: 28°48'
Mean diameter: 48,600 km (3.82 D$_\oplus$)
Mass: 1.03 × 10^{29} grams (17.2 M$_\oplus$)
Density: 1.7 grams/cm^3
Surface gravity: 1.08 earth gravities
Escape velocity: 24 km/sec
Surface temperature: 50 K
Albedo: 0.62
Satellites: 2

ASTRONOMICAL INSIGHT (14.1)

Science Politics, and the Discovery of Neptune.

There have been many cases throughout the history of science where a great discovery was just waiting to be made, as soon as the ncessary technology or preliminary information had been developed. If Galileo had not turned his telescope to the heavens, surely someone else would have done so very soon and found the wonders that he discovered. If Newton had not developed the laws of mechanics, some other student of the motions of the moon and planets would have.

The discovery of Neptune was such a case. The irregularities in the orbit of Uranus were well known by the mid-1800s, and the mathematical techniques that were needed to calculate the position of the eighth planet were also standard. It was only a matter of someone recognizing the possibility, and carrying out the computation. As it happened, two people did so, quite unbeknownst to each other, and the result was one of the more fascinating disputes in the history of science.

In England, a young college graduate named John C. Adams made the prediction in 1845 that an eighth planet should be at a certain location in the constellation Aries. Being young and unknown, Adams had difficulty convincing the established astronomers of England that an effort should be made to find the predicted planet. A contributing factor was the development of a minor tiff with the Astronomer Royal, with the result that the latter simply ignored the affair.

Within a few months, a French scientist named Urbain Leverrier carried out calculations that led him to the same conclusion, although he was totally unaware of Adams's work, which was not published.

Leverrier's work did appear in scientific journals, however, and this prompted the Astronomer Royal in England to take the idea of an eighth planet a bit more seriously. A search was begun.

The English astronomers were hindered by the lack of good star charts for the appropriate region of the sky, and substantial preliminary work was necessary before an earnest search could begin. Meanwhile, Leverrier contacted an acquaintance at the observatory in Berlin, who had excellent charts of Aries, and he discovered Neptune on the very first night he looked for it. This was in 1846.

England and France were international rivals, and years of contentious arguments followed as to who should be credited with the prediction of Neptune's existence. There never was an amicable settlement of this question; it just gradually faded from the forefront of French-English affairs. Today, the heat of the debate having cooled with the passage of time, Adams and Leverrier generally are equally credited with the discovery.

laws of motion and gravitation were so accurate and universal that they could actually be used to deduce the presence of unseen objects. The new planet was given the name Neptune, after the Greek god of the seas.

PROPERTIES OF NEPTUNE

This planet is difficult to examine in any great detail, even more so than Uranus. Ordinary photographs show little more than a spot of light, and only rather large-scale features can be detected. The fine resolution of the *Space Telescope* will allow smaller details to be seen on Neptune, if they exist. The angular diameter of Neptune, as seen from the earth at opposition, is 2.3 seconds of arc, and the *Space Telescope* will be capable of discerning features as small as about one-tenth of an arcsecond in size. Of course, this instrument will also provide better views of Uranus than any presently available, but presumably *Voyager 2* will be sending back close-ups of that planet by about the time the *Space Telescope* is launched.

For now, we surmise that Neptune closely resembles Uranus, since the two are so similar in all respects that can be measured from earth. The diameter of Neptune, like that of Uranus, was measured from a stellar occultation, and the mass was deduced from the application of Kepler's third law to the orbits of its satellites. It has proven difficult to measure the rotation period of Neptune, but recent observations indicate that it is 18.2 hours, and that its rotation axis is tilted by 29 degrees, similar to the tilts of several of the other planets.

The atmosphere may be even colder than that of Uranus, with a temperature of perhaps 50 K. Spectroscopy reveals the presence of hydrogen and methane, and again it is suspected that ammonia has precipitated out. It is likely that the internal structure is similar to that of Urnaus, with the probable exception that Neptune has a larger core, since it is known that its overall density is greater.

Surprising as it may seem, we have some data on atmospheric motions on Neptune. With the use of special filters that match the wavelengths of methane-gas absorption lines, it is possible to infer, from brightness variations measured through these filters (Fig. 14.6), the flow pattern of methane on Neptune. The information is not very detailed, but there are indications of a pattern of zonal winds like those on Jupiter and Saturn, with velocity differences of about 110 kilometers per second between zones. This tends to confirm our natural expectation that there strong similarities among all the giant, gaseous planets.

SATELLITES IN CHAOS, AND A POSSIBLE RING SYSTEM

Neptune has two known satellites (Fig. 14.7), one of which, Triton, is the largest in the entire solar system. This satellite, noted in the same year that Neptune itself was discovered, is larger than Mercury, and almost as large as Mars. Recent spectroscopic observations have shown that Triton has a gaseous atmosphere containing methane; thus it is the second satellite (along with Titan) known to have an atmosphere. The other satellite of Neptune, Nereid, is very small, with a diameter of roughly 600 kilometers (compared with Triton's diameter of over 7,200 kilometers), and was not detected until 1949.

Neptune has a fairly normal rotation, but the motions of its two satellites are extremely unusual. Triton orbits the planet backward, taking about six days to complete each retrograde orbit. Furthermore, the orbit is tilted by an unusually large angle, about 20 degrees with respect to the ecliptic. The orbit of Nereid is perhaps even more bizarre. It goes around in the normal direction (opposite of the motion of Triton), and its

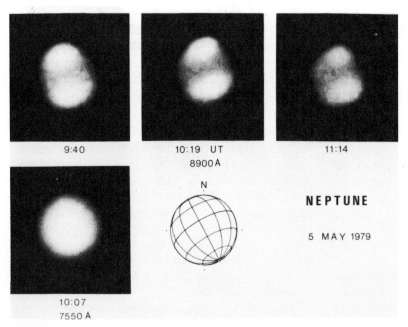

9:40   10:19 UT   11:14
              8900 A

N

NEPTUNE

5 MAY 1979

10:07

7550 A

FIG. 14.6.   NEPTUNE. This series of images, like the one of Uranus in Figure 14.3, was obtained in infrared light. The three upper images were made at a wavelength where methane absorbs light, so the dark areas are regions of methane concentration. The bright areas north and south of the equator are thought to have ice crystals and haze above the methane.

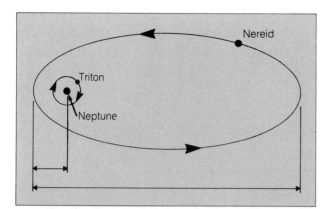

FIG. 14.7. THE SATELLITES OF NEPTUNE. This scale drawing illustrates the unusual orbital characteristics of the two moons, Triton and Nereid.

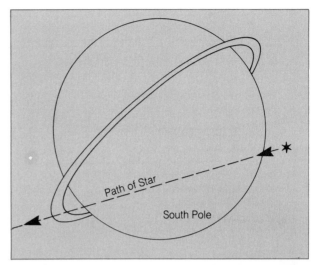

FIG. 14.8. RINGS OF NEPTUNE? This sketch roughly depicts the characteristics of Neptune's ring (or rings), whose existence was tentatively inferred from stellar-occultation data.

orbital plane is inclined 28 degrees to the ecliptic. Nereid is very distant from Neptune, and has a rather long period for a satellite, about one earth year. The strangest aspect of its orbit, however, is its shape, which is greatly elongated. Its distance from Neptune varies by a factor of five between its closest approach and the farthest reaches of its orbit. Newton's laws of motion allow for the possibility of such flattened orbits (and as we will see, the comets follow similar or even more elongated paths around the sun), but no planet nor any other satellite in the solar system has such an orbit.

The strange motions of Neptune's two satellites are not any sort of natural consequence of the formation of the planet, but rather, they are the result of the satellites being captured or the result of some immense gravitational disturbance that disrupted the system at some time in the past, after Neptune and its satellites had formed. Perhaps a near-collision with a massive body caused the upheaval. Some interesting speculations on this will be discussed in the next section, for it has been suggested that Pluto, the ninth planet, was a character in the drama.

It has been announced very recently that Neptune may have a ring system, which, if true, means that all four of the giants have rings. During a stellar-occultation observation, when Neptune was due to pass in front of a star, there was a dip in the star's brightness just before the planet itself blocked the star completely. This is analogous to the situation that occurred when the rings of Uranus were discovered, except there was no corresponding dip in the star's intensity as it emerged from behind Uranus on the other side. Analysis of the

geometry of the occultation in the case of Neptune, however, showed that the path of the star with respect to Neptune was such that it would have missed passing through the equatorial plane (where a ring system is most likely to lie), on its way out from behind the planet (Fig. 14.8). Therefore it is possible that the star was occulted by a ring system on one side of the planet, but missed being occulted on the other side. Until other observations can confirm the presence of the ring or rings, however, the discovery is being regarded as tentative.

Pluto

The ninth and last of the known planets is Pluto, orbiting at an average distance of nearly 40 AU from the sun. From that perspective, the sun would appear as a very bright star, but surely would not produce much warmth. This planet is the most obscure of them all, and we will raise many questions about its nature.

THE DISCOVERY:
AN INSPIRED ACCIDENT

After the triumph of Neptune's discovery, continued analysis of the orbits of both Uranus and Neptune seemed to indicate further, rather minor, discrepancies

TABLE 14.3   PLUTO

Orbital semimajor axis: 39.44 AU (5,900,000,000 km)
  Perihelion distance: 29.57 AU
  Aphelion distance: 49.31 AU
Orbital period: 248.8 years (90,700 days)
Orbital inclination: 17°10′12″
 	
Rotation period: 6.39 days
Tilt of axis: 65°
 	
Diameter: 3,000–3,600 km (0.16–0.47 $D_{\oplus}$)
Mass: 1.14 × 10^{25} grams (0.0019 $M_{\oplus}$)
Density: 0.5–0.8 grams/cm^3
Surface gravity: 0.024–0.034 earth gravity
Escape velocity: 0.9–1.0 km/sec
 	
Surface temperature: 40 K
Albedo: 0.25–0.36
 	
Satellites: 1

FIG. 14.9.   CLYDE TOMBAUGH AT THE BLINK COMPARATOR. This photo shows the discoverer of Pluto peering into the eyepiece of the device he used to find the planet on photographic plates. The blink comparator allows two images of the same field of stars to be examined alternately (with the use of a mirror that flips between two positions), so that moving objects stand out.

in their motions. The natural inclination of astronomers, given the recent experience with Neptune, was to speculate that yet another planet, beyond the orbit of Neptune, might be responsible. This was an example of a common human tendency to expect to make more fascinating new discoveries, and in this case the expectation was stronger than the evidence. In any case, several independent calculations, including most notably one carried out in 1915 by Percival Lowell (of Martian canal fame), led to predictions of the location of the supposed ninth planet, and the search was on.

Lowell set out to find his new planet, and spent the last years of his life searching photographic plates for it. The process was made difficult by the fact that the predicted position was in a portion of the Milky Way that is crowded with stars, so that a search for a dim object that slowly changed position among all the fixed stars was tedious and difficult. Lowell was unsuccessful.

In 1930 the American astronomer Clyde Tombaugh (Fig. 14.9), having taken up the search at the Lowell Observatory some time after Lowell died, found a tiny dot of light that displayed the expected motion (Fig. 14.10), and he found it only 7 degrees away from the position that Lowell had predicted. The new member of the sun's family was given the name Pluto, after the Roman god of the underworld.

Subsequent analysis of Pluto showed it to be far too small and of too low a mass to have created any

noticeable disturbance in the paths of Uranus and Neptune. Reexamination of the data that Lowell and others had used in predicting the existence of a ninth planet showed, furthermore, that there was no evidence for such a planet. The minor discrepancies between calculated and observed positions were all within the uncertainty of the measurements. Apparently Lowell and the others who made the calculations wanted to find evidence for a ninth planet, and so they found it, whereas dispassionate scientific analysis would not have shown that it was there. It is a remarkable coincidence, then, that Pluto was found, and especially that it was found not far from the position predicted by Lowell. If it had not been for that prediction, unsound though it may have been, it might have been a long time before Pluto was discovered.

PLANETARY MISFIT

The little we know about Pluto shows it to be a nonconformist. Throughout our discussions of the other planets, certain systematic regularities have shown up, and Pluto violates many of them.

FIG. 14.10.   THE DISCOVERY OF PLUTO. These are portions of the original 1930 photographs depicting the ninth planet. The position of the planet is indicated by arrows.

First, its orbit is highly irregular, compared to those of the other planets. It is both noncircular and tilted with respect to the ecliptic. At its greatest distance from the sun, Pluto is nearly 70 percent farther away than at its closest, when it actually moves inside of Neptune's orbit for a time. (In fact, Pluto is temporarily the eighth planet from the sun right now, having moved inside of Neptune's orbit in 1978, not to reemerge until 1998.) Pluto's orbit is tilted by 17 degrees to the ecliptic, whereas all the other planetary orbits are tilted by 7 degrees or less.

Pluto also does not seem to fit in with the other planets in terms of physical characteristics. The other outer members of the solar system are giant planets, with thick atmospheres; although it has been difficult to pinpoint the mass of Pluto, it has long been clear that this planet does not resemble the others. For a long time the only information on the mass of Pluto came from the lack of any significant effect on the motion of Neptune, which implied that Pluto's mass had to be less than about 10 percent of that of the earth. Its diameter was not established until rather recently, when the technique called interferometry (see chapter 6) was used to discover that it is between 3,000 and 3,600 kilometers.

Spectroscopy has shown that Pluto has methane gas and possibly methane ice as well, so there is a thin atmosphere and perhaps also a frost of methane ice on the surface. Most molecular compounds formed of common elements freeze out into solid form at the low temperature (perhaps 40 K) thought to characterize Pluto.

Brightness variations indicate that Pluto has some surface markings, and is rotating with a period of 6.39 days. The orientation of the rotation axis has been difficult to determine, but new information indicates a large tilt, probably about 65 degrees (with respect to the planet's orbital plane, which is itself tilted 17 degrees with respect to the ecliptic, as stated earlier).

The small size and apparent low mass of Pluto, combined with its unusual orbit, led to the speculation that it was once a satellite of one of the giant planets, and that it somehow escaped to follow its own path around the sun. Recall the disorder of the satellites of Neptune; it has been shown that if Pluto were once a satellite of Neptune, a three-way collision among Pluto, Triton, and Nereid could have altered the orbits of the latter two and allowed Pluto to escape into its present orbit around the sun. To do this would have required an extremely special set of circumstances, with the three meeting each other at just the right speed and direction. It could have happened, however; after all, some-

Text continues on page 225.

The Search for Planet X

The discovery of Pluto was by no means the end of the search for new planets. Quite to the contrary, this event sharpened the interest in continuing the search, and it was carried on for an extended period.

The technique by which Pluto was discovered was tedious but efficient, in that it was thorough. Observers carried out the task by comparing photographs taken at different times (usually a few days apart), to see whether any of the objects in the field of view moved. This would ordinarily be very difficult to do, particularly in areas of the sky that are crowded with stars, but was made easier with the aid of a special instrument called a *blink comparator*. This is a microscope for viewing two photographic plates simultaneously, with a small mirror that flips back and forth, providing, in rapid succession, alternating views of the two plates. If the plates are mounted so that the images of fixed stars are perfectly aligned, then an object that moves between the times of the two exposures will appear to blink on and off at two separate locations, whereas all the fixed ob-

jects will be steady. To examine a region of the sky using this technique therefore consists of mounting in the blink comparator two plates taken at different times, then systematically scanning them with the movable eyepiece so that each object can be checked, to see if it is blinking. The entire job could take days or weeks for each plate pair.

The blink-comparator technique, although time-consuming, has the capability of allowing thorough searches of large regions of the sky. The search for a tenth planet, often referred to as Planet X, continued at Lowell Observatory for thirteen years after the discovery of Pluto. Clyde Tombaugh, who found Pluto, was the principal worker in the extended search. Nothing was found.

During and since that time, occasional reports of evidence for a tenth planet have surfaced, but the evidence has always been indirect. Reanalysis of the motion of Neptune has convinced some astronomers that there really may be discrepancies attributable to gravitational perturbations caused by another planet. Other researchers have pointed to the motions of certain comets as indicative of gravitational influence exerted by Planet X. Most of these predictions have suggested a distance from the sun of 50 to 100 AU. The motion of Halley's comet, a bright and very famous object that visits the inner solar system every seventy-six years, has been anlyzed by several scientists, who find that its arrival near the sun is usually several days later

than it should be, as predicted by the laws of motion. Some have attributed this discrepancy to the gravitational influence of a tenth planet, but no one has found such an object at the predicted position. One of the predictions based on the motion of Halley's comet was rather spectacular: the inferred Planet X had a mass three times that of Saturn, and an orbit inclined by 120 degrees to the ecliptic!

Modern information on comets has explained the discrepancy in the motion of Halley's comet another way. As it nears the sun and heats up, volatile gases vaporize and stream outward from the heated side, acting like a rocket exhaust to slow the comet's motion. (This process is described in more detail in chapter 15.)

The routine search for new objects continues today, with occasional success, but no new planets have been found. There was a brief flurry of excitement in 1977, when a sun-orbiting body was discovered whose elliptical path carries it almost as far from the sun as Uranus, and as close in as Saturn. It is clearly not a planet, but the exact nature of Chiron, as it is called, is uncertain. It may be an asteroid in a highly unusual orbit, or an old cometary nucleus that never approaches the sun closely enough to develop a luminous tail (see chapter 15). Perhaps eventually Planet X will be found, but if so, it will undoubtedly be a dim, cold, distant object, much too faint to be seen without a large telescope.

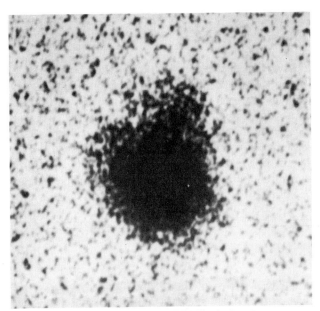

FIG. 14.11.   THE DISCOVERY OF CHARON. This is the photograph that revealed Pluto's satellite (the bulge at upper right).

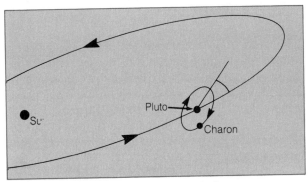

FIG. 14.12.   CHARON'S ORBIT. This sketch illustrates the orientation of the orbit of Pluto's satellite, with respect to the orbit of Pluto itself, which in turn is greatly tilted with respect to the ecliptic. The orbit of Charon is retrograde and tilted 65 degrees with respect to Pluto's orbital plane.

thing unusual must have occurred to create the motions of Triton and Nereid. Thus it seemed that a whole set of anomalous motions in the outer solar system could be explained by a single catastrophic event.

This theory has recently suffered a setback, however. Careful examination of photographs of the planet have shown that Pluto has a lump that moves back and forth from one side to the other (Fig. 14.11). Apparently this lump is a satellite very close to the planet, orbiting Pluto with a period of slightly more than six days. To create the distorted appearance of Pluto, the satellite must be relatively large and close to it. Pluto's satellite was given the name Charon. The idea that Pluto should have a satellite of its own does not square up with the theory that Pluto itself is an escaped satellite of one of the giant planets. We have to drop that theory, unless Charon split off from Pluto in the cataclysmic process that caused Pluto's escape from Neptune.

The orbital period of Charon is 6.39 days, precisely equal to the rotation period of Pluto. Thus, this is a case where both the satellite and the parent planet are locked in synchronous rotation. To an observer on the surface of Pluto, Charon would appear fixed in position overhead, while the background stars streamed past as a result of the planet's rotation.

Charon is unusually large compared with the planet it orbits. No doubt this has helped create sufficiently strong tidal forces to produce Pluto's synchronous rotation. Pluto might be accurately described as a double planet.

The orbital plane of Charon is tilted about 65 degrees with respect to the plane of Pluto's orbit about the sun, and Charon orbits in the retrograde direction (Fig. 14.12). The tentative conclusion that Pluto's rotation axis is tipped 65 degrees with respect to its orbital plane is based on the assumption that Charon orbits in the plane of Pluto's equator.

The discovery of Charon gave astronomers a chance to determine the mass of the planet, using Kepler's third law. The result shows that Pluto's mass is only about 0.0019 of the earth's mass, and this in turn leads to a rather low average density, in the range of 0.5 to 0.8 grams/cm^3. Evidently Pluto is not a rocky planet, as had previously been assumed, but rather is made largely of ices, much like the satellites of Jupiter and Saturn.

Perspective

We have thoroughly digested all the available information on the planets. We certainly have many unanswered questions about them, but on the other hand, we have seen a number of systematic trends. Two of the outer three planets fit the pattern established by the others, being gaseous giants that apparently are similar in many ways to Jupiter and Saturn. The last planet, Pluto, is mysterious, and remains an anomaly to be reckoned with.

Nearly all the pieces of the puzzle are now in place. We need only examine the nonplanetary bodies roving through the solar system, to see what they can tell us of its origins, and then we will visit the center of power, the sun itself, before finally attempting to unravel the sequence of events that led to the creation of the solar system as we see it today. We turn our attention now to the minor bodies: the asteroids, comets, and meteoroids.

Summary

1. Uranus is about fifteen time more massive than the earth, has an average density of 1.2 grams/cm^3, and probably resembles Jupiter and Saturn in structure and composition.

2. The spin axis of Uranus, as well as the orbital plane of its five satellites, are tilted 98 degrees, creating strange seasonal effects.

3. Uranus has five satellites and at least nine thin, dim rings.

4. Neptune was discovered because of its gravitational effects on Uranus, a discovery that vindicated the laws of motion and gravitation developed by Newton.

5. Neptune and Uranus are very similar in all observable properties, and probably resemble each other strongly in other aspects, such as atmospheric motions and internal structure.

6. Neptune's two satellites have unusual orbital properties, possibly created by a major gravitational disturbance at some time in the past.

7. Pluto, the ninth planet, was discovered fortuitously close to the position Percival Lowell predicted on the basis of suspected irregularities in the orbit of Neptune.

8. Pluto is unusual in that its orbit is highly eccentric and tilted, and it is much smaller and probably different in structure than the other outer planets.

9. Pluto has a relatively large satellite that provides data on the planet's mass and density; these data show that Pluto is probably composed largely of ice.

Review Questions

1. When Uranus occulted the light from a background star, it took about 1 hour, 40 minutes to pass in front of it. The apparent motion of Uranus as seen from the earth is about 0.0005 seconds of arc per second. Using these numbers, calculate the angular diameter of Uranus.

2. What is the length of the day at the north pole of Uranus? What is the length of day at the equator when the pole is pointed 90 degrees away from the direction of the sun?

3. Based on what we learned about Jupiter and Saturn,

how might we try to determine from remote observations whether Uranus has a magnetic field?

4. At the point in time when Uranus and Neptune are lined up on the same side of the sun, Uranus experiences a gravitational force in one direction from the sun, and another in the opposite direction from Neptune. Calculate how much (by what fraction) this reduces the sun's attraction for Uranus.

5. Summarize the similarities and differences between Uranus and Neptune.

6. Use Kepler's third law to calculate the semimajor

axis of Nereid's orbit about Neptune, given that its orbital period is one year. (Note: you have to use the form of Kepler's third law that includes the masses of the two bodies.)

7. Contrast the means by which Uranus, Neptune, and Pluto were discovered.

8. Calculate how much the intensity of sunlight on Pluto varies as the planet goes from perihelion to aphelion in its orbit. At perihelion, when Pluto is at its closest point to the sun, it is still about 30 AU away from it. How does the sun's intensity on Pluto at that time compare with its intensity on the earth?

9. Explain why Pluto has been forced into synchronous rotation with its satellite Charon, but the earth has not been similarly forced into synchronous rotation with the moon.

Additional Readings

Beatty, J. K., O'Leary, B., and Chaikin, A., eds. 1981. *The new solar system*. Cambridge England: Cambridge University Press.

Harrington, R. S., and Harrington, B. J. 1980. Pluto: still an enigma after fifty years. *Sky and Telescope* 59(6):452.

Hunten, D. M. 1975. The outer planets. *Scientific American* 233(3):130.

Newburn, R. L., and Gulkis, S. 1973. A survey of the outer planets—Jupiter, Saturn, Uranus, Neptune, Pluto, and their satellites. *Space Science Reviews* 3:179.

•CHAPTER 15•

Space Debris

The planets are the dominant objects among the inhabitants of the solar system (except, of course, for the sun), but they are not entirely alone as they follow their clockwork paths through space. The abundant craters on planetary and satellite surfaces have shown us that there must have been a time when interplanetary rocks and gravel were very plentiful, raining down continually on any exposed surface. Today the rate of cratering is much lower than it once was, but some vestiges of the space debris that caused it still remain, orbiting the sun and occasionally becoming obvious to us as they pass near the earth or enter its atmosphere.

There are at least four distinct forms of interplanetary matter, some of which are closely related. In this chapter we will discuss the **asteroids,** myriad rocky chunks up to several hundred kilometers in diameter that orbit the sun between Mars and Jupiter; the **comets,** whose ephemeral and striking appearances have caused us to pause and marvel since ancient times; **meteors,** the bright streaks often visible in our nighttime skies, and the objects that create them; and the

interplanetary dust, a collection of very fine particles that fill the void between the planets.

Bode's Law and the Asteroid Belt

As we have already seen, some of the early students of planetary motions were concerned with the distances of the planets from the sun. Kepler spent enormous amounts of his time and energy seeking a mathematical relationship that would describe the distances of the planets in terms of geometrical solids. What he finally did discover was something different, a relationship between the distances and orbital periods. Kepler's third law can be used to predict the period of a planet, given its distance from the sun; or its distance, if its period is known, but it did not provide any underlying basis for explaining why there are planets only at certain distances from the sun.

In 1766, a German astronomer named J. D. Titius found a simple mathematical relationship that seemed to accomplish what Kepler had set out to do. Titius discovered that if we start with the sequence of numbers 0, 3, 6, 12, 24, 48, and 96 (obtained by doubling each one in order), then add 4 to each and divide by 10, we end up with the numbers 0.4, 0.7, 1.0, 1.6, 2.8, 5.2, and 10.0, corresponding closely to the observed planetary distances in astronomical units. A few years after Titius found this numerological device, it was popularized by another German astronomer, Johann Bode, and eventually became known as Bode's law.

The sequence of numbers dictated by Bode's law includes one, 2.8 AU, where no planet was known to exist. After the discovery of Uranus in 1781, and the recognition that its distance fits the sequence (the next number is (192 + 4)/10 = 19.6, and the semimajor axis of Uranus's orbit is 19.2 AU), a great deal of interest arose in the idea that there ought to be a planet at 2.8 AU, since Bode's law was doing so well in predicting the positions of the others.

A deliberate search for such a planet was begun in 1800, but the discovery of the sought-after object came accidentally when an Italian astronomer named Piazzi noted a new object on the night of January 1, 1801, and within weeks found from its motion that it was probably a solar system body.

When the orbit of the new object was calculated, its semimajor axis turned out to be 2.77 AU, very close to the value predicted by Bode's law. The new fifth planet was named Ceres.

A little more than a year after the discovery of Ceres, a second object was found orbiting the sun at approximately the same distance, and was named Pallas. Because of their faintness, both Ceres and Pallas were obviously very small bodies and were not respectable planets. By 1807, two more of these **asteroids,** as they were called, had been found and designated Juno and Vesta. A fifth, Astrea, was discovered in 1845, and in the next decades, vast numbers of these objects began to turn up. The efficiency of finding them was improved greatly when photographic techniques began to be used. An asteroid will leave a trail on a long-exposure photograph, because of its orbital motion (Fig. 15.1).

Today the number of known asteroids is in the thousands, with about 2,000 of them sufficiently well observed to have had their orbits calculated and logged

FIG. 15.1. ASTEROID MOTION. The elongated image in this photo is the trail made by an asteroid, which moved relative to the fixed stars during the twenty-minute exposure. Most new asteroids are discovered through photos of this type, although the first few were spotted visually.

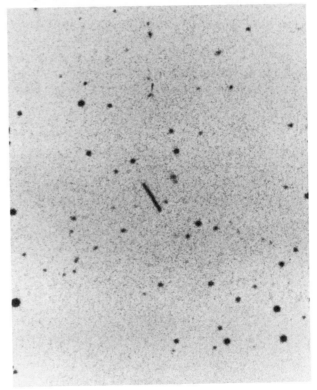

Trojans, Apollos, and Target Earth

Not all the asteroids are confined to the main belt between the orbits of Mars and Jupiter. There are exceptions, many of which make up two special categories.

The existence of one of these groups was actually predicted on the basis of mathematical calculations carried out in the late 1700's by the French scientist J. L. Lagrange. Using Newton's laws of motion, Lagrange showed that a system consisting of two orbiting bodies has certain points where additional objects may orbit in a stable fashion (ordinarily, when more than two objects are involved, the system becomes so complex that it is nearly impossible to describe mathematically all the possible motions). Lagrange showed that if one body orbits another, there are points 60° ahead and 60° behind it, in the same orbit, where additional bodies can stay. In chapter 13 we learned that there is a small satellite occupying the orbit of Dione, one of the larger satellites of Saturn, and that this small satellite stays in position 60° ahead of Dione. There are two such satellites sharing the orbit of Tethys, another moon of Saturn. One is 60° ahead of Tethys in its orbit, and the other is 60° behind. Similar points exist in the orbit of our moon, and there has even been discussion that these might be convenient places to locate permanent manned space stations. (The L–5 Society, a group advocating this idea, takes its name from one of these points, which is called the L–5 point, in honor of Lagrange.)

The same situation exists for the orbit of Jupiter, and after a few asteroids were found there in the early decades of this century, concerted searches were made, revealing a large number of asteroids in the Lagrange points, 60° in either direction from Jupiter. There may be hundreds of objects at these two locations, held in place by the combined gravitational pulls of Jupiter and the sun. These are called the **Trojan asteroids,** and they are individually named for classical Trojan and Greek heroes, such as Achilles and Hector.

Another distinct group of asteroids has been identified and named after the first to be discovered, Apollo. These are objects whose distinction is that their orbits bring them within one AU of the sun, so that their paths can cross the orbit of the earth. There are about thirty *Apollo asteroids* known, and a couple of them, Icarus and Eros, have provided useful information for scientists on earth. In 1968, Icarus, which of all the Apollo asteroids comes closest to the sun (0.19 AU), came within 16 million kilometers of Mercury, close enough for Mercury's gravitational pull to noticeably alter its direction, allowing astronomers to deduce the mass of Mercury. Eros passed the earth at a distance of about 23 million kilometers in 1931, giving astronomers a clear view of it through telescopes, so that its size and shape could be directly measured. It was seen to be an irregular chunk of rock, tumbling end over end with a period of about 5.3 hours. This same asteroid passed in front of a bright star in 1975, allowing its diameter to be measured precisely. It has an oblong shape, with dimensions of 7 × 19 × 30 kilometers.

Given a long enough time, some of the Apollo asteroids will probably eventually collide with the earth. They are so few in number, and the volume of space they occupy is so large, however, that the average time between such collisions is likely to be very long, measured in the millions of years. We can hope, therefore, to escape any truly major impacts for the foreseeable future.

in catalogues. The total number is probably much more, perhaps 100,000. The term **minor planet** is commonly adopted by modern astronomers, although we will use the traditional name for these objects.

Before we discuss the nature of the asteroids, let us return to our starting point: Bode's law. When the asteroids were first discovered, it seemed to lend a great deal of credence to this mathematical relationship among the planetary distances, and there were people who believed that it reflected some as-yet undiscovered physical principle that governed the layout of the solar system. Today the interpretation is rather different. Although there certainly is a regularity to the sequence of planetary distances from the sun (and this regularity must have been dictated by the laws of physics at the time of formation of the solar system), it is no longer believed that Bode's law itself represents a fundamental physical principle. Bode's law is now regarded as a mathematical coincidence; a numerical sequence that just happens to fit the observed planetary positions. There are, in fact, errors of a few tenths of an AU here and there, and the outermost planets, Neptune, and Pluto, do not fit the sequence at all.

The Nature of the Asteroids

It is possible to deduce some of the properties of the asteroids, from a variety of observational evidence. Measurements of their brightnesses can lead to size estimates; spectroscopy of light reflected from their surfaces provides information on their chemical makeup; and direct analysis of meteorites that may be remnants of asteroids adds data on their internal properties.

The largest asteroids were the first to have their diameters estimated; it was done by simply calculating what size they had to be to reflect the amount of light observed. Much more recently, infrared brightness measurements of both large and small asteroids have provided information on their sizes, since at temperatures of a few hundred degrees they glow at infrared wavelengths. The Stefan-Boltzmann law (chapter 5) can be used to determine the total surface area from the intensity of the emission.

A few direct measurements of asteroid sizes have also been possible, in cases where these objects have passed sufficiently close to earth that their angular sizes could be directly determined. In recent times, the angular diameters of some asteroids have been successfully measured by the use of interferometry. These assorted techniques led us to understand that asteroids have a wide range of diameters, a few as large as several hundred kilometers (Ceres, the largest, is about 1,000 kilometers in diameter), and most being rather small, with diameters of one or two hundred kilometers or less. The largest are apparently spherical in shape, whereas the smaller ones often are jagged, irregular chunks of rock (Fig. 15.2), varying in bright-

FIG. 15.2. A PORTRAIT OF AN ASTEROID? Many asteroids probably bear a general resemblance to this irregularly shaped object, which is Phobos, one of the tiny Martian moons.

ness as they tumble through space, reflecting light with different efficiencies on different sides. A few asteroids are binary, consisting of two chunks orbiting each other as they circle the sun. The total mass of all the asteroids together is small by planetary standards, amounting to only about 2 percent of the mass of the earth.

The compositions derived from spectroscopic analyses are highly varied. The principal ingredients range from metallic compounds to nearly metal-free ones, and from carbon-dominated to silicon-bearing minerals; and there are several classes of asteroids whose composition is not known. Among the biggest asteroids in the main belt between Mars and Jupiter, about three-quarters are carbonaceous, meaning that they contain carbon in complex molecular forms. Most of the rest are composed chiefly of silicon-bearing compounds, and about 5 percent are metal-rich, the primary metals being nickel-iron mixtures.

Kirkwood's Gaps: Orbital Resonances Revisited

As increasing numbers of asteroids were discovered and catalogued throughout the nineteenth century, calculations of their orbits showed remarkable gaps at certain distances from the sun. One such gap is at 3.28 AU, and another is at 2.50 AU.

The explanation for these gaps was offered by Daniel Kirkwood in 1866, when he realized that these distances correspond to orbital periods that are simple fractions of the period of Jupiter (Fig. 15.3). The giant planet, at its distance of 5.2 AU from the sun, takes 11.86 years to make a trip around it, whereas an asteroid at 3.28 AU, if one existed there, would have a period exactly half as long, 5.93 years. Thus this asteroid and Jupiter would be lined up in the same way frequently, every time Jupiter made one orbit. The asteroid would therefore be subjected to regular tugs by Jupiter's gravity, and would gradually alter its orbit, vacating the zone 3.28 AU from the sun. In this manner Jupiter has cleared out several gaps in the asteroid belt, at this and other distances where the orbital periods would result in regular alignments. The gap at 2.50 AU corresponds to orbits with a period exactly one-third that of Jupiter. These *orbital resonances* are exactly analogous to the ones created by the moons of Saturn, which are responsible for some of the gaps in the ring system of that planet (see chapter 13). Since Jupiter is the most massive and the nearest of the outer planets to the asteroid belt, it has by far the greatest effect in producing gaps, but in principle the other planets could do the same thing.

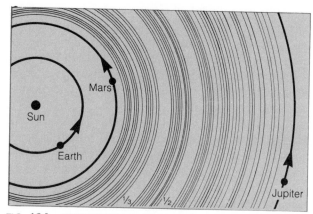

FIG. 15.3. KIRKWOOD'S GAPS. This schematically shows the orbits of the earth, Mars, and Jupiter, and a number of possible asteroid orbits. Gaps are shown at the distances from the sun where asteroids would have exactly one-half or one-third the orbital period of Jupiter.

The Origin of the Asteroids

For a long time the most natural explanation of asteroids was that a former planet broke apart, creating a swarm of fragments that continued to orbit the sun. One of the strongest arguments for this was the prevalent belief in Bode's law. This idea began to lose favor when it was accepted that Bode's law is not a fundamental physical principle, and it lost more ground when the total mass of the asteroids was estimated, and found to be much less than that of any ordinary planet.

Today an additional, rather strong argument against the planetary-remnant hypothesis is cited: it is difficult to comprehend how a planet could break apart once it had formed, whereas it is easy to understand why material orbiting the sun at the position of the asteroid belt could never have combined into a planet in the first place. It became simpler to accept the idea that the debris never was part of a planet, rather than to find a way to have it first form one, and then break apart.

The invisible hand of Jupiter's gravity is invoked again. If, as it is suspected, the planets formed from a swarm of debris orbiting the young sun in a disk, no planet could have formed in a location where the pieces of debris could not stick together. Calculations show that once Jupiter formed, its immense gravitational force stirred up the material near its orbit, so that collisions between particles occurred at speeds too great to allow them to stick together. It is as though Jupiter wielded a giant spatula, stirring up the debris near it and keeping it spread out as a loose collection of rocky fragments. Even today, Jupiter is still at work, keeping the asteroids mixed up, occasionally causing collisions between them that can sometimes break them up.

The study of meteorites, some of which probably originated as pieces of asteroids that broke apart in collisions, gives us a chance to examine material from the early solar system. It is interesting to note that some of the asteroids apparently have undergone differentiation, developing nickel-iron cores, because some meteorites are almost pure chunks of this material. There must have been sufficient heat inside some of the asteroids to create a molten state, allowing the heavy materials to segregate from the rest.

In contrast with the nickel-iron asteroids, some of the other types, particularly the carbonaceous ones, apparently have undergone almost no heating, since they contain high quantities of volatile elements that would have been easily cooked out.

Comets: Fateful Messengers

Among the most spectacular of all the celestial sights are the comets (Color Plate 13). With their brightly glowing heads and long, streaming tails, along with their infrequent and often unpredictable appearances, these objects have sparked the imagination (and often the fears) of people through the ages (Fig. 15.4). In antiquity, when astrological omens were taken very seriously, great import was attached to the occasion of a cometary appearance. Ancient descriptions of comets are numerous, and in many cases these objects were thought to be associated with catastrophe and suffering.

Included among the teachings of Aristotle was the notion that comets were phenomena in the earth's atmosphere. There was no good evidence for this idea, and it is not clear how Aristotle came upon it, but in any case it was accepted for centuries to come. However, Tycho Brahe in 1577 was able to prove that comets were too distant to be associated with the earth's atmosphere, because they do not exhibit any parallax when viewed from different positions on the earth. If a comet were really located only a few kilometers or even a few hundred kilometers above the surface of the earth, its position as seen from the earth would change from one location to another. Tycho was able to show that this was not the case, and that therefore comets belonged to the realm of space.

HALLEY, OORT, AND COMETARY ORBITS

A major advance in the understanding of comets was made by a contemporary and friend of Newton, Edmund Halley. Aware of the power of Newton's laws of motion and gravitation, Halley reviewed the records of cometary appearances, and noted one outstanding regularity. Particularly bright comets seen in 1531, 1607, and 1682 seemed to have similar properties, and Hal-

FIG. 15.4 CALAMITY ON EARTH ASSOCIATED WITH THE PASSAGE OF COMETS. This drawing is from a seventeenth century book describing the universe.

ley suggested that all three were appearances of the same comet, orbiting the sun with a 76-year period.

Calculations showed, through the application of Kepler's third law, that for a period of 76 years, this object must have a semimajor axis of nearly 18 AU. Halley realized, therefore, that in order to appear as dominant in our skies as the comet does, it must have a highly flattened orbit, so that it comes close to the sun at times, even though its average distance is well beyond the orbit of Saturn, nearly as far out as Uranus (Fig. 15.5). Such an eccentric orbit, as a very elongated ellipse is called, had not previously been observed, even though Newton's laws clearly allowed the possibility.

Since Halley's time, searches of ancient reports of comets have revealed that Halley's comet has been

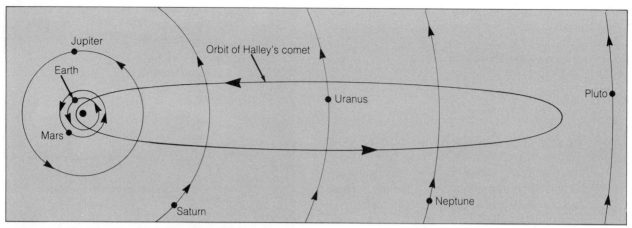

FIG. 15.5.   A COMETARY ORBIT. This is a rough scale drawing of the orbit of Halley's comet. Most comets actually have much more highly elongated orbits than this, and correspondingly longer periods.

making regular appearances for many centuries. The earliest records are from the ancient Chinese astronomers, who apparently observed its every appearance for more than 1,300 years, beginning in 467 B.C.

The most recent visit of Halley's comet was in 1910 (Fig. 15.6), when the earth actually passed through its tail (without any noticeable effects), and it will next be seen in 1986. Unfortunately for us, on its upcoming passage near the sun, Halley's comet will approach from the far side as we view it, and will not be in a favorable position for easy sighting from earth.

Other spectacular comets have been seen, and a number have rivaled Halley's comet in brightness (Comet Kohoutek 1973, for example; see Fig. 15.7 and

Color Plate 13). Traditionally, a comet is named after its discoverer, and there are astronomers around the world who spend long hours peering at the nighttime sky through telescopes, looking for a piece of immortality.

When a new comet is discovered, a few observations of its position are sufficient to allow computation of its orbit. The results of many years of comet-watching have shown that there are numerous comets whose orbits are so incredibly stretched out that their periods are measured in the thousands or even the millions of years. These comets, for all practical purposes, are only seen once. They return thereafter to the void of space well beyond the orbit of Pluto, there to spend millennia before visiting the inner solar system again.

FIG. 15.7.   COMET KOHOUTEK 1973. This was one of the brightest comets in recent years.

FIG. 15.6.   HALLEY'S COMET AS IT APPEARED IN 1910.

The orbits of comets, particularly these so-called long-period ones, are randomly oriented. Comets do not show any preference for orbits lying in the plane of the ecliptic, in strong contrast with the planets, and about half go around the sun in the retrograde direction.

Consideration of these orbital characteristics, especially the large orbital sizes, led the Dutch astronomer Jan Oort to suggest that all comets originate in a cloud of objects that surrounds the solar system. The **Oort cloud,** as it is now called, is envisioned to be a spherical shell with a radius of 50,000 to 150,000 AU, extending a significant fraction of the distance to the nearest star, which is almost 300,000 AU from the sun.

Occasionally a piece of debris from the Oort cloud is disturbed from its normal path, either by a collision with another object or by the gravitational tug of a nearby star, and it begins to fall inward toward the sun. Left to its own devices, a comet falling in from the Oort cloud would follow a highly elongated orbit with a period of millions of years, appearing to us as one of the long-period comets when it made its brief incandescent passage near the sun. In many cases, a comet is not left to its own devices, however. Instead it runs afoul of the gravitational pull of one of the giant planets, most often Jupiter. When this happens, the comet may be speeded up, so that it escapes the solar system entirely after it loops around the sun; or it may be slowed down, dropping into a smaller orbit with a shorter period, so that it becomes one of the numerous comets that are seen to reappear frequently.

By invoking the existence of the Oort cloud, we can account for all the observed cometary orbits, and this concept is widely accepted today. We must consider how the Oort cloud itself came into being, however. The best guess is that it consists of material left over from the formation of the solar system; if this is true, the comets are no doubt made of very old matter, representing the primordial stuff from which the planets and the sun were created. The formation of the Oort cloud will be discussed at greater length in chapter 17, where the formation and evolution of the solar system as a whole are described.

The Anatomy of a Comet

For most of its life, a comet is just a frozen chunk of icy material, probably consisting of small particles like gravel or larger rocks that are embedded in frozen gases.

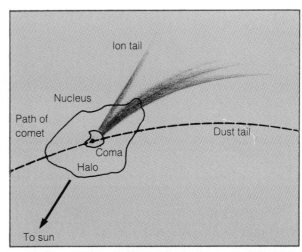

FIG. 15.8. ANATOMY OF A COMET. This sketch illustrates the principal features of a comet (not drawn to scale).

As it passes through the outer reaches of its orbit, far from the sun, it does not glow, has no tail, and is not visible from earth.

As a comet approaches the sun, however, it begins to warm up as it absorbs sunlight, and the added heat causes volatile gases to escape. A spherical cloud of glowing gas called the **coma** develops around the solid **nucleus** (Fig. 15.8). The gases that have been identified in the comae of comets by spectroscopic measurements are simple molecules such as H_2O, C_2, C_3, CH, CN, CO, and N_2. Other slightly more complex species, such as CO_2, ammonia (NH_3), and methane (CH_4), are probably present also, along with hydrogen molecules (H_2). The observed species glow by a process called **fluorescence.** The molecules absorb ultraviolet light from the sun, which causes them to be excited to high energy levels, and then they emit visible light as they return to low energy states. A cloud of hydrogen atoms, resulting from the breakup of the molecules in the coma, extends out to great distances from the nucleus. The visible coma may be as large as 100,000 kilometers in diameter, and the halo of hydrogen atoms may extend as much as ten times farther from the nucleus (Fig. 15.9). The solid nucleus is relatively tiny, having a diameter of perhaps a few kilometers.

Sometimes the release of gases from a cometary nucleus is so forceful that it actually alters the course of the comet. If gas is expelled from a specific location on the nucleus, such as the leading side, which is more directly exposed to heating from the sun, it can act like

FIG. 15.9. THE HALO OF COMET KOHOUTEK. This is an ultraviolet image, made by *Skylab* astronauts, showing the extent of the hydrogen cloud surrounding this famous 1973 comet. The halo was about a million km in diameter when this photo was taken.

a rocket exhaust. Erratic motions that apparently violate Newton's laws have been observed in several cases. This occurs most notably with Halley's comet, whose arrival in the inner solar system is often delayed by this effect.

As a comet nears the sun, the solar radiation and the solar wind force some of the gas from the coma to flow away from the sun, forming the tail, which in some instances is as long as 1 AU. There often are two distinct tails (Fig. 15.10): one formed of gas from the coma, usually containing molecules that have been ionized, such as CO^+, N_2^+, CO_2^+, and CH^+; and another formed of tiny solid particles released from the ice of the nucleus. The **ion tail,** the one formed of ionized gases, is shaped by the solar wind, and therefore points almost exactly straight away from the sun at all times. The other tail, called the **dust tail,** usually takes on a curving shape, as the dust particles are pushed away from the sun by the force of the light they absorb. This **radiation pressure** is not strong enough to force the dust particles into perfectly straight paths away from the sun, so they follow curved trajectories which are a combination of their orbital motion and the outward push caused by sunlight.

The gases that escape from the nucleus of a comet

FIG. 15.10. TWO TAILS. This photo of Comet West illustrates the distinction between the ion tail, the straight tail at right; and the dust tail, the broad, somewhat curved tail that points more nearly upward in this figure.

as it approaches the sun are highly volatile, and would not be present in the nucleus if it had ever undergone any significant heating. This tells us that comets must have formed and lived their entire lives in a very cold environment, probably never even getting as warm as 100 K before falling into orbits that bring them close to the sun. If a comet is so easily vaporized, then once it has begun to follow a path that regularly brings it close to the sun, its days are numbered. It may make many round trips, but eventually it will dissipate all of its volatile gases, leaving behind nothing but rocky debris. Several cases have been noted where a comet failed to reappear on schedule, but was replaced by a few pieces or perhaps a swarm of fragments (Fig. 15.11).

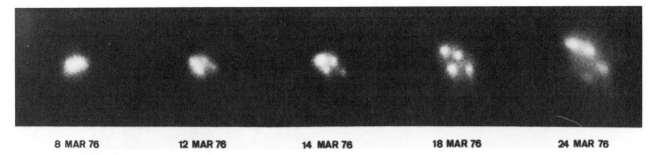

| 8 MAR 76 | 12 MAR 76 | 14 MAR 76 | 18 MAR 76 | 24 MAR 76 |

FIG. 15.11. THE BREAK-UP OF COMET WEST. This dramatic sequence shows the nucleus of Comet West fragmenting into four pieces.

In time, the remains of a dead comet will be dispersed all along the orbital path, so that each time the earth passes through this region, it encounters a vast number of tiny bits of gravel and dust, and we experience a meteor shower.

Meteors and Meteorites

Occasionally one of the countless pieces of debris floating through the solar system enters the earth's atmosphere, creating a momentary light display as it evaporates in a flash of heat created by the friction of its passage through the air. The streak that is seen in the sky is called a **meteor** (Fig. 15.12). Most of us are familiar with this phenomenon, commonly called a shooting star, since it is often possible to see one in just a few minutes of sky-gazing on a clear night. On

rare occasions an especially brilliant meteor is seen, possibly persisting for several seconds, and these spectacular events are called **fireballs** or **bolides.**

The piece of solid material that causes a meteor is called a **meteoroid.** Most are very small, amounting to nothing more than tiny grains of dust or perhaps fine gravel. A few, however, are larger solid chunks, which are responsible for the bright fireballs.

Occasionally one of the larger meteoroids survives the arduous trip through the atmosphere and reaches the ground intact. Such an object is called a **meteorite** (Fig. 15.13), and examples can be found in museums around the world. Meteorites have been the subject of intense scrutiny, for until the last fifteen years or so, they were the only samples of extraterrestrial material scientists could get their hands on.

FIG. 15.12. TWO BRIGHT METEORS. The streaks of light in this photo are created by tiny particles entering the earth's atmosphere from space.

FIG. 15.13. A METEORITE. This is a stony meteorite, the black coloring being caused by heating as the object passed through the atmosphere. The light-colored spots are breaks in the fusion crust where interior material is exposed.

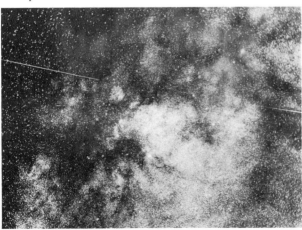

Throughout history, scientists scoffed at the notion that rocks could descend from the sky, until a meteorite was seen to fall near a French village in 1803, and was found and examined just after it dropped. Such falls, although rare, are occasionally observed, and even have been known to cause damage (but so far, few injuries).

PRIMORDIAL LEFTOVERS

Meteorites are old, much older than most surface rocks on the earth. This fact enhances the interest of scientists, for in studying these objects they get a glimpse of the early history of the solar system.

Meteorites generally can be grouped into three classes: the stony meteorites (Fig. 15.13), which represent about 93 percent of all meteorite falls; the iron meteorites, accounting for about 6 percent; and the stony-iron meteorites, which are the rarest. These relative abundances were determined indirectly, because the different types of meteorites are not equally easy to find on the ground. Most meteorites found are the iron ones, which, as just mentioned, represent only a small fraction of those which fall. The stony meteorites look so much like ordinary rocks that they are usually difficult to pick out, and some are burned up in their journey through the atmosphere. A particularly good place to search for meteorites is Antarctica, where a thick layer of ice conceals the native rock. Meteorites that fall there are relatively easy to find, and there is little chance for confusion with earth rocks.

The stony meteorites are mostly of a type called **chondrites,** so named because they contain small spherical inclusions called **chondrules** (Fig. 15.14). These are mineral deposits formed by rapid cooling, which most likely occurred at an early time in the history of the solar system, when the first solid material was condensing. A few of the stony meteorites are **carbonaceous chondrites** (Fig. 15.15), thought to be almost completely unprocessed since the solar system was formed, and therefore representative of the original stuff of which the planets were made. The primordial nature of the carbonaceous chondrites, like the carbonaceous asteroids mentioned in an earlier section, is deduced from their high volatile content, which indicates that they were never exposed to much heat. One particularly fascinating aspect of these meteorites is that in at least one case, complex organic molecules

FIG. 15.14. CROSS-SECTION OF A CHONDRITE. This shows the many chondrules (light patches) embedded within the structure of this type of stony meteorite.

called **amino acids** have been found inside a carbonaceous-chondrite meteorite, showing that some of the ingredients for the development of life were apparently available even before the earth formed.

The iron meteorites have varying nickel contents, and sometimes show an internal crystalline structure

Text continues on page 240.

FIG. 15.15. A CARBONACEOUS CHONDRITE. This example, not of the type in which amino acids have been found, shows a large chondrule (the light-colored spot, upper center), which is about 5 mm in diameter.

The Mysterious Tektites

In certain regions of the earth, small, glassy objects called **tektites** are found in large numbers. These are generally black in color, and have streamlined shapes as though they have been formed aerodynamically by wind. Their characteristics indicate that they have been heated to the melting point, hardening as they flew through the air. Their distribution over the surface of the earth is suggestive of an origin in giant showers of debris that fell to the ground.

Theories for terrestrial origins of the tektites have included ejection by impacts of large meteorites or creation in volcanic eruptions, when lava was spewed into the air, hardening into solid, glassy objects on the way down. Other theories suggest a lunar origin, by ejection in massive impacts or volcanic eruptions, in either case sending material upward so rapidly that some of it escaped the moon and fell to the earth, the fragments taking on the aerodynamic shapes characteristic of tektites as they passed through the earth's atmosphere.

The composition of tektites, an important clue in unraveling their origin, is quite unlike that of any volcanic rocks on earth, particularly with respect to their low content of volatile gases such as water vapor. Because of this, it is certain that they are not simply cinders from volcanic eruptions. Normal eruptions lack the energy required to eject objects above most of the earth's atmosphere, anyway. The composition of tektites may be consistent with the other suggested terrestrial origin, formation in giant meteorite impacts, but only if the impacts occurred in regions where the surface rocks are sedimentary, rather than volcanic. Thus if the tektites originated on the earth, it most likely was a result of bombardment from space.

There is another more likely explanation of the tektites; namely, that the tektites originated in lunar volcanoes, and were ejected from the moon with sufficient velocity to escape (recall that the moon's escape velocity is only 2.4 km/sec, less than one-fourth that of the earth). The composition of the tektites, although it fails to match that of lunar surface rocks, is very similar to the suspected composition of material under the moon's surface. A very interesting aspect of this theory is that it implies rather recent, major volcanic activity on the moon, for the tektites found in one area of the earth have been dated (by radioactive-isotope ratios) as being only 750,000 years old. If an eruption occurred that recently on the moon with sufficient power to eject material into space, there ought to be geological evidence of it on the moon. Perhaps the search for such evidence will be the objective of some future lunar exploration program.

This is an assortment of tektites, showing their aerodynamic shapes and characteristic black color.

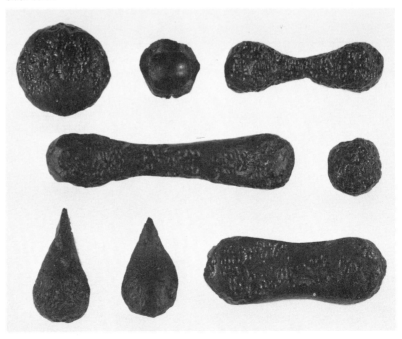

When It Rains Meteors

SHOWER	APPROXIMATE DATE	ASSOCIATED COMET
Quadrantids	January 3	—
Lyrids	April 21	Comet 1861 I
Eta Aquarids	May 4	Halley's Comet
Delta Aquarids	July 30	—
Perseids	August 11	Comet 1862 III
Draconids	October 9	Comet Giacobini-Zinner
Orionids	October 20	Halley's Comet
Taurids	October 31	Comet Encke
Andromedids	November 14	Comet Biela
Leonids	November 16	Comet 1866 I
Geminids	December 13	—

A meteor shower is an unforgettable experience, well worth the loss of sleep required to get the best view. There are a number of particularly dense showers that occur yearly as the earth passes through the paths of debris that create them. The best time of the night to see a meteor shower is after midnight, when you are on the side of the earth that faces forward as it moves along in its orbit. As the earth plows through the swarm of meteoroids, most of the entries into the earth's atmosphere occur on this leading side.

During a shower, the meteors all seem to approach from a single point, called the *radiant*. This is a simple effect of perspective: the meteoroids are traveling along in parallel paths, but they seem to diverge from a common point as we look in the direction from which they come.

Not all the meteor showers are associated with dead comets. Apparently even while a comet still lives, making regular appearances, a cloud of debris may be scattered along its orbit, and we are treated to the spectacle of a meteor shower whenever the earth passes through the orbit of one of these still-living comets. Halley's comet is an example: the earth passes through its orbit twice each year, and each of these occasions is marked by a meteor shower.

If the particles following a com-

FIG. 15.16.   CROSS-SECTION OF A NICKEL-IRON METEORITE. This example shows the characteristic crystalline structure indicative of a slow cooling process from a previous molten state. Such meteorites are thought to have been parts of larger bodies that differentiated.

(Fig. 15.16) that indicates a rather slow cooling process in their early histories. This has important implications for their origin, as we will see.

DEAD COMETS AND FRACTURED ASTEROIDS

The origins of the meteoroids that enter the earth's atmosphere can be inferred from what we know of the properties of meteorites, asteroids, and comets.

As mentioned earlier, most meteors are caused by

etary orbit are not scattered uniformly along it, but instead are concentrated in one region, we will not always see equally active meteor showers when the earth passes through the orbit. Instead the density of the shower will vary, depending on how close we pass to the part of the orbit having the greatest concentration of particles. The Leonid shower, for example, is especially brilliant and intense every thirty-three years, when the earth passes through the most densely populated part of the cometary leftovers that create this shower. In some cases, on the other hand, the particles seem to be uniformly distributed along the orbit, so we see about the same intensity of meteors each time.

The table that follows lists some of the more prominent meteor showers seen each year.

A meteor shower. Some 78 meteor trails appear in this exposure.

relatively tiny particles that do not survive their flaming entry into the earth's atmosphere. During **meteor showers,** when meteors can be seen as frequently as once per second, all seem to be of this type. As noted earlier, these showers are associated with the remains of comets that have disintegrated and left behind a scattering of gravel and dust. Therefore it is thought that the most common meteors, those created by small, fragile meteoroids, are a result of cometary debris.

The larger chunks that reach the ground as meteorites may have a different origin. It is likely that there are occasional collisions among the asteroids, sometimes sufficiently violent to destroy them, dispersing throughout the solar system the rubble that is left over. Most meteorites are probably fragments of asteroids. The iron meteorites apparently originated in asteroids that had undergone differentiation, whereas the stony ones came either from the outer portions of differen-

tiated asteroids or from smaller bodies that never underwent differentiation at all. The chondrites probably fall into the latter category, since the chondrules reflect a rapid cooling that would have characterized very small bodies. This is why the chondrites are thought to be the most primitive of the meteorites, having undergone no processing in the interiors of large bodies.

From our studies of the other planets and satellites, we know that there was a time long ago when frequent impacts occurred, forming most of the craters seen today. Certainly the earth was not immune, and no doubt it was also subjected to heavy bombardment. The difference, of course, is that the earth has an atmosphere, along with flowing water and glaciation, all of which combine to erase old craters in time. A few traces are still seen, however; there are several ancient craters still in existence (Fig. 15.17), including a very large basin under the Antarctic ice that is probably an ancient im-

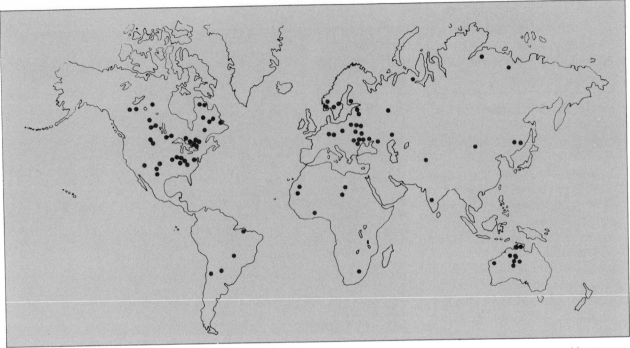

FIG. 15.17. IMPACT CRATERS ON THE EARTH. This map shows the locations of major craters thought to have been created by impacts of massive objects.

pact crater, and a portion of Hudson's Bay in Canada that shows a circular shape thought to have a similar origin.

Although the frequency of impacts has decreased, there are still rare occasions when major impacts occur. The Barringer crater, near Winslow, Arizona (Fig. 15.18), was formed only about 25,000 years ago, for example, and the possibility exists that other large bodies could hit the earth. Given a long enough time, it is almost inevitable.

Microscopic Particles: Interplanetary Dust and the Interstellar Wind

The empty space between the planets plays host to some very tiny particles, in addition to the larger ones we have just described. There is a general population of small solid particles, perhaps a millionth of a meter in diameter, called interplanetary dust grains. There is also a very tenuous stream of gas particles flowing through the solar system from interstellar space.

The presence of the dust has been known for some

time from two celestial phenomena, both of which can be observed with the unaided eye, although only with difficulty. The dust particles scatter sunlight, so that under the proper conditions a diffuse glow can be seen

FIG. 15.18. METEOR CRATER NEAR WINSLOW, ARIZONA. The impact that created this crater occurred about 25,000 years ago.

where the light from the sun hits the dust. This is analogous to seeing the beam of a searchlight stretching skyward; you only see the beam where there are small particles (either dust or water vapor) that scatter its light, so that some of it reaches your eye.

One of the phenomena created by the interplanetary dust is the **zodiacal light** (Fig. 15.19), a faintly illuminated belt of hazy light that can be seen stretching across the sky (along the ecliptic) on clear, dark nights, just after sunset or before sunrise. The second observable phenomenon created by the dust is a small bright spot seen on the ecliptic in the opposite direction from the sun. This diffuse spot, called the **gegenschein** (Fig. 15.20), is created by sunlight that is reflected straight back by the interplanetary dust, which is concentrated in the plane of the ecliptic. This is analogous to seeing a bright spot on a cloud bank or a low-lying mist when you look at it with the sun directly behind you; the bright spot is just the reflected image of the sun, and is the counterpart of the gegenschein.

FIG. 15.20. THE GEGENSCHEIN. The photo here shows the Milky Way (stretching across the upper portion) and a diffuse concentration of light (at lower center) which is the gegenschein. It is created by light reflected directly back to earth from interplanetary dust in the direction opposite from the sun.

FIG. 15.19. THE ZODIACAL LIGHT. This photo shows the diffuse band of light in the plane of the ecliptic that is caused by the scattering of sunlight from tiny interplanetary dust grains. The zodiacal light is most easily visible about an hour before sunrise or after sunset.

It has been possible to collect interplanetary dust particles for direct examination (Fig. 15.21). This is done most commonly by the use of high-altitude balloons, but recently a surprising new technique has been developed, which involves scooping sludge (which contains such dust particles) off of the ocean floor. The

FIG. 15.21. AN INTERPLANETARY DUST GRAIN. This is a microscopic view of a tiny particle from interplanetary space. The amorphous structure is highly variable from one grain to another.

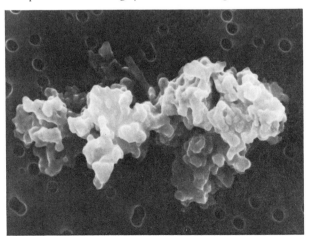

earth is constantly being pelted with dust particles (which add about eight tons per day to its mass!), and those which fall into the oceans can lie undisturbed on the seabed for long times. Studies of the grains show that they are probably of cometary origin, having been dispersed throughout space from the dead nuclei of old comets.

As we will learn in chapter 25, the space between stars in our galaxy is permeated by a rarefied gas medium. In the sun's vicinity, the average density of this gas is far below that of any man-made vacuum; it amounts to only about 0.1 particle/cm^3 (that is, there is one atom, on average, in every volume of 10 cm^3, cor-

responding to a cube about one inch on each side). Because of the motion of the sun in its orbit about the galaxy, the interstellar gas streams through the solar system with a velocity of about 20 kilometers per second. The presence of this ghostly breeze, consisting mostly of hydrogen and helium atoms and ions, was discovered in the early 1970s, when observations made from satellites revealed very faint ultraviolet emission from the hydrogen atoms in the gas. The interstellar wind, tenuous as it is, has very little effect on the other components of the solar system, but is nevertheless studied with some interest for what it may tell us about the interstellar medium.

Perspective

The interplanetary wanderers discussed in this chapter have given us insight into the history of the solar system, and have told us much about its present state as well. We have found two primary origins of the various objects: comets and asteroids. The former account for most of the meteors and for the interplanetary dust,

and the latter are responsible for the meteorites, including the massive bodies that formed the major impact craters in the solar system.

The pieces are nearly all in place. Next we turn our attention inward, to examine the engine that keeps all the machinery in operation.

Summary

1. The asteroids were accidentally and coinci-dentally discovered near the orbital distance predicted by Bode's law.

2. Thousands of asteroids have been catalogued, displaying a variety of sizes (up to 1,000 kilometers in diameter) and compositions (ranging from metals to rocky minerals).

3. Gaps in the asteroid belt are created by orbital resonances with Jupiter, whose gravitational influence was probably also responsible for preventing the asteroids from coalescing into a planet in the first place.

4. Comets are small, icy objects that develop their characteristic comae and tails only when in the inner part of the solar system.

5. Comets apparently originate in a cloud of debris very far from the sun. They occasionally fall inward, either to bypass the sun and return to the distant reaches of the solar system, to escape into interstellar space, or to be perturbed by the gravitational influence of one of the planets, becoming periodic comets.

6. When near the sun, a comet ejects gases that glow by fluorescence. Periodic comets eventually lose all of their icy substance in this process, and disintegrate into swarms of rocky debris.

7. A comet may have two tails, one created by ionized gas and the other made of fine dust particles.

8. A meteor is a flash of light created by a meteoroid entering the earth's atmosphere from space, and a meteorite is the solid remnant that reaches the ground in some cases.

9. Meteorites are either stony, stony-iron, or iron in composition, and are very old, providing information on the early solar system.

10. Most meteors are created by fine debris from comets, but most meteorites are fragments of asteroids.

11. Interplanetary space is permeated by fine dust particles and by an interstellar wind of hydrogen and helium atoms from the space between the stars.

Review Questions

1. What is the orbital period of an asteroid whose semimajor axis is 2.8 AU?

2. At what distance from the sun would an asteroid have exactly one-tenth the orbital period of Jupiter? Is there a gap in the asteroid belt at this distance?

3. How is the formation of the asteroid belt like the formation of the rings of Saturn?

4. How do the motions of comets differ from those of planets?

5. Summarize the effects of Jupiter on asteroids, comets, and meteoroids.

6. Summarize the life story of a typical comet.

7. Why are carbonaceous chondrites important clues to the early history of the solar system?

8. Would an observer on the surface of Mercury see meteors in the nighttime sky? Would he find meteorites on the surface of Mercury?

9. What does the concentration of the zodiacal light in the ecliptic tell us about the distribution of interplanetary dust in the solar system?

10. What are the orbital periods of the Trojan asteroids? How does the period of an Apollo asteroid compare with that of the earth?

Additional Readings

Chapman, C. R. 1975. The nature of Asteroids. *Scientific American* 232(1):24.

Hartmann, W. K. 1975. The smaller bodies of the solar system. *Scientific American* 233(3):142.

Morrison, D. 1976. Asteroids. *Astronomy,* June 1976, p. 6.

O'Keefe, J. A. 1978. The tektite problem. *Scientific American* 239(2):98.

Van Allen, J. A. 1975. Interplanetary particles and fields. *Scientific American* 233(3):160.

Whipple, F. L. 1974. The nature of comets. *Scientific American* 230(2):49.

Our star, the sun, is rather ordinary by galactic standards. It has modest mass and size, and there are stars as much as a few hundred times larger and a million times more luminous. Its temperature is also moderate, as stars go. In many respects the sun is entirely a run-of-the-mill entity.

Within our solar system, on the other hand, the sun is clearly the king. In mass it outranks even Jupiter by a factor of more than one thousand. It is the only body that glows under its own power (except for the excess infrared and radio emission from Jupiter and Saturn), and its light provides all the illumination by which we see the planets. The sun gives us the warmth necessary for biological activity on earth, and provides all the forms of energy available to us (except nuclear).

In this chapter we will consider the basic properties of our star, placing particular emphasis on its interaction with the planets.

Basic Properties and Internal Structure

The sun is a ball of hot gas (Fig. 16.1). Its density on average is 1.41 grams/cm^3, not much more than that of water, but its center is so highly compressed that the density there is about ten times greater than that of lead. The interior is gaseous, rather than solid, because

TABLE 16.1   THE SUN
Diameter: 1,319,980 km (109.3 $D_\oplus$)
Mass: 1.99 × 10^{33} grams (332,943 $M_\oplus$)
Density: 1.409 grams/cm^3
Surface gravity: 27.9 earth gravities
Escape velocity: 618 km/sec
Luminosity: 3.83 × 10^{33} ergs/sec
Surface temperature: 6,500 K (deepest visible layer)
Rotation period: 25.04 days (at equator)

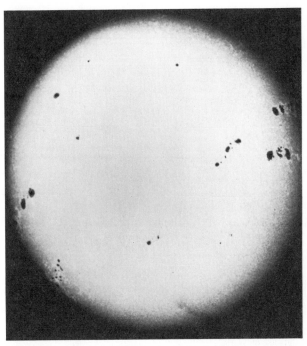

FIG. 16.1.   THE SUN. This is a white-light photo, showing a few sunspots.

the temperature is very high, around 10 million degrees at the center, diminishing to just under 6,000 K at the surface. At these temperatures, the gas is partially ionized in the outer layers of the sun, and completely ionized in the core, all electrons having been stripped free of their parent atoms.

The sun is held together by gravity. All of its constituent atoms and ions attract each other, and the net effect is for the solar substance to be held in a spherical shape. Gas that is hot exerts pressure on its surroundings, and this pressure, pushing outward, balances the force of gravity, which is pulling the matter inward. This balance is called **hydrostatic equilibrium,** with gravity and pressure equaling each other everywhere. The deeper the layer, the greater the weight of the overlying layers, and the more the gas is compressed. The higher the pressure, the greater the temperature required to maintain the pressure, so we find that the pressure and temperature both increase as we approach the center of the sun.

The composition of the sun is the same as that of most other stars: about 70 percent of its mass is hydrogen, 27 percent is helium, and the rest is made up of traces of other elements. In the outer layers, where no nuclear reactions have taken place, a greater fraction of the mass is in the form of hydrogen (about 79 percent). Thus the sun's chemical makeup is similar to that of the outer planets, and would resemble the terrestrial planets as well, except that these bodies have lost most of their volatile elements such as hydrogen and helium. It is apparent that all the components of the solar system formed from the same material.

The ultimate source of all the sun's energy is in its core (Fig. 16.2), within the innermost 10 percent or so of its radius. Here nuclear reactions create heat and photons of light at γ-ray wavelengths. It is this light that eventually reaches the surface and escapes into space, but it is a laborious journey. Each photon is absorbed and reemitted many times along the way, gradually losing energy (Fig. 16.3). In the process the photon becomes a visible-light photon, and the energy it loses

FIG. 16.2.   THE INTERNAL STRUCTURE OF THE SUN. This shows the relative extent of the major zones within the sun, except that the depth of the photosphere is greatly exaggerated.

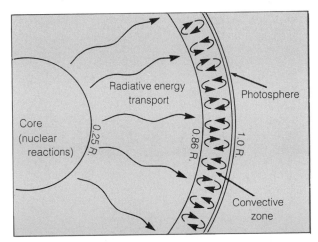

TABLE 16.2	THE COMPOSITION OF THE SUN'S PHOTOSPHERE		
Hydrogen	H	1.0000	0.792
Helium	He	0.0631	0.198
Lithium	Li	1.55×10^{-9}	8.45×10^{-9}
Beryllium	Be	1.41×10^{-11}	9.98×10^{-11}
Boron	B	2.00×10^{-10}	1.70×10^{-9}
Carbon	C	0.000372	0.00351
Nitrogen	N	0.000115	0.00127
Oxygen	O	0.000676	0.00850
Fluorine	F	3.63×10^{-8}	5.42×10^{-7}
Neon	Ne	3.72×10^{-5}	0.000590
Sodium	Na	1.74×10^{-6}	3.10×10^{-5}
Magnesium	Mg	3.47×10^{-5}	0.000662
Aluminum	Al	2.51×10^{-6}	5.30×10^{-5}
Silicon	Si	3.55×10^{-5}	0.000783
Phosphorus	P	3.16×10^{-7}	7.69×10^{-6}
Sulfur	S	1.62×10^{-5}	0.000408
Chlorine	Cl	2.00×10^{-7}	5.57×10^{-6}
Argon	Ar	4.47×10^{-6}	0.000140
Potassium	K	1.12×10^{-7}	3.44×10^{-6}
Calcium	Ca	2.14×10^{-6}	6.70×10^{-5}
Scandium	Sc	1.17×10^{-9}	4.13×10^{-8}
Titanium	Ti	5.50×10^{-8}	2.07×10^{-6}
Vanadium	V	1.26×10^{-8}	5.00×10^{-7}
Chromium	Cr	5.01×10^{-7}	2.00×10^{-5}
Manganese	Mn	2.63×10^{-7}	1.10×10^{-5}
Iron	Fe	2.51×10^{-5}	0.00175
Cobalt	Co	3.16×10^{-8}	1.46×10^{-6}
Nickel	Ni	1.91×10^{-6}	8.80×10^{-5}
Copper	Cu	2.82×10^{-8}	1.41×10^{-6}
Zinc	Zn	2.63×10^{-8}	1.35×10^{-6}
(All others combined)			(Less than 10^{-8} of total)

*The numbers given are relative to the number of hydrogen atoms.

FIG. 16.3. RANDOM WALK. A photon is continually being absorbed and reemitted as it travels through the sun's interior, and each time it is reemitted, it is in a random direction. Thus its progress from the core, where it is created, to the surface is very slow.

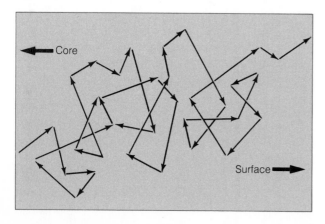

heats the surroundings. Because a photon only travels a short distance before it is absorbed, and because when it is reemitted it is in a random direction, progress toward the surface is very slow. It takes an individual photon as long as a million years to migrate from the center of the sun to the surface, even though the light travel time if it were unimpeded would be only two seconds. If the sun's energy source were suddenly turned off, we would not be aware of it for a million years!

Throughout most of the solar interior, the gas is quiescent, without any major large-scale flows or currents. The energy from the core is transported by the radiation wending its slow pace outward, except in the layers near the surface, where convection occurs and heat is transported by the overturning motions of the gas. As we will see, the bubbling, boiling action in the outer portions of the sun creates a wide variety of dynamic phenomena on the surface.

The sun rotates, and it does so differentially. This is apparent from observations of its surface features, which reveal that, like Jupiter and Saturn, the sun goes around faster at the equator than near the poles. The rotation period is 25 days at the equator, 28 days at middle latitudes, and even longer near the poles. The differential rotation probably plays an important role in governing variations in the solar magnetic field, which in turn have a lot to do with the behavior of the most prominent surface features, the sunspots.

There is evidence that the core of the sun rotates much more rapidly than the surface. The only clue is the presence of very subtle oscillations on the solar surface, which may be wave motions created by the rapid spin of the interior. The exact rate of internal rotation is not known, but the core probably spins with a period of only a few days, in contrast to the 25-day surface rotation period. The rapid spin of the solar core is probably a direct result of the collapse and accelerated rotation of the interstellar cloud from which the sun formed. (The origin of the sun's rotation is discussed in chapter 17.)

NUCLEAR REACTIONS

One of the biggest mysteries in astronomy in the early decades of this century involved the sun. The problem was how to account for the tremendous amount of energy it radiates, particularly perplexing since geological evidence shows that the sun has been able to produce this energy for at least four or five billion years. Two early ideas, that the sun is simply still hot from its formation, or that it is gradually contracting, releasing stored gravitational energy, were both ruled out, since neither situation could possibly supply the energy needed to run the sun for a long enough time.

The first hint at the solution came in the first decade of the 1900's, when Albert Einstein developed his theory of relativity. He showed that matter and energy are equivalent, and that one can be converted into the other by the famous formula $E = mc^2$, where E is the energy released in the conversion, m is the mass that is converted, and c is the speed of light. This mechanism can produce enormous amounts of energy, and physicists began to contemplate the possibility that somehow this energy was being released inside the sun and stars.

In the 1920's, following the pioneering work on atomic structure by Max Planck, Niels Bohr, and others, the concept of nuclear reactions began to emerge. Like chemical reactions, nuclear reactions are transformations, except that it is the nuclei of atoms, rather than the electrons in the outer orbits, that react with each other. There are **fusion** reactions, in which nuclei merge to create a larger nucleus, representing a new chemical element; and **fission** reactions, in which a single nucleus, usually of a heavy element with a large number of protons and neutrons, splits into two or more smaller nuclei. In either type of reaction, energy is released as some of the matter is converted according to Einstein's formula. In the 1920's, Enricho Fermi, Werner Heisenberg, and Wolfgang Pauli explored these possibilities, and by the 1930's, Hans Bethe had suggested a specific reaction sequence that might be operative in the sun's core.

Bethe envisioned a fusion reaction in which four hydrogen nuclei (each consisting of only a single proton) combine to form a helium nucleus, made up of two protons and two neutrons. The reaction occurs in several steps: (1) two protons combine to form **deuterium,** a type of hydrogen that has a proton and a neutron in its nucleus (one of the two protons undergoing the reaction converts itself into a neutron, by emitting a positively charged particle called a **positron**) and another particle called a **neutrino,** which has very unusual properties; (2) the deuterium combines with another proton to create an isotope of helium (^{3}He) consisting of two protons and one neutron; and (3) two of these ^{3}He nuclei combine, forming an ordinary helium nucleus (^{4}He, with two protons and two neutrons in the nucleus) and releasing two protons. At each step in this sequence, heat energy is imparted to the surroundings in the form of kinetic energy from the particles that are produced, and in step 2 a photon of γ-ray light is emitted as well.

The net result of this reaction, which is called the **proton-proton chain,** is that four hydrogen nuclei (protons) combine to create one helium nucleus. The end product has slightly less mass than the ingredients, 0.007 of the original amount having been converted into energy. Simple calculations (see chapter 20) show that this reaction can easily supply enough energy to keep the sun running at its present rate for many billions of years. The proton-proton chain is further discussed in Appendix 10, and outlined in Table 16.3.

Text continues on page 251.

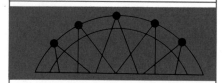

Imitating the Sun

Nuclear reactions can produce energy very efficiently, and therefore they have obvious potential benefits for human society, if methods are found for producing and controlling nuclear energy on earth. To this end, a great deal of effort has gone into developing nuclear power, but the only working reactors today are fission reactors. These have several disadvantages, particularly in that they produce radioactive waste products that are very difficult to store safely.

Fusion reactions, however, can also produce immense quantities of energy, without danger of runaway reactions that could catastrophically destroy the reactor. The sun, after all, runs on power produced by the fusion of hydrogen into helium, both of which are harmless gases. Thus if we could somehow produce and control fusion reactions, we might have a permanent solution to the problem of producing sufficient energy to run our society. Unfortunately, the only fusion reactions we have created to date have been the instantaneous ones that occurred in tests of hydrogen bombs.

The difficulty in controlling nuclear fusion reactions is obvious, in view of the conditions required for them to occur inside the sun. The problem is to somehow reproduce, in a controlled environment, the incredible temperatures and pressures of the sun's core. Two techniques for doing this are being developed.

The older of these ideas, dating back to the 1950's, is to contain the superheated gas in a sort of magnetic bottle. Recall that at sufficiently high temperatures, a gas is ionized, and consists entirely of charged particles. Such particles are subject to electromagnetic forces, and can be trapped within a fixed region by a properly shaped magnetic field. A great deal of theoretical work and considerable experimentation has gone into efforts to design a magnetic bottle capable of containing gas under the extreme conditions needed for fusion to occur. The most successful design thus far is a **torus** (a doughnut-shaped tube), which is twisted into a figure eight to help keep the particles away from the tube walls. The ionized gas circulates within the enclosed tube, and is kept away from the walls by immense magnetic fields created by electromagnets that surround the tube. This magnetic-confinement technique is being developed primarily at Princeton University. Tests conducted so far have succeeded, for very brief instants, in creating fusion reactions, but there is still a lengthy development program ahead, for it will

be some time before reactions can be sustained and can produce more energy than is required to generate the magnetic fields.

The second technique for controlling fusion is being studied primarily at the Lawrence Livermore Laboratories of the University of California. There a device called a **laser** is used. The laser, which was invented in the 1960's, produces a very narrow beam of light, with all the photons at precisely the same wavelength. The power in this beam of light can be immense; lasers have been developed for cutting metal and for performing surgery, for example. The use of lasers to produce fusion reactions involves subjecting small pellets of matter to intense laser beams which instantaneously vaporize the pellets, producing, for a brief instant, the conditions required for fusion to occur. When this technique has been developed to the point where more energy is produced than is required to power the laser, then it can become a useful means of producing energy.

The U.S. government has been supporting fusion research for some time, and, we can hope, will continue to do so. It will take several more years before all the problems and complications can be worked out and fusion becomes a viable source of energy. When that day does come, however, it could have major effects on human society, for in the long run it could eliminate our dependence on the earth's limited energy resources.

TABLE 16.3   THE PROTON-PROTON CHAIN

REACTION	EXPLANATION
$_1^1H + {}_1^1H \rightarrow {}_1^2H + e^+ + \nu$	Two protons ($_1^1H$) combine to produce a deuterium nucleus ($_1^2H$), a positron (e^+), and a neutrino (ν). The deuterium nucleus has excess kinetic energy, and this heats the gas.
$_1^2H + {}_1^1H \rightarrow {}_2^3He + \gamma$	A deuterium nucleus combines with another proton to produce a helium nucleus ($_2^3He$, with two protons and one neutron) and a γ-ray photon. Again, the product particle has excess kinetic energy.
$_2^3He + {}_2^3He \rightarrow {}_2^4He + 2{}_1^1H$	Two ^3He nuclei combine to produce an ordinary helium nucleus ($_2^4He$, with two protons and two neutrons) and two protons, each particle being left with excess kinetic energy.

Nuclear fusion reactions can take place only under conditions of extreme pressure and temperature, because of the electrical forces that normally would keep atomic nuclei from ever getting close enough together to react. Nuclei, which have positive charges because all their electrons have escaped, must collide at extremely high speeds in order to overcome the repulsion caused by their like electrical charges. The speed of particles in a gas is governed by the temperature, and only in the very center of the sun and other stars is it hot enough (around 10 million degrees) to allow the nuclei to collide fast enough to fuse. The high pressure in the sun's core causes nuclei to be crowded together very densely, and this means that collisions will take place very frequently, another requirement if a high reaction rate is to occur.

Structure of the Solar Atmosphere

Observations of the sun's appearance when viewed in different wavelengths of light make it clear that the outer layers are divided into several distinct zones (Fig. 16.4). The "surface" of the sun that we see in visible wavelengths is the **photosphere,** with a temperature ranging between 4,000 and 6,500 K. When we view the sun

at the wavelength of the strong line of hydrogen at 6,563 Å, we see the **chromosphere,** a layer above the photosphere whose temperature is 6,000 to 10,000 K. Outside of that is the very hot, rarefied **corona,** best observed at X-ray wavelengths, whose temperature is 1 to 2 million degrees. Overall, the tenuous gas within and above the photosphere is referred to as the solar atmosphere. If we consider the temperature throughout this region, we find that it decreases outward through the photosphere, reaching a minimum value of about 4,000 K. From there the trend reverses itself, and the temperature begins to rise as we go farther out. The chromosphere, immediately above the temperature minimum, is perhaps 2,000 kilometers thick. Above there the temperature rises very steeply within a few hundred kilometers, to the coronal value of more than a million degrees. Clearly something is creating extra heat at these levels; shortly we will consider where this heat comes from.

First let us discuss the photosphere, the "surface" of the sun as we look at it in visible light. It is here that the density becomes great enough for the gas to be opaque, making it impossible for us to see further into the interior. The sun's absorption lines are formed in the photosphere, as the atoms in this relatively cool layer absorb continuous radiation coming from the hot interior. A photograph of the photosphere reveals a cellular appearance called **granulation** (Fig. 16.5). Bright regions, representing areas where convection in the sun's outer layers causes hot gas to rise, are bordered by dark zones where cooler gas is descending back into the interior.

The temperature of the photosphere, roughly 6,000 K, is measured from the use of Wien's law and from the degree of ionization in the gas there; and the density, roughly 10^{17} particles per cm^3 in the lower photosphere, was found from the degree of excitation, as described in chapter 5. This is lower than the density of the earth's atmosphere, which is about 10^{19} particles per cm^3 at sea level.

Most of what we know about the sun's composition is based on the analysis of the solar Fraunhofer lines (the sun's absorption lines; Fig. 16.6), so strictly speaking, the derived abundances represent only the photosphere. We have no reason to expect strong differences in composition at other levels, however, except for the core, where a significant amount of the original hydrogen has been converted into helium by the proton-proton reaction.

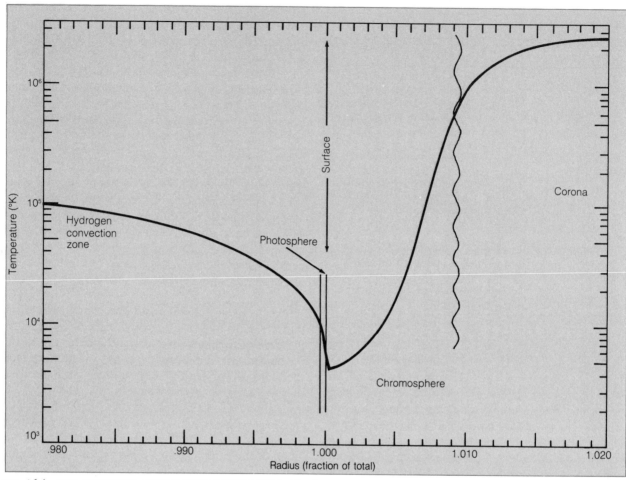

FIG. 16.4. THE STRUCTURE OF THE SUN'S OUTER LAYERS. This diagram shows the relative heights and temperatures of the convective zone, photosphere, chromosphere, and corona. Heights are expressed in terms of the solar radius, which is roughly 700,000 km.

The photosphere near the edge of the sun's disk looks darker than in the central portions (see Fig. 16.1). This effect, called **limb darkening,** is caused by the fact that we are looking obliquely at the photosphere when we look near the edge of the disk. We therefore do not see as deeply into the sun there as we do when looking near the center of the disk. Therefore at the limb the gas we are seeing is cooler than at the disk's center, and radiates less.

The chromosphere lies immediately above the temperature minimum. The fact that this region forms emission lines tell us, according to Kirchoff's laws, that the chromosphere is made of hot, rarefied gas, hotter than the photosphere behind it. When viewed through

a special filter that allows only light at the wavelength of the hydrogen emission line at 6,563 Å to pass through (Fig. 16.7), the chromosphere has a distinctive cellular appearance referred to as **supergranulation,** similar to the photospheric granulation, but with cells some 30,000 kilometers across instead of about 1,000 kilometers. There is also fine-scale structure in the chromosphere, in the form of spikes of glowing gas called **spicules** (Fig. 16.8). These come and go, probably at the whim of the magnetic forces that seem to control their structure.

The outermost layer of the sun's atmosphere is the corona, which extends a considerable distance above the photosphere and chromosphere. The corona is ir-

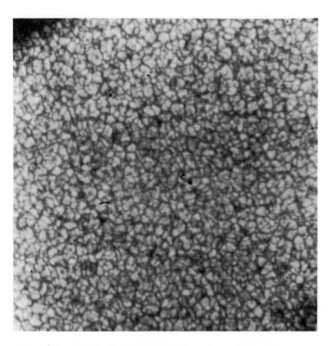

FIG. 16.5. SOLAR GRANULATION. This photo, obtained by a high-altitude balloon floating above much of the atmosphere's blurring effect, distinctly shows the granulation of the photosphere, which is a result of convective motions in the sun's outer layers.

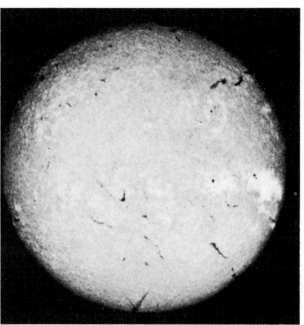

FIG. 16.7. THE CHROMOSPHERE. This photo, taken through a special filter that allows light to pass through only at the wavelength of the bright hydrogen emission line at 6,563 Å, reveals the locations of ionized hydrogen gas on the sun, which are primarily in the chromosphere. The light areas (such as at right) are active regions associated with sunspots, where the chromosphere glows especially brightly at the observed wavelength. The dark filaments are solar flares.

λλ 3300 – 4900 Å

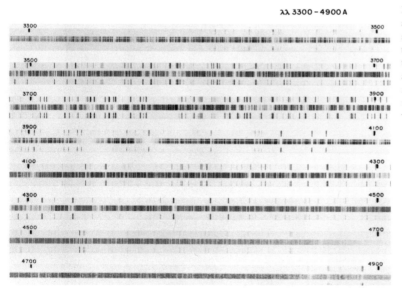

FIG. 16.6. FRAUNHOFER LINES. The major absorption lines in the sun's photospheric spectrum are called Fraunhofer lines, in honor of the German scientist who first catalogued many of them in the early nineteenth century. Here we see a photo of a portion of the solar spectrum, with matching emission lines just above and below, made by a special lamp in order to establish the wavelength scale.

FIG. 16.8.   SPICULES. This photo shows the chromosphere's transient features known as spicules. These spikes of glowing gas, which are apparently shaped by the sun's magnetic field, come and go irregularly.

FIG. 16.9.   THE CORONA. This photo, obtained during a total solar eclipse, shows the type of structure commonly seen in the corona. There are giant looplike features, and an overall appearance of outward streaming. Bits of more intense light from the chromosphere are seen around the edges of the moon's occulting disk. Visible at left is the planet Venus.

regular in form, patchy near the sun's surface, but with radial streaks at great heights suggestive of outflow from the sun (Fig. 16.9 and Color Plate 14). The density of the coronal gas is very low, only about 10^9 particles per cm^3. As we have already mentioned, the corona is very hot, containing highly ionized gas. The source of the energy that heats the corona to such extreme temperatures is not well understood, although a general picture has emerged.

X-ray observations reveal that the corona is not uniform, but instead has a patchy structure (Fig. 16.10). There are large regions that appear dark in an X-ray photograph of the sun, where the gas density is even lower than in the rest of the corona. These regions are called **coronal holes;** they are probably created and maintained by the sun's magnetic field. The coronal holes, as well as the overall shape of the corona, vary with time, which illustrates that the corona in general is a dynamic, active region (Fig. 16.11). Another impressive sign of the corona's dynamic nature is the existence of **prominences** (Figs. 16.12, 16.13, and Color Plate 14), great geysers of hot gas that spurt upward from the surface of the sun, taking on an arc-shaped appearance. These are usually associated with sunspots, and both phenomena are linked to the solar activity cycle, to be discussed shortly.

FIG. 16.10.   AN X-RAY PORTRAIT OF THE CORONA. This image, obtained by *Skylab* astronauts, shows the sun in X-ray light, which reveals only very hot regions. The bright regions are places where the corona is especially dense, and the dark regions, known as coronal holes, are places where it is much more rarefied. All the structure seen here changes with time.

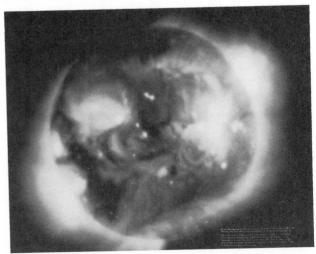

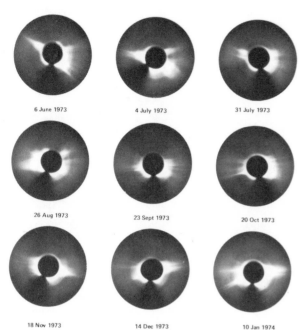

6 June 1973   4 July 1973   31 July 1973

26 Aug 1973   23 Sept 1973   20 Oct 1973

18 Nov 1973   14 Dec 1973   10 Jan 1974

FIG. 16.11. CHANGES IN THE CORONA. This series of photos was obtained through the use of a special shutter device that blocked out the solar disk, allowing the corona to be observed without a total solar eclipse. Here we see that the structure of the corona changes quite markedly over a period of a few months.

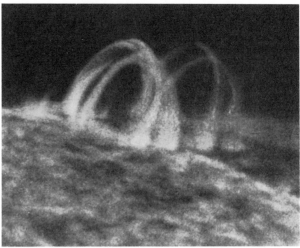

FIG. 16.12. A PROMINENCE. Here the looplike structures thought to be governed by the sun's magnetic field are readily seen. This photo was obtained by *Skylab* astronauts, using an ultraviolet filter.

The hot outer layers of the sun have provided astronomers with a second major mystery concerning the solar energy budget. In contrast to the mystery of the sun's internal energy source, which has been solved, the mystery regarding the mechanism for heating the chromosphere and corona has not. It is generally accepted that the heating must come from the boiling and churning of the outer convective layer of the solar interior, but it is not clear how this activity is translated into heat at great distances above the photosphere. For awhile it was assumed that sound waves were responsible, but today it is thought that various kinds of magnetic waves are accountable. There is certainly enough

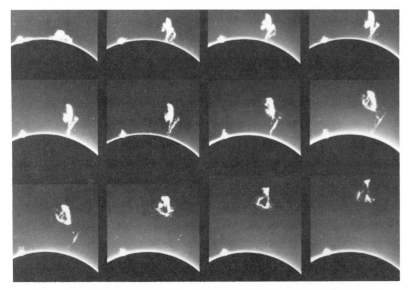

FIG. 16.13. AN ERUPTIVE PROMINENCE. This striking sequence shows an outburst of ionized gas from a prominence on the sun's limb. The photos were obtained through the use of a special device that blocked out the sun's disk.

energy in the convective motions in and below the photosphere; the question is how this energy is transported into the higher levels.

Recent satellite observations in ultraviolet and X-ray wavelengths have shown that stars similar to the sun also have chromospheric and coronal zones. Therefore if we can understand how the sun operates, we will also gain a deeper understanding of how other stars work.

The Solar Wind

The long, streaming tail of a comet always points away from the sun, regardless of the direction of the comet's motion. The significance of this was fully realized in the late 1950's, when the first U.S. satellites revealed the presence of the earth's radiation belts and the fact that they are shaped in part by a steady flow of charged particles from the sun. It is this river of charged particles that forces cometary tails to always point away from the sun.

Most of the direct information we have on the solar wind comes from satellite and spacecraft observations, because the earth's magnetosphere shields us from the wind particles. Solar-wind monitors are placed on board most spacecraft sent to the planets. One striking discovery has been the fact that the wind is not uniform in density, but rather seems to flow outward from the sun in sectors, as though it originates only from certain areas on the sun's surface. This is explained by X-ray data, which indicate that the wind emanates only from the coronal holes. Because the base of the wind is rotating with the sun, the wind sweeps out through space in a great curve, similar to the trajectory of water from a rotating lawn sprinkler (Fig. 16.14).

The existence of the solar wind is evidently a natural by-product of the same heating mechanisms that produce the hot corona of the sun. It was originally thought that particles in this high-temperature region move about with such great velocities that a steady trickle escapes the sun's gravity, flowing outward into space. However, X-ray observations of the sun have shown that the situation is not that simple. It is the solar magnetic field that governs the outward flow of charged particles. The coronal holes, mentioned earlier, are regions where the magnetic field lines open

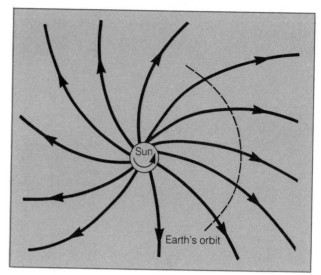

FIG. 16.14. THE SOLAR WIND. This schematic diagram illustrates how ionized gas from the sun spirals outward through the solar system in a steady stream. The solar magnetic field creates sectors of variable density in the wind.

out into space. Charged particles such as electrons and protons, constrained by electromagnetic forces to follow the magnetic field lines, therefore escape into space only from the coronal holes. The speed of the solar wind is relatively low close to the sun, but accelerates outward, quickly reaching a velocity of 300 to 400 kilometers per second, after which it is nearly constant. The wind nearly reaches its maximum velocity by the time it passes the earth's orbit, and beyond there it flows steadily outward, past the orbit of Saturn. It is thought that at some point in the outer solar system, the wind comes to an abrupt halt where it runs into an invisible and tenuous wall of matter swept up from the interstellar medium that surrounds the sun.

Occasional explosive activity occurring on the sun's surface releases unusual quantities of charged particles, and some three or four days later, when this burst of ions reaches the earth's orbit, we experience disturbances in the ionosphere that can interrupt shortwave radio communications and cause auroral displays. These magnetic storms, as they often are called, are outward manifestations of a much more complex overall interaction between the sun and the earth. We will discuss the so-called solar-terrestrial relations later in the chapter.

Sunspots, Solar Activity Cycles, and the Magnetic Field

The dark spots on the sun's disk were observed more than three hundred years ago, and were cited by Galileo as evidence that the sun is not a perfect, unchanging celestial object, but rather has occasional flaws (Figs. 16.15 and 16.16). During the centuries since then, observations of the spots (which individually may last for months), have revealed some very systematic behavior. The number of spots varies, reaching a peak every eleven years, and during the interval the spots move steadily from the sun's middle latitudes toward the equator. At the beginning of a cycle, when there is a maximum number of sunspots, most of them appear in activity bands about 30 degrees north or south of the solar equator. During the next eleven years, the spots tend to lie ever closer to the equator, and by the end of the cycle they are nearly on it. By this time the first spots of the next cycle may already be forming at middle latitudes. A plot of sunspot locations during a cycle clearly shows this effect, and is called a "butterfly diagram" because of the shape of the pattern (Fig. 16.17).

The sunspots are not totally black, although they appear so when seen against the bright background of the photosphere. Actually they glow rather intensely, but because they are somewhat cooler than their surroundings, they are not as bright. The typical temperature in a spot is about 4,000 K, compared with the 6,000 K or so temperature of the photosphere. Using Stefan's law, we see that the intensity of light emitted in a spot compared with the surroundings is $(4,000/6,000)^4 = 0.2$; that is, the brightness of the solar surface within a sunspot is only about one-fifth of the brightness in the photosphere outside.

A hint as to the origin of the spots was found when their magnetic properties were first measured. Astronomers accomplished this by using spectroscopy of the light from the spots. The energy levels in certain atoms are distorted by the presence of a magnetic field, which in turn causes the spectral lines formed by those levels to be split into two or more distinct, closely spaced lines. The degree of line-splitting, which is referred to as the **Zeemann effect,** depends on the strength of the magnetic field. Thus the field can be measured from afar simply by analyzing the spectral lines to see how widely split they are. This technique works for distant stars as well as for the sun.

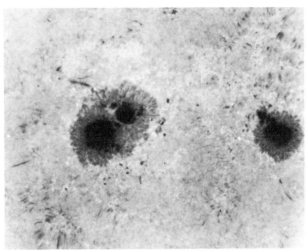

FIG. 16.15.   SUNSPOTS. This is a telescopic view of a group of spots showing their detailed structure. They appear dark only in comparison with their much hotter surroundings.

FIG. 16.16.   THE ACTIVE SUN. This is a composite image, showing the disk in the light of ionized hydrogen, such that sunspot groups appear white and flares dark. Outside the disk, solar flares and prominences are seen along the limb, as photographed with the disk occulted. Supergranulation in the chromosphere is especially clear in this image.

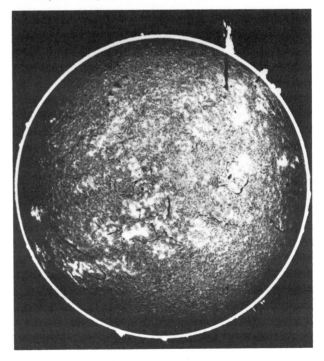

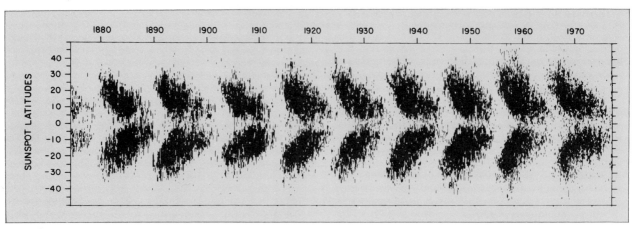

FIG. 16.17.   THE BUTTERFLY DIAGRAM. This is a plot of the latitudes of observed sunspots through several cycles of solar activity. During each cycle, the spots gradually shift their favored locations closer and closer to the solar equator.

When the initial measurements of the sun's field were made in the first decade of this century, it was found that the field is especially intense in the sunspots, about 1,000 times stronger than in the surrounding gas. A strong magnetic field creates pressure in a gas, just as high temperature does. Thus a sunspot maintains a balance with its hotter surroundings as the magnetic pressure within the spot counteracts the pressure of the hotter gas around it.

When the magnetic fields of sunspots were measured, they were found to act like either north or south magnetic poles; that is, each spot has a specific magnetic direction associated with it. Furthermore, pairs of spots often appear together, the two members of a pair usually having opposite magnetic polarities. During a given eleven-year cycle, in every sunspot pair the magnetic polarities always have the same orientation. For example, during one eleven-year cycle, the spot to the east in each pair generally will have a north magnetic polarity, and the one to the west a south magnetic polarity. During the next cycle, the polarity of all the pairs will reverse, with the south magnetic spot to the east, and the north magnetic spot to the west. Between cycles, when this arrangement is reversing itself, the sun's overall magnetic field also reverses, with the solar magnetic poles exchanging places. It is actually twenty-two years before the sun's magnetic field and sunspot patterns repeat themselves, so the solar magnetic cycle is truly twenty-two years long.

Sunspot groups are the scenes of the most violent forms of solar activity, the **solar flares.** These are gigantic outbursts of charged particles, as well as visible,

ultraviolet, and X-ray emission, created when extremely hot gas spouts upward from the surface of the sun (Fig. 16.18). Flares are most common during sunspot maximum, when the greatest density of spots is to be seen on the solar surface. Close examination of flare events shows that the trajectory of the ejected gas is shaped by the magnetic lines of force emanating from the spot where the flare occurred. Charged particles flow outward from a flare, some of them escaping into the solar wind. If the flare occurs at the place on the sun where the part of the wind arises that hits the earth,

FIG. 16.18.   A MAJOR FLARE. The gigantic looplike structure seen here, in an ultraviolet photograph obtained from *Skylab,* is one of the most energetic flares ever observed. Supergranulation in the chromosphere is easily seen on most of the disk.

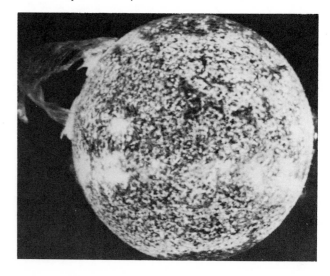

then the earth gets a heavy dosage of solar wind particles some three days later, affecting radio communications. The extra quantity of charged particles entering the earth's upper atmosphere also can cause unusually widespread and brilliant displays of aurorae. Apparently flares occur when twisted magnetic field lines suddenly reorganize themselves, releasing heat energy and allowing huge bursts of charged particles to escape into space.

The combination of all these bits of data on sunspots, magnetic fields, and solar activity cycles has led to the development of a complex theoretical picture, one that successfully accounts for many of the observed phenomena. This theory envisions ropes or tubes of magnetic field lines inside the sun, connecting its north and south magnetic poles. At some places these tubes become kinked, and loops break through the surface, creating pairs of sunspots with opposing magnetic polarities where they emerge and reenter. Early in the sunspot cycle the magnetic tubes break through the surface at middle latitudes, but later, as the sun's magnetic field is moving toward reversal of the poles, they do so near the equator. This accounts for the latitude-dependence of the spots during a cycle, as shown in the butterfly diagram. When the solar magnetic field reverses itself every eleven years, so do these magnetic ropes; thus when the new cycle begins, the sunspot pairs have their polarities reversed compared with the pairs of the previous cycle.

The origin of the sun's magnetic field and its periodic pole reversals is probably a dynamo, similar in nature to those thought to be at work in the interiors of the planets that have fields. Unlike the terrestrial planets, however, the sun is not rigid, and consequently differential rotation may play a role in creating the instability that causes the dynamo to reverse itself regularly, every eleven years. From the sun's behavior we might speculate that the magnetic fields of Jupiter and Saturn, which also rotate differentially, reverse themselves from time to time as well.

The solar activity cycle may have some indirect effects on the climate of the earth. Eleven-year patterns in the occurrence of droughts have been reported, and it seems possible that such patterns are related to the solar cycle, although it is not known how. In the late 1600's and early 1700's there was a prolonged period of very weak solar activity (called the *Maunder minimum,* after E. W. Maunder, who was also responsible for developing the butterfly diagram), and during this time the earth's climate was in chaos, with terrible droughts in many areas, and particularly severe winters in Europe and North America (Fig. 16.19).

There are other aspects of the relationship between solar activity and the earth's atmosphere. As the solar wind fluctuates in intensity, the earth's magnetosphere varies in extent. We have already noted that solar flares, which occur most often during sunspot maximum, have significant effects on the ionosphere. The chemistry of the upper atmosphere may also be strongly influenced by variations in the solar ultraviolet emission, which in turn are linked to the solar activity cycle. The entire question of solar-terrestrial relations is an important one that merits and is receiving greater attention.

FIG. 16.19.   THE SUN'S LONG-TERM ACTIVITY.   This diagram illustrates the relative level of activity (in terms of sunspot numbers) over three centuries. It is clear that the level varies. Note that there was a span of about fifty years (1650–1700), known as the Maunder minimum, when there was little activity. Evidence shows that the sun has long-term cycles that modulate the well-known 22-year period.

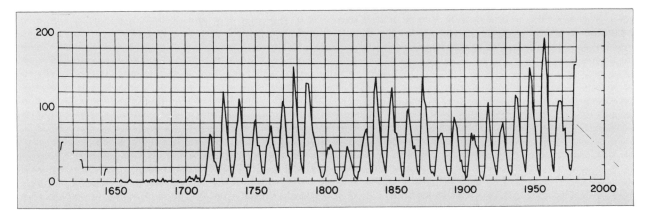

Perspective

Our sun, the source of nearly all our energy, is a very complex body. As stars go, it is apparently normal in all respects, so we imagine that other stars are just as complex, even though we cannot observe them in such detail.

We have explored the sun, both in its deep interior and in the outer layers that can be observed directly. We know that nuclear fusion is the source of all the energy, and that the size and shape of the sun are controlled by the balance between gravity and pressure throughout the interior. Somehow heat is transported above the surface, keeping the chromosphere and the corona hotter than the photosphere. Perhaps most intriguing of all is the solar activity cycle and its relationship to the sun's complex magnetic field.

With this examination of the sun, we have completed our survey of the solar system. We are ready to tie together all the diverse pieces of information that we have discussed, developing a coherent picture of the system as a whole and the manner in which it formed.

Summary

1. The sun is an ordinary star, one of billions in the galaxy.

2. The sun is gaseous throughout, and hydrostatic equilibrium causes it to be very hot and dense in the core.

3. Energy is produced in the core by nuclear fusion, and is slowly transported outward by radiation, except near the surface, where energy is transported by convection.

4. The nuclear reaction that powers the sun is the proton-proton chain, where hydrogen is fused into helium.

5. Observations of the sun through filters designed to pass various wavelengths allow us to see distinct levels of the sun's outer layers, each characterized by a different temperature.

6. The outer layers are the photosphere, with a temperature of about 6,000 K; the chromosphere, where the temperature ranges between 6,000 and 10,000 K; and the corona, where the temperature is higher than 1,000,000 K.

7. The photosphere, which is the visible surface, creates absorption lines, whereas the hotter chromosphere and corona create emission lines.

8. The excess heat in the outer layers is somehow transported there from the convective zone just below the surface; the mechanism for transporting the heat is not well understood.

9. The sun emits a steady outward flow of ionized gas called the solar wind. The wind originates from coronal holes, and is therefore controlled by the solar magnetic field.

10. Sunspots occur in 11-year cycles, which are a reflection of the 22-year cycle of the sun's magnetic field. The spots are regions of intense magnetic fields where flux tubes from the solar interior break through the surface.

11. The solar activity cycle may have important effects on the earth's climate.

Review Questions

1. Why is lead a solid material, whereas the sun, many times denser at its core than lead, is gaseous?

2. Why do nuclear reactions occur only in the innermost core of the sun? Explain in terms of hydrostatic equilibrium.

3. Explain why the use of special filters to isolate the

wavelengths of certain spectral lines allows us to separately examine distinct layers of the sun.

4. Why are the granules in the sun's photosphere bright, and the descending gas that surrounds them relatively dark?

5. Use Kirchhoff's laws to explain why the photosphere produces absorption lines, but the chromosphere and corona produce emission lines.

6. What would we conclude about the temperature just below the outermost layers of the photosphere if the sun's disk appeared brighter, instead of darker, at the edges?

7. Why does it take about three days for the effects of a solar flare to begin to occur in the earth's magnetosphere?

8. How would the appearance of sunspots be altered if they had weaker magnetic fields than their surroundings?

9. In previous chapters we discussed the effects of the solar wind on the planets. Summarize these effects.

10. Describe how energy produced in the sun's core by nuclear reactions heats the chromosphere and corona; in other words, describe how it gets from the core to the outer layers, step by step.

Additional Readings

Gibbon, E. G. 1973. *The Quiet Sun*. NASA Publication NAS 1.21. Washington, D.C.: NASA.

Gough, D. 1976. The shivering sun opens its heart. *New Scientist* 70:590.

Parker, E. N. 1975. The Sun. *Scientific American* 233(3):42.

Pasachoff, J. M. 1972. The fiery sun. *Natural History* 81(5):49.

————. 1973. The solar corona. *Scientific American* 229(4):68.

Wilson, O. C., Vaughan, A. H., and Mihalas, D. 1981. The activity cycle of stars. *Scientific American* 244(2):104.

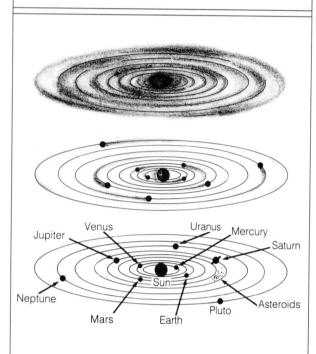

CHAPTER 17

Jupiter · Venus · Uranus · Mercury · Saturn · Neptune · Sun · Asteroids · Mars · Earth · Pluto

Adding It Up: Formation of the Solar System

We have collected a considerable quantity of information about the solar system and its contents. Along the way, we have uncovered various systematic trends, such as the similarities in internal structure and composition of the terrestrial planets, and the similarities among the outer planets, but so far we have said very little about how all of this arose. In discussing the histories of the individual planets, we have referred to some of the important processes thought to have been involved, such as the condensation of the first solid material from a cloud of gas surrounding the young sun, but it remains for us to put together a coherent story of the formation and evolution of the entire system.

Before we describe the modern theory, we will review the overall properties of the solar system that must be accounted for, and then we will have a look at the historical developments and alternative suggestions that scientists have considered over the years.

A Summary of the Evidence

A number of facts must be explained if a theory regarding the formation of the solar system is to be successful. In a way this is very much like a mystery story, where the detective (the scientist seeking the correct explanation) has certain clues that reveal isolated parts of the story, from which past events must be reconstructed.

One category of clues in our mystery has to do with the orbital and spin motions of the planets and satellites in the solar system. All the planetary orbits lie in a common plane (Fig. 17.1), all are nearly circular, and all go around the sun in the same direction (Fig. 17.2). The spins of nearly all the planets are in this same direction, as are the orbital and spin motions of most of the satellites in the solar system. The direction of these motions, counterclockwise when viewed from above the north pole, is said to be **direct,** or **prograde,** motion.

Although we do not believe in any physical basis for Bode's law, we still need to know what caused the planets to form where they did, and not elsewhere. A final clue having to do with the motions of the planets is that nearly all of them have small **obliquities,** meaning that their equatorial planes are nearly aligned

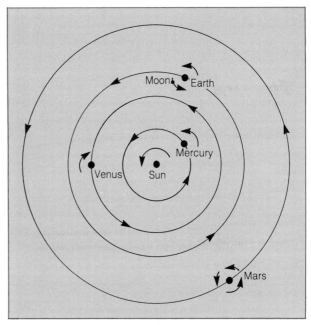

FIG. 17.2. PROGRADE MOTIONS IN THE INNER SOLAR SYSTEM. The rotations and orbital motions of the planets are generally in the same direction. This sketch of the orbits of the inner four planets shows that Venus is an exception, with retrograde rotation, but that the other three planets and their satellites obey the rule. All the outer planets and nearly all their satellites do also.

FIG. 17.1. THE CO-ALIGNMENT OF PLANETARY ORBITS. All nine planets have orbits that are nearly circular and whose planes are nearly parallel. An exception is Pluto, whose orbit is tilted 17 degrees with respect to the ecliptic, and is sufficiently eccentric (that is, elongated) that Pluto is actually closer to the sun than Neptune at times.

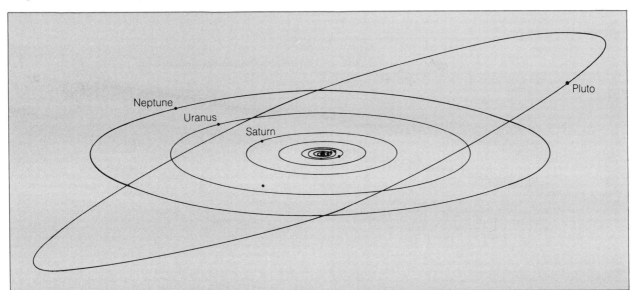

with the ecliptic plane. Even the sun has its axis of rotation nearly perpendicular to the ecliptic. Only Uranus, with its 98-degree tilt, does not fit the pattern (Pluto is also anomalous in this regard, as it is in most other respects).

A second general type of clue in the mystery is concerned with the natures of the individual planets, and the systematic trends from planet to planet that we have discussed. First, we must consider the planetary compositions, seeking an explanation for the fact that the inner four, the terrestrial planets, have high densities, which indicates that they are made mostly of rock and metallic elements; whereas the outer planets consist primarily of lightweight gases such as hydrogen and helium, as well as ices composed of these and similarly volatile gases. Within each of the two groupings of planets, there are relatively minor differences in chemical composition, such as the various isotope ratios we have discussed, which also must be accounted for.

A major mystery for a long time was the slow rotation of the sun; according to many early theories, the sun should be spinning much faster. The difficulty arises in theories that envision the solar system forming from the collapse of a cloud of gas and dust, and these theories, as we will soon learn, are the most successful ones. The laws of physics tell us that the angular momentum of an object (see chapter 4) must remain constant. This means that if a spinning object shrinks in size, it must spin faster to compensate, thereby maintaining constant angular momentum. Thus the sun, thought to have formed at the center of a collapsing

cloud, should have a very rapid rotation rate, rather than the leisurely 25-day period it has. The explanation for this was a long time coming, and for awhile astronomers were side-tracked in their efforts to understand the formation of the solar system.

A final category of information that bears on the origin of the solar system is the distribution and nature of the various kinds of interplanetary objects that were described in chapter 15. We learned there that asteroids, comets, meteoroids, and interplanetary dust inhabit the space between the planets, and we found that the motions and compositions of these objects reflect conditions that existed very early in the history of the solar system. Somehow this information must fit into our overall picture.

Catastrophe or Evolution?

In discussing theories of the formation of the solar system, we will confine ourselves to those which were suggested after the time of Copernicus, when it was established that the planets orbit the sun. Several primitive ideas that were based on the geocentric notion were outlined in chapter 2.

The first serious theory based on the heliocentric view of the solar system was developed by the French scientist René Descartes, who in 1644 advanced the idea that the solar system formed from a gigantic whirlpool, or vortex, in a universal fluid, with the planets and their satellites forming from smaller eddies. This theory was rather crude, without any clearly specified idea of the nature of the cosmic substance from which the sun and planets arose, but it did account for the fact that all the orbital motions are in the same direction.

The hypothesis of Descartes was the first of a general type known as **evolutionary theories,** theories in which the formation of the solar system occurred as a natural by-product of the sequence of events that produced the sun. Evolutionary theories state that no special circumstances were needed to create the planets in our solar system, other than the fact that the sun formed. (This leads us to suspect there may be planets orbiting other stars.)

Further elaboration on the idea of Descartes was developed by Immanuel Kant, who in 1755 applied the recently discovered Newtonian mechanics to the problem, and was able to show that a rotating gas cloud

---

TABLE 17.1   CLUES TO THE ORIGIN OF THE SOLAR SYSTEM.
ORBITAL AND SPIN MOTIONS
Planetary and satellite orbits lie in a common plane
Nearly all of the planetary and satellite orbital and spin motions are in the same direction
The rotation axes of nearly all the planets and satellites are roughly perpendicular to the ecliptic
PROPERTIES OF THE PLANETS
The terrestrial planets: small, high density, low volatile content
The giant planets: large, low density, high volatile content
THE SLOW ROTATION OF THE SUN
INTERPLANETARY MATERIAL
The existence and location of the asteroid belt
The Oort cloud and cometary orbits
The distribution of interplanetary dust

would flatten into a disk as it contracted (Fig. 17.3). Kant's theory was called the **nebular hypothesis,** because it invoked formation of the sun and planets from an interstellar cloud, or nebula. In 1796 Pierre Simon de Laplace, a French mathematician, added the notion that as the spinning cloud flattened into a disk, concentric rings of material broke off because of rotational forces, so that at one point the early solar system would have looked very much like the planet Saturn with its rings (Fig. 17.3). Each ring was supposed to have then condensed into a planet. Soon the problem of accounting for the slow solar rotation was recognized, and further development of the evolutionary theories was stymied.

Meanwhile an alternative group of ideas, called **catastrophic theories,** had been suggested. These concepts of the formation of the planetary system envision a solitary sun to begin with, but then invoke some singular, cataclysmic event that disrupted the sun and formed the planets. The first of these ideas was put forth by another Frenchman, Georges Louis de Buffon, who in 1745 suggested that a massive body (which he referred to as a comet, although it was much larger than any cometary nucleus) passed so near the sun that its gravitational pull forced material out of it, this gas then condensing to form the planets (Fig. 17.4).

Buffon's idea was largely ignored until the beginning of the twentieth century, when the difficulty with the sun's slow spin forced scientists to consider alternatives to the evolutionary theories. By 1905 the English astronomers T. C. Chamberlain and F. R. Moulton had suggested an elaboration of Buffon's original idea, in which another star was the object that passed near the sun, creating tidal forces that caused matter from inside the sun to stream out, thereafter condensing into planets.

Interestingly, in modern times, the idea of mass being transferred from one star to another in close double-star systems has been discovered to be very important under certain circumstances (see chapter 20), so the idea of gravitational forces extracting matter from the sun is not out of the question. Further refinements to the bypassing-star hypothesis were soon provided by the British astrophysicists H. Jeffreys and J. H. Jeans.

A disturbing problem with the catastrophic theories was the low probability that the sun could have collided with another star, given the great distances between stars in our part of the galaxy. Of course, this did not rule out the possibility, for even a low-proba-

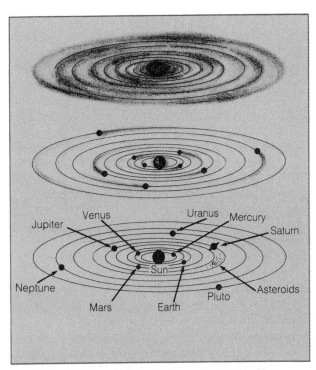

FIG. 17.3. THE HYPOTHESES OF KANT AND LAPLACE. Descartes's simple vision of a vortex was refined by Kant, who realized that rotation should cause a collapsing cloud to take on a disk-like shape (left); and by Laplace, who hypothesized that a rotating disk would form detached rings that could then condense into planets (right).

bility event can occur, but it was sufficiently worrisome to lead to the development of an alternative catastrophic theory, one in which the sun started out as a member of a triple system, but was then freed from

FIG. 17.4. THE CATASTROPHE HYPOTHESIS. Here a by-passing body gravitationally forces material out of the sun; this matter subsequently condenses to form the planets.

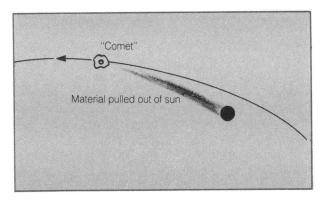

the gravitational bonds of the other two when the three nearly collided. In the process of expelling the sun from the triple-star system, the gravity of the other stars tore some of the solar material out, forming a tail, which later condensed to form the planets. (More recently, a similar three-body interaction has been suggested to explain the strange orbits of Neptune's two satellites and the expulsion of Pluto into its own solar orbit; see chapter 14.) This theory, developed by H. N. Russell, R. A. Lyttleton, and F. Hoyle, replaced the difficulty of forming the sun and the planets together with one of forming the sun and two other stars together. The advantage was that the slow rotation of the sun did not have to be produced during its formation, but could have developed later, when it nearly collided with the other stars and was ejected.

By the late 1930's the catastrophic theories were becoming quite unwieldy, not only because of the special circumstances that had to be assumed, but also because of several fundamental problems. For example, calculations showed that, even if some material really were pulled out of the sun's interior, it would have so much internal energy (that is, it would be so hot) that it would expand and dissipate into space, rather than condensing to form planets. Another difficulty was how to pull material out of the sun, giving it sufficient angular momentum for the planets, but without also giving it enough speed to escape entirely.

In the 1940's a new type of theory was proposed by the Soviet astronomer O. Y. Schmidt, and was developed in some detail by H. Alfven, the Swedish astrophysicist who was later to win a Nobel prize for his work on the behavior of ionized gas in the solar magnetic field. The idea was that once the sun had formed, its gravitational field trapped material from the surrounding interstellar medium, which then formed into planets. This general idea, called the **accretion theory** of solar-system formation, never gained much popularity, because the problems with the evolutionary theory began to be solved at about the time the accretion theory was being developed. Most research efforts after this time centered on the evolutionary theory.

Progress was being made in the 1940's concerning the general question of how a contracting cloud would flatten into a disk and then break up into eddies that could form planets. The German physicist C. F. von Weizsäcker showed that the disk would tend to rotate differentially, meaning that the inner parts would orbit the center faster than the outer regions, and that this would cause the disk to break up into eddies (Fig. 17.5).

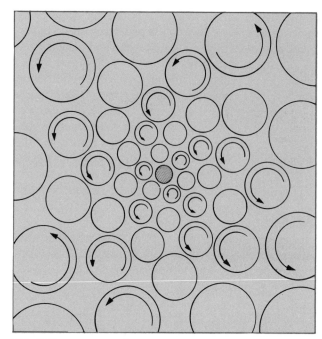

FIG. 17.5.   VON WEIZSÄCKER'S THEORY OF A TURBULENT DISK. This sketch shows how a rotating disk would form eddies, according to the theory of the German scientist C. F. von Weizsäcker.

The work of von Weizsäcker even showed how the relative sizes of the eddies would vary with distance from the center, accounting for both the relative distances of the planets from the sun and their varying sizes from one to the next. More detailed analyses of the stability of rotating disks were later carried out by a number of others, who showed that rather than regular eddies being formed, as envisioned by von Weizsäcker, a rotating disk would become clumpy, forming localized regions where the density was high enough to allow material to fall together gravitationally. This process would lead to the disk being broken up into a number of small solid bodies, which in turn could later merge to form the planets.

The breakthrough in understanding the sun's slow rotation came in the early 1960's, with the discovery of the solar wind. Because the sun continuously expels charged particles, interplanetary space is filled with them, and the sun's magnetic field tries to pull them along with it as the sun rotates (Fig. 17.6). This creates a constant drag on the sun (imagine trying to spin a pinwheel under water). This drag, given long enough to act, slows the sun's spin. If the sun was born with a rapid rotation, this **magnetic-braking** process could easily have slowed it to the present rate in the billions of years that have passed. Most likely the density of charged particles in space around the sun was much greater in the early days of the solar system, so the drag created as the sun's magnetic field tugged on the particles was probably greater than it is now, and the braking could have taken place in a relatively short time.

Once the magnetic-braking process was understood, most of the major pieces fell into place, and all that remained was further refinement of the theory, accounting for the details. There are still many unanswered questions, but the general idea of how the solar system formed is probably correct, as described in the next section.

A Modern Scenario

The first step in the formation of the solar system was the collapse of an interstellar cloud, one that must have been rotating before it began to fall in on itself. Our galaxy has a large quantity of interstellar material in it, containing both gas and small dust grains (about the size of the interplanetary grains described in chapter 15). The interstellar medium is not uniformly distributed throughout space, but rather is concentrated here and there in large amorphous regions called interstellar clouds, some of them sufficiently dense to block out starlight, appearing as dark regions in photographs of the sky (Fig. 17.7). Even in these areas of relatively high concentration, the material is so rarefied that a quantity of mass equal to that of the sun is spread over a volume up to ten or more light-years across.

The composition of the interstellar material is apparently quite uniform, consisting of about the same mixture of elements as the sun. Thus the sun and other stars are born with a standard composition of almost 80 percent hydrogen (by mass), more than 20 percent helium, and just traces of all the other elements. The planets must also have formed out of material with this composition, yet, as we have seen, their present makeup is quite different from this, especially in the case of the terrestrial planets.

Somehow the extended, tenuous gas in interstellar space had to become concentrated in a very small volume on its way to becoming a star. Gravitational forces were responsible for pulling the cloud together, but there must have been an initial push or a chance condensation to get the process started (see chapter 21 for

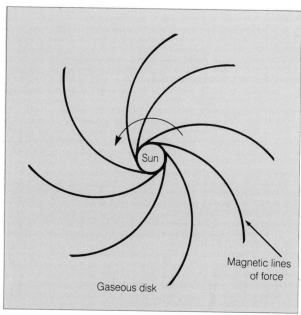

FIG. 17.6. MAGNETIC BRAKING. The young, rapidly spinning sun has a magnetic field structure as shown, the field lines rotating with the sun. The early solar system was permeated with large quantities of debris and gas, some of which was ionized and therefore subject to electromagnetic forces. The sun's magnetic field therefore exerted a force on the surrounding gas, which in turn created drag that slowed the sun's rotation.

FIG. 17.7. A REGION OF STAR FORMATION. The dark patches in this portion of the constellation Ophiuchus are dense interstellar clouds, the kind of environment where cloud collapse and star (and planet) formation take place.

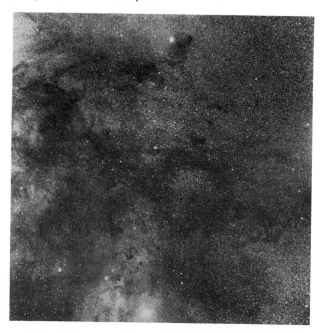

details). Once the cloud was falling in on itself, the rest of the process leading to the formation of the sun was inevitable, dictated by physical laws.

A combination of theory and observation tells us how the collapse proceeded (Fig. 17.8). The innermost portion of the cloud fell in on itself very quickly, leaving much of the outer material still suspended about the center. The rotation of the cloud sped up as its size diminished, and if the cloud had a magnetic field to begin with (most do), the field was intensified in the central part as a result of the condensation.

The core of the cloud began to heat up, because of the energy of impact as the material fell in. It eventually began to glow, first at infrared wavelengths, and finally, after a prolonged period of gradual shrinking, at visible wavelengths. Nuclear reactions began in the center when the temperature and pressure were sufficiently high, and the sun began its long lifetime as a star, powered by these reactions.

The steps involved in the formation of the sun that have been described up to this point are fairly well understood, and have been observed to be taking place today in many cloudy regions of the galaxy. In a num-

ber of stellar nurseries, infrared sources are found embedded inside dark interstellar clouds, indicating that newly formed stars are hidden there, still heating up. The details of planet formation are a little sketchier, however, because we have no way to observe the process as it occurs. All that we know about it has had to be inferred from observations of the end product: the present-day solar system.

As the central part of the interstellar cloud collapsed to form the sun, the outer portions were forced into a disk shape by rotation, just as Kant had shown. At this stage there was an embryonic sun surrounded by a flattened, rotating cloud called the **solar nebula.** The inner portions of the nebula were hot, and the outer regions were quite cold.

Throughout the solar nebula, the first solid particles began to form, probably by the growth of the interstellar grains that were mixed in with the gas. For every element or compound, there is a combined temperature and pressure at which it "freezes out" of the gaseous form (in direct analogy to the formation of frost on a cold night on earth). The volatile gases require a very low temperature in order to condense, so these

FIG. 17.8.   STEPS IN THE SUN'S FORMATION. An interstellar cloud, initially very extended and rotating slowly, collapses under its own gravitation. This happens most quickly at the center. As the collapse occurs, the internal temperature rises and the rotation rate increases. Eventually the central condensation becomes hot and dense enough to be a star, with nuclear reactions in its core (see chapter 21).

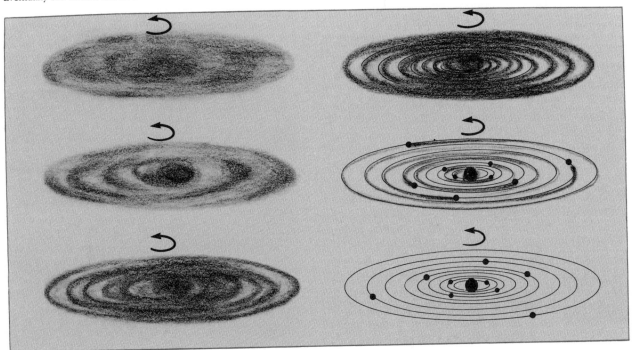

**TABLE 17.2.   STEPS IN THE EVOLUTION
OF THE SOLAR SYSTEM**

---

1. Gravitational collapse of an interstellar cloud.
2. Condensation of the young sun at the center, formation of a flattened disk by the surrounding nebula.
3. Condensation of refractory elements into solid objects in the inner portions of the nebula.
4. High-density concentrations formed in the disk by gravitational instabilities cause the buildup of planetesimals in the inner portion and the development of eddies in the outer portion.
5. Planetesimals coalesce to form the terrestrial planets; eddies in the outer part of the nebula collapse to form the giant planets and their ring and satellite systems.
6. Planetesimals orbiting the sun at about 2.8 AU from it are prevented from coalescing by the gravitational influence of Jupiter and instead form the asteroid belt.
7. Small, icy chunks of debris formed beyond Jupiter are gradually expelled to greater distances, forming the Oort cloud.
8. Extensive cratering occurs throughout the solar system because of impacts of planetesimals and other debris.
9. The sun's rotation is slowed by magnetic braking.
10. The remaining interplanetary debris is largely cleared out by a combination of the sun's T Tauri wind and impacts on planetary and satellite surfaces.

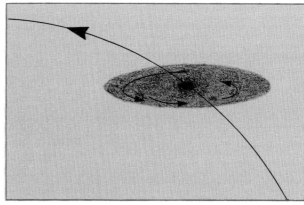

FIG. 17.9.   FORMATION OF A GIANT OUTER PLANET. Here a rotating disk has formed around a condensation in the outer solar system. Lumps in the disk grow to become satellites, except in the innermost portion where tidal forces prevent this, and instead a ring system forms. The planet that is forming at the center rotates rapidly as a result of its contraction from a much larger cloud. The entire system of planet, rings, and moons orbits the sun.

materials tended to stay in gaseous form in the inner portions of the solar nebula, but condensed to form ices in the outer portions. The elements that condense easily, even at high temperatures, are called **refractory elements,** and these were the ones that formed the first solid material in the warm inner portions of the solar nebula. This material therefore consisted of rocky debris containing only low abundances of the volatile species.

In due course, rather substantial objects built up, resembling asteroids in size and composition; these are referred to as **planetesimals.** By this time, gravitational instabilities had probably caused the solar nebula to build up eddies or condensations in some regions. Where these occurred, planetesimals had a greater chance of colliding, and in time these objects began to coalesce, forming the larger masses that became planets. In most cases, the planet that was formed rotated in the same direction as the overall rotation of the disk, but in two cases, Venus and Uranus, something happened to change the direction of rotation. Perhaps an unusually large planetesimal collided with

each planet at some point, altering the natural direction of spin.

The scenario just described apparently applies to the terrestrial planets, which therefore seem to have formed in two stages: (1) the condensation of refractory elements, leading to the development of planetesimals; and (2) the accretion of the planetesimals to form planets. The low quantity of volatile elements that characterizes the terrestrial planets was already established when the planets formed, and then was exaggerated by the release of volatile gases that occurred during their early histories, when they underwent molten periods.

For the outer planets, there is some uncertainty about the sequence of events that led to their formation. These planets contain a much higher proportion of volatile gases, which is to be expected, given the lower temperatures in the outer portions of the solar nebula. Thus planetesimals that formed there would have contained higher relative abundances of gases such as hydrogen and helium, and so would the planets that later formed through the coalescence of these planetesimals.

An alternative viewpoint has developed, however, in which the outer planets formed directly from the gas of the solar nebula, without an intermediate conden-

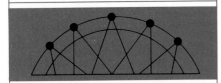

ASTRONOMICAL INSIGHT (17.1)

A Different Kind of Catastrophe

The catastrophic theories described in the text have as a common theme the formation of the planets from matter pulled out of the sun by a close encounter with another body. In the more modern versions of these theories, it is the differential gravitational force exerted by the other body that distorts the sun enough to cause some of its substance to break off. Consequently these theories are often called tidal theories. A radically different tidal theory was proposed in 1960.

The new tidal theory still invoked the disruption of one star by the gravitational attraction of another, but it was suggested that the sun disrupted a neighbor, rather than the reverse. The premise is that the sun formed as a member of a cluster of stars, and that after it had condensed to its present size, it passed near another cluster member that was just in the process of forming. This embryonic star was still rather extended and cool, so that it was easily torn apart by the sun's tidal force, and the material pulled out of it was not too hot to condense into planets.

This idea, first proposed by the astrophysicist M. Woolfson, has not been developed as completely as some other theories concerning the origin of the solar system, and may have faults that are not yet known. It does overcome the principal objections to the other catastrophic theories, however. It has failed to gain widespread favor in part because of the success of evolutionary theories.

One interesting aspect of Woolfson's hypothesis is the suggestion that the sun formed as part of a cluster. This was proposed for two reasons: it makes a near-collision between stars more likely; and it provides an opportunity for the young sun to be close to newly forming stars, creating the circumstances necessary for the rest of the scenario to be played out. There is, however, no evidence that the sun was ever part of a cluster of stars. On the other hand, it is possible that it was (some astronomers even claim that *all* stars form in clusters). A cluster of stars will naturally disperse, given enough time, partly because of the random motions of the stars within it, and partly because of the stretching and distortion it may suffer as it orbits the galaxy, occasionally passing through its spiral arms. The sun is roughly 4.5 billion years old, which is sufficient time for any cluster to be dissipated.

Although the Woolfson theory may have few obvious flaws, it is probably doomed to be largely ignored unless some serious problem is found with the evolutionary theory. The principle of Occam's Razor prevails, dictating that the simplest solution is most likely the correct one.

sation stage (Fig. 17.9). This idea is prompted in part by the extensive ring and satellite systems of the outer planets, which resemble miniature solar nebulae. The theory suggests that these planets simply formed like the sun did, from gravitational collapse of swirling gas clouds. The idea is that large eddies developed in the outer portions of the gaseous disk, and these became sufficiently dense to gravitationally fall in on themselves, forming their own small rotating disk systems in the process. The central portion of each disk then became a planet, whereas the outer portions underwent the same kinds of instabilities as the solar nebula, forming condensations that eventually became satellites (or rings, if the solid fragments that formed were inside the Roche limit, where coalescence into a satellite was prevented by tidal forces).

The fact that Uranus has a rotation axis that is tilted 98 degrees with respect to the ecliptic lends support to this idea. Apparently the rotating disk that was to become Uranus was tilted, so that not only the planet, but also its system of satellites and rings, were inclined equally, something that would be difficult to explain if Uranus was forced into its tilt at some time after the satellites had formed. An additional bit of evidence in favor of this theory regarding the formation of the outer planets is their rapid rotation rates, which would have developed naturally as the disks collapsed.

The asteroids probably formed as planetesimals, similar to those which eventually created the terrestrial planets, but were prevented from coalescing into a planet by the gravitational effects of Jupiter, as we have already noted (in chapter 15). The comets probably formed farther out, though not at the distance of the Oort cloud, where they now reside. At such a great distance (50,000 to 150,000 AU) from the center of the solar nebula, it is doubtful that any condensation could have occurred, because the density would have been too low. Therefore it is thought that the comets condensed at intermediate distances, probably near the orbits of Uranus and Neptune, and then were forced out to their present great distances by the gravitational effects of the major planets. Both Jupiter and Saturn, lying closer to the sun, would have forced the cometary bodies out farther, by speeding them up a little each time they passed by. Eventually most of the comets retreated to a great distance, forming the Oort cloud.

Once the planets and satellites were formed, the solar system was nearly in its present state, except that the solar nebula had not totally dissipated. There was still a lot of gas and dust swirling around the sun, along with numerous planetesimals that had not yet accreted onto planets. During this time the sun's magnetic field, pulling at ionized gas in the nebula, was very effective in slowing the sun's rotation, through the magnetic-braking process. Meanwhile the remaining planetesimals were floating about, now and then crashing into the planets and their satellites. Most of the cratering in the solar system occurred during this period, in the first billion years.

The leftover gas in the solar nebula was dispersed rather early, when the infant sun underwent a period of violent activity, developing a strong wind that swept the gas and tiny dust particles out into space (Fig. 17.10). This left behind only the larger solid bodies, the newly formed planets and the planetesimals. This wind phase has been observed in other newly formed objects called **T Tauri** stars, and is apparently a natural stage in the development of a star such as the sun.

Once most of the remaining interplanetary debris was eliminated, either by the T Tauri wind or by accretion onto planetary bodies, the solar system looked much like it does today. The planets each went their own ways, either continuing to evolve geologically, as

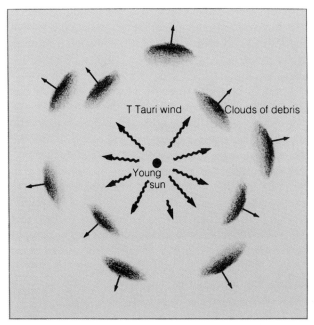

FIG. 17.10.   THE T TAURI PHASE. A star like the sun is thought to go through a phase very early in its lifetime when it violently ejects material in a high-velocity wind. This wind sweeps away matter, left over from the formation process, that is still drifting around the system.

most of the terrestrial planets have, or remaining perpetually frozen in their original condition, as the outer planets apparently have.

Are There Other Solar Systems?

As far as we know, the entire process by which our planetary system formed was a by-product of the sun's formation, requiring no singular, catastrophic event. We might expect, therefore, that the same processes have occurred countless times as other stars were born, and therefore there may be many planetary systems in the galaxy. Even if we rule out double- and multiple-star systems, in which perhaps all the material was either ejected or included in the stars, a vast number of candidates for having planetary systems remain.

To detect planets orbiting distant stars is not easy. For a long time astronomers suspected one star of having an orbiting planet. The gravitational pull of a very massive planet would cause its parent star to wobble a little in position as the planet orbited it. A star (labeled

Barnard's star) was thought to be wobbling in such a way, but a recent reexamination of the data has shown that there is no clear-cut indication of such a motion. At the moment the case for the existence of a massive planet circling Barnard's star is not well established.

It would be more direct if we could somehow see planets close to nearby stars, but this is very difficult. One reason is that the earth's atmosphere makes the image of a star appear extended and fuzzy, so that the star would drown out the feeble light from a close, dim object such as a planet. There is some hope of surmounting this problem with the *Space Telescope*, which will be situated above the earth's atmosphere, avoiding the fuzziness that it causes. It is expected that the *Space Telescope* will allow astronomers to see objects comparable to Jupiter, if they orbit nearby stars. There is little hope of directly detecting smaller objects such as the terrestrial planets, however. The clarity of

the *Space Telescope* images will also allow measurements of stellar motions to be more precise than those made from earth, so that tiny oscillations in the positions of stars will be more easily detected. Again, this probably will not allow the detection of small planets like the earth, but could locate Jupiter-like objects.

New techniques for observing from the ground are also being developed; perhaps eventually they will be accurate enough to detect small wobbles in stellar positions. One especially promising technique is **speckle interferometery;** by analyzing how the light waves are displaced and disorganized on the way through the earth's atmosphere, scientists are able to eliminate the distorting effects of our atmosphere. Astronomers are waiting eagerly for the chance to use these new instruments and techniques in a search for planets in other solar systems. It will be fascinating to learn the results in the coming years.

Perspective

We have outlined a natural, evolutionary process that seems to account for all the observed properties of the solar system. The orbital characteristics, as well as the compositions of the planets, the sun's slow rotation, and the nature and motions of the interplanetary material are now understood. Many details have yet to be worked out, but the overall scenario is probably in essence correct. We have completed our examination of the solar system, and we are ready to move on. The next logical step is to explore the realm of the stars.

Summary

1. The facts that must be explained by a successful theory of the formation of the solar system include the systematics of the orbital and spin motions of the planets and satellites, the contrasts between the terrestrial and giant planets, the slow rotation of the sun, and the existence and properties of the interplanetary bodies.
2. There are two general classes of hypotheses: catastrophic theories and evolutionary theories.
3. Catastrophic theories state that some singular event, such as a near-collision between stars, distorted the sun by tidal forces, pulling out matter that condensed to form the planets.
4. Evolutionary theories postulate that the planets formed as a natural by-product of the formation of the sun. These theories, although they are simpler, were not accepted for awhile because of their failure to explain the slow spin of the sun.

5. When magnetic braking was understood in the 1960's, the evolutionary theory became most widely accepted.
6. In the modern theory, the planets formed from condensations in the solar nebula, the flattened disk of gas and dust that formed around the young sun.
7. The terrestrial planets formed from the coalescence of planetesimals; solid, asteroid-like objects that condensed from the hot inner portions of the solar nebula.
8. The giant planets formed from small-scale eddies in the solar nebula that contracted and increased their spins, in parallel with the same processes that occurred in the solar nebula itself.
9. In the evolutionary theory, it is expected that many other solar systems should exist around other stars, but so far none has been detected because of the difficulty of observing planets at interstellar distances.

Review Questions

1. Why are the rotation of Venus and the retrograde orbits of some satellites not explained by the theory of solar-system formation that is described in the text?

2. We might have mentioned, as another fact that must be explained, the fraction of the total mass of the solar system that is contained in the sun. Using data from tables in the appendices (or from the tables in chapters 7 through 16), calculate the fraction of the total mass of the solar system that is contained in the sun.

3. How might the present-day solar system be different if the sun had been formed with no magnetic field?

4. The terrestrial planets have very low abundances of volatile elements. They also have lower masses than the giant planets, which have about the same composition as the sun. If you assume that 10 percent of the earth's mass is in the form of iron, what would the mass of our planet be if it were composed of all the elements in the same proportion to iron as in the sun? How would the earth compare with Jupiter in this case? (Hint: you will need to use data from table 16.2).

5. Summarize the differences between the formation of the terrestrial planets and that of the giant planets in terms of the evolutionary theory described in the text.

6. Compare the differences between the terrestrial and giant planets with the differences between the inner and outer of Jupiter's four Galilean satellites. Explain the contrasts among the Jovian satellites in terms of the evolutionary theory of the solar-system formation.

7. Why does the evolutionary theory make it doubtful that a major tenth planet exists?

8. The giant planets rotate rather rapidly, in comparison to the terrestrial planets. Why is this so? What forces might be acting to slow their rotations?

9. How much dimmer would Jupiter appear from the earth if it orbited alpha Centauri, the nearest star? The distance to alpha Centauri is about 4.3 light-years, or 270,000 AU.

Additional Readings

Beatty, J. K., O'Leary, B., and Chaikin, A., eds. 1981. *The new solar system*. Cambridge, England: Cambridge University Press.

Cameron, A. G. W. 1975. The origin and evolution of the solar system. *Scientific American* 233(3):32.

Head, J. W., Wood, C. A., and Mutch, T. A. 1977. Geologic evolution of the terrestrial planets. *American Scientist* 65:21.

Reeves, H. 1977. The origin of the solar system. *Mercury* 7(2):7.

Sagan, C. 1975. The solar system. *Scientific American* 233(3):23.

Wetherill, G. W. 1981. The formation of the earth from planetesimals. *Scientific American* 244(6):162.

Surveying the Solar System

Harold Masursky

Dr. Masursky is a senior scientist with the Branch of Astrogeological studies of the U.S. Geological Survey, and has been a leader in the exploration of landforms and the development of geological histories of the terrestrial planets and the moon. He was associated with the radar-mapping experiment on the Pioneer Venus *mission, and is actively advocating the launch of a higher-resolution radar mapper to better define the similarities and differences between the earth and its sister planet.*

We have made great progress in the exploration of the solar system by spacecraft. In the inner solar system we have returned samples from the moon and made many measurements by lander and orbiters of the geology, geophysics, and geochemistry of the lunar crust and interior. We have orbited Mercury and mapped half of this ancient and cratered heavenly body that so resembles our moon.

We have orbited Venus and mapped 93 percent of its surface by radar. The Soviets have put down five successful landers on the surface, taken pictures, and measured the chemistry. From these data we know that although Venus is by size and density almost a twin sister of the earth, its atmosphere and crustal evolution are very different.

Although the crust of Venus is broken by many fractures and there are volcanic materials ex-

truded along them, there is not a system of integrated plate tectonics present. The earth's crust is dominated by extrusions of lava along the mid-ocean rifts; these spread laterally and dive under the continents along subduction troughs. The ocean crust is all less than 200 million years old. The continents and plateaus of Venus are more extensive and higher than the earth's, and the largest mountain is higher than Mt. Everest. In contrast, the deep basins are only one-third as extensive as the earth's and have only one-fifth the maximum depth. Venus may once have had half the volume of oceans that the earth has, but the greenhouse effect dispersed the water; now the atmospheric pressure is ninety times that of the earth, and the surface tempera-

ture is 740 K—hot enough to melt lead.

It appears that microorganisms on the earth may have been responsible for pulling CO_2 from the atmosphere and depositing it as limestone. This not only moderated the environment but may also be controlling the crustal processes that result in plate tectonic motions. This means that the role of life on the earth, and the absence of that influence on Venus, have had a profound effect in creating the dichotomy that now exists between the two planets. In order to better study the present state of Venus, for further comparison with the early earth, plans are being made for a new Venus orbiter with a high-resolution radar imager that will map the surface terrain in far finer detail than has been previously possible.

We have investigated Mars with orbiters and landers. Although we did not find life on Mars, and therefore we assume that its influence on the evolution and crust of that planet is absent, we did find a complicated water story. It appears that, despite the fact that water cannot now survive on Mars in liquid form, it has carved great channels on the Martian surface at many times in the past. Local volcanic heat may have melted subsurface ice, releasing the water, or the climate may have been warmer in the past, allowing water to flow. We hope to bring back samples from Mars in the future, among other things to test whether glacial and interglacial periods on Mars coincided with those of the earth, which would indicate whether these climatic variations are due to changes in the solar output or are instead caused by fluctuations in the internal heat engines of the planets themselves.

In the outer solar system we have investigated the systems of Jupiter and Saturn, and will study Uranus, Neptune, and possibly Pluto in the future. The Jovian satellites vary from inner, hyperactive Io, with its nine active volcanoes at the time of the *Voyager* flyby, to the outer, totally inactive Callisto, whose only modifications over its multibillion-year lifetime have come from meteorite impacts on its surface. The activity on Io is governed by tidal interactions with Jupiter and the neighboring satellite Europa, which knead and heat the crust and interior of Io. The tiny icy satellites of Saturn also show dynamic internal activity that may be driven by tidal interaction; Enceladus is a case in point. We have seen strong electrostatic and/or electromagnetic interactions in the ring particles of Saturn, and similar effects may modify the small icy satellites also. The rich interplay of planetary magnetic fields, rings, satellites, and radiation belts is a subject for much more detailed study, and comprises one of the prime objectives of the upcoming *Galileo* mission, planned to orbit Jupiter and to drop a probe into its atmosphere within the decade.

These advances have caused us to look at the evolutionary history of the earth with new, powerful insights, and have modified our ideas of the fundamental processes that have affected the development of the earth and its companion planets.

The Stars

INTRODUCTION TO SECTION IV

This section is devoted to the stars and their properties. In its six chapters we learn about how stars are observed, how their fundamental properties are deduced from observations, and how stars work. We see what the processes are that control the structure of stars, and we learn how stars live, evolve, and die. We discover that stars are dynamic, changing entities, and only seem fixed and immutable because the time scales on which they evolve are vastly longer than the human lifetime.

The first of the six chapters describes the three basic types of stellar observations; measurements of positions, brightnesses, and spectra. The emphasis is on raw information that is gathered from observation, whereas discussion of how this information is used to derive the more basic stellar parameters is reserved for the next chapter. There we sys-tematically consider several stellar properties, discuss how each is determined, and summarize the typical values that are found.

Chapter 20 then initiates the discussion of how stars function and evolve by describing the physical processes that occur in stars, and how these processes govern the stellar parameters previously discussed. Here we draw upon knowledge of our own sun, whose well-observed properties provide vital information about other stars as well. (Although the present discussion is self-contained, and does not require knowledge of the contents of chapter 16, there are references to that chapter; thus it is recommended that it be read in conjunction with this section of the book.)

Chapter 21 illustrates how astronomers can learn about stellar evolution from observations, par-ticularly of star dusters. We will find that despite the immensity of the time scales involved, it is possible to deduce the life stories of stars.

The final two chapters, 22 and 23, describe the evolution of stars, from their births to their deaths. In chapter 22 we have to rely on the results of theoretical calculations, but we are guided by observations discussed earlier. In chapter 23 we discuss the remnants of stars that have finished their lives, and here we discover some of the most bizarre and exciting objects in the universe.

Having studied the stars, we will then be ready to consider large groupings such as our own galaxy, which contains many billions of stars, and whose own evolution is governed in part by that of the individual stars within it.

CHAPTER 18

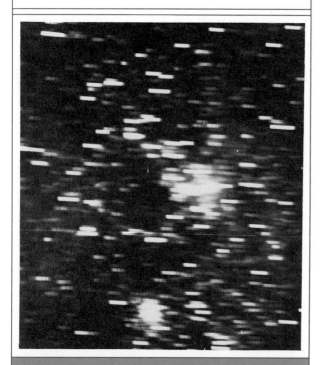

Stellar Observations: Positions, Magnitudes, and Spectra

Everything we can learn about a star is contained in the light that we receive from it. Fortunately, a lot of information is there, and astronomers have learned how to extract much of it. A wide variety of observations can be carried out, and they generally constitute three broad classes: (1) measurements of position; (2) measurements of brightness; and (3) measurements of spectra. In this chapter we will discuss these types of observations and what they tell us directly.

Positional Astronomy

For many centuries, positional measurements were practically the only kind of observation made, as we learned in chapter 2. Although some heed was paid to stellar brightnesses and especially to changes that occurred in the heavens, it was the positions of the stars and planets that were systematically observed and recorded. This was an inexact science until the first sextants (Fig. 18.1) and quadrants were available to provide accurate measurements of angular distances. In modern times highly precise positions can be measured.

The principle technique used today in **astrometry,** the science of measuring star positions, is to photograph the sky and to carefully measure the positions of the star images on the photographic plates. If a number of photographs of a given portion of the sky are taken and measured separately, the results can be averaged to produce a more accurate determination than is possible by measuring a single photograph. Today we can measure a stellar position to precision of much less than 0.01 second of arc (recall that a second of arc is one-sixtieth of a minute of arc, and that a minute is one-sixtieth of a degree; there are 1,296,000 arcseconds in a full circle). When the *Space Telescope* is in operation in the mid-1980's, even more accurate measurements will be possible, because the fuzziness of star images caused by the earth's atmosphere will be eliminated.

FIG. 18.1. A SEXTANT. Instruments of this type were used for centuries for the measurement of star positions relative to each other. Even in modern times, sextants are used as tools for navigation.

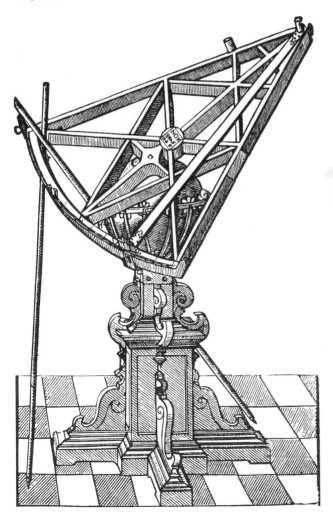

Nearby stars generally are found to be moving slowly across the sky (Fig. 18.2). All stars in the galaxy have individual, random motions, and in this sense they resemble bees in a swarm (although the galaxy has an overall rotation, whereas a swarm of bees does not). If a star is too far from the sun, even the most careful measurements will fail to reveal any motion. The apparent gradual changes in a star's position are called **proper motions**.

Astrometric measurements reveal another kind of motion, one referred to often in chapters 2 and 3. From ancient times through the Renaissance, astronomers were perplexed by the fact that no stellar parallax could be measured, even with the measurement precision available to Tycho Brahe. This failure to detect a parallax led many, including Brahe, to reject the notion that the earth was orbiting the sun, since it seemed obvious that if it were, the stars would appear to shift back and forth as a reflection of our motion. Eventually the heliocentric theory won out, but stellar parallax was still undetected.

Finally, in 1838, parallaxes were detected independently by F. W. Bessel, F. G. W. Struve, and T. Henderson, each observing a different nearby star. This was clearly a case where a great discovery awaited the development of suitable techniques for making the measurement. The reason stellar parallaxes had defied earlier observers became obvious: even for the closest stars, the maximum shift in position was much less than one second of arc! The stars are simply much farther away

FIG. 18.2. PROPER MOTIONS. Each star in the sky moves at its own velocity in its own direction. What we observe from earth are proper motions, which are measured in angular units, and are usually much smaller than 1 arcsecond per year. The measurement of such small motions requires that observations of star positions be made several years apart.

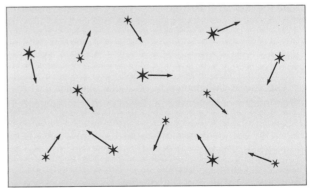

ASTRONOMICAL INSIGHT (18.1)

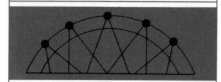

Star Names and Catalogues

From the earliest times when astronomers systematically measured star positions, they began to compile lists of these positions in catalogues. Hipparchus developed an extensive catalogue, as did early astronomers in China and other parts of the world, more than two thousand years ago (see chapter 2).

To list stars in a catalogue of positional measurements requires some kind of system for naming or numbering the stars, and a variety of such systems have been employed. The ancient Greeks designated stars by the constellation and the brightness rank within the constellation. For example, the brightest star in Orion is α Orionis, the next

brightest is β Orionis, the next is γ Orionis, and so on (the constellation names are spelled in their Latin versions, since this system was perpetuated by the Romans and the Catholic church after the time of the Greeks). Today many stars are still referred to by their constellation rankings, and the Greek alphabet is used to designate the rank.

After the time of Ptolemy, when western astronomy went into decline for some 1,300 years, Arab astronomers, occupying northern Africa and southern Europe, carried on astronomical traditions. These people assigned proper names to many of the brightest stars, and these names, such as Betelgeuse (α Orionis), Rigel (β Orionis), and Bellatrix (γ Orionis), are also still in use.

With the advent of the telescope, when many new stars were discovered that were too faint to be seen with the naked eye, catalogues rapidly outgrew the old naming systems. Generally each catalogue assigned numbers to the stars, usually in a sequence related to their positions. Thus modern catalogues,

such as the *Henry Draper Catalog*, the *Boss General Catalog*, the *Yale Bright Star Catalog*, and the *Smithsonian Astrophysical Observatory Catalog*, list stars by increasing right ascension (that is, from west to east in the sky, in the order in which the stars pass overhead at night). In some cases, separate listings are made for different zones of declination (for different strips of sky, separated in the north-south direction). Because each of these catalogues includes a different particular set of stars, each has its own numbering system, and often no cross-reference to other catalogues is provided. Hence a given star may have a constellation-ranked name, an Arabic name, and numbers in a variety of catalogues, and the astronomer must match up the coordinates to be certain of referring to the same star in different catalogues. Actually things are not quite so bad as that, since some modern catalogues do provide cross-references to others. Nevertheless, one of the principal tasks of a would-be astronomer is to become familiar with catalogues and their use.

than the ancient astronomers had dreamed possible. The annual parallax motion of the star alpha Centauri, the nearest to the sun, is comparable to the angular diameter of a dime as seen from a distance of about two miles, a very small angle indeed.

The successful detection of stellar parallax led to a direct means of determining distances to stars (Fig. 18.3). The amount of shift in a star's apparent position resulting from our own motion about the sun depends on how distant the star is; the closer it is, the bigger the shift. The parallax angle π is defined as one-half the total angular shift of a star during the course of a year.

This angle, once it is measured, tells us the distance to a star, in a unit of measure called the **parsec.** A star whose parallax is 1 arcsecond is, by definition, 1 parsec away (the word parsec, in fact, is a contraction of *parallax-second,* meaning a star whose parallax is 1 arcsecond). In mathematical terms, the distance to a star whose parallax is π is $d = 1/\pi$. Thus if a star has a parallax of 0.4 arcseconds, it is $1/0.4 = 2.5$ parsecs away. Recall that the closest star has a parallax of less than 1 arcsecond; we conclude therefore that even this star is more than 1 parsec away.

In more familiar terms, a parsec is equal to 3.26

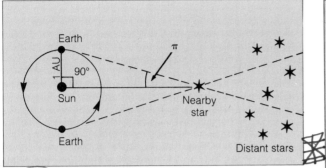

FIG. 18.3. STELLAR PARALLAX. As the earth orbits the sun, our line of sight toward a nearby star varies enough in direction to make that star appear to move back and forth with respect to more distant stars. The parallax angle π is defined as one-half of the total amount of apparent annual motion. If π is 1 second of arc, then the sun-star distance is 206,265 times the sun-earth distance. This defines the parsec, the distance to a star whose parallax angle is 1 arcsecond.

light-years, or 3.08×10^{18} centimeters, or 206,265 astronomical units. The use of parallax measurements to determine distances is a very powerful technique, and is the only direct means astronomers have for measuring how far away stars are. Unfortunately, parallaxes are large enough to be measured only for stars rather near us in the galaxy. The smallest parallax that can be measured is about 0.001 arcseconds, corresponding to a distance of 0/.001 = 1,000 parsecs, and accurate measurements can be made only for somewhat larger parallax angles, corresponding to distances within a few hundred parsecs. The galaxy, on the other hand, is more than 30,000 parsecs in diameter! Clearly, other distance-determination methods are needed if we are to probe the entire galaxy. One very powerful method will be described in the next chapter.

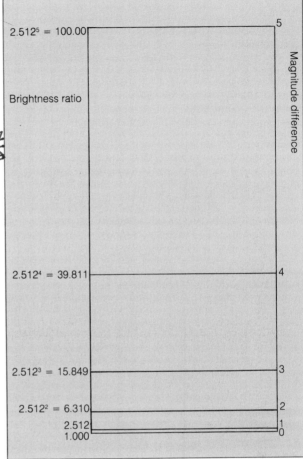

FIG. 18.4. STELLAR MAGNITUDES. This diagram shows schematically how brightness ratios and magnitude differences are related. To determine the relationship between brightness ratio and fractional magnitudes requires calculations that involve logarithms (see appendix 9).

Stellar Brightnesses

The second general type of stellar observation is the measurement of the brightnesses of stars. This was first attempted in a systematic way by Hipparchus, who, more than two thousand years ago, established a system of brightness rankings that is still with us today. Hipparchus ranked the stars in categories called **magnitudes,** from first magnitude (the brightest stars) to sixth (the faintest visible to the unaided eye). In his catalogue of stars, and in all since then, these magnitudes are listed along with the star positions.

By the mid-1800's, astronomers were beginning to make precise measurements of stellar brightnesses, rather than estimates made by the naked eye. It was discovered that what the eye perceives as a fixed **difference** in intensity from one magnitude to the next actually corresponds to a fixed intensity **ratio.** Measurements showed that a first-magnitude star is about 2.5 times brighter than a second-magnitude star, a second-magnitude star is 2.5 times brighter than a third-magnitude star, and so on (Fig. 18.4). The ratio between a first-magnitude star and a sixth-magnitude star was found to be nearly 100. In 1850, the system was formalized by the adoption of this ratio as exactly 100; thus, the ratio corresponding to a one-magnitude difference is the fifth root of 100, or $(100)^{1/5} = 2.512$.

Thus a precise, quantitative magnitude system was developed, and with the worldwide adoption of a few standard stars used for calibration, all astronomers could measure magnitudes in a comparable way. To do so no longer depended on rough estimates made by naked eye. Today photoelectric devices are used to make the measurements (Fig. 18.5). These are devices that produce an electric current when light strikes them; they are used in many familiar applications, such as door-openers in modern buildings. The amount of electrical current produced is determined by the intensity of light, so an astronomer need only measure the current to determine the brightness of a star.

Let us work out a few examples in order to gain familiarity with the magnitude system. If a first-magnitude star is 2.512 times brighter than a second-magnitude star, and a second-magnitude star is 2.512 times brighter than a third-magnitude star, then the first-magnitude star is $2.512 \times 2.512 = 6.3$ times brighter than the third-magnitude star. Similarly, a second-magnitude star is $2.512 \times 2.512 \times 2.512 = 15.9$ times brighter than a fifth-magnitude star. All we need to remember is that a *difference* of one magnitude corresponds to a brightness *ratio* of 2.512. It is sometimes confusing that the *brighter* a star is, the *smaller* the magnitude is.

Once magnitudes could be measured precisely, it was found that stars have a continuous range of brightnesses, and do not fall neatly into the various magnitude rankings. Therefore fractional magnitudes must be used; Deneb, for example, has a magnitude of 1.26 in the modern system, although it was classified simply as a first-magnitude star in the old days. Each of the former categories is found to include a range of stellar brightnesses. This is especially true for the first-magnitude stars, some of which turned out to be as much as two magnitudes brighter than others. To measure these especially bright stars in the modern system requires the adoption of magnitudes smaller than 1. Sirius, for example, which is the brightest star in the sky, has a magnitude of -1.42. By using negative magnitudes for very bright objects, astronomers can extend the system to include such objects as the moon, whose magnitude when full is about -12, and the sun, whose magnitude is -26 (thus the sun is 25 magnitudes, or a factor of $2.512^{25} = 10^{10}$, brighter than Sirius). A simple mathematical method for dealing with any magnitudes, even fractional or negative ones, is described in Appendix 9.

Of course, there are stars too faint to be seen by the human eye, so the magnitude scale must also extend beyond sixth magnitude. With moderately large telescopes, astronomers can measure stars as faint as fifteenth magnitude, and with long-exposure photographs taken with the largest telescopes, they can measure stars as faint as twenty-fifth magnitude. Such a star is nineteen magnitudes, or a factor of $2.512^{19} = 4 \times 10^7$, fainter than the faintest star visible to the unaided eye. The *Space Telescope* is expected to detect stars as faint as twenty-eighth magnitude, which is a factor of $2.512^3 = 15.9$ fainter yet.

The stellar magnitudes we have discussed so far all refer to visible light. Aas we learned in chapter 5, however, stars emit light over a much broader wavelength band than the eye can see. We also learned that the wavelength at which a star emits most strongly depends on its temperature. Therefore by measuring a star's brightness at two different wavelengths, astronomers can learn something about its temperature. Filters that allow only certain wavelengths of light to pass through are used for this purpose. Typically, a star's brightness is measured through two such filters, one that passes yellow light, and one that allows blue light to pass, resulting in the measurement of V (for visual, or yellow) and B (for blue) magnitudes. Since a hot star emits more light in blue wavelengths than in yellow, its B magnitude is *smaller* than its V magnitude.

FIG. 18.5. MEASURING STELLAR MAGNITUDES. This is a schematic illustration of photometry, the measurement of stellar brightnesses. Light from a star is focused (often through a filter that screens out all but a specific range of wavelengths) onto a photocell, a device that produces an electric current in proportion to the intensity of light. The current is measured, and, through comparison with standard stars, converted into a magnitude.

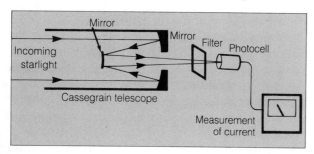

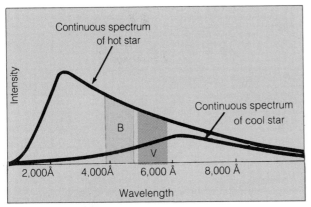

FIG. 18.6. THE COLOR INDEX. Here are continuous spectra of two stars, one much hotter than the other. The wavelength ranges over which the blue (B) and visual (V) magnitudes are measured are indicated. We see that the hot star is brighter in the B region than in the V region of the spectrum; therefore, its B magnitutde is *smaller* than its V magnitude, and it has a *negative* B-V color index. The opposite is true for the cool star.

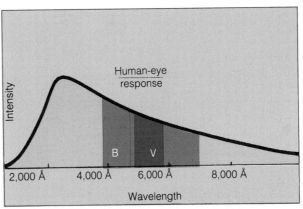

FIG. 18.7. THE BOLOMETRIC MAGNITUDE. This shows the continuous spectrum of a star, and the approximate portions of the spectrum that are covered by the human-eye estimate of the magnitude, and by the B and V magnitudes. None of these includes *all* of the light the star emits. The bolometric magnitude, however, is a measure of all the light, at all wavelengths, and usually requires ultraviolet or infrared observations.

For a cool star, the situation is reversed, and the B magnitude is larger than V (remember that the magnitude scale is backward in the sense that a smaller magnitude corresponds to greater brightness).

The difference between the B and V magnitudes is called the **color index** (Fig. 18.6). The exact value of this index is a function of the temperature of a star, so stellar temperatures can be estimated simply by measuring the V and B magnitudes. A very hot star might have a color index B-V = −0.3, whereas a very cool one might typically have B-V = +1.2.

One additional type of magnitude should be mentioned, although it is a very difficult one to measure directly. This is the **bolometric magnitude,** which includes all the light emitted by a star at all wavelengths. To determine a bolometric magnitude requires ultraviolet and infrared observations, as well as visible, and can best be done with the use of telescopes in space. This is particularly true for hot stars, which emit a large fraction of their light in ultraviolet wavelengths. For cool stars, which emit little ultraviolet light but a lot of infrared radiation, bolometric magnitudes can be measured from the ground with fair accuracy. The bolometric magnitude of a star is always smaller than the visual magnitude, because more light is included when all wavelengths are considered than when only the visual wavelengths are measured. We will see in the next chapter how bolometric magni-

tudes are related to other properties of stars, particularly the luminosity.

The Appearance of Stellar Spectra

Observations of stellar spectra began in the mid-1800's, well before scientists understood the physics of the atom and how it forms spectral lines. A great deal was learned about the appearance of the spectra of stars, even though the full significance was not appreciated until much later. For most of the nineteenth century, stellar spectra were examined through the use of a **spectroscope,** which is simply an eyepiece focused on the light emerging from a prism, which in turn is placed at the focus of a telescope. The first stellar spectra to be recorded, in the 1870's, were exposed on photographic plates, a technique that is still widely used. As we learned in chapter 6, however, modern astronomers are turning more and more to new electronic detectors to record the spectra of stars.

Regardless of how it is done, the basic principle is the same: the light is dispersed according to wavelength, and its intensity is recorded at each wavelength. The result is called a **spectrogram** if recorded photographically (Fig. 18.8), or simply a spectrum if recorded by other means.

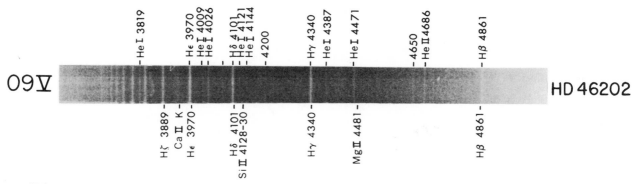

FIG. 18.8. A STELLAR SPECTRUM. This photograph shows the manner in which stellar spectra are usually displayed, as negative prints. Hence light features are absorption lines, wavelengths at which little or no light is emitted by the star.

NORMAL STARS AND SPECTRAL CLASSIFICATION

Among the first astronomers to systematically examine spectra of a large number of stars was the Roman Catholic priest Angelo Secchi, who in the 1860's catalogued hundreds of spectra, using a spectroscope. He found the appearance of the spectra to vary considerably from star to star, although they were consistent in one respect: they all showed continuous spectra with absorption lines. The work of Kirchhoff soon explained this to be the result of the relatively cool outer layers of a star, which absorb light from the hotter interior.

The primary differences between spectra of different stars lay in the differing strengths and patterns of the absorption lines. One star might have a particular series of strong absorption lines, whereas another might not have these lines at all, or they might be much weaker. Since Bunsen and Kirchhoff were able to identify many of the lines with known chemical elements, it was naturally assumed that stars with different patterns of lines were made of different elements. As we will see, this was not quite correct.

By the end of the nineteenth century, others had taken up and extended the work of Secchi; they used photographic plates to record the spectra, which allowed more detailed analyses (Fig. 18.9). Most notable were a group of astronomers at Harvard University, led by Edward C. Pickering. A sequence of spectral categories or classes was established, based on the different types of spectra that were being found, and Pickering's group set out to classify the spectra of all the stars in the sky brighter than ninth magnitude, some

FIG. 18.9. A MULTITUDE OF SPECTRA. The job of photographing numerous stellar spectra for classification and cataloguing is accomplished with the use of a thin prism in front of the telescope, so that each star image becomes a tiny spectrum. This is one of the hundreds of such objective prism photographs used by the Harvard group in their massive compilation of spectral classifications. The bright concentration near the center is Eta Carinae, a group of very hot, luminous stars and nebulosity.

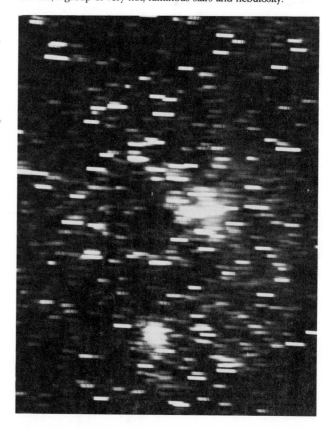

FIG. 18.10.   ANNIE J. CANNON. Cannon, a member of the Harvard College Observatory for almost fifty years, classified the spectra of several hundred thousand stars. At right is shown a page from one of her notebooks. Today she is recognized as the founder of modern spectral classification.   ⌐ *founder of Modern Spectral Classification*

200,000 in all. This immense task was eventually carried out by Annie J. Cannon (Fig. 18.10), who today is regarded as the founder of modern spectral classification (and as a pioneer of women's leadership in astronomy).

Stars were placed in categories according to which lines were prominent in their spectra, and the categories were simply labeled with letters of the alphabet. The sequence began with A stars, which showed the strongest lines of hydrogen, the simplest element. The B stars were those in which the next element, helium, dominated the spectrum. Other classes followed, in many cases not based on known chemical elements, but simply lumped together by reason of similar appearance and then assigned a letter of the alphabet. The spectral types were found, however, to represent a smooth sequence if arranged in the order O, B, A, F, G, K, M, (Fig. 18.11), although at that time the reason

Text continues on page 288.

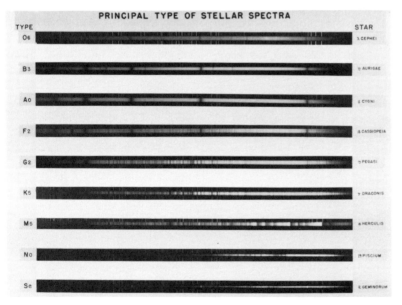

FIG. 18.11.   A COMPARISON OF SPECTRA. *Here we see several spectra representing different spectral classes, arranged in order of decreasing stellar temperature (top to bottom).*

Stellar Spectroscopy and the Harvard Women

There is some irony to be found in the fact that Harvard University, long a stronghold of the all-male tradition in American colleges, was also the institution that nurtured some of the nation's first leading women astronomers. Today, women are still underrepresented among professional astronomers, but were even more so at the turn of the century when the foundations of modern stellar spectroscopy were developed at the Harvard College Observatory.

In 1877, when Edward C. Pickering became director of the observatory, Secchi's work (described in the text), based on visual inspection of stellar spectra through a spectroscope, was the only attempt that had been made to classify stars according to the appearance of their spectra. Henry Draper, an American amateur astronomer, had in 1872 become the first to photograph the spectrum of a star. Upon Draper's death in 1882, his widow endowed a new department of stellar spectroscopy at Harvard. Pickering, as director, hired among his assistants a number of women, several of whom went to work on the problem of classifying spectra of stars.

An innovative technique was used to photograph spectra of large numbers of stars. A thin prism was placed in front of the telescope, so that each star image on the photographic plate at the focus was stretched, in one direction, into a spectrum (see Fig. 18.9). If color film had been used, each stellar image would have looked like a tiny rainbow. Pickering and his group refined this **objective prism** technique to the point where all stars in a field of view as faint as ninth or tenth magnitude would appear as spectra well-enough exposed for classification. Thus it became possible to amass stellar spectra in vast quantities.

The problem of sorting out and classifying the spectra fell to Pickering's associates. One of them, Williamina Fleming, published the first *Draper Catalog of Stellar Spectra* in 1890. In this catalogue, some 10,351 stars in the northern hemisphere were assigned spectral classes A through N, in a simple elaboration of a rudimentary classification scheme adopted earlier by Secchi. A number of Fleming's classes were later dropped.

While the first catalogue was being prepared, a niece of Henry Draper, Antonia Maury, joined the staff and set to work on the analysis of spectra of bright stars. For these objects, spectra could be photographed through the use of thicker prisms (actually, a series of two or three thin prisms), so that the spectra were more widely spread out according to wavelength, which allowed greater detail to be seen. Maury concluded, on the basis of these high-quality spectra, that stars should be grouped into three distinct sequences, rather than just one. She discovered that some stars had unusually narrow spectral lines, others were rather broad, and a third group was in between. It later was found that her sequence c, the thin-lined stars, are giant stars (the lower atmospheric pressure in these stars causes the lines to be less broadened than in main-sequence stars). Maury's discovery helped Einar Hertzsprung confirm the distinction he had recently found between giant and main-sequence stars, and he thought her work to be of fundamental importance. Unfortunately, Maury's separate sequences were not adopted in the subsequent work on classification at Harvard, which was thereafter based on a single sequence.

The most important of the Harvard workers in spectral classification arrived on the scene in 1896. Her name was Annie J. Cannon, and she gradually modified the classification system to the present sequence, finding that the arrangement O, B, A, F, G, K, M was a logical ordering, with smooth transitions from one type to the next (at the time, no one knew this was a temperature sequence). Cannon was also able to distinguish the gradations so finely that she established the ten subclasses for each major division that are in use today. In 1901 she published a catalogue of classifications for 1,122 stars, and then embarked on her major task: the classification of more than 200,000 stars whose spectra appeared on survey plates covering both hemispheres. By this time she was so skilled that she could reliably classify a star in a few moments, and most of the job was done in a four-year period between 1911 and 1914. The resulting *Henry Draper Catalog* appeared in nine volumes of the *Annals of the Harvard College Observatory,* the final one being published in 1924. Pickering died in 1919, before the cat-

alogue was complete, and was succeeded as director by Harlow Shapley. Cannon continued her work, later publishing a major *Extension* of the catalogue, along with a number of other specialized catalogues. She died in 1941, and the American Astronomical Society subsequently established the Annie J. Cannon prize for outstanding research by women in astronomy.

The successes of the Harvard women were not confined to stellar spectroscopy. Another major area of interest to Pickering and later to Shapley was the study of variable stars, and Henrietta Leavitt, who joined the group as a volunteer in 1894, played a leading role in this area. Following some early work in the establishment of standard stars for magnitude determinations, Leavitt by 1905 was at work identifying variables from comparisons of photographic plates (she was eventually to discover more than 2,000 of

them). In the process, she examined the Magellanic Clouds for variables, and noticed that their periods of pulsation correlated with their average brightnesses. This was a discovery of profound importance, for the period-luminosity relationship for variable stars was to become an essential tool for establishing both the galactic distance scale and the intergalactic scale (see discussions in chapters 24 and 27).

The Harvard women, including those discussed here and several whose names are now obscure, were a remarkable group, responsible for a number of major advances in the science of astronomy at a time when its basis in physics was just becoming clear. In this day of awakening recognition of the proper role of women in all areas, it is fitting to consider and appreciate the pioneering work done by this group.

A large group of Harvard astronomy assistants, about 1917. Sixth from the left is Henrietta Leavitt, who discovered the period-luminosity relationship for variable stars (see chapter 24); and fifth from the right is Annie J. Cannon, whose contributions to spectral classification are described in this chapter.

for this sequence was unknown. The Harvard classifications were collected in a massive compilation called the *Henry Draper Catalog* (named after an astronomer whose estate sponsored its publication; Draper had been the first, in 1872, to photograph a stellar spectrum), and today stars are commonly referred to by their HD numbers.

Through the work of Bohr, who showed how spectral lines are formed by atoms, and the work of later researchers such as Sir Arthur Eddington, Henry Norris Russell, and Sir James Jeans, it was later deduced that stars of different spectral classes do not differ significantly in chemical composition; they are all made of basically the same ingredients. What does differ from one spectral class to the next is temperature.

We learned in chapter 5 that the state of ionization of a gas determines which spectral lines it forms. When an atom has lost one of its electrons, the ion, as it is now called, has an entirely different set of energy levels through which the remaining electrons can jump, so it has an entirely different set of spectral lines. The number of electrons lost by the atoms in a gas depends on its temperature; thus, the hottest stars have spectra of highly ionized gases, whereas the coolest stars show little or no ionization (Fig. 18.12). The difference in ionization from one temperature to another is responsible for the differences in the appearance of the spectra from one class to the next. Hence the sequence of spectral types is really a sequence of stellar temperatures. Refinements in the classification guidelines, along with improvements in the quality of spectrograms, have led to the creation of subclasses, so that we have B0, B1, B2 stars, and so on. Each major division has up to ten subdivisions. The sun, in this modern system, is a G2 star.

PECULIAR SPECTRA

Not all stars fall into one of the standard spectral classes, and astronomers generally refer to any such misfits as "peculiar" stars. The fact is that these oddballs may be rare, but in many cases they probably represent normal stages in the evolution of stars.

Most of the peculiar stars are so designated because of unusual absorption lines in their spectra, or unusual relative strengths of the lines. These peculiarities, in most instances, do represent abnormal chemical abundances in the outer layers of the star, contrary to our

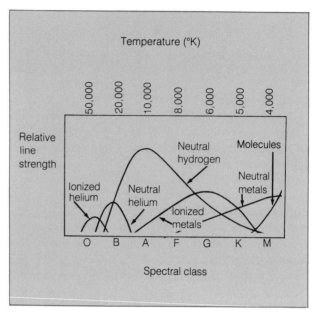

FIG. 18.12. IONIZATION FOR STARS OF VARIOUS SPECTRAL TYPES. This diagram shows which ions appear prominently in the spectra of stars of different classes. Note that the degree of ionization evident for the hot stars (left) is much greater than for the cool ones (right).

statement that most stars have about the same composition. Actually, the peculiar stars may have essentially normal element abundances, but for complex reasons, have undergone a kind of differentiation or sifting process that resulted in overabundances of certain elements at the surface. In other cases, elements produced by nuclear reactions inside the star (see chapter 20) are brought to the surface by internal currents, thereby altering the surface composition.

Some stars are referred to as peculiar because they have emission lines in their spectra. As we have seen, in a normal star the cool outer layers form absorption lines, so how do we understand these unusual cases? According to Kirchhoff's third law, a hot, rarefied gas forms emission lines, and this is how the emission lines in the spectra of some stars are formed. In these stars, hot gas extends above the surface, and it is in this region that the emission lines are produced (Fig. 18.13). In most cases, astronomers do not yet understand why this extra gas is present, although they have found some indications of its source. Many of the hottest and most luminous stars, for example, have steady "winds" of outflowing material, and these winds maintain a cloud of gas around the star, creating emission lines.

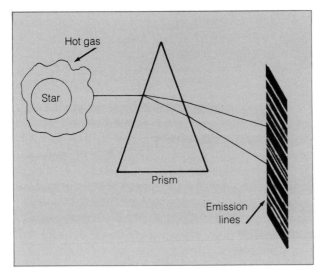

FIG. 18.13. STELLAR EMISSION LINES. Very few stars have emission lines in their visible-wavelength spectra (but many do in ultraviolet wavelengths; see chapter 20). When such lines are present, it usually signifies the existence of hot gas above the surface. According to Kirchhoff's second law, a rarefied, hot gas produces emission lines.

Other stars with gas swirling around them are found in double-star systems (see the next section), where one star may lose matter to the other. Such systems have received a lot of attention lately, because our best chance of detecting exotic remnants of stellar deaths is to find them in binary orbits with other stars.

Binary Stars

About half of the stars in the sky are members of double or **binary,** systems, where two stars orbit each other regularly. All types of stars can be found in binaries, and their orbits come in many sizes and shapes. In some systems the two stars are so close together that they literally are touching, and in others they are so far apart that it takes hundreds or thousands of years for them to complete one revolution.

Each of the three types of observations can be used to detect binary systems. Positional measurements, of course, tell us when two stars are very near each other in the sky. Sometimes this occurs by chance, when a nearby star happens to lie nearly in front of one that is in the background; this is called an **optical binary,** and is not a true binary system, since the two stars do not orbit each other (Fig. 18.14). When a pair of stars

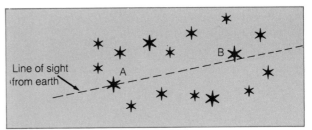

FIG. 18.14. AN OPTICAL BINARY. A chance alignment of two stars at different distances may give the appearance of a double star. Here stars A and B appear very close together, as seen from earth, but are in fact widely separated.

is seen close together, and measurements show that they are in motion about one another, this is called a **visual binary.** Accurate positional measurements are needed to reveal the orbital motion, because the two stars are very close together in the sky, and because they appear to move very slowly about each other. The distance between the two stars must be many AUs in order for both stars to appear separately (as seen from the earth), and therefore the orbital period is many years.

It is possible for binary systems to be recognized even when only one of the two stars can be seen. If positional measurements over a lengthy period of time reveal a wobbling motion of a star as it moves through space (Fig. 18.15), it may be inferred that the star is orbiting an unseen companion. Such systems, detected because of variations in position, are called **astrometric binaries.**

Simple brightness measurements can also tell us when a star, which may appear to be single, is actually part of a binary system. If we happen to be aligned with the plane of the orbit, so that the two stars alter-

FIG. 18.15. AN ASTROMETRIC BINARY. Careful observation of the motion of a star across the sky in some cases reveals a curved path such as the one shown here. Such a motion is caused by the presence of an unseen companion star, so that the visible star orbits the center of mass of the binary system as it moves across the sky.

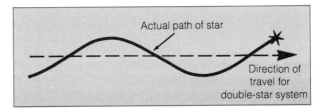

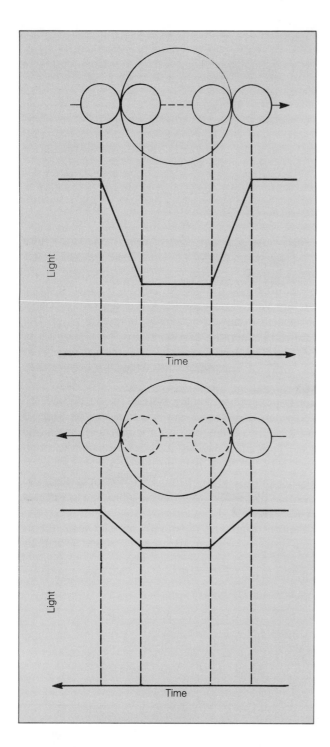

FIG. 18.16. AN ECLIPSING BINARY. If our line of sight happens to be aligned with the orbital plane of a double-star system, the two stars will alternately eclipse each other, and this causes brightness variations in the total light output from the system, as shown.

Finally, spectroscopic measurements can also be used to recognize binary systems, again even in the case where a star appears single. If two stars are actually present, and if they have nearly equal brightnesses, spectral lines from both will appear in the spectrum. If we find a star with line patterns representing two different spectral classes, we can be sure that two different stars are present; such a system is called a **spectrum binary.** Even if it is not clear that two different spectra are present (that is, if the two stars have similar spectral classes, or if one is too faint to contribute noticeably to the spectrum), the Doppler effect may reveal the fact that the star is double. Because the stars are orbiting each other, they move back and forth along our line of sight (as long as we are not viewing the system face-on), and this motion produces alternating blueshifts and redshifts of the spectral lines (Fig. 18.17). Binaries in which these periodic Doppler shifts are seen are called **spectroscopic binaries,** and they are more common than spectrum binaries.

A given double-star system may fall into more than one of these categories. For example, a relatively nearby system seen edge-on could be a visual binary, if both stars can be seen in the telescope; it may also be an eclipsing binary, if the two stars alternately pass in front of each other; and it almost surely will be a spectroscopic binary, since the motion back and forth along our line of sight is maximized when we view the orbit edge-on.

Binary stars are important, not just because there are so many of them, but also because they provide a direct means of determining many basic stellar properties, as we will learn in the next chapter.

Variable Stars

Although we tend to think of stars as eternal, steady beacons of light in the nighttime sky, close examination reveals that some of them are not so steady. There are stars that pulsate regularly, varying in brightness as they do so. A few of the brightest of these have been known since ancient times, but most have been found

nately pass in front of each other as we view the system, then the observed brightness will decrease each time one star is in front of the other (Fig. 18.16). This is called an **eclipsing binary.**

Variable stars classified according to period of pulsation

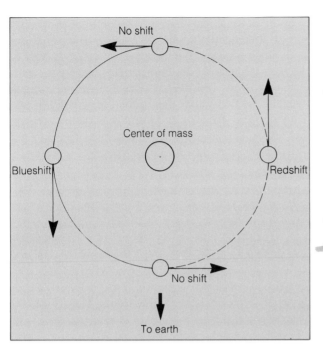

FIG. 18.17. A SPECTROSCOPIC BINARY. As a star orbits the center of mass of a binary system, its velocity relative to the earth varies. This produces alternating redshifts and blueshifts in its spectrum, as the star recedes from and approaches us. For simplicity, only one of the stars is shown here to be moving, although in reality both do.

by modern astronomers who made systematic searches for them, usually by comparing photographs of the same region of sky, taken at different times. Such a comparison will reveal any stars that change dramatically in brightness from one photograph to the next.

Like peculiar stars, variable stars have been classified according to their observed properties, primarily the period of pulsation. Most of these classes are named after the prototype, which is usually the first (and often the brightest) star discovered in that class. Examples include the δ **Cephei** (or **Cepheid**) stars, with periods of one to a hundred days; the **RR Lyrae** stars, with pulsation times of less than one day; and the Mira, or long-period, variables, which can take up to two years to complete one cycle.

*Cepher
RR Lyrae
Mira*

A pulsating variable star physically expands and contracts as it goes through its cycle (Fig. 18.18). This is known because the spectrum shows a periodic Doppler shift as the star's outer layers rise and fall in concert with the variations in brightness. It is as though the star is breathing in and out. The surface temperature also varies, and so does the spectral class. The brightness may change by as much as one magnitude (or much more, especially in the case of the long-period variables), or as little as a few tenths of a magnitude.

FIG. 18.18. A PULSATING VARIABLE STAR. The curved line shows how the brightness (in magnitude units) varies as the star expands and contracts. The sequence of sketches illustrates how the expansion and contraction phases are related in sequence to the variations in brightness. The surface temperature also varies.

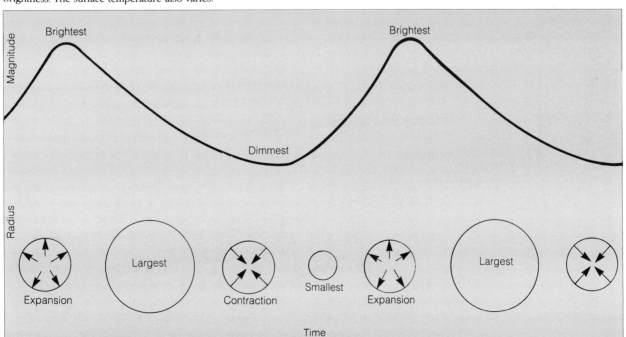

As in the case of most peculiar stars, the pulsating variables are normal stars passing through a particularly unstable point in their evolutions. A star has a natural vibration frequency, just like a bell, so that if it is disturbed at all, it will pulsate at this frequency for awhile before dying down to a steady state again. For certain stars, however, a kind of resonance is established, and once the vibrations start, they continue. These are the pulsating variables.

Some of these stars have certain uniform properties, such as average luminosities, that allow them to be used as distance indicators; these types of variables will be discussed in chapter 24. In other chapters (particularly chapter 23), we will discuss stars that vary in a highly irregular fashion rather than pulsating at a steady rate. Many of these irregular and eruptive variables are stars that are in late stages of their lifetimes. Thus we will describe them along with other types of dying stars.

Perspective

In this chapter we have learned the basics of stellar observations, by discussing the three fundamental types of observations and the characteristics of stars revealed by each. The stellar properties we have discussed; namely, positions, brightnesses, and spectra, represent the superficial information that comes directly from the observations. We are now ready to see how much more can be learned about stars by combining these directly observed properties with our knowledge of the laws of physics.

Summary

1. All that we know about stars is derived from the light we receive from them.

2. There are three basic types of stellar observation: measurements of position, brightness, and spectra.

3. Positional astronomy (that is, astrometry) has developed techniques capable of measuring position to an accuracy approaching 0.001 arcseconds, which is sufficient to detect proper motions, binary motions in some cases, and stellar parallaxes for stars up to 1,000 parsecs away.

4. Astronomers generally carry out stellar brightness measurements by using the stellar magnitude system, in which a difference of one magnitude corresponds to a brightness ratio of 2.512.

5. Magnitudes can be measured at different wavelengths, allowing the determination of color indices; or over all wavelengths, resulting in the determination of bolometric magnitudes.

6. Stellar spectra contain patterns of absorption lines that depend on the surface temperatures of stars, and which therefore can be used to assign stars to spectral classes that represent a sequence of temperatures.

7. Peculiar stars are those which do not conform to the usual spectral classes, most often because they have unusual surface compositions, but sometimes because they have emission lines.

8. Astronomers can detect binary-star systems on the basis of positional variations of one or both stars (astrometric binaries), brightness variations (eclipsing binaries), or composite or periodically Doppler-shifted spectral lines (spectrum or spectroscopic binaries).

9. Some stars are variable, in brightness and/or spectral appearance, in many cases because of physical pulsations.

Review Questions

1. How can the proper motion and the parallax motion of a star be distinguished from each other?

2. Does the sun have annual parallax motion, as seen from the earth?

3. What is the distance (in parsecs) to a star whose parallax angle π is (a) 0″25 (0.25 arcseconds), (b) 0″04, and (c) 0″005?

4. How much fainter is a twelfth-magnitude star than an eleventh-magnitude star? How does a fourth-magnitude star compare in brightness with a fifth-magnitude star?

5. Briefly discuss the role of stellar parallax in the development of the heliocentric theory.

6. Give the brightness ratio (star A compared with star B) for the following: (a) star A, with magnitude $m = 4$, compared with star B, with magnitude $m = 7$; (b) star A, with $m = 17$, compared with star B, with $m = 22$; and (c) star A, with $m = 6$, and star B with $m = -1$.

7. Which is cooler: star A, whose color index (B-V) is -0.15, or star B, with B-V $= -0.02$? Which is hotter, star C with B-V $= -0.2$, or star D, with B-V $= +0.2$?

8. Explain why the bolometric magnitude of a star is always smaller than the visual magnitude.

9. Explain, in terms of what you learned about ionization in chapter 5, why stars of different temperature have different absorption lines in their spectra.

10. Match the following color indices with the following spectral classes:

Color Index	Spectral Class
+0.5	B
−0.1	A
0.0	M
+0.2	F

Additional Readings

Probably the best sources of additional information on stars and their observational properties are general astronomy textbooks, of which there are a number readily available in nearly any library. A few more specific resources are listed here as well.

Aller, L. H. 1971. *Atoms, stars, and nebulae.* Cambridge, Ma.: Harvard University Press.

Mihalas, D. 1973. Interpreting early-type stellar spectra. *Sky and Telescope* 46(2): 79.

Strand, K. A., ed. 1965. *Basic astronomical data.* Chicago: University of Chicago Press.

Struve, O., and Zebergs, V. 1962. *Astronomy of the twentieth century.* New York: Crowell Collier and Macmillan.

Upgren, A. 1980. New parallaxes for old: a coming improvement in the distance scale of the universe. *Mercury* 9(6): 143.

CHAPTER 19

Fundamental Stellar Properties and the H-R Diagram

Before being able to understand how stars work, astronomers needed to determine their physical properties and how they are related to each other. A number of fundamental quantities characterize a star, and in this chapter we will see how the observations described previously can be used to determine those quantities. These include the luminosities, the temperatures, the radii, and above all, the masses of stars. In this chapter a new, very powerful distance-determination method will also be described.

Absolute Magnitudes and Stellar Luminosities

The luminosity of a star is the total amount of energy it emits from its surface, in all wavelengths; it is usually measured in units of ergs per second (see chapter 4). If the distance to a star is known, it seems an easy task to measure the apparent brightness and then allow for the distance by taking into account the inverse-square law of light propagation. It is, however, a very painstaking observational problem to measure the intensity of light from a star in absolute units, especially when it must be done over all wavelengths, including the ultraviolet and infrared (Fig. 19.1). Because of the complexities associated with such measurements, astronomers have done this only for a relatively small number of stars, and they have determined the luminosities of others by making comparisons with these standard stars.

To compare luminosities, however, astronomers must first eliminate the effects of differing distances of stars from the earth. Loyalty to the magnitude system led to the adoption of the **absolute magnitude,** defined as the magnitude a star would have if it were seen at a standard distance of 10 parsecs (Fig. 19.2). This distance was chosen arbitrarily, and is used by all astronomers. Thus a comparison of absolute magnitudes reveals differences in luminosities, because all the effects of distance have been canceled out.

It helps to consider a specific example. Suppose a certain star is 100 parsecs away, and has an **apparent magnitude** (the observed magnitude) of 7.3. To determine the absolute magnitude, we must find out what this star's magnitude would be if it were only 10 parsecs away. The inverse-square law tells us that since the star would be a factor of 10 closer, it would appear a factor of $10^2 = 100$ brighter. Thus if it were only 10 parsecs away rather than 100, this star would be a factor of 100 brighter, corresponding to exactly 5 magnitudes brighter. Therefore, its absolute magnitude is $7.3 - 5 = 2.3$.

It is also possible to derive the absolute magnitude from the measured apparent magnitude and the distance, by following a similar line of thought. Of course, the answer generally will not involve simple factors of 100, as did the example just cited, but the logic is the same. If we are doing these conversions mentally, it helps to remember that a factor of 10 in distance always corresponds to a difference of 5 magnitudes. Thus if in our example the star had been 1,000 parsecs away rather than 100, then to move it to a distance of 10 parsecs would have required bringing it a factor of 100 $= 10^2$ closer, so its absolute magnitude would have been $5 + 5 = 10$ magnitudes brighter than its apparent magnitude. Similarly, a star 10,000 parsecs away has an absolute magnitude 15 magnitudes brighter (smaller) than its apparent magnitude.

It should be clear that if we somehow knew the absolute magnitude of a star and measured its apparent magnitude, we could work these arguments backward

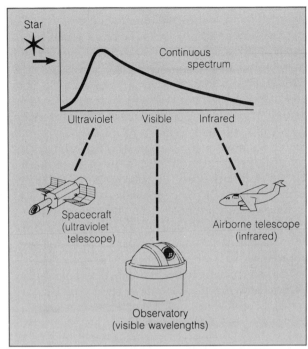

FIG. 19.1. MEASURING A STAR'S LUMINOSITY. A simple concept in principle, the determination of a stellar luminosity can be complex. The star's brightness must be measured at all wavelengths where it emits light (that is, its bolometric magnitude must be measured), and then allowance must be made for its distance, in order to calculate how much energy the star is emitting from its surface.

FIG. 19.2. THE ABSOLUTE MAGNITUDE. We imagine that we can move stars from their true positions to a uniform distance from us of 10 parsecs. The magnitude a star would have at this distance is called the absolute magnitude, and is directly related to the stellar luminosity, since the distance effect has been accounted for.

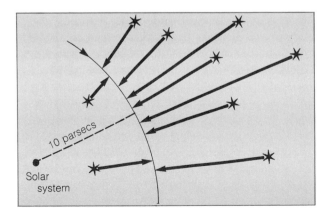

to determine the distance to the star. This is a very important point, and we will discuss it later.

Let us now return to the question of luminosities. By determining the absolute magnitudes of stars, we can compare their luminosities, since the distance effect has been removed. Thus if one star's absolute magnitude is 5 magnitudes smaller than another's, then we know that its luminosity is a factor of 100 greater. To actually express the luminosity in terms of ergs per second, we make a comparison with one of the standard stars for which a direct measurement of the luminosity has been made. Strictly speaking, since the luminosity refers to energy emitted over all wavelengths, we must use bolometric absolute magnitudes in order to determine luminosities (recall that a bolometric magnitude is simply a magnitude measured over all wavelengths).

What astronomers find when they determine stellar luminosities is that the values from star to star can vary over an incredible range. There are stars with luminosities as small as 10^{-4} that of the sun to as great as 10^6 that of the sun, a range of 10 billion from the faintest to the most luminous! The luminosity is by far the most highly variable parameter for stars; the others that we will discuss only cover ranges of a factor of 100 or so from one extreme to the other.

Stellar Temperatures

We have already learned something about how the temperatures of stars are determined, and we need not add much to that. In chapter 18 we saw that the color of a star depends on its temperature, as does the spectral class, so that by observing either, we can deduce the temperature. Recall that the color index, the difference between the blue (B) and visual (V) magnitudes, is a measured quantity that indicates temperature. A negative value of B-V means that the star is brighter in blue than in visual light, and therefore is a hot star. A large positive value indicates a cool star. A specific correlation of color index with temperature has been developed, and is used to determine temperatures from observed values of B-V.

More refined estimates of temperature can be made from a detailed analysis of the degree of ionization, which is done by measuring the strengths of spectral lines formed by different ions. This is basically the same as simply estimating the temperature from the spectral class, since in either case the point is that the strength of spectral lines of various ions depends on how abundant those ions are, which in turn depends on the temperature.

The temperature referred to here may be called the surface temperature, although stars do not have solid surfaces. We are really referring to the outermost layers of gas, where the absorption lines form. This region is called the photosphere of a star, and it actually has some depth (although it is very thin compared with the radius of the star).

Stellar temperatures range from about 2,000 K for the coolest M stars to 50,000 K or more for the hottest O stars. The temperature of the sun, a G2 star, is a little less than 6,000 K.

The Hertzsprung-Russell Diagram

We have seen that temperature and spectral class are closely related, and we have learned how astronomers deduce the luminosities of stars. In the first decade of this century, the Swedish astronomer Einar Hertzsprung and, independently, the American Henry Norris Russell (Fig. 19.3), began to consider how luminosity

FIG. 19.3. HENRY NORRIS RUSSELL. One of the leading astrophysicists of the era when a physical understanding of stars was first emerging, Russell made many major contributions in a variety of areas.

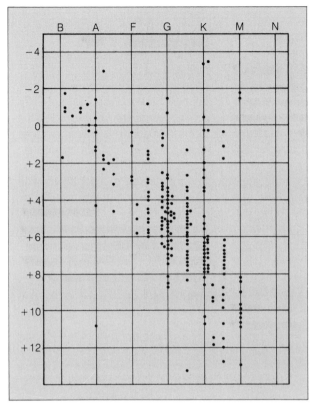

FIG. 19.4. THE FIRST H-R DIAGRAM. This is the first plot showing absolute magnitude versus spectral class, constructed in 1913 by Henry Norris Russell.

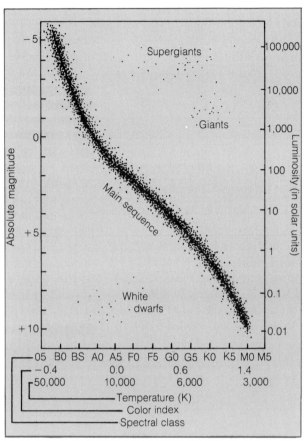

FIG. 19.5. A MODERN H-R DIAGRAM. This diagram shows the locations of a large number of stars, and gives alternative units on both axes for the two general parameters, luminosity and temperature.

and spectral class might be related to each other. Each gathered data on stars whose luminosities (or absolute magnitudes) were known, and found a close link between spectral class (that is, temperature) and absolute magnitude (or luminosity). This relationship is best seen in the diagram constructed by Russell in 1913 (Fig. 19.4), now called the **Hertzsprung-Russell,** or **H-R diagram.** In this plot of absolute magnitude (on the vertical scale) versus spectral class (on the horizontal axis), stars fall into narrowly defined regions rather than being randomly distributed. A star of a given spectral class cannot have just any absolute magnitude, and vice versa.

Most stars fall into a diagonal strip running from the upper left (high temperature, high luminosity) to the lower right (low temperature, low luminosity) (Fig. 19.5). This strip has been given the name the **main sequence.** A few stars are not in this sequence, but instead appear in the upper right (low temperature, high luminosity). Since the spectra of these stars indi-

cate that they are relatively cool, their high luminosities cannot be a result of the fact that they have greater temperatures than the main-sequence stars of the same type. The only way one star can be a lot more luminous than another of the same temperature is if it has a lot more surface area; recall that the two stars will emit the same amount of energy per square centimeter of surface. Hertzsprung and Russell realized that these extra-luminous stars located above the main sequence must be much larger than those on the main sequence, and they named these stars **giant** and **supergiants.**

The distinction among giants, supergiants, and main-sequence stars (commonly known as **dwarfs**) has been incorporated into the spectral classification system used by modern astronomers. A **luminosity class** has been added to the spectral type with which we are already

The Hertzsprung-Russell Diagram

H. N. Russell first constructed his famous diagram using absolute visual magnitudes and spectral classes as the two coordinates, and the diagram is often used in this way today. However, there are different (but equivalent) quantities frequently used instead.

As we saw in the text, the absolute magnitude is a measure of luminosity, with a precise correspondence when the bolometric absolute magnitude is used. Because the luminosity of a star is a more fundamental quantity than its visual absolute magnitude, astronomers sometimes use bolometric absolute magnitude or luminosity on the vertical axis of the H-R diagram. When luminosity is chosen, usually the logarithm of the luminosity (often expressed in units of the sun's luminosity) is used, to keep the scale similar to a magnitude scale; magnitudes are related logarithmically to actual intensities of light.

The spectral class is basically a measure of temperature, so any quantity that corresponds to temperature may be used as the horizontal coordinate on the H-R diagram. Sometimes the temperature itself (on the absolute scale) is shown on this axis, and often the B-V color index is used, this being a quantity that is rather simple to measure accurately. If B-V is the chosen temperature indicator, the diagram may be called a **color-magnitude diagram,** particularly when applied to star clusters (see chapter 21).

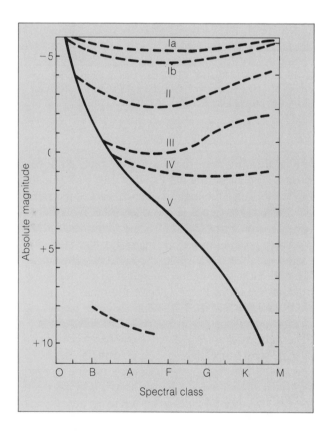

FIG. 19.6. LUMINOSITY CLASSES. This H-R diagram shows the locations of stars of the luminosity classes described in the text. A complete spectral classification for a star usually includes a luminosity-class designation, when it has been determined.

familiar. The luminosity classes, designated by Roman numerals following the spectral type, are I for supergiants (this group is further subdivided into classes Ia and Ib), II for extreme giants, III for giants, IV for stars just a bit above the main sequence, and V for main-sequence stars, or dwarfs (see Fig. 19.6). Thus the complete spectral classification for the bright summertime star Vega, for example, is AOV, meaning that it is an AO main-sequence star. The red supergiant in the shoulder of Orion, Betelgeuse, has the full classification M2Iab (it is intermediate between luminosity classes Ia and Ib). It is usually possible to assign a star to the proper luminosity class from examination of subtle details of its spectrum.

Another group of stars has become known (mostly since the time Russell first plotted the H-R diagram), that do not fall into any of the standard luminosity classes, but which instead appear in the lower left (high temperature, low luminosity) corner of the diagram.

Since these stars are hot, but not very luminous, they must be very small, and they have been given the name **white dwarfs.** These objects have some very bizarre properties, which will be discussed in chapter 23.

The H-R diagram is a fundamental tool for understanding stars and the relationships among their various properties, as well as how they evolve. We will be referring to it continually throughout this section of the book.

SPECTROSCOPIC PARALLAX

The H-R diagram can be used to find distances to stars, even stars that are very far away. The idea is really very simple: if we know how bright a star is intrinsically (that is, how much energy it is actually emitting from its surface), and we measure how bright it appears to be, we can determine how far away it is, because we know that the difference between its intrinsic brightness and its observed brightness is caused by the distance. The only problem lies in knowing the intrinsic brightness of the star (its luminosity), and this is where the H-R diagram comes in.

Once we determine the spectral class of a star, we can place it on the H-R diagram (as long as we are sure we know the luminosity class, so we know whether it is on the main sequence or is a giant or supergiant). After we have placed it on the diagram, we simply read off of the vertical axis the absolute magnitude of the star, which is a measure of its luminosity (Fig. 19.7). A comparison of the absolute magnitude with the apparent magnitude, therefore, amounts to the same thing as a comparison of the intrinsic and apparent brightnesses of the star, and from such a comparison the distance can be found.

Consider these examples. If the difference m-M (that is, the apparent minus the absolute magnitude), which is called the **distance modulus,** is 5, then the star appears 5 magnitudes, or a factor of 100, fainter than it would at the standard distance of 10 parsecs. A factor of 100 in brightness is created by a factor of 10 change in distance, so this star must be ten times farther away than it would be if it were at 10 parsecs distance; therefore, it is 10 × 10 = 100 parsecs away. Similarly, a star whose distance modulus m-M is 10 is 1,000 parsecs away. If m-M = 15, then the distance is 10,000 parsecs. It should be obvious that if $m = M$ (that is, m-M = 0), then the distance to the star must be 10 parsecs, because this is the distance that defines the ab-

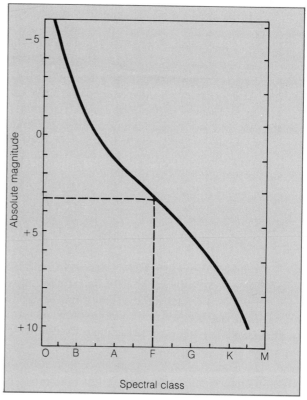

FIG. 19.7. SPECTROSCOPIC PARALLAX. Knowledge of a star's spectral class (including the luminosity class) allows the absolute magnitude, and hence the distance, to be determined. Here it is shown how the absolute magnitude of an F0 main-sequence star is read off of the diagram, leading to an absolute-magnitude estimate of +3.3. The distance can then be found by comparing the absolute and apparent magnitudes of the star (this is explained in the text and in appendix 9).

solute magnitude. As a rule of thumb, it helps to remember that for every 5 magnitudes of difference between the apparent and absolute magnitudes, the distance increases by a factor of 10.

This method is very powerful, because it can be used for very large distances. All that is needed is to be able to place a star on the H-R diagram so that its absolute magnitude can be determined, and to measure its apparent magnitude. Because this distance-determination technique requires that the spectrum of a star be classified so that it can be placed on the H-R diagram, it is called the **spectroscopic parallax** method (the word parallax is used by astronomers as a general word for distances, even though, technically speaking, no parallax is measured in this case).

It should be noted that any time it is possible to determine the absolute magnitude of an object, even if it is by some means other than placing it on the H-R diagram, its distance can be found by comparing the absolute and apparent magnitudes. Any object whose absolute magnitude can be determined so that it can be used as a distance indicator is called a **standard candle.** We will see in later sections how important these objects are in exploring the size scale both of the galaxy and of the universe.

Stellar Diameters

We have already seen that a star's position in the H-R diagram depends partly on its size, since the luminosity is related to the total surface area. If two stars have the same surface temperature (and therefore the same spectral class), but one is more luminous than the other, we know that it must also be larger. The Stefan-Boltzmann law (see chapter 5) specifically relates luminosity, temperature, and radius; use of this law allows the radius to be determined if the other two quantities are known.

Eclipsing binaries provide another means of determining stellar radius, one that is independent of other properties. Recall that these are double-star systems in which the two stars alternately pass in front of each other, as we view the orbit edge-on. The eclipsing bi-

nary is very likely also to be a spectroscopic binary, so that the speeds of the two stars in the orbits can be measured from the Doppler effect. We therefore know how fast the stars are moving, and from the duration of the eclipse we know how long it takes one to pass in front of the other; the simple formula distance equals speed times time thus gives us the diameter of the star that is being eclipsed (Fig. 19.8). Even if no information on the orbital velocity is available, the relative diameters of the two stars can be deduced from the relative durations of the alternating eclipses.

Eclipsing binaries provide the most direct means of measuring stellar sizes, but unfortunately, there are not many of them. In most cases the radii are estimated from the luminosity and temperature, as just described. In a few cases, stellar radii have been measured directly by use of speckle interferometry, a sophisticated technique for clarifying the image of a star by removing the blurring effects of the earth's atmo-

FIG. 19.9. THE EFFECT OF ROTATION ON STELLAR ABSORPTION LINES. Rotation of a star broadens its spectral lines because of the Doppler effect. Here the same line is shown as it would appear in the spectrum of a slowly rotating star (top) and one that is spinning rapidly (bottom).

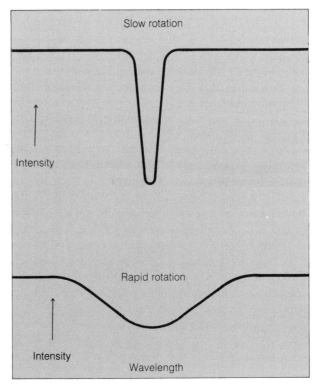

FIG. 19.8. THE MEASUREMENT OF A STELLAR DIAMETER IN AN ECLIPSING BINARY. The velocity of the eclipsing star is known from the Doppler effect (but cannot be measured at the time of the eclipse; it must be assumed that the velocity is constant), and the duration of the eclipse is known from observations of the brightness variations. The velocity times the eclipse duration gives the diameter of the eclipsed star.

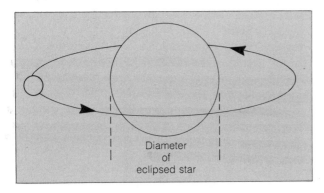

sphere. Only relatively nearby, large stars can be measured this way, however.

Stars on the main sequence do not vary greatly in radius, ranging from perhaps 0.1 times the sun's radius for the M stars at the lower right-hand end to 10 or 20 solar radii at the upper left. Of course, large variations in size occur as we go away from the main sequence, either toward the giants and supergiants, which may be 100 times the size of the sun, or toward the white dwarfs, which are as small as 0.01 times the size of the sun.

Stellar Rotation

To measure how rapidly a star is spinning is a fairly straightforward procedure, but there is one serious complication. Part of the light that reaches us from a spinning star is emitted from the portion of its surface that is moving toward us, and part is emitted from the portion that is moving away. Thus, because of the Doppler effect, spectral lines formed in the approaching portion are blueshifted, and those formed in the receding portion are redshifted. When we measure the spectrum of a star, the spectral lines are broadened, being made up of light from all parts of the surface. The more rapid the rotation, the greater the Doppler shift, and the broader the spectral lines (Fig. 19.9). Thus, in principle, a star's velocity of rotation can be determined from measurements of the width of its spectral lines.

The complication is that we do not know the orientation of the star's rotation axis. If we view a star from straight above its pole, there will be no Doppler shift or line broadening, no matter how fast it is spinning, because there will be no motion along our line of sight (recall that the Doppler effect occurs only for motions directed straight toward or away from the observer). If we view a star in the plane of its equator, on the other hand, then the Doppler shift does cause its spectral lines to be broadened, and we can make a direct measurement of its rotational velocity. Unfortunately, we rarely know whether we are viewing a star pole-on, equator-on, or, as is more likely, at some random angle in between. The result is that we usually cannot precisely determine the true rotational velocity of a particular star; instead, by sampling large numbers of stars, we can make general statements about the range of stellar rotation velocities that exist.

Many stars rotate slowly, with surface velocities of less than 10 km/sec (the sun's rotational velocity at the

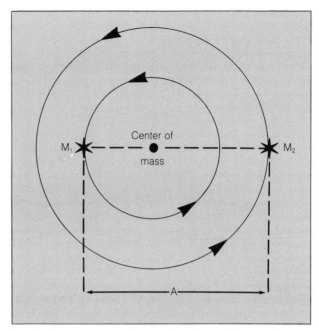

FIG. 19.10. BINARY-STAR ORBITS. This figure illustrates the terms used in Kepler's third law. Two stars of masses m_1 and m_2 (m_1 larger than m_2 in this case) orbit a common center of mass, each making one full orbit in period P. The semimajor axis a that appears in Kepler's third law is actually the sum of the semimajor axes of the two individual orbits about the center of mass; this sum corresponds to the average distance between the two stars.

equator is about 2 km/sec), and rotation periods of at least several days or a few weeks. There are exceptions, particularly among the hot stars (which, as we will see, are young objects that may still be spinning rapidly as a result of their formation). In these cases rotational velocities up to 450 km/sec are found, corresponding to rotation periods as short as one or two days.

Binary Stars and Stellar Masses

The mass of a star is the most important of all its fundamental properties, for it is the mass that governs most of the others. (This point will be discussed at some length in the next chapter.) Unfortunately, there is no direct way to see how much mass a star has. The only way to measure a star's mass is by observing its gravitational effect on other objects, and this is possible only in binary star systems, where the two stars hold each other in orbit by their gravitational fields. The fact that binary systems are common provides us with many op-

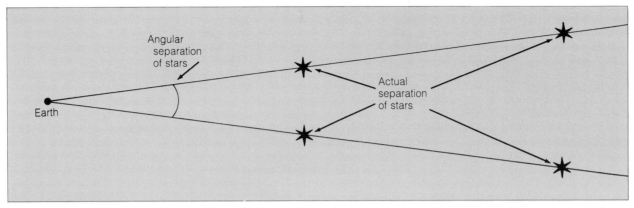

FIG. 19.11.   THE EFFECT OF DISTANCE ON MEASUREMENTS OF BINARY ORBITS. The angular separation of the two stars is all that we can measure directly. To determine the true separation, and hence the orbital semimajor axis, requires knowledge of the distance.

portunities to determine masses by analyzing binary orbits.

The basic idea is rather simple, although the application may be quite complex, depending on the type of binary system. Kepler's third law is used, in the form derived by Newton. Remember that if the period P is measured in years, the average separation of the two stars (the semimajor axis a) is measured in astronomical units, and the masses m_1 and m_2 in units of solar masses, then Kepler's third law is

$$(m_1 + m_2)P^2 = a^3$$

We need only observe the period and the semimajor axis in order to solve for the sum of the two masses.

Careful observation of the sizes of the individual orbits (actually, of the relative distances of the two stars from the center of mass; see Fig. 19.10) also yields the ratio of the masses; when both the sum and the ratio are known, it is simple to solve for the individual masses.

Complications arise when some of the needed observational data are difficult to obtain. The period is almost always easy to measure with some precision, but not so the semimajor axis. The main problems are that the apparent size of the orbit is affected by our distance from the binary (Fig. 19.11), and so the distance must be well known if a is to be accurately determined; and that the orbital plane is inclined at a random, unknown angle to our line of sight, so that

FIG. 19.12.   DETERMINATION OF STELLAR COMPOSITION. This diagram illustrates a modern technique for measuring the chemical composition of a star. A theoretical spectrum (solid line) is compared with two observed spectra (dotted and dashed line). The abundances of elements assumed present are altered in the computed spectrum until a good match is achieved.

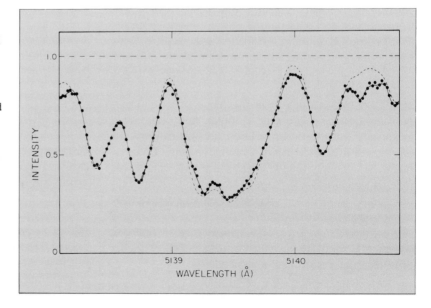

the apparent size of the orbit is foreshortened by an unknown amount. In some cases it is possible to unravel these confusing effects by carefully analyzing the observations, and in other cases it is not. Even so, some information about stellar masses can be gained, but usually only in terms of broad ranges of possible values, rather than precise answers.

The masses of stars vary along the main sequence from the least massive stars in the lower right to the most massive in the upper left. The M stars on the main sequence have masses as low as 0.05 solar masses, whereas the O stars reach values as great as 60 solar masses. It is likely that stars occasionally form with even greater masses (perhaps up to 100 solar masses), but as we will see in the next chapter, such massive stars have very short lifetimes, so it is rare to find one.

The giants, supergiants, and white dwarfs have masses comparable to those of main-sequence stars. Hence their obvious differences from main-sequence stars in other properties such as luminosity and radius have to be the result of something other than extreme or unusual mass. (This is discussed in the next chapter.)

For main-sequence stars, there is a smooth progression of all stellar properties from one end to the other. The mass and the radius vary by similar factors, whereas the luminosity changes much more rapidly along the sequence. The mass of an O star is perhaps 100 times that of an M star, whereas the luminosity is greater by a factor of 10^9 or more.

Stellar Composition

The next property we should consider is the chemical makeup of a star. We have already learned that this does not vary much from one star to another; that the major observed differences among the spectral types are a result of temperature effects. It is important, however, to be able to measure the elemental abundances in stars, both to see what they are made of and to understand the variations that do exist.

We learned previously that each chemical element has its own set of spectral lines. It would therefore seem simple to determine the composition of a star by seeing which elements are represented in its spectrum. This is complicated, however, by the fact that the temperature and density, which control the ionization and excitation, have a very profound, dominant influence on the spectrum.

To measure chemical abundances, therefore, re-

quires a more subtle analysis. It is necessary to know the temperature and density fairly well to begin with, so that these effects may be taken into account, and then to analyze the strengths of the spectral lines to see what the abundances are. This process is carried out most accurately by using complex computer programs that can calculate a simulated spectrum for comparison with the observed spectrum (Fig. 19.12). In the calculation of the theoretical spectrum, the temperature, density, and chemical composition are all varied until the best match with the observed spectrum is found. This is a costly and time-consuming process, and is not done in such detail for many stars.

In general terms, the result of stellar composition studies shows that most stars consist almost entirely of hydrogen and helium, with everything else constituting at most a few percent of the total mass. We have already learned that the sun and indeed the entire solar system were formed out of material with similar composition; we now find this to be a nearly universal effect. This provides an important clue regarding how the universe itself formed, as we shall see in section V.

Magnetic Fields

An often overlooked, but undeniably fundamental, property of a star is its magnetic field. We expect that stars should have fields; after all, the sun and many of

Text continues on page 305.

FIG. 19.13. THE MEASUREMENT OF A MAGNETIC FIELD. The presence of a magnetic field splits certain absorption lines of some elements, in a process called Zeeman splitting. The amount of separation is a measure of the strength of the magnetic field.

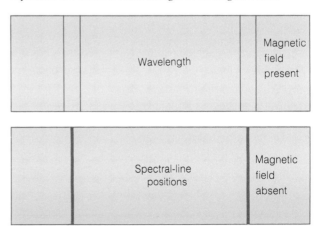

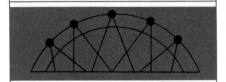

ASTRONOMICAL INSIGHT (19.2)

An Authentic Superstar

The word *superstar* is, no doubt, overused in our society. It is not used at all in astronomy, however, but perhaps it will be soon, in view of the incredible properties of an object recently discovered.

In our discussion of the properties of stars, we have given the typical range of values that are found for such parameters as luminosity, mass, and radius. Recently an object was found that may far exceed these usual ranges. Although there is still some uncertainty about the true nature of this object, the weight of the evidence seems to indicate that it is a gigantic star, something not even thought previously to be possible.

In the Large Magellanic Cloud, one of the two small galaxies that orbits the Milky Way, there is a very large and bright region of glowing gas clouds called the Tarantulus Nebula (sometimes referred to as 30 Doradus, its name in a particular catalogue). Lying at a distance of 55,000 parsecs from the sun, this nebula is so bright that it is visible to the unaided eye as a bright spot within the Large Magellanic Cloud.

Recent observations, particularly those made in ultraviolet wavelengths with the *International Ultraviolet Explorer* satellite (see chapter 6), show that at the heart of the nebula is a very compact, brightly glowing spot that may be the source of the energy by which the nebula glows. From the estimates of the intensity of radiation filling the nebula, it is guessed that this central object has a luminosity of 50 million times that of the sun. This is equivalent to about 1,000 O and B stars (the most luminous of normal stars), yet the diameter of the object is less than 0.7 parsec. Ultraviolet spectra of the object reveal lines similar to those normally seen in the spectra of hot stars, except that the degree of ionization (and hence the surface temperature) seems to be higher, and there appears to be a very strong stellar wind (stellar winds are discussed in chapter 20).

Two possible explanations of this object have been suggested: (1) it is a very compact, dense cluster of hot stars; or (2) it is a single, gigantic star, far exceeding the ordinary stellar parameters. The evidence seems to favor the latter hypothesis, because the spectrum does not exactly match that expected for a group of ordinary O and B stars, and because the cluster would have to be much denser than any ordinary one, in order to fit enough stars into a small enough volume.

The second possibility, that the object is a single star, has tremendous implications. To produce the observed luminosity, the star would

have to have a mass of about 3,000 solar masses, a surface temperature of around 65,000 K, and a radius of roughly 80 solar radii. This would be a superstar indeed!

There are reasons to wonder if such an object is possible. It was calculated long ago (by Sir Arthur Eddington, a leader in the early development of stellar-structure theory, whose major work was done during the first three decades of this century) that stars with masses greater than about 100 solar masses should not exist. Such massive stars would be unstable, because the radiation pressure created by their enormous luminosities would exceed gravity, and their outer layers would be blown off. Although that prediction is still valid, today we know of stars that have dense, rapid winds, apparently driven by radiation pressure. The strange object in the Tarantulus Nebula has a strong wind; perhaps stars can exist that exceed Eddington's limit, but are required to lose mass in this fashion.

We may also ask why only one such object has been found. Clearly, its lifetime must be short (about one million years, according to calculations), but other stars with similarly short lifetimes are found in greater numbers. Perhaps some special conditions are required for the formation of a 3,000-solar-mass star, thus very few are ever born. It will be fascinating to see the results of continued research into the nature of the superstar in the Large Magellanic Cloud.

the planets do, and we suspect that magnetism in these bodies originates in fluid motions in their interiors. There is every reason to expect the same processes to be at work in other stars.

To determine whether a distant star has a magnetic field, and furthermore to measure its strength, is very difficult. It was found some years ago that some spectral lines of certain elements are split into two or more closely spaced lines in the presence of a magnetic field (Fig. 19.13). In a process called **Zeeman splitting,** electromagnetic forces split certain energy levels of an atom, so that an electron can be in one of two or more different levels whereas there would only be one if there were no magnetic field. Thus a spectral line involving one of these split levels is itself split into two lines because the electron that absorbs a photon of light may be in either level. The amount of splitting is determined by the strength of the magnetic field, so in principle it is possible to determine the field strength by measuring how widely the lines are split.

In practice such measurements are very difficult, however, for a number of reasons. The splitting of spectral lines is normally very small, so very precise measurements are needed. Furthermore, stellar rotation or gas motions in the outer layers can cause the spectral lines to be broadened by the Doppler effect, so that the separate lines are merged into one, and their separation cannot be measured. Because of these difficulties, we do not have much information on magnetic-field strengths in typical stars. The fields have been measured in a number of stars having especially strong ones, and values up to hundreds or thousands of times greater than the sun's magnetic-field strength have been found.

The lack of more complete data on magnetic-field strengths in normal stars is frustrating, because it is likely that the fields play important roles in stellar formation and evolution.

Perspective

We now know how all the basic properties of stars are derived from observational data. We can categorize stars, classify them; describe them in any way we wish. Now we are ready to see how they work; *why* the various quantities are related the way they are, and no other way.

Summary

1. We determine stellar luminosity through knowledge of the distance to a star and the star's apparent magnitude. The absolute magnitude, a measure of luminosity, is the magnitude a star would have if it were seen from a distance of 10 parsecs.

2. Stellar temperatures can be inferred from the *B-V* color index, estimated from the spectral class, or determined from the degree of ionization in the star's outer layers.

3. The Hertzsprung-Russell diagram shows that the luminosities and temperatures of stars are closely related, and that stars which do not fall on the main sequence are either larger (as in the case of the red giants) or smaller (white dwarfs) than those on the main sequence.

4. We can measure the distance to a star by first determining the star's spectral class, then using the H-R diagram to infer its absolute magnitude, and finally comparing the absolute magnitude with its apparent magnitude to yield the distance. This technique is called spectroscopic parallax.

5. We can determine stellar diameters directly in eclipsing binaries through knowledge of the orbital speed and the duration of the eclipses.

6. The rotational velocities and periods of stars are determined from the Doppler-shifting of the spectral lines, which causes the lines to be broadened. However, these velocities are almost always ambiguous because of the unknown orientation of the rotation axis.

7. We can derive stellar masses in binary systems through the use of Kepler's third law, when the period and the orbital semimajor axis are observed.

8. The chemical composition of a star can be deduced from the strengths of the absorption lines in its spectrum, and is most commonly and accurately done by comparison with theoretically computed spectra.

9. The magnetic-field strengths of stars are probably important in their structure and evolution, and in some cases can be inferred from the Zeeman splitting of spectral lines. However, these strengths are not easily measured for most stars.

Review Questions

1. Why is it difficult to directly measure a star's brightness at all wavelengths?

2. Determine the absolute magnitude of the following: (a) a star 1,000 parsecs from the sun with apparent magnitude $m = 12.3$; (b) a star 10,000 parsecs from the sun with $m = 12.3$; (c) a star 10 parsecs away with $m = 6.8$; and (d) a star 1 parsec away with $m = 2.1$

3. If a certain star has absolute bolometric magnitude $M_{bol} = 1.5$, and another has $M_{bol} = 4.5$, which is more luminous, and by how much?

4. If one star is four times hotter than another, and has twice as large a radius, how does its luminosity compare with that of the other star? (Note: it is necessary for you to review some material in chapter 5.)

5. Explain why a G2 star that lies well above the position of the sun on the H-R diagram must have a larger radius. If its luminosity is a hundred times greater than that of the sun, how much larger is its radius?

6. Using the spectroscopic-parallax method, determine the distances to the following stars: (a) a star with apparent magnitude $m = 7.2$ and absolute magnitude -2.2; (b) a star with $m = 18.6$ and $M = 8.6$; (c) a star with $m = 9.0$ and $M = 9$; and (d) a star with $m = 5.8$ and $M = -4.2$.

7. Suppose, in an eclipsing binary system, that the orbital speed of star A is 10 km/sec and its eclipse of star B lasts 27.78 minutes. What is the diameter of star B?

8. Summarize the various reasons it is difficult to determine the masses of a pair of stars in a binary system, even though in the ideal case this can be done.

9. Explain why the spectral lines of a particular element (say, hydrogen) can have quite different strengths from one star to another, yet the abundance of hydrogen may be the same for the two stars.

10. Using information from chapter 17, describe the role a star's magnetic field might play in establishing its rotation rate.

Additional Readings

Aller, L. H. 1971. *Atoms, stars, and nebulae*. Cambridge, Ma.: Harvard University Press.

De Vorkin, D. 1978. Steps towards the Hertzsprung-Russell diagram. *Physics Today* 31 (3): 32.

Hack, M. 1966. The Hertzsprung-Russell diagram today. *Sky and Telescope*, May, p. 260 (pt. 1); June, p. 332 (pt. 2).

Smith, E. P., and Jacobs, K. C. 1973. *Introductory astronomy and astrophysics*. Philadelphia: Saunders.

Struve, O., and Zebergs, V. 1962. *Astronomy of the twentieth century*. New York: Crowell Collier and Macmillan.

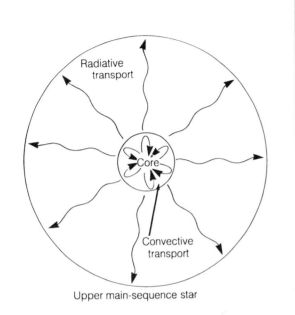

Radiative
transport

Core

Convective
transport

Upper main-sequence star

CHAPTER 20

Stellar Structure: What Makes a Star Run?

We have learned how astronomers determine from observations the physical characteristics of stars. From this we have discovered a great deal about the stars' surface properties, but very little about what goes on inside them, or how they evolve. To understand these secrets, we must apply the laws of physics and calculate theoretically the internal structure and evolution of stars, and the calculations must reproduce the observable properties. If this is done successfully, we can be somewhat confident that the calculations also accurately describe the internal conditions.

What Is a Star, Anyway?

We have discussed the surfaces of stars as hot, gaseous regions, often sufficiently hot that the gas is ionized. Because only absorption lines form in this region, we know that deeper in the star it must be hotter, so that continuous radiation is created there and flows out through cooler layers before escaping into space. The surface layer where the lines form (called its **photosphere**) has a low density, well below that of the earth's atmosphere at sea level. The density must be very high in the interior, however, because the average density (the total mass divided by the volume of the star) is close to that of water, or roughly 10^8 times greater than the surface density. To achieve such a high average density when the outer layers are so rarefied requires a still higher interior density, perhaps 100 times that of water, or about 30 times that of rock.

Despite this, a star is not solid inside. Although the density increases toward the center, so does the temperature, keeping the star in a gaseous state. Temperatures range from 10 to 100 million degrees absolute in the cores of stars. Under such extreme conditions, the gas particles whiz around at very high velocities (typically 10^8 cm/sec, or 1,000 km/sec), and they collide very frequently and with great impact. The result is that *all* electrons are knocked loose, and the gas is fully ionized, meaning that it consists only of bare nuclei and free electrons. This situation is different from that of the surface layer, where the atoms may have lost a few electrons, but most elements still have some electrons orbiting their nuclei.

A star, then, is a spherical ball of gas with density and temperature increasing toward the center. Most stars are made primarily of hydrogen, although as we will see, the composition changes gradually over a star's lifetime.

Hydrostatic Equilibrium and the Central Role of Mass

We may wonder what keeps a star in the state we have described. Why is it spherical? Why do density and temperature increase inward? The answer is that all the gas particles exert gravitational forces on each other, so that the star is held together by its own gravity. This force is always directed toward the center; thus, the star is forced into a symmetric, spherical shape. The fact that gas is compressible explains how gravity creates a state of high density inside.

A gas that is compressed heats up, causing it to exert greater pressure on its surroundings. Thus a star's interior is very hot and the internal pressure is high. If it were not for this pressure, gravity would cause a star to keep shrinking. A balance is struck between gravity,

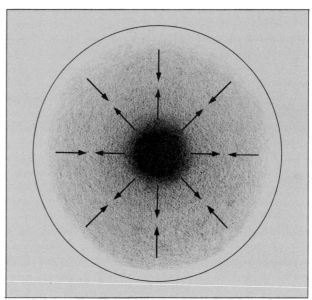

FIG. 20.1. HYDROSTATIC EQUILIBRIUM. This is a cutaway sketch of a star, showing its spherical shape. The arrows represent the balanced forces of gravity inward and pressure outward, and the shading indicates that the density increases greatly toward the center. Equilibrium is reached when the core becomes sufficiently hot to attain the pressure necessary to counterbalance gravity.

which is always trying to squeeze a star inward, and pressure, which pushes outward. This balance is called **hydrostatic equilibrium** (Fig. 20.1), and it plays a dominant role in determining the internal structure.

But what determines the balance; that is, what causes a star to reach a certain state of internal compression and no other? The answer is the mass of the star. The amount of gravitational force is set by the total mass, and this force in turn determines how much pressure is needed to balance gravity. The star will be compressed until this pressure is reached, and then it will become stable. The pressure required to balance gravity dictates the temperature inside the star. Hence a star's mass determines the internal density and temperature, as well as the overall size of the star, since this is a function of mass and density.

Luminosity is also governed largely by the mass. The luminosity of a star is simply the amount of energy generated inside it that eventually reaches the surface and escapes into space, and it is determined by the temperature in the interior. As we have just learned, the temperature is set primarily by the mass.

Thus virtually all the observed properties of a star depend on its mass (Fig. 20.2). This explains why the

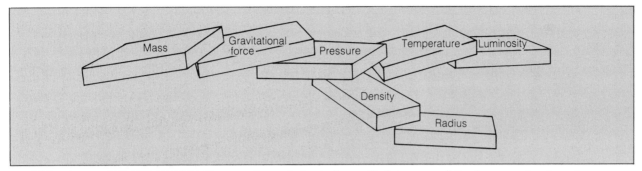

FIG. 20.2. THE IMPORTANCE OF MASS. The mass of a star is the single quantity that governs all its other properties, for a given composition. The sequence shown here is the same in general, but the details vary for different chemical compositions.

main sequence represents a smooth run of masses increasing from the lower right to the upper left in the H-R diagram: the luminosity and temperature, which define a star's place in the diagram, both depend on mass.

Of course, there are stars that do not fall on the main sequence, yet their masses are not different from those of main sequence stars. This tells us there must be something other than mass that can influence a star's properties. This other parameter is the chemical composition. As we will see, this varies as a star ages. The supergiants, giants, and white dwarfs have chemical makeups different from those of main-sequence stars (in the next section we will see how these differences arise).

The amount of internal compression, and hence the temperature and luminosity of a star, depend on the average mass of the individual nuclei in the star's core. The heavier the particles that make up the gas, the more tightly they are compressed by gravity, the hotter it gets, and the greater the luminosity. When a star is formed, it consists mostly of hydrogen, but as it ages, its core material at first is converted to helium, and later possibly to other, even heavier elements. In the process the core heats up and the star becomes more luminous.

The fact that all properties of a star depend on just its mass and composition was recognized several decades ago, and is usually referred to as the **Russell-Vogt theorem**, after the astrophysicists who first stated it. In modern times, it is known that other influences, such as magnetic fields and rotation, must play roles in governing a star's properties as well. Just what these roles are is not yet well understood, however.

Nuclear Reactions and Energy Transport

We have seen that the core of a star must be very hot, in order to maintain the pressure required to counterbalance gravity. Since energy flows outward from the core, there must be some source of heat in the interior. If there were not, the star would gradually shrink.

Nuclear fusion reactions provide the only source of heat capable of maintaining the required temperature over a sufficiently long period of time. A particular type of reaction was described in chapter 16, where we discussed the sun's interior. Recall that in atomic fusion reactions, nuclei of light elements such as hydrogen combine to form nuclei of heavier elements (Fig. 20.3). In the process, a small fraction of the mass is con-

FIG. 20.3. NUCLEAR FUSION. The core of a star is a sea of rapidly moving atomic nuclei. Occasionally a pair of these nuclei, most of which are simple protons (hydrogen nuclei) collide with sufficient velocity to merge together, forming a new kind of nucleus (deuterium in this case, composed of one proton and one neutron). Energy is released in the process.

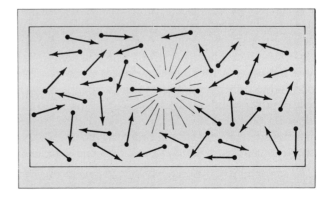

ASTRONOMICAL INSIGHT (20.1)

Henry Norris Russell

During the first decades of the twentieth century, physics underwent a period of extraordinary change and ferment. In what could be called a second Renaissance, the structure of the atom was revealed, the secrets of the photon and the emission and absorption processes in atoms were unraveled, Einstein developed his theories of general and special relativity, and, by the late 1930's, the production of energy by nuclear reactions was understood. All these discoveries had applications to

stars, and among the scientists who led the way into these new territories, none showed greater insight and breadth than the American astrophysicist Henry Norris Russell. This was the era when astronomy truly became astrophysics, and Russell played a key role in the transition.

Educated at Princeton, Russell received a Ph.D. there in 1900, and was at first quite inactive because of illness. By 1905, following a three-year stay at the Cambridge University in England, Russell took a position at Princeton, resuming what was to become a lifelong association with that university. In 1912 he was appointed director of the Princeton University Observatory, a post he was to hold for the next thirty-five years. During these years he not only made numerous fundamental breakthroughs in understanding the physics of stars, but also traveled and lectured exten-

sively, keeping scientists at widespread locations abreast of new developments, in a sense acting as mentor to a whole generation of astronomers. His seemingly limitless energy also led him into endeavors that brought him before the public eye (he wrote more than five hundred popular articles, along with a textbook that became a standard for more than thirty years). In a day before television and instant celebrity, he was known to the public by name and by sight in every corner of the United States.

Russell worked in nearly every area of astronomy, in each instance quickly developing new insights and techniques. He was both an observer and a theorist. His first work involved the development of astrometric methods and the measurement of stellar distances through the trigonometric parallax technique, and his later efforts were devoted to the structure and evolution of

verted into energy according to Einstein's famous formula $E = mc^2$: it is this energy, in the form of heat and radiation, that maintains the internal pressure in a star.

Fusion reactions can only occur under conditions of extremely high temperature and density. The nuclear force that holds the protons and neutrons together in a nucleus and which causes fusion to occur only acts over very short distances. The electromagnetic force, which causes particles of like electrical charge to repel each other, acts over a much greater distance; for this reason it is difficult for nuclei in a gas to undergo nuclear fusion reactions. The nuclei must therefore collide at very high velocities in order to combine, so that they can get close enough together despite their electromagnetic repulsion for each other. The high temperature of a stellar core imparts high speeds to the nuclei, so they collide with great energy, and the high

density means that collisions will be frequent. Even so, only occasionally do two nuclei combine in a fusion reaction; a single particle may typically collide and bounce around inside a star for millions or even billions of years before it reacts with another. There are so many particles, however, that reactions are constantly occurring.

The amount of energy released in a single reaction between two particles is small, about 10^{-5} erg. An erg (defined in chapter 4) is itself a small quantity; a 100-watt light bulb radiates 10^9 ergs per second. Hence a light bulb would require 10^{14} reactions per second to keep glowing, if nuclear reactions were its energy source. A star like the sun emits more than 10^{33} ergs per second, so a tremendously large number of reactions must be occurring in its interior at all times.

The reaction that takes place inside all stars on the main sequence converts hydrogen nuclei into helium

stars, with a good deal of research on stellar spectra and the structure of atoms thrown in for good measure.

His principal work can be classified into four general areas. First, he studied double stars, and developed very sophisticated techniques for analyzing the light variations in eclipsing binaries to determine not only stellar radii, but also stellar densities and internal structure. To do this required the development of new mathematical methods, as well as a physical theory as to how light is emitted from the surface of a star.

Russell's second major area of research was the study of stellar evolution, an interest that was inspired by the recognition that there exist giant and dwarf stars, and which led to his development of the H-R diagram. He was led into this work by his earlier studies of stellar distances, which enabled him to determine absolute magnitudes for

a number of nearby stars. Along the way, Russell developed several theories about stellar evolution, each of which was consistent with what was known at the time, but each of which later had to be modified.

Russell's work on stellar spectra was inspired by research on ionization balance done by others, and his own recognition that it should be possible to measure a star's chemical composition through the analysis of its spectral lines. Having developed techniques for doing this, Russell is credited with ascertaining that the primary constituent of the sun (and, by inference, the stars in general) is hydrogen. This was a profound discovery about the nature of the universe, one that later played a major role in studies of its evolution.

Finally, Russell's fourth major area of work, the study of atomic structure (specifically, electron energy levels and the spectra they pro-

duce) was an outgrowth of his interest in stellar composition. He found a great lack of laboratory data when he sought to analyze stellar spectra, and accordingly he developed both theoretical and experimental means to supply the needed information.

Although it is true that many of Russell's advances were made through collaboration with other scientists and not in isolation, it is equally true that he was quick to rise to a role of leadership in every instance. If his era was a new beginning in astronomy, then surely he must be regarded as a true Renaissance man of his time.

nuclei. A hydrogen nucleus consists solely of a proton, whereas a helium nucleus contains two protons and two neutrons. The result of the reaction is that four hydrogen nuclei are combined into one helium nucleus; two of the protons must be converted into neutrons in the process. The helium nucleus has less mass than the total of the four hydrogen nuclei that went into the reaction. The difference, 0.007 of the original mass of the four protons, is the amount that is converted into energy.

We have said that all main-sequence stars are converting hydrogen into helium in their cores. The details of the reactions differ from one part of the main sequence to another, however. In stars on the lower portion of the main sequence, including the sun, the dominant process is a rather simple sequence of reactions called the **proton-proton chain.** The more massive stars on the upper main sequence, with higher

internal temperatures, undergo a more complex sequence called the **CNO cycle.** In this sequence, carbon acts as a **catalyst,** meaning that it is necessary to get the reactions going, but it is not used up in the reactions. The name CNO cycle is adopted because oxygen and nitrogen also appear in the reaction sequence. The result is the same as before: four hydrogen nuclei are combined into one helium nucleus, and 0.007 of the original mass is converted into energy.

Once it is produced in the core, energy must somehow make its way outward to the stellar surface. For most stars, two energy transport mechanisms are at work at different levels (Fig. 20.4). One of these is **radiative transport,** meaning that the energy is carried outward by photons of light. In the core, where it is very hot, the photons are primarily γ-rays, but as they slowly move outward, being continually absorbed and reemitted on the way, they are gradually converted to longer wave-

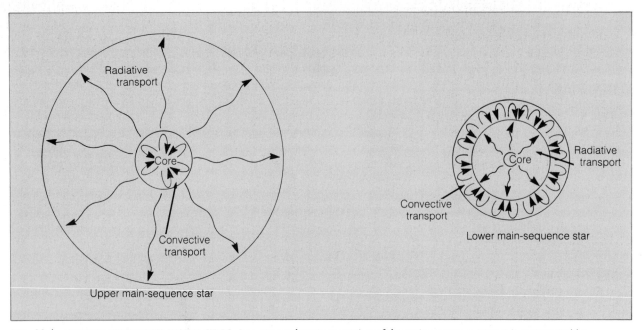

FIG. 20.4. ENERGY TRANSPORT INSIDE STARS. In a star on the upper portion of the main sequence, energy is transported by convection in the inner zone, and by radiation in the outer regions. The opposite is true of lower main-sequence stars.

lengths. When the light emerges from the stellar surface, it is primarily in the visible-wavelength region (or the ultraviolet or infrared, if it is a very hot or very cool star).

The second means of energy transport inside stars is **convection** (a process discussed briefly in chapter 16). Convection is an overturning of the gas, as heated material rises and cooled material sinks. The same process causes warm air to rise toward the ceiling of a room, and is responsible for the overturning of water in a pot that is being heated from the bottom.

Whether radiative transport or convection will be the dominant energy transport mechanism in a star depends on the temperature structure (specifically, on how rapidly the temperature decreases with distance from the center). Most stars have both a convective zone and a radiative region. In stars like the sun (that is, for stars on the lower half of the main sequence, of spectral types F, G, K, and M), convection occurs in the outer layers (see Color Plate 15), whereas radiative transport is the principal means of energy transport in the interior. For stars on the upper main sequence, the situation is reversed: there is convection in the central core, but not in the rest of the star, where radiative transport is responsible for conveying the energy to the surface.

Stellar Life Expectancies

The lifetime of a star is measured by the amount of energy it can produce in nuclear reactions, and the rate at which the energy is radiated away into space. When all the available nuclear fuel has been used up, the star undergoes major changes in its structure and properties, and it leaves the main sequence (this will be discussed in detail in the next chapter). We can estimate the hydrogen-burning lifetime of a star simply by determining how much mass it can convert to energy in the reactions, then using the formula $E = mc^2$ to see how much energy will be produced.

Let us consider the sun first. It has a mass of 2×10^{33} grams. Only the innermost 10 percent or so of this mass will ever undergo nuclear reactions, because only in the core are the temperature and density sufficiently high. Therefore we expect only 2×10^{32} grams of the sun to be available for reactions. Only 0.007 of this, or 1.4×10^{30} grams, will actually be converted into energy. From Einstein's formula, we find that $E = mc^2 = (1.4 \times 10^{30} \text{ grams}) \times (3 \times 10^{10} \text{cm/sec})^2 = 1.26 \times 10^{51}$ ergs. This is all the energy the sun can ever produce in its lifetime on the main sequence.

To estimate how long it will take the sun to use up all of this energy, we need only to take into account its

The rings of Saturn ▲

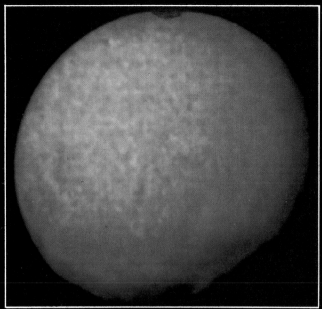

Enceladus ▲

◄ Saturn and some
of the major satellites

Color Plate 12

◀ A false-color image of Saturn

Structure in
Saturn's atmosphere ▼

Ring structure reconstructed
from *Voyager* photopolarimeter data ▼

An ultraviolet false-color image of Comet Kohoutek ▲

False color representing ▲
brightness structure of Comet Bennett

Comet Kohoutek ▼

Color Plate 14

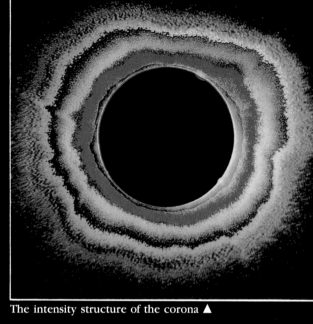

Dynamics of the sun's corona ▲

The intensity structure of the corona ▲

A solar prominence ▼

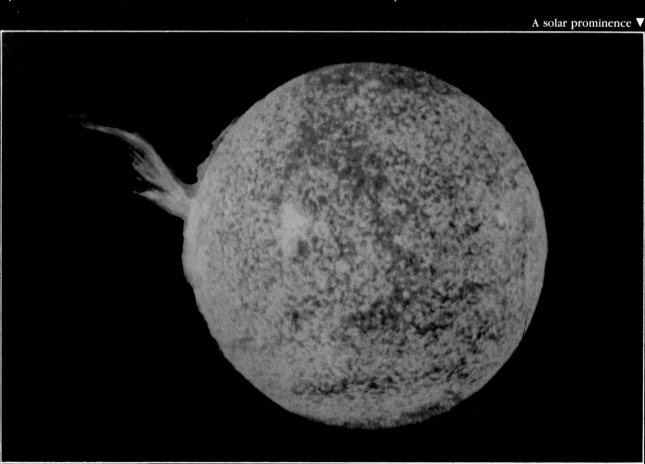

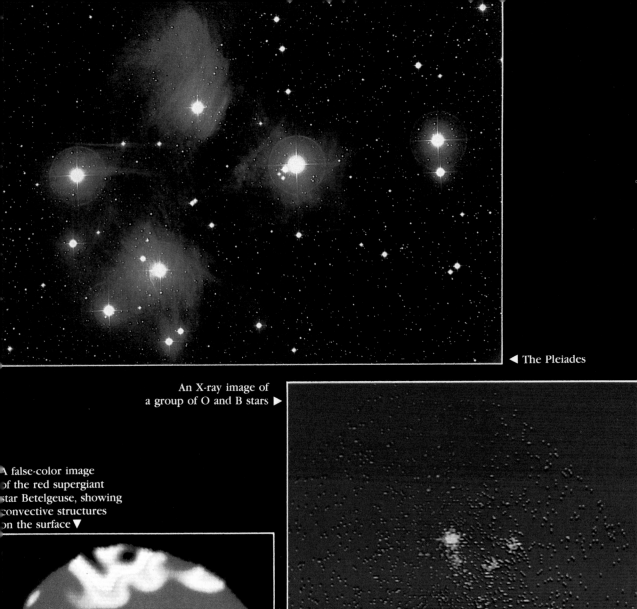

◄ The Pleiades

An X-ray image of
a group of O and B stars ►

A false-color image
of the red supergiant
star Betelgeuse, showing
convective structures
on the surface ▼

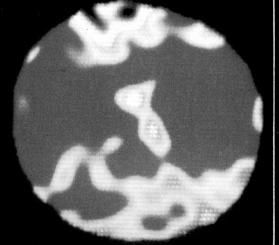

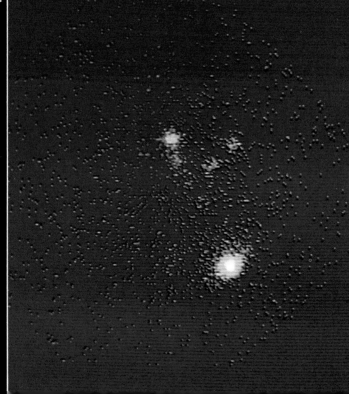

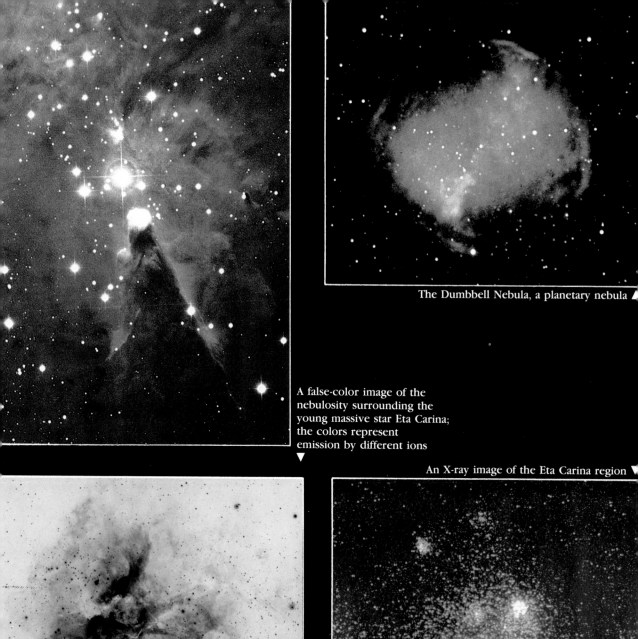

The Dumbbell Nebula, a planetary nebula ▲

A false-color image of the nebulosity surrounding the young massive star Eta Carina; the colors represent emission by different ions ▼

An X-ray image of the Eta Carina region ▼

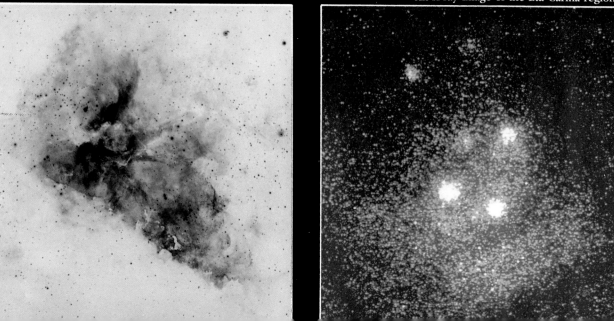

luminosity, which is the rate at which the energy is produced and radiated away. For the sun, this is roughly 4×10^{33} ergs/sec, so the lifetime is $(1.26 \times 10^{51}$ ergs)/ $(4 \times 10^{33}$ ergs/sec) $= 3.15 \times 10^{17}$ sec. This is just about 10^{10} years. Therefore we can expect the sun to run out of hydrogen fuel 10 billion years after it began burning it. As we learned in chapter 16, the present age of the sun is roughly 4.5 billion years, so there are still more than 5 billion years to go before the fuel is expended.

What of other stars, with greater or smaller masses? A star near the top of the main sequence may have 50 times the mass of the sun, but at the same time it uses its energy as much as 10^6 times more rapidly. These numbers lead to an estimated lifetime which is $50/10^6 = 5 \times 10^{-5}$ times that of the sun. This star will last only half a million years! (Actually, in such a massive star, more than 10 percent of the total mass can undergo reactions, and the lifetime is accordingly longer than this estimate; it is a few million years.) By astronomical standards, such massive stars exist only for an instant before using up all their fuel and dying (Fig. 20.5). This explains why stars of this type are very rare;

only a few may be around at any given moment, even though many may have formed in the past.

Let us consider a lower main-sequence star, for example an M star with mass 0.05 solar masses, and luminosity 10^{-4} times that of the sun. Its lifetime is $0.05/10^{-4} = 500$ times greater than the sun's lifetime, or about 5×10^{12} years. This is a very long time, probably greater than the age of the universe. No star with such a low mass has yet had time to use up all of its fuel; all such stars ever born are still with us. For this reason, the majority of all stars in existence today are low-mass stars (it is also true that these stars form in greater numbers).

Heavy-Element Enrichment

Clearly nuclear reactions have an effect on the chemical composition of a star, because they change one element into another. We have stressed that all stars consist of about the same mixture of hydrogen and helium, with a trace amount of other elements, but now

FIG. 20.5. STELLAR LIFETIMES. The large bucket on the left represents the large quantity of energy produced by a massive star during its lifetime, and the water gushing out of the hole at its bottom depicts its high luminosity, or energy-loss rate. At right is a small bucket representing a low-mass star, which produces much less energy over its lifetime, but which loses it so slowly (that is, it has such a low luminosity) that it outlives the more massive star by a wide margin.

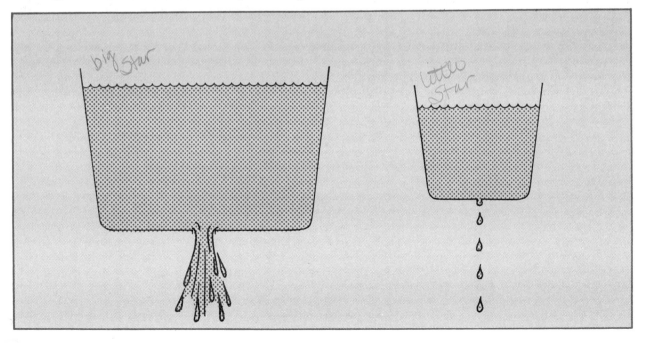

we find that in the core of a star, hydrogen is gradually converted into helium. Although this may not immediately affect the surface composition, which is what astronomers can measure directly from spectral analysis, it does change the internal composition. It is this change that causes a star to evolve, because the core density is altered as hydrogen nuclei combine into the heavier helium nuclei, and because when the hydrogen runs out, the star must make major structural adjustments as its primary source of internal pressure disappears.

These adjustments are discussed in chapter 22; for now, let us consider only the change in abundances of the elements. There are other nuclear reactions that can take place in stars, after the hydrogen-burning state has ended, if the core temperature reaches sufficiently high levels. One such reaction is the conversion of helium into carbon, by a sequence called the **triple-alpha reaction** (helium nuclei are called alpha particles, and in this reaction, three of these combine to form one carbon nucleus). When the triple-alpha reaction takes place, the composition of the star's core changes from helium to carbon.

There are many additional reactions that can occur, if in later stages the core temperature goes even higher (Fig. 20.6). These reactions produce ever-heavier products, so that as a star goes through successive stages of nuclear burning, its internal composition changes from what had originally been predominantly hydrogen to

elements as heavy as iron (which has 26 protons and 26 neutrons in its nucleus).

These reactions require ever-higher temperatures, because as the particles become more massive, it takes more energy to keep them moving fast enough to react. Furthermore the heavier nuclei have greater electrical charges and therefore stronger repulsive forces that tend to keep them apart. For both of these reasons, reactions involving the fusion of heavy nuclei require higher temperatures than those in which lightweight nuclei such as hydrogen undergo fusion.

It is the most massive stars that go through the greatest number of reaction stages and, as we have just seen, these stars live only a short time. In later chapters (23, and especially 26) we will discuss the role played by these massive stars in the enrichment of chemical abundances in the galaxy.

Stellar Chromospheres and Coronae

Let us now consider other processes that may affect a star's structure and evolution. Some very important ones take place near the surface, rather than deep inside.

As we have discussed, in most stars convection occurs at some level in the interior. For stars on the upper portion of the main sequence, it takes place near the center, and there are no directly observable con-

FIG. 20.6.   THE ENRICHMENT OF HEAVY ELEMENTS. This diagram schematically illustrates the sequence of element formation that occurs in a stellar core, as one nuclear fuel after another is exhausted. The first step, the conversion of hydrogen into helium, occurs in all stars, but the number of subsequent steps that a star goes through depends on its mass. The heaviest element that can be formed by stable reactions inside stars is iron.

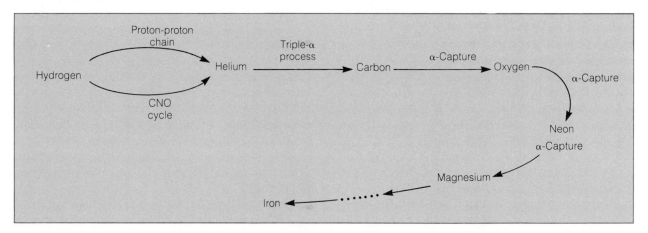

sequences. For stars on the lower portion of the main sequence, however, convection occurs in the outer layers, and there are important effects. Stars of spectral types F, G, K, and M, which are thought to undergo surface convection, have faint emission lines in their visible-wavelength spectra. These lines are usually almost drowned out by the intense light from the stellar photosphere, but there are ultraviolet emission lines that are easy to observe because the stellar surface emits little ultraviolet light (remember, these are relatively cool stars, so they emit most strongly at longer wavelengths).

The sun has hot regions above its surface that create emission lines, and these zones are called the chromosphere and the corona (Fig. 20.7; see chapter 16). The fact that other cool stars have the same kinds of emission lines (Fig. 20.8) indicates that they, too, have chromospheres and coronae. The sun's corona is extremely hot, exceeding 10^6 K, some two hundred times hotter than the photosphere. Apparently very high temperatures also exist in the coronae of other stars, although as yet the observations have not been sufficiently extensive to allow precise determinations. The emission lines that best indicate the temperature of a corona lie at very short ultraviolet wavelengths not accessible to any telescopes yet launched.

The presence of chromospheres and coronae seems to be linked with the presence of convection in the outer layers. Stars on the lower portion of the main sequence generally have both phenomena (Fig. 20.9). Somehow, the kinetic energy of the turbulent motions in the convective layer is transported into higher zones, where it causes heating. As we learned in our discussion of the sun, the precise mechanism for converting the energy of convection into heat is not understood.

What of the red giants and supergiants? These stars can have surface temperatures comparable to the lower main-sequence stars, but obviously their internal structure is rather different, since they are so much larger. These stars also have convection in their outer layers, however, and indeed they also have chromospheres, although it is not certain that the extremely high temperatures characteristic of coronae are present.

A very important modern area of research, made feasible by the development of ultraviolet telescopes for use in space, has to do with the study of phenomena in stars that also occur in the sun. These phenomena include chromospheres and coronae, and

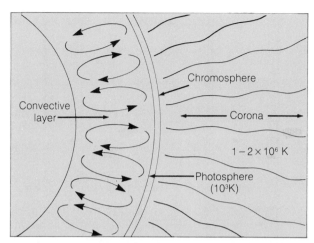

FIG. 20.7. A STELLAR CHROMOSPHERE AND CORONA. Stars in the lower portion of the main sequence all have chromospheres and coronae, analogous to the sun (see chapter 16). The photosphere is the region where the star's continuous spectrum and absorption lines are formed; the chromosphere is a thin, somewhat hotter region just above; and the corona is a very hot, extended region outside of that. The source of heat for the chromosphere and corona is probably related to convective motions in the star's outer layers.

probably activity cycles like the sun's 22-year cycle (see chapter 16). Interest in the connection between the sun and stars has inspired intensive new observing programs.

Stellar Winds and Mass Loss

So far we have said nothing about the possibility of the hot stars having chromospheres or coronae. Some of the upper main-sequence stars are known to have emission lines in their spectra, indicating that they have some hot gas above their surfaces. Visible-wavelength emission lines are found in only a few extremely hot or luminous stars, however.

In the late 1960's, with the first observations of ultraviolet spectra, it was immediately found that many hot stars (nearly all the O stars and many of the hotter B stars) have ultraviolet emission lines. Furthermore there are absorption features that show enormous Doppler shifts, indicating that the stars are ejecting material at speeds as great as 3,000 kilometers per second! The O stars have the most extreme **stellar winds,**

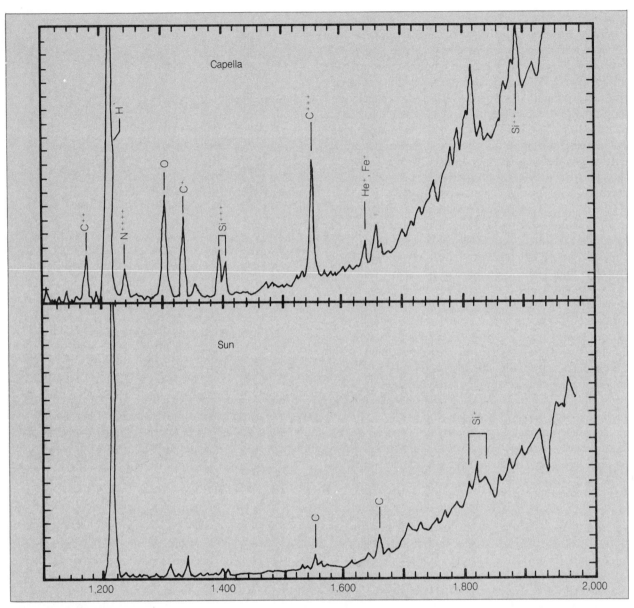

FIG. 20.8. ULTRAVIOLET EMISSION LINES IN THE SUN AND CAPELLA. This star, a binary consisting of a G giant and an F giant, has a much richer ultraviolet emission-line spectrum than the sun, a G main-sequence star. Some of the emission lines are labeled with the identities of the ions that produce them (the number of pluses represents the number of electrons missing).

as these outflows are called, but most B stars also have winds (see fig. 20.9), usually with lower velocities.

The cause of the winds is not known, although there are indications of how the high velocities are reached. Once gas begins to move outward, light from the star can exert sufficient force to accelerate it to high speeds. This force, called **radiation pressure,** is very weak (we certainly cannot feel the breeze from a light bulb,

for example), but O and B stars are so luminous that strong acceleration of the wind is possible. The mystery is how the outflow gets started in the first place.

Very high temperatures are observed in the winds from O and B stars. Recent X-ray measurements (for example, see Color Plate 15) show that temperatures of 10^6 degrees or more are common, which implies that these stars have coronae. This is rather surprising,

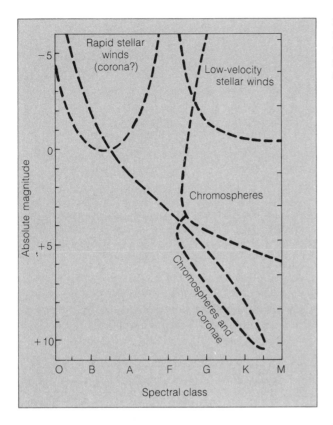

FIG. 20.9. CHROMOSPHERES, CORONAE, AND STELLAR WINDS IN THE H-R DIAGRAM. Here we see that chromospheres exist in most stars of type F or cooler, but coronae may be confined to the main sequence. Very luminous stars, hot and cool alike, have winds strong enouth to cause significant loss of mass.

because the O and B stars are not thought to have convection in the outer layers, and no other process is known that could create coronae.

Whatever the cause of the winds, they have important consequences. Analysis of the ultraviolet emission lines shows that in some cases stars are losing matter at such a great rate that a large fraction of the initial mass may be lost during their lifetimes. An O star might lose as much as 1 solar mass every 100,000 years; if such a star begins life with 20 or 30 solar masses and lives 1,000,000 years, it could lose 10 solar masses, a significant fraction of what it had to begin with. This has important effects on how such a star evolves, as we will see.

The supergiants in the upper right of the H-R diagram also lose mass through stellar winds, but these winds have a distinctly different nature (Fig. 20.10). The red supergiants are so large that their surface gravity is very low, so that gas in the outer layers is not tightly

FIG. 20.10. STELLAR WINDS. A luminous hot star (left) ejects gas at a very high velocity; the lengths of the arrows indicate that the gas accelerates as it moves away from the star. A luminous cool star (especially a K or M supergiant) is so extended in size that the surface gravity is very low, and material drifts away at relatively low speeds. In both types of stars, radiation pressure probably helps accelerate the gas outward.

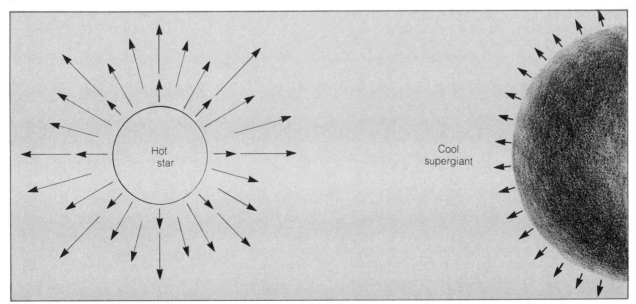

bonded to the star. Radiation pressure can easily push the gas outward at the relatively low speed of 10 or 20 kilometers per second. The amount of mass lost can be just as great as in the hot stars, however, because these low-velocity winds are much denser than the high-speed winds from the O and B stars.

Because the red supergiants are relatively cool objects, the gas in their outer layers is not ionized, and even molecular species form there. In addition, small solid particles called dust grains can condense, so that the star becomes shrouded in a cloud. In extreme cases the dust cloud becomes so thick that little or no visible light escapes to the outside. The dust grains become heated, however, and emit infrared radiation, so the star can still be detected with the use of an infrared telescope.

Some of the dust that forms in the outer layers of red supergiants escapes into the surrounding void, and interstellar space gets contaminated in this way with a kind of pollution, an interstellar haze (the interstellar dust is discussed at greater length in chapter 25).

Mass Exchange in Binary Systems

All of the processes discussed so far can occur in single stars or in stars that are part of binary systems. There are additional factors that come into play in certain kinds of binaries, but which do not affect single stars.

We have seen that stars can lose matter into space. If such a star has a companion, then the companion may catch some of the cast-off material and gain mass in the process. In come cases the companion actually helps its neighbor lose matter. There is a point between the two stars where their gravitational forces just balance, and a particle of gas that reaches that point can fall either way, into either star. If one of the stars is swelling up on its way to becoming a red giant, when its outer layers reach this balance point, gas will begin to flow down onto the other star (Fig. 20.11). It is actually possible for a pair of stars with unequal initial masses to reverse themselves, so that the one that started out with most of the mass ends up with the least.

A number of binary systems have been observed where this mass-exchange phenomenon is taking place. These systems are characterized by flowing streams of gas swirling around the two stars, creating emission lines with Doppler shifts indicating the motions of the

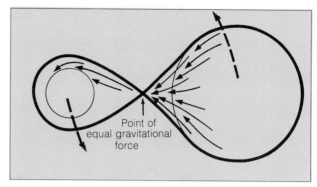

FIG. 20.11. MASS TRANSFER IN A BINARY SYSTEM. The more massive star in a binary will be the first to evolve and swell up as a red giant. In the process, its outer layers may come sufficiently close to the companion star to be pulled to it gravitationally. Such a transfer of mass can radically affect both stars.

gas. In some of these systems the star that is receiving mass has finished its evolution and has become a stellar remnant such as a white dwarf, neutron star, or black hole (see chapter 23), and the gas falling in creates spectacular effects ranging from nova explosions to intense X-ray emission.

No matter how it happens, the loss or gain of mass by a star will have important effects on how it evolves. We have already learned that the mass of a star determines all its other properties; hence it follows that the other properties will vary if the mass does.

Stellar Models: Tying It All Together

Once astrophysicists have understood all the phenomena that govern a star's internal structure and its outward appearance, they must combine their knowledge of all these processes into a theoretical calculation that will tell them what goes on inside the star, and what will happen to it in time (Fig. 20.12). When the calculations correctly reproduce the observable quantities, it is assumed that they also correctly illustrate the non-observable aspects, such as the star's interior conditions. We can have reasonable confidence in such calculations, because it can be demonstrated that the resulting solutions are unique; that is, no other solution to the mathematical equations can exist. The most difficult question is whether we know all the correct equations to begin with.

The calculation of a model star consists of simultaneously solving a set of basic mathematical relationships describing the physical processes, and solving them repeatedly for different depths inside the star. In this way the calculations may proceed from the observed surface conditions to the center, or from assumed central conditions to the surface, with adjustments made until the calculations match the observed properties of the star. To carry out such a calculation requires a substantial amount of time on a large computer.

In order to see how a star evolves, we must compute the model many times over, each time taking into account changes produced in the star that occurred in the previous step in the calculation. For example, a certain amount of hydrogen is converted into helium at each step, and this change in the chemical composition of the core must be considered during the next step. The conversion of hydrogen into helium uses up the hydrogen fuel, so that the star will eventually exhaust its supply, and it causes the density of the core to increase, because helium nuclei are more massive then hydrogen nuclei. These factors cause the star's overall structure to change, and the calculations keep track of the changes, allowing us to trace the evolution. In the next chapters we will discuss the evolution of stars, and how it is determined from a combination of theory and observation.

FIG. 20.12. BUILDING A MODEL STAR. Modern astrophysicists reconstruct what is happening inside stars by using complex computer programs that simultaneously solve large numbers of equations describing the physical conditions. Often the result is plotted, as a convenient way of showing the implications of these data.

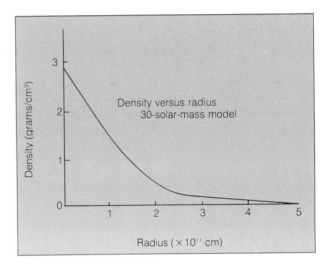

FINAL MODEL

POINT	RADIUS	DENSITY	TEMP	PRESSURE	ENERGY GEN		LUMIN	OPACITY	DELAD	DELRAD	BETA
1	0.	2.8526E+00	3.5459E+07	1.7651E+16	1.9682E+05	1	0.	3.4602E-01	2.8868E-01	3.0013E+00	7.7421E-01
2	5.4736E+08	2.8526E+00	3.5459E+07	1.7651E+16	1.9680E+05	1	3.8569E+32	3.4602E-01	2.8868E-01	3.0013E+00	7.7421E-01
3	7.3124E+08	2.8525E+00	3.5459E+07	1.7650E+16	1.9679E+05	1	9.1397E+32	3.4602E-01	2.8868E-01	3.0011E+00	7.7421E-01
4	8.8736E+08	2.8525E+00	3.5459E+07	1.7650E+16	1.9677E+05	1	1.6376E+33	3.4602E-01	2.8868E-01	3.0009E+00	7.7421E-01
5	1.0365E+09	2.8525E+00	3.5459E+07	1.7650E+16	1.9675E+05	1	2.6288E+33	3.4602E-01	2.8868E-01	3.0008E+00	7.7421E-01
6	1.1927E+09	2.8524E+00	3.5458E+07	1.7649E+16	1.9673E+05	1	3.9864E+33	3.4602E-01	2.8868E-01	3.0006E+00	7.7421E-01
7	1.3548E+09	2.8524E+00	3.5458E+07	1.7649E+16	1.9671E+05	1	5.8460E+33	3.4602E-01	2.8869E-01	3.0004E+00	7.7421E-01
8	1.5283E+09	2.8523E+00	3.5458E+07	1.7648E+16	1.9668E+05	1	8.3930E+33	3.4602E-01	2.8869E-01	3.0001E+00	7.7421E-01
9	1.7159E+09	2.8523E+00	3.5457E+07	1.7648E+16	1.9664E+05	1	1.1881E+34	3.4602E-01	2.8869E-01	2.9998E+00	7.7421E-01
10	1.9200E+09	2.8522E+00	3.5456E+07	1.7647E+16	1.9659E+05	1	1.6659E+34	3.4602E-01	2.8869E-01	2.9994E+00	7.7421E-01
11	2.1449E+09	2.8521E+00	3.5456E+07	1.7646E+16	1.9654E+05	1	2.3202E+34	3.4602E-01	2.8869E-01	2.9989E+00	7.7422E-01
12	2.3916E+09	2.8519E+00	3.5455E+07	1.7645E+16	1.9647E+05	1	3.2162E+34	3.4602E-01	2.8869E-01	2.9983E+00	7.7422E-01
13	2.6639E+09	2.8518E+00	3.5454E+07	1.7643E+16	1.9639E+05	1	4.4432E+34	3.4602E-01	2.8869E-01	2.9976E+00	7.7422E-01
14	2.9647E+09	2.8515E+00	3.5453E+07	1.7641E+16	1.9628E+05	1	6.1231E+34	3.4602E-01	2.8869E-01	2.9967E+00	7.7423E-01
15	3.2977E+09	2.8512E+00	3.5452E+07	1.7639E+16	1.9615E+05	1	8.4230E+34	3.4602E-01	2.8869E-01	2.9956E+00	7.7423E-01
16	3.6664E+09	2.8510E+00	3.5450E+07	1.7636E+16	1.9600E+05	1	1.1571E+35	3.4602E-01	2.8869E-01	2.9942E+00	7.7424E-01
17	4.0755E+09	2.8506E+00	3.5448E+07	1.7632E+16	1.9580E+05	1	1.5880E+35	3.4602E-01	2.8869E-01	2.9925E+00	7.7424E-01
18	4.5291E+09	2.8501E+00	3.5446E+07	1.7628E+16	1.9556E+05	1	2.1770E+35	3.4602E-01	2.8869E-01	2.9905E+00	7.7425E-01
19	5.0324E+09	2.8496E+00	3.5443E+07	1.7623E+16	1.9527E+05	1	2.9841E+35	3.4602E-01	2.8869E-01	2.9878E+00	7.7426E-01
20	5.5910E+09	2.8488E+00	3.5439E+07	1.7616E+16	1.9491E+05	1	4.0871E+35	3.4602E-01	2.8870E-01	2.9847E+00	7.7427E-01
21	6.2112E+09	2.8480E+00	3.5434E+07	1.7608E+16	1.9446E+05	1	5.5949E+35	3.4602E-01	2.8870E-01	2.9808E+00	7.7429E-01
22	6.8998E+09	2.8469E+00	3.5428E+07	1.7598E+16	1.9391E+05	1	7.6550E+35	3.4602E-01	2.8870E-01	2.9760E+00	7.7431E-01
23	7.6646E+09	2.8455E+00	3.5421E+07	1.7585E+16	1.9324E+05	1	1.0468E+36	3.4602E-01	2.8871E-01	2.9701E+00	7.7434E-01
24	8.5141E+09	2.8439E+00	3.5412E+07	1.7570E+16	1.9241E+05	1	1.4307E+36	3.4603E-01	2.8872E-01	2.9628E+00	7.7437E-01

Perspective

We now are acquainted with all the physical processes that affect a star; we know what a star is and what makes it run. We have seen that mass plays a dominant role in dictating all the other properties, although compo-sition, which changes with time, is also important. We are prepared to see what the life story of a star is; how it is born, how it lives, and how it ends its days.

Summary

1. A star is gaseous throughout, with temperature and density increasing toward the center.

2. The balance between gravity and pressure within a star, called hydrostatic equilibrium, governs the internal structure.

3. The mass of a star, along with its chemical composition, governs its central temperature and pressure (through hydrostatic equilibrium), its luminosity, its radius, and its internal structure.

4. Nuclear fusion reactions take place in the core of a star, where the temperature and density are high enough to allow nuclei to collide with sufficient velocity and frequency.

5. In most stars (those on the main sequence) hydrogen is converted into helium in the reactions in the core.

6. The energy inside a star is transported by convection or by radiation. On the lower main sequence (including the sun), radiative transport dominates in the inner parts of the star, and convection operates in the outer layers. On the upper main sequence, convection is dominant in the core, whereas radiative transport dominates in the outer layers.

7. The lifetime of a star depends on its mass and how rapidly it uses up its nuclear fuel, and decreases dramatically from the lower main sequence to the upper main sequence.

8. Nuclear reactions in stars, and the recycling of matter between stars and interstellar space, result in a gradual increase in the abundance of heavy elements in the universe.

9. Stars in the cool half of the H-R diagram have chromospheres and coronae, in analogy with the sun.

10. Stars in the hot portion of the H-R diagram, and very luminous cool supergiants as well, have stellar winds that can cause the loss of large fractions of the initial stellar masses. The loss of mass can have significant effects on how the stars evolve.

11. In certain binary star systems, matter is transferred from one star to the other, with important effects on the structure and evolution of both.

12. To determine the interior conditions in a star, astronomers must compute them from known laws of physics and observations of the surface conditions.

Review Questions

1. How do we know that a star must be in a state of balance between pressure and gravity? What would happen if this balance did not exist?

2. Explain in your own words why stellar mass and composition control all other basic properties of a star, such as temperature, luminosity, and radius.

3. Explain how energy is produced by nuclear reactions, and why these reactions occur only in the central core of a star.

4. Using information from chapter 16, explain why radiative energy transport is a very slow process that takes millions of years for a single photon of light to travel from the center to the surface of a star.

5. Estimate the lifetimes of the following stars: (a) a B star, with 15 times the mass of the sun and 4,500 times the sun's luminosity; and (b) a K star, with half of the sun's mass and one-eighth of its luminosity.

6. Why are very luminous stars very rare?

7. Explain why each nuclear reaction stage that can occur in a star as heavier and heavier elements are created in its core requires a higher temperature than the preceding stage.

8. If a star has 20 solar masses to begin with and a lifetime of 5 million years, how much of its mass will it lose if it has a stellar wind that removes mass at a rate of 1 solar mass every 500,000 years?

9. Summarize the differences in energy generation, internal structure, and overall properties, between stars on the lower main sequence and those near the top of the main sequence.

Additional Readings

Aller, L. H. 1971. *Atoms, stars, and nebulae*. Cambridge, Ma.: Harvard University Press.

Cox, A. N., and Cox, J. P. 1967. Cepheid pulsations. *Sky and Telescope,* May, p. 278.

Hoyle, F. 1975. *Astronomy and cosmology: a modern course*. San Francisco: W. H. Freeman.

Percy, J. R. 1975. Pulsating stars. *Scientific American* 232(6): 66.

Shu, F. 1982. *The physical universe*. San Francisco: W. H. Freeman.

Smith, E. P., and Jacobs, K. C. 1973. *Introductory astronomy and astrophysics*. Philadelphia: Saunders.

Weymann, R. J. 1978. Stellar winds. *Scientific American* 239 (2): 34.

CHAPTER 21

Star Clusters and Observations of Stellar Evolution

How can we watch a star evolve? After all, even the most short-lived ones are around for hundreds of thousands of years, and human studies of astronomy date back only three or four millennia. Occasionally we catch a star in the act of change, as it makes a transition from one stage in its evolution to another, but clearly we cannot hope to see a star through its whole lifetime, watching it form, live, and die.

We can, however, piece together a story of stellar evolution by examining many stars of different ages, and deducing from this the sequence of events that occurs in a single star. An apt analogy is the deduction of the life story of a tree from the examination of a forest, where seeds, young saplings, mature trees, and dead logs are all found. For stars matters are complicated by the fact that the manner of evolution depends on the mass of the star, whereas for a group of trees we expect all to have about the same properties. As we shall see, however, stars, particularly groups of them, provide certain clues for the detective trying to reconstruct their life stories.

Properties of Clusters

Within our galaxy, not all stars are uniformly distributed, but instead many are located in concentrated regions called clusters. A few of these are sufficiently prominent to be visible to the unaided eye, and others

Galactic (open clusters) contain modest # of stars (up to few 100) found in disk of galaxy

FIG. 21.1. THE PLEIADES. This is a relatively young galactic cluster, and is a prominent object in the late fall and winter.

FIG. 21.2. THE DOUBLE CLUSTER IN PERSEUS. Known as *h* and χ Persei, these are young galactic clusters.

can be viewed with a small telescope or a good pair of binoculars. The Pleiades (Fig. 21.1), a bright group to the west of Orion, is perhaps the best-known cluster for northern-hemisphere observers, being easily visible in the evening sky throughout the late fall and early winter. Other clusters easy to find with a small telescope are the Hyades, the double cluster in Perseus (Fig. 21.2), and M13, a fantastic, spherically shaped collection of hundreds of thousands of stars. It is apparent that the stars in a cluster are gravitationally bound to it, each one orbiting the common center of mass.

There are several distinct types of star clusters. Those containing a modest number of stars (up to a few hundred), found in the disk of a galaxy, are referred to as **galactic,** or **open** clusters (Fig. 21.3). The Pleiades (Color Plate 15) and the Hyades are examples. Loose groupings of hot, luminous stars are also found in the plane of the galaxy, and are called **OB associations,** being dominated by stars of spectral types O and B. These clusters, in contrast with the other types mentioned here, are probably not gravitationally bound together, but appear grouped simply because the stars

FIG. 21.3. THE VERY SPARSE GALACTIC CLUSTER NGC 7510.

ASTRONOMICAL INSIGHT (21.1)

Main-Sequence Fitting

The technique of determining the distance to a cluster by plotting its H-R diagram can best be explained by example. Suppose the B and V magnitudes are measured for all the members of a particular cluster of stars. A plot is then made of V (on the vertical axis) versus B-V (on the horizontal axis). Such a plot, often called a color-magnitude diagram, is really an H-R diagram for the cluster, since, as explained in the

text, the relative positions of the individual stars along the vertical direction depend only on their relative luminosities, just as if an absolute-magnitude scale were used.

Suppose on the left is the color-magnitude diagram of our sample cluster, and on the right is the standard H-R diagram, plotted in the usual absolute-magnitude units. We see that each diagram has a main sequence. It is also clear that if the two diagrams could be superimposed, the main sequences would match each other, except that the numbers on the vertical scale would be different. In this example, we see that if 15 were subtracted from each number on the vertical axis of the cluster diagram, the scale would be the same as that of the standard H-R diagram. This means that the difference between the apparent and

absolute magnitudes of the stars in the cluster is 15; that is, the distance modulus is $m - M = 15$, and the distance to the cluster is 10,000 parsecs (see chapter 19 to review the use of the distance modulus).

The principle of main-sequence fitting is precisely the same as that of the spectroscopic-parallax method: the absolute and apparent magnitudes are compared in order for us to determine the distance. The advantage of main-sequence fitting is that a number of stars can be used, so there is no ambiguity about whether a star is on the main sequence. When applying the spectroscopic-parallax technique to a single star, we cannot always be sure that the star is on the main sequence, so the absolute magnitude can be ambiguous.

FIG. 21.4. A GLOBULAR CLUSTER. This is M92, a prominent example.

have recently formed together. The giant **globular clusters** such as M92 (Fig. 21.4) are found in a spherical volume about the galactic center, and are not confined to the plane of the galaxy (Fig. 21.5). In later chapters (especially chapters 24 and 26) we will dis-

FIG. 21.5. THE LOCATIONS OF CLUSTERS IN THE GALAXY. The open or galactic clusters are found in the disk of the Milky Way, as are the OB associations, which are always located in spiral arms (see chapter 24). The globular clusters, to the contrary, inhabit a large spherical volume, and are not confined to the plane of the disk.

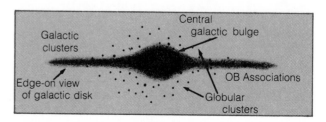

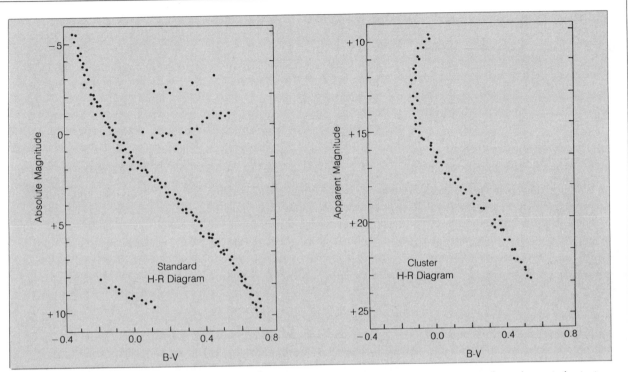

H-R DIAGRAM. This is a standard H-R diagram, with a cluster H-R diagram next to it using apparent magnitude on the vertical axis. A comparison of the two diagrams leads to an estimate of the distance to the cluster.

cuss how the various types of clusters fit into the story of the formation and evolution of the galaxy; for now we will exploit certain properties of the clusters that help us see how stars evolve.

There are two important assumptions astronomers make about a cluster of stars: (1) the stars in a cluster are all at the same distance from us; and (2) they formed together so that all are the same age and have the same composition initially.

THE ASSUMPTION OF A COMMON DISTANCE

The first of these assumptions is, strictly speaking, only approximately true. Clearly a cluster of stars has some spatial extent, so that the stars on the near side are closer to us than those on the far side. The diameter

of a cluster, however, is rarely more than a few parsecs, and is usually small compared with the distance from the earth to the cluster. Therefore, for all practical purposes the stars in a cluster are all at the same distance from us. This means that any observed differences from star to star within a cluster must be real differences, and cannot be a false effect created by differing distances from us. Brightness differences, for example, have to be the result of variations in the luminosities of stars within the cluster.

Because distance effects are eliminated when we compare stars within a cluster, it is possible to plot an H-R diagram for a cluster without first determining the absolute magnitudes of the stars. We simply plot apparent magnitude against spectral type, or, more commonly, against the color index B-V, which is easier to measure for a large number of stars (Fig. 21.6). To do this requires only the determination of the visual (V) and blue (B) magnitudes of the cluster stars, a rather straightforward observational procedure, and then the construction of a plot of V versus B-V. The result, called

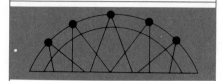

Star Tracks

We have seen that as a star evolves, its position in the H-R diagram changes. The correct deduction of *how* a star's position changes was not easily made, however. During the years following the development of the H-R diagram, great intellectual effort was concentrated on discovering how stars evolve, but because the necessary information was not yet complete, many of the first ideas were wrong. This resulted in an assortment of views on the direction of motion in the H-R diagram, and evolutionary tracks

thought to be made by stars during their lifetimes changed directions a few times.

By about 1910 it was well established that the red giants formed a distinct grouping of stars in the H-R diagram, above the main sequence. Nothing was known of the true energy source that powers stars, and little was known of stars' chemical composition. It was thought that stars glowed by virtue of energy released in a slow gravitational contraction, which caused the stars to shrink and heat, at least in their early stages. It was also suspected that within the main sequence, the hot stars were youngest and the cool stars the oldest. In 1913 Henry Norris Russell, in an address to the American Astronomical Society, put these ideas together when he suggested a comprehensive theory of stellar evolution. He proposed that a star begins life as a red giant, and for

awhile it contracts and heats, moving across the top of the H-R diagram from right to left. Eventually the temperature gets so hot inside the star that the gas there becomes highly ionized (now we know this is true from the start). Because ionized gas (it was supposed incorrectly) cannot compress and heat further, the star begins to cool while still contracting, and it moves down the main sequence from upper left to lower right. The evolutionary track of a star was therefore thought to resemble a backward figure seven, and the main sequence was thought to be just an age sequence.

In the decade following Russell's hypothesis, several astrophysicists, including most notably Sir Arthur Eddington and Sir James Jeans, developed theoretical models of stellar structure. These models, which of necessity incorporated numerous approximations, appeared to be consistent with the evolution-

---

a **color-magnitude** diagram, looks just like an H-R diagram, except that the vertical scale is an apparent-magnitude scale. This is because the *differences* in magnitude of the cluster stars are the same, whether we use apparent or absolute magnitudes.

One advantage of plotting cluster H-R diagrams is that these diagrams can be used to determine the distances to clusters. A comparison of a cluster H-R diagram with a standard one (that is, one with an absolute-magnitude scale) allows us to see what the absolute-magnitude scale for the cluster diagram should be. This in turn tells us the difference between the absolute and apparent magnitudes for the cluster, and from this difference we deduce the distance. This procedure, known as **main-sequence fitting,** depends only on the assumption that the cluster has a main sequence that is identical to that of the standard H-R diagram, a safe assumption as long as the chemical composition of the cluster is not unusual.

THE COMMON–AGE ASSUMPTION

The second major assumption made about clusters, that all the stars are the same age, is very important in studying stellar evolution. The basic premise is that the stars formed together, out of a common cloud of interstellar material, and did not just happen to come together. The basis for this assumption is the extremely low probability that a number of stars could ever encounter each other and form a cluster by chance, as a result of random stellar motions. A close encounter between stars is very rare indeed, and furthermore even when one does occur, it is very unlikely to result in gravitational bonding. Binary systems as well as clusters must *form* as binaries or clusters. A corollary is that all stars in a cluster must begin with the same chemical composition, since they form from a common source of material.

The knowledge that all stars in a cluster have the

ary scenario of Russell, but by the early 1920's, discrepancies began to arise. The chief problem was that there appeared to be no reason for the ionized gas in a star's interior to become incompressible, and therefore no reasonable explanation of why a contracting star should stop heating up and start cooling when its interior became ionized. The new theoretical work, along with developing evidence for a close relation between a star's mass and its luminosity, led to the first suggestion (again by Russell) that the main sequence represented a mass sequence, where stars are stable and spend a long time with no movement in the H-R diagram. Russell suggested that the difference between the main-sequence stars and the red giants was that the two groups had different internal energy sources, although his ideas of what those sources might be were rather vague. He still thought that the

red giants were young stars, and that they moved down to the main sequence as their first energy source was exhausted and the second took over. Thus in 1923 (when this was hypothesized), the evolutionary tracks attributed to stars were much like those known today, but with the motion in the exact opposite direction.

The proper sequence of events in a star's lifetime finally began to be sorted out in the late 1920's and early 1930's, with Russell once again leading the way. Two major clues were developed: (1) cluster H-R diagrams (studied by Hertzsprung, among others) showed that as a cluster ages, it loses stars from its upper main sequence and gains red giants; and (2) the Russell-Vogt theorem was developed, which demonstrated that a star's properties are all determined specifically by its mass and composition. It finally became clear that a star

moves from the main sequence to the red-giant region, and that this movement must be caused by a change in its internal composition. It was even suggested that the change of composition involves the conversion of hydrogen into helium, although it was not until the late 1930's that the exact nature of this process was understood.

The study of star tracks in the H-R diagram demonstrates how plausible, but incorrect, theories can be deduced from incomplete information, yet it also shows how patient and careful collection of data will eventually lead to the correct picture. The incorrect hypotheses developed along the way were important, for they helped inspire the continued efforts to solve the problem.

same age and initial composition gives astronomers tremendous leverage in deducing the evolutionary histories of stars. The only major parameter that differs from one star to another within a cluster is mass, so the observed differences between stars in a cluster give us clues as to how the properties of stars depend on mass. Comparison of H-R diagrams of clusters of different ages then gives us a picture of how these different masses evolve.

The Main-Sequence Turn-Off and Cluster Ages

Striking differences are found between H-R diagrams of different clusters. The Pleiades, for example, has an almost complete main sequence, but no red giants. By contrast a globular cluster has no stars on the upper

portion of the main sequence, but it has a large number of red giants. A wide variety of intermediate cases can be found.

From considerations discussed in the last chapter, we expect massive stars to evolve most rapidly. These are the stars that start out near the top of the main sequence, and they have such tremendous luminosities that they use up their nuclear fuel rapidly. Tying this together with the observed differences between cluster H-R diagrams leads to the conclusion that the upper main-sequence stars must evolve into red giants when they have used up all their hydrogen fuel. In a very young cluster, even the most massive stars might still be on the main sequence, whereas in an older cluster, these massive stars will have had time to use up their core hydrogen and become red giants. Hence the Pleiades is relatively young, whereas a globular cluster is very old.

As a cluster ages, more and more of its stars use up

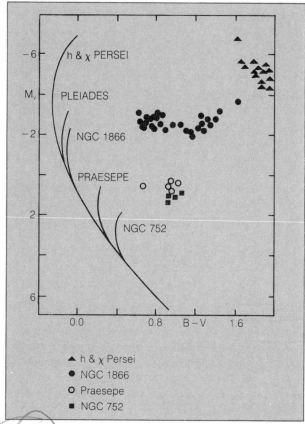

FIG. 21.6. H-R DIAGRAMS FOR STAR CLUSTERS. Here, plotted on the same axes, are H-R diagrams (using color index rather than spectral type) of several clusters. Different symbols are used in the right-hand portion to distinguish among the giant stars in different clusters.

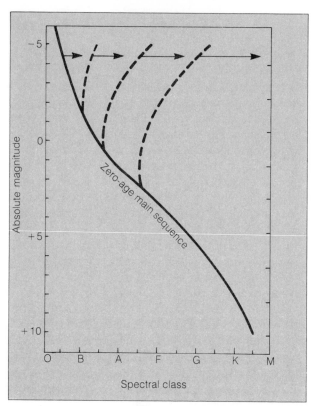

FIG. 21.7. THE EVOLUTION OF THE MAIN SEQUENCE. As a cluster ages, the stars on the upper main sequence move to the right. The farther down the main sequence this has happened, the older the cluster.

their hydrogen and leave the main sequence. The upper main sequence gradually erodes away as its stars move to the right in the H-R diagram, toward the red-giant region (Fig. 21.7). This process creates a definite cutoff point on the main sequence, above which there are no stars. The location of this cutoff depends entirely on the age of the cluster. The position of this **main-sequence turn-off,** as it is commonly called, indicates the age of a cluster. If, for example, we find a cluster whose main sequence stops at the position of stars like the sun (that is, it has no stars above G2, or $B-V = 0.6$, on the main sequence), then we know that all stars more massive than one solar mass have burned their hydrogen and evolved toward the red-giant region. In the last chapter we learned that the sun's lifetime for hydrogen burning is about 10 billion years;

we conclude, therefore, that such a cluster is just about 10 billion years old. If it were younger, its main sequence would extend farther up, and if it were older, even the sun-like stars would have had time to evolve into red giants, and the main-sequence turn-off would be farther down.

Young Associations and Stellar Infancy

Earlier in this chapter we mentioned OB associations, loosely bound groups of O and B stars. We now know that these are very young clusters, probably the youngest in the galaxy, because of the fact that they contain stars on the extreme upper portion of the main sequence. In many cases the stars are moving away from the center and escaping the association, so within a few million years no concentration of stars will remain.

If the OB associations are the youngest ensembles of stars in the galaxy, we might expect to find star formation actually occurring in or near them. Some young clusters show remnants of the process of star birth, in the form of gas and dust surrounding the stars (Fig. 21.9 and Color Plates 16 and 17). The Pleiades provide a good example of this; photographs show bluish-colored clouds of dust, shining by the reflected light of the stars in the cluster (Color Plate 15). In other cases not only these **reflection nebulae,** but also hot, glowing gas clouds called **HII regions** may be found. The notation *HII* refers to hydrogen in its ionized form; these hot gas clouds are called HII regions because they glow most strongly in emission lines of hydrogen (these lines form when protons and electrons combine to form hydrogen atoms, which can only occur in a region that is ionized). HII regions appear red in color because hydrogen has a very strong emission line at a wavelength of 6,563 Å, in the red portion of the spectrum.

Since a number of young clusters are permeated by

FIG. 21.8. A REGION OF YOUNG STARS AND NEBULOSITY. This is a portion of the Rosette Nebula, showing numerous dark patches and globules where star formation takes place.

FIG. 21.9. A VERY YOUNG CLUSTER WITH NEBULOSITY. This is the cluster M16, a group of stars associated with the gas and dust from which it formed.

gas and dust left over from star formation, it is natural to expect that in even younger, newly forming clusters, the gas and dust would be much denser; so dense, in fact, that the embryonic stars might be hidden from the view of observers. Hence the search for stars in the process of forming takes us into dense interstellar clouds (Figs. 21.10 and 21.11).

How do we probe these dark and dusty regions? For a couple of reasons, this is best done by observing in infrared wavelengths. One reason is that infrared radiation penetrates much farther through clouds of gas and dust than does visible light; another is that the dust around a newborn star is hot, and it glows at infrared wavelengths. Therefore, to see infant stars, we must look for infrared sources buried in dark clouds (Fig. 21.12 and Color Plate 17).

We have found many regions in our galaxy where star formation is apparently taking place. One of the best observed of these is in the sword of Orion, in the Orion Nebula, a great, glowing HII region (Color Plate 17). A very young cluster of stars is located there, and

FIG. 21.10. A DARK CLOUD. The regions without stars in this photo are portions of the Coalsack nebula, a well-known concentration of interstellar gas and dust in the southern Milky Way.

FIG. 21.11. A NEWBORN STAR LEAVES THE WOMB. This remarkable photo of a dark cloud shows a wispy, glowing trail left behind by a newly formed star that has recently emerged from the cloud.

FIG. 21.12. EMBEDDED INFRARED SOURCES. This sketch shows a dark interstellar cloud with a few young, infrared-emitting stars inside. The infrared light is able to escape the cloud more readily than visible light, so the best technique for finding young stars is to use infrared telescopes.

infrared observations reveal a number of infrared sources enbedded within the dark cloud associated with the cluster (Figs. 21.13 and 21.14).

In other, similar regions faint variable stars called **T Tauri** stars are found. These appear to be newborn stars just in the process of shedding the excess gas and dust from which they formed; in fact, spectroscopic measurements indicate that some material may still be falling into these stars, as though the collapse of the cloud from which they are forming is not yet complete. We learned in chapter 16 that the sun probably went through a phase as a T Tauri star early in its evolution.

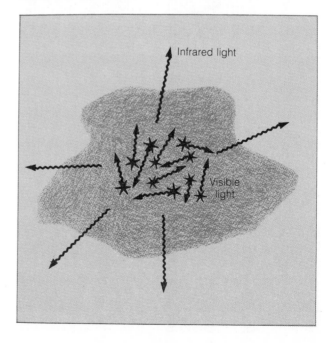

Star Formation

The process of star formation deduced from the types of observations just described begins with the gravitational collapse of an interstellar cloud (Fig. 21.15). It is not certain what causes a cloud to begin to fall in on itself, but once it starts, gravity takes care of the rest. Calculations show that the innermost portion collapses most quickly, while the outer parts of the cloud are still slowly picking up speed. The temperature in the core builds up as the density increases, but for a long time the heat escapes as the dust grains in the interior radiate infrared light. Eventually, however, the cloud becomes opaque in the center, and radiation no longer can escape directly into space. After this the heat builds up much more rapidly, and the resulting pressure causes the collapse to slow to a very gradual shrinking. Material still falling in crashes into the dense core, creating a violent shock front at its surface. The luminosity of the core, which by this time may have a temperature

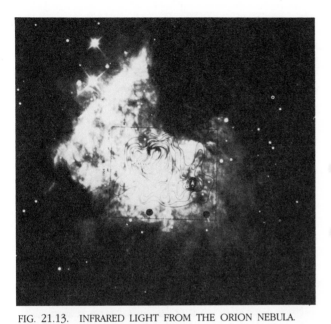

FIG. 21.13.   INFRARED LIGHT FROM THE ORION NEBULA. Here we see intensity contours showing the locations of sources of infrared light within the great nebula in the sword of Orion. Several young stars are thought to lie within this complex cloud of gas and dust.

FIG. 21.14.   THE ORION NEBULA. At left is a normal, long-exposure photograph, showing the familiar nebula. At right is a short exposure of the central region, taken in infrared light. Here little of the glowing gas can be seen, but the young, hot stars that are embedded within the outer portions are clearly visible.

ASTRONOMICAL INSIGHT (21.3)

Triggers of Star Formation

An interstellar cloud can collapse under its own gravity only if its density exceeds a certain value for its size, such that the inward gravitational force overcomes the ordinary gas pressure as a result of random motions of the cloud particles. There is a specific relationship among density, temperature, and cloud size called the **Jeans criterion** (named after Sir James Jeans, the British astrophysicist who derived it in the early decades of the twentieth century), which predicts the mass of cloud that can collapse under a given set of physical conditions. This criterion states that, as long as there are normal interstellar cloud densities (several thousand to perhaps a million atoms per cubic centimeter

in dark clouds), only very massive clouds can collapse gravitationally. This poses the dilemma of explaining how stars with masses like that of the sun or smaller can form.

One answer to this question has been to assume that as a massive cloud collapses, it breaks up into smaller fragments that can then individually collapse into stars. This is possible because as the overall density increases during collapse, the Jeans criterion allows smaller and smaller masses to condense individually. Furthermore, this picture provides a nice explanation of how a cluster of stars may form, starting with a very massive cloud that creates fragments as it collapses, the fragments in turn forming individual stars.

But what of individual, isolated stars? There is no evidence, for example, that the sun was ever a member of a cluster. Evidently it is possible for small-mass clouds to collapse and form stars, contrary to the Jeans criterion.

A hint at the answer to this problem is found in regions of active star formation, such as the Orion Nebula. Here observations show that the

interstellar material is chaotic; it is stirred up by violent shock fronts, probably themselves triggered by supernova explosions or strong stellar winds. The entire interstellar medium is in a constant state of turmoil (see chapter 25), so that a cloud of gas and dust is likely to be ravaged sooner or later by a passing shock front. Theoretical calculations show that in some circumstances this event will cause the cloud to compress enough to make it collapse, even though the cloud's initial density may not have been high enough to satisfy the Jeans criterion.

Supernova explosions occur in massive stars when all the nuclear fuel in the core is depleted (see the next chapter for a more complete discussion). These are the same stars that have very short lifetimes. Hence when a cluster of stars forms, its most massive members will evolve and explode as supernovae rather quickly. The shock waves from these explosions can then trigger the collapse of interstellar material remaining from the original formation of the cluster, creating a new generation of stars. The most

---

of 1,000 degrees or more, depends on its mass. If the mass is large; that is, several times the mass of the sun, then the luminosity may be sufficient to blow away the remaining material by radiation pressure. For a lower-mass **protostar,** as the central dense object is called, material continues to fall in. Eventually the density of this material becomes low enough that the protostar can be seen through it. Regardless of mass, a point is reached where a starlike object becomes visible to an observer; this stage may be identified with the T Tauri stars mentioned earlier. If the obscuring matter is swept away violently, the object may appear to brighten up

dramatically in a very short time. There are a few cases where a previously rather faint star was found to have brightened suddenly by several magnitudes.

In any case, the protostar continues to shrink slowly, growing hotter in its core. The shrinking stops when the temperature becomes high enough for nuclear reactions to begin to occur in the center. This is a major landmark in the process of star formation, for once the reactions have started, the star is on its way to becoming stable (that is, no further shrinking), and it lies on the main sequence, where it will spend most of its life converting hydrogen into helium in its core.

massive of these new stars will also evolve quickly, themselves dying in supernova explosions and creating shock waves that can give birth to yet another generation of stars. A sort of chain reaction of star formation is set off, gradually consuming an entire interstellar cloud complex.

There is some evidence that the sun's formation was associated with a nearby supernova explosion. Certain atomic isotopes found in meteorites show evidence of having formed out of an isolated interstellar cloud that was compressed by shocks created in a supernova outburst. This was not part of a chain-reaction process like that just described, presumably because the sun's predecessor cloud was not in a dense region of gas and dust.

This schematically illustrates the sequence of events that occurs as massive stars that have formed in a large, dense interstellar cloud region go through their lifetimes and explode as supernovae, creating shock waves that travel through the cloud. These shocks, in turn, compress other portions of the cloud, leading to the formation of new stars. The process repeats, as more and more of the cloud is consumed.

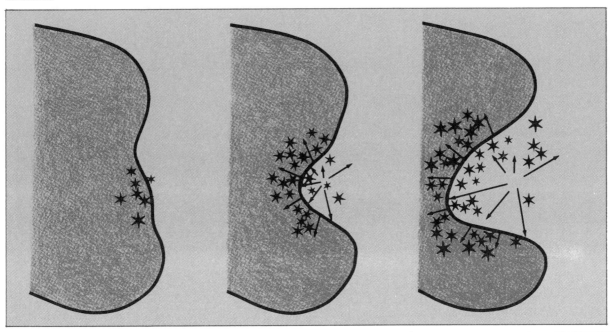

It is interesting to trace on the H-R diagram the path the star takes as it forms (Fig. 21.16). When it is a cold, dark cloud, its position is far to the right and down, well off of the scales of temperature and luminosity appropriate for stars. As the cloud heats up in the core and begins to glow in infrared wavelengths, it moves up and to the left, eventually attaining sufficient luminosity to fall onto the standard H-R diagram, but still off to the right of the main sequence. When it reaches the phase of slow contraction, it moves gradually to the left and down, finally reaching the main sequence when the reactions begin.

The entire process, from the beginning of cloud collapse to the point where the newly formed star is on the main sequence, takes many millions or even billions of years for the least massive stars, and only a few hundred thousand years for the most massive. In a cluster with stars of various masses, it can happen that the most massive stars will form, live, and die before the least massive ones even reach the main sequence. The H-R diagram for such a cluster shows stars located to the right of the main sequence at the lower end (Fig. 21.17).

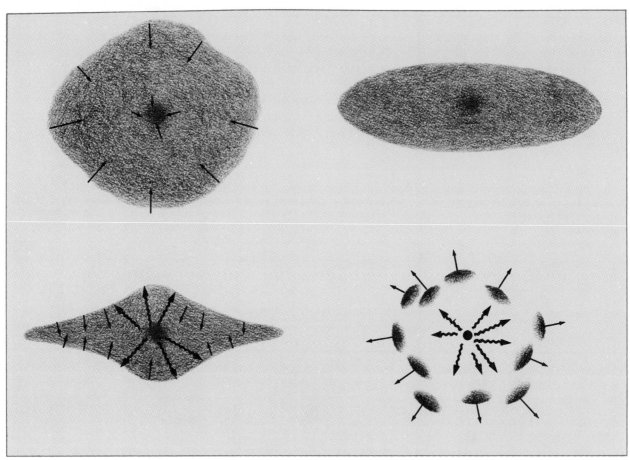

FIG. 21.15.   STEPS IN STAR FORMATION. A rotating interstellar cloud collapses, most rapidly at the center. At first, infrared radiation escapes easily and carries away heat, but eventually, the central condensation becomes sufficiently dense that this radiation is unable to escape, and the core becomes hot. In time, after a lengthy period of slow contraction, nuclear reactions begin in the protostar. At some stage, in a process not well understood, the young star develops a strong wind that helps clear away remaining debris.

Perspective

The observational evidence tells us that stars must form out of interstellar material, and that they spend most of their lives on the main sequence. H-R diagrams of clusters show that, upon completion of the main-sequence lifetime, a star moves up and to the right, be-coming a red giant. By combining these observational deductions with the results of stellar model calculations, we can deduce the detailed life stories of individual stars.

Summary

1. Steps in stellar evolution are deduced from observations of stars in various stages, from newly formed ones to those in the late stages of their lives.

2. Clusters are groups of stars that formed together out of the same material.

3. The relative brightnesses of stars in a cluster re-

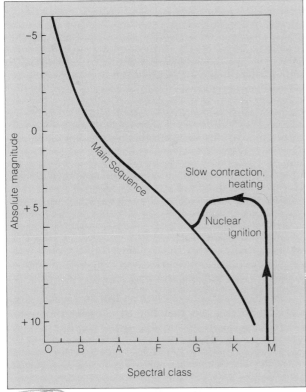

FIG. 21.16. THE PATH OF A NEWLY FORMING STAR ON THE H-R DIAGRAM. As a protostar heats up, it eventually becomes hot and luminous enough to appear in the extreme lower right-hand corner of the H-R diagram. When the core becomes opaque, the rapid collapse slows to a gradual shrinking, and the protostar moves across to the left as it heats. Eventually nuclear reactions begin, and the star is on the main sequence.

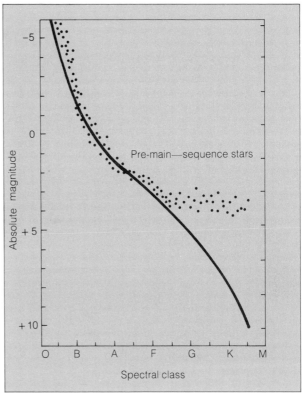

FIG. 21.17. PRE-MAIN-SEQUENCE STARS. H-R diagrams of very young clusters often show stars that have not yet reached the main sequence, but are still moving toward it. On the other hand, some of the massive, short-lived stars at the top of the main sequence may have already evolved away from it, having used up their hydrogen fuel.

flect real differences in luminosity, because all the stars in a cluster are essentially at the same distance from us.

4. The distances to clusters can be determined by plotting a color-magnitude diagram and comparing the location of the main sequence with that on a standard H-R diagram.

5. The fact that all stars in a cluster have the same age and initial composition means that the observed differences from star to star within the cluster are entirely a result of differences in stellar mass. This allows us to determine how stars of different mass evolve.

6. As stars evolve, they move off of the main sequence, becoming cooler and more luminous, and move into the red-giant region of the H-R diagram.

7. The age of a cluster can be inferred from the location of its main-sequence turn-off point.

8. In associations of young, hot stars there often is interstellar material left over from the process of star formation.

9. Stars in the process of forming are often embedded within dense clouds, and are best observed in infrared wavelengths.

10. A star forms from the gravitational collapse of an interstellar cloud, which condenses most rapidly at the center. The new star begins to heat up only after the gas in the core of the cloud becomes opaque, and the resulting protostar continues to slowly shrink until its interior becomes hot enough for nuclear reactions to begin.

Review Questions

1. How do we know that a cluster containing hot, luminous stars is a young cluster?

2. Is it possible for a cluster to contain both hot O stars and cool G stars at the same time?

3. Suppose that comparison of the main sequence in a cluster color-magnitude diagram with that in the standard H-R diagram shows that the apparent magnitudes of the stars on the cluster main sequence are five magnitudes greater than the absolute magnitudes of stars in the corresponding positions on the standard H-R diagram main sequence. How far away is the cluster?

4. Suppose two stars in the same cluster have identical apparent magnitudes. How do their masses and other properties compare? Could you reach the same conclusion for a pair of stars that are not in the same cluster, but which also have identical apparent magnitudes?

5. Cluster *A* has no stars of spectral types O, B, and A on its main sequence, whereas cluster *B* has stars of all types but O on its main sequence. Which cluster is older? Explain how you decided this.

6. Give two reasons why an association of O and B stars can exist for only a few million years at most.

7. Why are infrared observations needed in order for astronomers to observe stars in the process of forming?

8. Why does a protostar not heat up significantly until it becomes opaque?

9. Why can a young star never reach hydrostatic equilibrium until nuclear reactions start in its core?

10. It has been suggested, because many stars exist that are not in clusters, that isolated interstellar clouds can be caused to collapse by shock waves created in supernova explosions or stellar winds. Can you think of an alternative explanation for the fact that many stars are isolated individuals, not members of clusters? Hint: very few of the isolated stars in the galaxy are hot, young objects.

Additional Readings

Bok, B. J. 1972. The birth of stars. *Scientific American* 227(2):48.

Cohen, M. 1975. Star formation and early evolution. *Mercury* 4(5):10.

Herbst, W., and Assousa, G. E. 1979. Supernovas and star formation. *Scientific American* 241(2):138.

Werner, M. W., Becklin, E. E., and Neugebauer, G. 1977. Infrared studies of star formation. *Science* 197:723.

Zeilik, M. 1978. The birth of massive stars. *Scientific American* 238(4):110.

CHAPTER 22

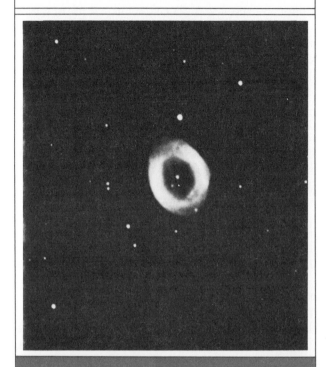

Life Stories of Stars

Combining our knowledge of stellar structure with observations showing how stars evolve, we can develop a picture of the story of an individual star, tracing it from its birth to its death. Because so much depends on the mass of a star, we will discuss three separate examples: a star like the sun, a star of a few times the sun's mass, and a very massive star. In the process we will see many important differences among the three cases.

We have already discussed the formation of stars, finding that the process is rather similar for all masses, except for the time required. We will not reiterate this discussion here; we will begin our story for each star on the main sequence.

Stars Like the Sun

A star with the mass of the sun, when it arrives on the main sequence, is similar to the sun in all its properties. Thus it is a G2 star, with a surface temperature of around 6,000 K and a luminosity of about 10^{33} erg/sec. Its composition initially is nearly 80 percent hydrogen by mass, and more than 20 percent helium, with only about 1 percent of the mass composed of other elements. It has convection in the outer layers, and it has a chromosphere and a corona.

In the core, the proton-proton chain converts hydrogen into helium. As we saw in chapter 19, this process lasts some 10 billion years for a star of 1 solar mass. During this time, the core gradually shrinks as the hydrogen nuclei are replaced by a smaller number of the heavier helium nuclei, and the internal temperature rises. This causes an increase in luminosity, and the star gradually moves up in the H-R diagram. The main sequence, therefore, is not a perfectly narrow strip, but has some breadth since stars on it slowly move upward as their luminosities increase (Fig. 22.1). The starting point, the lower edge of the main sequence, is called the **zero-age main sequence (ZAMS),** because this is where newly formed stars are found.

When the hydrogen in the core is gone, reactions there cease. By this time, however, the temperature in the zone just outside has nearly reached the point where reactions can take place there, and with a little more shrinking and heating of the core, the proton-proton chain begins again, in a spherical shell surrounding the core (Fig. 22.2). At this point the star has an inert helium core, and it is producing all its energy in the hydrogen-burning shell, which steadily moves outward inside the star, eating its way through the available hydrogen.

The core continues to shrink and heat. This heating enhances the nuclear reactions in the shell, causing it to produce more and more energy. As the source of energy is heated and moves toward the surface, the outer layers of the star are forced to expand, and as this occurs, the surface cools. The luminosity of the star increases because of its increased surface area, so it moves upward on the H-R diagram (Fig. 22.3). It also moves to the right, because the surface temperature is decreasing. The star rapidly becomes a red giant, reaching a size of 10 to 100 times its main-sequence radius (Fig. 22.4). The outer layers are constantly overturning as convection occurs to a great depth (Fig. 22.5).

The helium core becomes extremely dense. At first the temperature is not high enough for any new nuclear reactions to start there (the triple-alpha reaction, in which helium nuclei combine to form carbon, requires a temperature of about 100 million degrees). As the core continues to shrink, the matter there takes on a very strange form. A new kind of pressure gradually takes over from the ordinary gas pressure that has been

FIG. 22.1. THE WIDTH OF THE MAIN SEQUENCE. As stars on the main sequence gradually convert hydrogen into helium in their interiors, the cores shrink a little and get hotter. This in turn increases the luminosity, and the stars gradually move upward on the H-R diagram. As a result, the main sequence, consisting of stars of a variety of ages, is not a narrow strip, but has some breadth.

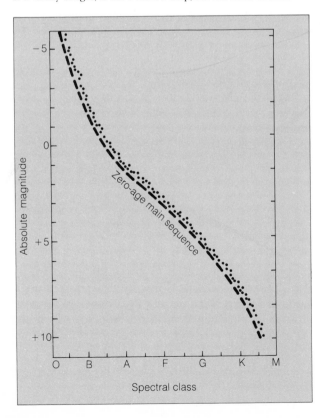

FIG. 22.2. THE HYDROGEN SHELL SOURCE. After the hydrogen in the core of a star is completely used up, the core, now composed of helium, continues to shrink and heat. This causes a layer of gas outside the core to reach the temperature required for nuclear reactions. A shell source, in which hydrogen is converted into helium, is ignited. The star's outer layers expand and cool, and the star becomes a red giant.

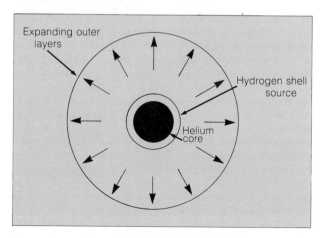

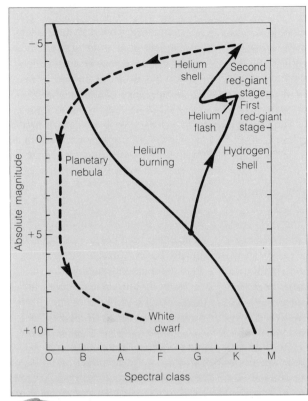

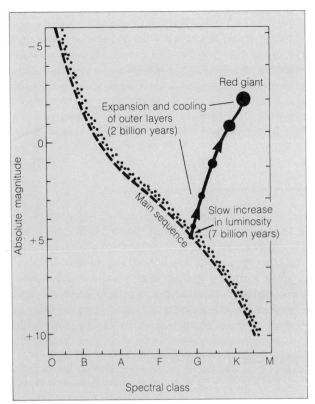

FIG. 22.3. THE EVOLUTIONARY TRACK OF A STAR LIKE THE SUN. This H-R diagram illustrates the path a star of 1 solar mass is thought to follow as it completes its evolution. The path indicated by dashes, following the second red-giant stage, is less certain than the earlier stages.

FIG. 22.4. THE DEVELOPMENT OF A RED GIANT. As a star's outer layers expand, they cool. At the same time, the surface area increases, raising the star's luminosity. The star therefore moves up and to the right on the H-R diagram.

FIG. 22.5. THE INTERNAL STRUCTURE OF A RED GIANT. The stellar core, which actually contains most of the mass of the star, is very small, compared with the huge extent of the outer layers. Most of the volume of the star consists of relatively rarefied gas, constantly overturning because of convection.

supporting the core. Electrons have a property that prevents them from being squeezed too close together, and this creates the new pressure. The gas in the stellar core contains many free electrons, and eventually their resistance to being compressed becomes the dominant pressure that supports the core against further collapse. When this happens, the gas is said to be **degenerate.**

A degenerate gas has many unusual properties. One of them is that the pressure no longer depends on the temperature, and vice versa. If the gas is heated further, it will not expand to compensate, as an ordinary gas would. As we will soon learn, there are important consequences if nuclear reactions start in the degenerate region of a star.

We now have a red giant star, with highly expanded outer layers. Near the center is a spherical shell in which hydrogen is burning in nuclear reactions, and inside of this is a degenerate core containing helium nuclei.

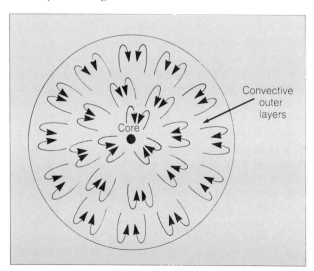

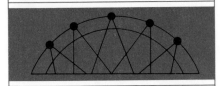

ASTRONOMICAL INSIGHT (22.1)

Electron Degeneracy

In the text a degenerate gas was described in very crude terms. There it was stated simply that electrons cannot be forced any closer together once a certain state of compression has been reached. It is interesting to examine a little more closely the reasons for this.

We learned, when discussing atomic structure, that an electron in orbit around the nucleus of an atom can only be in certain very specific energy levels. In the core of a star, however, the electrons are all free of atoms, since the extremely high temperature has fully ionized the gas. Ordinarily, this means that the electrons are free to fly around randomly, without regard for fixed energy levels.

This is only an illusion, in a way; under most conditions it only seems that there are no restrictions on the allowed energy state of a free electron. When a certain combination of high density and only moderately high temperature is reached, however, the electrons experience certain limitations on their freedom. These limitations result from a combination of two laws of **quantum mechanics,** the branch of physics that describes the motions and forces acting upon subatomic particles.

One of these axioms states that even in a gaseous state, electrons (or any other particles, for that matter) are confined to be in specific energy levels, regardless of the fact that they are not bound to a nucleus. An energy level in this case means a certain combination of speed, direction, and spin of the electron. The second axiom holds that no two electrons can occupy the same energy state. The law of physics that makes this statement is called the **Pauli Exclusion Principle,** after the scientist, Wolfgang Pauli, who discovered it.

As a star compresses itself after the nuclear fuel is all gone, the number of available energy states for the electrons to be in is reduced. A point is reached where all the states have electrons in them. The star cannot collapse any further because to do so would require squeezing more than one electron into the same energy state. This inability of the electrons to be forced any more tightly together creates a physical pressure that is sufficient to hold the star up against gravity, as long as the star's mass does not exceed a certain limit.

Neutron stars, whose cores are made entirely of neutrons, are also subject to the Pauli Exclusion Principle. The neutrons in these stars can be very close together, as close as particles in an atomic nucleus. In a way, a neutron star is a single, gigantic atomic nucleus, with a density as high as that of nuclear material.

The core is small, but it may contain as much as a third of the star's total mass.

THE HELIUM FLASH

During the red giant phase, the helium core continues to be heated by the reactions going on around it. Eventually the temperature becomes sufficiently high for the triple-alpha reaction to begin. When it does, there are spectacular consequences.

Ordinarily when a gas is heated, it expands, and this limits how hot it can get, because an expanding gas tends to cool. In a degenerate gas, however, no such expansion occurs, because pressure does not depend on temperature in the usual way. Hence when the reactions begin in the degenerate core of a red giant, the temperature goes up quickly, and the core retains the same density and pressure. The increased temperature speeds up the reactions, producing more heat, in turn accelerating the reactions even further. There is a rapid snowball effect, and in an instant (literally seconds), a large fraction of the core is consumed in the reaction. This spontaneous runaway reaction is called the **he-**

lium flash. Although it creates dramatic consequences for the star's interior, calculations show that there is little or no immediate effect visible to an observer. The overall direction of the star's evolution is changed, however, which would be apparent after thousands of years.

The helium flash quickly disrupts the core, destroying its degeneracy. The core then returns to a more normal state where temperature and pressure are linked, and the triple-alpha reaction continues, now in a more stable, steady fashion. The outer layers of the star begin to retract as the star reverts to a more uniform internal structure, and as they do so, they become hotter. The star moves back to the left on the H-R diagram (Fig. 22.3). Old clusters, particularly globular clusters,

FIG. 22.6. THE H-R DIAGRAM FOR A GLOBULAR CLUSTER. These star clusters are very old, and all the upper main-sequence stars have evolved to later stages. Here we see a number of stars in or approaching the red-giant stage, and a number on the horizontal branch, where they move after the helium flash occurs. Only very old clusters have a horizontal branch. (Besides their great ages, globular clusters also have a relatively low heavy-element content, which affects how the stars evolve. See chapter 26).

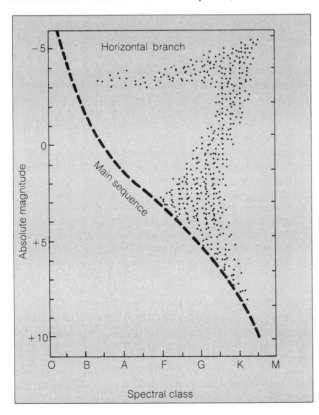

are found to have a number of stars in this stage of evolution, forming a sequence called the **horizontal branch** (Fig. 22.6). This is seen only in clusters old enough for stars as small as one solar mass to have evolved this far; as we learned in previous chapters, this requires an age of 10 billion years or more.

In due course, the helium in the core becomes exhausted, leaving an inner core of carbon (Fig. 22.7). Helium still burns in a shell around the core; there could even be an active hydrogen-burning shell farther out in the star at the same time. The star expands and cools, becoming a red giant for the second time. This second red-giant stage is short-lived, however, lasting perhaps one million years. A degenerate core again develops, but there is not another dramatic flare-up in its interior. In fact, no more nuclear reactions ever occur in the core of this low-mass star, because the temperature never reaches the extremely high levels needed to cause heavy elements like carbon to react.

PLANETARY NEBULA TO WHITE DWARF

The star in its second red-giant stage may be viewed as having two distinct zones: the relatively dense interior, consisting of a carbon core inside a helium-burning shell which in turn is surrounded by a hydrogen-

FIG. 22.7. A STAR LATE IN ITS LIFETIME. Here the core has completed both the hydrogen-burning and helium-burning stages, and is composed of carbon. Outside the core, there may still be an active hydrogen shell source as well as a helium-burning shell, and the star is becoming a red giant for the second time.

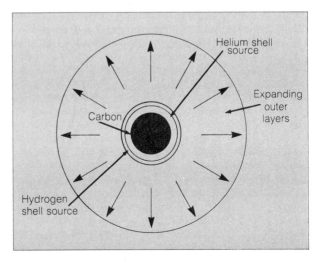

ASTRONOMICAL INSIGHT (22.2)

Planetary Nebulae and Forbidden Emission Lines

When the first spectra of the light from planetary nebulae were taken, a number of emission lines were found that had never been seen in laboratory spectra. These lines defied identification for a long time, and were attributed to a mysterious element called *nebulium*. One of the strongest of the mystery lines lay at a wavelength of 5,007 Å, in the green portion of the spectrum, lending planetary nebulae their characteristic color. (Usually there are red-colored portions as well, resulting from emission of the well-known strong line of hydrogen at 6,563 Å.)

Several decades ago, the science of quantum mechanics became sufficiently sophisticated to allow theoretical calculations to be made, showing the precise energy-level structure of an atom. When these calculations were done for twice-ionized oxygen (that is, oxy-

gen atoms that have lost two of their usual eight electrons), a pair of energy levels were found whose separation corresponds precisely to the wavelength of the unidentified line at 5,007 Å. The only problem was that certain rules of quantum mechanics showed that an electron was very unlikely to ever undergo this transition, which was said to be "forbidden." Thus, under ordinary conditions, an oxygen ion could never emit a photon with a wavelength of 5,007 Å, and this was the reason the line was never seen in the laboratory.

How, then, could oxygen ions in space, in planetary nebulae, emit such photons? The answer lies in the particular combination of physical conditions that prevails in a planetary nebula. The density is very low, so low that an ion can float around for years without colliding with another. At the same time the radiation field is intense, because of the nearby presence of the nebula's central star, which is very hot and emits mostly in ultraviolet wavelengths. Thus an ion in the nebula is constantly bombarded with energetic photons, and is likely to have its electrons excited into high energy states.

Now the stage is set. Imagine an oxygen ion in this situation. It is flying around, minding its own business, when an ultraviolet photon hits it and is absorbed, causing one of the

electrons orbiting the nucleus to jump to an excited state. Normally the electron would jump right back down to the ground state, but if it finds itself in the upper level of the transition that creates the 5,007 Å emission, it is stuck. It cannot readily make this jump downward, according to quantum mechanics. Under most conditions, the ion would stay in this state so long that eventually a collision with another atom or ion would occur, and the electron would be jarred out of that level. If the density is low enough, however, such a collision is unlikely, and the electron continues to sit there. If this is allowed to go on long enough, in due course (perhaps in ten years), the electron may spontaneously jump to the lower level, emitting a 5,007 Å photon as it does so. Given sufficient time, an electron can make a low-probability transition, despite the rules against it.

The emission lines created in this way are called **forbidden emission lines,** and there are a number of them in the spectra of planetary nebulae, in addition to the 5,007 Å line of twice-ionized oxygen. Their identification was both a triumph for quantum mechanics and a major discovery for astronomers, for it immediately revealed a great deal about the conditions in planetary nebulae, until that time very mysterious objects.

burning shell; and the very extended, diffuse outer layers. The inner portion may contain up to 70 percent of the mass, but occupies only a small fraction of the star's volume.

As the nuclear fuel in the interior runs out, the core

shrinks. The details of what happens to the outer layers are not clear, but apparently the star ejects portions of this material in a series of minor outbursts that could be likened to blowing smoke rings. The star gently gets rid of its outer layers.

The evidence that this occurs is primarily observational, since a successful theory of how it happens has not yet been developed. A number of objects have been observed that seem to consist of a hot, compact star surrounded by a shell of expanding gas. These shells often have a bluish-green color because of the emission lines by which they glow, and through a telescope or on a photograph they bear some resemblance to the planets Uranus and Neptune. Because of this resemblance, they are called **planetary nebulae.** Some, such as the Dumbbell Nebula (Color Plate 16) or the Ring Nebula (Fig. 22.8), are well-known objects visible with a small telescope.

The star that remains in the center of a planetary nebula lies far to the left in the H-R diagram (Fig. 22.3), having a surface temperature that may be as high as 100,000 K or more, far hotter than any main-sequence star. Evidently when the shell is being ejected, the star's surface temperature increases as hot inner gas becomes exposed on the surface. In essence such a star is a naked core, left behind when the outer layers are cast off.

Some nuclear reactions may still be going on in a shell inside the star, but they do not last much longer, since the fuel is depleted. The density, already very high, increases as the stellar remnant condenses under the force of gravity. A larger and larger fraction of the star's interior becomes degenerate. As the star shrinks, it stays hot but its luminosity decreases because of its diminishing surface area, and it moves downward on the H-R diagram (Fig. 22.3). Eventually the shrinking stops, and the star becomes stable again, but now it is a very small object, so dim that it lies well below the main sequence, in the lower left-hand corner of the H-R diagram. It is called a **white dwarf.**

A white dwarf is a bizarre object in many ways. It is made of degenerate matter, whose peculiar properties govern its internal structure. It has a mass as great as that of the sun (some white dwarfs are even slightly more massive), yet it is approximately the size of the earth. This means that its density is incredibly high, roughly a million times that of water. A cubic centimeter of white-dwarf material would weigh a ton at the surface of the earth!

A white dwarf does not do much. It has no more nuclear fuel, so it does not evolve further. It just sits there, radiating away the heat energy still contained inside, and slowly cooling off. A white dwarf can be stirred into action under certain circumstances (as described

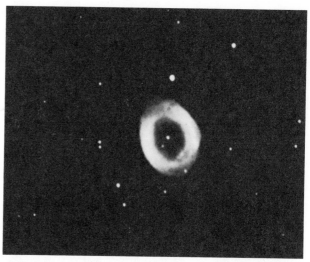

FIG. 22.8.   A PLANETARY NEBULA. This is the Ring Nebula, whose striking symmetry makes it a favorite object for viewing through telescopes. The star that ejected the nebula is visible at its center.

in the next chapter), but in most cases, this is the end of the line for a star like the sun.

The Middleweights

Let us now consider a star of significantly greater mass than the sun, say 5 to 10 solar masses. For this star, much of the evolution is similar to that of the 1-solar-mass star, although there are some major differences. One of the contrasts, of course, is the time scale, because this star will have perhaps 100 times the solar luminosity, and will use up its fuel accordingly faster. Let us quickly escort this star through its development, starting again on the main sequence. The path in the H-R diagram is shown in Figure 22.9.

Our middleweight star lands on the middle to upper portion of the zero-age main sequence when it has completed its formation, and nuclear reactions begin. Its spectral type is B8 if its mass is 5 solar masses, and B2 if it is 10 solar masses. The nuclear reaction in the core is the CNO cycle, producing helium from hydrogen. There is no convection in the outer layers, but there is in the core. The surface temperature is roughly between 12,000 K and 20,000 K, depending on the mass.

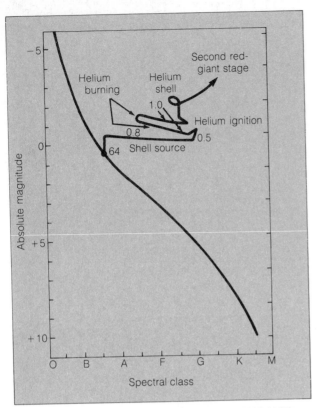

FIG. 22.9. THE EVOLUTIONARY PATH OF A 5-SOLAR MASS STAR. This star, with its multiple red-giant loops, goes through a more complicated series of stages than a less massive star. There are probably additional red-giant stages following those indicated here, but calculations have not yet been carried that far. The numbers indicate times in units of 10^6 years.

MULTIPLE RED GIANT LOOPS

As in the case of the sunlike star, the more massive star gradually moves upward on the main sequence as its core becomes more compressed. The hydrogen in its core is gone (after about a hundred million years) before a shell outside the core is ignited, and for a brief period the star simply contracts and heats, with no reactions taking place. During this time, it turns sharply to the left in the H-R diagram. After only one or two million years, however, hydrogen begins burning again, in a shell outside the core, and the star reverses itself and heads for the red giant region. Within a few hundred thousand years, it reaches the upper right-hand portion of the H-R diagram. In this case, the core is not degenerate because the temperature is high enough to prevent a collapse to the necessary density, so no helium flash occurs. When the core is hot enough, the triple-alpha reaction quietly begins, and the star contracts and moves to the left in the H-R diagram. This star does not move far to the left, however, and never really leaves the red-giant region before it develops a new reaction shell outside the core, and expands and cools once again, reversing itself and heading back to the right.

What happens next is not clear, but a sketchy picture is emerging. The core temperature continues to increase, and it is possible for new nuclear reactions to begin. These reactions, called **alpha-capture reactions,** start with carbon and helium nuclei (remember, a helium nucleus, consisting of two protons and two neutrons, is called an alpha particle). A carbon nucleus can capture an alpha particle, forming an oxygen nucleus. This in turn can capture another alpha particle, forming neon, and so on. If the core temperature reaches high enough levels, a complex sequence of reactions can occur, gradually building up a supply of heavy elements in the star's core. Each time one fuel source is exhausted and shell burning begins, the star expands and moves upward and to the right in the H-R diagram; each time a new reaction starts in the core, the star settles down and moves to the left. Thus a middleweight star can execute a number of left-to-right loops in the red-giant region of the H-R diagram. The number of loops depends on the mass. The situation is very complicated, however, and theoretical calculations do not give definitive answers.

One thing that complicates matters is that extremely luminous red giants lose mass through strong, dense winds (as described in chapter 20). The rate of mass loss can be sufficient to significantly alter the mass of the star in a few hundred thousand years. If a star's mass is reduced, its evolution is modified. The central density and temperature are lower, which in turn reduces the number of reaction stages the star goes through. At some difficult-to-determine point, all reactions cease. Then the red giant contracts and moves down and to the left in the H-R diagram, possibly passing through a phase as a planetary nebula.

DEATH: TO GO OUT QUIETLY OR WITH A BANG?

Whether these middleweight stars end up as white dwarfs is uncertain. It has been shown theoretically, and confirmed from observations, that a white dwarf cannot exist if the star's mass is greater than a certain

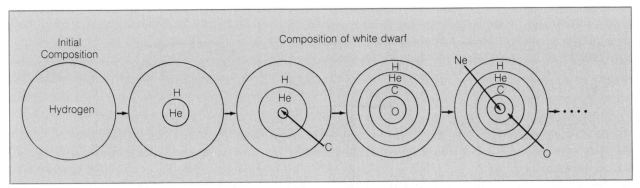

FIG. 22.10. WHITE-DWARF COMPOSITION. The more nuclear-burning stages a star goes through, the greater the variety of elements produced in its interior. The white dwarf that results, therefore, has a composition that depends on the number of nuclear-burning stages, and this, in turn, depends on the mass of the star.

limit, about 1.4 times the mass of the sun. If a mass is any larger it creates sufficient gravitational force to overcome the pressure of the degenerate electron gas in the core, and the star collapses catastrophically. As we will see, this violent ending certainly occurs in the most massive stars; for the 5- to 10-solar-mass star, however, a white dwarf may form if enough mass is lost to bring the star below the 1.4-solar-mass limit. Stellar wind and planetary-nebula ejection might reduce the mass far enough, so it is possible that many stars that start out in the middleweight division end up as white dwarfs.

The white dwarfs formed from the middleweight stars are different from those formed by lower-mass stars, however. Recall that for a star of 1 solar mass, the most advanced nuclear reaction that occurs is the triple-alpha process, converting helium into carbon. The 5- to 10-solar-mass star goes well beyond this point, however, so heavier elements are built up in its core. The white dwarf that results, then, may have an interior made of oxygen, neon, or even heavier elements (Fig. 22.10), whereas one produced by a lower-mass star consists of carbon or perhaps only helium.

What of the relatively massive stars that do not shed enough material to become white dwarfs? These stars contract beyond the white-dwarf stage and therefore exceed the stupendous densities achieved by the white dwarfs. There seem to be two possible end results: the star may contract to another kind of degenerate state, where it can stabilize; or it may collapse indefinitely, never again reaching stability. In either case, the collapse may be violent, with the in-falling matter creating an outburst called a **supernova** explosion.

In the first of these two possibilities, the collapse of

the interior is stopped when a new kind of degenerate pressure is created. As the collapse occurs, the protons and electrons are forced so close together that a certain kind of nuclear reaction takes place, in which neutrons are formed (Fig. 22.11). Neutrons, like electrons, can reach a state where they cannot be squeezed any further; this creates a gas pressure that may be sufficient to balance the force of gravity, creating a kind of super white dwarf called a **neutron star.** Like a white

FIG. 22.11. INVERSE BETA DECAY. In this process, which only occurs at a significant rate under conditions of very high pressure and temperature, a proton and an electron are forced together, forming a neutron and a neutrino. Energy is absorbed in the process, rather than being released.

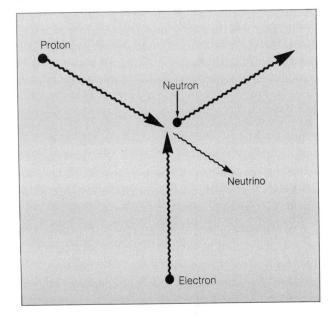

dwarf, a neutron star has a strict limit on its mass, around 2 or 3 solar masses. If the stellar core has more mass than this, the second possibility just mentioned occurs: the collapse never stops. (We will return to this case in the next section.)

The neutron star has properties even more fantastic than a white dwarf. Its diameter is about 10 kilometers, yet it contains more than 1.4 solar masses, and its density far exceeds that of the white dwarf. A cubic centimeter of neutron-star material would weigh a billion tons at the earth's surface!

Observations have confirmed that neutron stars can form during supernova explosions, because there are cases where a neutron star was found in the center of the expanding cloud of debris left over from such an event. One of these apparently formed in historical times, when the supernova that created the present-day Crab Nebula occurred (this was the "guest star" observed by Chinese astronomers in A.D. 1054).

Neutron stars are exceedingly dim objects, very difficult to see directly, and were for a long time thought to be unobservable. The ways in which they have been observed will be described in the next chapter. For now, we have finished our life story of the middle-weight star, and we are ready to see how the most massive stars evolve.

The Heavyweights

The most massive stars form in basically the same manner as their less massive relatives, but they do so more quickly. All phases of their evolution are speeded up. As we learned in the previous chapter, the massive stars in a cluster can go through the entire evolution from birth to death before the less massive ones even reach the main sequence. The evolutionary track of a massive star is shown in the H-R diagram in Figure 22.12.

Stars containing 20 or more solar masses start their main-sequence lifetimes as O stars. They have surface temperatures of 30,000 K or more, and luminosities of 10^5 to 10^6 times that of the sun. The dominant nuclear reaction during the main-sequence phase is the CNO cycle. These stars have some excess surface heating, which causes coronae to form, but it is not thought likely that convection occurs in the surface layers. There is convection in the cores of these stars, however.

As in the other cases, these stars undergo gradual core contraction as the hydrogen there is converted

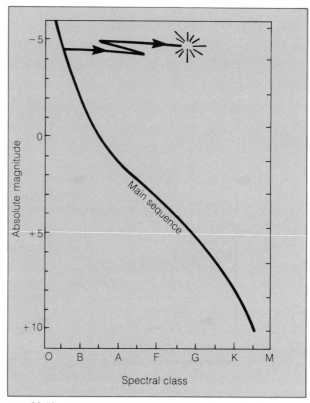

FIG. 22.12. EVOLUTION OF AN O STAR. The evolutionary track of a very massive star is quite simple: it moves to the right (not up, because mass loss caused by a stellar wind counteracts the effects of expansion), briefly moves to the left during helium burning, then goes back to the right. It may go back and forth (rapidly) a few times, but within a total lifetime of around one million years, it exhausts all nuclear fuel and explodes in a supernova.

into helium. By contrast with the less massive stars, however, the O stars do not move upward in the H-R diagram; instead, they move straight across to the right. The reason is that O stars steadily lose matter through high-velocity winds (described in chapter 20), so that the tendency to increase their core temperature and hence the luminosity is offset by the decreased mass of the overlying layers.

The details of the later stages of evolution for these massive stars are not very clear. Most likely, several stages of nuclear burning take place as the star goes through numerous loops in the red-giant region of the H-R diagram, leading ultimately to the formation of iron in the core. At the same time, the wind may rip away so much of the star's substance that the outer portion of the core is exposed, creating a kind of "peculiar" object called a **Wolf-Rayet star** (Fig. 22.13). These are

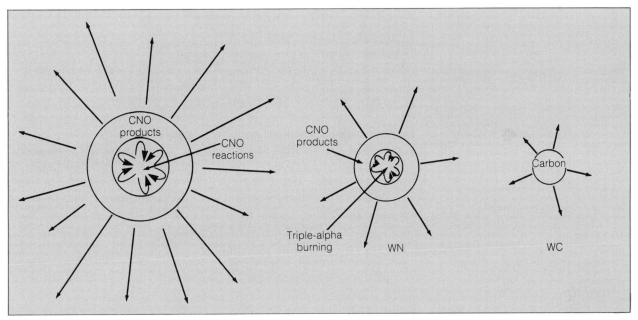

FIG. 22.13. FORMATION OF A WOLF-RAYET STAR. A massive star loses its outer layers through a stellar wind; meanwhile, its core is undergoing CNO-cycle reactions, gradually being transformed into helium. When the outer layers have been blown away, the exposed inner region of the star, containing enhanced quantities of nitrogen and oxygen as by-products of the CNO cycle, is a type of Wolf-Rayet star known as a WN star. If the star has progressed to helium burning by the triple-alpha process, then it is a WC star, with an abundance of carbon.

hot stars with very strong winds that have excess quantities of carbon (as result of the triple-alpha reaction) or nitrogen and oxygen (the products of the CNO cycle) at their surfaces. When the outer layers have been stripped off, we get a direct view of the innards of the star, where material that has recently undergone nuclear reactions is seen. The convection that is taking place in the core helps to bring this processed matter up to the surface.

THE ULTIMATE ENERGY CRISIS: THE SUPERNOVA

When iron has been formed in the core of a star, no further nuclear reactions can support the star against the inward force of gravity. Indeed, any additional reactions that could occur would help gravity, because reactions involving elements heavier than iron are all **endothermic,** meaning that more energy is required to create the reaction than is produced. Hence if these reactions start, they actually cool the star's interior, thereby reducing the pressure and allowing gravity to dominate further.

Apparently the following sequence of events can occur (Fig. 22.14). When the core is composed of iron, the reactions stop, and the star begins to collapse, having no means of support. As the collapse begins, the compression of the core causes endothermic reactions to start, and suddenly there is an effective vacuum at the center, as the heat there goes into the reactions. The star now collapses very quickly, in free-fall, and it implodes. It is as though the rug were suddenly pulled out from under the outer layers, and they come crashing down into the center of the star. Nuclear reactions occur at a tremendous rate during this implosion, and they produce a very large quantity of the tiny subatomic particles called neutrinos (discussed in chapter 16). These particles carry away great amounts of energy, and create a strong outwards pressure force. This causes a massive shock, and the material bounces back violently, in a supernova explosion. A good deal of matter from the star is dispersed into space as a result. (A second supernova mechanism, creating a so-called Type I supernova, is briefly mentioned in chapter 23. The kind of explosion caused by the collapse of a massive star, as described here, is called a Type II supernova.)

During the supernova explosion, densities and tem-

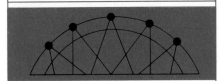

The Wolf-Rayet Stars

The so-called Wolf-Rayet stars were identified as a class by their discoverers in the late 1800's. They are recognizable by their numerous emission lines, particularly those of carbon and nitrogen. Other lines that were seen (for example, those of ionized helium), showed that these are hot stars, and they were assigned a sequence of numbers corresponding to the subclasses of O stars. Thus, a W7 star was thought to be roughly comparable to an O7 star in temperature, a W5 corresponded to an O5, and so on.

Additional classification was based on whether carbon lines or nitrogen lines are dominant among the emission lines, because it is never both. Those with strong carbon lines are called WC stars, whereas the others, with strong nitrogen lines, are WN stars (the numerical suffixes are still used to denote the temperature class). Apparently WC stars are those in which the triple-alpha process has produced carbon in the core, and the WNs are stars in which the CNO cycle has taken place, but not the subsequent triple-alpha reaction.

For a long time, though, it was not understood how the material that had been enhanced by nuclear reactions reached the surfaces of the Wolf-Rayet stars, where it could produce the observed emission lines. Finally, in the last decade, the universal existence of stellar winds in O stars was recognized, and this offered an explanation. If an O star

has a sufficiently strong wind that substantial portions of its outer layers are blown away, the "surface" of the star will consist of matter that was originally buried deep inside. Apparently in the Wolf-Rayet stars, material that has been involved in nuclear reactions in the stellar core is revealed. It is unlikely we are actually seeing the core itself; what has probably happened is that convection in the center of the star has carried processed material some distance outward, away from the core, and we are seeing the upper portion of the region where convection has transported the enriched matter.

The Wolf-Rayet stars themselves have very strong winds, losing mass at an even greater rate than the O stars. It is fascinating to speculate on how these stars will end up, because it is not clear how much mass they can lose before they run out of nuclear fuel and die.

FIG. 22.14. A SUPERNOVA EXPLOSION. When the last possible reaction stage is completed, the star's outer layers collapse inward, because the source of internal pressure is gone. After the collapse, the star "bounces back," and explodes as a supernova.

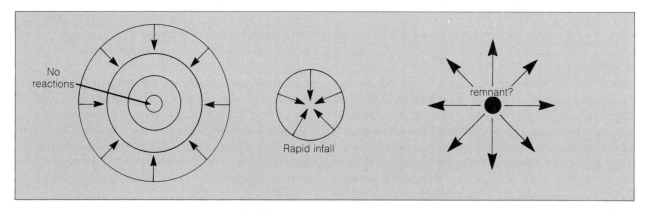

peratures reach incredible levels for a short time, sufficient to cause the formation of very heavy elements. Most of the atoms heavier than iron in the universe were created in supernova explosions. Of course, some or all of the intermediate-weight elements that had been created in the star's interior during its more placid stages of evolution are also dispersed into space by the explosion, so the process enriches the interstellar gas with a wide spectrum of elements.

What remains of the original star? Apparently in many cases nothing at all is left, except a cloud of cosmic debris. In other cases, a fraction of the star's final mass remains in the form of a highly condensed remnant. This may be a neutron star, if the remnant mass is below the limit of 2 or 3 solar masses.

If too much matter is contained in this remnant, however, a neutron star cannot form, because the force of gravity is too strong. The star simply continues to collapse. Nothing can ever stop it. The resulting object, first envisioned mathematically nearly seventy years ago, is called a **black hole,** and today there is observational evidence that such things exist.

Although a black hole literally cannot be seen, there are indirect means by which its presence may be detected. These means, and the information about black holes that can be obtained from observations, are described in the next chapter. For now, we return to the question of the evolution of massive stars.

It seems almost certain that all stars starting out with masses in excess of 10 solar masses or so end their lives in supernova explosions. The race between mass loss and the nuclear reactions in the core determines how many nuclear reaction stages the star undergoes, as well as how much mass it will have left when all the possible reactions are finished. The outcome of the race is difficult to call theoretically. Hence it is difficult to predict exactly which stars will end up as neutron stars and which will form black holes. Some may simply explode, leaving no remnant behind.

Evolution in Binary Systems

It was pointed out in chapter 20 that some special effects may occur in certain binary systems. If the two stars are sufficiently far apart, say, several astronomical units or more, then most likely each evolves on its own, as though the companion were not there. If, however, they are very close together, then significant interaction takes place which can drastically alter the normal course of events.

If either star is massive enough to have a strong wind, then matter can be transferred to the other star. Even if no strong wind is present initially, as soon as either star reaches the red-giant stage, it can swell up sufficiently to dump material onto the other through the point between them where their gravitational forces are equal (see chapter 20).

Regardless of how it happens, once mass begins to transfer from one star to the other, the evolution of one or both is modified (Fig. 22.15). The one that gains mass speeds up its evolution, whereas the one that loses it may slow down. The most spectacular observational effects occur when one star has gone all the way through its evolution and has become a white dwarf, neutron star, or black hole, and then it gains new material from the companion star. Special effects, often including X-ray emission, are then observed, and it is under these circumstances that stellar remnants are most easily detected. This will be discussed at some length in the next chapter.

Perspective

We have now seen how stars are born, live their lives, and die, and we know how the sequence of events depends on mass. The least massive stars, those with much less than 1 solar mass, have not been discussed much simply because they evolve so slowly that none has yet had time, since the universe began, to complete its main-sequence lifetime. The evolution of these stars has had little effect on the rest of the galaxy, whereas the much more rare heavyweight stars are profoundly important, because it is only through their rapid evolution that the galaxy has been enriched with heavy elements.

Before we finish discussing stars and turn our attention to larger-scale objects in the universe, we must consider the properties of the stellar remnants left behind as stars die.

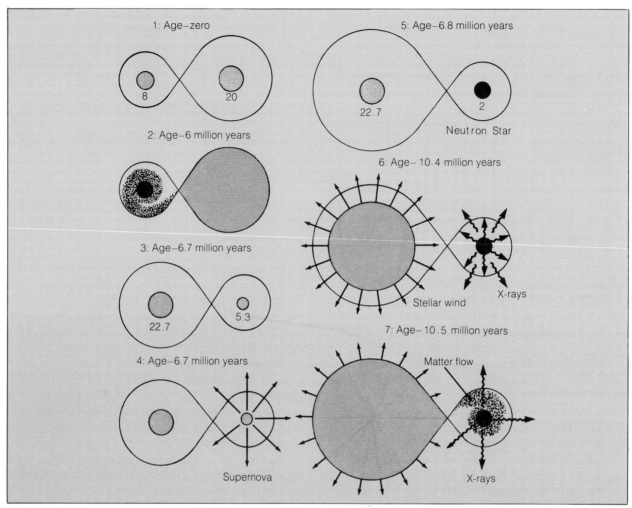

FIG. 22.15.   MASS EXCHANGE IN A CLOSE BINARY SYSTEM.   The star that is initially more massive evolves faster, and as a red giant it deposits material on the companion. This speeds up the evolution of the second star, so that it expands and returns matter to the first star. If the first one is by this time a compact object such as a neutron star, the consequences of this stage of mass-transfer can be violent (see chapter 23).

Summary

1. A star with the mass of the sun spends about 10^{10} years on the main sequence, producing its energy by nuclear reactions that convert hydrogen to helium in its core.

2. When the core hydrogen is gone, the reactions stop there, but continue in a spherical shell outside the core. The outer layers expand and cool, and the star becomes a red giant.

3. The core continues to shrink and heat until it becomes degenerate. When it eventually gets hot enough, a helium flash occurs, after which the star is powered by the triple-alpha reaction, converting helium into carbon.

4. Following the helium-burning phase, the star may undergo instabilities that cause the ejection of one or more planetary nebulae.

5. The star eventually becomes degenerate throughout, and ends its evolution as a white dwarf.

6. A more massive star, in the 5- to 10-solar-mass range, undergoes many of the same evolutionary steps as the 1-solar-mass star, but does so much more quickly.

7. The moderate-mass star may go through several red-giant phases, as successive nuclear reaction stages produce ever-heavier elements in the stellar core.

8. This star will become a white dwarf only if it loses enough mass, through a stellar wind or other processes, to bring its mass below the 1.4-solar-mass limit for white dwarfs.

9. If the star is too massive at the end of its evolution to become a white dwarf, it may explode in a supernova, leaving a neutron star or possibly a black hole.

10. The most massive stars also undergo the same evolutionary steps as the less massive stars, but do so even more quickly.

11. There will be several red-giant stages, producing elements as heavy as iron in the stellar core.

12. When all possible reaction stages are completed, the star will implode and then explode in a supernova, dispersing some of its material into space.

13. The mass of the remnant that is left determines whether it will be a neutron star or a black hole.

14. In close binary systems, mass can be transferred from one star to the other, speeding the evolution of the one that gains mass.

Review Questions

1. Explain why the main sequence is not a very thin strip on the H-R diagram, but instead has some width. Where does the sun lie within the broad strip of the main sequence?

2. Why is there not a "hydrogen flash" in the core of a star when it forms and the proton-proton chain (or the CNO cycle) begins in its core for the first time?

3. If a star loses one-tenth of a solar mass each time it ejects a planetary nebula, how many times would a star of initial mass 2.4 solar masses have to eject planetary nebulae in order to become a white dwarf?

4. Using Wien's law (see chapter 5), calculate the wavelength of maximum emission for a planetary nebula central star whose temperature is 100,000 K. What kind of telescope would be needed to best observe such a star?

5. Knowing that a 10-solar-mass star has 10 times more total nuclear energy available to it than the sun does, but that the more massive star has about 1,000 times

the sun's luminosity, estimate how much shorter the lifetime of the 10-solar-mass star is.

6. Discuss the role of stellar winds in the evolution of massive stars.

7. If a neutron star has only 1/500 the radius of a white dwarf of the same mass, compare the average densities of the two.

8. Summarize the differences between the evolution of a 1-solar-mass star and one that starts with 10 solar masses.

9. It has been discovered that iron is a relatively abundant element in the galaxy, more abundant than the elements that are a little heavier or a little lighter. In view of what you have learned in this chapter about nuclear reactions in stars, explain why iron is so abundant in the galaxy.

10. Compare the evolution of a 30-solar-mass star with that of a 5-solar-mass star.

Additional Readings

Flannery, B. P. 1977. Stellar evolution in double stars. *American Scientist* 65:737.

Kaler, J. 1981. Planetary nebulae and stellar evolution. *Mercury* 10(4):114.

Page, T., and Page, L. W., eds. 1968. *The evolution of stars.* New York: Macmillan.

Shklovskii, I. S. 1978. *Stars: their birth, life, and death.* San Francisco: W. H. Freeman.

Sweigart, A. V. 1976. The evolution of red giant stars. *Physics Today* 29(1):25.

Wyckoff, S. 1979. Red giants: the inside scoop. *Mercury* 8(1):7.

•CHAPTER 23•

Stellar Remnants

Our story of stellar evolution is almost complete. We have seen how the properties of stars are measured, and we have learned how stars form, live, and die. The only missing link is what is left of a star after its life cycle is completed.

As we saw in the previous chapter, the form of remnant that remains at the end of a star's life depends on the mass the star had when it finally ran out of nuclear fuel. Three possibilities were mentioned: white dwarf, neutron star, and black hole. In this chapter each of these three bizarre objects is discussed, with emphasis on the observational properties.

White Dwarfs, Black Dwarfs

An isolated white dwarf is not a very exciting object, from the observational point of view (Fig. 23.1). It is rather dim, and its spectrum is nearly featureless, except for a few very broad absorption lines created by the limited range of chemical elements it contains (Fig. 23.2). The great width of the lines is caused by the immense pressure in the star's outer atmosphere; pressure tends to smear out the atomic energy levels, because of collisions between atoms. In effect, the energy levels become broadened, and this results in broadened spectral lines when transitions of electrons occur between the levels.

Additional broadening or shifting of spectral lines may occur because of magnetic effects. If the star had a magnetic field before its contraction to the white-dwarf state, that field is not only preserved but also intensified in the process of contraction. Thus a white dwarf may have a very strong magnetic field, and as we learned previously (chapters 16 and 19), a magnetic field causes certain spectral lines to be split into two or more parts. In a white dwarf, where the lines are already very broad, this splitting (called the Zeeman effect) usually makes the lines appear even broader.

The lines in the spectrum of a white dwarf are shifted, always toward longer wavelengths. This is not caused by the star's motion; if it were, it would be telling us that somehow all the white dwarfs in the sky are receding from us. A different type of redshift occurs here, one caused by gravitational forces. One of the predictions of Einstein's theory of general relativity is that photons of light are affected by gravitational fields. White dwarfs, with their immensely strong surface gravities, provide a confirmation of this prediction. In later sections, we will discuss even more extreme examples of these **gravitational redshifts.**

Left to its own devices, a white dwarf will simply cool off, eventually becoming so cold that it no longer emits visible light. At this point it cannot be seen, and simply persists as a burnt-out cinder that may be referred to as a black dwarf. The cooling process takes a long time, however; several billion years may go by before the star becomes cold and dark. This may seem surprising, since the white dwarf has no source of energy and only stays hot as long as it can retain the heat that it contained when it was formed. The outer skin of the white dwarf acts like a very efficient thermal blanket, however, since it consists of gas that is nearly opaque (Fig. 23.3). Radiation cannot penetrate it easily, and because there is no other way of transporting heat away from the interior into space, it takes a long time for the heat to filter out.

The thermal blanket created by the outer atmosphere is very thin; most of the volume of the star is filled with degenerate electron gas, as discussed in the last chapter. The atmosphere, consisting of ordinary gas, may be only about 50 kilometers thick (remember, the white dwarf itself is about the size of the earth, with a radius of some 5,000 to 6,000 kilometers). The com-

FIG. 23.1.   SIRIUS B. The dim companion to Sirius was the first white dwarf to be discovered; the analysis of its mass, as well as its temperature and luminosity, led to the realization that it is very small and dense.

FIG. 23.2.   A WHITE DWARF SPECTRUM.
Here we see that white dwarf spectra are without many strong features, and that the spectral lines tend to be rather broad. This white dwarf has a large abundance of carbon, as a result of the triple-alpha reaction.

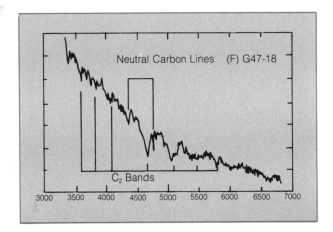

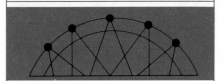

The Story of Sirius B.

Sirius, also known as α Canis Majoris, is the brightest star visible in the heavens. It was discovered in 1844 to be a binary, because of its wobbling motion as it moves through space (recall the discussion of astrometric binaries in chapter 18). Its period is about fifty years, and the orbital size about 20 astronomical units; if we analyze these data using Kepler's third law we come to the conclusion that the companion must have a mass similar to that of the sun. At the distance of Sirius (only 2.7 parsecs), a star like the sun should be easily visible, yet the companion to Sirius defied detection for some time.

Part of the difficulty lay in the overwhelming brightness of Sirius itself, whose light would tend to drown out that of a lesser, very close star. Astronomers were sufficiently intrigued by the mysterious nature of the companion, however, that they made intensive efforts to find it, and finally, in 1863, it was seen with an exceptionally fine, newly developed telescope.

It was immediately confirmed that this star was unusually dim for its mass, having an apparent magnitude of 8.7 (the apparent magnitude of the primary star is −1.4) and an absolute magnitude of 11.5, nearly 7 magnitudes fainter than the sun. Spectra obtained with great difficulty (again because of the brightness of the companion) showed that this object, called Sirius B, has a high temperature, similar to that of an O star.

When Russell constructed his first H-R diagram in 1913, he included Sirius B, finding it to lie all by itself in the lower left-hand corner. The only way for the star to have such a low luminosity while its temperature was so high was for it to have a very small radius, and it was deduced at that time that Sirius B was about the size of the earth. The incredible density was estimated immediately, and it was realized that a totally new form of matter was involved. The term **white dwarf** was coined to describe Sirius B and other stars of this type, which were discovered soon afterward in other binary systems.

The physics of this new kind of matter took some time to be understood, partly because the theory of relativity had to be applied (the electrons in a degenerate gas move at speeds comparable to that of light). The Indian-American astrophysicist S. Chandrasekhar was the first to solve the problem of the behavior of this kind of gas, and by the 1930's he was able to construct realistic models of the structure of a white dwarf. From this work came the 1.4-solar-mass limit we referred to in the text.

Modern observational techniques offer the possibility of detecting white dwarfs more easily. For example, even though Sirius B is about 10 magnitudes (or a factor of 10,000) fainter than Sirius A in visible wavelengths, its greater temperature makes it relatively brighter in the ultraviolet. There is even a wavelength, about 1,100 Å, below which the white dwarf is actually brighter than its companion. Hence Sirius B and other white dwarfs havebeen detected with ultraviolet tele-scopes. The development of more sophisticated space instruments will make it possible to study the properties of many white dwarfs in this way.

position of the interior may be primarily helium, if the star never progressed beyond the hydrogen-burning stage; it may be carbon, if the triple-alpha reaction was as far as the star got in its nuclear evolution before dying; or it may be some heavier element, if the original star was sufficiently massive to have undergone several reaction stages. In any case, the mass of the white dwarf must be less than the limit of 1.4 solar masses.

WHITE DWARFS AND NOVA EXPLOSIONS

Ordinarily, a white dwarf cools off and is never heard from again. There are circumstances, however, in which it can be resurrected briefly for another role in the cosmic drama. **Nova** outbursts have been observed for centuries, and it has recently become apparent that white dwarfs are central characters in producing these spectacular events.

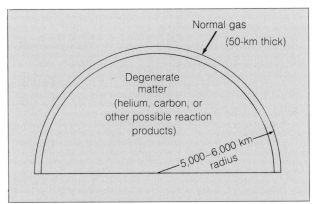

FIG. 23.3. THE INTERNAL STRUCTURE OF A WHITE DWARF. The star's interior is degenerate, and would cool very rapidly, except that the outer layer of normal gas acts as a very effective insulator, trapping radiation so that heat escapes only very slowly.

The word nova means *new,* and is applied here because a star that becomes a nova increases its brightness so much that it might be visible only for the brief time that it takes to flare up (Fig. 23.4). Typically a nova brightens up to its peak in less than a day, then takes months or even years to return to its former dimness.

At maximum brightness, the absolute magnitude is usually between -6 and -9, comparable to or even brighter than the brightest supergiants. Apparently material is ejected by a star during the nova outburst, because blueshifted absorption lines created by matter moving away from the star have appeared on spectra, and in some cases expanding nebulae surrounding the star that flared up have been visible on photographs (Fig. 23.5).

Some stars become novae more than once; several examples are known where two or more nova outbursts have occurred in the same star, separated by several decades. These **recurrent novae** are reminiscent of another phenomenon, the **dwarf novae.** These stars undergo minor flare-ups, and are called dwarf novae because they do not approach the brightness of a genuine nova; in many ways they appear to be scaled-down versions. The dwarf novae always occur in binary systems where one star is a white dwarf. This appears to be the case for the true novae as well.

To understand how the flare-up occurs, we recall the helium flash that takes place in the degenerate core of certain red giants. When nuclear reactions start in a degenerate gas, they quickly create a violent runaway

FIG. 23.4. NOVA CYGNI 1975. This is one of the brightest novae seen in recent years. The photo at left shows the region without the nova; at right is the same region near the time of peak brightness of the object, when the apparent magnitude was near $+1$.

FIG. 23.5. AN EXPANDING SHELL AROUND AN OLD
NOVA. This is Nova Persei (1901), surrounded by an expanding
cloud of gas that was ejected during the outburst.

explosion, because the degenerate gas cannot respond
as ordinary gas would. It cannot expand and cool,
thereby holding the reactions down to a moderate,
steady rate. Because a white dwarf consists of degenerate gas, it will experience the same kind of runaway

FIG. 23.6. THE NOVA PROCESS. Matter is transferred onto a
white dwarf by a red-giant companion. Theoretical calculations
show that the new material will orbit the white dwarf in a disk, but
will eventually become unstable and fall down onto the white
dwarf. When it does so, nuclear reactions flare up, creating a nova
outburst.

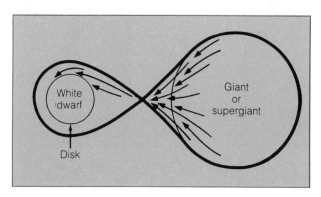

explosion as a helium flash, if somehow nuclear reactions are started.

In an isolated white dwarf, there is no way to start
new nuclear reactions; all the possible fuel is used up
earlier in the star's evolution. In a binary system, however, the companion may shed material when it expands into the red-giant stage, or as a result of a stellar
wind, and some of this material can fall onto the white
dwarf (Fig. 23.6). This matter forms the fuel for new
nuclear reactions, and the impact of its descent onto
the surface of the white dwarf can provide the necessary heat. When the temperature gets high enough, reactions begin in the newly accreted matter, and the nature of the underlying degenerate gas ensures that the
reactions will quickly flare up, consuming all the new
fuel in a very short time. This is the nova outburst, and
its magnitude depends on how much material is burned
in the brief nuclear conflagration. Apparently in the
dwarf novae the outburst occurs after only a small
amount of material has accumulated, whereas in a standard nova a great deal more must build up before the
explosion occurs; when it does, however, the outburst
is correspondingly more energetic. The explosions re-
cur as more material is subsequently supplied to the
white dwarf by its binary companion.

White dwarfs, then, are most prominent in binary
systems where mass transfer from a normal companion takes place. We will see later in this chapter that
other forms of stellar remnants are also most easily
observed under these circumstances.

Supernova Remnants

We learned in the last chapter that stars too massive at
the end of their lifetimes to become white dwarfs will,
in most cases, explode in supernovae. This violent de-
mise may leave a stellar remnant (a neutron star or
black hole), and it will certainly leave an expanding
gaseous cloud called a **supernova remnant** (see Color
Plates 19 and 20).

A number of supernova remnants are known to astronomers. A few, like the prominent Crab Nebula (Fig.
23.7 and Color Plate 19), are detectable in visible light,
but many are most easily observed at radio wavelengths. This is partly because visible light is affected
by interstellar dust, which can hide our view of a remnant, whereas radio waves are largely unaffected; and

ASTRONOMICAL INSIGHT (23.2)

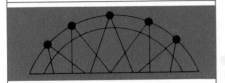

The Celestial Crustacean

In the eighteenth century, a French amateur astronomer named Charles Messier, who was fond of searching the skies for comets, collected and published a list of fuzzy-looking objects that never moved and were therefore not comets. His interest was to provide a reference list for other comet hunters, so that they would not be confused by these objects, but his list is best known today as a convenient reference for those who like to view some of the most conspicuous nonstellar objects in the skies, ranging from giant gas clouds to distant galaxies.

The very first object on Messier's list, and therefore designated MI, is a brightly glowing gas cloud in the constellation Taurus. Photographs show it to have a filamentary structure and a peculiar overall shape reminiscent of certain crustaceans, and it became known as the Crab Nebula. Spectroscopic analysis showed that the filaments glow largely because of emission lines of hydrogen; and that the continuous radiation from throughout the nebula is synchrotron radiation, which implies that within the nebula there is both a strong magnetic field and a source of energetic, fast-moving electrons.

Checks of ancient Chinese astronomical records showed that the supernova observed in A.D. 1054 coincided in position with the Crab Nebula, and it was realized that the Crab must be the remains of the star that exploded some nine hundred years ago. Careful comparison of photographs taken of the Crab at intervals of several years shows that it is still expanding. The rate of expansion can be measured, and from this it is seen that the expansion began from a single point about nine hundred years ago, further corroborating the conclusion that it began with the supernova of 1054.

The expansion also provides a direct estimate of the distance to the Crab. The apparent velocity of expansion is determined from the angular growth of the nebula, and the true expansion rate, in units of kilometers per second, is found from the Doppler shift of the gas that is moving directly toward us (this velocity is about 600 kilometers per second). Since the apparent angular motion of the expanding gas depends on both the distance and its true velocity, the distance could be determined. (This is analogous to estimating how far away a jet plane is by looking at its apparent speed across the sky, knowing that in reality it is moving at several hundred miles per hour.) The distance to the Crab Nebula is approximately 1,000 parsecs.

Even though the origin and essential nature of the nebula were known, mysteries remained. What was the energy source that produced the fast-moving electrons, and was there, buried somewhere inside, a remnant of the original star that had exploded? The answers to these questions were not found until the late 1960's, with the discovery of the pulsars.

---

partly because the remnants are intrinsically strong radio emitters. The reason is not that they are cold, which would cause their peak thermal emission to occur at long wavelengths, as in the case of the outer planets. Here a completely different process, called **synchrotron emission,** produces the radio waves.

Synchrotron emission occurs when electrons move rapidly through a magnetic field (Fig. 23.8). The electrons must have speeds near that of light; as they travel through the magnetic field, they are forced to move in a spiraling path, and they emit photons as they do so.

The emission occurs over a very broad range of wavelengths, including some visible, ultraviolet, and even X-ray radiation, but these other wavelengths are often not as easily detected as radio, because the radio emission is the strongest.

Supernova remnants that are detected in visible wavelengths usually glow in the light of several strong emission lines, most notably the bright line of atomic hydrogen at a wavelength of 6,563 Å, in the red portion of the spectrum. These lines often show large Doppler shifts, indicating that the gas is still moving rapidly as

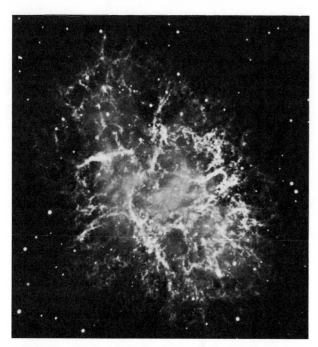

FIG. 23.7. THE CRAB NEBULA. This is a well-studied supernova remnant, the result of the explosion observed by Chinese astronomers in A.D. 1054. The filamentary structure stands out most clearly in photos such as this, taken through a red filter which reveals directly the regions of ionized hydrogen.

FIG. 23.8. THE SYNCHROTRON PROCESS. Rapidly moving electrons spiral along magnetic-field lines, and emit photons as they do so. The radiation that results is polarized, and its spectrum is continuous but lacks the peaked shape of a thermal spectrum. The electrons must be moving very rapidly (at speeds near that of light), so whenever synchrotron radiation is detected, it is known that a source of large quantities of energy must be present.

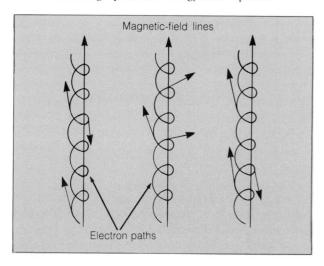

Magnetic-field lines

Electron paths

the entire remnant expands outward from the site of the explosion that gave it birth. Often a filamentary structure is seen, suggestive of turbulence, but probably modified in shape by magnetic fields.

Remnants are visible at the locations of several famous supernova explosions. The most prominent is the Crab Nebula, which was created in the supernova observed by Chinese astronomers in A.D. 1054. Other supernovae seen by Tycho Brahe (in 1572) and by Kepler and Galileo (in 1604) also left detectable remnants, although neither is as bright in visible wavelenghts as the Crab. Apparently a supernova remnant can persist for 10,000 years or more before becoming too dissipated to be recognizable.

The energy of a supernova explosion is immense, comparable to the total amount of radiant energy the sun will emit over its entire lifetime. Much of this energy takes the form of mass motions as the remnant expands. This kinetic energy heats the expanding gas to very high temperatures. The entire process has a profound effect on the interstellar gas and dust that permeates the galaxy, as we will learn in chapter 25.

Neutron Stars

We have referred to a neutron star as a stellar remnant composed entirely of neutrons that are in a degenerate state, similar to that of the electrons in a white dwarf. The pressure created by the degenerate neutron gas is greater than that of a degenerate electron gas, so a star slightly too massive to become a white dwarf can be supported by the neutrons. Again, however, the mass of one of these objects is limited. This limit, which depends on other factors such as the rate of rotation, is between 2 and 3 solar masses. Thus there is a class of stars, whose masses at the end of their nuclear lifetimes are between 1.4 and about 2 or 3 solar masses, that become neutron stars.

The structure of a neutron star is even more extreme than that of a white dwarf (Fig. 23.9). All the mass is compressed into an even smaller volume (with a radius of about 10 kilometers), and the gravitational field at the surface is immensely strong. The layer of normal gas that constitutes its atmosphere is only centimeters thick, and beneath it may be zones of different chemical composition resulting from previous shell-

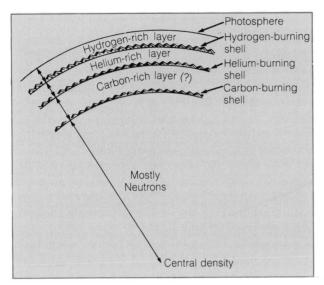

FIG. 23.9. STRUCTURE OF A NEUTRON STAR. This diagram shows that the outer regions of a neutron star may consist of thin layers of various elements that were produced by nuclear reactions during the star's lifetime. Because of the intense gravitational field of the neutron star, these outer layers are thought to have a rigid crystalline structure.

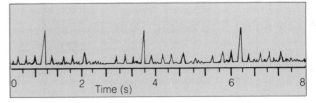

FIG. 23.10. THE RADIO EMISSION FROM A PULSAR. Here, plotted against time, is the radio intensity from a pulsar. The radio emission is weak or nonexistent except for a very brief flash once each cycle (in some cases there is a weaker flash between each adjacent pair of strong flashes).

burning episodes, each zone only a few meters thick at the most. Inside these surface layers is the incredibly dense degenerate neutron-gas core, which takes the form of a crystalline lattice. The temperature throughout is very high, but because the surface area is so small, a neutron star is very dim indeed. It may have a magnetic field that is also much more intense than that of a white dwarf, resulting from the compression of the star's original field.

PULSARS: COSMIC CLOCKS

The properties of neutrons stars were predicted theoretically several decades ago, but until 1967 it was not thought that they could be observed, because of their low luminosities. In that year, however, an accidental discovery by a radio astronomer led to the establishment of a new class of objects called **pulsars,** which are now believed to be neutron stars. As it turned out, observation of pulsars is only one of two ways to detect neutron stars (the other is described in the next section).

Pulsars are radio sources that flash on and off very

regularly, and with a high frequency (Fig. 23.10). The most rapid of them, the one located in the Crab Nebula, repeats itself thirty times a second, whereas the slowest known pulsars have cycle times of more than a minute. In every case, the pulsar is only "on" for a small fraction of each cycle.

When pulsars were discovered, there was a great deal of excitement and a lot of speculation, including the suggestion that they were beacons operated by an alien civilization. Once the initial shock of discovery wore off, however, a number of more natural explanations were offered by astronomers who sought to establish the identity of the pulsars. It was well known that a variety of stars pulsate regularly, alternately expanding and contracting, but none were known to do it so rapidly. Theoretical studies showed that such quick variations as those exhibited by the pulsars should occur only in very dense objects, denser even than white dwarfs. This led astronomers to think of neutron stars.

Enough was known from theoretical calculations, however, to rule out rapid expansion and contraction of neutron stars as the cause of the observed pulsations. The calculations showed that the vibration period of such an object would be even shorter than the observed periods of the pulsars. A second possibility, that the rapid periods were produced by rotation of the objects, was considered. This hypothesis further ruled out ordinary stars and white dwarfs. In order to rotate several times per second, a normal star or white dwarf would have to have a surface velocity in excess of the speed of light, a physical impossibility. Such an object would be torn apart by rotational forces before approaching such a rotational velocity. A neutron star, on the other hand, could rotate several times per second and remain intact.

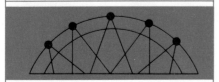

Discovery of the Pulsars, and the Crab Revisited

The radio observatory of Cambridge University, in England, has been used for the last three decades to map the skies in radio wavelengths. This kind of survey creates a vast amount of data, which are routinely processed and which are often not examined in great detail, at least not immediately as they come in.

In 1967, a graduate student named Jocelyn Bell was perusing data from the radio survey when she found an astonishing thing: a radio source that seemed to be blinking on and off very regularly, at a rapid rate. Careful checking showed that this source was always at the same position in the sky, and hence was not some earthly object or a moving vehicle. The regularity of the pulses was so perfect that there arose a great deal of speculation as to whether this was an interstellar beacon left by some intelligent race.

Several additional pulsars, as they were soon dubbed, were found, and in scientific circles, the search was on for a natural explanation. The concept of the neutron star, first devised several decades earlier, was revived, because it was soon realized that such rapid pulsations could not be produced by anything less dense, including a white dwarf. When a pulsar was then discovered in the midst of the maelstrom of gas known as the Crab Nebula, where it was already known that a star had exploded recently, the suspicion that neutron stars were responsible for the pulsars approached certainty. The question was settled when the slowdown rate of the pulsar in the Crab was found to be precisely what was needed to provide the energy of the synchrotron emission, until then a mystery. The Crab pulsar rotates at a rate of some thirty times per second, making it the fastest of all known pulsars. Apparently this is because it is also the youngest; older ones have had more time to slow down their spin rates as they lose energy to their surroundings.

The Crab pulsar was identified in visible light by a group of astronomers who built a shutter device and placed it in their telescope in front of a photometer (an instrument for measuring star brightnesses). The shutter was set to open and shut rapidly, at the same rate as the radio pulses from the Crab pulsar. When the astronomers pointed the telescope at a particular dim star near the center of the Crab, and adjusted the shutter, the star suddenly disappeared. This proved that the star was turning itself off and on at the same rate the shutter was opening and closing, and so showed that the star was definitely the pulsar. It pulses not just in radio, but also in visible and, it was found much later, in X-ray wavelengths as well.

When the Crab pulsar was discovered, a great deal of additional information became available. This pulsar was found to be gradually slowing its pulsations, something that could best be explained if the pulses were linked to the rotation of the object. The rotation could slow as the pulsar gave up some of its energy to its surroundings.

This discovery cleared up another mystery. The source of energy that powers the synchrotron emission from supernova remnants had been unknown; now it was suggested that the slowdown of the rotating neutron star could provide the necessary energy, probably by magnetic forces exerted on the surrounding ionized gas. This process is analogous to magnetic braking, which apparently slowed the sun's rotation during the early history of the solar system (see chapter 17).

The remaining question was how the pulses were created by the rotation. Evidently a pulsar acts like a lighthouse, with a beam of radiation sweeping through space as it spins. The question was why a rotating neutron star should emit a beam from just one point on its surface. The most probable explanation has to do with the strong magnetic fields that neutron stars are likely to have. If a neutron star has a strong field, then electrons falling onto it from the surrounding gas are forced to follow the lines of the field, hitting the sur-

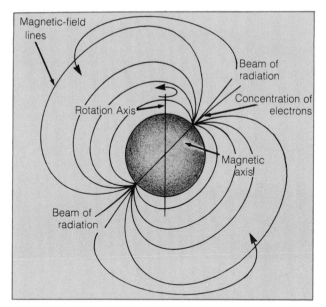

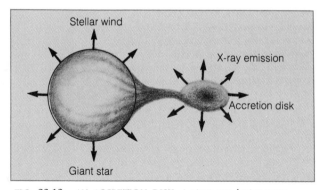

FIG. 23.12. AN ACCRETION DISK. A giant star, losing mass through a stellar wind, has a neutron-star companion. Material that is trapped by the neutron star's gravitational field swirls around it in a disk, which is so hot because of compression that it glows at X-ray wavelengths. A nearly identical situation can arise when the compact companion is a black hole.

FIG. 23.11. THE PULSAR MECHANISM. Here we see a rapidly rotating neutron star, with its magnetic axis out of alignment with the rotation axis. Synchrotron radiation is emitted in narrow beams from above the magnetic poles, where charged particles, constrained to move along the magnetic-field lines, are concentrated. These beams sweep the sky as the star rotates, and if the earth happens to lie in a direction covered by one of the beams, we observe the star as a pulsar.

face only at the magnetic poles. Similarly, electrons ejected from the surface by electrical forces will escape only at the magnetic poles, creating synchrotron emission as they do so. This results in narrow beams of radiation emitted from the star in opposite directions from the two poles (Fig. 23.11). If the magnetic axis of the star is not aligned with the rotation axis, these beams will sweep across the sky in a conical pattern as the star rotates. If the earth happens to lie in the direction intersected by one of these beams, then we see a flash of radiation every time the beam sweeps by us.

Although the details are still somewhat vague, this seems to be the best explanation of the pulsars. Since special conditions (nonalignment of the magnetic and rotation axes, and the need for the beam to sweep across the direction toward the earth) are required for a neutron star to be seen from earth as a pulsar, it follows that there should be many neutron stars that do not manifest themselves as pulsars. In the next section we will see how some of these non-pulsar neutron stars are detected.

NEUTRON STARS IN BINARY SYSTEMS

Earlier we made a general statement that stellar remnants are often most easily observed when they are in binary systems. We have already seen that a white dwarf in a binary can flare up violently if it receives new matter from its companion star. A neutron star reacts similarly in the same circumstances.

If a neutron star is in a binary system where the companion object is either a hot star with a rapid wind or a cool giant losing matter because of its great size, some of the ejected mass will reach the surface of the neutron star. Very little of it falls directly down onto the surface; instead, much of it swirls around the neutron star, forming a disk of gas called an **accretion disk** (Fig. 23.12). The individual gas particles orbit the neutron star like microscopic planets, and only fall inward as they lose energy in collisions with other particles. The accretion disk acts as a reservoir, slowly feeding the gas inward, toward the neutron star.

The disk is very hot because of the gigantic gravitational forces exerted on it by the neutron star. The gas in the disk is highly compressed, and reaches temperatures of several million degrees. A gas at such a high temperature emits X rays. Hence a neutron star in a mass-exchange binary system is likely to be an X-ray source; a number of such systems have been found in the last decade, since the advent of X-ray telescopes launched on rockets and satellites (Fig. 23.13). Often it is known that an X-ray object is part of a binary system

FIG. 23.13. AN X-RAY MAP. This shows the locations of X-ray sources discovered by the *Uhuru* satellite, the first major X-ray survey instrument. Most of the sources of X rays shown here are binary systems containing evolved, compact companions such as neutron stars or black holes.

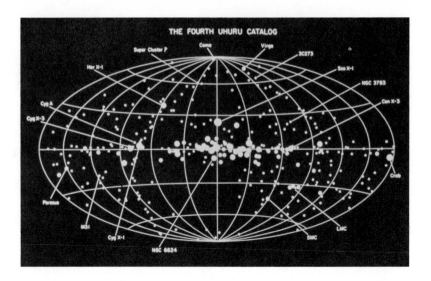

because the X-rays are periodically eclipsed by the companion star. Eclipses are made likely by the proximity of the two stars (mass exchange would not occur unless it were a close binary) and the fact that the mass-losing companion is likely to be a large star, either an upper main-sequence star or a supergiant.

The so-called **binary X-ray sources** are among the strongest X-ray emitting objects known. Many of them are, most likely, neutron stars, although the possibility also exists that some are black holes, which would similarly produce X-ray emission as material falls inward.

There is a slightly different way in which neutron stars can emit X rays, which also occurs in mass-exchange binaries. If the in-falling material trickles down onto the neutron star in a steady fashion, there is continuous radiation of X rays, as we have just learned. If, however, the material falling in does so sporadically, and arrives in substantial quantities every now and then, a major nuclear outburst occurs each time, in close analogy to the nova process involving a white dwarf. This apparently happens in some cases, producing random but frequent X-ray outbursts (Fig. 23.14). The intensity of the outburst, as in the case of a nova, depends on how much matter has fallen in and been consumed in the reactions. The neutron-star binary systems where this occurs are called **bursters,** and were discovered accidentally by scientists at Massachusetts Institute of Technology who were examining data acquired with an X-ray satellite. One important contrast with novae is that the flare-up of a burster occurs much more rapidly, lasting only a few seconds. Another is that most of the emission occurs only in very energetic

X rays, so these objects do not show up in visible light.

Neutron stars in binary systems may also be responsible for some supernova explosions. If a neutron star accumulates enough extra material through mass exchange, it may exceed the mass limit for neutron stars, and explode. This process is thought to be responsible for the so-called Type I supernova (the other kind, created by the collapse of very massive stars at the end of their nuclear-burning lifetime, is called Type II).

Black Holes: Gravity's Final Victory

In the last chapter we saw what happens to a massive star at the end of its life. It falls in on itself, and there is no barrier, neither electron degeneracy nor neutron degeneracy, that can stop it.

FIG. 23.14. THE X-RAY LIGHT CURVES OF A BURSTER. Here we see the very brief flashes of X rays that occur at erratic intervals in these sources. The data shown here are from the so-called rapid burster, which is unique among bursters because of the rapidity with which the bursts reoccur. These outbursts, in analogy with the nova process, are thought to be caused by material falling onto the surface of a neutron star and igniting brief episodes of nuclear reactions.

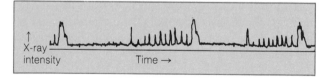

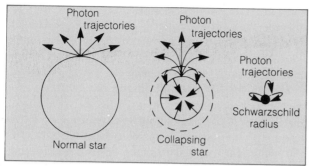

FIG. 23.15. PHOTON TRAJECTORIES FROM A COLLAPSING STAR. Light escapes in essentially straight lines in all directions from a normal star (left), whose gravitational field is not sufficient to cause large deflections. At an intermediate stage of collapse (center), photons emitted in a cone nearly perpendicular to the surface can escape, whereas others cannot. Those emitted at just the right angle go into orbit around the star, whereas those emitted at greater angles fall back onto the stellar surface. After collapse has proceeded to within the Schwarzschild radius, no photons can escape.

The gravitational field near the surface of a collapsing star grows in strength as the mass of the star becomes concentrated in an ever-smaller volume. This gravitational field has important effects in the near vicinity of the star, although at a distance it remains unchanged from what it was before the collapse. Close to the star, though, the structure of space itself is distorted, according to Einstein's theory of general relativity. Einstein discovered that accelerations caused by changing motion and those caused by gravitational fields are equivalent, and from this it follows that space must be curved in the presence of a gravitational field, so that moving particles follow the same path they would follow if they were being accelerated. Not only physical particles, but also photons of light are affected (Fig. 23.15).

Usually the effects of the curvature of space are not noticeable, except when we engage in very careful observation, or consider very large distances. Later in the text, when we discuss the universe as a whole and its overall structure, we will go into this in more detail. For now, we confine ourselves to local regions where space can be distorted by very strong gravitational fields.

Let us consider a photon emitted from the surface of a star as it falls in on itself (Fig. 23.15). If the photon is emitted at any angle away from the vertical, its path will be bent over further. If the gravitational field is strong enough, the photon's path may be bent over so far that it falls back onto the stellar surface. Photons emitted straight upward follow a straight path, but they lose energy to the gravitational field, which causes their wavelengths to be redshifted (we already discussed this phenomenon, in the case of white dwarfs). When the gravitational field becomes strong enough, the photon loses all its energy, and cannot escape. Another way of stating this is that the velocity required for the photon to escape from the star exceeds that of light. At the point when this happens the star becomes invisible to an outside observer, because no light from the star can reach the earth.

The radius of the star at the time when its gravitational field becomes strong enough to trap photons is called the **Schwarzschild radius,** after the German astrophysicist who first calculated its properties some sixty years ago. The Schwarzschild radius depends only on the mass of the star that has collapsed. For a star of 10 solar masses, it is 30 kilometers. A star of twice that mass would have twice the Schwarzschild radius, and so on.

When a collapsing star has shrunk inside of its Schwarzschild radius, it is said to have crossed its **event horizon,** because an outside observer cannot see it or

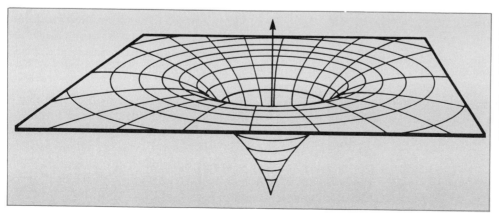

FIG. 23.16. GEOMETRY OF SPACE NEAR A BLACK HOLE. Einstein's theory of general relativity may be interpreted in terms of a curvature of space in the presence of a gravitational field. Here we see how this curvature varies near a black hole.

anything that happens to it after that point. We have no hope of ever seeing what happens inside the event horizon, but since no force is known that could stop the collapse, we assume that it continues. The mass becomes concentrated in an infinitesimally small region at the center, which is called a **singularity,** because mathematically it is a single point (Fig. 23.16).

What happens to the matter that falls into a singularity is a subject for speculation, for no concrete theories have been developed. Among the more fantastic ideas that have been suggested is the possibility that the matter, having disappeared from one place in the universe, reappears somewhere else, having tunneled through to a different point in space and time. The path it would take has been dubbed a **worm hole,** and the place where it reappears a **white hole.**

Let us now return to considering our collapsed star from the outsider's point of view. The time sense is distorted by the extreme gravitational field and acceleration in such a way that to an outside observer, the collapse seems to slow gradually, coming to a halt just as the star reaches its event horizon. This slowdown, however, is most significant in the last moments before the star's disappearance, and by then the star is essen-

FIG. 23.17. CYGNUS X–1. This X-ray image obtained by the orbiting *Einstein Observatory* shows the intense source of X rays at the location of a dim, hot star which is thought to be the normal star that is losing mass to its invisible black hole companion.

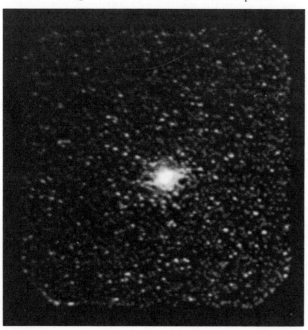

tially invisible anyway, as most escaping photons are diverted into curved paths or redshifted into the infrared or beyond. The star seems to disappear rather quickly, despite the stretching of the collapse in time resulting from relativistic effects.

Mathematically, a black hole can be described completely by three quantities: its mass, its electrical charge, and its spin. The mass, of course, is determined by the amount of matter that collapsed to form the hole, plus any additional material that may have fallen in later. The electrical charge, similarly, depends on the charge of the material from which the black hole formed; if it contained more protons than electrons, for example, it would end up with a net positive charge. Because particles with opposite charges attract each other, and those with like charges repel, it is thought that electrical forces would maintain a fairly even mixture of particles during and after the formation of the black hole, so that the overall charge would be nearly zero. To illustrate this, let us imagine that a black hole was formed with a net negative charge. If there were ionized gas around it afterward (as there likely would be, with some of the matter from the original star still drifting inward), the negative charge of the hole would repel additional electrons, preventing them from falling in, while protons would be accelerated inward. In time, enough protons would be gobbled up to neutralize the negative charge.

The spin of a black hole is not so easily dismissed, however. It stands to reason that if the star were spinning before it collapsed, and most likely it would have been, then the rotation would speed up greatly as the star shrank. A high spin rate actually shrinks the event horizon, allowing an outside observer to see closer in to the singularity residing at the center. It is even possible mathematically, although it presents a physical dilemma, for the black hole to have so much spin that there is no event horizon, which means that whatever is in the center is exposed to view. A **naked singularity,** as this has been dubbed, would not produce the usual gravitational effects of a black hole, and it would be possible to blunder into one without any forewarning.

For the most part, it is assumed that the spin rate is never so large that a naked singularity can form, and in fact black hole properties are usually specified by mass alone, and the effects of charge and spin are neglected. It is assumed that our main hope of detecting a black hole is by its gravitational effects, determined entirely by the mass.

ASTRONOMICAL INSIGHT (23.4)

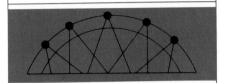

A Black Hole in Cygnus X–1?

One of the brightest X-ray sources in the sky was found in the constellation Cygnus by rocket experiments, even before the launch of the *Uhuru* satellite in 1972. At that time so few X-ray objects were known that they were given simple names based on the constellation name, followed by "X" and the rank in the constellation. Since then, more sophisticated X-ray telescopes have discovered thousands of X-ray sources, and this naming system has become impractical. The brightest sources are still known by their early names, however.

X-ray telescopes, particularly the first ones developed, could not pinpoint the location of a source very accurately. Thus it took awhile for astronomers to identify the visible star associated with Cygnus X–1, but finally they found that a giant B star, known from catalogues as HDE226868, coincided with the position of the X-ray emitting object. Very weak variations in the light from the visible star were detected, with the same period as the eclipses of the X rays. This proved that the X-ray source and the star were orbiting each other.

Very intensive efforts have been made to determine the mass of the unseen companion. Without knowing the exact orientation of the orbit nor the orbital velocity of the dark object (nor even an accurate velocity of the B star, which has broad spectal lines, difficult to measure precisely), astronomers have not been able to deduce exactly what the mass is. Fairly stringent limits have been set, however, indicating that the unseen object must contain at least 8 solar masses. This makes it a black hole, if the theory of stellar collapse is at all correct.

Several similar systems are known, in which it is thought that a black hole might exist, but Cygnus X–1 is still the most likely candidate. The only way out is to hypothesize a third star in the system, whose gravitational force would produce the observed orbital characteristics of the visible star without requiring that the X-ray source be so massive. This is a contrived situation for which there is no supporting evidence, however, and although it remains a possibility, it is considered extremely unlikely. Cygnus X–1 is very likely a black hole.

Before turning to a discussion of how to find a black hole, we should mention that there may be additional kinds of black holes (discussed in later chapters), formed by processes other than the collapse of individual massive stars. These include the so-called mini black holes, very small ones postulated to have formed under extreme density conditions early in the history of the universe; and supermassive black holes, thought possibly to inhabit the cores of large galaxies, having formed from thousands or millions of stellar masses coalescing in the center.

DO BLACK HOLES EXIST?

The mass that goes into a black hole during its formation still exists there, hiding inside its event horizon. Even though no light can escape, gravitational effects persist. The gravitational force of the star is exactly the same as it was before the collapse, except at distances from the hole that are less than the radius of the original star. Our best chance of detecting a black hole, then, is to look for an invisible object whose mass is too great to be anything else. Even if we do find such a thing, it is really only a circumstantial argument, because the conclusion that it is a black hole relies on the theory that neither a white dwarf nor a neutron star can survive if the mass is sufficiently great.

The best opportunity for determining the mass of an object occurs when the object is in orbit around a companion star; we can apply Kepler's third law to find the masses. The search for black holes, then, leads us to examine binary systems, where we look for invisible, but massive, objects. This search has been facilitated by another property of black holes; if new material is pulled into black holes, the trapped matter gets so highly compressed and heated on the way in that it emits X rays. We have already seen that this happens in the case of neutron stars accreting new material, so the mere emission of X rays does not prove that a black hole is present. We still must determine the mass of the object, to see whether it is too great to be a neutron star.

The skies were first scanned at X-ray wavelengths in the early 1970s by the *Uhuru* satellite, and later by a number of other X-ray telescopes aboard spacecraft. Many sources of X-ray emission have been found, and the most intense are the X-ray binaries. These are usually known to be binaries because the X-ray emission periodically dips in intensity owing to eclipses by the companion star as it moves in front of the X-ray emitting object. The orbital period is always only a few hours, and the star that emits the X rays is always invisible. Only the normal companion can be seen. This is often an O or B star with a strong wind, so apparently the matter that falls into the collapsed companion and emits the X rays comes from this wind.

The determination of the masses in a binary system is difficult, if not impossible, if only one of the two stars can be seen, because the information on their velocities and on the inclination angle of the orbits is incomplete. Sometimes enough information can be derived to at least place limits on the mass of the invisible companion, however. In one such case, the X-ray binary called Cygnus X–1 (Fig. 23.17), it appears that the collapsed object has to contain at least 8 solar masses. This object is the leading candidate for being a black hole.

Although the jury is still out on the question of the existence of black holes, the circumstantial evidence in favor is strong. Furthermore, if we follow a principle called **Occam's Razor,** which to the scientist means that the simplest explanation of the observed facts is most likely the correct one, acceptance of the existence of black holes is natural, there being no simpler way to explain what happens when a massive star collapses. Most astronomers today have adopted the concept of black holes, and not only believe they exist, but also consider them an integral part of our universe.

Perspective

We have hounded the stars into their graves. We have examined their corpses to see how they decay, and have come away with our minds filled with wonder at the novel forms matter can take. We have seen that the nature of stellar remnants depends entirely on how much mass is left when stars run out of nuclear fuel and collapse, the three possibilities being white dwarfs, neutron stars, and black holes. Stars that die alone, no matter what final form they take, are not likely to be detected again (except for the closest white dwarfs and the neutron stars that happen to appear to us as pulsars), whereas those which exist in binary systems may be reincarnated in spectacular fashion if mass exchange takes place.

Having learned all we can about stars as individuals, we now are ready to move on to larger scales in the universe, to examine galaxies and ultimately the universe itself.

Summary

1. A white dwarf gradually cools off, taking billions of years to become a cold cinder.

2. If new matter falls onto the surface of a white dwarf, for example in a binary system where the companion star loses mass, then a nova can occur as the degenerate matter of the white dwarf causes a rapid nuclear reaction. Nova outbursts can occur many times in the same white-dwarf binary system.

3. Massive stars are likely to explode as supernovae when all possible nuclear reaction stages have ceased. The supernova explosion creates an expanding cloud of hot, chemically enriched gas known as a supernova remnant.

4. In some cases a remnant of 2 to 3 solar masses is left behind in the form of a neutron star, which consists of degenerate neutron gas.

5. A neutron star is too dim to be seen directly in most cases, but may be observed as a pulsar (depending on

the alignment of its magnetic and rotation axes, and our line of sight), and in a close binary system it may become a source of X-ray emission.

6. Some neutron stars that receive new material in clumps flare up occasionally as X-ray sources called burster. Others may exceed the mass limit for neutron stars as a result of mass exchanges and explode as Type I supernovae.

7. If the final mass of a star exceeds 2 or 3 solar masses, it will become a black hole at the end of its nuclear-reaction lifetime.

8. The immensely strong gravitational field near a black hole traps photons of light, rendering the black hole invisible.

9. A black hole may be detected by its gravitational influence on a binary companion, or by the X rays it emits if new matter falls in, as in a close binary system where mass transfer takes place. A black hole binary X-ray source can be detected only by analysis of the orbits to determine the mass of the unseen object.

Review Questions

1. Why is the final mass of a star, rather than the mass it begins with, the important criterion for determining what form of remnant the star will leave? Why should the final mass be different from the initial mass for a given star?

2. Using information from chapter 4, compare the surface gravity of the sun with that of a white dwarf having the same mass but only 1/100 of the sun's radius. Make the same calculation for a neutron star that has twice the mass of the sun but only 10^{-6} of the sun's radius.

3. Summarize the techniques by which white dwarfs can be detected from the earth.

4. Calculate (using Wien's law) the temperature at which a cooling white dwarf becomes essentially invisible to the human eye, assuming that this occurs when the wavelength of maximum emission has shifted to the infrared wavelength of 10,000 Å.

5. Novae and supernovae, despite the similarity of their names, are completely different phenomena. Explain the differences.

6. Discuss the effect of supernova explosions on the chemical composition of the galaxy.

7. Compare the internal structure of a neutron star with that of a white dwarf.

8. If a certain pulsar has a much longer period than another, which is more likely to be surrounded by a detectable supernova remnant?

9. When a star collapses to become a black hole, the gravitational field close to it becomes very strong. Why does this not affect the orbit of the companion star, in the case where the black hole is a member of a binary system?

10. Both neutron stars and black holes can be detected as X-ray sources under certain circumstances (that is, in mass-transfer binary systems). How can we expect to distinguish between these two kinds of stellar remnants in such cases?

Additonal Readings

Gursky, H., and van den Heuvel, E. P. J. 1975. X-Ray emitting double stars. *Scientific American* 232(3):24.

Helfand, D. J. 1978. Recent observations of pulsars. *American Scientist* 66:332.

Kaufmann, W. 1974. Black holes, worm holes, and white holes. *Mercury* 3(3):26.

Lewin, W. H. G. 1981. The sources of celestial X-ray bursts. *Scientific American* 244(5):72.

Schramm, D., and Arnett, W. 1975. Supernovae. *Mercury* 4(3):16.

Smarr. L., and Press, W. H. Spacetime: black holes and gravitational waves *American Scientist* 66:72.

Thorne, K. S. 1974. The search for black holes. *Scientific American* 233(6):32.

Van Horn, H. M. 1973. The physics of white dwarfs. *Physics Today* 32(1):23.

Wheeler, J. C. 1973. After the supernova, what? *American Scientist* 61:42.

The Secrets of the Stars

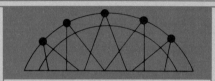

J. Craig Wheeler

Dr. Wheeler is a professor of astronomy at the University of Texas, having served earlier in his career on the faculties of Harvard and the California Institute of Technology. His research centers on active galaxies and the final stages of stellar evolution, and he has developed new understandings of the supernova process and the strange objects that can be left behind. He is a noted writer on astronomical subjects, having won competitions for essays on astronomy, and he is the coauthor of a new book on extreme and peculiar objects in the universe, entitled Astronomy Bizarre.

Thanks to observational and theoretical constructions of the Hertzsprung-Russell diagram, most of the behavior of most of the stars we see is comprehensible. This overwhelming achievement of modern astronomy cannot be underestimated. Nevertheless, many problems remain. Stated briefly, we understand fairly well that which has been observed, but when new techniques lead to new observations, whole new puzzles arise. The solar neutrino experiment (described in chapter 16 of this text) is a prime example of a new technique leading to new problems.

We have little direct information concerning the details of the interiors of stars, but the stars themselves give us one way to probe the insides from the out-

side. If heavy elements processed on the inside are somehow swept to the surface, they carry clues about the conditions under which they were formed. This apparently is happening in cases where processed CNO abundances are seen in stars low on the red-giant branch in the H-R diagram. Some of the internal mixing that brought these elements to the stellar surface must have occurred while the stars were still on the main sequence. The reason for this mixing, in the face of theoretical predictions to the opposite, is one of the main problems in the study of stellar evolution.

An important area that will benefit from future studies of stellar evolution is that of the advanced stages of stellar lifetimes.

Here the problem lies with the extended red-giant envelope. The processes that control the evolution of a star are in the central region, but the outer layers that can be directly observed look the same regardless of what is going on down in the core. Checking on the evolution in the core is therefore difficult because we must rely on indirect means.

One of the first crucial stages a star reaches after the main sequence, if its mass is moderate, is the helium flash. We have no observations to help us understand the helium flash. Until recently, the theory predicted that the red-giant envelope would remain unperturbed during the core flash. Recent, more sophisticated calculations have suggested, to the contrary, that the helium flash could be considerably more violent than we previously predicted. If this is the case, the helium flash may be observable. The problem then is that it is a quick phenomenon, so the chance of catching a given star in the act is low. If it could be done, the payoff of a direct observation to compare with theory would be immense.

Another major problem area concerns the final stages of evolution. We are fairly confident that low-mass stars eject their red-giant envelopes to leave white-dwarf remnants. We do not know how massive a star can be and still shed its envelope. Presumably if a star cannot eject the envelope, it must evolve to some catastrophic end, but no star has ever been observed immediately prior to exploding, so in no case do we know directly what kind of star exploded.

Most supernovae probably come from stars of about ten to twenty times the mass of the sun. These stars are predicted to trap most of their heavy elements in cores that collapse to make neutron stars, and so they are not expected to contribute much to the synthesis of the heavy elements available to the outside environment. Recent observations have mapped the change in some of the most populous heavy elements, such as carbon, oxygen, magnesium, and iron, as the galaxy ages. Tight constraints are placed as a result on stars that can contribute to the processes of nucleosynthesis. Either the predictions for the elements produced in stars are wrong, or the stars with masses from twenty to forty times the sun do not explode. An alternative exists. Rather than exploding, these stars could collapse to make black holes. The observed heavy elements could come from even more massive stars, or from an odd class of supernovae that do not involve massive stars at all.

A major uncertainty in stellar evolution is rotation. Stars do rotate, but none, even the final compact products, are observed to rotate particularly fast. Perhaps rotation is not so crucial. On the other hand, some calculations that have included rotation show that the late stages of evolution could be altered severely. Rotation may play a role in the mixing processes we have already discussed, and may change the amount of heavy elements generated in a model star of a given mass.

What we need most dearly is a technique that has not even been invented. The root of all this uncertainty is the same. We do not have direct observations of the sort that will guide our theories in the right direction. We need observations to help us decide whether we are calculating the evolution of moderately massive stars incorrectly, and to learn more specifically which stars explode and how.

We need a way to see directly to the heart of a star. We need to observe the mixing in action, to detect the buildup of heavy elements deeply buried beneath red-giant envelopes, and we need a way to watch a star as its core collapses to form a neutron star. The most likely technique is to build

a true neutrino telescope, an instrument that would monitor the elusive particles that are escaping from the hearts of all the stars. At each stage, neutrinos escape to carry information concerning the precise nuclear reactions that are going on. In the final collapse, a gigantic burst of neutrinos is emitted that would tell precisely what was happening in the collapse and succeeding explosion, if any.

The solar neutrino experiment is only the first hint of the power of such a device to launch the next gigantic leap in our understanding of the stars. Already experiments exist that could detect the neutrinos from a supernova in our galaxy. Such events are rare, so we must be patient and wait. Someday, perhaps hundreds or even thousands of years from now, we will have a device, probably gigantic in scale, that can sweep from star to star, wresting secrets from their innards that must remain hidden from our current observations of the stellar surfaces.

The Milky Way

INTRODUCTION TO SECTION V.

In this section we explore the great conglomeration of stars to which our sun belongs. The Milky Way, a spiral galaxy, is itself a dynamic entity, with its own structure and evolution governed by the complex interactions of stars with each other and with the interstellar gas and dust. Although this system of some 10^{11} individual stars is infinitely more complicated than a single star, and its life story is correspondingly more difficult to unravel, great progress has nevertheless been made in piecing together the puzzle.

In the first of the three chapters of this section, we examine the overall properties of our galaxy. We learn how its size and shape, and the sun's location within it, were deduced from observations carried out from our position within the great disk, where our view is obscured by interstellar haze. We discuss how the various parameters describing the galaxy were derived from observation, and we learn of some immense and fascinating mysteries that remain.

Chapter 25 focuses on the diffuse material between the stars. The interstellar medium is so rarefied that a superficial consideration might lead us to deem it unworthy of our time and effort, but we find, to the contrary, that the interstellar medium is a vital part of the galactic life story. Stars form from this material, and in evolving and dying they return their substance to it. The interstellar medium itself is a dynamic, active component of the galaxy, one that has in recent years provided astronomers with several major surprises. In this chapter we survey the general properties of the gas and dust that pervade space, and pay particular attention to their role in the evolution of the galaxy.

Finally, chapter 26 ties together the disparate data gathered and examined in the preceding two chapters, and tells the story of the formation and evolution of our galaxy. We see how the major features of the Milky Way arose from a sequence of developments dictated by simple laws of physics previously learned. Although there remain some mysteries, such as the quantity of mass in the far reaches of the galactic halo and the traces of violent activity at the core of the galaxy, most of the story can be told.

Having probed the Milky Way, we will then be ready to move outward, to examine and understand the countless galaxies in the universe beyond our own.

•CHAPTER 24•

Structure and Organization of the Galaxy

We have discussed stars as individuals, as though they lived in a vacuum, isolated from the rest of the universe. For the purposes of analyzing the structure and evolution of stars, this is a suitable approach, but if we want to understand the full stellar ecology, the properties of individuals must be discussed in the context of the larger environment.

Even a casual glance at the nighttime sky shows that the stars tend to be grouped, rather than being randomly distributed. The most obvious concentration is the Milky Way, the diffuse band of light stretching from horizon to horizon, clearly visible only in areas well away from city lights (Fig. 24.1). To the ancients the Milky Way was merely a cloud; Galileo with his primitive telescope recognized it as a region of great concentration of stars. In modern times we know the Milky

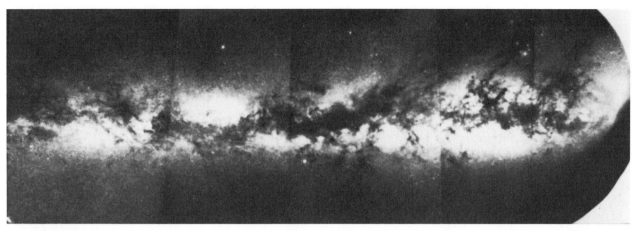

FIG. 24.1. THE MILKY WAY. This mosaic shows a major portion of the cross-sectional view of our galaxy that we see from earth.

Way as a galaxy, the great pinwheel of billions of stars to which the sun belongs. The hazy streak across our sky is a cross-sectional, or edge-on view of the galaxy, seen from a point within its disk.

FIG. 24.2. THE STRUCTURE OF OUR GALAXY. These sketches illustrate the modern view of the Milky Way.

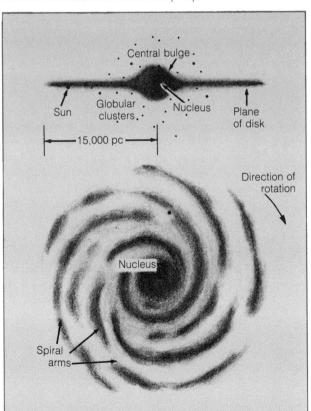

The overall structure of the Milky Way is like a phonograph record, except that it has a large central bulge, the center of which is called the **nucleus** (Fig. 24.2 and 24.3). Nearly all the visible light is emitted by stars in the plane of the disk, although the galaxy also has a **halo**, a distribution of stars and star clusters cen-

FIG. 24.3. A SPIRAL GALAXY SIMILAR TO THE MILKY WAY. This galaxy (NGC 2997) probably resembles our own, as seen from afar.

tered on the nucleus but extending well above and below the disk. The most prominent objects in the halo
are the **globular clusters,** very old, dense clusters of
stars characterized by their distinctly spherical shape.

To envision the size of the Milky Way requires a
new unit of distance. In chapter 19 we discussed stellar
distances in terms of parsecs, and we found that the
nearest star is about 1.3 parsecs from the sun. To expand to the scale of the galaxy, we speak in terms of
kiloparsecs, or thousands of parsecs. The visible disk
of the Milky Way is roughly 30 kiloparsecs (abbreviated
kpc) in diameter, and the disk is a few hundred parsecs thick. Light from one edge of the galaxy takes about
100,000 years to travel across to the far edge.

Our galaxy is so large that we must discuss new
methods of measuring distance, for even the main-sequence-fitting technique, the most powerful we have
mentioned so far, fails when star clusters are too distant and faint to allow determination of the individual
stellar magnitudes and spectral types or color indices.

Variable Stars as Distance Indicators

It was stressed in chapter 19 that it is always possible
to determine the distance to an object if both its absolute and apparent magnitudes are known. In that chapter we talked about spectroscopic parallaxes, where the
absolute magnitude of a star is derived from its spectral class, and we discussed main-sequence fitting, where
a cluster of stars is observed and its main sequence
determined, so that it can be fitted to the main sequence in the standard H-R diagram. Now we will learn
about a special type of star whose absolute magnitude
can be determined from its other properties, allowing
it to be used to measure distances.

In chapter 18 we briefly mentioned variable stars,
including those which pulsate regularly. These stars
physically expand and contract, changing in brightness
as they do so. One of the first such stars to be discussed is δ Cephei, which is sufficiently bright to be
seen with the unaided eye. Following the discovery of
δ Cephei in the mid-1700's, other similar stars were
found, and these as a class became known as **Cepheid
variables,** named after the prototype.

The usefulness of pulsating variables as distance indicators was discovered in 1912, when an astronomer
named Henrietta Leavitt (Fig. 24.4) at the Harvard Col-

FIG. 24.4.   HENRIETTA LEAVITT. Leavitt was part of the group of
astronomers at Harvard who pioneered the development of
spectral classification. In addition, she discovered the period-
luminosity relationship for Cepheid variables in the course of
studying stars in the Magellanic Clouds.

lege Observatory carried out a study of variable stars
located in the Magellanic Clouds, two small galaxies
near the Milky Way. Leavitt noticed that the average
brightnesses of these stars were correlated with the period of pulsation; that is, the longer the time between
brightenings, the brighter the star. The Magellanic
Clouds are so far away (about 55 kpc) relative to their
own dimensions that all the stars are, for practical purposes, at the same distance from us. This meant that
the correlation of period with apparent magnitude was
actually a correlation with absolute magnitude, sug-
gesting the possibility that the absolute magnitude could
be determined from the period of pulsation.

Subsequent analysis established the numerical relationship between the period and absolute magnitude
(Fig. 24.5), so that the measurement of distance to a
variable star becomes a simple matter of determining
the period of pulsation by counting the days between
times of maximum brightness, then using the established correlation to determine the absolute magnitude. The distance is then found by comparing the absolute and apparent magnitudes, using the standard
technique discussed in chapter 19. Because the Ce-

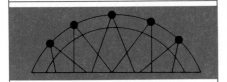

What Makes a Cepheid Run?

Every star has a natural vibration frequency, a period at which it will pulsate if distrubed by an outside force. Ordinarily, however, the pulsations, if started, will die out very quickly, so most stars are quiescent and stable. The pulsations will continue only if some force or energy input is made in synchronization with the vibrations, giving the star a little boost each cycle. A weight on a string behaves the same way; if it is given an initial push and then left alone, it will oscillate for a while, but gradually stop. If, on the other hand, it is given a push each time it reaches a certain point, it will keep oscillating indefinitely, with a regular frequency. This is how a pendulum clock works; a weight or a spring is used to give the pendulum a boost each cycle, and the pendulum in turn maintains a steady tempo.

But what provides the boost to keep a pulsating star expanding and contracting? The answer has to do with the star's outer layers and the way they allow light to pass through. When a Cepheid contracts, the density in the atmosphere increases, and helium ions combine with electrons to form helium atoms. These atoms in turn absorb light very efficiently, and in this state the atmosphere acts as a closed valve, keeping radiant energy bottled up inside the star. Heat energy builds up, causing the star to expand again. As it does so, the density in the atmosphere decreases, and the helium atoms become ionized by the emerging radiation. Light is now free to escape from the star's interior, since there are fewer helium atoms to hold it in, and therefore the pressure that caused the expansion diminishes. Soon enough, the outer layers begin to fall back in, and the star enters a new contraction phase. Helium atoms form again, and again they block the light from the interior, causing a new buildup of pressure and creating a new expansion phase. The cycle continues indefinitely, with the helium acting as a valve, letting light out or keeping it bottled up.

Similar valve mechanisms apply to some of the other types of variable stars, although rather than helium, some other element may play the role of the valve. All stars with just the right combination of temperature and pressure in their outer layers will pulsate because of the mechanism just described. The variable stars occupy a narrow region in the H-R diagram, where stars have the correct conditions. This area is called the **instability strip,** because all the stars in it are unstable and therefore pulsate.

The period of pulsation is determined by the average density of a star. The denser it is, the more rapidly it vibrates. Thus the range of periods of Cepheid variables represents a range of densities.

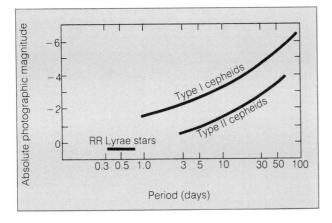

FIG. 24.5. THE PERIOD-LUMINOSITY RELATIONSHIP FOR VARIABLE STARS. This diagram shows how the pulsation periods for Cepheid and RR Lyrae variables are related to their absolute magnitudes. The fact that there are two types of Cepheids, with somewhat different relationships, was not recognized at first, and this led to some early confusion regarding distance scales.

pheid variables are giant stars, they are quite luminous, and can be observed at great distances, even beyond the Milky Way. Therefore these stars are very useful in measuring the scale of our galaxy.

Another type of pulsating star, the **RR Lyrae variables,** also were found to be reliable distance indicators. These stars have periods of only a few hours, and they all have the same absolute magnitude. Any time we identify an RR Lyrae star by measuring its period, we immediately know its absolute magnitude and hence its distance. The RR Lyrae stars are not as luminous as the Cepheids, but they are still bright enough to be observed at large distances. These stars are often found in globular clusters, and are sometimes called **cluster variables.** They played a very important role in the early measurements of the size and shape of our galaxy, as we will learn later in this chapter.

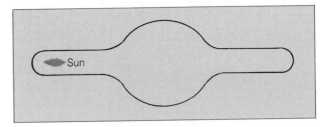

FIG. 24.6.   THE KAPTEYN UNIVERSE. The darkly shaded area here illustrates roughly the extent and shape of the galaxy as inferred by J. C. Kapteyn on the basis of star counts. Because of obscuration by interstellar matter, Kapteyn was unable to include stars very far from the sun in the plane of the disk.

The Structure of the Galaxy and the Location of the Sun.

Because the solar system is located within the disk of the Milky Way, we have no easy way of getting a clear view of where we are in relation to the rest of the galaxy. All we see is a band of stars across the sky, which tells us that we are in the plane of the disk. It is not so easy to determine where we are within the disk with respect to its edge and center.

The first serious attempts to solve this problem were based on determinations of the density of stars in space around the sun. The method was to choose random regions of the sky and count the number of stars of each magnitude in these regions. The number of stars increases with decreasing brightness, because faraway stars appear faint, and in a given direcion, there are many more faraway stars than nearby ones. Therefore we can determine the distribution of stars in space by seeing how rapidly their number increases with decreasing brightness (that is, with increasing distance).

In the first decade of the twentieth century, the Dutch astronomer J. C. Kapteyn employed this method, and got a surprising result. The density of stars appeared to fall off in all directions from the sun, implying that our solar system is in the densest portion of the Milky Way (Fig. 24.6). The most logical interpretation was that the sun is at the very center of the galaxy, for it seemed likely that it should be densest at the core. The thought that the sun was at the center of the known universe made many astronomers uneasy, because in the past, every assumption that the earth held a special place in the universe had been proved incorrect. It turns out there was indeed a flaw in the work of Kapteyn, although he could hardly have been aware of it. Before the error in Kapteyn's method was found, however, two

other studies showed that the sun was probably not at the center of the galaxy.

One of these studies, carried out by Harlow Shapley (Fig. 24.7), made use of the globular clusters, the spherical star clusters found to lie outside the confines of the galactic disk. As noted earlier, these clusters tend to contain RR lyrae variables, so it was possible for Shapley to determine their distances and hence their locations with respect to the sun and the disk of the Milky Way (Fig. 24.8). He found that the globular clusters are arranged in a spherical volume centered on a point several thousand parsecs from the sun, and he argued

FIG. 24.7.   HARLOW SHAPLEY. His work on the distances to globular clusters was a key step in the determination of the size of the Milky Way. He also played a prominent role in the discovery of other galaxies (see chapter 27). Much of his observational work was done before 1920, when Shapley was a staff member at the Mt. Wilson Observatory. He later became director of the Harvard College Observatory.

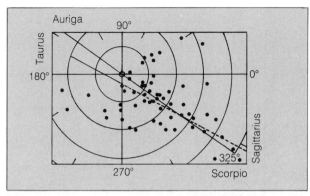

FIG. 24.8. SHAPLEY'S MEASUREMENTS OF GLOBULAR CLUSTERS. This is one of Shapley's original figures illustrating the distribution of globular clusters in the galaxy. The sun is at the point where the straight lines intersect, and each circle centered on that point represents an increase in distance of 10,000 parsecs. Note that the distribution of globulars is centered some 10 to 20 kiloparsecs from the sun.

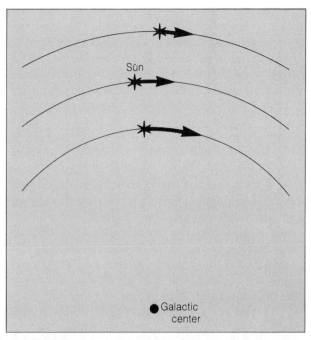

FIG. 24.9. STELLAR MOTIONS NEAR THE SUN. Stars just inside the sun's orbit move faster than the sun, whereas those farther out move more slowly. Analysis of the relative speeds (as inferred from measurements of Doppler shifts) and distances of stars like these led to the realization that the sun and stars near it are orbiting a distant galactic center.

that this point must represent the center of the galaxy. It would not make physical sense for the globular clusters to be concentrated around any location other than the center of the entire galactic system.

Shapley's conclusion, first published in 1917, was not widely accepted initially, but other supporting evidence came to the fore in the 1920's. Two scientists, Jan Oort of Holland (already mentioned in connection with the origin of comets; see chapter 15), and the Swede Bertil Lindblad, carried out careful studies of the motions of stars in the vicinity of the sun. What they found was that these motions could best be understood if the sun and the stars around it were assumed to be orbiting a distant point; that is, there are systematic, small velocity differences between stars, similar to those between runners on a track who are in the inside and outside lanes (Fig. 24.9). It appeared from these studies that the sun is following a more-or-less circular path about a point several thousand parsecs away, indicating that the center of the galaxy is located at that distant center of rotation. This supported Shapley's view of the galaxy, although there were still uncertainties about how far the sun was from the center.

Let us now return to the work of Kapteyn, for the next major discovery was the one that revealed the flaw in his conclusions. In 1930 a study carried out by the American R. Trumpler showed that the galaxy is filled with an interstellar haze that makes stars appear fainter than they otherwise would. Trumpler made this dis-

covery by examining the apparent brightnesses of star clusters whose distances he knew from their apparent sizes; what he found was that distant clusters appear much fainter than they should, and that the effect increases with increasing distance. This was the first direct evidence that there is a pervasive interstellar medium in the space between the stars; until then, it was known only that interstellar clouds and nebulae existed, as seen on photographs showing dark regions and brightly glowing gas clouds.

Trumpler's discovery explained the contradiction between Kapteyn's results based on star counts, and the picture developed by Shapley and by Lindblad and Oort. Kapteyn's star counts failed to take the interstellar haze into account, so his estimates of stellar distances were erroneous. It was the obscuration by the interstellar medium that made the density of stars appear to fall off with distance from the sun.

Consideration of the effects of interstellar material, along with refinements in the assumed relation between pulsation period and absolute magnitude for variable stars, led eventually to a consensus on the size of the galaxy and the sun's location within it. As mentioned earlier, the disk is some 30 kpc in diameter. The sun is located about 10 kpc from the center, or about two-thirds of the way to the edge. As we will see in later sections, there is still some uncertainty about how far the galaxy extends beyond the sun's position.

Galactic Rotation and Stellar Motions

An important result of the work of Lindblad and Oort was the development of an understanding of the overall motions in the galaxy. Oort's analysis especially was useful in this regard, for he showed not only that the sun and the stars near it are orbiting the distant galactic center, but also that the rotation of the galaxy is differential, meaning that each star follows its own orbit at its own speed. Thus the galaxy in the vicinity of the sun does not act like a rigid disk but like a fluid, with each star moving as an independent particle.

This is not true of the inner portion of the galaxy. There the entire system does rotate like a rigid object; like a record on a turntable. The stars in the inner part of the galaxy are firmly held in place by the combined gravitational forces of all the stars around them, and they are not free to follow Keplerian orbits about the galactic center. In the region where rigid-body rotation is the rule, the speeds of individual stars increase with distance from the center, whereas in the outer portion they decrease with distance (Fig. 24.10). In the outer reaches of the galaxy, each star orbits the central portion of the galaxy with little influence from its neighbors, and the stellar orbits are approximately described by Kepler's laws. This is why the orbital speeds decrease with distance from the center in this part of the galaxy. There is an intermediate distance, just inside the sun's orbit, where a transition between rigid-body and Keplerian orbits occurs, and it is here that stars have the greatest orbital velocities. The sun, near this peak position, travels at about 250 kilometers per second, taking roughly 250 million years to make 1 complete circuit about the galaxy. In its 4- to 5-billion-year lifetime, the sun has completed 15 to 20 orbits.

Individual stellar motions do not necessarily follow precise circular orbits about the galactic center. What we have described so far is the overall picture that develops from looking at the composite motions of large numbers of stars. If we look at the individual trees in-

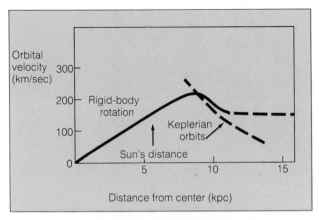

FIG. 24.10.   THE ROTATION CURVE FOR THE MILKY WAY. This diagram shows how the stellar orbital velocities vary with distance from the center of the galaxy. The fact that the curve does not simply drop off to lower and lower velocities beyond the sun's orbit, as it would if all the mass of the galaxy were concentrated at its center, indicates that there is a lot of mass in the outermost portions of the galaxy.

stead of the forest, we find that each star in the great disk has its own particular motion, which may deviate slightly from the ideal circular orbit. These individual motions are comparable to the paths of cars on a freeway, where the overall direction of motion is uniform, but a bit of lane-changing occurs here and there.

In the galaxy, a star's deviation in motion from a perfect circular orbit is called the **peculiar velocity;** in the case of the sun, it is the **solar motion.** The sun has a velocity of about 20 kilometers per second with respect to a circular orbit, in a direction about 45 degrees from the galactic center and slightly out of the plane of the disk. Most peculiar velocities of stars near the sun are comparable, amounting to only minor departures from the overall orbital velocity of about 250 kilometers per second. In chapter 26 we will discuss the **high-velocity** stars, which greatly deviate from the circular orbits followed by most stars in the sun's vicinity.

Spiral Structure and the 21-cm Line

So far we have spoken of the galactic disk as though it were a uniform, featureless object, but we know that this is not a completely accurate picture. The Milky Way is a spiral galaxy, and if we could see it face-on, we

ASTRONOMICAL INSIGHT (24.2)

21-cm Emission from Hydrogen

To understand how a hydrogen atom can emit radiation at a wavelength of 21 centimeters, we need to take a closer look at the structure of the atom. A hydrogen atom consists only of a proton, which forms the nucleus, and a single electron in orbit about it. We have already learned that the electron can occupy a variety of possible orbits, each corresponding to a different energy state. We have also learned that, if a photon of light is absorbed by the electron, the electron moves from a low energy level to a higher one, and a photon is emitted if the electron drops from a high energy level to a lower one. The wavelength of the photon is related to the energy difference between the two electron energy levels; the greater the difference, the shorter the wavelength, and vice versa.

It happens that the energy-level structure of the electron is more complicated than previously described. The electron and the proton are both spinning, and the energy of the electron depends on whether it is spinning in the same or the opposite direction as the proton. If both spin in the same (parallel) direction, then the energy state is slightly greater than in the case where the electron and proton spin in opposite (antiparallel) directions. As before, the electron can change from one state to the other by either emitting or absorbing a photon. The energy difference is so small, however, that the wavelength of the photon is very much longer than that of visible light. It is 21.1 centimeters.

Hydrogen atoms in space tend to have the electron in the lowest possible energy state, with its spin antiparallel to that of the proton. Occasionally an atom will collide with another, however, causing the electron to jump to the higher state, and its spin is then parallel to that of the proton. Following this, the electron spontaneously reverses itself, seeking to return to the lowest energy state, and when it does so, it emits a photon of 21.1-cm wavelength. The probability that the electron will make the downward transition is very low, and the electron may remain in the upper state for as long as 10 million years before spontaneously dropping back down. There are so many hydrogen atoms in space, however, that at any given instant, many photons are being emitted. Hence radio telescopes capable of receiving this wavelength can trace the locations of hydrogen clouds throughout the galaxy and beyond.

---

would see the characteristic pinwheel shape normally found in galaxies of this type.

There is a common misconception about the spiral structure in the Milky Way (and other spiral galaxies); namely, that there are few stars between the visible spiral arms. In reality, the density of stars between the arms is nearly the same as it is in the arms. The most luminous stars, however, the young, hot O and B stars, tend to be found almost exclusively in the arms. Because these are the brightest of stars, their residence in the arms makes the spiral structure stand out.

The fact that we live in a spiral galaxy was not easily discovered, again because of our location within it where we see only a cross-sectional view. It was not until 1951 that investigations of the distribution of luminous stars revealed traces of spiral structure in the Milky Way, and even those studies were limited to a small portion of the galaxy. Because of the obscuration caused by interstellar material, our view of even the brightest stars is limited to a local region, about a thousand parsecs from the sun at most.

A major advance in measuring the structure of the galaxy occurred in the same year, when the 21-cm radio emission of interstellar hydrogen atoms was first detected. It had been predicted that hydrogen should emit radiation at this wavelength (see chapter 6), and the Americans E. M. Purcell and H. I. Ewen were the first to detect this emission, using a specially built radio telescope.

The great advantage of the hydrogen 21-cm emission for measuring galactic structure is its ability to penetrate the interstellar medium to great distances. Whereas the brightest stars can be detected at distances of a few hundred parsecs at most, hydrogen clouds in space can be "seen" by radio telescopes from all the way across the galaxy, at distances of several thousand parsecs. Hydrogen gas is the principal component of the interstellar medium, and it tends to be concentrated along the spiral arms, so observations of the 21-cm radiation can be used to trace out the spiral structure throughout the entire Milky Way.

When a radio telescope is pointed in a given direction in the plane of the galaxy, it receives 21-cm emission from each segment of spiral arm in that direction (Fig. 24.11). Because of differential rotation, each arm has distinct velocity from the others that are closer to the center or farther out. Therefore what is seen, instead of a single emission peak at exactly 21.1-cm wavelength, is a cluster of emission lines near this wavelength, separated from each other by the Doppler effect. By combining measurements such as these with Oort's mathematical analysis of differential rotation, astronomers were able to reconstruct the spiral pattern of the entire galaxy.

The pattern is quite complex, much more so than in some galaxies, where two arms are seen elegantly spiraling out from the nucleus. The Milky Way consists of bits and pieces of a large number of arms, giving it a definite overall spiral form, but not a smoothly co-

herent one. As we will learn in chapter 26, the differences between the type of spiral structure seen in our galaxy and the regular appearance seen in others may reflect different origins for the arms.

The Mass of the Galaxy

Once the true size of the Milky Way was determined, it became possible to estimate its total mass. This could be achieved by measuring the star density in the vicinity of the sun and then assuming that the entire galaxy has about the same average density, but a much simpler and more accurate technique became possible with the discovery that the stars in our region of the galaxy obey Kepler's laws.

Kepler's third law, in the more complete form developed by Newton, expresses a relationship among the period, the size of the orbit, and the sum of the masses of the two objects in orbit about each other:

$$(m_1 + m_2)P^2 = a^3,$$

where m_1 and m_2 are the two masses (in units of the sun's mass), P is the orbital period (in years), and a is

FIG. 24.11. 21-CM OBSERVATIONS OF SPIRAL ARMS. At left is a schematic diagram of the galaxy, showing the direction in which a radio telescope might be pointed to produce a 21-cm emission-line profile like the one sketched at right. There are many components of the 21-cm line, each corresponding to a distinct spiral arm, and each at a wavelength reflecting the Doppler shift between the velocity of that arm and the earth's velocity.

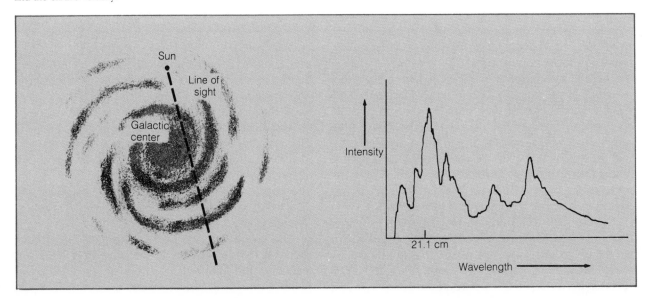

the semimajor axis (in astronomical units). If we consider the sun to be one of the two objects, and the galaxy itself to be the other, we can use this equation to determine the mass of the galaxy. As we have already seen, the orbital period of the sun is roughly 250 million years; that is, $P = 2.5 \times 10^8$ years. The orbit is nearly circular, with the galactic nucleus at the center, so the semimajor axis is approximately equal to the orbital radius; $a = 10$ kpc $= 2 \times 10^9$ AU. Now we can solve Kepler's third law for the sum of the masses:

$$m_1 + m_2 = a^3/P^2 = 1.3 \times 10^{11} \text{ solar masses.}$$

Since the mass of the sun (1 solar mass) is inconsequential compared with this total, we can say that the mass of the galaxy itself is about 1.3×10^{11} solar masses. The sun is slightly above average in terms of mass, so we conclude that the total number of stars in the galaxy must be $3-4 \times 10^{11}$; that is, several hundred billion.

This method, based on Kepler's third law, refers only to the mass inside the orbit of the sun; the matter that is farther out has no effect on the sun's orbit. Thus when we estimate the mass of the galaxy by applying Kepler's third law in this way, we neglect all the mass that lies farther out. So that this problem could be overcome, radio measurements of interstellar gas in the outer portions of the galaxy have recently been used to determine the orbital velocity (and therefore the orbital period) of material in the outer reaches of the galaxy. It was found that the velocity does not decrease as rapidly with distance as had previously been thought, and this in turn led to the conclusion that quite a bit of mass lies beyond the sun's orbit. In a later section, we will discuss the possibility that most of the mass of the galaxy lies in the halo.

The Galactic Center: Where the Action Is

The center of our galaxy is a mysterious region, forever blocked from our view by the intervening interstellar medium (Fig. 24.12 and Color Plate 20). From Shapley's work on the distribution of globular clusters, as well as Oort's analysis of stellar motions, it was known by the 1920's that the center of our great pinwheel lies in the direction of the constellation Sagittarius. There we find immense clouds of interstellar matter and a

FIG. 24.12.   THE GALACTIC CENTER. The dark regions are dust clouds in nearby spiral arms. Because of obscuration, photos such as this do not reveal the true galactic center, but only nearby stars and interstellar matter in the plane of the disk.

great concentration of stars. It is one of the richest regions of the sky to photograph, although the best observations can be made only from the southern hemisphere.

The central portion of the galaxy consists of a more-or-less spherical bulge, populated primarily by relatively cool stars. The absence of hot, young stars implies that the central region has had relatively little recent star formation, something we will have to deal with in our discussion of the history of the galaxy (chapter 26).

Although we cannot see into the central portion of the nucleus in visible light, this region can be probed at longer wavelengths where the interstellar extinction is not such a problem (Fig. 24.13). Radio and, more recently, infrared observations have revealed some interesting features of the galactic core. The first clue that something unusual was taking place there came from

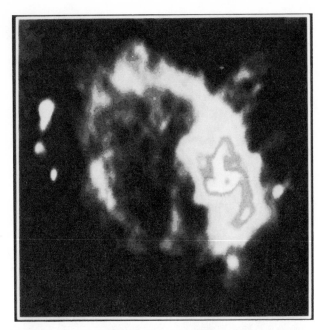

FIG. 24.13. A RADIO MAP OF THE GALACTIC CENTER. This image, made at a wavelength of 6 cm (where thermal-continuum emission is measured, rather than a spectral feature such as the 21-cm line), shows a tiny, intense spot at the precise location of the galactic nucleus.

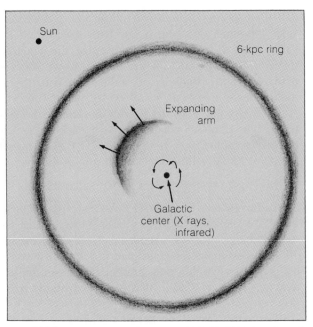

FIG. 24.14. ACTIVITY AT THE GALACTIC CENTER. A variety of evidence for energetic motions associated with the center of our galaxy is indicated here.

21-cm observations of hydrogen, which revealed a turbulent mixture of clouds moving about at high speed, and which showed that one of the inner spiral arms, about 3 kpc from the center, is expanding outward at a velocity of more than 100 kilometers per second (Fig. 24.14). This is suggestive of some sort of explosive activity at the center that forced matter to move outward. More recently, infrared observations of emission lines from hot gas clouds have shown that interstellar clouds near the center of the nucleus are moving quite rapidly in their orbits, which in turn implies (through the use of Kepler's third law) that they are orbiting a very massive central object.

Additional evidence of violent activity is seen in another portion of the radio spectrum, at the wavelength where interstellar carbon monoxide molecules emit (see chapter 25 for a discussion of molecules in space). Observations of CO have revealed a gigantic ring of interstellar clouds circling the galaxy at a distance of some 6 kpc from the center. It looks as though the ring could have been built up by the expansion of matter away from the galactic center, perhaps as the result of an ancient explosion. The concentration of matter at 6 kpc

may be the end result of an earlier episode of the same mysterious activity that is responsible for the current expansion of the 3 kpc arm.

Finally, space observations show that a very small object precisely at the center of the galaxy is emitting enormous amount of energy in the form of X rays. This object, whatever it is, is smaller than one parsec in diameter, yet it appears to be responsible for all the violent and energetic activity just described from the observed cloud velocities, astronomers have estimated the mass of the central object to be roughly 10^6 solar masses. This is only a small fraction of the total mass of the galaxy, but is much greater than the mass of any known object within it, and must therefore represent some new kind of astronomical entity. The best explanation offered by astronomers is that a very massive black hole resides at the core of our galaxy. Its gravitational influence would be responsible for the rapid orbital motions of nearby objects, and matter falling in would be compressed and heated, accounting for the X-ray emission. We do not yet understand how such an object formed, but its presence would imply that at some time during the formation and evolution of the Milky Way, great quantities of matter were compressed into a very small volume at its center. Such an intense buildup of matter may have resulted from frequent collisions among stars in the early days of the galaxy, causing vast numbers of them to gradually coalesce into a single, massive object at the center. The prospect that such a beast inhabits the core of our galaxy is quite a bizarre one, yet it may tie in directly to some of the strange happenings that have been observed in the nuclei of other galaxies (see chapter 30).

Globular Clusters Revisited

We have not yet said much about the globular clusters, the gigantic spherical conglomerations of stars that orbit the galaxy (Fig. 24.15). A large globular cluster can be an impressive sight when viewed with a small telescope, and a number of them are popular objects for astronomical photography. More than a hundred of these clusters have been catalogued; no doubt many more are obscured from our view by the disk of the Milky Way.

A single cluster may contain several hundreds of thousands of stars, and have a diameter of ten to twenty parsecs. The mass is usually in the range of several hundred thousand to a few million solar masses. To contain so many stars in a relatively small volume, globular clusters must be very dense, compared with the galactic disk. The average distance between stars in a globular cluster is only about one-tenth of a parsec (recall that the nearest star to the sun is more than one parsec away). If the earth orbited a star in a globular cluster, the nighttime sky would be a spectacular sight, with hundreds of stars brighter than first magnitude.

An H-R diagram for a globular cluster is rather peculiar looking, compared with the familiar diagram for stars in the galactic disk (Fig. 24.16). The main sequence is almost nonexistent, having stars only on the extreme lower portion. On the other hand, there are many red giants, and a number of blue stars that lie on a horizontal sequence extending from the red-giant region across to the left in the diagram. These facts together point to a very great age for globular clusters. As we learned in chapter 21, in a cluster H-R diagram, the point where the main sequence turns off toward the red-giant region indicates the age of the cluster; the lower the turn-off point, the older the cluster. Us-

FIG. 24.15. THE GLOBULAR CLUSTER M13.

FIG. 24.16. THE H-R DIAGRAM FOR A TYPICAL GLOBULAR CLUSTER. The main sequence branches off at a point near the bottom, indicating the great age of the cluster.

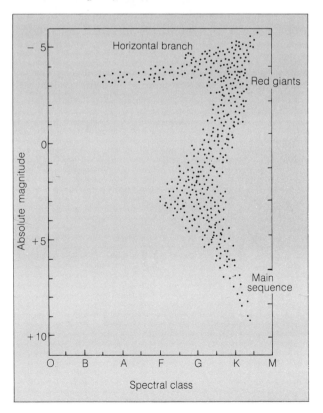

ing this technique to date globular clusters leads typically to age estimates of 14 to 16 billion years, comparable to the accepted age of the galaxy itself. Thus globular clusters are among the oldest objects that exist, and, as we will see, their presence in the galactic halo provides important data on the early history of the galaxy.

The sequence of stars extending across the globular cluster H-R diagram from the red-giant region to the left is called the **horizontal branch** (also mentioned in chapter 22). The evolutionary status of these stars has been the subject of a good deal of uncertainty; theoretical models have not yet been developed that can explain how stars evolve to this region of the H-R diagram. It is not even known for certain whether these stars are moving to the left (getting hotter, perhaps following a red-giant phase), or to the right. Most likely these stars have completed their red-giant stages and are moving to the left on the diagram, perhaps on their way to becoming white dwarfs.

Some globular clusters have been found to contain X-ray sources in their centers (Fig. 24.17). In a few cases these are X-ray *bursters,* described in chapter 23 as being neutron stars that are slowly accreting new matter on their surfaces, probably in binary systems where the companion star is losing mass. In other cases, however, the X ray does not have the recognizable properties of a binary system, and may instead be caused by a single object at the center of the cluster. This could be a giant black hole, formed as stars in the cluster collided and fell together in the core.

A Massive Halo?

The halo of the Milky Way galaxy has traditionally been envisioned as a very diffuse region, populated by a scattering of dim stars and dominated by the giant globular clusters. There was little or no evidence of any substantial amount of interstellar material, and the halo was assumed not to contain a significant fraction of the galaxy's mass.

Some of these ideas are changing as a result of very recent discoveries. We mentioned earlier that radio observations revealed an unexpectedly high amount of mass in the outer portions of the disk, indicating that the galaxy is more extensive than previously thought (Fig. 24.18). At the same time, other evidence (to be described in later chapters) led to a general expectation that many galaxies have large quantities of matter in their halos. Finally, in the early 1980's, ultraviolet observations revealed a large amount of very hot, rarefied interstellar gas in the halo of our galaxy, probably extending to distances of several thousand parsecs above and below the plane of the disk. This gas is very turbulent, with clouds traveling at speeds of several hundred kilometers per second, and is highly ionized, indicating a temperature of at least 100,000 K or more.

FIG. 24.17. A GLOBULAR CLUSTER X-RAY SOURCE. Several globular clusters have been found to contain intense X-ray sources at their centers. Here is an X-ray image of a globular, obtained with the *Einstein Observatory,* clearly showing the central source.

FIG. 24.18. THE EXTENSIVE HALO OF THE MILKY WAY. This illustrates the shape and scale of the very extended galactic halo whose presence has been inferred from the shape of the galactic rotation curve (see figure 24.10) and from absorption lines formed by interstellar gas.

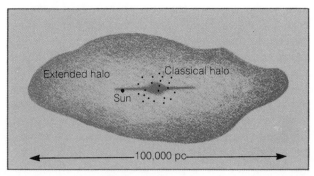

At the present time we cannot say much about the origin or total quantity of material in the halo of our galaxy. It is quite possible that in addition to the gas, there are enough stars, too dim to be detected, to add significantly to the total mass of the galaxy. Some as-tronomers believe that as much as 90 percent of the galaxy's mass may be in the halo. The *Space Telescope,* to be launched later in the 1980's, should be able to settle this question by making more sensitive measure-ments of both the gas and the stars in the halo.

Perspective

We have discussed the anatomy of the galaxy, particu-larly its overall structure and motions. The solar system is located some 10 kpc from the center of a flattened, rotating disk containing a few hundred billion stars. With the help of radio and infrared observations, as-tronomers have unraveled the structure of the disk, which consists of spiral arms (delineated by bright, hot stars and interstellar gas), and a central nucleus that contains relatively dim, red stars along with a mysteri-ous, energetic gremlin that stirs up the core region. Many questions remain about the structure and extent of our galaxy; some will be answered with the launch of the *Space Telescope.*

We have yet to discuss the workings of the galaxy; we have seen what it is like, but have said little about why. Before we endeavor to do so, we will discuss the interstellar medium, a vital part of the galactic ecology.

Summary

1. The Milky Way is a spiral galaxy, consisting of a disk with a central nucleus and a spherical halo, where the globular clusters reside. The disk is about 30 kpc in diameter.

2. Distances within the Milky Way can be determined from the measurement of variable-star periods and ap-plication of the period-luminosity relation.

3. The true size of the Milky Way and the sun's loca-tion within it were difficult to determine because the view from our location within the disk is obscured by interstellar obscuration.

4. Star counts seem to indicate that the sun is in the densest part of the Milky Way, but measurements of the distribution of globular clusters and analysis of stellar motions show that the sun is some 10 kpc from the center. The discrepancy was resolved when it was dis-covered that interstellar extinction affects the star counts.

5. The inner part of the galactic disk rotates rigidly, whereas the outer parts are fluid, with each star follow-ing its own individual orbit, approximately described by Kepler's laws.

6. Individual stellar orbits in the sun's vicinity are generally in the plane of the disk, and are nearly cir-cular. Deviations from perfect circular motion by stars are called peculiar velocities, and in the case of the sun, the solar motion. The high-velocity stars are those which pass through the disk in the perpendicular di-rection.

7. The spiral structure of our galaxy is most easily and directly measured through radio observations of the 21-cm line of hydrogen atoms in space. The spiral pattern is complex, with many segments of spiral arms.

8. The mass of the galaxy, determined by the appli-cation of Kepler's third law to the sun's orbit, is roughly 10^{11} solar masses. This technique does not take into account any mass that resides in the halo.

9. A variety of evidence indicates there is chaotic, en-ergetic activity associated with the central core of our galaxy. The data show a compact, massive object exist-ing there, and the best explanation is that it is a mas-sive black hole.

10. The globular clusters that inhabit the halo of the galaxy are very old, and therefore provide information on the early history of the galaxy.

11. There may be large quantities of mass in the galac-tic halo, in the form of interstellar gas or dim, cool stars, or both. It appears possible that as much as 90 percent of the mass of the galaxy is in the halo.

Review Questions

1. Recalling what you learned in chapter 18 about stellar parallaxes as distance determinators, calculate the distance to a star whose parallax angle is $\pi = 0''.001$ (this is about the smallest angle that can be measured with current techniques). Compare the distance to such a star with the diameter of the galactic disk, and discuss the practicality of using parallax measurements to probe the structure of our galaxy.

2. Suppose a Cepheid variable is found to have a period of ten days. From the period-luminosity relation, its average absolute magnitude is found to be $M = -4$. Its average apparent magnitude is $m = +16$. What is its distance? How does this compare with the diameter of the galactic disk?

3. Why wasn't the period-luminosity relation for Cepheid variables discovered from observations of these stars in our own galaxy?

4. What would have been the result of Kapteyn's star-count method if there had not been any interstellar extinction?

5. Compare Shapley's reasoning, that the center of the galaxy should be the place about which the globular clusters orbit, with that of Aristarchus, who determined, for similar reasons, that the sun, and not the earth, is at the center of the solar system.

6. Why do the most rapid circular orbits around the galaxy occur at the distance from the center where rigid-body rotation gives way to fluid rotation?

7. Summarize the reasons that the 21-cm line of hydrogen is the best tool we have for tracing the spiral structure of the galaxy.

8. Suppose the orbit of a star lying 20 kpc from the galactic center were analyzed to determine the mass of the galaxy. How would you expect the result to differ from what was found from analysis of the sun's orbit?

9. Summarize the evidence that there may be a massive black hole at the center of the Milky Way.

10. We have seen that the globular clusters are very old objects, comparable in age to the galaxy itself. What does this imply about the chemical composition of the stars in these clusters?

Additional Readings

Bok, B. J. 1972. Updating galactic spiral structure. *American Scientist* 60:708.

————. 1981. The Milky Way galaxy. *Scientific American* 244 (3):92.

Bok, B. J., and Bok, P. 1981. *The Milky Way.* Cambridge, Ma.: Harvard University Press.

Geballe, T. R. 1979. The central parsec of the galaxy. *Scientific American* 241(1):52.

Kraft, R. P. 1959. Pulsating stars and cosmic distances. *Scientific American* 201(1):48.

Sanders, R. H., and Wrixon, G. T. 1974. The center of the galaxy. *Scientific American* 230(4):66.

Seeley, D., and Berendzen, R. 1978. Astronomy's great debate. *Mercury* 7(3):67.

Weaver, H. 1975 and 1976. Steps towards understanding the large-scale structure of the Milky Way. *Mercury* 4(5):18, Part 1; 5(6):19.

CHAPTER 25

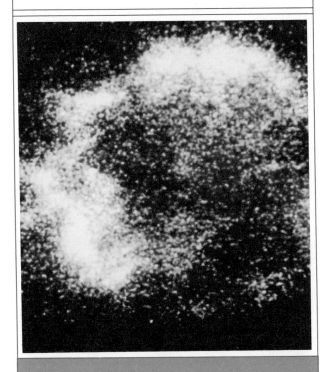

The Interstellar Medium

The very rarefied matter between the stars, diffuse though it is, constitutes one of the most important elements in the galactic environment. Stars form from it, and as they age and die, they return their substance to it. As we have seen, the interstellar haze limits our observations of the galactic disk to the nearest few hundred parsecs, allowing us to view only a very small fraction of the entire galactic volume.

Despite its obvious importance, the interstellar medium, by ordinary earthly standards, is so tenuous that it verges on nonexistence. Even in the densest interstellar clouds, the density of particles is less than one-trillionth of the earth's sea-level atmospheric density. Only the most sophisticated artificial vacuum pumps can even approach the natural vacuum of space.

The medium is so pervasive, however, and the volume of space so large, that some 10 percent of the total mass of the galactic disk is contained in this form. There are two distinct types of interstellar material, which are thoroughly mixed together throughout space: tiny solid particles called interstellar dust grains; and interstellar gas particles, which may be atoms, ions, or molecules, depending on the temperature and density. The grains are relatively few in number, and constitute only about 1 percent of the total mass in the interstellar medium.

The Interstellar Dust

The fact that there are interstellar dust grains was recognized long ago; photographs of the Milky Way show extensive, irregular dark regions (Fig. 25.1 and Color Plates 16-19) where dense dust-bearing clouds completely hide the stars behind them. Knowledge of the presence of these clouds is ancient, but it was not recognized until Trumpler's work in 1930 that there is a general distribution of dust grains throughout the spaces in between the obvious clouds. Today the dust is studied by a number of techniques, and a great deal has been learned about its nature.

Even though the dust grains constitute only a tiny fraction of the total mass in the interstellar medium, they have a very important influence on the starlight that passes through space. There are two basic effects: (1) the obscuration of light from distant stars, commonly referred to as **interstellar extinction;** and (2)

FIG. 25.1. A REGION OF INTERSTELLAR CLOUDS. This photograph shows the region near the star rho Ophiuchi (itself buried in nebulosity at the center of the photo). This region contains a rich assortment of interstellar clouds and nebulae. At bottom is a globular cluster, much more distant than the interstellar clouds seen here.

the polarization of starlight. Both effects provide important information about the nature of the dust grains themselves.

The extinction of starlight by interstellar dust particles is the result of a combination of absorption and **scattering,** a process in which the light essentially bounces off of the dust grains. When a photon is scattered, its wavelength remains fixed, but its direction is altered. The scattering process is more effective for short wavelengths of light than for longer wavelengths, so a distant star appears redder in color than it otherwise would, because the red light that it emits reaches us more easily than the blue (Fig. 25.2). For a similar reason, the sun appears red in color when it is near the horizon so that its light must traverse a long pathlength through the earth's atmosphere; the fine particles in the air scatter the blue light but allow the red to come through relatively unhindered. Because the extinction of starlight is so much more severe at short wavelengths, ultraviolet telescopes cannot probe the great distances that visible-light telescopes can.

It is possible to determine how much dust lies in the direction of a given star by measuring how much redder the star appears than it should. Recall the color index B-V, the difference between the blue and visual magnitudes of a star (see chapter 18). The redder a star is, the greater the value of B-V (do not forget: if a star is red in color, it is *brighter* in visual than in blue light, so the V magnitude is *smaller* than the B magnitude). The effect of interstellar extinction is to make the value of B-V greater than it would otherwise be.

FIG. 25.2. THE SCATTERING OF STARLIGHT BY INTERSTELLAR GRAINS. The shorter-wavelength photons are more readily scattered by grains; thus it is the longer wavelengths that more easily pass through a cloud. This means that stars seen through interstellar clouds appear reddened.

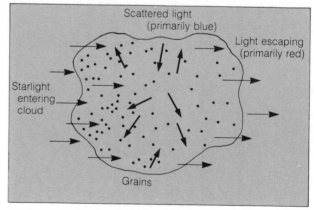

To estimate the amount of dust in the line of sight to a distant star, therefore, simply requires a comparison of the star's *B-V* color index with the value it would have without extinction. The latter quantity is determined from the spectral class of the star. Remember that *B-V* and spectral class both measure the same stellar property, namely the temperature; therefore all stars of a given spectral class have the same intrinsic value of *B-V*.

Theoretical considerations allow astronomers to determine the average size of the grains, from the variation of extinction with wavelength (known as the *extinction curve;* Fig. 25.3). The fact that the grains scatter blue light more effectively than red, and the way in which the scattering efficiency varies as a function of wavelength leads to the conclusion that there are two distinct types of grains, both very small: (1) those with an average diameter of about 5×10^{-5} cm (or about 5,000 Å, comparable in size to the wavelength of visible light); and (2) a group of even smaller grains, whose diameters are only about 10^{-6} cm (or about 100 Å), which create extremely strong ultraviolet extinction.

The polarization of starlight tells us something about the shape of the grains. In chapter 5 we learned that when light is polarized, the waves have a preferred orientation. The interstellar dust creates this effect by selectively absorbing light of one orientation, so that what

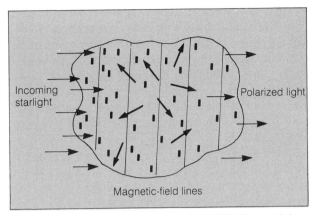

FIG. 25.4.   THE POLARIZATION OF STARLIGHT. Elongated dust grains, aligned by a galactic magnetic field, tend to scatter photons whose orientations are not parallel to the grains. Thus the light that emerges from an interstellar cloud is polarized.

gets through to us is the light with the orientation perpendicular to that (Fig. 25.4). The dust has the same effect as the polarizing lenses commonly used in sunglasses.

The only way the interstellar dust grains can produce such an effect is if they are not spherical in shape, but elongated (Fig. 25.5). They must also be aligned in some organized fashion, so that a majority of the grains

FIG. 25.5.   AN INTERSTELLAR GRAIN? No direct information is yet available on the composition and the detailed shapes of interstellar grains. This sketch illustrates a popular model, but for a sobering look at what the grain shapes might really be like, see the photo of an interplanetary grain in chapter 15.

FIG. 25.3.   THE INTERSTELLAR EXTINCTION CURVE. This diagram shows how the extinction (obscuration) of starlight varies with wavelength. The extinction increases steadily toward short wavelengths, and there is a pronounced peak centered near 2,200 Å, in the ultraviolet.

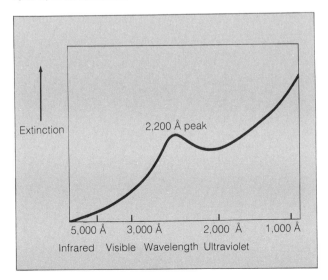

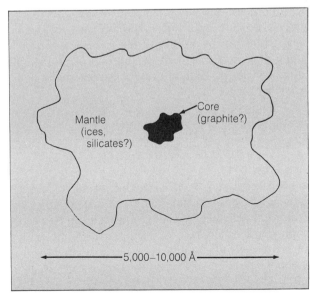

between the earth and a given star are parallel to each other. Their alignment is probably caused by an interstellar magnetic field, which has a constant orientation over large regions of space. Thus the fact that the grains produce polarization tells us something not only about their properties, but also about the galaxy; namely, that it has an overall magnetic field. It appears that the field tends to parallel the spiral arms, at least in the vicinity of the sun.

Little has been learned about the chemical composition of the grains. Theoretical work indicates that a variety of common substances, including silicates, oxides, graphite, and iron-bearing minerals, could cause the observed extinction. There is a pronounced peak of extinction at the ultraviolet wavelength of 2,200 Å, which is best explained as being a result of absorption by graphite (a form of carbon), so it is probable that some of the grains contain this substance.

The origin of the interstellar grains is not well understood; there seem to be several possible processes, all of which may contribute to their existence. Tiny solid particles can form from a gas through condensation, if the proper combination of pressure and temperature is reached. As we learned in chapter 17, this probably occurred early in the formation of the solar system, and therefore we expect that some interstellar grains are born in the vicinity of newly formed stars, then expelled into space, perhaps by a stellar wind. It also appears likely that some grains form by condensation in the outer layers of cool supergiant stars, where the proper conditions apparently exist. In addition, it is clear that grains can form in the expanding material in nova and supernova explosions, and in matter expanding outward in planetary nebulae. The formation of interstellar grains is a by-product of the formation and evolution of stars, and is therefore an important part of the large-scale recycling of matter between stellar and interstellar forms.

Observation of Interstellar Gas

Interstellar gas can be detected by several different techniques. As in the case of dust, one manifestation is in the form of rather obvious clouds that can be seen on photographs of the sky. Rather than being dark clouds, however, the visible gas regions are bright concentrations of material that is hot enough to glow. Interstellar gas is also observed in the radio portion of the spectrum, as we learned in the previous chapter when discussing the 21-cm line of hydrogen. As we will see, there are other types of radio emission lines from the interstellar gas as well.

Perhaps the best method for measuring the properties of the interstellar gas, however, is to observe the absorption lines this gas forms in the spectra of stars (Fig. 25.6). As light from a distant star travels through space on its way to the earth, it encounters atoms, ions, and molecules along the way, each of which can absorb light at specific wavelengths. The result is that the spectrum of the star has extra absorption lines in it, in addition to the ones created in its own atmosphere. It is possible to distinguish the interstellar from the stellar absorption lines by several criteria: the interstellar lines are usually much narrower than those formed in the star's atmosphere; they may occur at a different Doppler shift; and they may represent a different degree of ionization than occurs in the gas in the stellar atmosphere. The first interstellar absorption lines were discovered accidentally just after the turn of the century, but it was not until the 1920's that their interstellar origin was firmly established.

Because of the low density of the interstellar gas, collisions between particles are very rare, and as a result, nearly all the ions and atoms have their electrons in the lowest possible energy level (Fig. 25.7). For most elements, the absorption lines that can be formed when the electrons are in this lowest state lie at ultraviolet

FIG. 25.6. INTERSTELLAR ABSORPTION LINES. This sketch shows the contrast in width between absorption lines formed in a star's atmosphere and those formed in interstellar gas.

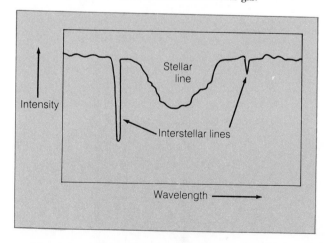

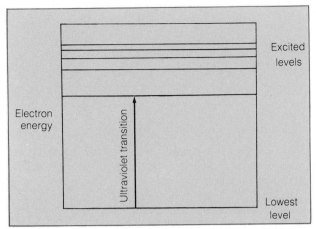

FIG. 25.7. THE FORMATION OF INTERSTELLAR ABSORPTION LINES. This energy-level diagram illustrates why most elements have their interstellar absorption lines at ultraviolet wavelengths. In the rarefied interstellar gas, collisions between atoms are rare, and the electrons nearly always stay in their lowest possible energy levels. For most elements, a relatively large amount of energy is needed to cause a transition from the lowest level to a higher one; therefore, only ultraviolet photons can be absorbed, because only they have sufficient energy.

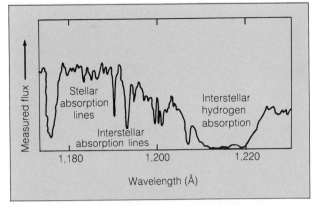

FIG. 25.8. ULTRAVIOLET INTERSTELLAR ABSORPTION LINES. This is a portion of a spectrum obtained with the *Copernicus* satellite. Several stellar lines and a few interstellar lines are seen here. The interstellar hydrogen feature is so wide because of the great abundance of this element in space.

wavelengths; only a few relatively rare elements, such as calcium and sodium, have interstellar absorption lines in visible wavelengths. Therefore it was not until the 1970's, when ultraviolet telescopes were launched on satellites, that the most common species could be directly observed in space (Fig. 25.8).

The chemical composition of the interstellar medium is nearly identical to that of the sun and other stars, which is no surprise, since the stars form from this material. As we will see in the next section, however, some elements tend to be in the form of grains rather than gas, so the interstellar gas itself has a somewhat different distribution of the elements than does the sun.

CLOUDS AND NEBULAE

The interstellar material is not uniformly distributed, but tends to be patchy, with most of the matter contained in relatively dense concentrations called **interstellar clouds.** In the line of sight to a faraway star, there are likely to be several clouds, their presence being revealed to us by the extra redness they impart to the color of the star and by the interstellar absorp-

tion lines they form in its spectrum. These most common of interstellar clouds are not usually hot enough to glow nor thick enough to show up as dark patches in the sky like the dense clouds found in some regions, and for this reason they are called **diffuse clouds.**

In comparing the properties of different types of interstellar clouds, we find it convenient to think in terms of the density; the number of particles per cubic centimeter. At sea level on the earth, the atmospheric density is about 2×10^{19} molecules/cm^3. In a typical diffuse interstellar cloud, the number is more like 10 to 100/cm^3, a factor of 10^{17} or 10^{18} less! The densest clouds, the dark ones that allow no light to pass through, have densities of about 10^6 particles/cm^3 at most, still a factor of 10^{13} less than the density of the earth's atmosphere. The temperature of the gas in a diffuse cloud is typically in the range of 50 to 100 K, and is as low as 10 to 20 K in dark clouds.

Diffuse clouds are thought to be ordinarily less than a parsec thick, although in many cases they may stretch for may parsecs in one dimension, as though they were thin sheets of matter, rather than being more or less spherical. The total mass of a diffuse cloud is difficult to assess because of its unknown extent, but is often comparable to the mass of the sun. Thus we conclude that these clouds are not the type from which massive stars or star clusters are formed.

The composition of diffuse clouds is dominated by hydrogen, with other elements present in much smaller

FIG. 25.9. AN EMISSION NEBULA. This is the North American Nebula, so named because of its distinctive shape.

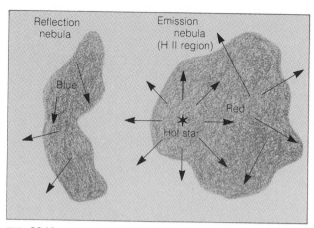

FIG. 25.10. BRIGHT NEBULAE. This shows how reflection and emission nebulae are formed. Light from a hot star is reflected off of a background cloud containing dust. The cloud has a bluish color, because short-wavelength photons are scattered most effectively. The stellar radiation also ionizes gas that is very close to the star, creating a region around it that glows at the wavelengths of certain emission lines. This type of emission nebula is also known as an H II region.

quantities. The hydrogen is primarily in atomic form, except in the central portions of the clouds, where the density may be sufficient to allow the formation of molecular hydrogen (H_2). The abundances of some of the elements are lower than the accepted "cosmic" values. It is as though some of the elements, among them iron, calcium, titanium, and other metallic species, are missing from the interstellar gas. The likely explanation is that these materials are preferentially in the solid dust grains, which gives us some indirect indication of what the grains are made of.

As we have pointed out, diffuse clouds are essentially transparent, but there are conditions under which they show up in photographs. If a hot star is embedded within a cloud, its radiation ionizes and heats the gas to the point where it glows. Such a cloud is known as an **emission nebula** (Fig. 25.9 and Color Plates 16–19), nebula being a general word for any interstellar cloud that is dense or bright enough to show up in photographs. In the case of a hot star embedded within a diffuse cloud, the emission occurs when atoms are ionized and then combine again with electrons to form new atoms (Fig. 25.10). The electron usually starts out in a high energy state, and then quickly drops down to

the lowest level, emitting a photon of light each time it makes a downward jump. Because most of the light from an emission nebula is from hydrogen, these nebulae are called **H II regions,** the *H* standing for hydrogen, and the Roman numeral *II* indicating the ionized state (astronomers generally use this type of notation to indicate the degree of ionization of a gas). The strongest line of hydrogen has a wavelength of 6,563 Å, in the red portion of the spectrum, so H II regions stand out especially well in photographs taken with red-sensitive film. A particularly prominent H II region surrounds the stars of the Trapezium, in the sword of Orion, and is commonly called the Orion Nebula (Color Plate 17).

If a diffuse cloud happens to lie just behind a hot star, it may be visible to us as a **reflection nebula** (Fig. 25.11). In this case it is the dust, rather than the gas, that causes the cloud to glow. What we see is the light that has been scattered in our direction by dust grains (Fig. 25.10). This light is always blue, for the same reason the daytime sky is blue: the tiny particles scatter blue light more effectively than longer wavelengths. The best-known reflection nebula is the one in the young star cluster called the Pleiades, and it lends photographs of this group of stars an ethereal quality.

FIG. 25.11. A REFLECTION NEBULA (NGC 7129).

FIG. 25.12. A MOLECULAR CLOUD. The dark regions seen here near the bright star omicron Persei (left center) emit a rich spectrum of radio lines produced by interstellar molecules.

Dark Clouds and the Molecular Zoo

In many ways the most fascinating of the interstellar objects are the dark clouds, whose densities and dimensions are sufficient to block out all the light of stars behind them. Dark-cloud regions are highly concentrated in the plane of the Milky Way, and have a variety of amorphous sizes and shapes (Fig. 25.12). Often bright H II regions are located within and among them, for the regions where dark clouds are common are also the sites of star formation, and therefore many young, hot stars are mixed in.

The masses of dark clouds are comparable to, or, in the cases of large cloud complexes, far in excess of a single star mass, ranging up to hundreds of thousands of solar masses in some instances. Thus it is quite possible for massive stars and star clusters to form from material in this type of cloud, and it is in dark-cloud regions that the most active star formation occurs.

As we have already noted, the temperature in the interior of a dark cloud is extremely low, perhaps only

10 to 20 K, and the density is in the range of 10^4 to 10^6 particles/cm^3. Under these conditions, atoms can combine into molecular form, and a wide variety of species have been detected. The discovery of the molecular nature of the gas in dark clouds was relatively recent, the first species (the simple molecule OH) having been detected in 1963. The reason these observations were not made sooner is that sophisticated new radio astronomy techniques had to be developed.

Because these clouds are so dense, it is not possible to observe them in the same manner as the diffuse clouds. No stars can be seen through the dark clouds, so there is no way to measure absorption lines formed there. Molecules have special properties, however, that produce radio emission lines, and these lines can be observed.

A molecule is a combination of atoms in a single particle, held together by chemical bonds. These bonds may be viewed as a merging of energy states, so that some of the electrons orbit the entire combination of atoms, acting as the glue that holds them together. The electrons still have fixed energy levels, and there is more complex structure as well. The atoms act as though they are held together with springs, in the sense that they can vibrate, and different frequencies of vibration correspond to different energy states. A molecule can also rotate on its axis, and different rotation speeds also correspond to different energy levels. The vibration and

ASTRONOMICAL INSIGHT (25.1)

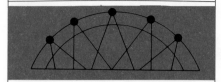

The Confusion Limit

A very interesting yet confounding problem has been recognized recently by researchers involved with radio observations of interstellar molecules. As receiver technology improves and it becomes possible to detect ever weaker radio emission lines, more and more of these lines are found. The reason is that complex molecules tend to have very complicated energy-level structure, so that rather than having just a few strong radio emission lines, they have hundreds or thousands of very weak ones. As a result, the entire portion of the radio spectrum where molecular lines occur is crowded with weak features, and it has already begun to happen that lines from different molecules coincide so closely that they cannot be distinguished from one another.

This sort of coincidence will occur more and more often as the technology continues to improve. Eventually, as extremely weak lines become detectable, the point may be reached where the spectrum is so crowded that the characteristic patterns of individual species will be lost in the bewildering array of overlapping lines, and it will be impossible to identify any new molecules. Hence the true limits on the size and complexity of interstellar molecules may never be known, because for practical reasons we will not be able to identify them, no matter how sensitive the radio receivers.

This potential limit on our ability to detect the largest molecules is fundamental and insurmountable, having to do with the physical nature of the universe out there, rather than our abilities to develop technology here. The term *confusion limit* has been applied to this situation, and there are other examples (see chapter 30).

rotation states are close together, so the wavelengths of light corresponding to transitions between them are long. In most cases the rotational levels are so closely spaced that the differences between them correspond to radio wavelengths. Thus if a molecule jumps from one rotational state to a lower one (that is, if it slows its rotation), it emits a photon at a radio wavelength. Each kind of molecule has its own characteristic set of rotational energy levels, so each has its own particular set of radio emission lines. Molecules have the same sort of fingerprints in the radio spectrum that atoms and ions do in visible and ultraviolet light. (Molecules, having such complex structure, also have spectral lines at visible and ultraviolet wavelengths, but these cannot be easily observed in dark clouds for the reasons already stated.)

In a dark cloud, the density is sufficiently high that molecules collide every now and then, and when they do they can be excited to high rotational energy levels. Once a molecule has been kicked into a rapid rotation state, it will soon slow (in a single, instantaneous jump) to a lower state, emitting a radio photon in the process. Thus a dark cloud is constantly releasing radio emission at a variety of fixed wavelengths corresponding to the types of molecules inside. As we learned in discussing the 21-cm line, radio lines are not affected by interstellar extinction, so the emission from a dark cloud escapes easily into space.

Most of the molecules that exist in dark clouds emit at wavelengths of a few millimeters or centimeters. Since the emission from a distant cloud is greatly attenuated (because of the inverse-square law) by the time it reaches the earth, very sensitive receivers are required to detect it, and the technology for doing this has only been developed in the last two decades. Every time a new advance is made in receiver sophistication, new molecules are immdeiately detected.

The number of molecules discovered in space has grown to more than seventy (see Appendix 11). Most are rather simple species, the most common being diatomic molecules, which consist of only two atoms. Of these, the dominant one is molecular hydrogen which, for a subtle reason, does not emit radio waves, and has been detected in dark clouds only in active regions where it is excited to emit infrared radiation. The next most common molecule is carbon monoxide (CO), and following this the abundant species include OH (called the hydroxyl radical), water vapor (H_2O), and an assortment of other simple combinations of abundant elements.

Some rather complex large molecules have begun to be detected as well, the current grand champion being $HC_{11}N$, a species consisting of a total of thirteen atoms, eleven of them carbon. There may be even larger molecules in space, most likely other species containing long chains of carbon atoms, which form easily under the conditions that prevail inside dark clouds.

All the molecules detected so far are made up of combinations of just a few of the most common elements, primarily hydrogen, carbon, nitrogen, oxygen, and sulfur. There probably are molecules containing other elements, but because of their lower abundances, these species are difficult to detect. The second most common element after hydrogen is helium, but this is an inert element, meaning that it cannot easily combine with other atoms to form molecules. Helium is certainly present in interstellar clouds, but only in atomic form.

Perhaps the most interesting of all the species detected in dark clouds are the **organic molecules,** those containing certain combinations of carbon atoms that are also found in living material. Their presence shows that when stars form in dark clouds, the ingredients for life are already present. We have seen (in chapter 15) that amino acids have been found in meteorites, indicating that these molecules were present before the planets formed. We can speculate that perhaps these relatively complex organic molecules formed even before the first solid matter in the solar system.

Interstellar Violence and the Role of Supernovae

Now that we have examined the principal components of the interstellar medium, let us take another look at its overall properties. We have found a diverse collection of ingredients, which can be characterized by a wide variety of temperatures and densities.

When the first spectroscopic observations of the interstellar gas were made more than fifty years ago, it was noticed that some diffuse clouds are moving through space at rather large velocities, up to 100 kilometers per second or even more (Fig. 25.13). Since

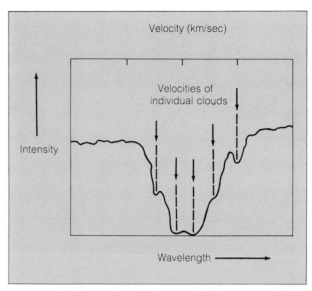

FIG. 25.13. THE EFFECT OF MOTIONS IN THE INTERSTELLAR MEDIUM. Interstellar absorption lines often show structure such as this, caused by Doppler shifts of clouds moving at different velocities.

then it has been found that the motions are often directed away from groups of bright stars, as though the gas were being expelled by some force originating in the stars.

When 21-cm observations of hydrogen gas became possible, similar motions were found. A variety of rapidly moving clouds was discovered, and in some regions there appeared to be expanding loops or shells of gas (Fig. 25.14 and 25.15). Some of these regions were associated with known supernova remnants (Fig. 25.16 and Color Plates 19 and 20), where it was clear that the gas was still expanding from the original explosion. In other cases, however, nothing was found at the center of expansion except perhaps a group of young stars. Whatever the cause, the picture developed by these observations was one of an interstellar medium in turmoil, with random, high-speed motions throughout.

In the 1970's new evidence of energetic phenomena in space was uncovered, in two forms. Data obtained from X-ray satellite and rocket experiments showed that the interstellar medium in the plane of the galaxy emits X rays. At about the same time, ultraviolet spectroscopic observations revealed previously undetected ions in space that form only at very high temperatures. The combination of these X-ray and ultraviolet data led to

FIG. 25.14. BARNARD'S LOOP. This photo of Orion shows a tenuous loop of glowing gas encircling the region of the Belt and Sword, where interstellar clouds and young stars are found. Evidently the loop, named for its discoverer (E. E. Barnard), is composed of gas ejected long ago by the stars in the nebulosity near the center.

the conclusion that the space between cool clouds is filled with a very hot gas, whose temperature is a million degrees or more. The density of the hot gas between the clouds is very low even by interstellar standards; it is 10^{-4} to 10^{-3} particle/cm^3, the equivalent of one particle in every cubic volume 10 to 20 centimeters on a side!

We now see that the interstellar medium not only is diversified, but also is being constantly disturbed, which creates the observed high-speed motions and high temperatures. All this activity requires a substantial amount of energy input; a quiet pond does not spontaneously form ripples and eddies unless something stirs it up.

The agitation of the interstellar medium, it is now understood, is primarily the result of supernova explosions. These occur in the galaxy at a frequency of one every few decades, and each time, a vast amount of

FIG. 25.15. A 21-CM PICTURE OF THE INTERSTELLAR MEDIUM. This cross-section of the disk of the Milky Way shows many looplike structures of hydrogen gas, probably created by old supernova explosions.

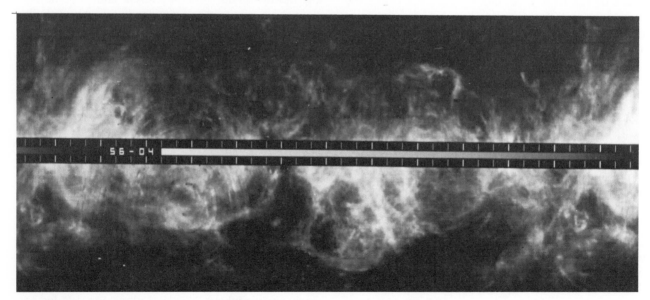

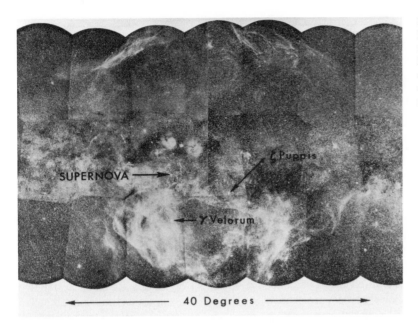

FIG. 25.16. AN EXTENSIVE SUPERNOVA REMNANT. Here we see, spread over a large region of the sky, a large supernova remnant known as the Gum Nebula. Filamentary structures are obvious, and measurements of interstellar absorption lines show the presence of rapidly moving clouds.

energy is released into space. Material expands outward from the site of the explosion, sweeping up a shell of gas that continues to expand, and creating a cavity of very rarefied, extremely hot gas inside the shell (Fig. 25.17). As we have seen, interstellar clouds can be compressed by the outburst, triggering star formation, and clouds can be accelerated to high velocities as well. The net effect of repeated supernova explosions in the galaxy is that a large fraction of its volume has become filled with the hot gas that is left behind by the expanding shells (Fig. 25.18), and much of the cloud material has been accelerated, so that high-velocity clouds are seen here and there, particularly in loops or fragments of spherical shells.

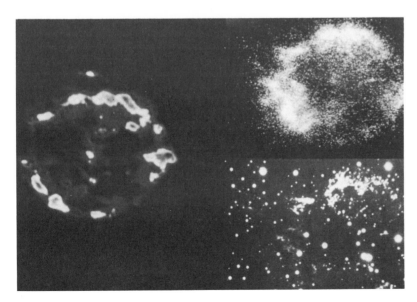

FIG. 25.17. X-RAY, OPTICAL, AND RADIO IMAGES OF A SUPERNOVA REMNANT. At upper right is an X-ray image of Cas A, a supernova remnant. Below that is an optical photo, and at left is a radio map. The X-ray and radio images are similar, because both kinds of radiation are produced by the synchrotron process. The visible light seen here appears different, as it is in the form of emission lines.

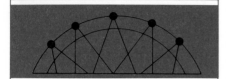

The Mysterious Cosmic Rays

It has been known for a long time that the earth is constantly being bombarded by rapidly moving sub-atomic particles from space, which have been given the name **cosmic rays.** Consisting of electrons or atomic nuclei, these particles can be detected in several ways, the most common of which is to trace their tracks in special liquid-filled containers called bubble chambers. The cosmic-ray particles are electrically charged, and their passage through such a container creates a trail of fine bubbles. From the length and shape of this track, we can infer the charge and speed of the particles. Cosmic-ray velocities are usually near the speed of light.

The earth's magnetic field deflects most cosmic rays, particularly the less massive and energetic ones, so that they reach the ground most easily near the poles. From the point of view of life on earth, this is probably a good thing, for cosmic rays are suspected of causing genetic mutations, and could be harmful in other ways if allowed to reach the surface unimpeded. (On the other hand, mutations caused by cosmic rays are thought to have assisted evolutionary processes in the development of life on earth.)

Extensive observations of cosmic rays show that most are either electrons or nuclei of atoms, with all of the electrons stripped free. Most of the latter are single protons, the nuclei of hydrogen atoms, but some are nuclei of heavy elements, including metallic species known to be formed only in supernova explosions. The majority of the cosmic rays that reach the earth originate in the sun, but many, particularly the more massive and energetic ones, come from all directions in the plane of the galaxy, and seem to fill its volume more or less uniformly. Apparently they are trapped by the

galactic magnetic field and cannot escape into intergalactic space. Because their paths are curved and deflected by the magnetic field, it is impossible to tell where they come from. Therefore the source of the galactic cosmic rays is not understood, but a strong possibility is that they are released in supernova explosions. The total energy contained in cosmic rays darting about the galaxy is very large, and only supernovae seem capable of producing the necessary energy.

The distribution of cosmic-ray nuclei of various elements is generally similar to the chemical abundances found in stars and the interstellar material, although there are some differences, most of which can be attributed to nuclear reactions that can occur when the rapidly moving nuclei strike interstellar dust grains. Such reactions are responsible for the production of certain elements in space, such as the lightweight species, beryllium.

Cosmic rays have little effect on our everyday lives, but they have the potential of telling us quite a bit about the energetic activities that take place in interstellar space.

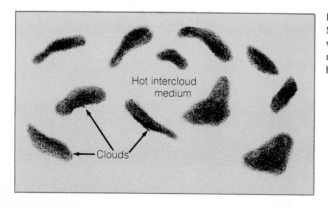

FIG. 25.18. A MODERN VIEW OF THE INTERSTELLAR MEDIUM. Recent X-ray and ultraviolet data have led to a picture in which the interstellar clouds are embedded in a very hot (1 to 2 million degree) intercloud gas. The hot gas is created and maintained by expanding supernovae and stellar-wind bubbles.

Another mechanism that has the same effect, and which contributes to this picture, is the expansion of stellar winds. We have learned in chapter 20 that hot, blue stars as a rule have high-velocity winds; these are streams of gas flowing outward at speeds as great as 3,000 kilometers per second. Such a wind will act in

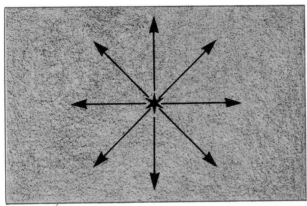 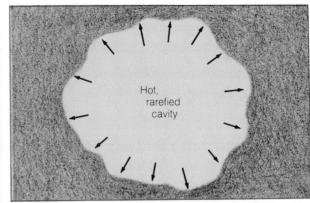

FIG. 25.19. FORMATION OF AN INTERSTELLAR BUBBLE. A supernova explosion or a strong stellar wind occurs in the midst of a cloudy interstellar medium (left). As the resultant shock waves expand, a spherical cavity is formed, with a concentration of swept-up gas surrounding it. The cavity, or bubble, is filled with very hot, rarefied gas. In real situations, the medium is probably not so uniform initially, and the resulting cavity is not so symmetrically shaped as shown here.

some ways like a supernova explosion, sweeping up material around the star, and creating an expanding shell (Figs. 25.19 and 25.20). Although a supernova releases far more energy in a brief moment, a stellar wind may last several million years, eventually injecting a

comparable amount of energy into the surrounding gas. In a cluster of young stars, the winds can combine to evacuate a large bubble around the entire cluster, closely simulating the effect of a supernova explosion.

In one recently discovered case, a gigantic shell was found surrounding a vast volume in the constellation Cygnus (Fig. 25.21); the total energy required to create

FIG. 25.20. CIRCUMSTELLAR BUBBLES. Two Wolf-Rayet stars are shown here, each surrounded by a roughly spherical shell of glowing gas created by the strong winds from the stars. This is a negative print, so stars and glowing gas appear dark against a light background.

FIG. 25.21. THE SUPERBUBBLE IN CYGNUS. This is an X-ray map of a large portion of the constellation Cygnus, showing a gigantic loop of gas that is so hot it emits X rays. This structure has been interpreted as a huge bubble created by numerous supernovae that occurred in the same general location. The superbubble is roughly 600 by 450 parsecs in size.

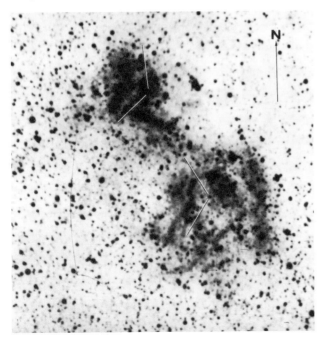 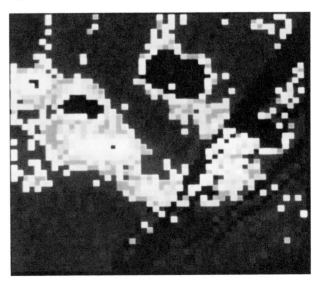

that structure is estimated to be comparable to that of several hundreds or even thousands of supernovae. It is not clear how this "superbubble" formed. Possibly a very large star cluster at its center experienced a number of supernovae whose expanding shells combined to form the single huge structure seen today.

The hot, tenuous interstellar gas is known to extend well above and below the plane of the galactic disk. Perhaps the galaxy as a whole has a wind emanating outward, a composite of the expansion of all the supernova explosions and stellar winds that have occurred within it. In any case, we can no longer think of the interstellar void as a cold, desolate place where nothing ever happens.

Perspective

If this chapter had been written ten years ago, it would have had a much different flavor. The interstellar medium would have been viewed as a quiescent background against which all the activities of stellar evolution are played out. Instead, we now know that the material between the stars plays a dynamic role in the affairs of the galaxy, with the stars providing the necessary energy as they live and die, and the interstellar material in turn creating the conditions necessary for the births of new generations. The rich interplay between stars and interstellar matter is the dominant theme of galactic ecology, and its role will be brought to the fore again in the next chapter, as we discuss the formation and evolution of the Milky Way.

Summary

1. The interstellar medium, although extremely tenuous, contains about 10 percent of the mass of the galactic disk.

2. Interstellar grains cause extinction and reddening of starlight, because they scatter and absorb photons, and do so most efficiently at short wavelengths.

3. Analysis of the interstellar extinction created by dust grains leads to the conclusion that many of the grains are about 5,000 Å in diameter, and that there is a second group, responsible for the strong extinction of ultraviolet wavelengths, that are much smaller in size, with diameters of about 100 Å.

4. The grains also cause polarization of starlight, implying that they are elongated in shape and that they are aligned by a galactic magnetic field.

5. The gas in space was first detected through absorption lines in the spectra of distant stars. Most of these lines lie at ultraviolet wavelengths.

6. The interstellar medium is patchy, containing cold clouds embedded in a hot, more rarefield, intercloud medium. Most of the mass in space is in the clouds, and most of the volume is filled with the hot gas.

7. Interstellar clouds may have a variety of densities and temperatures. The so-called diffuse clouds have densities of about 1 to 100 particles/cm^3 and temperatures of 50 to 100 K. We can observe these clouds in absorption-line measurements. Also, if near a hot star, they may appear as emission or reflection nebulae.

8. The dark clouds, with temperatures of 10 to 20 K and densities up to 10^6 particles/cm^3, consist of gas that is primarily molecular in form. The most common molecules are hydrogen (H_2) and carbon monoxide (CO). A large number of different species have been detected, primarily through their radio emission lines.

9. The interstellar medium displays various forms of energetic activity: there are random, often rapid, cloud motions; the gas between clouds is extremely hot; and here and there are gigantic loop and ring structures suggestive of expanding shells. All this activity can be attributed to energy injected into the interstellar medium by supernova explosions and stellar winds.

Review Questions

1. Explain how absorption lines formed in the interstellar medium can be distinguished from those formed in a star's atmosphere.

2. Explain why ultraviolet telescopes are required to observe the interstellar absorption lines of most elements. Why does the same argument not apply to absorption lines formed in stellar atmospheres?

3. How is it that most of the mass in interstellar space is in clouds, but the clouds fill only a small fraction of the volume?

4. Suppose a BO star is observed to have a *B-V* color index of $B\text{-}V = 0.20$. From measurements of nearby stars, it is known that the normal *B-V* color index for a star of this type is $B\text{-}V = -0.30$. What is the color excess for this star?

5. How can astronomers distinguish between a cool star that is intrinsically red in color and a hotter star that looks just as red because of interstellar dust?

6. Compare the composition of interstellar grains with that of the terrestrial planets, and discuss the similarities in the formation processes for both (see chapter 17).

7. Explain why it is primarily the diffuse clouds, and not the intercloud gas or the dark clouds, that produce the 21-cm emission with which the galaxy is mapped (hint: you will have to refer to information on the formation of the 21-cm line that is in chapter 24).

8. Why are the dark clouds not observed through interstellar absorption lines?

9. The overall chemical composition of the interstellar medium is thought to be the same everywhere, yet the material is observed in quite different forms in diffuse clouds, dark clouds, and in the hot area between clouds. Summarize the distribution of the elements in each of these components of the interstellar medium.

10. Explain why most of the evidence for violent, energetic activity in the interstellar medium was not discovered until recently.

Additional Readings

Buhl, D. 1972. Light molecules and dark clouds. *Mercury* 1(5):4.

Chaisson, E. J. 1978. Gaseous nebulas. *Scientific American* 239(6):164.

Chevalier, R. A. 1978. Supernova remnants. *American Scientist* 66:712.

Heiles, C. 1978. The structure of the interstellar medium. *Scientific American* 238(1):74.

Herbig, G. H. 1974. Interstellar smog. *American Scientist* 62:200.

Herbst, E., and Klemperer, W. 1976. The formation of interstellar molecules. *Physics Today* 29(6):32.

McCray, R. A., and Snow, T. P. 1979. The violent interstellar medium. *Annual Review of Astronomy and Astrophysics* 17:213.

Miller, J. S. 1974. The structure of emission nebulas. *Scientific American* 231(4):34.

CHAPTER 26

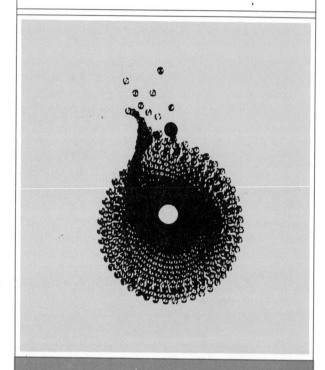

The Formation and Evolution of the Galaxy

The galaxy we observe today is the end product of some 10 to 20 billion years of evolution. A variety of processes have shaped it, and many have left unmistakable evidence of their action. In this chapter we discuss the major influences on the galaxy, along with the developments they have brought about. We have laid the groundwork that will enable us to understand one of the most elegant concepts in the study of astronomy: the interplay between stars and interstellar matter and the long-term effects of this cosmic recycling on the evolution of the galaxy.

Stellar Populations and Elemental Gradients

To begin this story, we focus on the overall structure of the galaxy (Fig. 26.1), and the types of stars found in various regions. We have already learned that most of the young stars in the galaxy tend to lie along the spiral arms; indeed, it is the brilliance of the hot, massive O and B stars and their associated H II regions that delineates the arms, making them stand out from the rest of the galactic disk. We have also noted that in the nuclear bulge of the galaxy, as well as in the globular clusters and the halo in general, there are few young stars and relatively little interstellar material.

Careful scrutiny has revealed a number of distinctions

between the stars that lie in the disk, particularly those in spiral arms, and the stars in the nucleus and halo. Analysis of chemical compositions has shown that in the halo and nucleus, stars tend to have relatively low abundances of heavy elements such as metals, whereas in the vicinity of the sun (and in spiral arms in general), stars have such elements in greater quantity. Nearly all stars are dominated by hydrogen, of course, but in the halo stars the heavy elements represent an even smaller trace than in the spiral-arm stars. The relative abundance of heavy elements in a halo star may typically be a factor of 100 below that found in the sun. If iron, for example, is 10^{-5} as abundant as hydrogen in the sun, it may be only 10^{-7} as abundant in a halo star.

A picture has emerged in which the galaxy has two distinct groups of stars, those in the halo and nuclear bulge, and those in the spiral arms. They have been designated **Population II** (Pop. II) for the stars in the halo, and **Population I** (Pop. I) for those in the spiral arms (Fig. 26.2). The sun is thought to be typical of Pop. I, and its chemical composition is usually adopted as representative of the entire group.

Almost every time an attempt is made to classify astronomical objects into distinct groups, the boundaries between the objects turn out to be a little indistinct. There are usually intermediate objects whose properties fall between categories, and the stellar populations are no exception. For example, the stars in the disk of the galaxy that do not fall strictly within the spiral arms have intermediate properties between those of Pop. I and Pop. II, and are often called simply the **disk population.** Furthermore, there are gradations within the two principal groups, and we therefore speak of extreme or intermediate Pop. I or Pop. II objects. It is much more accurate to view the stellar populations as a smooth sequence of properties represented by very low heavy-element abundances at one end, and sunlike abundances at the other.

The differences between chemical compositions from one part of the galaxy to another provide astronomers with very important clues in their efforts to piece together the history of the galaxy. Differences that occur gradually over substantial distances are referred to as **gradients,** and the variations of stellar composition from one part of the galaxy to another are called **abundance gradients** (Fig. 26.3). There is a gradient of increasing heavy-element abundances from the halo to the disk. Within the disk itself, there is a similar gradient from the outer to the inner portions (this does

FIG. 26.1. A SPIRAL GALAXY SIMILAR TO THE MILKY WAY. This vast conglomeration of stars and interstellar matter is following its own evolutionary course, just as individual stars do.

FIG. 26.2. THE DISTRIBUTION OF POPULATIONS I AND II. This cross-section of the galaxy shows that Pop. II stars lie in the halo and central bulge, whereas Pop. I stars inhabit the spiral arms in the plane of the disk. Intermediate population stars are distributed throughout the disk.

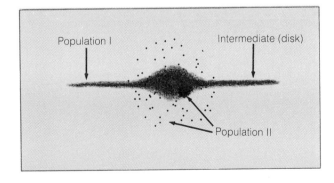

not include the stars in the nuclear bulge, which, as we pointed out, tend to have low abundances of heavy elements).

Most of the stars near the sun belong to Pop. I; they

FIG. 26.3. AN ABUNDANCE GRADIENT. This diagram shows how the abundance of a heavy element (relative to hydrogen) varies with distance from the center of the galaxy. More stellar generations have lived and died in the dense regions near the center than farther out, so nuclear processing is more advanced in the central region.

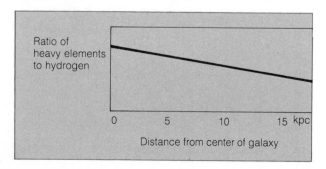

are disk stars orbiting the galactic center in approximately circular paths, like the sun. There are a few nearby Pop. II stars here and there, and they are distinguished by a number of properties in addition to their compositions. The most easily recognized is that their motions depart drastically from those of Pop. I stars. As a rule, Pop. II stars do not follow circular paths in the plane of the disk, but instead follow highly elliptical orbits that are randomly oriented, much like cometary orbits in the solar system. Thus, most of the Pop. II stars that are seen near the sun are just passing through the disk from above or below it. The sun and the other Pop. I stars in its vicinity move in their orbits at speeds of around 250 kilometers per second, so the sun moves very rapidly with respect to the Pop. II stars that pass through its neightborhood in a perpendicular direction. We classify these Pop. II objects as **high-velocity stars** (Fig. 26.4).

These stars were discovered and recognized as a distinct class even before the differences between Pop. I and Pop. II were enumerated. It was only later that

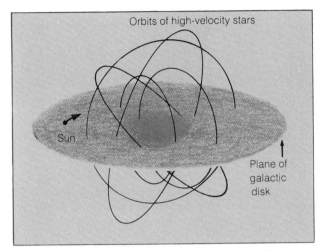

FIG. 26.4.   THE ORBITS OF HIGH-VELOCITY STARS. These stars follow orbits that intersect the plane of the galaxy. When such a star passes near the sun, it has a high velocity relative to us, because of the sun's rapid motion along its own orbit.

they were equated with Pop. II stars following randomly oriented orbits in the galactic halo.

The systematic differences between the stellar populations arose from differing conditions at the times when they formed. In this regard it is important to recall that globular clusters (representative of Pop. II), are among the oldest objects in the galaxy; Pop. I, on the other hand, includes young, newly formed stars. We will follow this discussion further in a later section.

Stellar Cycles and Chemical Enrichment

The fact that stars in different parts of the Milky Way have different chemical makeups is explained in terms of stellar evolution. Recall (from discussions in chapter 20) that as a star lives its life, nuclear reactions gradually convert light elements into heavier ones. The first step occurs while the star is on the main sequence, during which hydrogen nuclei in its deep interior are fused into helium. Later stages, depending on the mass of the star, may include the fusion of helium into carbon, and possibly the formation of even heavier elements by the addition of more helium nuclei. The most massive stars, which undergo the greatest number of reaction stages, also form heavy elements in the fiery instants of their deaths in supernova explosions.

All these processes work in the same direction: they act together to gradually enrich the heavy-element abundances in the galaxy. Material is cycled back and forth between stars and the interstellar medium, and with each passing generation a greater supply of heavy elements is available. As a result, stars formed where there has been a lot of stellar cycling in the past are born with higher quantities of heavy elements than those formed where little previous cycling has occurred. Stars that formed before much cycling occurred therefore tend to have low abundances of heavy elements. This explains why Pop. II stars are very old; they formed out of material that had not yet been chemically enriched.

The Pop. I stars, such as the sun, are those which condensed from interstellar material that had previously been processed in stellar interiors and supernova explosions. Therefore, as a rule, they formed at moderate to recent times in the history of the galaxy, and are not as ancient as Pop. II objects.

Besides elemental-abundance variations with age, there can also be variations with location. As noted in the preceding section, there is a distinct abundance gradient within the disk of the galaxy, the stars nearer the center having higher abundances of heavy elements than those farther out. This is a reflection of enhanced stellar cycling near the center, rather than an age difference. The central portion of the disk is where the density of stars and interstellar material is highest, and in this region there has been relatively active stellar processing, because star formation has proceeded at a greater rate than in the less dense outer portions of the disk. Over the lifetime of the galaxy, more generations of stars have lived and died in the central region than in the rarefied outer reaches.

The Care and Feeding of Spiral Arms

In chapter 24 we described the structure of the galaxy and how its spiral nature was discovered and mapped. But we said nothing about how the spiral arms formed, nor why they persist. Both questions are important, for it is clear that if the arms were simply streamers trailing along behind the galaxy as it rotates, they would wrap themselves up tightly around the nucleus, like string being wound into a ball (Fig. 26.5). Because the galaxy is old enough to have rotated at least forty to

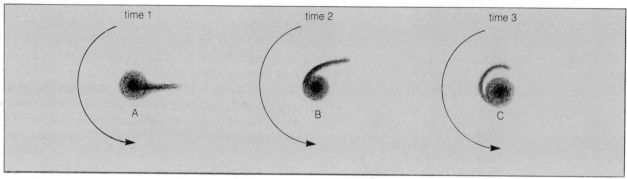

FIG. 26.5.   THE WIND-UP OF SPIRAL ARMS. If spiral arms were simple streamers of material attached to a rotating galaxy, they would, within a few hundred million years, wind up tightly around the nucleus. The galaxy is much older than that, so some other explanation of the arms is needed.

fifty times since it formed, something must be preventing the arms from winding up, or they would have done so long ago.

Maintaining the spiral arms, a distinct question from that of forming them in the first place, is a very complex business, and is not yet fully understood. The first successful theory on this appeared in 1960, and is still being refined. The essence of this hypothesis is that the large-scale organization of the galaxy is imposed on it by wave motions. We understand waves to be oscillatory motions created by disturbances, and we know that waves can be transmitted through a medium over long distances while individual particles in the medium move very little. In the case of water waves, for example, a floating object simply bobs up and down as a wave passes by, whereas the wave itself may travel a great distance. Here we are distinguishing between the wave and the medium through which it moves.

The waves that apparently govern the spiral structure in our galaxy are not transverse waves, like those in water, but are compressional waves, similar to sound waves and certain seismic waves (the P waves; see chapter 7 for an elaboration on the different types of waves). In this case the wave pattern consists of alternating regions of high and low density. When a compressional wave passes through a medium, the individual particles vibrate back and forth along the direction of the wave motion. There is no motion perpendicular to that direction, in contrast with water waves.

The theory of galactic spiral structure that invokes waves as a means of maintaining the spiral arms is called the **density-wave theory.** This hypothesis supposes that there is a spiral-shaped wave pattern centered on the galactic nucleus, creating a pinwheel shape of al-

ternating dense and relatively empty regions (Fig. 26.6). The density waves have more effect on the interstellar medium than on stars, so the spiral arms are characterized primarily by concentrations of gas and dust. These in turn lead to concentrations of young stars, because star formation is enhanced in regions where the interstellar material is compressed.

FIG. 26.6.   THE DENSITY-WAVE THEORY. This depicts the manner in which circular orbits are deformed into slightly elliptical ones by an outside gravitational force. The nested elliptical orbits are aligned in such a way that there are density enhancements in a spiral-shaped pattern. This pattern rotates at a steady rate, and does not wind up more tightly.

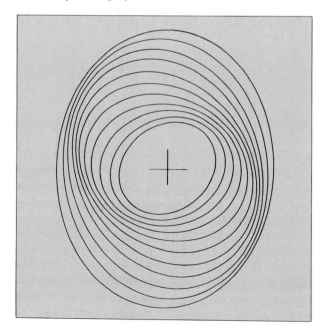

In its simplest form, the wave pattern is double; that is, there are just two spiral arms emanating from the nucleus, on opposite sides. There are galaxies that have such a simple spiral structure, but many, including the Milky Way, are more complicated. The density-wave theory allows the possibility of more arms, if the waves have shorter "wavelength," or distance between them.

The waves rotate about the galaxy at a fixed rate that is constant from inner portions to the outer edge. Thus, although the outer portions of the arms appear to trail the rotation, in fact they move all the way around the galaxy in the same time period that the inner portions do. The waves are essentially rigid, in strong contrast with the motions of the stars. Just as in the case of water waves, the motion of the spiral density waves is quite distinct from the motions of individual particles (that is, stars) in the medium through which the waves travel. In the Milky Way, individual stars each orbit the galaxy at their own speed, which is faster than that of the density waves. This means that as a star circles the galaxy, every so often it will overtake and pass through a region of high density, as it penetrates a spiral arm. Because their motion is slightly slower when they are in a density wave, stars tend to become concentrated in the arms, just as cars on a highway become jammed together at a point where traffic flow is constricted.

We have noted that the main reason the arms stand out is that hot, luminous young stars are found exclusively in the arms because star formation is enhanced there, where the interstellar gas is compressed. According to the density-wave theory, the most active stellar nurseries should be located on the inside edges of the arms, where stars and gas catch up with and enter the compressed region (Fig. 26.7). This is difficult to check observationally for our own galaxy, but seems to be true of others.

The most luminous stars have such short lifetimes that they evolve and die before having sufficient time to pull ahead of the arm where they were born; therefore, we find no bright, blue stars between the arms. Less luminous stars, those which live billions instead of only millions of years, can survive long enough to orbit the galaxy several times, and these stars become spread almost uniformly throughout the galactic disk, between the arms as well as in them. The sun, for example, is old enough to have circled the galaxy some fifteen to twenty times, and therefore has passed through spiral arms and the intervening gaps on several occasions. It is only coincidence that the sun is presently

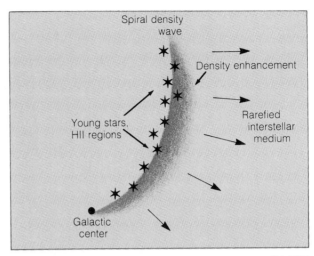

FIG. 26.7. THE EFFECT OF A SPIRAL DENSITY WAVE ON THE INTERSTELLAR MEDIUM. As the wave moves through the medium, material is compressed, leading to enhanced star formation. The young stars and H II regions and nebulosity along the density wave are what we see as a spiral arm.

in a spiral arm; this is not necessarily where it formed.

The initial formation of the density wave that is the cause of the spiral arms is a separate question, one for which there are several possible answers. One idea is that gravitational effects of nearby galaxies can disturb a disk-shaped galaxy in such a way that the wave motion is started, in some manner comparable to dropping a rock into a pond and initiating wave motions (Fig. 26.8). In this case, the factor that creates the density waves would be the differential gravitational force, which tends to stretch and distort the galaxy that is subjected to it. The part of the disk nearest the disturbing galaxy would feel a stronger force than the portions farther away, and as a result, stresses could be placed on the disk that cause oscillatory wave motion. Another possible mechanism for creating spiral density waves, with no help from neighboring galaxies, occurs in certain galaxies that have noncircular disks. In these **barred spiral galaxies** (see chapter 27), the asymmetric shape of the disk can create the gravitational disturbance needed to initiate spiral density waves.

Aside from the problem of explaining how the arms form in the first place, another area of current research related to spiral arms is aimed at understanding the effects of a massive galactic halo. As we learned in chapter 24, evidence is mounting that the halo of our galaxy may contain as much as 90 percent of the total

FIG. 26.8. CREATION OF SPIRAL
STRUCTURE. This sequence of computer-
generated models shows how a rotating disk
forms spiral density waves when subjected to a
gravitational force.

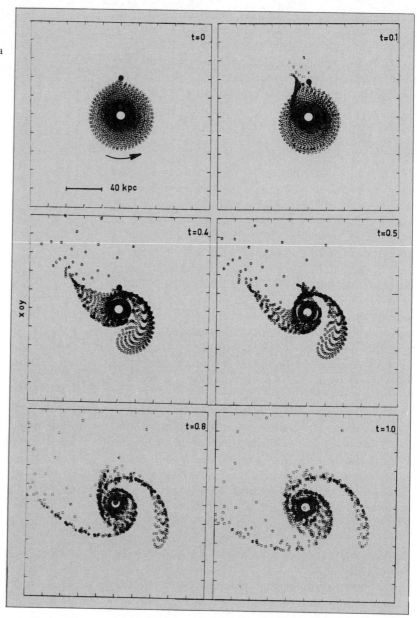

Galactic History

mass; if so, the gravitational effects of all this material would have some influence on the nature of the spiral arms.

We are now in a position to tie together all the diverse information on the nature of the Milky Way, and to develop from that a picture of its formation and evo-lution. The pertinent facts that must be explained in-clude the size and shape of the galaxy, its rotation, the distribution of interstellar material, elemental-abun-dance gradients, and the dichotomy between Pop. I and Pop. II stars in terms of composition, distribution, and motions. The task of fitting all the pieces of the puzzle together is made easier by the fact that we can recon-struct the time sequence, knowing that as a rule stars with high abundances of heavy elements were formed more recently than those with lower abundances.

ASTRONOMICAL INSIGHT (26.2)

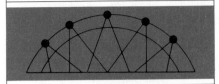

A Different Kind of Spiral Arm

Recently an alternative to the density-wave theory has been proposed to explain the existence and persistence of the spiral arms. In this picture, the differential rotation of the galaxy, along with the great extent of some regions of star formation, play key roles.

There are regions in the galaxy where very large complexes of interstellar gas and dust are found. Here star formation seems to take place by the sequential process described in chapter 21, in which the stellar winds and supernova explosions of massive, short-lived stars give rise to new generations in the near vicinity, as shock waves compress the adjacent clouds. In several cases these hotbeds of star formation are many tens of parsecs in extent. On this size scale, the differential rotation of the galaxy distorts the star-forming region. The part of the vast cloudy region nearer the galactic center is pulled ahead of the outer part, and the entire complex is stretched into a curved segment that resembles a portion of a spiral arm. In addition to its shape, such a region has other characteristics of spiral arms, most notably a concentration of interstellar matter and hot, young stars.

In due course a giant dark cloud is consumed by star formation, the luminous stars in it die out, and the entire region fades into obscurity. A spiral-arm segment created in this way is therefore only a temporary feature of the galaxy. If such segments were continually being formed and dissipated, at any given moment a galaxy would have plenty of spiral structure. The arms would not be permanent, but instead would be constantly coming and going. This is especially likely in the cases of galaxies with rather chaotic spiral structure, consisting of many bits and pieces of arms, rather than a small number of smooth, complete ones in a simple overall pattern. From what we know of the structure of the Milky Way, this process may be occurring in our galaxy.

The oldest objects in the galaxy are in the halo, which is dominated by the globular clusters but which contains a large (but unknown) number of isolated, dim, red stars as well. Estimates based on the main sequence turn-off in the H-R diagrams for globular clusters indicate ages of between 14 and 16 billion years, and we conclude that the age of the galaxy itself is comparable.

The spherical distribution of the halo objects about the galactic center demonstrates that when they formed, the galaxy itself was round (Fig. 26.9). Evidently the progenitor of the Milky Way was a gigantic gas cloud, spherical in shape, consisting almost exclusively of hydrogen and helium. Very early, perhaps even before the cloud began to contract, the first stars and globular clusters formed in regions where localized condensations occurred. These stars contained few heavy elements, and were distributed throughout a spherical volume, with randomly oriented orbits about the galactic center. Eventually the entire cloud began to fall in on itself, and as it did so, star formation continued to occur, so that many stars were born with motions directed toward the galactic center. These stars assumed highly elongated orbits, accounting for the motions observed today in Pop. II stars.

Apparently the pregalactic cloud was originally rotating, for we know that as the collapse proceeded, a disk-like shape resulted. Rotation forced this to happen, just as it did in the case of the contracting cloud that was to form the solar system (see chapter 17). Rotational forces slowed the contraction in the equatorial plane but not in the polar regions, so material continued to fall in there. The result was a highly flattened disk. Stars that had already formed before the disk took shape retained the orbits in which they were born; stars are unaffected by the fluid forces (such as viscosity) that caused the gas to continue collapsing to form a disk.

Stars that were formed after the disk had developed had characteristics unlike those of their predecessors, in at least two respects: they contained greater abundances of heavy elements, because by this time some

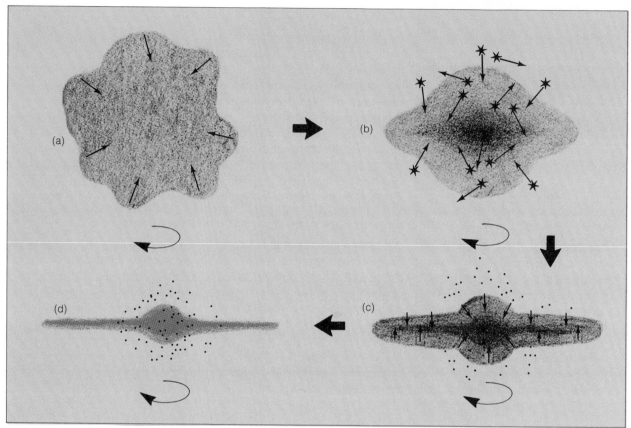

FIG. 26.9. FORMATION OF THE GALAXY. The pregalactic cloud, rotating slowly, begins to collapse (a). The stars formed before and during the collapse have a spherical distribution and noncircular, randomly oriented orbits (b). The collapse leads to a disk with a large central bulge, surrounded by a spherical halo of old stars and clusters (c). The disk flattens further, and eventually forms spiral arms (d). Stars in the disk have relatively high heavy-element abundances, having formed from material that had been through stellar nuclear processing.

stellar cycling had occurred, enriching the interstellar gas; and they were born with circular orbits lying in the plane of the disk. These are the primary traits of Pop. I stars.

Since the time of the formation of the disk, there have been additional, but relatively gradual, changes. Other generations of stars have lived and died, continuing the chemical-enrichment process (particularly in the inner regions of the disk) and creating the chemical-abundance gradient mentioned earlier. Apparently the enrichment process was once more rapid than it is today, because the present rate of star formation is too slow to have built up the quantities of heavy elements that are observed in Pop. I stars. At some time in the past, probably just when the disk was forming, there must have been a period of intense star formation,

during which the abundances of heavy elements in the galaxy jumped from almost zero to nearly the present level. A large fraction of the galactic mass must have been cooked in stellar interiors and returned to space in a brief episode of stellar cycling whose intensity has not been matched since.

We have just about recounted all the events that led up to the present-day Milky Way, except for the formation of the spiral arms. It is not known when this took place, but it was probably soon after the formation of the disk itself. We see relatively few disk-shaped galaxies without spiral structure, which leads to the conclusion that a disk galaxy does not exist long in an armless state.

It is difficult to guess exactly what may become of the galaxy in the future. It appears that some sort of balance has been reached in the grand recycling process between stars and the interstellar material, so that the interstellar medium is being replenished by evolving and dying stars about as rapidly as it is being consumed by stellar births. Therefore we do not expect the interstellar material to gradually disappear, terminating star formation, as it apparently has in certain other types of galaxies.

Perspective

We have now seen our galaxy in a new light, as an active, dynamic entity. Individual stars orbit its center in individual paths, yet the overall machinery is systematically organized. The majestic spiral arms rotate at a stately pace, while stars and interstellar matter pass through them. We have come to understand the active processes of stellar cycling and chemical enrichment, which still occur today. In a constant turnover of stellar generations, the deaths of the old give rise to the births of the new, while the violence of their death throes energizes a chaotic interstellar medium. The lessons we have learned by examining our Milky Way will be remembered as we move out into the void and probe the distant galaxies.

Summary

1. Stars in the galactic halo have low abundances of heavy elements, and are referred to as Population II stars, whereas those in the spiral arms of the disk have "normal" compositions, and are called Population I stars.
2. There are gradients of increasing heavy-element abundance from halo to disk, and from the outer to the inner portions within the disk.
3. Population II stars have randomly oriented, highly elliptical orbits, whereas Population I stars have nearly circular orbits that lie in the plane of the disk. Population II stars, when passing through the disk near the sun's location, are seen as high-velocity stars.
4. The variations in heavy-element abundance from place to place within the galaxy reflect variations in age: very old stars formed before stellar nuclear reactions produced a significant quantity of heavy elements, and these stars therefore have low abundances of heavy elements; younger stars formed after the galactic composition was enriched by stellar evolution.
5. Spiral arms are probably density enhancements produced by spiral density waves, which rotate about the galaxy while stars and interstellar material pass through them. Because interstellar gas is compressed in these density waves, star formation tends to occur there, and this in turn explains why young stars are found predominantly in the spiral arms.
6. The existence and characteristics of Population I and II stars can be explained in a picture of galactic evolution that begins with a spherical cloud of gas that has little or no heavy elements at first. Population II stars formed while the cloud was still spherical or just beginning to collapse, but while it still had no significant quantities of heavy elements. Population I stars formed later, after the galaxy collapsed to a disk, and after stellar evolution produced some heavy elements.
7. The spiral arms formed after the disk was created by the collapse of the original cloud. The spiral density waves that maintain the arms probably started as a result of gravitational disturbances, possibly caused by nearby galaxies.

Review Questions

1. Explain why the globular clusters have the lowest heavy-element abundances of any objects in the galaxy.
2. The "intermediate disk" population of stars, those which uniformly fill the disk of the galaxy without being confined to the spiral arms, have lower heavy-element abundances than extreme Population I stars (those which

have formed recently in the spiral arms). Explain how this has come about.

3. Suppose a star is observed to have its spectral lines shifted to the red. The strong line of hydrogen whose rest wavelength is 6,563 Å is found to lie at a wavelength of 6,567.38 Å. What is the line-of-sight velocity of this star (see chapter 5), and is it a Population I or Population II star?

4. Explain why the evolution of massive stars, rather than that of the much more common low-mass stars, has been the principal contributor to the enrichment of heavy elements in the galaxy.

5. Why are O and B stars in the galaxy found only in the spiral arms?

6. Explain the contrast between the orbital motions of stars circling the galaxy in the disk and the motions of the spiral density waves that are thought to be responsible for the spiral arms.

7. If the mass of the galaxy is 2×10^{11} solar masses, use Kepler's third law to determine the orbital period of a globular cluster whose semimajor axis is 20 kpc (remember to convert this into astronomical units).

8. How was the formation of our galaxy similar to the formation of the solar system (as described in chapter 17)? Does the solar system have the equivalent of a halo?

9. How do we know that at some time in the past, the interstellar material that pervades space has been completely processed through stellar interiors?

10. Summarize the various roles played by supernovae in the evolution of the galaxy.

Additional Readings

Arp, H. 1969. On the origin of arms in spiral galaxies. *Sky and Telescope,* Dec., p. 385.

Bok, B. J. 1981. Our bigger and better galaxy. *Mercury* 10(5):130.

Bok, B. J., and Bok, P. 1981. *The Milky Way.* Cambridge, Ma.: Harvard University Press.

Burbidge, G., and Burbidge, E. M. 1958. Stellar populations. *Scientific American* 199(2):44.

Iben, I. 1970. Globular cluster stars. *Scientific American* 223(1):26.

Shu, F. H. 1973. Spiral structure, dust clouds, and star formation. *American Scientist* 61:524.

———. 1982. *The physical universe.* San Francisco: W. H. Freeman.

A New View of Our Home Galaxy

Bart J. Bok

Dr. Bok has a very impressive and distinguished record, both as a researcher and as a moving force in the administration of our science. A native of the Netherlands, Dr. Bok has held important posts in many countries, including a lengthy term as associate director of the Harvard College observatory, and directorships of two major observatories (Mt. Stromlo, in Australia, and the Steward Observatory of the University of Arizona, where he continues to hold a faculty position). He has also won awards for his contributions to research in astronomy, which have centered on the structure of the Milky Way and in particular on dark clouds where star formation takes place. With his late wife (and fellow astronomer) Priscilla Bok, he authored a book called The Milky Way *which has been a standard text for several years. The fifth edition has recently been published by Harvard University Press.*

I was asked to write a few words about the future of Milky Way research. I had a chance to review the subject carefully before I wrote the text for the fith edition of *The Milky Way*. By 1979, the fourth edition (1973) seemed out of date, so it was then decided, earlier than had been anticipated, to proceed with a new edition. It took twenty-nine lengthy inserts and the re-writing of two chapters to bring the text up to date. Milky Way research is clearly on the move!

The most important change was that our home galaxy, the Milky Way system, extends much further out than we thought possible in 1973. The evidence for the "Bigger and Better Milky Way" began to accumulate about 1976. Purely dynamical studies by Lynden-Bell, Ostriker, and others indicated that our thin central galactic disk might literally dissolve if it did not have the oppressive support of the thinly populated but very massive outer halo, referred to by some as the galactic corona. Its radius was estimated at 60,000 pc or more, and its total mass was supposed to be at least 750 billion solar masses, five times the earlier estimated total mass of our galaxy.

Additional evidence for the massive corona has come from three sides: (1) the rotation curve of our galaxy does not drop off at a "Keplerian" rate, as it would if the galaxy were to be limited by the conventional inner halo. It goes up and up, and at 20 kpc from the center the hydrogen clouds of our galaxy move at somewhere between 250 and 300 km/sec; (2) Einasto and his colleagues in Estonia found evidence for a strong attraction that our galaxy apparently exerts on our neighbor, the Andromeda galaxy; and (3) eleven outlying globular star clusters (up to 60 kpc from the center of our galaxy) plus seven dwarf galaxies have sufficiently high velocities that there must be much unseen matter to the limit of 60 kpc, possibly more beyond.

The big problem now is what form of matter constitutes the tremendous new-found mass? Is it dead stars or interstellar matter? White dwarfs cannot possibly do the trick. Modest black holes in terrific numbers have been suggested, but we have no proof of their existence, certainly not of their presence in large numbers. Some physicists have talked in terms of "heavy" neutrinos, possibly left over from the big bang, but this is for the present sheer speculation. We have as yet no definite indications as to the mass of a neutrino, leave alone its numbers.

What we have today is simply good solid evidence for much

mass in the corona. We also have some meager evidence for the presence of atomic hydrogen clouds in this region, and then there is the good evidence from the eleven globular clusters and the seven dwarf galaxies. But at best all of these only represent the tip of the iceberg. We had better go to work and find out what's going on. The Space Telescope and the Space Schmidt (a wide-angle photographic survey telescope that has been proposed for operation in earth orbit, so that ultraviolet pictures of the entire sky can be obtained) may give the answer.

A totally different area of research in which much work lies ahead is the study of atomic abundances, especially with reference to globular star clusters. Which are the metal-poorest ones (presumably the oldest), where are they in the galaxy, and what are their ages? Are there truly gradients of abundances? And what are we to do if the atomic ages of some globular clusters come out

greater than the age of our universe as derived from its expansion rate (for which we will soon have a firm and improved value from the Space Telescope)?

Another, totally different, problem faces us with regard to star births in our galaxy. Here we are dealing at present with a new and very young field of research, about which all sorts of suggestions and new evidence are appearing in the literature. Do supernova remnants indicate places of star birth about to happen, or are they simply strings of filamentary nebulosity that will merge into the interstellar medium, enriching it in gaseous matter that contains, percentage-wise, more heavy atoms than are now present in the general interstellar gas? What are we to make of the frequently explosive gas outflows that are being discovered deep inside dense dark nebulae, places where infrared research indicates that some of the youngest stars are found? And, finally, what are the conditions for the formation of the plentiful

young stars in many parts of the Magellanic Clouds where there is a great paucity of cosmic dust, but where atomic hydrogen is present in astounding amounts? Obviously, there is much work ahead in the field of star formation.

I could go on and on, writing about the prospects for spiral-structure studies, or on research in astrometry (stellar parallaxes, proper motions, binary-star studies), a field that the Space Telescope and its cousins promise to revive and change completely. I have only touched on the problems relating to the distant past of our galaxy, on the theories for the dynamics of our local parts, and the more comprehensive theories for the dynamical status of the several components of our galaxy. How permanent or nonpermanent is the spiral structure, and how did it originate? And, in a way above all, what is happening near the center of our galaxy? We might as well put up road signs all over our galaxy: Men and Women at Work!

Extragalactic Astronomy

INTRODUCTION TO SECTION VI.

We are ready now to move out of the confines of our own galaxy, to explore the universe beyond. The Milky Way is just one of billions of galaxies in the cosmos, and we will find that although our own system is typical of a certain class of these objects, there is a wide variety of shapes, sizes, and peculiarities associated with other galaxies.

The first chapter of this section describes both the history of the discovery of galaxies beyond the Milky Way and the observational properties of these galaxies. We rely here on what was learned in the previous section on our own galaxy, for many aspects of its structure and evolution are characteristic of other galaxies, particularly spirals. We do not discuss the evolution of individual galaxies in great detail; we assume that the forces that have shaped the Milky Way are at work in other situations as well.

Chapter 28 begins to bridge the gap between our discussion of individual objects (that is, galaxies) and the structure and organization of the universe as a whole. In this chapter we examine the distribution of galaxies, paying particular attention to groupings and clusters and the possibility that these are in turn organized into larger associations. In the process we learn that galaxies in clusters interact with each other in ways that modify their individual properties, and they also influence the nature of the clusters in which they reside. One question addressed in this chapter, which is of critical importance to our later discussions of the universe as a whole, has to do with the uniformity of the distribution of matter. The implications of this question are not discussed until chapter 31, but the observational evidence is cited in chapters 28 and 29.

Chapter 29 describes two major aspects of the universe, and for the time being, we direct our attention away from individual objects and focus on the universe itself. The two profound observational discoveries are the expansion of the universe and the cosmic background radiation that fills it. Both are legacies of the fiery origin of the universe, and their observed properties teach us something of its early history. Here we see clearly for the first time that the universe itself is a dynamic, evolving entity. It had a beginning, has been changing since, and will continue to develop.

With the perspective gained from chapter 29, in the next chapter we turn our attention back to individual objects, ones whose properties can best be appreciated in the context of an expanding universe. A variety of peculiar galaxies, and especially the quasistellar objects, have fantastic properties that were only discovered by astronomers because the nature of the universal expansion was already known. These objects, on the frontiers of the observable universe, can potentially tell us a great deal about its history. An important theme here is the realization that very distant objects are only seen as they were long ago, because of the light-travel time. This is useful because it allows us to probe the early history of the universe, but is also a hindrance because of the difficulty of comparing distant objects with nearby ones. The quasistellar objects are a fundamental component of the early universe, and therefore the mysteries they present for astronomers are among the most important now under study.

Finally, the last chapter of this section describes the current state of cosmology, the science of the universe as a whole. Here we tie together all the diverse information gained from the preceding chapters and assess the overall structure of the universe, and we especially emphasize the question of its future. If there is a single premier challenge to modern astronomy, it is this, and we are on the verge of knowing the answer.

Having completed this section, we will be nearly finished with our introduction to astronomy. We are left only with the question of life in the universe, and that is reserved for the final, brief section.

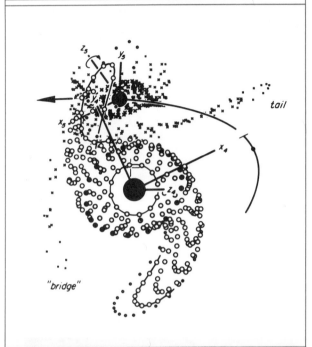

tail

"bridge"

The Nature of The Nebulae

We have spoken of our galaxy as one of many, a single member of a vast population that fills the universe. Having traced our painful progression from the geocentric view to the realization that we occupy an insignificant planet orbiting an ordinary star in an obscure corner of the galaxy, we should be surprised if we found that our galaxy held any kind of unique status in the larger environment of the universe as a whole. It does not. Despite the seeming inevitability of this idea, the actual proof that our galaxy is not alone was some time in coming, and only arrived after considerable debate and controversy.

Island Universes and the Shapley-Curtis Debate

The fact that the heavens are sprinkled with fuzzy objects (Color Plates 21 and 22), some of them spiral-shaped (Figs. 27.1, 27.2, and 27.3), others spherical or spheroidal (Fig. 27.4), has been known for a long time, but there was uncertainty concerning the nature of these objects. They originally were considered to be gas clouds, and were called **nebulae,** along with all the dark and bright interstellar clouds described in chap-

FIG. 27.1. A SPIRAL NEBULA. This is NGC 6946, an example of a face-on spiral galaxy. During the early years of the twentieth century, there was considerable controversy among astronomers over the true nature of objects like this.

FIG. 27.3. AN EDGE-ON SPIRAL. This view clearly shows the thin layer of interstellar material that resides in the plane of the disk.

ter 25. Recall that the size and nature of our own galaxy was not well established until the 1920's; before then, there was little basis for even considering the possibility that some of the nebulae were other galaxies. Thus the question of the nature of these objects only became a serious one in the early decades of the twentieth century.

The nebulae inspired a deep controversy, which reached its peak around 1920. By this time, substantial evidence had been collected supporting the opposing

FIG. 27.4. AN ELLIPTICAL GALAXY. These smooth, featureless galaxies are probably more common than spirals.

FIG. 27.2. A BARRED SPIRAL. This is M83, an example of a spiral galaxy with a barlike structure through the nucleus.

viewpoints; that nebulae were local objects within our galaxy on one hand, and that they were distant galaxies or "island universes" on the other. One of the strongest arguments for the former hypothesis was based on an observation that later turned out to be erroneous. There were reports, based on photographs, that some of the spiral nebulae were rotating, showing slightly different orientations on photographs taken at different times (Fig. 27.5). Although it is no surprise that nebulae rotate, rotation would not be measurable in an object the size of a galaxy, which would take many millions of years to spin once. There would be no hope of detecting differences of orientation in times of a few months or years.

The view that nebulae were local objects was supported by Harlow Shapley, who became the chief spokesman for it. Recall that Shapley was a key figure in establishing the size of the Milky Way and the sun's location in it (see chapter 24). He initially overestimated the size of our galaxy by a substantial amount, and concluded that it was large enough to include the Magellanic Clouds within it. These nearby galaxies were, in his view, local objects. This was important, for the Magellanic Clouds clearly were conglomerates of individual stars rather than gas clouds. If they were proven to lie outside our galaxy, this would lend weight to the hypothesis that the nebulae were more distant, but similar, objects also consisting of vast collections of stars.

In the face of the evidence favoring this hypothesis, the proponents of the island-universe view had difficulty developing strong counterarguments. There was disagreement as to the size of the Milky Way, and those who favored a smaller size than that proposed by Shapley tended to believe that the nebulae, like the Magellanic Clouds, were external objects. Another central point in the controversy hinged on the nature of the bright points of light that occasionally flared up within nebulae; some people argued that these were ordinary novae, in which case they could be near enough to earth to still be within our galaxy, while others argued that these were supernovae, whose brilliance was so great that they had to be at extragalactic distances to appear as faint as they did.

The controversy reached a peak in 1920, when Shapley publicly debated H. D. Curtis, a leading proponent of the island-universe hypothesis, before the National Academy of Sciences. The debate was separated into two parts: a discussion of the size of the Milky Way; and an exchange of arguments on the na-

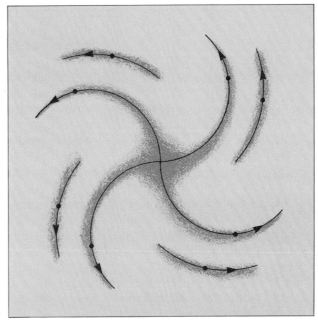

FIG. 27.5. SPURIOUS MEASUREMENTS OF ROTATION. Observations of bright knots (now known to be H II regions) in the spiral arms of some nebulae seemed to show that they were moving outward, along the arms, which implied that all the nebulae were rotating, with the arms leading. Only relatively small objects can rotate fast enough for such an effect to be seen over a period of a few years. Therefore, these erroneous measurements led some astronomers to believe that the nebulae were small, local objects.

ture of nebulae. The outcome of the debate was not a clear-cut victory for either point of view, although there appears to have been a consensus that Shapley's arguments were better presented and more thoroughly backed up by observational evidence.

Finally, in 1924, the question was settled. Edwin Hubble (Fig. 27.6), working with the recently completed 100-inch telescope at Mt. Wilson, was able to make out individual stars in a few of the most prominent nebulae, and furthermore identified some Cepheid variables within them (Fig. 27.7). By applying the standard period-luminosity relation to these variables, Hubble demonstrated beyond any doubt that these nebulae were well outside of our galaxy, even if Shapley's size estimate was accepted. For once and for all it was established that nebulae are distant galaxies, true island universes.

Later examination of the counterevidence showed the error in the earlier conclusion that some nebulae could be seen to rotate. The reported differences in

ASTRONOMICAL INSIGHT (27.1)

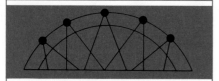

A Classic Telescope

The pioneering observations of nebulae, described in this chapter, were carried out with the 100-inch telescope at the Mt. Wilson Observatory, which overlooks Pasadena, California. This telescope has a remarkable history of involvement with major discoveries, and has ushered in many of the fundamental changes in our perspective on the universe that have taken place in this century.

Named the Hooker Telescope after a Los Angeles businessman who donated funds for the purchase of the primary mirror, the 100-inch instrument has a history dating back to 1908, when the mirror blank was acquired from a French glassworks. Techniques for pouring and annealing such a large mirror were at that time not well developed, and the blank arrived in Pasadena with many small bubbles embedded in it. It was deemed unsuitable, and the glassworks agreed to try to make a better one, but war in Europe intervened. The astronomers at Mt. Wilson therefore reexamined the existing glass, and decided after some debate that it would be adequate; most of the bubbles were buried deeply enough not to distort the shape of the surface or to weaken the mirror. Grinding and polishing operations began in 1910, and were to take five years.

The shaping of the mirror was undertaken by the leading designer and builder of large telescopes of the day, G. W. Ritchey, and the design and construction of the dome were overseen by George Ellery Hale, director of the Mt. Wilson Observatory. Hale, a solar astronomer, was a pioneer in the development of large telescopes, and on no less than four occasions he oversaw the establishment of the then-largest telescope in the world. The 200-inch telescope on Mt. Palomar is named in his honor.

The 100-inch telescope finally went into operation in 1918, a full decade after the arrival of the mirror blank from Europe. New and profound discoveries were immediately made, as it became possible for the first time to observe individual objects within the numerous nebulae. Among the early users of the telescope was Henry Norris

FIG. 27.6. EDWIN HUBBLE. His discovery of Cepheid variables in the Andromeda nebula led to the unambiguous conclusion that this object lies well beyond the limits of the Milky Way, and must therefore be a separate galaxy. Hubble later made important discoveries about the properties of galaxies and what they tell us about the universe as a whole (see chapter 29).

orientation had actually been smaller than the uncertainties of the measurements, so the perceived rotation was not real. As for Shapley's overestimate of the size of the Milky Way, this was found to have been caused by confusion over the period-luminosity relationship for Cepheids. It was eventually recognized that there are two distinct types of Cepheid variables, with different period-luminosity relations, and when the ambiguity was straightened out, Shapley's estimated size for the galaxy shrank to the point where the Magellanic Clouds were definitely outside, and they were recognized as external galaxies in their own right.

Russell, whose ground-breaking work on double stars and on stellar structure and evolution was based largely on observations made with it. Harlow Shapley's work also commenced in the early years of the telescope's operations; Shapley was a staff astronomer at Mt. Wilson before assuming the directorship of the Harvard College Observatory in the late 1920's. All his work on globular clusters and the structure of the galaxy was done with the 100-inch telescope.

If that had been all that was accomplished with this telescope, it would have been sufficient to ensure its place in astronomical history. It was really only the beginning, however. Another project that began early in the lifetime of the 100-inch instrument was the observation of nebulae, and Hubble's great discovery and analysis of variable stars in the Andromeda galaxy were accomplished with this telescope. Thus, the same instrument played a key role both in the determination of the properties of our own galaxy and in the confirmation of the existence of other galaxies, a truly fundamental change in our outlook on the universe.

Hubble continued his work on nebulae and, in 1929, announced his discovery of the expansion of the universe (see chapter 29). Again, the 100-inch telescope had ushered in a new era in astronomy. Later breakthroughs achieved with this instrument included Baade's discovery of stellar populations (described in Chapter 26), based on observations of the Andromeda galaxy; pioneering work on the nature of interstellar gas by T. R. Dunham and P. Merrill; analysis of interstellar extinction by J. Stebbins and A. E. Whitford; and, as long ago as 1920, interferometric measurements of stellar angular diameters by A. A. Michelson and F. G. Pease.

The 100-inch telescope was finally surpassed in size in 1948, when the 200-inch Hale Telescope went into operation. By that time industrial and residential expansion of Los Angeles had created both atmospheric pollution and a bright nighttime sky at Mt. Wilson, and the 100-inch telescope no longer enjoyed such a fine observing site. Today the Hooker Telescope is still in operation, but is generally used for observations of relatively bright stars, although the addition of sophisticated electronic detectors has helped to preserve its capability for analyzing the light of faint objects.

The Hubble Classification System

Having established that nebulae are truly extragalactic objects, Hubble began a systematic study of their properties. The most obvious basis for establishing patterns among the various types was to categorize them according to shape. Hubble did this, designating the spheroidal nebulae **elliptical galaxies,** as distinct from the spiral galaxies. Within each of the two general types, Hubble established subcategories on the basis of less dramatic gradations in appearance. The ellipticals displayed varying degrees of flattening, and were sorted out according to the ratio of the long axis to the short axis, with designations from E0 (spherical in shape) to E7 (the most highly flattened). The number following the letter E is determined from the formula $10(1-b/a)$, where a is the long axis and b is the short axis, as measured on a photograph (Fig. 27.8).

Hubble's classification of the spirals was based on the tightness of the arms and the compactness of the nucleus (Fig. 27.9 and Color Plate 22). The types ranged from Sa (tight spiral, large nucleus) to Sc (open arms, small nucleus). The Milky Way is probably an Sb in this system, intermediate in both characteristics, although it is difficult to determine, because we cannot get an outsider's view of our galaxy.

Hubble recognized a variation of the spiral galaxies in which the nucleus has extensions on opposing sides, with the spiral arms emanating from the ends of the extensions. These he called **barred spirals,** and assigned them the designations SBa through SBc, using the same criteria as before in establishing the a, b, and c subclasses (Fig. 27.10 and Color Plate 22). Following Hubble's original work on galaxy classification, astronomers recognized an intermediate class called the SO galaxies. These appear to have a disk shape, but no trace of spiral arms.

Although Hubble himself did not draw an evolutionary connection among the various types of galaxies, he assembled an organization chart that was for a time thought by many astronomers to represent an age sequence. Because there are two types of spirals,

FIG. 27.9.   AN ASSORTMENT OF SPIRALS. This sequence shows spiral galaxies of several subclasses.

FIG. 27.7.   CEPHEID VARIABLES IN ANDROMEDA. This is a portion of the Andromeda galaxy, with the locations of two variable stars indicated. By comparing the photos at right and left, we can see that both of these stars vary in brightness.

FIG. 27.8.   THE SHAPES OF ELLIPTICALS. The numerical designation (following the letter E) is given by the formula $(1 - b/a) \times 10$, where b is the short axis and a the long axis of the galaxy image. Here are three examples: the E0 galaxy has $a = b$ (that is, it is circular in shape); the E3 galaxy has $a = 1.43b$; and the E7 has $a = 3.33b$. E7 is the most highly elongated of the elliptical galaxies.

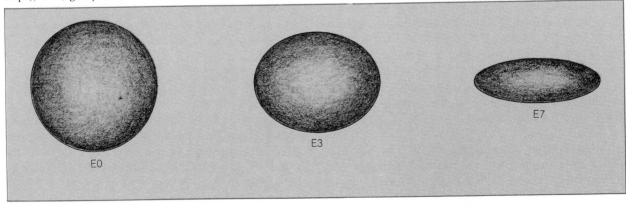

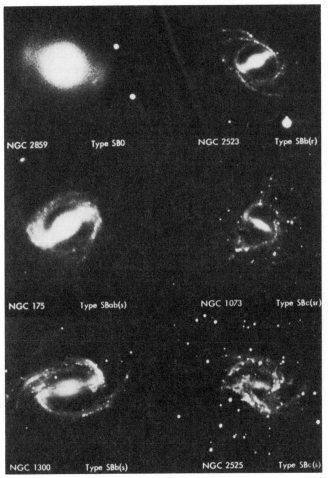

NGC 2859   Type SB0   NGC 2523   Type SBb(r)

NGC 175   Type SBab(s)   NGC 1073   Type SBc(sr)

NGC 1300   Type SBb(s)   NGC 2525   Type SBc(s)

FIG. 27.10. BARRED SPIRALS. Roughly half of all spiral galaxies have central elongations, or bars, from which the spiral arms emanate.

Hubble chose not to force all the types into a single sequence, but instead split the sequence into two branches, and the result became known as the tuning-fork diagram (Fig. 27.11). It was the prevailing belief that elliptical galaxies evolve into spirals, the originally spherical distribution of gas gradually flattening into a disk (as in SO galaxies) and then developing spiral arms. This view was later reversed, after the discovery of stellar populations (see chapter 26). Since elliptical galaxies consist primarily of old stars, with little interstellar matter and therefore little active star formation; and spirals contain prominent young stars and interstellar gas and dust, it seemed reasonable to assume that spirals evolve into ellipticals. This was the accepted view as recently as the early 1960's. It was eventually found, however, that even spiral galaxies contain old populations of stars and clusters, and are just as old as the

FIG. 27.11. THE TUNING-FORK DIAGRAM. This is the traditional manner of displaying the galaxy types, originally devised by Hubble. For quite some time, this was thought to be an evolutionary sequence, although the imagined direction of evolution was reversed at least once. Now it is known that galaxies do not, in the normal course of events, evolve from one type to another.

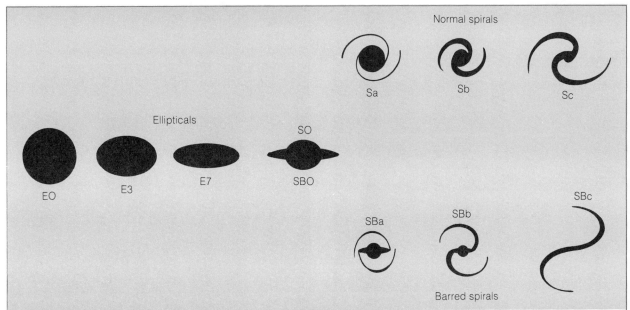

Normal spirals

Sa   Sb   Sc

Ellipticals   SO

EO   E3   E7   SBO

SBa   SBb   SBc

Barred spirals

ellipticals. The tuning-fork diagram is not an age sequence after all, and the differences in galactic type must be explained in some other way.

Most of the galaxies listed in catalogues are spirals, which are about evenly divided between normal and barred spirals. Only about 15 percent of the listed galaxies are ellipticals, and a comparable number are SO galaxies. The remaining few percent are called irregular galaxies, which do not fit into the normal classification scheme. Because there probably are many small, dim elliptical galaxies that are usually not sufficiently prominent to appear in catalogues, it seems likely that ellipticals actually outnumber spirals in the universe. This certainly is the case in dense clusters of galaxies, as we will see.

Although the irregular galaxies are, by definition, misfits, it has proved possible to find some systematic characteristics even in their case. Most have a hint of spiral structure, although they lack a clear overall pattern, and these have been designated as **type I irregulars.** The rest, a small group, simply do not conform in any way to the normal standards, and are assigned the classification of **type II irregulars** or peculiar galaxies (Fig. 27.12). The Magellanic Clouds (Color Plate 21) are irregulars of type I.

Some of the type II irregulars have the appearance of pairs of galaxies in collision (Fig. 27.13). In these cases very unusual overall shapes are seen, but careful examination reveals hints of two galaxies in the process of merging, or at least passing through each other (Fig. 27.14). Since galaxies are mostly empty space, in such a collision there are few direct impacts of stars with each other. The galaxies can pass through each other like ghosts and go on their separate ways, except that their gravitational effects cause major distortions of their shapes. Even the Milky Way suffers from such an effect; the gravitational forces of the Magellanic Clouds have caused a slight warping of the galactic disk. As we will see in the next chapter, collisions among galaxies, particularly in clusters, can have very important effects on the evolution of both the individual galaxies and the clusters in which they are found.

Galactic Distances

In order to probe the physical nature of the galaxies, we must first know galactic distances, because without this information, we cannot deduce such fundamental parameters as masses and luminosities. We have already mentioned one technique for distance determination that can be applied to some galaxies: the use of Cepheid variables. These stars are sufficiently luminous to be identified as far away as a few million parsecs (that is, a few **megaparsecs,** abbreviated **Mpc**), which is sufficient to reach the Andromeda Nebula and

FIG. 27.12.   A PECULIAR GALAXY. Not all galaxies fit the standard classifications. This is M82, a well-known example of a galaxy with an unusual appearance. For a time, it was thought that the nucleus of this galaxy is exploding, but more recent analysis has indicated that the odd appearance is simply the result of a very extensive region of interstellar gas and dust surrounding the center of a spiral galaxy that is being viewed edge-on.

FIG. 27.13.   GALAXIES IN COLLISION? This unusual object is thought to be a pair of galaxies running into each other.

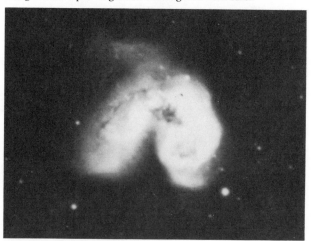

several other neighbors of the Milky Way. This technique is not adequate, however, for probing the distances of most galaxies; so other techniques had to be developed.

Recall once again, from our discussions of stellar distance determinations (chapter 19), that we can always find the distance to an object if we know both its apparent and absolute magnitudes. This is the basis of

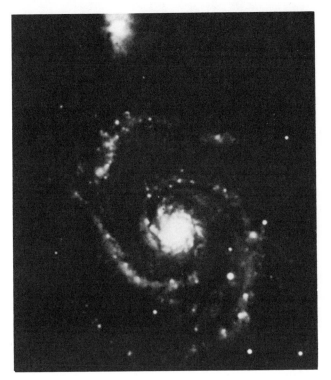

FIG. 27.14. THE EFFECT OF A NEAR-COLLISION BETWEEN GALAXIES. This is M51 (also known as the Whirlpool galaxy), with its smaller companion. Below is a computer simulation of the correct orientation of these two galaxies as seen from different angles.

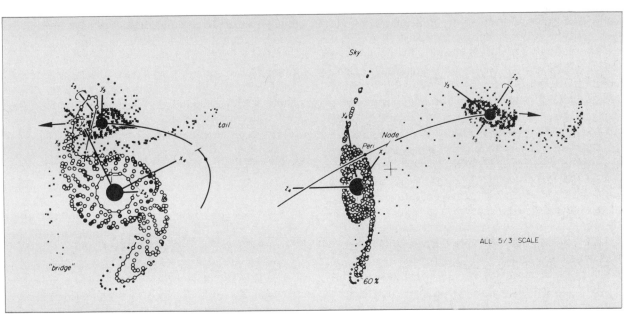

the spectroscopic-parallax method, the main-sequence-fitting technique, and even the Cepheid variable period-luminosity relation. A general term for any object whose absolute magnitude is known from its observed characteristics is **standard candle,** and an assortment of these are used in extending the distance scale to faraway galaxies.

The most luminous of stars are the red and blue supergiants, those which occupy the extreme upper regions of the H-R diagram. These can be seen at much greater distances than Cepheid variables, and therefore are important links to distant galaxies. The absolute magnitudes of these stars are inferred from their spectral classes, just as in the spectroscopic-parallax technique, although they are so rare that there is substantial uncertainty in assuming that they conform to a standard relationship between spectral class and luminosity. In a variation on this technique, it is simply assumed that there is a fundamental limit on how luminous a star can be, and that in a collection of stars as large as a galaxy, there will always be at least one star at this limit. Then to determine the distance to a galaxy, we need only to measure the apparent magnitude of the brightest star in it, and to assume that its absolute magnitude is at the limit, which is about $M = -8$. This technique extends the distance scale by about a factor of 10 beyond what is possible with Cepheid variables; that is, to 10 or more megaparsecs.

Other standard candles come into play at greater distances. These include supernovae, which, at peak brightness, always reach the same absolute magnitude (although care must be exercised, because there are at least two distinct types of supernovae, with different absolute magnitudes). A supernova is fantastically bright, and can even outshine the entire galaxy in which it resides (Fig. 27.15), reaching an absolute magnitude as bright as $M = -21$. By adopting supernovae as standard candles, we can measure distances of hundreds of millions of parsecs. One drawback is that we can only use this technique when a supernova happens to occur in a galaxy, which may be only once every ten to a hundred years. Therefore, if we wish to estimate the distance to a specific galaxy, and we do not want to wait that long, we must use some other standard candle.

Other objects useful for this purpose include certain types of star clusters and bright H II regions. In the latter case, it is not the absolute magnitude but the diameter of the H II region that is assumed to be known. Then the apparent diameter (that is, the angular diameter) can be used to infer the distance. The ultimate

JUNE 9, 1950   FEB. 7, 1951

FIG. 27.15. A SUPERNOVA IN A DISTANT GALAXY. Before the distinction between novae and supernovae became clear, these flare-ups contributed to the controversy over the nature of nebulae. Now that these occurrences in galaxies are known to be supernova explosions, comparisons of their apparent and assumed absolute magnitudes provide distance estimates.

standard candle is the brightest galaxy in a cluster of galaxies, in which case we adopt the assumption that this object should have the same absolute magnitude in every instance. This assumption makes it possible to measure distances to objects up to several billion parsecs away.

It is important to keep in mind how uncertain these techniques are. To assume that all objects in a given class are identical in basic properties such as luminosity is always a risky business, especially when such assumptions are applied to objects as distant as external galaxies or as rare as the brightest star in a galaxy. There is little else that can be done, however, so we must simply recognize the inherent limitations in accuracy and take them into account.

The Masses of Galaxies

The masses of nearby galaxies can be measured in the same manner as the mass of our own galaxy; by applying Kepler's third law to the orbital motions of stars or gas clouds in the outer portions. All that is required is to measure the orbital velocity at some point well out from the center, and to determine how far from the center that point is (which in turn requires knowledge

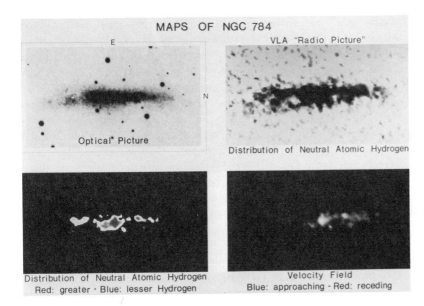

MAPS OF NGC 784

Optical Picture

VLA "Radio Picture"

Distribution of Neutral Atomic Hydrogen

Distribution of Neutral Atomic Hydrogen
Red: greater - Blue: lesser Hydrogen

Velocity Field
Blue: approaching - Red: receding

FIG. 27.16. DETERMINATION OF A GALACTIC ROTATION CURVE FROM 21-CM OBSERVATIONS. These are images of NGC 7784, a spiral galaxy seen nearly edge-on. At upper left is an optical photo, and at upper right is a radio map showing the distribution of interstellar hydrogen gas. The image at lower left illustrates the density of gas implied by the radio emission, and the image at lower right shows the relative velocities of gas in different parts of the galaxy. Here the lightest areas are receding from us, and the intermediate gray areas are approaching. From data such as these, the variation of velocity with distance from the center of a galaxy is measured. This rotation curve, in turn, is used to estimate the mass of the galaxy.

of the distance to the galaxy). Then Kepler's third law leads to

$$M = a^3/P^2,$$

where M is the mass of the galaxy (in solar masses), and a and P are the semimajor axis (in AU) and the period (in years) of the orbiting material at the observed point (see chapter 24).

The difficulties of this technique are that it is hard to isolate individual stars in distant galaxies and then to measure their velocities (using the Doppler shift), and that both the distance and the orientation of the galaxy must be known before the true orbital velocity and semimajor axis can be determined. The orientation can usually be deduced for a spiral galaxy, since it has a disk shape whose tilt can be seen, and the distance can be estimated through the use of one of the methods just outlined. By allowing light from many stars to enter the telescope and spectrograph, and then measuring the aggregate velocity of a large area of the galaxy, we can avoid the problem of isolating individual stars. This is somewhat imprecise, because there is always some internal motion within the portion of the galaxy that is observed, but it leads to useful estimates of the orbital velocity in the outer regions.

Radio astronomy techniques are especially useful, because the gas that produces the radio emission (usually the 21-cm line of hydrogen is used in this case) often extends farther out in the galaxy than the visible stars, and it is important to know the orbital velocities at the greatest possible distance from the galactic cen-

ter (Fig. 27.16). Furthermore, velocity measurements can be made very accurately with radio emission-line data, and interstellar extinction is not a problem.

In most cases the orbital velocities are measured at several points within a galaxy, from the center out as far as possible, and the data are plotted on a **velocity curve,** which is simply a diagram showing the variation of orbital velocity with distance from the center (Fig. 27.17). This is useful for several reasons: it pro-

FIG. 27.17. A SCHEMATIC VELOCITY CURVE FOR A SPIRAL GALAXY. The dashed curve shows the idealized case where the stars in the outer portions of the galaxy move as individuals orbiting a central object, whose mass can therefore be determined through the use of Kepler's third law. The solid line indicates what actual rotation curves for spiral galaxies usually look like; the failure of the curve to drop off indicates that the galaxy has a great deal of mass beyond its visible limits.

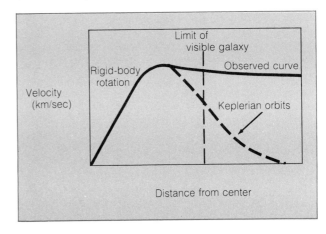

Limit of visible galaxy

Observed curve

Rigid-body rotation

Keplerian orbits

Velocity (km/sec)

Distance from center

vides a means of checking whether the observations go sufficiently far out in the galaxy to reach the region where the orbiting material follows Kepler's third law; it helps in determining the orientation of the galaxy; and it ensures that the velocities have been measured as far out as possible from the center.

The technique of measuring rotation curves as just described can best be applied to spiral galaxies, where there is a disk with stars and interstellar gas orbiting in a coherent fashion. In elliptical galaxies there is no such clear-cut overall motion, and a slightly different technique must be used. The individual stellar orbits are randomly oriented within an elliptical galaxy, so there is a significant range of velocities within any portion of the galaxy's volume. This range of velocities, called the **velocity dispersion** (Fig. 27.18), is greatest near the center of the galaxy, where the stars move fastest in their orbits, and is smaller in the outer re-

FIG. 27.18. THE VELOCITY DISPERSION FOR AN ELLIPTICAL GALAXY. There is no easily measured overall rotation of an elliptical galaxy, so the mass cannot be estimated from a rotation curve. Instead the average velocities of stars at a known distance from the center are used. Light from a small area of the galaxy's image is measured spectroscopically. The random motions of the stars in the part of the galaxy that is observed broaden the spectral lines by the Doppler effect, and the amount of broadening is a measure of the average velocity of the stars in the observed region. At a given distance from the center of the galaxy, the higher the average velocity, the greater the mass contained inside that distance.

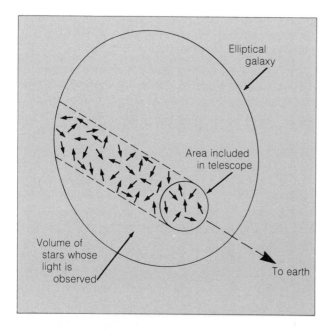

gions. Furthermore, the greater the mass of the galaxy, the greater the velocity dispersion (at any distance from the center), so a measurement of this parameter can lead to an estimate of the mass of a galaxy. The velocity dispersion is deduced from the widths of spectral lines formed by groups of stars in different portions of a galaxy; the greater the internal motion within the region that is observed, the greater the widths of the spectral lines, owing to the Doppler effect (caused by the fact that some stars are moving toward the earth and others away from it).

Kepler's third law can sometimes be used in an entirely different way in determining galactic masses. There are double galaxies here and there in the cosmos, orbiting each other exactly like stars in binary systems. In these cases, Kepler's third law can be applied, leading to an estimate of the combined mass of the two galaxies. The uncertainties are even more severe than in the case of a double star, however, because the orbital period of a pair of galaxies is measured in hundreds of millions of years. Thus the usual problems of not knowing the orbital inclination or the distance to the system are compounded by inaccuracies in estimating the orbital period, something that is usually very well known for a double star. Still, this technique is useful, and has one major advantage: it takes into account *all* the mass of a galaxy, including whatever part of it is in the outer portions, beyond the reach of the standard velocity curve or velocity-dispersion measurements. Galactic masses estimated from double systems are generally much larger than those based on measurements of internal motions within galaxies, possibly indicating that most galaxies have extensive halos containing large quantities of matter. As noted in chapter 24, there is independent evidence that our own galaxy has a massive halo, perhaps containing as much as 90 percent of the total mass.

Both the mass-to-light ratios and the colors of galaxies are indicators of the relative content of Population I and Population II stars. Recall that Pop. I stars tend to be younger, and included in this group are all the bright blue O and B stars. Population II, by contrast, includes old, only red, relatively dim stars. Therefore a red overall color, along with a high mass-to-light ratio, implies that Pop. II stars are dominant, whereas a low value of M/L and a bluer color means that some Pop. I stars are mixed in. Thus, elliptical galaxies seem to consist almost entirely of Pop. II objects, whereas spirals contain a mixture of the two populations.

This dichotomy between the two types of galaxies is

found also when we compare the interstellar-matter content of spirals and ellipticals. Photographs of spirals, especially edge-on views, often clearly show the presence of dark dust clouds, and face-on views typically reveal a number of bright H II regions. Neither shows up on photographs of elliptical galaxies. Radio observations of the 21-cm line of hydrogen bear this out; emission is usually present in spirals, but is rarely seen in ellipticals.

The Origins of Spirals and Ellipticals

In attempting to explain the differences between spiral and elliptical galaxies, astronomers have suggested several theories, none of which is fully developed at present. All evidence shows that spiral galaxies are dynamic, living entities, with active star formation and recycling of material between stellar and interstellar forms, whereas elliptical galaxies have reached some sort of equilibrium in which these processes are not taking place at a significant rate. We know that there is no significant age difference between the two types of galaxies; we cannot simply conclude that the ellipticals are older and have run out of gas.

One of the earliest suggestions was that rotation is responsible for the differences between the two types. We learned in the previous chapter that our own galaxy is thought to have formed from a rotating cloud of gas that flattened into a disk as it contracted. The rotation of the cloud was the cause of the disk formation, so perhaps elliptical galaxies are the result of contracting gas clouds that did not rotate rapidly enough to form disks. It is not clear how this would account for the lack of both interstellar material and star formation, however.

Luminosities, Colors, and Diameters

Once the distance to a galaxy has been established, its luminosity and size can be deduced directly from the apparent magnitude and apparent diameter. Both quantities are found to vary over wide ranges, with luminosities as low as 10^6 and as high as 10^{12} times that of the sun, and diameters ranging from about 1 to 100 kpc.

Generally the elliptical galaxies display a wider range of luminosities and sizes than the spirals. The latter tend to be more uniform, with luminosities usually between 10^{10} and 10^{12} solar luminosities and diameters between 10 and 100 kpc. The smallest elliptical galaxies are called **dwarf ellipticals.** These may be very common, but because they are too dim to be seen at great distances, we can only say for sure that there are many of them near our own galaxy.

It appears that galaxies of a given type tend to be fairly uniform in other properties, just as stars of a given spectral type are the same in other ways. One quantity often used by astronomers to characterize galaxies is the **mass-to-light** (M/L) ratio, which is simply the mass of a galaxy divided by its luminosity, in solar units. Values of the mass-to-light ratio typically range from 50 to 200. Any value larger than 1 means that the galaxy emits less light per solar mass than the sun; that is, such a galaxy is dominated in mass by stars that are dimmer than the sun. Even the smaller values of M/L that are observed for galaxies are much larger than 1; a value of M/L = 50, for example, means that 50 solar masses are required to produce the luminosity of one sun.

Elliptical galaxies tend to have the largest mass-to-light ratios, consistent with our earlier statement that these galaxies are relatively deficient in hot, luminous stars. Spirals, on the other hand, which contain some of these stars, have the lower M/L values. It is worth stressing again, however, that even the low values of M/L for these galaxies are much greater than 1, indicating that they too are dominated in mass by low-luminosity stars. Hence a spiral galaxy, with all its glorious bright blue stars, actually has far more dim red ones.

The colors of galaxies also can be measured, with the use of filters to determine the brightness at different wavelengths. In general, the spirals are not as red as the ellipticals, again indicating that the latter galaxies contain a higher fraction of cool, red stars. There are also color variations within galaxies; for example, in a spiral, the central bulge is usually redder than the outer portions of the disk where most of the young, hot stars reside.

One problem with this idea is that elliptical galaxies can and do rotate, yet they have not formed disks. Rotation alone therefore cannot be the full explanation. Apparently the key is whether or not a disk forms before most of the gas in the contracting cloud has been consumed by star formation. If all the gas is converted into stars quickly, before the collapse has proceeded

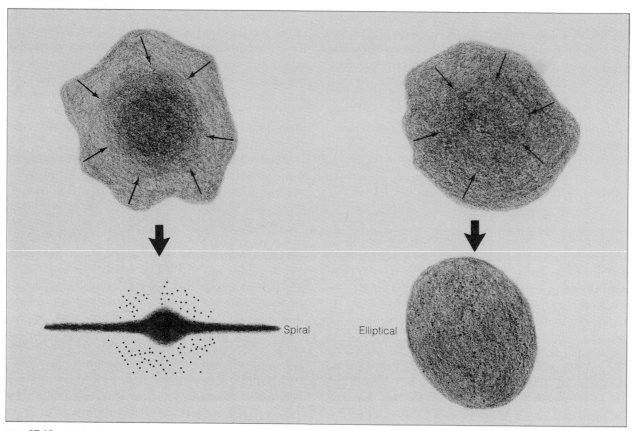

FIG. 27.19. THE ORIGINS OF SPIRALS AND ELLIPTICALS. Both types of galaxies begin as giant gas clouds that collapse gravitationally. If collapse to a disk occurs before all the gas is converted into stars, the result is a spiral galaxy. If, on the other hand, the gas is entirely used up in star formation before a disk forms, then the result is an elliptical. The rate of star formation relative to the rate of collapse to a disk is probably determined by the initial density, and perhaps is influenced by the rotation rate of the cloud.

very far, then the result is an elliptical galaxy (Fig. 27.19). If, on the other hand, the initial rate of star formation is not so great, then the cloud has time to form a disk while it still contains a large quantity of interstellar gas and dust. The reason timing is so important is that stars, once formed, will act as individual particles and will continue to orbit the galaxy without forming a disk. In contrast, gas acts as a fluid and will settle into a disk. (The essence of this is whether or not there is internal friction that will dissipate energy and allow the material to sink into the plane of a disk; stars do not exert friction forces on each other, but gas particles do.)

Although these ideas represent progress toward understanding why some galaxies are ellipticals and others are spirals, there remain substantial questions. It is not clear why the initial rate of star formation should differ from one galaxy to another, for example, although it has

been suggested that it may have to do with the density of the cloud from which the galaxy contracts.

Other as-yet unexplained questions regarding the two types of galaxies include why ellipticals have a wider range of masses and sizes than spirals, and why many spirals have bars. Some progress has been made on the latter question; calculations show that instabilities that can develop naturally in a rotating disk may lead to the formation of a symmetric bar like those observed. It has turned out, in fact, to be more difficult to explain why not all spirals have bars, than to account for the fact that some of them do, because the responsible instabilities should be effective in all cases. It appears that the presence of a massive halo may inhibit the formation of a bar, however, so perhaps the difference between barred spirals and those without bars is that the latter have extensive halos and the former do not.

Perspective

We have expanded our horizons almost to the ultimate limit, by examining objects that inhabit the full volume of the universe. In this chapter we have confined ourselves to a discussion of individual galaxies and their properties, a necessary step before moving on to the large-scale organization of the universe. We are prepared now to examine groups and clusters of galaxies, to see what they can tell us about the origin and evolution of the universe itself.

Summary

1. Nebulae were observed by astronomers for centuries before it was known whether they were local objects, such as gas clouds, or other galaxies, comparable to the Milky Way.

2. The controversy over the nature of nebulae peaked in 1920, and was settled in 1924, when Hubble's analysis of Cepheid variables in the Andromeda nebula proved that this object is too distant to be part of the Milky Way.

3. Galaxies are categorized by shape, and fall into two general classes: spirals and ellipticals. There are also a substantial number of SO galaxies (disk-shaped, but with no spiral structure) and irregular galaxies.

4. Distances to galaxies are measured using a variety of standard candles, such as Cepheid variables, extremely luminous stars, supernovae, and galaxies of standard types.

5. Masses of spiral galaxies are determined through the application of Kepler's third law to the outer portions, where the orbital velocity and hence the period are determined from a rotation curve. For elliptical galaxies, the internal velocity dispersion is used. Galactic masses can also be determined from the application of Kepler's third law to binary galaxies.

6. Determinations of mass based on external gravitational influences of galaxies usually yield higher values than those based on internal diagnostics such as rotation curves or velocity dispersions. This leads to the conclusion that many galaxies have massive halos.

7. Galactic luminosities and diameters vary quite widely among elliptical galaxies, but not so widely among spirals.

8. Although all galaxies are dominated by Population II stars, spirals tend to contain a greater proportion of Population I stars, have substantial quantities of interstellar matter, and generally seem to be in a state of continuous evolution and stellar cycling. Ellipticals, on the other hand, have a few or no Population I stars, contain little or no interstellar matter, and generally do not seem to have active stellar cycling at the present time.

9. Spiral galaxies appear to originate from rotating gas clouds that flatten into disks before all the gas is used up in star formation, whereas ellipticals seem to result when star formation consumes all of the gas before collapse to a disk occurs.

Review Questions

1. To help demonstrate the difficulty of observing individual stars in other galaxies, calculate the apparent magnitude of a supergiant star whose absolute magnitude is $M = -8$, if such a star were observed in a galaxy that is 100,000 pc away (that is, a bit beyond the Magellanic Clouds), and in a galaxy 1,000,000 pc away (a little farther than Andromeda).

2. Suppose an elliptical galaxy is found that is twice as long as it is wide. What type of elliptical is it?

3. A supernova of one type has an absolute magnitude of $M = -21$ at peak brilliance. Assuming such objects can be observed to apparent magnitudes as faint as $m = +24$, how far away can supernovae of this type be used as distance indicators?

4. Explain how our knowledge of the distance to a faraway galaxy depends on how well we know the sun-earth distance.

5. Explain, in terms of Newton's laws of motion and the properties of orbits (discussed in chapter 4), why the average velocity of stars at a given distance from the center of a galaxy depends on how much mass the galaxy contains within that distance.

6. In a spiral galaxy with a mass-to-light ratio of 50, it takes 50 solar masses to produce each solar luminosity of energy that the galaxy emits. What does this tell us about the type of star that is most common in such a galaxy?

7. Explain why, to an outside observer, the Milky Way would appear to have a color gradient; that is, it would appear bluer in the outer portions than in the central bulge and halo.

8. Would you expect elliptical galaxies to have the same proportion of heavy elements as spiral galaxies? Explain.

9. What would our solar system look like if all the interstellar gas and dust from which it formed had been converted into planets before the cloud that it formed from had collapsed into a disk?

10. Contrast the evolution of an elliptical galaxy with that of the Milky Way.

Additional Readings

Hodge, P. W. 1966. *Galaxies and cosmology*. New York: McGraw-Hill.

Mitton, S. 1976. *Exploring the galaxies*. New York: Scribner's.

Page, T., and Page L. W., eds. 1969. *Beyond the Milky Way*. New York: Macmillan.

Sandage, A., Sandage, M., and Kristian J., eds. 1976. *Galaxies and the universe*. Chicago: University of Chicago Press.

Shapley, H., and Hodge, P. W. 1972. *Galaxies*. Cambridge, Ma.: Harvard University Press.

Strom, S. E., and Strom, K. M. 1977. The evolution of disk galaxies. *Scientific American* 240(4):56.

Talbot, R. J., Jensen, E. B., and Dufour, R. J. 1980. Anatomy of a spiral galaxy. *Sky and Telescope* 60(1):23.

van den Bergh, S. 1976. Golden anniversary of Hubble's classification system. *Sky and Telescope* 52:410.

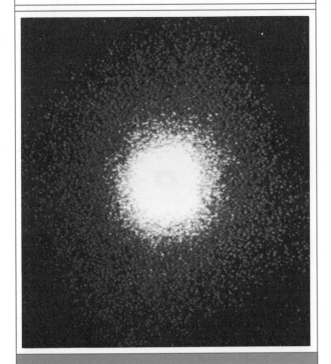

• CHAPTER 28 •

Clusters and Superclusters: The Distribution of Galaxies

Although galaxies may be considered the largest single objects in the universe (if indeed an assemblage of stars orbiting a common center can be viewed as a single object), there are yet larger scales on which matter is organized. Galaxies tend to be located in clusters (Fig. 28.1) rather than being distributed uniformly throughout the cosmos, and these in turn have an uneven distribution, with concentrations of clusters referred to as **superclusters.** It may be that superclusters are the largest things in the universe.

Clusters of galaxies range in membership from a few (perhaps half a dozen) to many hundreds or even thousands. Just as stars in a cluster orbit a common center, so are galaxies in a group or cluster gravitationally bound together, following orbital paths about a central point. In a large, rich cluster, the frequent encounters between galaxies have, over time, caused the members to take on a smooth, spherical distribution; whereas in a small group like the one to which the Milky Way belongs, the arrangement of individual galaxies is more haphazard, creating an amorphous overall appearance.

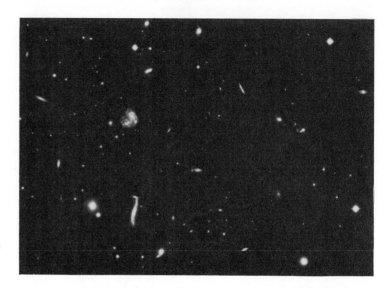

FIG. 28.1. A PORTION OF A CLUSTER OF GALAXIES IN THE CONSTELLATION HERCULES. Many different galaxy types are seen here.

The Local Group

We can begin by discussing the **Local Group** of galaxies, the cluster that we live in. This group consists of about thirty members arranged in a random distribution. Despite the relative proximity of these galaxies to us, it has been difficult to ascertain their properties in some cases, because of obscuration by our own galactic disk. It is not even possible to say with certainty how many members belong to the Local Group.

Among the member galaxies are three spirals, two of which, the Andromeda nebula (Color Plate 21) and the Milky Way, are rather large and luminous. These are probably the brightest and most massive galaxies in the entire cluster. Most of the other members are ellipticals, many of them dwarf ellipticals (Fig. 28.2). There may be additional members of this type, undetected so far because of their faintness. Four irregular galaxies (Fig. 28.3), two of them being the Large and Small Magellanic Clouds (Color Plate 21), are also found within the Local Group. In addition, there are globular clusters, which are probably distant members of our galaxy, but which lie so far away from the main body of the Milky Way that they appear isolated.

The Local Group is about 800 kpc in diameter, and has a roughly disk-like overall shape, with the Milky Way located a little off-center. Beyond the outermost portions of the Local Group there are no conspicuous external galaxies for a distance of some 1,100 kpc.

The Magellanic Clouds and the Andromeda nebula

FIG. 28.2. THE SCULPTOR DWARF GALAXY. This loose conglomeration of stars lies about 83 kpc from the sun. Such a small, dim galaxy would not be noticed if it were much farther away. The number of systems of this type in the Local Group leads to the assumption that they may be very common in the universe.

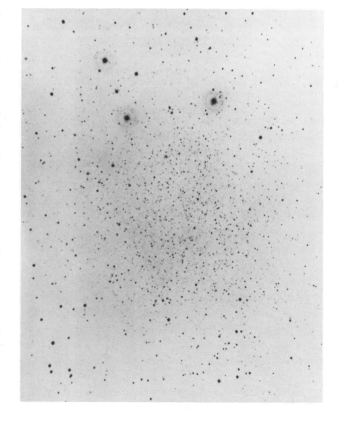

FIG. 28.3. IC1613. This is an irregular galaxy, a member of the Local Group.

FIG. 28.4. THE LARGE MAGELLANIC CLOUD. This is our nearest neighbor galaxy. The bright concentration at upper left is the region known as the Tarantulus Nebula, also listed in catalogues as 30 Doradus. In its center is a bright spot tentatively identified as a supermassive young star (see chapter 19).

(also known commonly as M31, its designation in the widely cited Messier Catalog) have been particularly well studied, because of their proximity and prominence, and because of what they can tell us about galactic evolution and stellar processing. The Large and Small Magellanic Clouds appear to the unaided eye as fuzzy patches, easily visible only on dark, moonless nights (Figs. 28.4 and 28.5). They lie near the south celestial pole, and can therefore be seen only from the southern hemisphere. Their name originated from the fact that the first Europeans to see them were Ferdinand Magellan and his crew, who made the first voyage around the world in the early sixteenth century.

The Magellanic Clouds are considered to be satellites of the Milky Way, having lesser masses and following orbits about it, taking several hundred million years to make each circuit. Lying between 50 and 60 kpc from the sun, both are irregulars of type I. They contain substantial quantities of interstellar matter and are quite obviously the sites of active star formation, with many bright nebulae and clusters of hot, young stars. Measurements of the colors of these galaxies, and of the spectra of some of their brighter stars, indicate that they have somewhat lower heavy-element abundances than Pop. I stars in our galaxy. This seems to indicate that

the Magellanic Clouds have not undergone as much stellar cycling and recycling as the Milky Way. These, and other type I irregular galaxies, may generally be viewed as galaxies in extended adolescence, not yet having settled down into mature disks. In the case of

FIG. 28.5. BOTH MAGELLANIC CLOUDS. The Small Cloud is at left and the Large Cloud is at right.

the Magellanic Clouds, the reason for the unrest is probably the gravitational tidal forces exerted by the Milky Way.

The great spiral of M31, the Andromeda nebula, (Fig. 28.6) is the most distant object visible to the unaided eye, lying some 700 kpc from our position in the Milky Way. All that the eye can see is a fuzzy patch of light, even if a telescope is used, but when a time-exposure photograph is taken, then the awesome disk and spiral arms stand out. The Andromeda nebula is so large, extending over a full degree across the sky, that full portraits can only be obtained by using relatively wide-angle telescope optics (most large telescopes have extremely narrow fields of view).

The Andromeda nebula is probably very similar to our own galaxy, and thus has taught us quite a bit about the nature of the Milky Way. The two stellar populations were first discovered through studies of stars in Andromeda. In fact, the effects of stellar processing on the chemical makeup of different portions of a galaxy are better determined for Andromeda than for the Milky Way.

Another very important by-product of observations of the Andromeda galaxy, as well as those of some other members of the Local Group, is the calibration of our distance scales for measuring objects farther away. Knowledge of distances to galaxies within the Local Group, measured through the use of the period-lumi-

nosity relationship for Cepheids, has allowed astronomers to establish the absolute magnitudes of various standard candles which in turn are used to find distances to more faraway galaxies. Each step in extending the distance scale must be built upon previous steps, and the Local Group galaxies represent an important stage in the process of reaching out to the greater distances of other clusters of galaxies. It would have been very difficult indeed to establish the distances of extragalactic objects if we did not live within a cluster of galaxies where this calibration process is possible.

Before leaving the Local Group, we return to the question of its membership, and the rank of the Andromeda galaxy and the Milky Way within it. New galaxies are occasionally discovered and found to belong to the Local Group. These are most often dwarf ellipticals (one was named Snickers by its discoverers, to provide an alternative to the Milky Way). In the early 1970's, however, infrared observations revealed two new galaxies (Fig. 28.7), both lying behind the disk of the Milky Way, and therefore obscured from our direct view. These galaxies, called Maffei 1 and Maffei 2 after their discoverer, are both rather large, one being an elliptical and the other a spiral. Their distances were not well known for awhile, and it appeared possible that they were members of the Local Group. If this had been the case, they would have displaced the Andromeda galaxy and the Milky Way as the dominant members, and this would have altered our perception of the cluster as a whole. Recent evidence has shown, however, that two galaxies are not members of the Local Group.

FIG. 28.6.   THE ANDROMEDA GALAXY. This is a large spiral, comparable to the Milky Way, and the most distant object visible to the unaided eye. Studies of this galaxy have been very useful in helping us better understand the structure and evolution of the Milky Way.

Rich Clusters: Dominant Ellipticals and Galactic Mergers

In contrast with small clusters such as the Local Group, many clusters of galaxies are much larger and contain hundreds or even thousands of members (Fig. 28.8). The density of galaxies in such a cluster is relatively high, and therefore there are numerous close encounters between galaxies as they follow their individual orbits about the center. When two galaxies pass close by each other, they exert mutual gravitational forces that can have profound effects. One cumulative result of many such encounters is that the overall distribution

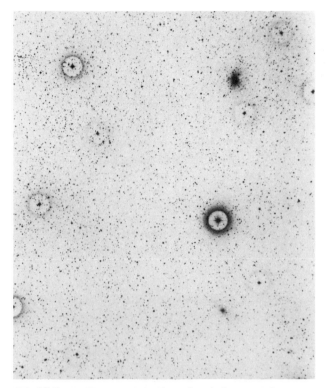

FIG. 28.7.   MAFFEI 1 AND 2. This infrared photo reveals two large galaxies (the fuzzy images, upper right and lower right), which were for awhile considered possible members of the Local Group. More extensive analysis has shown, however, that they are not.

Fig. 28.8.   A PORTION OF A RICH CLUSTER OF GALAXIES. Most of the objects in this photograph are galaxies. This cluster lies in the constellation Coma Berenices.

of galaxies in the cluster becomes smooth and more or less spherical, with the greatest density being toward the center. By contrast, small clusters rarely reach this state, but instead retain a more irregular appearance.

The central regions of large, rich clusters are highly dominated by elliptical and SO galaxies, more so than in small groups or isolated galaxies. Near the center of a rich cluster, some 90 percent of the galaxies may be ellipticals or SOs, whereas among noncluster galaxies about 60 percent are spirals. This contrast is probably a direct result of the frequent near-collisions between galaxies in dense clusters. When two galaxies have a close encounter, the tidal forces they exert on each other stretch and distort them (Fig. 28.9). Under some circumstances any interstellar matter in them can be pulled out and dispersed. The outer regions, such as the halos, can also be stripped away. The effect is similar to what happens to rocks in a tumbler: the galaxies gradually are ground down into smoothly shaped remnants. A spiral galaxy subjected to these cosmic upheavals may assume the form of an elliptical. In time, most of the spirals in an entire cluster, particularly those in the dense central region, may be converted into ellipticals and SOs (Fig. 28.10).

The frequent gravitational encounters between galaxies in a rich cluster have another interesting effect: they cause a buildup of galaxies right at the center of the cluster. When two galaxies orbiting within a cluster come together, one will always gain speed and move to a larger orbit, while the other (always the more massive of the two) will lose speed and drop closer to the center. Thus a gradual sifting process, analogous to the differentiation that occurs inside some planets as the heavy elements sink toward the core, gradually builds up a dense central conglomeration of galaxies at the heart of the cluster. There these galaxies may acutally merge, the end result being a single gigantic elliptical galaxy, which continues to grow larger as new galaxies fall in.

The scenario just described was developed rather recently, as astronomers sought to explain why elliptical galaxies are so prevalent in many rich clusters, and why a single, giant elliptical is found at the center of many of these clusters (Fig. 28.11). These giants, designated cD galaxies, are at the extreme upper limit of known galactic sizes, masses, and luminosities, and are often peculiar in certain ways, as we will learn in chapter 30.

FIG. 28.9.   A COLLISION BETWEEN GALAXIES. This shows a pair of galaxies undergoing a near-collision. At right is a computer simulation of their interaction, showing how the present appearance was created.

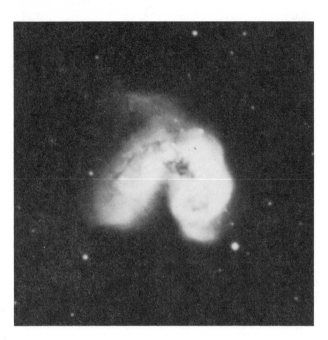

FIG. 28.10.   GALAXY TYPES IN A RICH CLUSTER. In dense, highly populous clusters of galaxies, nearly all the members in the central portion are ellipticals, and there is often a giant, dominant elliptical at the center. Only in the outer portion are many spirals seen.

Fig. 28.11.   A GIANT ELLIPTICAL GALAXY. This is M87, a well-known example of a dominant central galaxy in a rich cluster. This particular galaxy is discussed further in chapter 30.

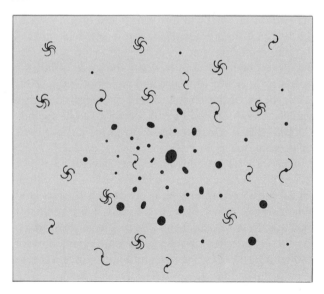

Masses of Clusters

There are two methods for measuring the mass of a cluster of galaxies, and both are quite uncertain. This is a crucial problem, as we will see in chapter 31, because of the importance of knowing how much mass the universe contains.

The simpler and more straightforward of the two methods is to estimate the masses of the individual galaxies in a cluster, using techniques described in the previous chapter, and then add up these masses. In many cases, particularly for distant clusters where it is impossible to measure the rotation curves or velocity dispersions of individual galaxies, the only way to estimate their masses is to measure their brightnesses and then use a standard mass-to-light ratio to derive their masses. This technique is inaccurate because it depends on our knowing the distance, and because it assumes that the galaxies adhere to the usual mass-to-light ratios for their types. It also does not take into account any matter in the cluster that may lie between the galaxies.

The second method is similar to the velocity-dispersion technique used to estimate masses of elliptical galaxies (Fig. 28.12). The mass of a cluster is estimated from the orbital speeds of galaxies in its outer portions; the faster they move, the greater the mass of the cluster. This method has the advantage that it measures all the mass of the cluster, whether the mass is in galaxies or between them, but it has the disadvantage that the necessary velocity measurements, particularly for a very distant cluster, are difficult. Furthermore, the technique is only valid if the galaxies are in stable orbits about the cluster; if the cluster is expanding or some of the galaxies are not really gravitationally bound to it, then the results are incorrect. In rich clusters at least, the smooth overall shape and distribution gives the appearance of a bound system, and this technique is probably valid. It may not be valid for small clusters such as the Local Group, however.

Whatever the reason, this method always leads to an estimated cluster mass that is much greater than that derived by adding up the masses of the visible galaxies. It is as though a large fraction of the cluster's mass is either in extensive galactic halos, which are ignored when the individual galaxy masses are estimated, or in intergalactic space.

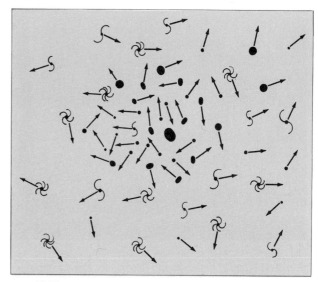

FIG. 28.12. GALAXY MOTIONS WITHIN A CLUSTER. Galaxies move randomly within their parent cluster. In a large cluster, where the overall distribution of galaxies is uniform, analysis of the average galaxy velocities leads to an estimate of the total mass of the cluster, in analogy with the velocity-dispersion method for measuring masses of elliptical galaxies (see chapter 27).

Intracluster Gas and X-ray Emission

For a variety of reasons, astronomers have sought to know whether the space between galaxies is truly empty or whether there is an intergalactic medium analogous to the interstellar material that exists within galaxies. The quantity of intergalactic matter could play a vital role in the future evolution of the universe (see chapter 31), and could have a profound effect on the fate of individual galaxies, in terms of their motions and their shapes. If galaxies move through a gaseous medium, drag will be created, which will help them congregate toward the centers of rich clusters and aid in stripping off their halos and interstellar gas; in this way spirals will be converted to ellipticals.

Searches for intergalactic gas that utilize techniques similar to those traditionally employed to study the interstellar matter in our own galaxy have generally yielded negative or ambiguous results. The lack of strong evidence for an intergalactic medium provides useful information, however. It tells us there is very little gas there, and what is there must be so hot that it is essentially invisible, because a very highly ionized gas does not create visible-wavelength absorption lines.

VIRGO CLUSTER

FIG. 28.13. INTRACLUSTER GAS IN THE VIRGO CLUSTER. This is an X-ray image from the *Einstein Observatory,* showing emission from the entire central portion of this cluster of galaxies. The X rays are being emitted by very hot gas that fills the space between galaxies.

The possibility that there is a very hot intergalactic medium seemed to be supported by recent X-ray observations of clusters of galaxies, which showed that many clusters are filled with a diffuse gas that glows at X-ray wavelengths (Fig. 28.13 and Color Plate 23). The spectrum of the X-ray emission implies that this gas is indeed very hot; its temperature is estimated to be as high as a hundred million degrees, much hotter than even the highly ionized gas in the interstellar medium in our galaxy. This observation raised the possibility that the general intergalactic void is filled with such gas, although if it is, the density outside of clusters must be still rather low, or the X-ray data would have revealed its presence.

A different type of X-ray measurement, however, carried out very recently, indicated that the hot intracluster gas is not part of a general intergalactic medium, but instead originates in the galaxies themselves. Spectroscopic measurements made at X-ray wavelengths revealed that the gas contains iron, a heavy element, in nearly the same quantity (relative to hydrogen) as in the sun and other Pop. I stars. Such a high abundance of iron could only have been produced in nuclear reactions inside of stars. Therefore this intracluster gas must once have been involved in part of the cosmic recycling that goes on in galaxies, as stars gradually enrich matter with heavy elements before returning it to space. How the gas was then expelled into

the regions between galaxies is not clear, but it may have been swept out during near-collisions between galaxies, or it may have been ejected in galactic winds created by the cumulative effect of supernova explosions and stellar winds.

Superclusters

We turn now to consider the overall organization of the matter in the universe. We noted at the beginning of this chapter that clusters of galaxies may represent the largest scale on which matter in the universe is organized, but that there is some evidence for a higher-order grouping. It appears that clusters of galaxies are concentrated in certain regions; these congregations of clusters are commonly referred to as **superclusters.** The Local Group appears to be a part of a supercluster (Fig. 28.14).

The reality of superclusters has been difficult to establish, and for awhile many astronomers were not convinced. Today, however, there are few doubts that clusters of galaxies tend to be grouped, although there is still disagreement on the significance. The best evidence lies in the distribution of rich clusters (some 1,500 of these have been catalogued), which clearly tend to congregate in certain regions, with relatively empty space in between.

The uncertainty that remains has to do with whether the groupings are random occurrences, or whether they reflect a fundamental unevenness in the distribution of matter in the universe. This is usually tested by comparing the apparent clustering of clusters with what would be expected from coincidence if they were actually distributed in a random fashion (after all, in any random scattering of objects, there will always be occasional accidental groupings). Mathematical experiments of this type appear to show that the observed superclusters could be random concentrations of clusters, and therefore may not have profound significance in terms of the universe as a whole.

The Origins of Clusters

The question of the formation of galaxies obviously must involve the formation of clusters of galaxies, since they are so common. In chapter 24 we discussed a scenario

The Orion Nebula ▶

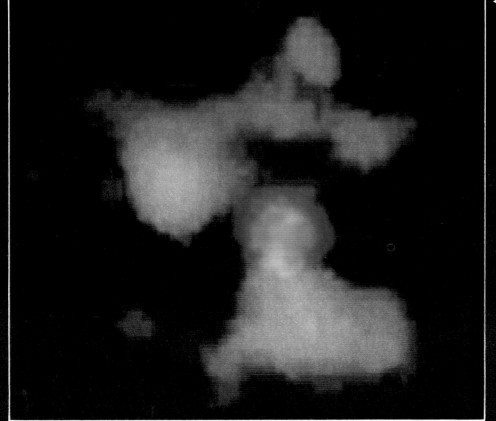

◀An infrared image
of two protostars
embedded in a dark cloud

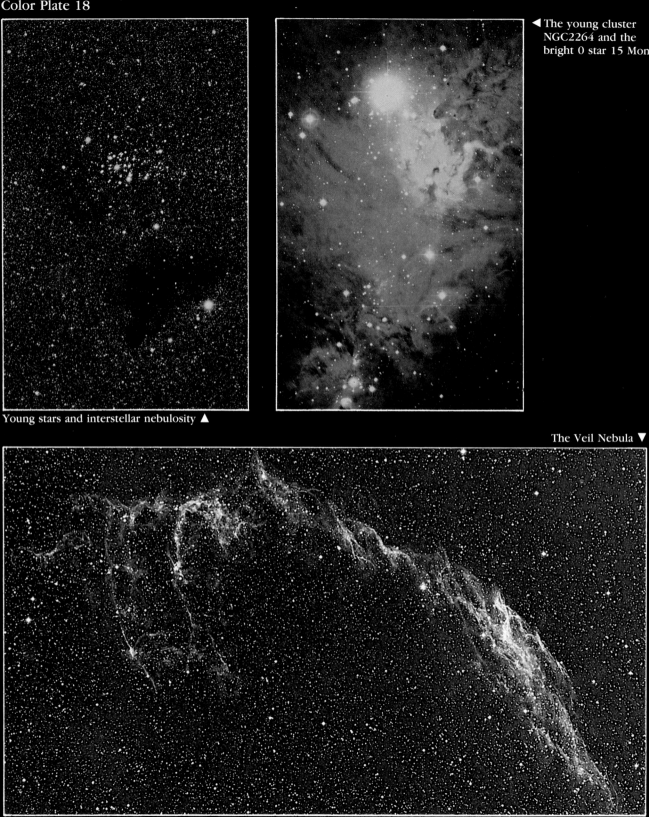

Color Plate 18

◄ The young cluster NGC2264 and the bright 0 star 15 Mon

Young stars and interstellar nebulosity ▲

The Veil Nebula ▼

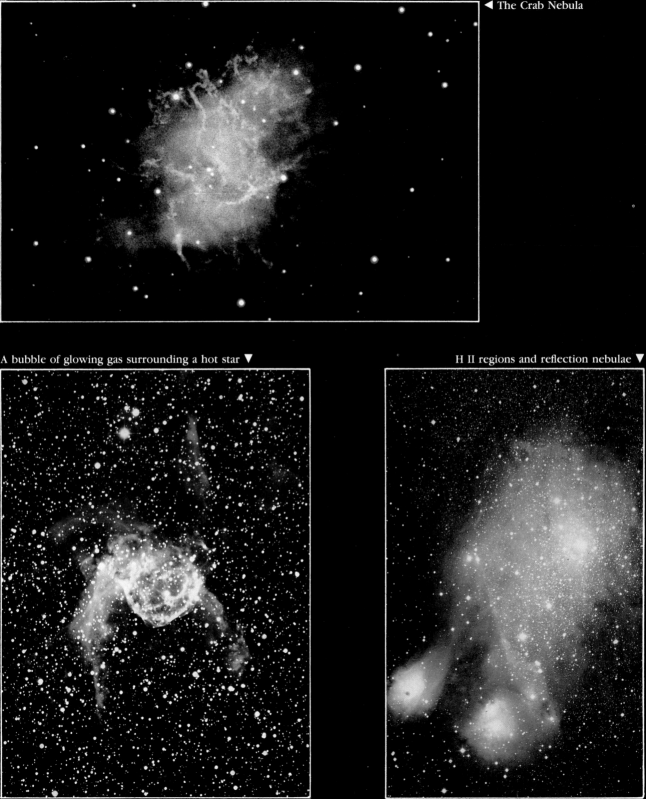

◄ The Crab Nebula

A bubble of glowing gas surrounding a hot star ▼

H II regions and reflection nebulae ▼

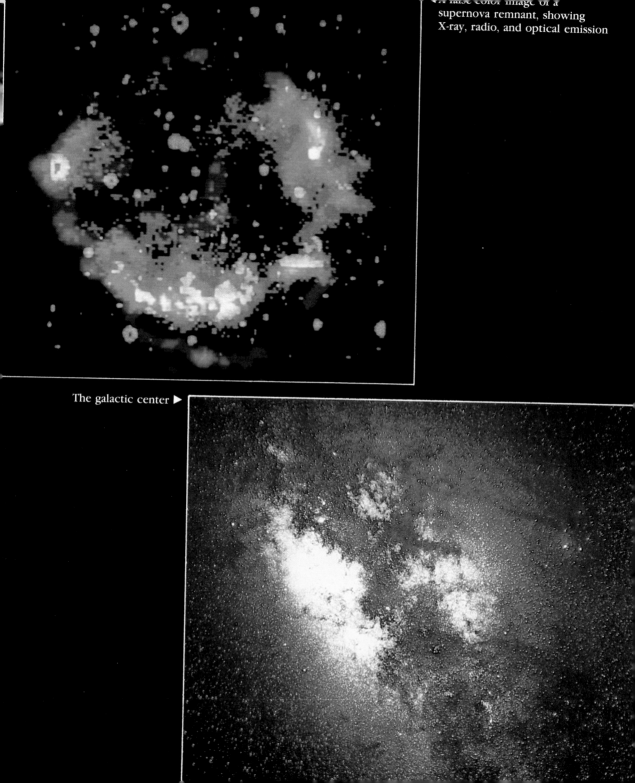

A false color image of a
supernova remnant, showing
X-ray, radio, and optical emission

The galactic center ▶

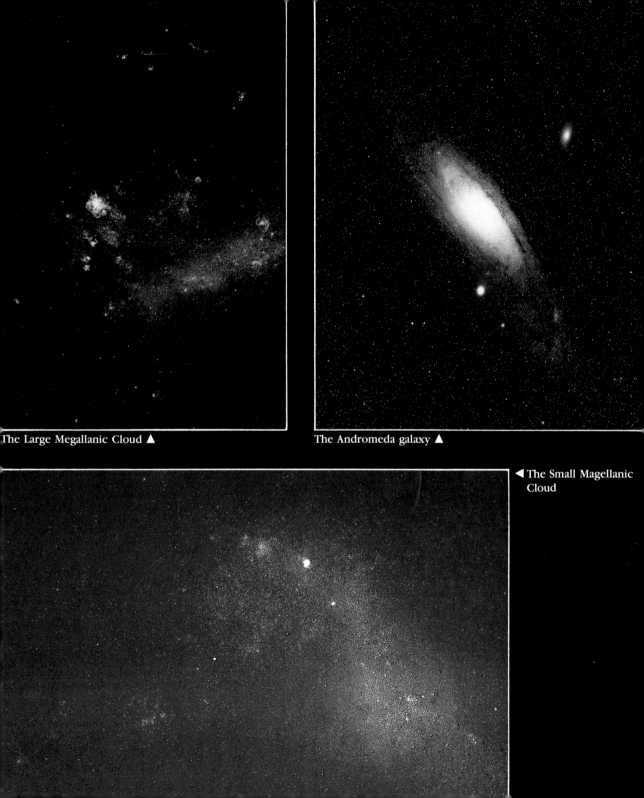

The Large Megallanic Cloud ▲

The Andromeda galaxy ▲

◄ The Small Magellanic Cloud

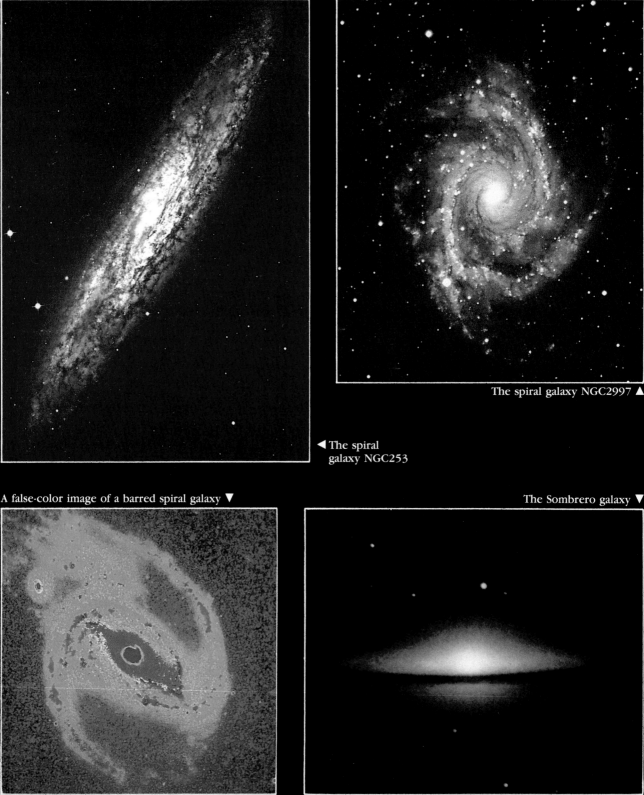

The spiral galaxy NGC2997 ▲

◄ The spiral
galaxy NGC253

A false-color image of a barred spiral galaxy ▼

The Sombrero galaxy ▼

Centaurus A ▲

A false-color radio image of a
radio galaxy with trailing side lobes ▼

An X-ray image of the Virgo cluster
of galaxies, showing the hot intracluster gas ▼

VIRGO CLUSTER

10 ARC-MIN:

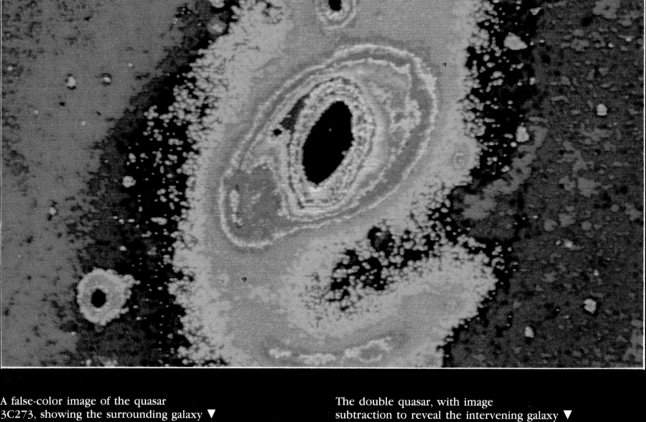

A false-color image of the quasar
3C273, showing the surrounding galaxy ▼

The double quasar, with image
subtraction to reveal the intervening galaxy ▼

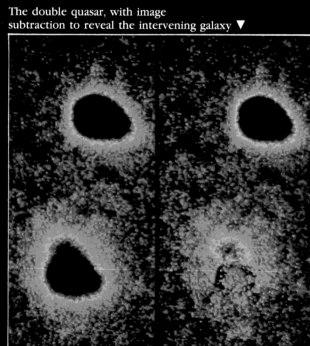

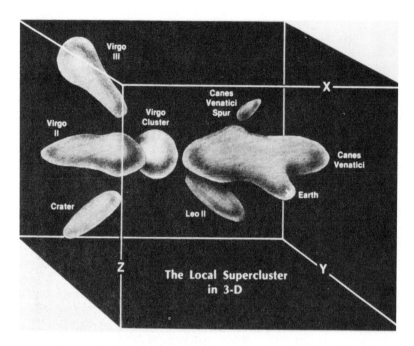

FIG. 28.14. THE LOCAL SUPERCLUSTER. This is an artist's concept of the cluster of galaxy clusters to which the Local Group belongs.

for the formation of the Milky Way, and in chapter 27 we outlined some of the theories contrasting the formation of spiral and elliptical galaxies in general. In both cases the starting assumption was that a galaxy condenses from a giant gas cloud. Now we must look a little farther into the past, to try to understand how the clouds developed in the first place, and why many of them appeared in groups and clusters.

The basic problem of galaxy formation bears a great deal of resemblance to that of star formation: we know that collapse will occur only if a sufficiently high density of matter exists in a volume sufficiently small that the mutual gravitational attraction of the gas particles overcomes the particles' natural tendency to fly about as free individuals. The Jeans criterion for collapse, discussed in chapter 21, comes into play, and it tells us that in the early days of the universe, matter was probably not dense enough (or, conversely, was too hot) to spontaneously form clouds that could collapse. We encountered a similar problem in discussing the formation of individual stars, and in that case we found a solution in interstellar turbulence, which can squeeze a cloud and force it to collapse when it otherwise would not do so. Astronomers have suggested a similar solution to the problem of galaxy formation, but there is considerable controversy over the question of whether the early universe was turbulent.

Another suggestion is that massive stars formed early in the history of the universe (calculations show that relatively small condensations could have led to star formation even before galaxies formed), and that the supernova explosions resulting from their deaths could have created sufficiently strong ripples in the universal gas to cause it to fragment and collapse into galaxies. This idea is similar to the theory of sequential star formation discussed in chapter 21.

Whatever the initial cause, the clouds that began to collapse were in many cases much more massive than individual galaxies. As in the case of star formation, when these overly massive condensations developed, they soon fragmented, because as their densities increased, smaller volumes of the gas were able to collapse on their own (Fig. 28.15). The initial giant cloud (whose mass was probably about 10^{15} solar masses) broke apart into many small ones, each on its way to becoming an individual galaxy. The entire group was still held together gravitationally, so the galaxies that formed remained in orbit about a common center, and the result was a cluster of galaxies.

One alternative theory attempts to show that the preponderance of elliptical and SO galaxies in rich clusters is the result of the cluster-formation process, rather than the subsequent conversion of spirals into ellipticals and SOs through collisions, as described earlier in this chapter. The suggestion is that when the initial large cloud fragmented, the individual clouds that

Searching for Holes in Space.

One of the fundamental properties of the universe is its homogeneity, or uniformity from one location to another. We have discussed in this chapter the hierarchy of galaxies, clusters, and superclusters, but with the tacit assumption that the universe as a whole is more or less uniformly filled with these objects. As we will see in Chapter 31, the assumption of homogeneity is commonly made in the construction of models of the universe, but it is an assumption that is sometimes questioned. If the cosmos is not uniformly filled with matter, if the distribution is somehow unbalanced or uneven, it may have important implications for the formation and history of the universe.

A great deal of effort is made by astronomers to measure the distribution of matter. The technique most often used is to choose one or more specific, representative regions of the sky and carefully search for all galaxies in those regions brighter than some pre-determined limit. Distances to all the detected galaxies are then estimated (usually using the Hubble expansion law described in Chapter 29), so that a three-dimensional picture can be developed of each sampled region

of the sky. From this it is possible to determine the overall distribution of galaxies, and to discover whether any large-scale clumpiness or unevenness exists.

From a study of this type, the astronomical world recently received news of a large portion of space that was apparently devoid of visible matter. In the course of an extensive survey of faint galaxies, a group of astronomers found a region on the sky some 35° across where there appeared to be no galaxies at distances between 240 million parsecs (Mpc) and 360 Mpc. At that distance, 35° of angular diameter corresponds to a linear dimension of 180 Mpc, so it appeared that there was a void over 100 Mpc across in all dimensions, centered some 300 Mpc away.

Statistical studies show that random distributions of galaxies should produce voids, or empty zones, no larger than about 20 Mpc across. Gravitational effects, in which clumps in the distribution of matter are enhanced by the attraction of galaxies for each other, are thought capable of producing voids up to 35 Mpc in diameter. An empty space as large as 100 Mpc across would have profound implications, for at the present time there is no known way to produce such a hole in space, unless the universe is fundamentally inhomogeneous. Therefore the announcement of the discovery of a hole in space immediately provoked a great deal of interest, and a number of astronomers began independent studies of the region of the sky where the void was thought to exist.

The original announcement of the

existence of the void was based on the discovery of three smaller regions in the same distance range that were empty of galaxies, and the assumption that these three small voids were connected. This had seemed a reasonable assumption, because it seemed quite unlikely that three unrelated voids in the same general portion of the sky should lie at the same distance from us. Several astronomers set out to search more of the region, however, to test the assumption that no galaxies would be found in the volume of space between the three small voids.

It was not long before a few faint galaxies began to be found in the region of the proposed void, and in the distance range previously thought to be empty of galaxies. Apparently there are some galaxies between the voids originally discovered. The evidence for the presence of a hole in space large enough to imply a fundamental unevenness of the distribution of matter has been considerably weakened by these discoveries. At the present time, therefore, there is no evidence to contradict the assumption that the universe is homogeneous.

To search for faint galaxies and accurately determine their distances is not a simple task, but the importance of the question of the overall distribution of matter is so significant that the work will continue. The first announcement of the discovery of a major void, even though it now appears to have been ruled out by more complete data, is helping to provoke interest in the search for holes in space.

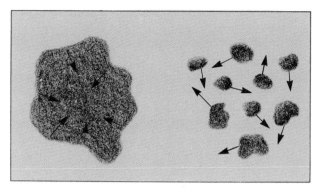

FIG. 28.15. FRAGMENTATION OF A PRIMORDIAL CLOUD. At left is a gigantic cloud, containing the mass of many galaxies. As it collapses, it breaks up into fragments, which then collapse individually, forming galaxies. This is one scenario for the origin of galaxies; in another, the individual clouds that result from fragmentation then merge to form galaxies of varying size and mass.

gether, building up masses comparable to those of galaxies. When a large number of clouds merged together, the result most often was a smooth elliptical or SO galaxy. When a small number, perhaps only two or three, combined, the result was usually a flattened, rotating cloud that became a spiral galaxy. This happened much more often in the outer parts of clusters and in small ones such as the Local Group, where the density of clouds was low.

This picture explains many of the observed characteristics of galaxy clusters, and reduces the need for subsequent galactic cannibalism to explain why there is often a gigantic elliptical at the center of a cluster. In this view, the dominant galaxy is simply the one that resulted from a merging of the greatest number of clouds. This would naturally occur at the center, where the density of clouds would be highest.

It is quite likely that the situation is actually more complex, and that a variety of processes has been involved in the formation of clusters of galaxies and their evolution to the present state. This is a central part of the problem of understanding the early development of the universe, and one that will continue to receive a great deal of attention.

formed were actually smaller than galaxies. These small clouds then orbited a common center, so for awhile there existed a cluster of gas clouds. These clouds, especially in the dense central portions of the cluster, occasionally collided with each other and merged to-

Perspective

We have, at last, finished our tour of the universe, having discussed nearly all the forms and types of organization that matter can take. We still must deal with a few specific kinds of objects that have not yet been described, but the overall picture of the universe in its present state is now more or less complete. We are ready to tackle questions having to do with the nature of the universe itself, its overall properties and its dynamic nature, and we will do so before examining some of the peculiar objects that are clues to the past.

Summary

1. Many galaxies are members of clusters, rather than being randomly distributed throughout the universe.

2. The Milky Way is a member of a cluster called the Local Group, which contains about thirty galaxies.

3. The nearest neighbors to the Milky Way are the Magellanic Clouds, both irregulars of type I.

4. The Andromeda galaxy, a huge spiral of type Sb, is similar in size and general properties to the Milky Way, and the two are among the most prominent members of the Local Group.

5. Rich clusters of galaxies may contain thousands of members. In contrast to the general situation for isolated galaxies and small groups, in rich clusters elliptical and SO galaxies are the most common, and there is a relatively small fraction of spirals.

6. The dominance of elliptical galaxies in rich clusters is probably the result of the conversion of spirals into ellipticals by tidal forces exerted by other galaxies and by drag created by intracluster gas.

7. In many rich clusters there is a giant elliptical galaxy at the center that probably formed from the merger of several galaxies that settled there as a result of collisions.

8. Masses of clusters can be determined from the sum of the masses of individual galaxies (measured by techniques described in chapter 27), or from the internal velocity dispersion of the galaxies in the cluster.

9. Although there is no evidence for a general intergalactic medium, there is a very hot, rarefied gas filling the space between galaxies in some rich clusters. Observed best at X-ray wavelengths, this gas appears to have been created by the galaxies themselves (through mass loss in galactic winds that result from stellar winds and supernova explosions).

10. Clusters of galaxies tend to be grouped into aggregates called superclusters, but these may be random concentrations, rather than fundamental inhomogeneities of the universe.

11. Clusters of galaxies formed because of an uneven distribution of matter at some point early in the history of the universe, but it is not known how this clumpiness originally developed. Theories range from a turbulent early universe to one that fragmented into clouds which then merged to form galaxies.

Review Questions

1. Why does a small cluster of galaxies such as the Local Group have an amorphous shape rather than a spherical, smooth shape such as is seen in many rich clusters?

2. If a dwarf elliptical galaxy has an absolute magnitude of $M = -15$, and could be detected with an apparent magnitude as faint as $m = +20$, how far away can these galaxies be found? Compare this distance with the diameter of the Local Group, and with the distance to the Virgo cluster, a moderately large cluster some 15 Mpc distant.

3. Why does our membership in the Local Group enable us to measure distances to galaxies farther away?

4. Discuss the similarities between a rich cluster of galaxies and a globular cluster of stars.

5. Why would it be surprising if a giant elliptical galaxy were found to be a member of the Local Group?

6. Why is it inaccurate to estimate the mass of a cluster of galaxies by using standard mass-to-light ratios for the individual members?

7. Suppose the temperature of the intracluster gas in a rich cluster of galaxies is 100 million degrees (10^8 K). At what wavelength does this gas emit most strongly? (Review the discussion of Wien's law in chapter 5.)

8. Based on what you learned about the Jeans criterion for gravitational collapse (in chapter 21), discuss how the problem of the formation of individual stars parallels that of the formation of individual galaxies.

Additional Readings

Bok, B. J. 1964. The large cloud of Magellan. *Scientific American* 210(1):32.

Geller, M. J. 1978. Large-scale structure of the universe. *American Scientist* 66:176.

Gorenstein, P., and Tucker, W. 1978. Rich clusters of galaxies. *Scientific American* 239(5):98.

Groth, E. J.; Peebles, P. J. E.; Seldner, M.; and Soneira, R. M. 1977. The clustering of galaxies. *Scientific American* 237(5):76.

Hodge, P. W. 1964. The Sculptor and Fornax dwarf galaxies. *Sky and Telescope,* Dec., p. 336.

———. 1981. The Andromeda galaxy. *Scientific American* 244(1):88.

Larson, R. B. 1977. The origin of galaxies. *American Scientist* 65:188.

Toomre, A., and Toomre, J. 1973. Violent tides between galaxies. *Scientific American* 229(6):38.

CHAPTER 29

Universal Expansion and the Cosmic Background

A recurrent theme throughout our study of astronomy has been the dynamic nature of celestial objects. Everything we have studied in detail, from planets to stars to galaxies, has turned out to be in a constant state of change. Now that we are prepared to examine the entire universe as a single entity, we should not be surprised to see this theme maintained. That the universe itself is evolving, with a life story of its own, has been accepted by most astronomers, but there was doubt in the minds of some until very recently. In this chapter we will study the evidence.

Hubble's Great Discovery

Even before it was established that nebulae were definitely galaxies, a great deal of effort went into observing them. Their shapes were studied carefully, and their spectra were analyzed in detail. Typically the spectrum of a galaxy resembles that of a moderately cool star, with many absorption lines (Fig. 29.1). Because the spectrum represents the light from a huge number of stars, each moving at its own velocity, the Doppler effect causes the spectral lines to be rather broad and indistinct. Nevertheless, it is possible to analyze these lines in some detail, and one of the early workers in this field, V. M. Slipher, discovered during the decade

FIG. 29.1. SPECTRA OF GALAXIES. These examples illustrate the broad absorption lines characteristic of spectra of large groups of stars such as galaxies. It was noticed before 1920 that the spectral lines tend to be shifted toward the red in galaxy spectra, indicating that galaxies as a rule are receding from us.

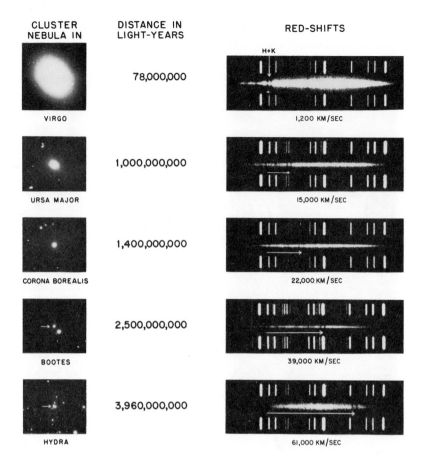

CLUSTER NEBULA IN	DISTANCE IN LIGHT-YEARS	RED-SHIFTS
VIRGO	78,000,000	1,200 KM/SEC
URSA MAJOR	1,000,000,000	15,000 KM/SEC
CORONA BOREALIS	1,400,000,000	22,000 KM/SEC
BOOTES	2,500,000,000	39,000 KM/SEC
HYDRA	3,960,000,000	61,000 KM/SEC

before the Shapley-Curtis debate in 1920 that nebulae tend to have large velocities, almost always directed away from the solar system. The spectral lines nearly always were found to be shifted toward the red, and this tendency was most pronounced for the faintest of the nebulae, Slipher measured velocities as great as 1,800 kilometers per second.

Following his work, others continued to study spectra of nebulae, with heightened interest after Hubble's demonstration in 1924 that these objects were undoubtedly distant galaxies, comparable to the Milky Way in size and complexity. Hubble and others estimated distances to as many nebulae as possible, by use of the Cepheid variable period-luminosity relation for the nearby ones, and other standard candles such as novae and bright stars for more distant nebulae.

In 1929 Hubble made a dramatic announcement: the speed with which a galaxy moves away from the earth is directly proportional to its distance. If one galaxy is twice as far away as another, for example, its velocity

is twice as great. If it is ten times farther away, it is moving ten times faster. The implications of this relationship between distance and velocity are enormous. It means that the universe itself is expanding, its contents rushing outward at a fantastic pace (Fig. 29.2). All the galaxies in the cosmos are moving away from each other.

To envision why the velocity increases with increasing distance, we find it useful to resort to a commonly used analogy. Imagine a loaf of bread that is rising. If the dough has raisins sprinkled uniformly through it, they will move farther apart as the dough expands. Suppose that the raisins are one centimeter apart before the dough begins to rise. One hour later, the dough has risen to the point where adjacent raisins are two centimeters apart. A given raisin is now 2 cm from its nearest neighbor, but 4 cm away from the next one over, and 6 cm from the next one over, and so on. The distance between any pair of raisins has doubled. From the point of view of any one of the raisins, its nearest neighbor had to move away from

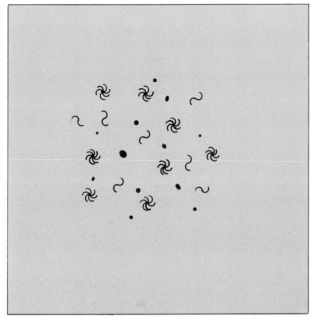 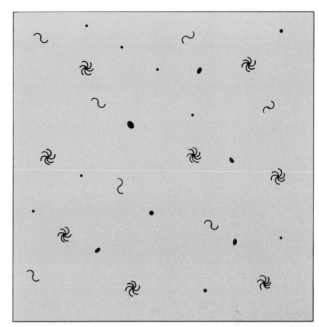

FIG. 29.2.   THE EXPANDING UNIVERSE. This shows a number of galaxies at one time (left), and again at a later time (right). All the spacings between galaxies have increased because of the expansion of the universe.

it at a speed of 1 cm/hr, the next raisin farther away had to move at 2 cm/hr, the next one had a speed of 3 cm/hr, and so on. If we let the dough continue to rise to the point where the adjacent raisins are 3 cm apart, then we find that all the distances between pairs have tripled over what they were to start with, and the speeds needed to accomplish this are again directly proportional to the distance between a given pair. The farther away a raisin is to begin with, the farther it must move in order to maintain the regular spacing during the expansion, and therefore the greater its velocity. In the same way, the galaxies in the universe must increase their separations from each other at a rate proportional to the distance between them, or else they would become bunched up (Fig. 29.3). By observing the velocities of galaxies, astronomers are keeping watch on the raisins in order to see how the loaf of bread is coming along. It is important to realize that it does not matter which raisin we choose to watch; from any point in the loaf, all other raisins appear to move away with speeds proportional to their distances from that point. Thus we do not conclude that our galaxy is at the center of the universe; from *any* galaxy, it would appear that all others are receding.

Following Hubble, a number of other astronomers

extended the study of galaxy motions to greater and greater distances, with the same result: as far as the telescopes can probe, the galaxies are moving away from each other at speeds proportional to their distances. Expansion is a major feature of the universe we live in; one that must be taken into account as we seek to understand its origins and its fate. Hubble and others

FIG. 29.3.   VELOCITY OF EXPANSION AS A FUNCTION OF DISTANCE. Above is a row of galaxies, and below is the same row one billion years later, when the distances between adjacent galaxies have doubled from 1 to 2 Mpc. From the viewpoint of an observer in *any* galaxy in the row, the recession velocity of its nearest neighbor is 1 Mpc/billion years; of its next nearest neighbor, 2 Mpc/billion years; of the next, 3 Mpc/billion years; and so on. The velocity is proportional to the distance in all cases, for any pair of galaxies.

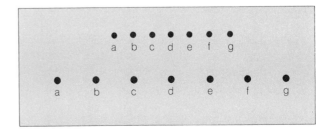

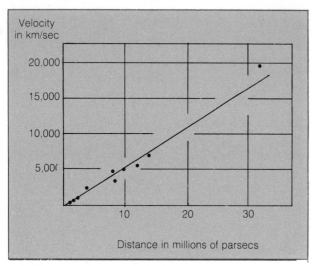

working since his time have found it useful to display the relationship between distance and velocity in the form of a plot (Figs. 29.4 and 29.5).

The galaxies within a cluster have their own orbital motions, which are separate from the overall expansion (Fig. 29.6). These motions are relatively unimportant for very distant galaxies that are moving away from us at speeds of thousands of km/sec, but for nearby ones, the orbital motion can rival or exceed the motion caused by the expansion of the universe. The Andromeda nebula, for example, is actually moving *toward* the Milky Way at a speed of about 100 km/sec, whereas at its distance of 700 kpc from us, the universal expansion should give it a velocity away from us of about 40 km/sec.

FIG. 29.4.   THE HUBBLE LAW. This early diagram prepared by Hubble and M. Humason shows the relationship between galaxy velocity and distance. The slope of the relation has been altered since, with the addition of data for more galaxies, and with improvements in the measurement of distances, but the general appearance of this figure is the same today.

FIG. 29.5.   A MODERN VERSION OF THE HUBBLE LAW. Often apparent magnitude is used on the horizontal axis instead of distance, because the two quantities are related, particularly if the diagram is limited to galaxies of the same type, as this one is. At lower left the small rectangle indicates the extent of the relationship as Hubble first discovered it; today many more galaxies, much dimmer, have been included.

FIG. 29.6.   LOCAL MOTIONS. Although there is a systematic overall expansion of the universe, individual galaxies within clusters, and even clusters within superclusters, have random individual motions. Hence within the Local Group the galaxies are not uniformly receding from each other.

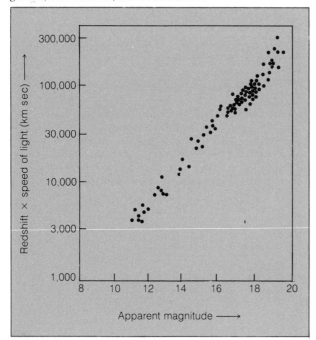

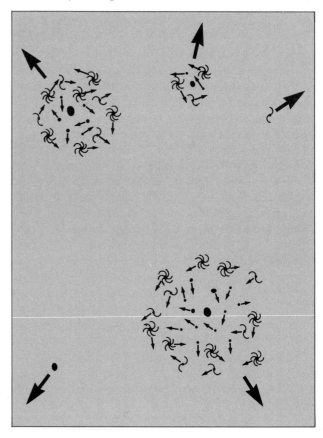

Hubble's Constant and the Age of the Universe

The relation discovered by Hubble can be written in simple mathematical form:

$$v = Hd,$$

where v is a galaxy's velocity of recession and d is its distance, in megaparsecs. The Hubble constant, H, is then given in units of km/sec/Mpc.

The value of H is difficult to establish. It is done by collecting as large a body of data as possible on galactic velocities and distances, and then deducing the value of H that best represents the relationship between distance and velocity. Hubble did this first, finding $H = 500$ km/sec/Mpc, meaning that for every megaparsec of distance, the velocity increased by 500 km/sec. The standard candles were not very well established in Hubble's day, when it was still news that nebulae were distant galaxies, and this turned out to be an overestimate. The best modern values for H are between 50 and 100 km/sec/Mpc. Until very recently, a consensus had been developing that the best value is 55 km/sec/Mpc, but a new distance determination for a number of galaxies has now suggested a value close to 90 km/sec/Mpc. The precise value of H has extremely important implications for our understanding of the universe and its expansion, and an intense research effort is being devoted to refining the estimates.

If the universe is expanding, it follows that all the matter in it used to be closer together than it is today (Fig. 29.7). If we follow this logic to its obvious conclusion, then we find that the universe was once concentrated in a single point, from which it has been expanding ever since.

From the rate of expansion, we can calculate how long ago the galaxies were all together in a single point. From the simple expression

$$\text{Time} = \text{distance/velocity},$$

we find

$$\text{age} = d/Hd = 1/H.$$

The age of the universe is equal to $1/H$, if the expansion has been proceeding at a constant rate since it began. As we will see in chapter 31, this assumption of a constant expansion rate is not strictly true (it was more rapid in the beginning), but it gives a useful estimate of the age for now. If we accept a value for H of 55 km/sec/Mpc,

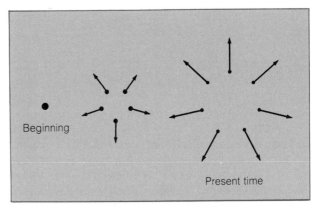

FIG. 29.7. BACKTRACKING THE EXPANSION. If the universe originated from a single point, then the present rate of expansion provides information on how long ago the expansion began. Although there is considerable uncertainty in calculations owing to the fact that the expansion rate has probably not been constant, we arrive at an estimated age for the universe of between 15 and 20 billion years.

then the age of the universe is 5.6×10^{17} sec $= 1.8 \times 10^{10}$ years (to carry out this calculation, first we must convert the value of a megaparsec into kilometers).

A simple calculation based on the observed expansion rate has led to a profound conclusion: our universe is about 18 billion years old. This fits in with what we have learned about galactic ages. The sun, for example, is thought to be about 4.5 billion years old, and the most aged globular clusters are apparently some 16 billion years old.

It is interesting to consider what happens to our estimate of the age of the universe if we choose other values of H. If Hubble's value of 500 km/sec/Mpc had been correct, then the age would have been calculated as only 2 billion years, and if the recently suggested value of about 90 km/sec/Mpc is found to be correct, then the age is 11 billion years. It is important to keep in mind that these values are somewhat overestimated, because they do not take into account the fact that the expansion rate at the earliest times was somewhat more rapid than it is today.

There are many other important questions about the universe that depend on the value of H, and this is the reason so much effort is being expended to define its value as accurately as possible. One of the principal reasons for development of the *Space Telescope* is to probe the most distant galaxies, to determine their distances and velocities so the information can be used to help us find the correct value of *H*.

Redshifts as Yardsticks

The expansion of the universe has a very practical advantage for astronomers concerned with the properties of distant galaxies: it provides a means of determining their distances. If we know the value of *H*, then we can find the distance to a galaxy simply by measuring its velocity as indicated by the Doppler shift of its spectral lines (Fig. 29.8). The distance is given by

$$d = v/H.$$

Thus, if we adopt $H = 55$ km/sec/Mpc, then a galaxy found to be receding at 5,500 km/sec, for example, has the distance

$$d = 5,500/55 = 100 \text{ Mpc}.$$

FIG. 29.8. FINDING DISTANCES FROM THE HUBBLE LAW. This diagram shows how the distance to a galaxy can be deduced directly from the Hubble law, if the redshift, hence the velocity, is measured.

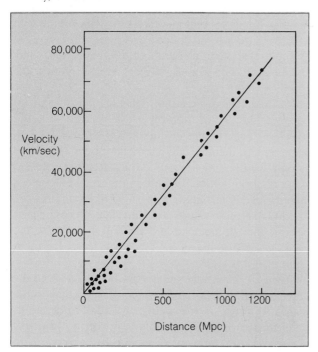

This technique can be applied to the most distant and faint galaxies, where it is nearly impossible to observe any standard candles. It is important to remember, however, that the use of Hubble's relation to find distances is only as accurate as our value of *H*, and the value that we use is derived in turn from distance determinations that depend on standard candles. Furthermore, because the expansion rate of the universe has not been constant, for very distant galaxies we need to know what the rate was in the past in order to accurately determine the distance.

A Cosmic Artifact: The Microwave Background

Since Hubble's discovery of the universal expansion, many astronomers and physicists have explored its significance, both for what it may tell us about the future and for the information it provides about the past. A number of scientists, beginning in the 1940's with George Gamow and his associates, have carried out extensive studies of the nature of the universe in its earliest days. It was quickly realized that if all the matter were originally compressed into a small volume, it must have been very hot, for the same reason that any gas heats up when compressed. The fiery conditions inferred for the early stages of the expansion have led to the term **big bang,** which is the commonly accepted title for all theories of the universal origin that start with an expansion from a single point.

Gamow's primary interest was in analyzing nuclear reactions and element production during the big bang (the modern understanding of this is described in chapter 31), but in the course of his work he came to another important realization: the universe must have been filled with radiation when it was highly compressed and very hot, and this radiation should still be with us. At first, when the temperature was in the billions of degrees, γ-ray radiation dominated, but later, as the universe cooled, X rays, and then ultraviolet light filled the universe. During the first million years or so after the expansion started, the radiation was constantly being absorbed and reemitted (that is, scattered) by the matter in the universe, but eventually (when the temperature had dropped to around 4,000 K), protons and electrons combined to form hydrogen atoms, and from that time on there was little interac-

ASTRONOMICAL INSIGHT (29.1)

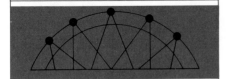

The Distance Pyramid

Having reached the ultimate distance scales, we can pause to look back at the progression of steps that got us here. As noted in the text, the use of Hubble's relation to estimate distances to the farthest galaxies depends critically on the value of H, which can only be determined from knowledge of the distances to a representative sample of galaxies. These distances in turn depend on various standard candles, themselves calibrated through the use of other methods, such as the Cepheid variable period-luminosity relation, which are applicable only to the most nearby galaxies. Our knowledge of these close extragalactic distances rests on the known distances to objects within the Milky Way, such as star clusters containing variable stars, which are used to calibrate the period-luminosity relation. The distances to these clusters, in their turn, are known from more local techniques, such as spectroscopic parallaxes and main-sequence fitting, and these ultimately depend on the distances to the most nearby stars, derived from trigonometric parallaxes.

Thus the entire progression of distance scales, all the way out to the known limits of the universe, depends on our knowledge of the distance from the earth to the sun, the basis of trigonometric parallax measurements. At every step of the way outward, our ability to measure distances rests on the previous step. If we revise our measurement of local distances, we must accordingly alter all our estimates of the larger scales, which affects our perception of the universe as a whole. This elaborate and complex interdependence of distance-determination methods is known as the *distance pyramid,* and it is sometimes depicted in graphic form, as follows.

A very important step in the distance pyramid is represented by the Hyades cluster, a galactic star cluster some 43 parsecs away. The distance to this cluster is determined from the **moving-cluster method,** in which a combination of Doppler-shift measurements and proper-motion measurements (see chapter 18) reveal the true direction of cluster motion. A comparison of angular motion and true velocity then gives the distance to the cluster. Because much of the rest of the distance pyramid depends on the application of this technique to the Hyades cluster, great care has been taken in the analysis, and every so often it is redone. Whenever the distance to the Hyades is altered slightly, the impact spreads as astronomers judge how this affects other distance scales in the universe.

DISTANCE	KEY OBJECT	METHOD
10^{-6} pc	Sun	↓ Radar, planetary motions
10^{-5}		
10^{-4}		
10^{-3}		Trigonometric parallax
10^{-2}		
10^{-1}		
1	α Centauri	
10^{1}		↓ Moving-cluster method
10^{2}	Hyades cluster	
10^{3}		
10^{4}	Limits of Milky Way	Main-sequence fitting
10^{5}	Magellanic Clouds	
10^{6}	Andromeda galaxy	Cepheid variables
10^{7}		↓ Brightest stars
10^{8}	Virgo cluster	
10^{9}		
10^{10}		↓ Brightest galaxies

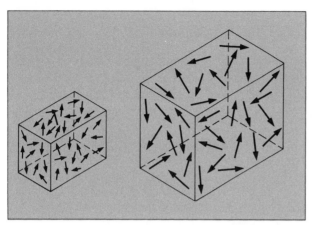

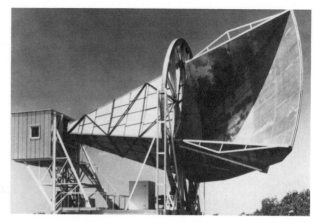

FIG. 29.9.   THE COOLING OF THE UNIVERSAL RADIATION. At left is a box representing the early universe, filled with intense gamma-ray radiation. At right is the same box, greatly expanded. The radiation is still there, but its wavelength has increased, and the temperature represented by its thermal spectrum has decreased.

FIG. 29.10.   THE RADIO RECEIVER THAT DISCOVERED THE 3-DEGREE BACKGROUND RADIATION. This is the antenna at Bell Laboratories with which Penzias and Wilson found a persistent noise that turned out to be the remnant radiation from the big bang.

tion between the matter and the radiation. The matter, in due course, organized itself into galaxies and stars, whereas the radiation has simply continued to cool as the universe has continued to expand (Fig. 29.9). The intensity and spectrum of the radiation are dependent only on the temperature of the universe, and are described by the laws of thermal radiation (see chapter 5).

Gamow and others made rough calculations that showed the present temperature of the radiation filling the universe to be very low indeed, perhaps 25 K or less. More recently, R. H. Dicke and colleagues quite independently carried out new calculations with similar results. Using Wien's law, we can easily calculate the most intense wavelength of such radiation: if the temperature is 5 K, for example, then the value of λ_{max} is 0.06 cm, in the microwave part of the radio spectrum.

Gamow's principal interest was the element production during the big bang, rather than the remnant radiation. Furthermore, in the late 1940's the technology needed to detect the radiation was years away from being ready. Hence no attempt was made in Gamow's time to search for it, and the radiation was forgotten until the work of Dicke and his colleagues. By the mid-1960's, the necessary technology was available, and Dicke and his co-workers set out to build a radio telescope capable of detecting radiation at the appropriate wavelength.

Because the expected signal would be very weak,

and because the earth's atmosphere blocks out much of the microwave radiation that enters from space, the development of such a device was difficult. In 1965, while Dicke's group was still at work on the task, a pair of radio astronomers named Arno Penzias and Robert Wilson, working at Bell Laboratories (only a few miles away from Dicke's group at Princeton University), were puzzled by a persistent static in a new radio receiver (Fig. 29.10). Try as they might, they could not eliminate this noise, and they were forced to conclude that it was of cosmic origin.

The implication soon became clear: Penzias and Wilson had discovered the universal radiation that the theorists had predicted should exist. This was one of the most significant astronomical observations ever made, for it provided a direct link to the origins of the universe. More than a piece of circumstantial evidence, this was a real artifact, the kind of hard evidence that carries weight in a court of law.

The Crucial Question of the Spectrum

There remained questions about the interpretation of the radiation detected by Penzias and Wilson, however, and doubt in the minds of some who did not accept the big-bang theory of the origin of the universe. If the radiation was really the remnant of the primeval fireball, then it was expected to fulfill certain conditions, and intensive efforts were made to see whether it did.

One important prediction was that the spectrum should be that of a simple glowing object whose emission of radiation is the result only of its temperature (that is, it should be a thermal spectrum, as explained in chapter 5). If, on the other hand, the source of the radiation were something other than the remnant of the initial universe, such as distant galaxies or intergalactic gas, it could have a different spectrum. Thus the shape of the spectrum, its intensity as a function of wavelength, was a very important test of the interpretation of its origin.

Some of the early results were confusing because of unforeseen complications, but now the situation seems to have been resolved. The spectrum does indeed follow the shape expected (Fig. 29.11), and the radiation is considered to be a relic of the big bang. The most difficult part of the task of measuring the spectrum was observing the intensity at short wavelengths, around 1 millimeter or less, where the earth's atmosphere is especially impenetrable, but this was finally accomplished with high-altitude balloon payloads. The result of all the effort expended in determining the spectrum is that the temperature of the radiation is 2.7 K, and its peak emission occurs at a wavelength of 1.1 millimeter. The radiation is commonly referred to as the **microwave background,** or the **3-degree background radiation.**

The care that went into establishing the true nature of the background radiation reflects the importance of what it has to tell us about the universe. As we will see in chapter 31, there are alternatives to the big-bang theory, and those who support these other theories have sought different explanations of the microwave radiation. Its close adherence to the spectrum envisioned in the big-bang theory has presented grave difficulties for these alternative points of view.

Isotropy and Daily Variations

Another expectation of the background radiation, in addition to the nature of its spectrum, is that it should fill the universe uniformly, with no preference for any particular location or direction. A medium or radiation field that has no preferred orientation, but instead looks the same in all directions, is said to be **isotropic.** Some of the early observations of the 3-degree background

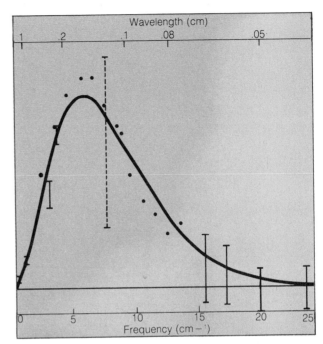

FIG. 29.11. THE MEASURED BACKGROUND SPECTRUM. This shows the spectrum of the cosmic background radiation, plotted as a function of frequency (a wavelength scale is given at the top as well). The solid line represents a thermal spectrum for a temperature of just under 3 K.

radiation were designed to test its isotropy, for again, its failure to meet this expectation could imply some origin other than the big-bang.

A test of isotropy, in simple terms, is just a measurement of the radiation's intensity in different directions, to see whether it varies. From the time it was discovered, the microwave background showed a high degree of isotropy, as expected. The issue has been pressed, however, for a variety of reasons. One possible alternative explanation of the radiation, for example, was that it arises from a vast number of individual objects, such as very distant galaxies, so closely spaced in the sky that their combined emission appears to come uniformly from all over. To test this idea, a group of astronomers have attempted to see whether the radiation is patchy on very fine scales, as it would be if it came from a number of point sources. So far no evidence of any clumpiness has been found, and the big bang theory has not been threatened.

There is a kind of subtle nonuniformity that is expected of the microwave radiation, even if it is the remnant of the big bang. Because of the Doppler shift,

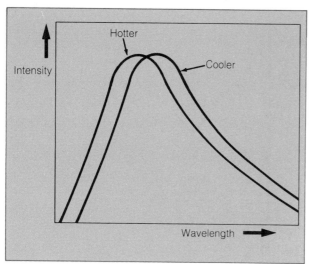

Fig. 29.12. THE EFFECT OF A DOPPLER SHIFT ON THE
COSMIC BACKGROUND RADIATION. If the observer on the earth
has a velocity with respect to the radiation, this will shift the peak
of the spectrum a little, and affect the temperature that the
observer deduces from his measurements. Motion of the earth
results in a daily cycle of tiny fluctuations in the observed
temperature, known as the 24-hour anisotropy.

FIG. 29.13. THE *COSMIC BACKGROUND EXPLORER.* This
satellite, to be launched in the late 1980's, will make the most
precise and extensive measurements yet of the microwave
background radiation.

the intensity of the radiation as seen by a moving observer will vary with direction, depending on whether the observer is looking toward or away from the direction of motion (Fig. 29.12). If the observer looks ahead, toward the direction of motion, then a blueshift occurs, the peak of the spectrum being shifted slightly toward shorter wavelengths. In terms of temperature, this means that the radiation looks a little hotter when viewed in this direction. If the observer looks the other way, there is a slight redshift, and the temperature looks cooler.

The earth is moving in its orbit about the sun. The sun, in addition, is orbiting the galaxy, and the galaxy is moving along its own path through the Local Group. All of these motions combined represent a velocity of the earth with respect to background radiation, a velocity that should produce slight differences in the radiation temperature if it is viewed in different directions. Because of the earth's rotation, if we simply point our radio telescope straight up, we should alternately see high and low temperatures, as our telescope points toward, and then twelve hours later, away from the direction of motion. This means that there should be a daily variation cycle in the radiation, referred to as the **24-hour anisotropy** (anisotropic is the term for a nonuniform medium).

The observed temperature difference from the for-

ward to the rearward direction is only a few thousandths of a degree, and very sophisticated technology was required in order for this to be measured. The data show that the earth is moving with respect to the background radiation at a speed of about 350 km/sec. When the sun's known orbital velocity about the galaxy is taken into account, this implies that the galaxy itself is moving at 520 km/sec with respect to the background radiation. This motion must be a combination of the motion of the galaxy in its orbit within the Local Group and possibly the motion of the Local Group within the supercluster to which it belongs.

The use of the 3-degree background radiation as a tool for deducing the local motions with respect to the radiation is based on the assumption that the radiation itself is fundamentally isotropic. At present we have no reason to doubt this is true. The possibility remains, however, that the radiation is not isotropic; that the universe is somehow lopsided. If it is, we will have much difficulty measuring its asymmetry, because of the complex motions of the earth, our observing platform.

Astronomers are planning future observations to improve upon the measurements of the 3-degree background radiation. A satellite called the *Cosmic Background Explorer* (Fig. 29.13) is due to be launched in the late 1980's, and will provide, from its vantage point above the earth's atmosphere, the best data yet on the spectrum and isotropy of the background radiation.

Perspective

We have learned now that the universe, like all the subordinate objects within it, is a dynamic, evolving entity. One of the grandest stories in astronomy has been the unfolding of the concept of universal expansion. An idea forced on astronomers by the evidence of the redshifts, it has led directly to the big bang picture, which in turn tells us the age of the universe, explains the origin of some of the elements, and is verified through the presence of the microwave background radiation.

Before we finish the story, by studying modern theories of the universe and their implications for its future, we must take a closer look at some of the objects that inhabit the universe. There is a breed of strange and bizarre entities whose very nature seems to be tied up with the evolution of the universe, and which therefore will provide us with a few additional bits of evidence for our final discussion.

Summary

1. It was discovered early in this century that galaxies tend to have redshifted spectra, which implies that they are moving away from us.

2. Hubble discovered in 1929 that the velocity of recession is proportional to distance, demonstrating that the universe is expanding.

3. The rate of expansion, expressed in terms of Hubble's constant H, is between 50 km/sec/Mpc and 100 km/sec/Mpc.

4. The fact that the universe is expanding implies that it originated from a single point, and the rate of expansion, if constant, tells us that it began some 10 to 20 billion years ago. This may be considered to be the age of the universe.

5. The relationship between velocity and distance allows the distance to a galaxy to be estimated from its velocity, through the Doppler effect. This method of determining distance extends to the farthest limits of the observable universe.

6. The fact that the universe originated from a single point implies that it was very hot and dense initially, and this in turn implies that the early universe was filled with radiation. As the expansion has proceeded, this radiation has been transformed into microwave radiation, with a thermal spectrum corresponding to a tem-

perature of about 3 K. The cosmic background radiation was discovered in 1965.

7. To distinguish a primordial origin for the background radiation from other possible origins (such as numerous galaxies spread throughout the universe), we must measure the spectrum to see whether it truly is a thermal spectrum. Observations to date are consistent with the assumption that it is.

8. Another important test is to determine whether the radiation is isotropic, or whether it might be unevenly intense in different directions. Careful observations have revealed no evidence of patchiness or unevenness that might imply that the radiation arises in a large number of individual sources, or that the early universe was nonuniform in any way.

9. There is a 24-hour anisotropy in the radiation, however, created by the Doppler effect resulting from the earth's motion. By comparing the radiation intensity in the forward and backward directions, we have been able to determine the earth's velocity with respect to the radiation. The earth's motion is a combination of its orbital motion about the sun, the sun's orbital motion about the galaxy, and the galaxy's motion within the Local Group.

Review Questions

1. From the information on the Doppler shift in chapter 5, determine the speed of recession of a galaxy whose ionized calcium line (rest wavelength 3,933 Å) is observed at a wavelength of 3,936.93 Å.

2. Suppose three galaxies are situated initially so that the distance from galaxy A to galaxy C is twice the distance from A to B. After a period of time, the distance between galaxies A and B has doubled. How does the distance between A and C now compare with that between A and B?

3. To help understand how the age of the universe is determined from the rate of expansion, determine the age of a supernova remnant whose outermost portions are expanding away from the center at a rate of 1,000 km/sec, if its radius is 10 parsecs. (You will have to convert parsecs to kilometers to do this. Express your answer in years.)

4. If the past expansion of the universe was more rapid than the present expansion, the correct age is less than the value calculated by assuming that H has been constant at the present value. Suppose the correct present value of H is 55 km/sec/Mpc, but the average value over the entire history of the universe has been 70 km/sec/Mpc. Calculate the age of the universe, using this value.

5. Using information from chapter 28, summarize the kinds of observations needed in order to determine the value of H, the Hubble constant.

6. Suppose three galaxies are observed, and in their spectra the position of the line of ionized calcium is observed. The rest wavelength of this line is 3,933 Å, and in the three galaxies it is observed at wavelengths of 3,936 Å, 4,028 Å, and 3,942 Å. Rank the three galaxies in order of increasing distance. (As an additional exercise, you may want to calculate the distances to the galaxies, but you need not do that to simply rank them.)

7. If the Hubble constant is $H = 55$ km/sec/Mpc, calculate the distance to a galaxy whose ionized calcium line is observed at a wavelength of 4,028 Å, instead of the rest wavelength of 3,933 Å.

8. What was the wavelength of maximum emission for the universal radiation field when hydrogen atoms formed from protons and electrons, when the temperature was 4,000 K?

9. Explain why it is necessary, when attempting to observe the cosmic background radiation, to cool the telescope and instruments to a low temperature.

10. Discuss the possible ways in which the cosmic background radiation could be anisotropic (apart from the 24-hour anisotropy), and the implications if it is.

Additional Readings

Abell, G. 1978. Cosmology—the origin and evolution of the universe. *Mercury* 7(3):45.

Layzer, D. 1975. The arrow of time. *Scientific American,* 233(6):56.

Muller, R. A. 1978. The cosmic background radiation and the new aether drift. *Scientific American* 238(5):64.

Shu, F. H. 1982. *The physical universe.* San Francisco: W. H. Freeman.

Silk, J. 1980. *The big bang: the creation and evolution of the universe.* San Francisco: W. H. Freeman.

Webster, A. 1974. The cosmic background radiation. *Scientific American* 231(2):26.

Weinberg, S. 1977. *The first three minutes.* New York: Basic Books.

In discussing the characteristics of galaxies in the preceding chapters, we have overlooked a variety of objects, some of them galaxies and some possibly not, that have unusual traits. As in many other situations in astronomy, the so-called peculiar objects, once understood, will have quite a bit to tell us about the more normal ones.

So that we could fully appreciate the bizarre nature of some of these astronomical oddities, it was best to delay their introduction until this point, where we have essentially completed our survey of the universe. We know about the expansion and the big bang, and we are just in the process of tying it all together. In this chapter we will uncover a number of vital clues in the cosmic puzzle.

The Radio Galaxies

Most of the first astronomical sources of radio emission to be discovered, other than the sun, were galaxies. Most of these, when examined optically, have

FIG. 30.1. CENTAURUS A. This is a giant elliptical radio galaxy, showing a dense lane of interstellar gas and dust across the central region. Many radio galaxies are giant ellipticals with peculiarities in visual appearance.

FIG. 30.2. A RADIO IMAGE OF A RADIO GALAXY. Here the lighter areas represent maximum radio brightness. The galaxy as seen in visible wavelengths does not even appear here; this image is completely dominated by the double side lobes.

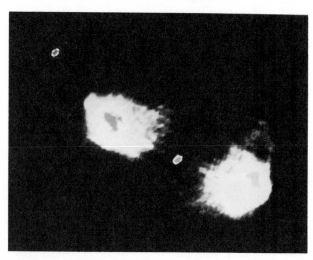

turned out to be large ellipticals, often with some un-usual-appearing structure (Fig. 30.1). The first of these objects to be detected, and one of the brightest, is called Cygnus A (it was named under a preliminary cataloguing system in which the ranking radio sources in constellations are listed alphabetically). This galaxy has a strange double appearance, and astronomers eventually found, after sufficient refinement of radio-observing techniques (see the discussion of radio interferometry in chapter 6) that the radio emission comes from two locations on opposite sides of the visible galaxy and well separated from it (Figs. 30.2 and 30.3).

This double-lobed structure is a common feature of **radio galaxies,** those with unusually strong radio emission. Most, if not all, ordinary galaxies emit radio radiation, but the term *radio galaxy* is only applied to the special cases where the radio intensity is many times greater than the norm. The core of the Milky Way is a strong radio source as viewed from the earth, but is not in a league with the true radio galaxies.

Although the double-lobed structure is standard among elliptical radio galaxies (Color Plate 23), the visual appearance varies quite a bit. We have already noted the double appearance of Cygnus A; another bright source, the giant elliptical Centaurus A, appears to have a dense band of interstellar matter bisecting it (Fig. 30.1 and Color Plate 23), and it has not one, but two pairs of radio lobes, one much farther out from the visible galaxy than the other. Other giant elliptical radio galaxies have other kinds of strange appearances. One of the most famous is M87, also known as Virgo A, which has a jet protruding from one side that is aligned with one of the radio lobes (Fig. 30.4). Careful examination shows that this jet actually consists of a series of blobs that appear to have been ejected from the core of the galaxy in sequence. Many galaxies have been discovered to have radio-emitting jets. These include galaxies already known to be peculiar, such as Centaurus A (Fig. 30.5), and at least one spiral galaxy (Fig. 30.6). In some cases double jets are seen on opposite sides of a galaxy (Fig. 30.7).

The size scale of the radio galaxies can be enormous. In some cases the radio lobes or jets extend as far as a million parsecs from the central galaxy. Recall that the diameter of a large galaxy is only one-tenth of this distance, and that the Andromeda galaxy is less than a million parsecs from our position in the Milky Way.

The first radio galaxies were detected in the 1940's, and by the 1950's studies of the radio spectra of these

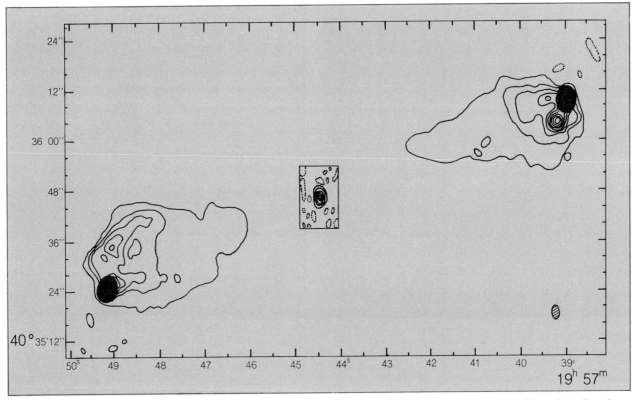

FIG. 30.3. A MAP OF A RADIO GALAXY. The intensity contours here represent radio emission from the two side lobes of a radio galaxy, Cygnus A. The image at the center illustrates the size of the visible galaxy compared with the gigantic side lobes.

FIG. 30.4. A GIANT ELLIPTICAL RADIO GALAXY WITH A JET. This is M87, also known as Virgo A, with its remarkable linear jet of hot gas. This is a short-exposure photo, specially processed to maximize visibility of the jet and its lumpy structure. On a longer-exposure photo, (see Fig. 28.11), this galaxy looks like a normal elliptical, the light from the jet being drowned out by the intensity of light from the galaxy.

objects were in progress. It was discovered that they are emitting by the synchrotron process (mentioned in chapters 12 and 23), which requires a strong magnetic field and a supply of rapidly moving electrons. The electrons are forced to follow spiral paths around the magnetic field lines, and as they do so, they emit radiation over a broad range of wavelengths. The characteristic signature of synchrotron radiation, in contrast with the thermal radiation from hot objects such as stars, is a sloped spectrum with no strong peak at any particular wavelength. The radiation from a synchrotron source is also polarized. Whenever an object is found to be emitting by the synchrotron process, it is immediately concluded that some highly energetic activity is taking place, producing the rapid electrons that are required to create the emission.

In some cases, it appears that a different process is responsible for the radio emission, but one that also requires a supply of rapidly moving electrons. If there is ordinary thermal radio emission from a galaxy, then the presence of electrons moving near the speed of

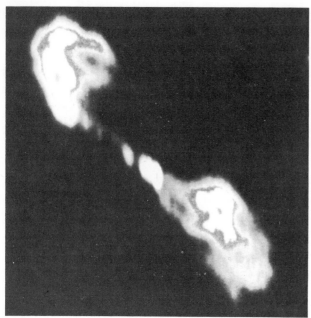

FIG. 30.5. A RADIO JET AND INNER LOBES IN CENTAURUS A. This is a radio image of the galaxy shown in figure 30.1, in the same orientation. It shows a pair of radio lobes close to the galaxy (there is more structure much farther out as well), and a section of a jet that is aligned with the lobes.

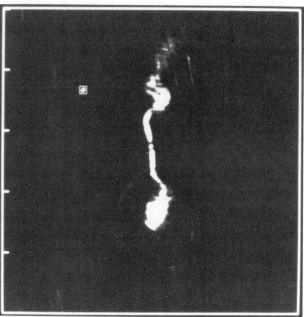

FIG. 30.7. DUAL RADIO JETS. Although visible jets are rare, radio images often reveal jetlike structures. Here is a radio image of a galaxy with a pair of jets emanating from opposite sides, aligned with the normal double radio-emitting lobes. It appears that jets such as these are responsible for the formation of the radio lobes.

FIG. 30.6. JETS IN AN UNUSUAL GALAXY. At left is a normal barred spiral galaxy, a type of object not normally found to have jets from its center. At right, however, is a computer-enhanced image of an object called 3C120, which may be the nucleus of a spiral galaxy, and which clearly has two linear jets extending through its center.

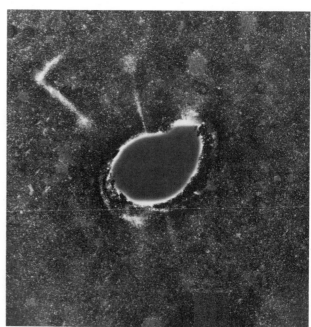

light can alter the spectrum of the radio emission, as energy is transferred from the electrons to the radiation. This process is called **inverse Compton scattering,** and it has the same implication as synchrotron radiation: there must be a source of vast amounts of energy, to produce the rapidly moving electrons required to create the radiation.

The source of this energy in radio galaxies is a mystery, although some interesting speculation has been fueled by certain characteristics of the galaxies. The double-lobed structure indicates that two clouds of fast-moving electrons are located on either side of the visible galaxy, and this leads us naturally to the impression that the clouds of hot gas may have been ejected from the galactic core, perhaps by an explosive event that expelled material symmetrically in opposite directions. This impression is heightened by the cases where jets are seen protruding from the core, as in M87, and by instances where more than one pair of lobes, aligned in the same direction but with one farther out than the other, are seen. In these galaxies it seems that successive outbursts have taken place, each one ejecting a pair of electron clouds.

One method envisioned by astronomers for producing prodigious amounts of energy from a small volume such as the core of a galaxy is to have a giant black hole there, surrounded by an accretion disk, and this has naturally been a suggested source for the energy in the radio galaxies. Recently it was reported that evidence for such an object has been found in the heart of M87, the radio galaxy with the jet. Measurements of the velocity dispersion near the center of the galaxy led astronomers to deduce that a large amount of mass must be confined in a very small volume at the core, and a black hole seemed the best explanation. Subsequent observations, however, have failed to confirm the reportedly high velocities of stars in the inner portions of the galaxy, and the case for a black hole in M87 has been weakened. The source of the energy and of the explosions, if they really do occur, remains unknown.

Seyfert Galaxies and Explosive Nuclei

Although it is true that the strongest radio emitters among galaxies are giant ellipticals, they are by no means the only ones with evidence of explosive events in their cores. Some spiral galaxies also display such behavior.

We learned in chapter 24 that even the Milky Way is not immune; there is evidence of a compact, massive object existing at its center. There are other spiral galaxies with much more pronounced violence in their nuclei. These galaxies as a class are called **Seyfert galaxies,** after the astronomer who catalogued many of them in the 1930's.

Seyfert galaxies have the appearance of ordinary spirals, except that the nucleus is unusually bright and blue in color, in contrast with the red color of most normal spiral-galaxy nuclei (Fig. 30.8). About 10 percent of them are radio emitters, with spectra indicating that the synchrotron process is at work. The radio emission usually comes from the nucleus, rather than from double lobes. The spectrum of the visible light from a Seyfert nucleus typically shows emission lines, something completely out of character for normal galaxies. The emission lines, formed in an ionized gas, are sometimes very broad, indicating velocities of several thousand kilometers per second in the gas that produces the emission. Evidently the cores of these galaxies are in extreme turmoil, with hot gas swirling about and tremendous amounts of energy being generated.

Few spirals show such effects, and it is not clear whether the Seyfert behavior is a phase they all pass through at some time, or whether only a few act up in this manner. Later in this chapter we will discuss evidence supporting the first of these possibilities.

FIG. 30.8. A SEYFERT GALAXY. This photo of NGC 4151, a well-studied example, shows the enormous intensity of light from the nucleus compared with the rest of the galaxy.

The Discovery of Quasars

In 1960, spectra were obtained of two starlike, bluish-colored objects that had been found to be sources of radio emission. No radio stars were known, and astronomers were very interested in the two objects, called 3C48 and 3C273 (their designations in a catalogue of radio sources that had recently been compiled by the radio observatory of Cambridge University, in England).

The spectra of the two objects had a completely unprecedented appearance: they contained several strong emission lines at totally unrecognizable wavelengths. The lines defied identification, and the objects became known as **quasistellar objects,** soon shortened to **quasars** or **QSO's,** because of their starlike appearance (Fig. 30.9 and Color Plate 24) but nonstellar spectra. The mysterious spectra of 3C48 and 3C273 continued to confound astronomers until finally, in 1963,

FIG. 30.9.   QUASISTELLAR OBJECTS. These starlike objects are quasars, whose spectra are quite unlike those of normal stars. These objects are probably more distant, and therefore more luminous, than normal galaxies.

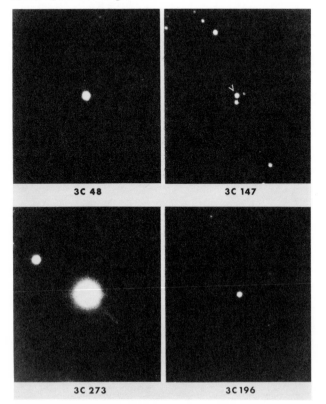

Maarten Schmidt of the Hale Observatories found the answer to the puzzle. Schmidt realized that the spacing of some of the most prominent emission lines in 3C273 coincided perfectly with the separations of the bright lines of hydrogen, except that the entire pattern was displaced by hundreds of Ångstroms. The hydrogen line whose rest wavelength is 4,861 Å, for example, was found at a wavelength of 5,639 Å, and the line that is normally located at 4,340 Å was shifted to 5,034 Å. This indicated a redshift of 16 percent; that is, the object emitting these lines must be moving away from the earth at 16 percent of the speed of light, or 48,000 km/sec! In the case of 3C48, the shift was even greater, about 37 percent, corresponding to a velocity of 111,000 km/sec.

Since the early 1960's, hundreds of additional quasars have been discovered. Only a small fraction are radio sources, and in other ways they may differ from the first two quasars, but they invariably have highly redshifted emission lines, and in very many cases they also have weak absorption lines. In a number of quasars the redshift is so huge that spectral lines whose rest wavelengths are in the ultraviolet portion of the spectrum are shifted all the way into the visible region. The grand champion today is a quasar with redshift of 378 percent, so that the strongest of all the hydrogen lines, whose rest wavelength is 1,216 Å, is shifted all the way to 4,292 Å. This quasar is *not* traveling away from us at 3.78 times the speed of light, however; for very large velocities, the Doppler-shift formula has to be modified in accordance with relativistic effects, as explained in Astronomical Insight 30.1. The velocity of this quasar is 92 percent of the speed of light, still an enormous speed.

The Origin of the Redshifts

To understand the physical properties of the quasars, we must first discover the reason for the high redshifts. We have already tacitly adopted the most obvious explanation, that they are a result of the Doppler effect in objects that are moving very rapidly, but we still must ascertain the nature of the motions. Furthermore, one alternative explanation was suggested that has nothing to do with motions at all.

One consequence of Einstein's theory of general relativity, verified by experiment, is that light can be

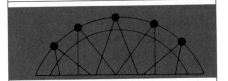

ASTRONOMICAL INSIGHT (30.1)

The Relativistic Doppler Effect

In the text we made the fantastic assertion that the largest redshift observed so far in any quasar is 378 percent; that is, the wavelengths of the spectral lines have been shifted toward the red by a factor of 3.78.

Recall from the discussion of the Doppler effect in chapter 5 that the speed v of an object is related to the shift in wavelength ($\Delta\lambda$) and to the rest wavelength (λ), by

$$v = (\Delta\lambda/\lambda)c$$

where c is the speed of light. A red shift of 378 percent means that $\Delta\lambda/\lambda = 3.78$, and this appears to imply that the quasar is receding at more than 3¾ times the speed of light! The laws of physics tell us this is not possible.

Fortunately, Einstein's theory of relativity had already provided a way out of this dilemma, long before the quasars were discovered. The simple form of the Doppler relation that we have used is actually only an approximation of the more complete form developed by Einstein, which takes into account time-dilation effects that occur when objects travel at speeds near that of light. The correct formula is:

$$\frac{\Delta\lambda}{\lambda} = \sqrt{\frac{1 + v/c}{1 - v/c}} - 1$$

which leads to the following solution for the velocity:

$$v = c\left[\frac{(z+1)^2 - 1}{(z+1)^2 + 1}\right]$$

where $z = \Delta\lambda/\lambda$ and is the most common form in which quasar redshifts are expressed.

If we use this expression to find the velocity of the most rapid quasar, which has $z = 3.78$, we get

$$v = 0.92\ c.$$

This quasar is moving at 92 percent

of the speed of light, a very high velocity indeed, but not in violation of the known laws of physics.

It is interesting to carry this idea one step further, and calculate the distance to this object, assuming that Hubble's constant has the value $H = 55$ km/sec/Mpc. In this case we find

$$d = v/H = .92 \times 300,000/55$$
$$= 5{,}018 \text{ Mpc.}$$

This is a fantastic distance, corresponding to more than 16 billion light years, meaning that we are looking back almost to the beginning of the universe. We have overestimated the distance, however, because the correct value of H in this case should be one that corresponds to an earlier time. The initial stages of the expansion were more rapid, so a larger value of H would be more appropriate, and the distance derived would be correspondingly less. Nevertheless, this is a very faraway object, seen as it was in the infancy of the universe.

---

redshifted by a gravitational field. Photons struggling to escape an intense field lose some of their energy in the process, and as this happens, their wavelengths are shifted toward the red. We have already been exposed to this concept in chapter 23 when we discussed the behavior of light near black holes; the black holes have such strong gravitational fields that no light can escape at all. For awhile it was considered possible that quasars are stationary objects sufficiently massive and compact to have large gravitational redshifts.

This suggestion has now largely been ruled out, for two reasons. One is that there is no known way for an

object to be compressed enough to produce such strong gravitational redshifts without falling in on itself and becoming a black hole. If enough matter is squeezed into such a small volume that its gravity produces redshifts as large as those in quasars, no known force could prevent this object from collapsing further. A neutron star does not have as large a gravitational redshift as those found in quasars, and we already know that the only possible object with a stronger gravitational field than that of a neutron star is a black hole.

The second objection is that even if such a massive, yet compact object could exist, its spectral lines would be very much broader than those observed in quasar spectra. The high pressure would distort electron energy levels so that the spectral lines would be smeared out (as in a white dwarf, but more extreme), and light emitted from slightly different levels in the object would have different gravitational redshifts, again causing spectral lines to be broadened.

For both of these reasons, we are forced to accept the Doppler-shift explanation for the redshifts. The problem then is to explain how such large velocities can arise. One possibility is that the quasars are relatively nearby (by intergalactic standards), and are simply moving away from us with very high speeds, perhaps as the result of some explosive event. This "local" explanation requires some care: to accept it, we must explain why no quasars have ever been found to have blueshifts. In other words, if quasars are nearby objects moving very rapidly, it is difficult to understand why none happen to be approaching us, but rather are all receding. We could argue they originated in a nearby explosion (at the galactic center, perhaps) a long time ago, so any that happened to be aimed toward us have had time enough to pass by, and are now receding (Fig. 30.10). There are serious difficulties with this picture, though, primarily in the amount of energy that would be required to get all these objects moving at the observed

velocities, an amount that would dwarf the total light output of the galaxy over its entire lifetime.

We are left with one other alternative, one that still poses problems but perhaps is more acceptable than the others. Let us consider the possibility that the quasars are very distant objects, moving away from us with the expansion of the universe. In this view they are said to be at "cosmological" distances, meaning that they obey Hubble's relation between distance and velocity, just as galaxies do. We find, if we adopt this assumption, that 3C273 is some 870 Mpc away, and 3C48 is more than 2,000 Mpc distant. (These distances are based on the assumption that the rate of expansion of the universe has been constant, something that is probably not true; this will be discussed in chapter 31.) The best evidence that the quasars are at cosmological distances is the fact that some have been found in clusters of galaxies, with the same redshift as the galaxies (Fig. 30.11). This was an extremely difficult observation to make, because at the distances where the quasars exist, the much fainter galaxies are very hard to see, even with the largest telescopes.

Supporting evidence for the cosmological-redshift interpretation is provided by the **BL Lac objects**. Named for the prototype, an object called BL Lacertae that was

FIG. 30.10. A "LOCAL" HYPOTHESIS FOR QUASAR REDSHIFTS. If quasars were very rapidly moving objects ejected from our galactic nucleus some time ago, they could all be receding from our position in the disk. This would explain the fact that only redshifts are found, but there is no known mechanism for providing the energy required to produce such great speeds.

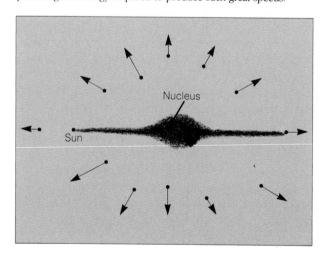

FIG. 30.11. A QUASAR IN A CLUSTER OF GALAXIES. The image at the center of this very long-exposure photo is a quasar. The much fainter objects around it are galaxies, forming a cluster of which the quasar is apparently a member. The galaxies and the quasar have the same redshift, indicating that they lie together at the same distance.

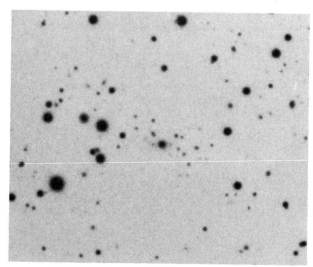

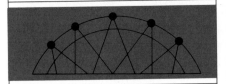

The Redshift Controversy

In the text we presented a standard set of arguments demonstrating that quasar redshifts are cosmological, because of the expansion of the universe, and implying that the quasars must be very distant objects. Although most astronomers accept this viewpoint, it is not universally adopted, and there are those who favor other explanations of the redshifts.

The evidence cited by the opponents of the cosmological interpretation consists primarily of cases where quasars are found apparently associated with galaxies or clusters of galaxies, but do not have the same redshift as the galaxies (an example is seen in Color Plate 24). It is argued in these cases that the quasar is physically associated with the galaxy or cluster of galaxies, and is therefore at the same distance, so that its redshift cannot be cosmological. The evidence favoring these arguments can be quite striking. When a quasar is found within a cluster of galaxies, there is a natural tendency to think of it as physically a member of the group. It can be argued in terms of statistics that the chances of accidental alignments between galaxies and quasars are so low that the observed associations between these objects cannot be coincidental. There are even cases where long-exposure photographs appear to show gaseous filaments connecting a galaxy with a quasar that has a different redshift. This would seem to prove that the redshift cannot be cosmological.

This radical view of quasars leaves many important questions open. The most difficult of these is how to explain the redshifts, if they really are not cosmological, for the arguments cited in the text against gravitational and local Doppler redshifts are still valid. No satisfactory explanation of the quasar redshifts has been offered by the opponents of the cosmological viewpoint.

The counterarguments center on statistical calculations. One point is that the calculations showing that the quasar-galaxy associations have low probability of occurring by chance are made after the fact. For example, after a few such connections were found, some scientists argued that these associations were very unlikely to occur as a result of chance alignments of objects randomly distributed over the sky. This is somewhat akin to arguing that the chances of being dealt a certain combination of cards in a game are very low. This is true before the deal, but meaningless after the fact. Thus it is thought by many scientists incorrect to argue that chance associations of quasars and galaxies are improbable. More to the point, it is argued, quasars associated with galaxies are more likely to be discovered than those which are not; therefore, it is natural to find a disproportionately high number of quasars that happen to be aligned in the sky with galaxies. This is an example of a **selection effect,** because the process by which quasars are found is not perfectly random and thorough; the attention of astronomers is naturally concentrated on regions where there are many galaxies, so those regions are more thoroughly searched for quasars.

Probably the most telling argument against the noncosmological redshifts for quasars arises from the fact (noted in the text) that a number of quasars have now been found to be embedded in galaxies or within clusters of galaxies that share the same redshift. Recent observations show, in fact, that *every* quasar that is sufficiently nearby for an associated galaxy to be detected is indeed embedded within a galaxy that has the same redshift as the quasar. There has not yet been an exception to this, and the data amount to virtual proof that quasars are the cores of galaxies and that their redshifts are cosmological.

The controversy is by no means over, however. Adherents of the noncosmological interpretation of the redshifts continue to find remarkable cases of apparent galaxy-quasar connections where the redshifts do not match. Perhaps observations made in the near future, with the *Space Telescope* or other planned instruments, will provide unequivocal evidence supporting one side or the other.

once thought to be a variable star, these are elliptical galaxies with very bright central cores. The nucleus of a BL Lac object displays many of the properties of a quasar, with radio synchrotron emission, variable brightness, and enormous luminosity. The surrounding galaxy shows a normal absorption-line spectrum with a redshift that is consistent with its apparent distance and which is undoubtedly cosmological. Generally, BL Lac objects have relatively small redshifts (by quasar standards), and they apparently are simply nearby quasars. The implications of the fact that they are embedded in galaxies will be discussed later in this chapter.

Distances of hundreds or thousands of megaparsecs, inferred for the quasars with very large redshifts, have enormous implications. One is that the light that we receive from quasars has traveled on its way to us for many billions of years. When we look at quasars, we are looking far into the history of the universe, and we must keep this in mind as we attempt to interpret them. The fact that quasars are seen only at high redshifts, meaning great distances, indicates that they were more common in the early days of the universe than they are now. This will prove to be a very important clue to their true nature. It has been speculated that there may be a limit to quasar redshifts; that is, no quasars can be found with a redshift greater than this limit. The implication is fantastic, for it means that we would be looking back to a time before the universe had organized itself into such objects as galaxies and quasars.

Another important implication of the great distances to quasars is that they are the most luminous objects known. Their apparent magnitudes (the brightest are thirteenth-magnitude objects) combined with their tremendous distances, imply that their luminosities are far greater then even the brightest galaxies, by factors of hundreds or even thousands. As we will see, explaining this astounding energy output is one of the central problems of modern astronomy.

The Properties of Quasars

More than 1,500 quasars have been catalogued, and more are being discovered continually. Many are being studied with care, primarily by means of spectroscopic observations. Within the last few years, it has even become possible to observe quasars at ultraviolet and X-ray wavelengths, with the use of satellite observatories such as the *International Ultraviolet Explorer* and the *Einstein Observatory*. The faintness of the quasars makes all these observations difficult, but their importance makes the effort worthwhile. The result of all the intensive work being done on quasars is that a great deal is now known about their external properties, even though their origin and especially their source of energy remain mysterious.

The blue color that characterized the first quasars discovered is a general property of quasars (except for extremely red-shifted ones, where the blue light has been shifted all the way into the red part of the spectrum), but the radio emission is not. Only about 10 percent are radio sources, contrary to the earlier impression that all are. This type of misunderstanding is known as a **selection effect**, because the process of selecting quasars was based at first on a certain assumption about their characteristics. For awhile, the only method used to look for quasars was to search for objects with radio emission, so naturally all those which were found were radio sources. The radio emission, and usually the continuous radiation of visible light as well, show the characteristic synchrotron spectrum.

Apparently *all* quasars emit X rays (again by the synchrotron process), according to the data collected by the *Einstein Observatory*, the most sensitive X-ray sat-

FIG. 30.12. A QUASAR WITH A JET. This is 3C273, one of the first two quasars discovered. This visible-light photo reveals a linear jet very much like those seen in many radio galaxies. The radio structure of quasars is usually double-lobed, also similar to radio galaxies.

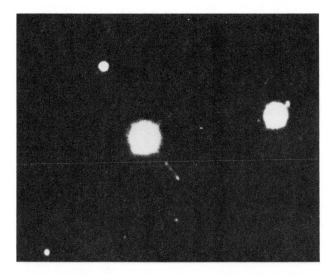

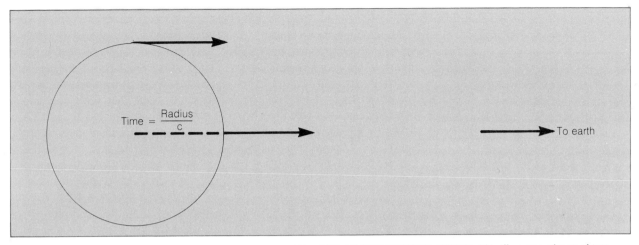

FIG. 30.13. THE IMPLICATION OF TIME VARIABILITY FOR THE SIZE OF THE EMITTING OBJECT. This illustrates why an object cannot appear to vary in less time than it takes for light to travel across it. As a simple case, this spherical object is assumed to instantaneously change its luminosity. At earth, we observe the first hint of this change when light from the nearest part of the object reaches us, but we continue to see the brightness changing gradually as light from more distant portions reaches us.

ellite yet launched. Hence X-ray observations may be the most reliable technique for finding new quasars.

In some cases, photographs of quasars reveal evidence of structure, instead of a single point of light. The most notable of these is 3C273, which played such a key role in the initial discovery of quasars. This object shows a linear jet extending from one side, closely resembling the one emanating from the giant radio galaxy M87 (Fig. 30.12). Perhaps this means that similar processes are occurring in these two rather different objects.

Many quasars vary in brightness, usually over times of several days to months or years (although in one case, variations were seen over just a few hours). This variability is very important, for it provides information on the size of the region in the quasar that is emitting the light. This region cannot be any larger than the distance light travels in the time over which the intensity varies (Fig. 30.13). There is no way an entire object that might be hundreds or thousands of light years across can vary in brightness in a time of a month or so. The light travel-time across the object would guarantee that we would only see changes over times of hundreds or thousands of years, as the light from different parts of the object reached us. Therefore the observed short-term variations in quasars tell us that the fantastic energy emitted by these objects is produced within a volume no more than a light-month, or about 0.03 parsecs, in diameter. This obviously places stringent limitations on the nature of the emitting object.

The spectra of quasars have already been described in broad outline, but there is a great deal of detail as well. All quasars have emission lines, generally of common elements such as hydrogen, helium, and often carbon, nitrogen, and oxygen. (Some of the latter elements only have strong lines in the ultraviolet, and therefore are best observed in cases where the redshift is sufficiently large to move these lines into the visible portion of the spectrum.) In many cases, the emission lines are very broad, showing that the gas that forms them has internal motions of thousands of kilometers per second. The degree of ionization tells us that the gas is subjected to an intense radiation field, which continually ionizes the gas by the absorption of energetic photons.

Many quasars have absorption lines in addition to the emission features (Fig. 30.14). Strangely enough, the absorption lines are usually at a different redshift (always smaller) than the emission lines, indicating that the gas which creates the absorption is not moving away from us as rapidly as the gas that produces the emission. Further confounding the issue is the fact that many quasars have multiple absorption redshifts; that is, they have several distinct sets of absorption lines, each with its own redshift, and each therefore representing a distinct velocity. This shows that there are several absorbing clouds in the line of sight. There are two possible origins of these clouds (Fig. 30.15). One is that the quasar itself, which is moving at the high speed indi-

ASTRONOMICAL INSIGHT (30.3)

The Double Quasar and a New Confusion Limit

In chapter 25, in our discussion of radio emission lines produced by molecules in interstellar clouds, we introduced the concept of a **confusion limit.** This is a point beyond which, because of the complexity of the universe, it is simply impossible to learn anything more by further observation. It is a fundamental limit on how deeply we can probe.

So far in the development of astronomy, all the limitations have been imposed by technological shortcomings, by the earth's atmosphere, or by other factors that in principle can be overcome. Now, however, we find the first ominous signs of a confusion limit that could prevent us from ever extending our view to the limits of the universe.

Recently a pair of quasars was discovered, very near each other in the sky, with properties so nearly identical that it has been concluded that they are in fact two images of the same quasar. If so, then astronomers have found the first example of a **gravitational lens,** something predicted to be a possibility on the basis of Einstein's theory of general relativity. Since a gravitational field can bend light rays, it is possible for such a field to act as a lens. In particular, calculations have shown

that the gravitational field of a galaxy can bend the light from a distant object around it. If the galaxy were perfectly aligned with the more distant object, then we would see either a single-point image, as if the galaxy were not there (this would happen only if we were precisely at the focal point of the lens), or, more likely, a circular ring-shaped image. In the more probable event that the galaxy is slightly off-center between us and the distant light source, we will see two or more distinct images, on either side of the intervening galaxy. The double quasar is thought to be such a case, the two images reaching us by coming around opposite sides of the gravitational lens formed by a galaxy between us and the quasar. Careful photographic studies have revealed what appears to be a galaxy between the two images of the quasar

FIG. 30.14. A SCHEMATIC QUASAR SPECTURM. All quasars apparently have emission lines, usually of great breadth. Many also have absorption lines, usually at different (always smaller) redshifts than the emission lines. The absorption lines are formed in material that is not receding from us as rapidly as the quasar.

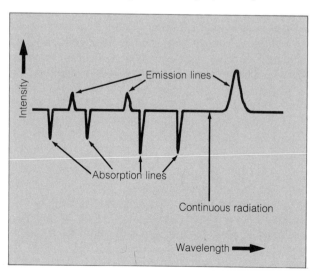

cated by the redshift of the emission lines, is ejecting clouds of gas, some of which happen to be aimed toward the earth. These clouds therefore have lower velocities of recession than the quasar itself, so the redshift of the absorption lines they form is less than the redshift of the emission lines formed in the quasar.

The alternative view, which is favored by most astronomers, is that the absorption lines arise in the gaseous halos of galaxies that happen to lie between us and the more distant quasar. Each galaxy has a recession velocity determined by the expansion of the universe, and because the intervening galaxies are not as far away as the quasars, they are not moving as fast and therefore have lower redshifts. If this interpretation is correct, then analysis of the absorption lines should provide important information on the nature of the extensive halos thought to surround many galaxies.

In some cases the absorption spectra of quasars show only lines created by hydrogen, and none created by heavier elements. These lines apparently arise in gas

(Color Plate 24), which lends support to this interpretation.

If what we are observing is in fact a gravitational lens, it follows that if we extend our observations to more and more distant quasars, the likelihood of encountering other intervening galaxies will increase, so that our view is more and more likely to be distorted by gravitational lenses (in fact, several other possible gravitational lenses have already been found). If we look far enough away, it will be like looking through a shower-stall glass, and everything in the distance will be distorted. At that point we will have reached a fundamental limit; it will be impossible to unambiguously study the farthest frontiers of the universe, because we will not be able to see clearly. We will not know the true locations of the objects we see, nor will we be able to say how many objects there are, because of the multiplicity of images.

This diagram illustrates how an intervening galaxy can act as a gravitational lens, forming two images of a distant quasar.

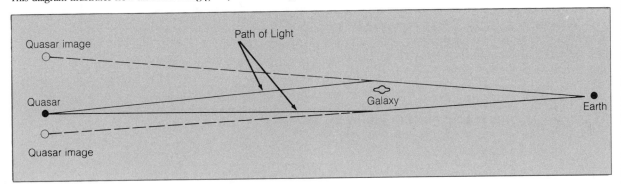

that has not been enriched at all by stellar processing, and it is thought that these might be intergalactic clouds that have never formed into galaxies, so that no stars have formed in them. These may be material remnants of the big bang; if so, they can tell us much about the exact amount of nuclear processing that occurred in the early stages of the expansion.

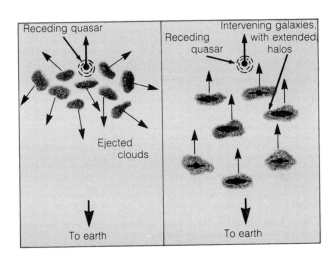

FIG. 30.15. TWO EXPLANATIONS OF QUASAR ABSORPTION LINES. In one scenario (left), the absorption lines are formed by clouds of gas ejected from the quasar. Clouds ejected directly toward the earth have lower velocities of recession than the quasar itself. In the other more likely view (right), the absorption lines are formed in the extended halos of galaxies that happen to lie between us and the quasar. The galaxies are closer to us than the quasar, and therefore have lower velocities of recession and smaller redshifts.

Galaxies in Infancy?

A variety of theories, some of them rather fanciful, have been proposed to explain the quasars. One idea has gradually become widely accepted, however, and we will restrict ourselves to discussing only that one hypothesis, keeping in mind that there are other suggestions.

The prevailing interpretation of the nature of quasars was inspired by the fact that they existed only in the long-ago past, by their association with the BL Lac objects (which clearly are galaxies), and by their resemblance to the nuclei of Seyfert galaxies. The statement that they only existed in the past is based on the fact that all are very far away; the light travel-time ensures that we are seeing things only as they were billions of years ago, not as they are today. The resemblance to Seyfert nuclei is striking: both are blue; both are radio sources in about 10 percent of the cases; both vary on similar time scales; and both have very similar emission-line spectra. Seyferts lack the complex absorption lines often seen in the spectra of quasars, but this would be expected if the quasar absorption lines are formed in the halos of intevening galaxies, since Seyfert galaxies are not so far away that there are likely to be many other galaxies along their lines of sight. The nuclei of Seyferts differ from quasars also in the amount of energy they emit, being considerably less luminous.

The picture that is developing is that quasars are very young galaxies, with some sort of youthful activity taking place in their centers. In this view, the Seyfert galaxies are descendants of quasars, still showing activity in their nuclei, but with diminished intensity. If we carry this idea a step further, we are led to the suggestion that normal galaxies like the Milky Way are later stages of the same phenomenon; recall the mildly energetic (by quasar standards) activity in the core of our galaxy.

The notion that quasars may be infant galaxies is supported by some rather direct evidence. As noted earlier, some quasars have been found associated with clusters of galaxies, showing that they can be physically located in the same region of space at the same time. In addition, there are a few examples where careful observation has revealed a fuzzy, dimly glowing region surrounding a quasar (Fig. 30.16 and Color Plate 24). This is probably a galaxy, at the center of which the quasar is embedded. It is likely that all quasars are located in the nuclei of galaxies, which are so much fainter that they usually cannot be seen. A short-exposure

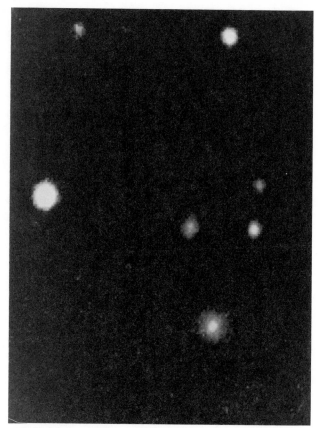

FIG. 30.16. A QUASAR EMBEDDED IN A GALAXY. This photo, obtained with an electronic detector, contains several star images (top, sides), a galaxy (slightly elongated object just to the right of center), and a quasar (lower right). The quasar image is extended, just as the galaxy image is, and the fuzzy region surrounding it is indeed a galaxy. In every case where a quasar is close enough to us that a galaxy would be bright enough to be detected, one has been detected, indicating that all quasars lie at the centers of galaxies.

photograph of a Seyfert galaxy (Fig. 30.17) shows only the nucleus, which is very much brighter than the surrounding galaxy. It is clear that if we observed one of these galaxies from so far away that it was near the limit of detectability, all we would see would be a blue starlike object resembling a quasar.

Although we can make a strong case that quasars are very young galaxies, we are still far from understanding all their properties. The main mystery remaining is the source of the tremendous energy of quasars and active galactic nuclei. A number of possibilities have been raised, but so far none is certain.

One idea invokes the presence in quasars of **antimatter,** which combines with normal matter, convert-

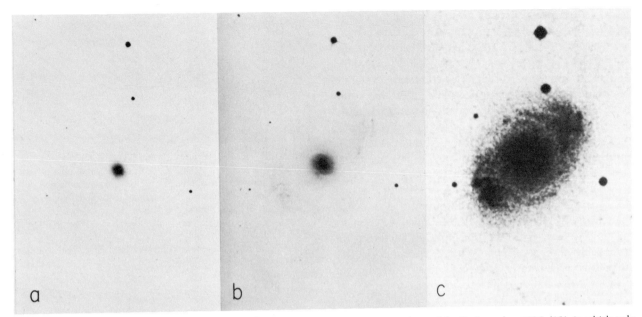

FIG. 30.17.   THREE EXPOSURES OF A SEYFERT GALAXY. At left is a short-exposure photo of the Seyfert galaxy NGC 4151, in which only the nucleus is seen, resembling the image of a star. The center and right-hand images are longer exposures of the same object, revealing more of its outer, fainter structure. This sequence illustrates how a quasar, which is even more luminous than a Seyfert galaxy nucleus, can appear as a starlike image even though it is embedded in a galaxy.

ing all the mass of both into energy. Every subatomic particle has an analogous **antiparticle,** similar in mass but opposite in other quantities such as electrical charge. We have mentioned, for example (in chapter 20), the positron, which is the antiparticle of the electron. The matter of which we and our surroundings are made is "normal" matter, but it is not known how much antimatter might have been produced in the big bang, nor whether much of it exists today in other parts of the universe. If a large amount does exist, or did in the distant past, then it is possible that it combined with normal matter in dense cores of young galaxies, releasing vast amounts of energy in the process. The rate at which energy is produced is given by Einstein's famous equation $E = mc^2$. Since the matter and antimatter that combine are totally annihilated, this is the most efficient mechanism for producing energy, easily capable of accounting for the tremendous luminosities of quasars.

Another suggestion, made shortly after the quasars were discovered, is that their power arises in multiple supernova explosions. The idea is that in the dense inner regions of a young galaxy, many massive, short-lived stars could form and quickly (by astronomical standards) evolve to the point where they explode in supernovae. If we imagine that these explosions are taking place at a sufficiently great rate, they could maintain a steady luminosity comparable to that of a quasar. The observed variability could then be a result of random fluctuations in the frequency of supernovae. A mechanism of this sort could help explain how a galaxy such as ours could have undergone a rapid buildup of heavy elements early in its history, as evidence suggests (see chapter 26). The supernova rate in the galactic core would then decrease as the galaxy aged and the gas was gradually dispersed, slowing the rate of star formation.

A third possibility, one that keeps coming up in situations where large amounts of energy seem to be emanating from a small volume, is that quasars are powered by massive black holes. In the chaotic early days of collapse, when a galaxy was just forming out of a condensing cloud of primordial gas, it may be that a great deal of material collected at the center. If many stars formed there, frequent collisions could have caused them to settle in more and more tightly, until they coalesced to form a black hole. The gravitational influence of such a massive object would have then stirred up the surroundings, creating the high electron velocities required to fuel the synchrotron emission. Infalling matter would have formed an accretion disk around the black hole, from which X rays would be emitted, in analogy with stellar black holes and neutron stars in binary systems. It has even been suggested that somehow very hot gas can be ejected along the poles of the accretion disk, accounting for the jets and double-lobed radio sources that are often observed (Fig. 30.18). This mechanism, like the others just cited, seems capable of producing the required amounts of energy.

Whatever the origin and power source of the quasars, we will certainly learn a lot about the nature of matter and the early history of the universe when we are able to answer all the questions about these fundamentally important objects. The *Space Telescope,* with its broad wavelength coverage and great sensitivity, will provide invaluable information on quasars.

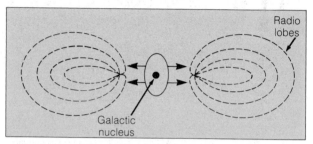

FIG. 30.18. A GEOMETRICAL MODEL FOR A QUASAR, SEYFERT GALAXY, OR RADIO GALAXY. All of these objects have certain features in common that fit the picture shown here. A central object (most likely a supermassive black hole with an accretion disk) ejects opposing jets of energetic charged particles. These jets produce synchrotron radiation, and build up double radio-emitting lobes on either side of the central object. These lobes extend well beyond the confines of the galaxy in which the central black hole is embedded.

Perspective

In this chapter we have learned of new and wondrous things, the mighty radio galaxies and the enigmatic quasars. All of the fantastic objects discussed here may be related, since each is characterized by explosive activity at the core and the apparent ejection of hot gas in opposing directions. Along the way we have gained several important bits of information on the universe itself. It is time to tackle the fundamental question of the origin and fate of the cosmos.

Summary

1. Radio galaxies emit vast amounts of energy in the radio portion of the spectrum. These galaxies, often giant ellipticals, commonly show structural peculiarities. The radio emission is nonthermal (usually synchrotron radiation), and in most cases is produced from two lobes on opposite sides of the visible galaxy.
2. The source of energy in a radio galaxy is unknown, but appears to be concentrated at the core.
3. Seyfert galaxies are spiral galaxies with compact, bright blue nuclei that produce emission lines characteristic of high temperatures and rapid motions.
4. Several point sources of radio emission were found in the early 1960's to look like blue stars, and were therefore called quasistellar objects, or quasars.
5. Quasars have emission-line spectra with very large redshifts, corresponding to velocities that are a significant fraction of the speed of light. If these redshifts are cosmological, then quasars are on the frontier of the universe, are extremely luminous, and are seen as they were billions of years ago.
6. Quasars, which are compact, blue objects, are sometimes radio sources and they usually emit X rays; in some cases they vary over times of only a few days. This implies that their tremendous energy output arises in a volume only a few light-days across.
7. Many quasars have absorption lines at a variety of redshifts (always smaller than the redshift of the emission lines), which may be created by matter being ejected from the quasars, or, more likely, by intervening galactic halos.
8. The most satisfactory explanation of the quasars is that they are very young galaxies. This interpretation is

supported by the fact that they are only seen at great distances (that is, they only existed long ago), and by their resemblance to the nuclei of Seyfert galaxies. In several cases photographs have revealed galaxies surrounding quasars, which confirms this suggestion.

9. The source of the energy that powers quasars is a premier mystery of modern astronomy. The most probable explanation is that a quasar has a massive black hole at its core; this can produce enough energy (from the infall of matter) to account for the quasar's great luminosity as well as its time variability. By inference, similar objects may exist in the nuclei of galaxies such as Seyfert galaxies, radio galaxies, and the Milky Way.

Review Questions

1. Summarize the differences between radio galaxies and the radio emission from the Milky Way.

2. In radio galaxies with more than one pair of radio-emitting lobes on either side, the alignment between the inner and outer lobes is generally quite precise. What does this tell us about the source of the lobes?

3. Suppose that emission lines are found in the spectrum of a quasar at wavelengths of 2,189 Å, 2,790 Å, and 5,040 Å. If these are identified as the Lyman-alpha line of hydrogen (rest wavelength 1,216 Å), the three-times-ionized carbon line (rest wavelength 1,550 Å), and the strong line of ionized magnesium (2,800 Å), what is the redshift of this quasar? How far away is it, if Hubble's constant has the value $H = 55$ km/sec/Mpc? (Note: You must use the relativistic redshift equation.) If the apparent magnitude of this quasar is $m = +16$, what is its absolute magnitude?

4. Explain, in your own words and using your own sketch, why the time scale of variations in light from a source such as a quasar places a limit on the diameter of the emitting region.

5. Compile a list of similarities and differences between quasars and Seyfert galaxies.

6. Why are quasar absorption lines *always* observed at smaller redshifts than the emission lines? Would this necessarily be true of the redshifts if the emission lines were not cosmological?

7. From what you have learned about quasars in this chapter and about the *Space Telescope* in chapter 6, describe some important observations of quasars that might be made with the *Space Telescope*.

8. If quasars are young galaxies, how does this help demonstrate the difficulty of using faraway galaxies as standard candles in estimating large distances in the universe?

9. Explain why nearly all the first quasars detected are radio sources, but only about 10 percent of *all* quasars are.

10. Explain why the material in intergalactic clouds may potentially provide clues about conditions very early in the life of the universe.

Additional Readings

Blandford, R. D., Begelman, M. C., and Rees, M. J. 1982. Cosmic jets. *Scientific American* 246(5):124.

Chaffee, F. H. 1980. The discovery of a gravitational lens. *Scientific American* 243(5):60.

Disney, M. J., and Veron P. 1977. BL Lac objects. *Scientific American* 237(2):32.

Hazard, C., and Mitton S., eds. 1979. *Active galactic nuclei.* Cambridge, England: Cambridge University Press.

Kellerman, K. I. 1973. Extragalactic radio sources. *Physics Today* 26(10):38.

Sandage, A., Sandage, M., and Kristian J., eds. 1976. *Galaxies and the universe.* Chicago: University of Chicago Press.

Shipman, H. L. 1976. *Black holes, quasars, and the universe.* Boston: Houghton-Mifflin.

Silk, J. 1980. *The big bang: the creation and evolution of the universe.* San Francisco: W. H. Freeman.

Strom, R. G., Miley, G. M., and Oort, J. 1975. Giant radio galaxies. *Scientific American* 233(2):26.

CHAPTER 31

Cosmology: Past, Present, and Future of the Universe

Given the time and opportunity, humans have, through the ages, devoted themselves to speculation on the grandest scale of all. Poets, philosophers, and theologians have approached the question of the origin and future of the universe in a countless variety of ways. So have astronomers, with the exception that a somewhat restrictive set of rules is followed: the answers that are accepted must not violate known laws of physics. By retaining this requirement, scientists attempt to approach the problem in an objective, verifiable manner.

There are difficulties in maintaining this idealized posture, however, and we shall try to clearly point them out. Because the universe in which we live is, as far as we know, unique, we have no opportunity to check our hypotheses by comparison with other examples. Furthermore, no matter how thoroughly and rigorously we trace the evolution of the universe by application of known physical laws, there will always be fundamental questions that are beyond the scope of physics. As a result, even the most careful and objective scientists reach a point where they have to make certain unverifiable assumptions, and at that point, they become philosophers or theologians. In this chapter we restrict ourselves to questions that in principle have objective, verifiable answers.

Technically, the study of the universe as it now appears is **cosmology,** which is really what the preceding thirty chapters have been all about. The study of its origins is **cosmogony,** and this word applies to the big bang as well as to the earlier theories on the origin of the solar system, once thought to be the entire universe. In practice, the general subject of the nature of the universe and its evolution is lumped under the heading of cosmology, and so it is on a pursuit of this subject that we now embark.

Underlying Assumptions

To even begin to study the universe as a whole, we need to make certain assumptions. These can in principle be tested, although it is not clear that there is a practical way to do so. Therefore in making these assumptions, the astronomer is on the verge of acting as philosopher.

The central rule that the cosmologist sets forth is that the universe must look like the same at all points within it. This does not mean that the appearance of the heavens should be literally identical everywhere, but it means that the general structure, the density and distribution of galaxies and clusters of galaxies, should be constant (Fig. 31.1). This assumption states that the universe is **homogeneous.**

A related, but slightly different assumption has to do with the appearance of the universe when viewed in different directions. The assumed property in this case is **isotropy;** that is, the universe must look the same to an observer, no matter in which direction the telescope is pointed. We have already encountered this concept in connection with the microwave background radiation. Again, the assumption of isotropy does not imply identical constellations of galaxies and stars in all directions, just comparable ones.

Both of these assumptions are thought to apply only

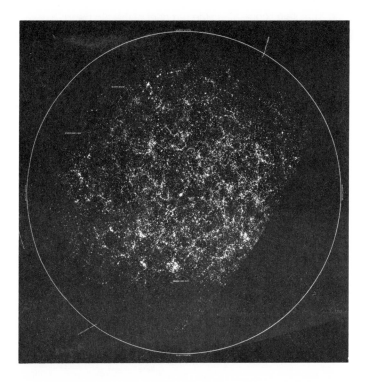

FIG. 31.1. THE COSMOLOGICAL PRINCIPLE. The universe is assumed to be homogeneous and isotropic, meaning that it looks the same to all observers in all directions. Here is a segment of the universe, filled with galaxies. Their distribution, although not identical from place to place, is similar throughout.

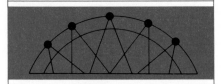

The Steady-State Universe

Despite the nearly overwhelming evidence favoring the big-bang theory, alternative ideas have persisted almost to the present time. One of the most widely regarded of these is called the **steady state universe,** championed for many years by the British astrophysicist Fred Hoyle and his colleagues.

The steady-state model of the universe is based on a modification of the Cosmological Principle, which states that the universe is homogeneous and isotropic, and *does not change with time*. This statement, called the **Perfect Cosmological Principle,** says, in other words, that the universe looks the same in all directions to all observers every-

where and at every time. This allows the universe to have no beginning and no end, and it forbids any changes to occur in its overall properties.

The observed fact that expansion is occurring would seem to immediately rule out the steady state, because it would lower the density of the universe with time, but this is not necessarily the case. A constant density could be maintained if matter were continuously created at just the right rate to compensate for the expansion. This may seem outlandish, and therefore impossible, but it is no more impossible than the fact that the universe exists at all. No matter which theory you prefer, there is a fundamental, unanswerable question about where the matter came from. The steady-state theory simply shifts this mystery from a single event at a long-ago time to continuous creation throughout the universe. The rate of creation of matter required to maintain constant density despite the expansion would be exceedingly small, averaging one new hydrogen atom per year in a cubical volume one kilo-

meter on each side. This would never be detected if it were occurring. The steady-state theory cannot be ruled out on that basis.

In the last fifteen years or so, however, some rather compelling evidence against it has been found. The requirement that the universe cannot change with time means that galaxies should have been just as numerous long ago as they are today, yet careful counts of distant radio galaxies show that they were more abundant in the past. Even more striking is the disparity between the number of quasars that existed long ago and the lack of quasars today.

The advocates of the steady-state theory were already burdened with these problems in 1965, when the 3 degree microwave background was discovered. This has turned out to be the death knell for the steady-state theory, for no satisfactory explanation of this radiation, other than as the remnant of the big bang, has been found. Today acceptance of the big-bang theory is nearly unanimous.

on the largest scales, larger than any obvious groupings in the universe, such as clusters or superclusters of galaxies.

The statement that the universe is homogeneous and isotropic is often referred to as the **Cosmological Principle,** and is the primary underlying assumption made by most cosmologists. This principle is often stated in terms of how the universe looks to observers; that is, the universe looks the same to all observers everywhere. To violate this principle would mean that there are special locations in the universe that are distinctly different from others, and this would fly in the face of

all the lessons we have learned over the ages since the geocentric theory was overturned.

It is difficult to verify the Cosmological Principle. We can test the assumption of isotropy by looking in all directions from earth, and indeed we do not find any deviations. On the other hand, we cannot test the homogeneity of the universe by traveling to various other locations to see whether things look any different. What we can do is count very dim, distant galaxies, and observe whether their densities appear to be any greater or less in some regions than in others, but even this is complicated by the fact that we are looking back in time to an era when the universe was more compressed and dense than it is now. The 3-degree background radiation gives us another tool for testing both homogeneity and isotropy, and it also appears to satisfy the Cosmological Principle. The best we can say for now is that the available evidence supports this principle, but does not completely rule out the possibility of anisotropies or inhomogeneities.

Einstein's Relativity: Mathematical Description of the Universe

The simple act of making assumptions about the general nature of the universe by itself tells us very little about its past or its future. It serves to elucidate certain properties of the universe as it is today, but to tie this into a quantitative description, to develop a theory capable of making predictions that can be tested, requires a mathematical framework. In developing such a mechanism for describing the universe, the astronomer seeks to reduce the universe to a set of equations and then to solve them, much as a stellar-structure theorist studies the structure and evolution of stars by constructing numerical models.

The most powerful mathematical tool for describing the universe was developed by Einstein (Fig. 31.2). His theory of general relativity represents the properties of matter and its relationship to gravitational fields. Within the context of his theory, Einstein developed a set of relations, called the **field equations** (Fig. 31.3), that express in mathematical terms the interaction of matter, radiation, and gravitational forces in the universe. Although alternatives to general relativity have been developed and their consequences explored, most research in cosmology today involves finding solutions to Einstein's field equations, and testing these solutions with observational data.

The basic premise of general relativity is that acceleration caused by gravitational field is indistinguishable from acceleration caused by a changing rate or direction of motion. One way to visualize this is to imagine we are inside of a compartment with no windows (Fig. 31.4). If this compartment is on the earth's surface, our weight feels normal because of the earth's gravitational attraction. If, however, we are in space and the compartment is being accelerated at a rate equivalent to one earth gravity, our weight will feel normal

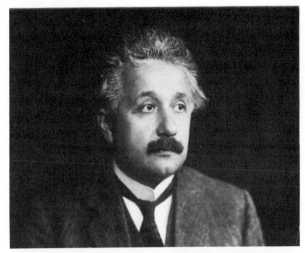

FIG. 31.2. ALBERT EINSTEIN. Among his many great contributions was the development of a mathematical formalism to describe the interaction of gravity, matter, and energy in the universe. This framework, general relativity, has withstood all observational and experimental tests applied to it so far.

here, too. There is no experimental way for us to tell the difference between the two situations, short of opening the door and looking out.

One consequence of the equivalence of gravity and acceleration is that an object passing near a source of gravitational pull (that is, any other object) undergoes acceleration, and therefore follows a curved path. In a

FIG. 31.3. THE FIELD EQUATIONS. Here, in an early German publication, are the equations that describe the physical state of the universe. To a large extent, the science of cosmology involves finding solutions to these equations.

EINSTEIN: Einheitliche Feldtheorie von Gravitation und Elektrizität   415

Jnabhängig von diesem affinen Zusammenhang führen wir eine kontravariante Tensordichte $\mathfrak{g}^{\mu\nu}$ ein, deren Symmetrieeigenschaften wir ebenfalls offen lassen. Aus beiden bilden wir die skalare Dichte

$$\mathfrak{H} = \mathfrak{g}^{\mu\nu} R_{\mu\nu} \qquad (3)$$

und postulieren, daß sämtliche Variationen des Integrals

$$\mathfrak{J} = \int \mathfrak{H} dx_1 dx_2 dx_3 dx_4$$

nach den $\mathfrak{g}^{\mu\nu}$ und $\Gamma^{\tau}_{\mu\nu}$ als unabhängigen (an den Grenzen nicht varierten) Variabeln verschwinden.
    Die Variation nach den $\mathfrak{g}^{\mu\nu}$ liefert die 16 Gleichen

$$R_{\mu\nu} = 0, \qquad (4)$$

die Variation nach den $\Gamma^{\tau}_{\mu\nu}$ zunächst die 64 Gleichungen

$$\frac{\partial \mathfrak{g}^{\mu\nu}}{\partial x_\alpha} + \mathfrak{g}^{\sigma\nu} \Gamma^{\mu}_{\sigma\alpha} + \mathfrak{g}^{\mu\sigma} \Gamma^{\nu}_{\sigma\alpha} - \delta^{\nu}_\alpha \left(\frac{\partial \mathfrak{g}^{\mu\sigma}}{\partial x_\sigma} + \mathfrak{g}^{\tau\sigma} \Gamma^{\mu}_{\tau\sigma} \right) - \mathfrak{g}^{\mu\nu} \Gamma^{\sigma}_{\sigma\alpha} = 0. \qquad (5)$$

Wir wollen nun einige Betrachtungen anstellen, die uns die Gleichungen (5) durch einfachere zu ersetzen gestatten. Verjüngen wir die linke Seite von (5) nach den Indizes ν, α bzw. μ, α, so erhalten wir die Gleichungen

Testing Relativity

In the framework of Einstein's theories of special and general relativity, most departures from the classical laws of physics do not become significant except under rather extreme conditions, such as velocities near the speed of light, or the presence of very intense gravitational fields. Therefore, under normal circumstances it is difficult or impossible to tell whether the theory of relativity is correct or not. There are means of testing the theory, however, and the importance of doing so has inspired many experiments.

One of the earliest tests of general relativity was attempted in 1919, when star positions near the sun's limb were measured during a total solar eclipse, to see whether light from the stars was deflected by the sun's gravity, as the theory predicted. The stellar positions were found to be shifted by the expected amount, scoring a resounding victory for the then-new theory. Much more recently, this experiment has been carried out using radio interferometry techniques to observe the positions of quasars near the sun, and the agreement with Einstein's theory is so good that many of its competitiors have been ruled out.

It has also been possible to detect the expected, very small, gravitational redshift of sunlight, and also to confirm Einstein's prediction that the highly elongated orbit of Mercury should be gradually changing its orientation. These and other, more complex, tests have never yet shown any measurable deviation from the expectations of general relativity theory, and there are few alternative theories still in the running. Many experiments being planned, such as those to be launched on the satellite called the *Relativity Explorer,* are designed to test further, more subtle predictions.

Long before he developed the concepts of general relativity, Einstein published his theory of **special relativity,** in which he first stated the equivalence of mass and energy, and delved into the effects of velocities near that of light. One consequence of special relativity is that time should advance at a different rate for a person traveling near

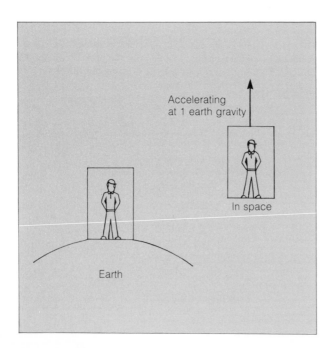

FIG. 31.4. GENERAL RELATIVITY. The person in the enclosed room has no experimental or intuitive means of distinguishing whether he is motionless on the surface of the earth, or in space, accelerating at a rate corresponding to one earth gravity. This implies that acceleration caused by motion and acceleration caused by gravity are equivalent, which in turn implies that space is curved in the presence of a gravitational field.

universe containing matter, this means that all trajectories of moving objects are curved, and it is often said that space itself is curved. The degree of curvature is especially high close to massive objects (Fig. 31.5), but there is also an overall curvature of the universe, owing to its total mass content. The solutions to the field equations specify, among other things, the degree and type of curvature. We will return to this point shortly.

Einstein's solution to the equations, developed in 1917, had a serious flaw, in his view: It did not allow for a static, nonexpanding universe. In what he later admitted was the biggest mistake he ever made, Ein-

the speed of light than for someone who is at rest. This leads to the famous paradox, often exploited in science fiction novels, in which a space-faring traveler returns to earth after a high-speed voyage and finds himself younger than his twin.

Although we do not yet have the ability to travel at speeds near that of light, we can develop clocks so accurate that the time-dilation effect can be detected at lower speeds. A few years ago, a pair of very high-precision atomic clocks were used in an experiment; one was taken around the world by airplane, at an average speed of a few hundred miles per hour, while the other remained stationary. The two clocks were synchronized at the start, and were found to disagree by exactly the expected amount after the flight, providing another confirmation of Einstein's work.

One of the best laboratories for testing relativistic theory was discovered in space, in the form of a binary pulsar. This is a pair of neutron stars, one of them a pulsar, in mutual orbit. Because the gravitational fields involved are so much stronger than that of the earth or even the sun, many of the effects of general relativity should be much greater and easier to measure than in the experiments just described. It will still take several years of observation to answer all the questions that may be asked of the binary pulsar, but the data that are in so far are all in agreement with Einstein's predictions.

One of the most interesting consequences of general relativity is that there should be a form of radiation completely distinct from electromagnetic radiation, that is emitted whenever massive objects undergo strong acceleration. The radiation, referred to as **gravity waves,** might be detectable by means of the minute vibrations it would set off in a large object such as a rigid metal bar. Experiments have not yet reached the level of sensitivity that is thought necessary, but within a very few years it is likely that meaningful searches for gravitational radiation will be possible. The cosmic events that might produce gravity waves include the nearly instantaneous collapse of massive stars as they become supernovae, the rapid motions in close binary systems, and perhaps the mysterious energetic phenomena that occur in many galactic nuclei.

stein added an arbitrary force called the **cosmological constant** to the field equations, solely for the purpose of allowing the universe to be stationary, neither expanding nor contracting.

Others developed different solutions, always by making certain assumptions about the universe. Some of these assumptions were necessary in order to simplify the field equations so that they could be solved.

FIG. 31.5. CURVATURE OF SPACE NEAR A MASSIVE STAR. The curved surface represents the shape of space very close to a massive star. The best way to envision this curvature is to imagine photons of light as marbles rolling on a surface of this shape.

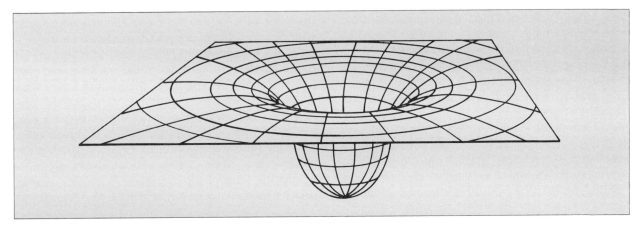

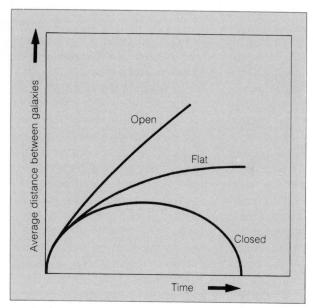

FIG. 31.6. THE THREE POSSIBLE FATES OF THE UNIVERSE. This diagram shows how the average distance between galaxies changes with time for the open, flat, and closed universe. In the first case, galaxies continue to separate forever, although the rate of separation is slow. In the second case, in an infinite time the rate of separation slows to a halt, but will not reverse. In the third case, galaxies eventually begin to approach each other, and the universe returns to a single point.

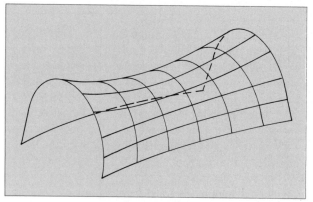

FIG. 31.7. A SADDLE SURFACE. This is a representation of the geometry of an open universe, one that has negative curvature, is infinite in extent, and has no boundaries.

For example, W. de Sitter in 1917 developed a solution that corresponded to an empty universe, one with no matter in it. By the 1920's, solutions for an expanding universe were found, primarily by the Soviet physicist A. Friedmann and later by the Belgian, G. LeMaitre, who went so far as to propose an origin for the universe in a hot, dense state from which it has been expanding ever since. This was the true beginning of the big-bang idea, and it was developed some three years before Hubble's observational discovery of the expansion. Of course, once it was found that the universe is not static, but is actually in a dynamic state, the original need for Einstein's cosmological constant disappeared. Nevertheless, it is usually included by modern cosmologists in the field equations, but its value is assumed to be zero.

The general relativistic field equations allow for three possibilities regarding the curvature of the universe, which correspond to three different possible futures (Fig. 31.6). A central question of modern studies of cosmology has to do with deciding which possibility is correct. One is referred to as "negative curvature," and an analogy to this is a saddle-shaped surface, which is curved everywhere, has no boundaries, and is infinite in extent (Fig. 31.7). This type of curvature corresponds to what is called an "open" universe, a solution to the field equations in which the expansion continues forever, never stopping (it does slow down, however, because the gravitational pull of all the matter in it tends to hold back the expansion).

A second possibility is that the curvature is "positive," corresponding to the surface of a sphere, which is curved everywhere, has no boundaries, but is finite in extent (Fig. 31.8). This solution to the field equations is called the "closed" universe, and it implies that the expansion will eventually be halted by gravitational forces and will reverse itself, leading to a contraction back to a single point.

The third and last possibility is that the universe is flat, with no curvature. This corresponds to the case where the outward expansion is precisely balanced by the inward gravitational pull of the matter in the universe, so that the expansion will eventually come to a stop, but will not reverse itself. This balanced, static state requires a perfect coincidence between the momentum of expansion and the inward gravitational pull of the combined total mass of the universe, and may therefore seem very unlikely. We do not know what set of rules governed the amount of matter the universe was born with, however, so there is no logical reason to rule out this possibility. Most astronomers generally pose the question in terms of whether the universe is open or closed, without specifically mentioning the third possibility, that it is flat. If a perfect balance has been achieved, it will be very difficult to

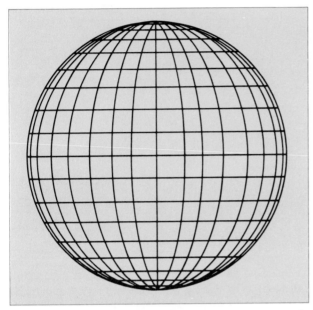

FIG. 31.8.   THE SURFACE OF A SPHERE. This surface represents a closed universe; one that has positive curvature, has a finite extent, but has no boundaries.

determine this from the observational evidence, which has large uncertainties.

Open or Closed: The Observational Evidence

Substantial intellectual and technological resources are devoted today to the question of determining the fate of the universe. Theorists are at work developing and refining the solutions to the field equations, or seeking alternatives to general relativity that might provide equally or more valid representations. Observers are busily attempting to test the theoretical possibilities by finding situations where competing theories should lead to different observational consequences. This is a difficult job, because most of the differences will show up only on the largest scales, and therefore to detect them requires observations of the farthest reaches of the universe.

Within the context of general relativity, the premier question is whether the universe is open or closed. There are two general observational approaches to answering this: determine whether there is enough matter in the universe to produce sufficient gravitational attraction to close it; or measure the rate of deceleration of the expansion, to see whether it is slowing rapidly enough to eventually stop and reverse itself.

Total Mass Content

The total mass in the universe is the quantity that determines whether gravity will halt the expansion or not. The field equations express the mass content of the universe in terms of the density; the amount of mass per cubic centimeter. This is convenient for observers, because it is obviously simpler to measure the density in our vicinity than to try to observe the total mass everywhere in the universe. The field equations can be solved for the value of the density that would produce an exact balance between expansion and gravitational attraction; that is, the density corresponding to a flat universe. If the actual density is greater than this **critical density,** then there is sufficient mass to close the universe, and the expansion will stop. The critical density depends on the value of Hubble's constant H; for a value near 55 km/sec/Mpc, it is calculated to be about 5×10^{-29} grams/cm^3, or about 1 proton per cubic meter, a very low value by earthly standards.

The most straightforward way to measure the density of the universe is to simply count galaxies in some randomly selected volume of space, add up their masses, and divide the total by the volume. Care must be taken to choose a very large sample volume, so that clumpiness owing to clusters of galaxies is not important. This technique yields very low values for the density, around 10^{-30} grams/cm^3, less than about 2 percent of the critical density. It may seem that we have already answered the question, and that the universe is open, but this method overlooks substantial quantities of mass.

One clue to this arises from the determination of cluster mass based on the velocity dispersion of the galaxies in the cluster (see chapter 28). These measurements of mass, for reasons still not entirely clear, always yield much higher values than the estimated total mass of the visible galaxies in the cluster. The disparity can be as great as a factor of ten or more, so if the larger values are correct, the average density of the universe comes closer to the critical value. Even the larger values based on velocity-dispersion measurements, however, fall short of the amount needed to close the universe. If the mass density is to exceed the

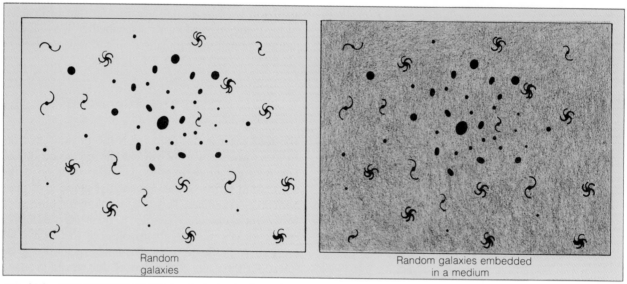

Random galaxies | Random galaxies embedded in a medium

FIG. 31.9.   THE "MISSING MASS." Are the isolated galaxies that we see all there is in the universe? Or are they merely points that happen to glow, embedded in a universal sea of unknown substance, whose mass density overwhelms that represented by the galaxies?

critical value, there must be large quantities of matter in some form that has not yet been detected (Fig. 31.9). The hidden matter cannot be inside clusters of galaxies, because its presence would have been detected by the velocity-dispersion measurements.

If there is a lot of invisible matter in the universe, it could take several different forms. One possibility that is obvious to us by now is that there may be many black holes in space between clusters. The astrophysicist Stephen Hawking hypothesized the existence of countless numbers of "mini" black holes in intergalactic space, formed during the early stages of the big bang. These would have very small masses, much smaller even than the mass of the earth, but would be so numerous that they could easily exceed the critical density. Current observational evidence argues against the existence of such objects in great numbers, however.

Another possibility is that neutrinos, the elusive subatomic particles produced in nuclear reactions, have mass. Recall (from chapter 16) that these particles permeate space, freely traveling through matter and vacuum alike, but that standard theory says they have no mass. A recent controversial experiment indicates that this last assertion may not be correct; that neutrinos may contain miniscule quantities of mass after all. If so, they are sufficiently plentiful to provide more than the critical density, thereby closing the universe.

For now, however, we are forced to conclude from the available observational evidence that the mass density of the universe is less than the critical density, and therefore that the universe is open.

The Deceleration of the Expansion

The second approach to answering the question of whether the universe is open or closed is perhaps being pursued more vigorously today. The objective is to compare the present expansion rate with what it was early in the history of the big bang, to see how much slowing, or deceleration, has occurred (Fig. 31.10). The expansion has certainly slowed; the question is how much. If it has only decelerated a little, then we infer that it is not going to slow down enough to stop and reverse itself, but if it has already decelerated a lot, then we conclude that the expansion is coming to a halt, and that the universe is closed.

To establish the deceleration rate requires knowing both the present expansion rate and the expansion rate at a time long ago, shortly after the big bang began. Neither quantity is easy to determine: we already learned that the value of the Hubble constant H, which tells us the present expansion rate, is quite uncertain to begin with. To measure the expansion rate that occurred at early times in the history of the universe is even more difficult, for it involves determining the distances and

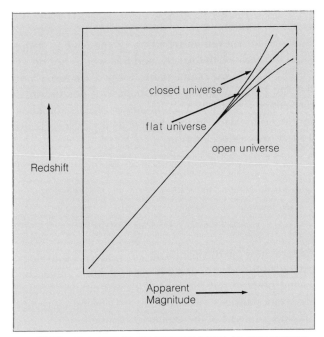

FIG. 31.10. THE EFFECT OF DECELERATION IN THE HUBBLE DIAGRAM. This shows the relationship between redshift and apparent magnitude for the three possible cases: closed (upper curve), flat (middle), and open (lower). Present data seem to support the idea of an open universe, although there is substantial uncertainty, largely because of problems in applying standard-candle techniques to galaxies so far away that they are being seen as they were at a young age. The distinction is also made difficult by the subtlety of the contrast between the shapes of the curves.

velocities of very faraway galaxies, so distant that we see them as they were when the universe was young. Such objects are now at the very limits of detectability, and it is hoped that the *Space Telescope* will push the frontier back far enough to permit observations of velocities in the early days of the expansion.

We still must rely on standard candles to establish the distance scale, and for such distant objects this procedure becomes even more uncertain than usual. When we look so far back in time, we are seeing galaxies that are much younger than those near us, and which therefore may have quite different properties than mature galaxies. For example, their content of stellar populations may be rather different than in nearby galaxies, their brightest stars and nebulae may be different, and even total galactic luminosities may be different. It may not be valid to assume that the brightest galaxy in a cluster of galaxies has the same absolute magnitude as in more nearby clusters.

For these reasons, it has been difficult to determine the relationship between distance and velocity for very faraway galaxies. The present evidence on deceleration is that the universe has not slowed very much, however, and therefore is not on its way to stopping and beginning to contract.

Astronomers have recently employed an indirect technique for measuring the deceleration. Some light elements were created in the early stages of the expansion, and the amounts that were produced depend on how long the conditions were appropriate for nuclear reactions to occur. If the early expansion was very rapid, then relatively low quantities of these elements were produced, and the large difference between the expansion velocity then and the present expansion rate tells us that there has been a great deal of deceleration. If, on the other hand, the early expansion was relatively slow, then there was time for more of the light elements to be produced, in which case there has been relatively little deceleration. Thus, the abundances of these elements that were produced in the big bang can tell us how much deceleration has occurred.

The most abundant element (other than hydrogen) produced in the big bang is helium, so if we could measure how much helium was created, then this would be a good indicator of the deceleration. The problem with this element, however, is that it is also produced in stellar interiors, so it is difficult to determine how much of what we see in the universe today is really left over from the big bang.

A better candidate is deuterium, the form of hydrogen that has one proton and one neutron in the nucleus. This was also produced in the big bang, and, as far as we know, is not made in any other way. Although deuterium can be destroyed by nuclear processing in stars, it is not produced in that way (even if it were, it would not survive the high temperatures of stellar interiors without undergoing further reactions). The present deuterium abundance in the universe should therefore represent an upper limit on the quantity created in the big bang. Direct measurements of the amount of deuterium in space became possible in the 1970's, with the launch of the *Copernicus* satellite, which made ultraviolet spectroscopic measurements and was able to observe absorption lines of interstellar deuterium atoms (Fig. 31.11). The abundance of deuterium that was found is sufficiently high that it implies a slow early expansion rate, in turn pointing to a small amount of deceleration. Thus this test, in agreement with others,

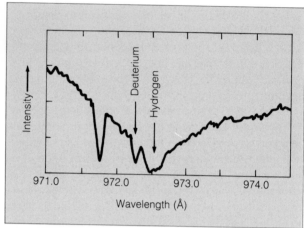

FIG. 31.11. AN ULTRAVIOLET ABSORPTION LINE OF INTERSTELLAR DEUTERIUM. The abundance of deuterium in space is determined from the analysis of absorption lines it forms in the spectra of background stars. Here we see a weak absorption line due to deuterium, close to a strong absorption feature caused by normal hydrogen. The spectrum shown here was obtained with the *Copernicus* satellite.

indicates that the universe is open. There is some uncertainty, though, because the *Copernicus* data seem to indicate an uneven distribution of interstellar deuterium throughout the galaxy, and this would not be expected if the deuterium really formed exclusively in the big bang. A clumpy distribution of deuterium in the galaxy could indicate that some of it is somehow produced by stars, in which case its abundance is not a strict test of deceleration.

The History of the Universe

We are now at the forefront of modern cosmology, having outlined the present state of knowledge, and having pointed out the basis for current and planned observational tests. At this point it is useful to review the development of the universe, highlighting some of

FIG. 31.12. ELEMENT FORMATION IN THE BIG BANG. This diagram illustrates the relative rates of formation of various light elements by nuclear reactions during the early stages of the big-bang expansion. The production rates are shown as functions of time and temperature.

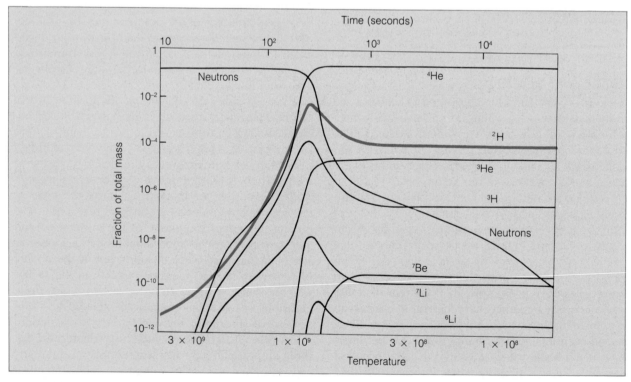

the significant events. The very fact that we can do this, that we can say with any degree of certainty what conditions were like at the beginning and at points along the way, is a triumph of modern science. It is impossible to physically describe the universe at the precise moment that the expansion began; it is physical nonsense to deal with infinitely high temperature and density. It is possible, however, to calculate the conditions immediately after the expansion started, and at any later time (Figs. 31.12 and 31.13).

At first the universe was entirely filled with radiation and elementary subatomic particles. By 0.01 seconds after the expansion started, however, when the temperature was perhaps 100 billion degrees, some more familiar particles, primarily electrons, positrons, and neutrinos, had appeared. The temperature dropped to 10 billion degrees by 1.09 seconds after the start, and by then protons and neutrons were appearing. The first atomic nuclei, formed of combinations of these particles, began to form 10 to 15 seconds after the beginning, although even at this point, most of the energy of the universe was still in the form of radiation. Within a few minutes, conditions became better suited for nuclear reactions to take place efficiently, and the most active stage of element creation began. The principal products, in addition to helium and deuterium, were tritium (another form of hydrogen, with one proton and two neutrons in the nucleus), lithium (three protons and four neutrons) and beryllium (four protons and five neutrons). Nearly all the available neutrons combined with protons to form helium nuclei, a process that was complete within four minutes after the beginning of the expansion. At this point, with some 22 to 28 percent of the mass in the form of helium, the reactions were essentially over, except for some production of lithium and beryllium over the next half hour.

The expansion and cooling continued, but nothing significant happened for a long time after the nuclear reactions stopped. Eventually the density of the radiation had become reduced sufficiently that its energy was less than that contained in the mass; that is, the energy derived from $E = mc^2$ became greater than that contained in the radiation. At that point, it is said, the universe became matter-dominated rather than radiation-dominated.

Matter and radiation continued to interact, however, because free electrons scatter photons of light very efficiently. The strong interplay between matter and ra-

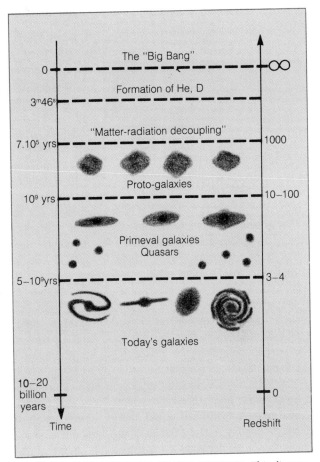

FIG. 31.13. THE EVOLUTION OF THE UNIVERSE. This diagram shows, as a function of time since the start of the big bang, significant stages in the evolution of the universe. It also shows the types of objects that existed at each stage, and the redshift at which they are (or would be) observed at the present time.

diation finally ended nearly a million years after the start of the big bang, when the temperature became low enough to allow electrons and protons to combine into hydrogen atoms. The atoms still absorbed and reemitted light, but much less effectively, because they could only do so at a few specific wavelengths. From this time on, matter and radiation went separate ways. The radiation simply continued to cool as the universe expanded, reaching its present temperature of 2.7 K some 10 to 20 billion years after the expansion began.

Sometime in the first billion years or so, the matter

ASTRONOMICAL INSIGHT (31.3)

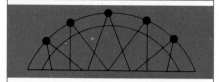

The Mystery of the Nighttime Sky

A very simple question, first asked more than two centuries ago and discussed at length in the early 1800's by the German astronomer W. Olbers, leads to profound consequences. The question is why the sky is dark at night. In an infinite universe filled with stars, every line of sight, regardless of direction, should intersect a stellar surface. The nighttime sky should therefore be uniformly bright, not dark, as the star images in the sky would literally overlap each other. The fact that this is not so presents a paradox, one worth pondering.

One suggested solution is that interstellar extinction can diminish the light from distant stars sufficiently to make the sky appear black. Careful consideration shows,

however, that this is not so. Light that is absorbed by dust in space causes the grains to heat up. If the universe were really filled with starlight as suggested by Olbers' paradox, then the grains would become so hot that they would glow, and we would still have a uniformly bright nighttime sky.

One potential solution to Olbers' paradox is to realize that the universe is not really infinite, at least from a practical point of view. At sufficiently great distances we expect to see no galaxies or stars, because we are looking back to a time before the universe began. Thus, even though the universe has no edge, there is a horizon beyond which stars and galaxies do not exist. Therefore, this argument goes, all lines of sight do not have to intersect stellar surfaces; many reach the darkness beyond the horizon of the universe, and we have a dark nighttime sky.

Interestingly, it now appears that this horizon is being approached by observations. As noted in chapter 29, very recent data hint that there may be a limit on quasar redshifts, implying that no quasars exist beyond a certain distance. Hence we

may already be nearing the point where we are looking back to a time before quasars had formed.

Given that the universe began in a very hot phase filled with energetic radiation, it seems, however, that if we look far enough into the past, we should see the brightness of the primeval fireball. We must therefore look for another solution to the dilemma.

The ultimate explanation of Olbers' paradox has to do with the expansion of the universe. The redshift caused by the expansion, if we look sufficiently far away, becomes so great that the light never gets here in visible form. It becomes redshifted to extremely large wavelengths, losing nearly all its energy. This solution to Olbers' paradox works also in a steady-state universe, so that even if there is no beginning to the universe, there is still a horizon.

It is interesting that this simple question, if analyzed fully, could have led to the conclusion that the universe has a finite age or is expanding, a full century before these things were learned from other observational and theoretical developments.

in the expanding universe became clumpy, and fragmented into clouds and groups of clouds that eventually collapsed to form galaxies and clusters of galaxies. As we noted in chapter 28, the cause of the clumpiness and fragmentation is as yet unknown. Because there is so much uncertainty about how the initial stages of fragmentation took place, there is a great deal of interest in probing far enough into the past to get some information on this process. To do so will be another of the goals of the *Space Telescope*.

What Next?

To discuss the future of the universe is obviously a speculative venture. We cannot even answer with absolute certainty the basic question of whether it is open or closed (although it seems likely it is open).

If the universe is open, then it will have no definite end; it will just gradually run down. The radiation background will continue to decline in temperature, approaching absolute zero. As stellar processing continues in galaxies, the fraction of matter that is in the form of heavy elements will continue to grow, and the supply of hydrogen, the basic nuclear fuel, will diminish. The recycling process between stars and the interstellar medium will continue, but gradually matter will become locked up in black holes, neutron stars, and white dwarfs, from which there is no return. Perhaps in the distant future the universe will be totally dark, populated only by black holes and black dwarfs, the cold cinders of neutron stars and white dwarfs.

If the universe is closed, then someday, billions of years from now, the expansion will stop, and will be replaced by contraction. The deterioration will still take place, until the final moments when the universe once again becomes hot and compressed, entering a new singularity. In some views, purely conjectural and without possibility of verification, such a contraction would be followed by a new big bang, and the universe would be reborn. This concept of an oscillating universe, pleasing to the minds of many, will not be fulfilled unless the present weight of the evidence favoring an open universe is found to be in error.

Perspective

Our story is essentially complete. We have discussed the universe and all the various objects and structures in it, and we have described what is known of its evolution and its fate.

We have, however, omitted consideration of what is perhaps the most important ingredient of all: life. Although much of what we might say about this is beyond the scope of astronomy, it is appropriate to assess what can be learned from objective scientific consideration.

Summary

1. In cosmological studies, astronomers usually adopt the Cosmological Principle, which states that the universe is both homogeneous and isotropic. Existing observational data tend to support this assumption.

2. Einstein's theory of general relativity, which describes gravitation and its equivalence to acceleration, is used to mathematically describe the universe as a whole. In the context of this theory, field equations are written that represent the interaction of matter, radiation, and energy in the universe. Solutions of the field equations amount to definitions of the properties of the universe.

3. There are three general solutions to the field equations that are considered by modern cosmologists. They correspond to a closed universe (positive curvature), an open universe (negative curvature), and a flat universe (no curvature).

4. A major question in modern astrophysics is whether the universe is closed (expansion eventually to be reversed) or open (expansion never to stop).

5. There are two ways to test whether the universe is open or closed: (1) ascertain whether there is enough mass in the universe to gravitationally halt the expansion; or (2) determine the rate of deceleration of the expansion, to see whether it is slowing enough to eventually stop.

6. The total mass content is measured in terms of the average density of the universe. If the density is estimated by adding up the masses of visible galaxies, it is not sufficient to close the universe. Various suggestions have been made concerning other forms in which the necessary mass could exist.

7. The deceleration is measured in two ways: (1) by comparing past and present expansion rates, through observations of very distant galaxies; and (2) by inferring the early expansion rate from the present-day abundances of elements that formed only during the big bang.

8. Both the total mass content that is observed and the inferred deceleration of the expansion indicate that the universe is open. This conclusion is not universally accepted, and observational tests are continuing.

9. The early stages of the universal expansion, up to the time when matter and radiation decoupled, can be described quite precisely and with certainty by modern physics. Following the initial moment of infinite

density and temperature, the first atomic nuclei formed just over a second later, and all the early element production was finished within a few minutes. Matter and radiation decoupled almost a million years later; it is not so well understood how the universe subsequently organized itself into stars and galaxies.

The future of the universe appears to have two possibilities. If it is open, it will gradually become cold and disorganized. If it is closed, it will eventually contract, perhaps to a new beginning in another big bang.

Review Questions

1. In chapter 28 we discussed the fact that galaxies often are grouped into clusters. In this chapter we have asserted that the universe appears to be homogeneous; that it has a uniform distribution of matter. Explain how galaxies can be grouped into clusters and at the same time the universe can be homogeneous.

2. Recall, from chapter 29, that there is a 24-hour anisotropy in the 3-degree background radiation. Why does this not violate the assumption adopted here that the universe is isotropic?

3. Are Einstein's field equations more like Kepler's laws of planetary motion or are they more akin to Newton's derivation of Kepler's laws?

4. Suppose a star ejects a spherical shell of gas as it forms a planetary nebula. The shell of gas, depending on the mass of the star and the initial velocity of ejection, will either escape completely or fall back onto the star. What determines this, and how is it analogous to the question of whether the universe is open or closed?

5. Why do astronomers not simply measure the density of matter within the well-observed solar neighborhood, within a few hundred parsecs of the sun, in de-

termining whether there is enough mass to close the universe? What do you think the result would be if this local density were used? (You may want to review chapter 25.)

6. How do we know that the existence of massive black holes in the nuclei of all galaxies would not be sufficient to close the universe?

7. Summarize the difficulties in measuring very distant galaxies in order to infer the early expansion rate of the universe.

8. Why is deuterium a better species to use in inferring the early expansion rate of the universe than helium? Why would either be better than iron, for this purpose?

9. What would it tell us about the early expansion of the universe if it was discovered that substantial quantities of elements such as carbon, nitrogen, and oxygen had been formed in the big bang? Would this imply that the universe is open or closed?

10. Summarize the ways in which the *Space Telescope* should help answer cosmological questions raised in this chapter.

Additional Readings

Barrow, J. D., and Silk, J. 1980. The structure of the early universe. *Scientific American* 242(4):118.

Ferris, T. 1977. *The red limit: the search for the edge of the universe.* New York: William Morrow.

Gott, J. R.; Gunn, J. E.; Schramm, D. N.; and Tinsley, B. M. 1976. Will the universe expand forever? *Scientific American* 234(3):62.

Margon, B. 1975. The missing mass. *Mercury* 4(1):2.

Penzias, A. A. 1978. The riddle of cosmic deuterium. *American Scientist* 66:291.

Schramm, D. N. 1974. The age of the elements. *Scientific American* 230(1):69.

Shipman, H. L. 1976. *Black holes, quasars, and the universe.* Boston: Houghton-Mifflin.

Tinsley, B. M. 1977. The cosmological constant and cosmological change. *Physics Today* 30(6):32.

Trefil, J. S. 1978. Einstein's theory of general relativity is put to the test. *Smithsonian* 11(1):74.

Weinberg, S. 1977. *The first three minutes.* New York: Basic Books.

Will, C. M. 1974. Gravitation Theory. *Scientific American* 231(5):24.

The Elusive Content of the Universe

P. J. E. Peebles

Dr. Peebles, a professor of physics at Princeton University, is among the world's foremost authorities on observational cosmology. More than anyone else, he has pursued and answered questions having to do with the distribution of matter in the universe, and its implications for the evolution of the cosmos. He was also a part of the Princeton team that carried out the first detailed analyses of the 3-degree microwave background radiation. Dr. Peebles has written two books on the structure and content of the universe, both of which are standard texts in the field, and he was recently awarded the distinguished Heineman Prize by the American Astronomical Society.

The astronomer's lot can be compared to Tantalus: we can see and admire but never touch the subject of our attention. Unlike our colleagues in laboratory science who are accustomed to controlled experiments that test many aspects of a hypothesis, we must rely on the relatively scarce and ambiguous bits of evidence nature happens to send our way. That is why there are such heated and prolonged debates over questions like the nature of quasars, the origin of redshifts of galaxies, and the geometry of the universe. But still extragalactic astronomy does make progress because we find effects whose significance seems clear or at least highly suggestive. A particularly important example is the observation that the universe is isotropic about us to high accuracy. That is seen in the smooth distribution of galaxies across the sky (after we remove minor fluctuations like the great clusters of galaxies) and it is seen also in the very isotropic distribution of radio and X-ray sources. As distant galaxies seem to be no poorer as possible homes for observers than our Milky Way galaxy, I have no trouble arguing for the next step, that it would be highly contrived to imagine that the universe is inhomogeneous but accurately isotropic about our particular galaxy. Thus it seems clear that, as Einstein guessed, the universe is simple: it looks much the same when viewed from any spot or direction, with no center and no edge. This remarkable simplicity is of great importance because it means that we can speak of the large-scale structure of the universe and analyze its behavior in terms of just a few variables, as has been described in chapter 31 of this book.

Of course, apparently clear indications can be misleading and a good example of that concerns the question, what are galaxies made of? A few years ago the answer seemed clear: the main component is stars, there are trace amounts of gas and dust and possibly planets, and there is a light sprinkling of cosmic-ray particles and magnetic fields. The light from a galaxy is thought to be mainly starlight because the spectrum looks like what would be produced by a mixture of stars moving relative to each other at a few hundred kilometers per second. Also, as is discussed in chapter 27, the mass in the central parts of a galaxy is comparable to the mass in stars wanted to produce the light, which again is evidence that galaxies are made of stars. However, more detailed measurements have shown that in large isolated spirals the mass is distinctly more spread out, the starlight more concentrated to the center of the galaxy. Thus we conclude that the seen stars could not provide the dominant part of the mass in the outskirts of the galaxy. A point to be borne in mind but invoked only if we are forced to it is that this inference is based on Kepler's laws. It may be that the inverse-square law of Newtonian gravity fails on the scale of galaxies, but I think a more conservative bet is that much of the matter is in a form that happens not to be readily observed by us.

What might this hypothetical unseen matter be? White dwarfs and neutron stars are not attractive candidates because we observe that the formation of these objects at the deaths of normal stars is attended by the release of considerable amounts of gas and dust. If the unseen matter were star remnants we would have expected to see this debris, the gas and dust, and we do not. Stars with very low masses are exceedingly faint and so are acceptable candidates for the unseen matter. These very low-mass stars emit infrared radiation. Indeed even Jupiter, which is like a star except that the central pressure is much too low to permit sustained thermonuclear reactions, is an infrared source: if placed well away from any star it would glow in the infrared like any thermal source at 200 K, still cooling off from the heat of its formation. Tests for infrared radiation from galaxies are possible, though infrared technology still is in a state of rapid development, and the most sensitive tests will be costly because they must be done from space to avoid infrared emission from our atmosphere. I will be eagerly following the progress of this difficult but important program of infrared studies of galaxies.

Black holes certainly are acceptable forms for the unseen matter but I find the idea unappealing because it is ad hoc. Where did all these black holes come from? I am more attracted to the thought that the unseen matter may be neutrinos or some other sort of elementary particle. Neutrinos are known to exist, detected in the laboratory in nuclear reactions. They obey the Pauli Exclusion Principle, and just as that principle explains why electrons cannot all be stuffed toward the center of an atom we find that neutrinos cannot be stuffed into a galaxy to the wanted density unless the neutrino mass exceeds a critical value (about one ten-thousandth of the mass of an electron). This wanted mass seems not unreasonable according to current elementary-particle theories. Laboratory measurements do not provide useful constraints but the subject is under intense study and I will be watching the newspapers for developments. Neutrinos would be produced in the hot big bang, and knowing the temperature of the universe and borrowing some current ideas from elementary-particle physics we can compute the expected density of these neutrinos: it amounts to a mean density of about 100 per cm^3. I find it an amusing and tantalizing coincidence that this number density multiplied by the above-mentioned critical neutrino mass yields a mass density that agrees with the critical density for a closed universe (chapter 31). And there would be ample mass in

neutrinos to provide the matter in galaxies. This picture is just as romantic as the black-hole hypothesis but I prefer it because it is more meaty. We have a definite idea of where the neutrinos came from: they ought to be present as thermal by-products of the big bang, just like the 3-degree background (chapter 29). Laboratory physics should be capable of narrowing the range of masses of these neutrinos. And if the neutrinos are present in our galaxy at a local density of about $10^8/cm^3$ (the mean density averaged over all space would then be around $100/cm^3$; the local density in mass concentrations would be higher) some clever person may well find a way of detecting them. One way, that would be feasible but a very costly experiment, is to look for ultraviolet radiation from the predicted occasional annihilation of a neutrino. Again, we will watch the newspapers for developments.

The unseen-matter puzzle is an example of the excitement of astronomy. We are presented with a definite question; we can think of observations to test possible answers, and we must expect that in the process of testing we will stumble on some pretty fundamental discoveries.

Another definite question is, what happened before the big bang? The 3-degree background radiation is thought to offer tangible evidence that the universe expanded away from a dense hot state, because that is the only way anyone has thought of to produce such radiation (chapter 29). According to the conventional theory of gravity (general relativity theory) this expansion can be traced all the way back to an instant at which the density of the universe was infinite. Then the equations break down and offer no information on what (if anything) happened before that time. Singularities of this sort have been encountered before in physics. The prominent example is the prediction in classical mechanics that all the electrons in an atom should quickly pile up on the nucleus, quite contrary to the true state of affairs. By the third decade of this century it had become clear that the fault was in classical mechanics, which becomes inaccurate on a scale of atoms and must be replaced by the more accurate quantum mechanics. Perhaps the singular behavior of general-relativity theory at the moment of the big bang is only a symptom of the inaccuracy of this theory at extremely high densities.

Comparable symptoms in more easily studied events may be provided by black holes. We observe many stars and star clusters massive and dense enough to be in danger of collapsing to black holes. The predicted collapse occurs for densities and masses modest enough that moderate adjustments of the gravity theory would not prevent the collapse. The end point, a singular black hole which is much like the time reversal of a singular big bang, is a prediction of general relativity theory and may be altered in detail or substance by an improved gravity theory. We may hope for clues to such an improved theory in the behavior of the collapsed objects in our neighborhood. As is discussed in chapter 30 of this book, there is circumstantial evidence that black holes with masses around 10^9 solar masses are present in the nuclei of some galaxies, perhaps including the Milky Way. We will continue to pay a lot of attention to these nuclei.

Is there really a compact massive object at the center of our galaxy? If not, how did the galaxy avoid the expected accumulation of mass as viscosity caused matter to rain down on its center? If so, does this object really act just as a singular hole in the curvature of space? Or does it on occasion give more than it receives? We know a good deal about white dwarfs and neutron stars because these objects are seen in handy nearby systems such as binary stars. Black holes are not so accommodating: the only candidates so far are in messy systems where the observations are hard to interpret. Apparently relativistic collapse to a black hole is rare. Conceivably there is something wrong with our picture of collapse.

I do not look for firm answers to the puzzles of black holes and big bangs anytime soon, but I also see no reason to think we are near the end of the rope. We will continue as people have done since science got organized, looking for simple patterns that can be fitted into the structure we are pleased to call mathematics. There never was any guarantee that this is the way to do science, but it has proved to be a remarkably useful course and I suppose will continue to do so.

Life in the Universe

INTRODUCTION TO SECTION VII. LIFE IN THE UNIVERSE.

We finish our survey of the dynamic universe with a discussion of living organisms and the prospects for finding that we are not alone in the cosmos. To many, this is the most central question we can ask. In the sole chapter of this section, we will examine the astronomer's attempt to answer it.

In discussing the question of life elsewhere in the universe, we will begin by looking into the origins of life here on earth. In doing so, we will learn of biological evolution and the evidence that has been unearthed to tell us of our own beginnings. To do this, we will step outside the normal arena of astronomy, but we will then return to it in assessing whether the conditions that led to life on earth might exist on other planets in our solar system and in the galaxy at large. We will discuss the probability that technological civilizations might be thriving here and there in the Milky Way.

We will complete our brief treatment of life in the universe by discussing the strategy for searching for other civilizations. It seems most likely that radio communications will be our initial means of contact, and there is an interesting exercise in logic involved with choosing the wavelength at which to search for or to send signals.

CHAPTER 32

The Question of Companionship

We have attempted to answer all the fundamental questions about the physical universe that can be treated scientifically. Having done this, we know our place in the cosmos: we know something of its age, and we realize how insignificant our habitat is.

In this chapter we contemplate whether we as living creatures are unique in the universe, or whether even that distinction must be shared. Nothing that we have learned so far leads us to rule out the possibility that other life forms, some of them intelligent, may exist. We believe that the earth and the other planets are a natural by-product of the formation of the sun, and we have evidence that some of the essential ingredients for life were present on the earth from the time it formed. Similar conditions must have been met countless times in the history of the universe, and will occur countless more times in the future. Science cannot yet tell the full story of how life began, however, and we have not found any evidence that it actually does exist elsewhere. The mystery remains.

Life on Earth

We can start by discussing the origins of the only life we know. Besides giving us some insight into the processes thought to have been at work on the earth, this will help us later, when we are speculating about whether the same processes have occurred elsewhere.

Before the time of Charles Darwin (Fig. 32.1), in the mid-1800's, the view was widely held that life could arise spontaneously from nonliving matter. Darwin's work in the study of evolution, showing how species develop gradually as a result of environmental pressures, made such an idea seem improbable.

An alternative to the idea that life arose from spontaneous formation was proposed in 1907 by the Swedish chemist S. Arrhenius, who suggested that life on the earth was introduced billions of years ago from space, originally in the form of microscopic spores that floated through the cosmos, landing here and there to act as seeds for new biological systems. This idea, called the **panspermia** hypothesis, cannot be ruled out, but several arguments make it seem unlikely. It would take a very long time for such spores to permeate the galaxy, and there would have to be a very high density of them in space in order for one or more of them to reach the earth by chance. More important, it seems very unlikely that the spores could survive the hazards of space, such as ultraviolet light and cosmic rays. Even if the panspermia concept is correct, the question of the ultimate origin of life remains, although it is transferred to some other location. In view of what is known today about the evolution of life and the early conditions on the earth, scientists generally agree that life arose through natural processes occurring here, and was not introduced from elsewhere.

It is believed that the early atmosphere of the earth was composed chiefly of hydrogen and hydrogen-bearing molecules such as ammonia (NH_3) and methane (CH_4), as well as water (see chapter 7). Therefore the first organisms must have developed in the presence of these ingredients.

In the 1950's, scientists performed experiments that involved reproducing the conditions of the early earth. The starting point of these experiments was to place water in containers filled with the type of atmosphere just described. Water was introduced because it is apparent that the earth had oceans from very early times, and because it is thought that life started in the oceans, where the liquid environment provided a medium in

FIG. 32.1. CHARLES DARWIN. The scientific inquiries by Darwin led to an understanding of evolution, one of the most profound concepts of human intellectual development.

which complex chemical reactions could take place. Reactions occur much more slowly in solids and in gases, in the former case because the atoms are not free to move about easily and interact, and in the latter case because the density of gas is low, and particles are relatively unlikely to encounter each other. Water is the most stable and abundant liquid that can form from the common elements thought to have been present when the earth was young.

The first of these experiments (Fig. 32.2) was carried out in 1953 by the American scientists H. Urey and S. Miller, who concocted a mixture of methane, ammonia, water, and hydrogen, and exposed it to electrical discharges, a possible source of energy on the primitive earth (ultraviolet light from the sun is another, but was more difficult to work with in the laboratory). After a week, the mixture turned a dark brown, and Urey and Miller analyzed its composition. What they found were large quantities of amino acids, complex molecules that form the basis of proteins, which are the fundamental substance of living matter. Other scientists later showed that exposure to ultraviolet light produced the same results. All these experiments demonstrated that at least some of the precursors of life probably existed in the primitive oceans almost immediately after the earth had cooled enough to support liquid water. Other similar experiments produced more complex molecules, including sugars and larger fragments of proteins.

FIG. 32.2. SIMULATING THE EARLY
EARTH. Urey and Miller constructed this
apparatus to reproduce conditions on the
primitive earth, in hopes of learning how life
forms developed.

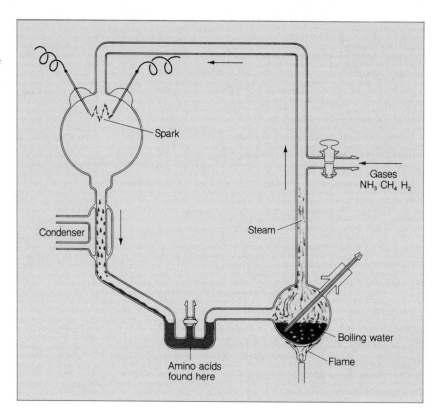

FIG. 32.3. PRIMORDIAL AMINO ACIDS. This is a section of the
Murchison meteorite, which fell in Australia. Amino acids found in
this carbonaceous chondrite were undoubtedly present in it when
it fell.

As noted in chapter 15, it may be that amino acids were
present in the solar system even before the earth formed,
because traces of them have been found in some meteorites (Fig. 32.3) and we know that meteorites are very
old, representing the first solid material in the solar system. We also know that several kinds of complex molecules, including organic (carbon-bearing) molecules,
exist inside dense interstellar clouds (see chapter 25), and
we may speculate that amino acids formed in these regions. In order to have survived on the earth, these
primordial amino acids would have had to reach our
planet sometime *after* its molten period.

It is not so clear what direction things took, once
amino acids and other organic molecules existed.
Somehow these building blocks had to combine to form
ribonculeic acid (RNA) and **deoxyribonucleic acid**
(DNA; Fig. 32.4). These very complex molecules carry
the genetic codes that allow living creatures to reproduce themselves. Experiments have successfully produced molecules that are fragments of RNA and DNA
from conditions like those which prevailed on the early
earth, but not the complete forms required. Maybe it

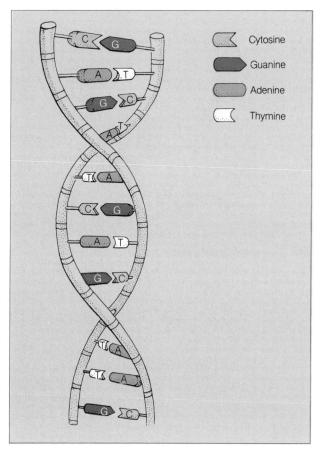

FIG. 32.4. THE DNA MOLECULE. This is a schematic diagram of a section of a DNA molecule. DNA carries the genetic code that allows organisms to reproduce themselves. A critical question in understanding the development of life on earth is how DNA arose.

FIG. 32.5. A FOSSIL MICROORGANISM. This is evidence for the fact that primitive life forms existed on the earth millions of years ago.

is a simple matter of time; if such experiments could be carried out for years or millennia, perhaps the vital forms of these proteins would appear. This is one of the areas of greatest uncertainty in our present knowledge of how life began.

Fossil records tell us that the first microorganisms (Fig. 32.5) appeared some 3 to 3.5 billion years ago, when the earth was barely 1 billion years old. The evolution of increasingly complex species seems to have followed naturally. At first the development was very slow, only reaching the level of simple plants such as algae a billion years later. Increasingly elaborate multicellular plant forms followed, and gradually altered the earth's atmosphere by introducing free oxygen. Meanwhile, the gases hydrogen and helium, light and fast-moving enough to escape the earth's gravity, essen-

tially disappeared. Nitrogen, always present from outgassing and volcanic activity, became more predominant through the decay of dead organisms. By about one billion years ago, the earth's atmosphere had reached its present composition.

The first broad proliferation of animal life occurred about 600 million years ago, and the great reptiles arose some 350 million years later. The dinosaurs died out after about 200 million years, and mammals came to dominance about 65 million years ago. Our primitive ancestors appeared only in the last 3 or 4 million years. Once the development of intelligence had provided the ability to control the environment, the entire world became our ecological home, and our physical evolution essentially stopped.

Could Life Develop Elsewhere?

The scenario just described, if at all accurate, should occur almost inevitably, given the proper conditions. If this is so, the question of whether life exists elsewhere amounts to asking whether the conditions that existed on the primitive earth could have arisen elsewhere.

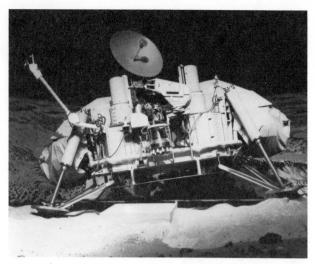

FIG. 32.6.   SEARCHING FOR LIFE IN THE SOLAR SYSTEM. Mars was long thought to be the most likely home in the solar system for extraterrestrial life. Here we see a *Viking* lander in a simulated Martian environment (left), and an artist's concept of a planned return spacecraft, lifting off of the Martian surface with a cargo of soil and rocks (right). The *Viking* mission reached Mars in 1976, and found no evidence of life forms.

It is clear that no other planet in the solar system could have provided an environment exactly like that of the early earth. It therefore seems unlikely that earthlike organisms will be found anywhere else in the solar system (but recall the argument in chapter 12 that the atmosphere of Jupiter may come close to reproducing the conditions of the early earth). Mars is the only planet where life forms have been sought so far (Fig. 32.6), but even there, the conditions are quite unlike those on earth. Given the billions of stars in the galaxy, however, and the vast number that are very similar to the sun, it seems highly probable that the proper conditions must have been reproduced many times in the history of the galaxy.

So far we have worked from the tacit assumption that life on other planets, if it exists, must be similar to life on earth, and we have considered only the question of whether other earthlike environments may exist. We might question the premise that life could only have developed in the form that we are familiar with, however. Here, obviously, we must indulge in speculation, having no examples of other types of life at hand for examination.

Life as we know it is based on carbon-bearing molecules, and it has been argued that only carbon has the capability of combining chemically in a sufficiently wide variety of ways with other common elements to produce the complexity of molecules thought to be nec-

essary for life. After all, it is clear that the basic elements available to begin with must be the same everywhere, given the homogeneous composition of the universe. This may seem to rule out life forms based on anything other than carbon, but it has been pointed out that another common element, silicon, also has a very complex chemistry, and therefore might provide a basis for a radically different type of life. If so, we cannot begin to speculate on the conditions necessary for such life forms to arise.

Another assumption that might be subject to question is that water is a necessary medium. As noted earlier, it is the most abundant liquid that can form under the temperature and pressure conditions of the early earth, and it is thought that only a liquid medium could support the required level of chemical activity. Other liquids can exist under other conditions, however, and it is interesting to consider whether life forms of a wholly different type than we are familiar with might arise in oceans of strange composition. It is interesting to speculate, for example, about what goes on in the lakes of liquid methane that are thought to exist on Titan, the mysterious giant satellite of Saturn.

If we are satisfied that life probably has formed naturally in many places in the universe, we can address a related, and to many a more important, question: given the existence of life, how likely is it that intelligence will follow? Here we have no means of answering, except to reiterate that as far as we know, the evolution of our species on earth was the natural product of environmental pressures.

The Probability of Detection

In view of the limitations that prohibit faster-than-light travel, it is exceedingly unlikely that we will be able to visit other solar systems, seeking out life forms that may reside there. We will continue to explore our own system, so there is a reasonable chance that if life exists on any of the other planets of our sun, we will someday discover it. It seems, however, that our best hope of finding other intelligent races in the galaxy will be to make long-range contact with them, through radio or light signals. Since this requires both a transmitter and a receiver, we can only hope to contact other civilizations as advanced as ours, with the capability of constructing the necessary devices for interstellar communication.

A mathematical exercise in probabilities was developed some years ago by Frank Drake of Cornell University as a means of assessing, as objectively as possible, the chances for making contact with an extraterrestrial civilization. The aim was to separate the question into several distinct steps, each of which could then be treated independently. The underlying assumption is that the number of technological civilizations in our galaxy today with the capability for interstellar communication is the product of the number of planets that exist with appropriate conditions, the probability that life arose on those planets, the probability that such life developed intelligence that gave rise to a technological civilization, and finally, the likelihood that the civilization was not killed off, through evolution or catastrophe.

Mathematically, Drake's statement of the problem is written:

$$N = R_* f_p n_e f_l f_i f_c L,$$

where N is the number of technological civilizations presently in existence, R_* is the number of stars of appropriate spectral type formed per year in the galaxy, f_p is the fraction of these that have planets, n_e is the number of earthlike planets per star, f_l is the fraction of these on which life arose, f_i is the fraction of those planets on which intelligence developed, f_c is the fraction of planets with intelligence on which a technolog-

ical civilization evolved to the point where interstellar communications would be possible, and, finally, L is the average lifetime of such a civilization.

This equation allows us to isolate the factors for which we can make educated guesses from those we are more ignorant about. It is an interesting exercise to go through the terms in the equation one by one, to see what conclusions we reach under various assumptions. People who do this have to make sheer guesses for some of the terms, and the result is a variety of answers ranging from very optimistic to very pessimistic ones. We will adopt middle-of-the-road numbers for most of the unknown terms.

The first two factors, R_* and f_p, are quantities that in principle can be known with some certainty from observations. For the rate of star formation to be relevant in this exercise, stars must be of a spectral type similar to the sun's. A much cooler star would not have a temperate zone around it where a planet could have the moderate temperatures needed for life to begin (we are, throughout this exercise, limiting ourselves to life forms similar to our own), and a very hot star would be short-lived, so there would be insufficient time for life to develop on its planets, before the parent star blew up in a supernova explosion. Taking these considerations into account, some people estimate that up to 10 suitable stars form in our galaxy per year; for the sake of discussion, we will be more cautious and adopt $R_* = 1$/year. This may even be a little high for the present epoch in galactic history, but the star formation rate was surely much higher early in the lifetime of the galaxy, and low-mass stars of solar type which formed that long ago are still in their prime. Thus, the adoption of an average formation rate of one sun-like star per year is probably reasonable.

From what we know of the formation of our solar system, it seems that the formation of planets is almost inevitable, except perhaps in double or multiple star systems. Let us assume that $f_p = 1$; that is, all stars of solar type have planets. Observations made with future instruments such as the *Space Telescope* may soon provide real information on this term.

The number of earthlike planets, n_e, is highly uncertain, and depends on how wide the zone is where the appropriate temperature conditions could exist. Recent studies show that this zone might be rather small; that the earth would not have been able to support life if its distance from the sun were only about 5 percent closer or farther than it is. Estimates of n_e vary from

10^{-6} to 1. Let us be moderate and assume $n_e = 0.1$; that in one out of ten planetary systems around solar-type stars, there is a planet within the temperate zone where life can arise.

Now we get to the *really* speculative terms in the equation. We have no way of estimating how likely it is that life should begin, given the right conditions. It can be argued that because of the seeming naturalness of its development on earth, life would always begin if given the chance. Let us be optimistic here and agree with this, adopting $f_l = 1$.

Again, the chance of this life developing intelligence, and furthermore advancing to the point of being capable of interstellar communication, are complete unknowns. All we know is that in the only example that has been observed, both things happened. For the sake of argument, we therefore set both f_i and f_c equal to 1.

At this point it is instructive to put the values adopted so far into the equation. We find:

$$N = (1/\text{year}) \ (1) \ (0.1) \ (1) \ (1) \ (1) \ L$$
$$= 0.1L$$

Having taken our chances and guessed at the values for all the other terms, we now face a critical question: how long can a technological civilization last? Ours has been sufficiently advanced to send or receive interstellar radio signals for only about fifty years, and the instability in our society leads some pessimists to think we will not last many more decades. If we take this viewpoint and adopt fifty years as the average lifetime, then we find:

$$N = 5,$$

meaning that we should expect the total number of technological civilizations present in the galaxy at one moment to be very small, about 5. If this is correct, then the average distance between these outposts of civilization is nearly 20,000 light years (Fig. 32.7). The time it would take for communications to travel between civilizations would therefore be very much longer than their lifetimes, and there would be no hope of our establishing a dialogue with anyone out there. If this estimate is correct, it is no surprise that we have not heard from anybody yet.

We can be more optimistic, though, and assume that a technological civilization solves its internal problems and lives much longer than fifty years. Extremely optimistic people would argue that a civilization is immortal; that it colonizes star systems other than its own, so

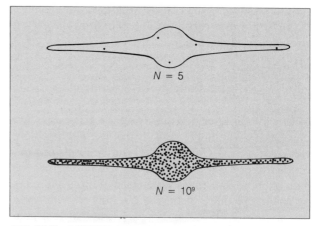

FIG. 32.7.   POSSIBLE VALUES OF N. The number of technological civilizations in the galaxy may be very small (upper), in which case the average distance between them is very large; or N may be large, so that the distance between civilizations is relatively small. The chances for communication with alien races are much higher in the latter case, where the distances are only a few tens of light years.

FIG. 32.8.   EARTH'S MESSAGE TO THE COSMOS. As our entertainment and communications broadcasts travel out into space, they provide a history of our culture for anyone who may be receiving the signals. At the present time, the growing sphere that is filled with our broadcasts has a radius of more than fifty light years.

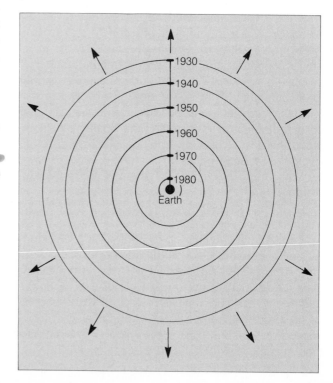

ASTRONOMICAL INSIGHT (32.1)

The Case for a Small Value of N

Although is is an interesting exercise to write down an equation for N, the number of technological civilizations in the galaxy, no two astronomers can agree on the values of the various terms.

There is one very pessimistic argument, developed recently, that concludes that we are alone in the galaxy. This is based on the likely fact that if there were other civilizations, a significant fraction of them would have arisen long before ours. This is a straightforward consequence of the great age of the galaxy, some 5 or 10 billion years greater than the age of the earth. This argument postulates further that if technological civilizations, once begun, live a long time, then the galaxy must be inhabited by a number of very advanced races, much older and more mature than our own. Up to this point, there is

little controversy about any of these assertions.

The next logical step in the argument, however, is hotly debateable; it says that any advanced civilization that has survived for millions or even billions of years will necessarily have done so by perfecting some form of interstellar travel, and colonizing other planets. Once this started, the argument goes, the spread of a given civilization throughout the galaxy would accelerate rapidly, and by this time in galactic history, no habitable planet such as the earth would remain uncolonized. The first successful galactic civilization, in this view, would quickly rise to complete dominance.

The only logical corollary, if the argument is accepted up to this point, is that no other civilizations exist, because if they did, the earth would have been colonized long ago. The absence of interstellar visitors on the earth is taken as proof that there are no other civilizations in the galaxy.

Those who adhere to this line of reasoning believe that N = 1. Therefore at least one of the terms in Drake's equation must be very much smaller than the more optimistic values discussed in the text. One suggestion is that the temperate zone around a sun-like star where planets can have moderate

conditions is really very much narrower than usually supposed, perhaps because of the more ready development of a Venus-like greenhouse effect than is normally thought to be the case. If the temperate zone is very small, then the term n_e, representing the number of habitable, earthlike planets per star, would be very much smaller than the value of 0.1 adopted in the text.

Another term that has recently been singled out is f_l, the fraction of earthlike planets on which life begins. As we learned in the text, somehow the amino acids that were present on the primitive earth had to arrange themselves into special combinations to produce the long chainlike protein molecules RNA and DNA. From a purely statistical point of view, the probability that the proper combination would come together by chance is very small. This apparently happened on the earth, but it may be so unlikely that it has not occurred anywhere else in the galaxy.

Whatever the correct values of the terms in the equation, the debate will rage on until our civilization reaches its life expectancy L and dies; or we discover another civilization; or we ourselves expand to colonize the galaxy and find no one else out there.

that it is immune to any local crises such as planetary wars or suns expanding to become red giants. In that case, allowing a few billion years for the development of such civilizations, we can set $L = 10^{10}$ years (nearly equal to the age of the galaxy), and we find

$$N = 10^9,$$

where the average distance between civilizations is only

about thirty light years, coincidentally just a little less than the distance our own radio signals have traveled since the early days of radio and television (Fig. 32.8). If this estimate of N is correct, we should be hearing from somebody very soon.

We have presented two extreme views of the likelihood that other civilizations exist in our part of the galaxy. As we mentioned earlier, opinions among the scientists who seriously study this question vary throughout this large range. Those who favor the optimistic viewpoint advocate that we should make deliberate attempts to seek out other civilizations.

The Strategy for Searching

The probability arguments just outlined are amusing and perhaps somewhat instructive, but obviously not very accurate. There are entirely too many unknowns in the equation for us to develop a reliable estimate of the chances for galactic companionship. Perhaps we will not know for certain what the answer is until we make contact with another civilization.

The problem of developing an experiment to search for or to send interstellar signals is that we do not know the ground rules. There is an infinite number of ways in which a distant civilization might choose to try to communicate, and it is impossible to search for them all. We must try to guess what the most probable technique would be.

In view of the power that is transmitted by radio signals, it has often been assumed that radio communications are most likely to succeed, although other techniques have been tried (Fig. 32.9). In the early 1960's, the U.S. National Radio Astronomy Observatory (Fig. 32.10) "listened" for transmissions from two nearby sunlike stars, tau Ceti and epsilon Eridani, without success. More recently, larger surveys of neighboring stars have been carried out, in the United States and in the Soviet Union, with similar results. The wavelengh chosen for most of these observations was 21 centimeters, selected partly for practical reasons (radio telescopes designed for observing at this wavelengh already existed), and partly because it was hypothesized that if a civilization out there wanted deliberately to send signals, it might choose to do so at a wavelength that other astronomers around the galaxy might be observing anyway, even if they were not trying to find distant civilizations. Another wavelength that has been considered, but tried only in a very limited search, lies in the microwave region, between the emission lines of interstellar water vapor and OH. Dubbed the "water hole," this region has the advantages that it, too, might be a target of extraterrestrial astronomers, and it is in the quietest part of the radio spectrum (various objects, ranging from supernova remnants to radio galaxies, create radio noise that fills the galaxy, and the water hole lies in a region of the spectrum where this noise is minimized; see Fig. 32.11).

One pessimistic viewpoint is that we are all listen-

FIG. 32.9. ANOTHER MESSAGE FROM EARTH. This recording of music and other sounds from earth is traveling beyond the solar system aboard the *Voyager* spacecraft. Only an advanced race of beings would have the technological skills needed to learn how to listen to and decode the recording.

FIG. 32.10. A RADIO TELESCOPE USED IN THE SEARCH FOR LIFE. These are radio antennae at the U.S. National Radio Astronomy Observatory, in Green Bank, West Virginia, which has been used in concerted efforts to detect signals from alien civilizations. Most of these attempts were made at the 21-cm wavelength of atomic hydrogen emission.

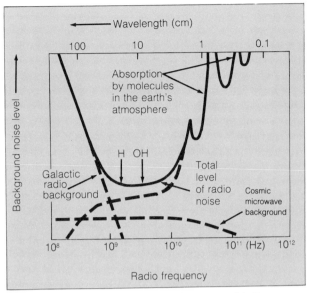

FIG. 32.11. THE GALACTIC NOISE SPECTRUM. This diagram shows the relative intensity of the background noise in the galaxy as a function of wavelength. The quietest portion of the spectrum is in the vicinity of the so-called water hole, between the natural emission wavelengths of H_2O and OH molecules.

ing, and no one is sending. In that case, we can only hope to pick up accidental emissions, such as entertainment broadcasts on radio or television, and these would be much weaker and more difficult to detect. It has been estimated that at the present level of technol-

ogy we could deliberately send radio signals that could be received by a similarly advanced civilization up to several thousand parsecs away, whereas our accidental radio and television signals could only be detected at distances of one or two parsecs. Therefore, it makes a big difference whether someone out there is trying to send a message or not.

Although most thinking about means of communication with other civilizations has been based on the assumption that radio signals are the best choice, there have been other suggestions. One is that a **laser** might be used. This is a device that can emit a powerful, narrow beam of visible light at a single wavelength; lasers today are being used for many purposes ranging from microsurgery to weapons to telephone communications. The lasers developed so far, however, do not have the range of radio transmissions, so at this point most scientists favor a search for radio signals. Laser technology is improving, but there is a fundamental limitation on the use of visible light: the universe is far "noisier," more filled with confusing natural emissions, at visible wavelengths than in the radio portion of the spectrum.

It is not clear what the future will bring, in terms of deliberate searches for extraterrestrial intelligence. The most ambitious project so far is called *Project Cyclops* (Fig. 32.12), and it is probably several years or decades away from being implemented. The plan is to build a veritable forest of small radio antennae, along with a very sophisticated data-processing system, so that up to

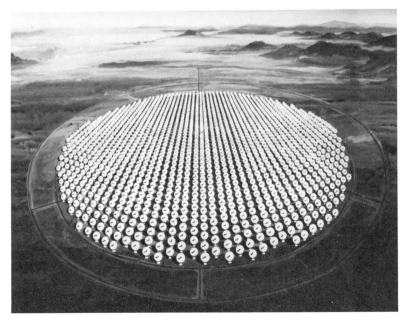

FIG. 32.12. *PROJECT CYCLOPS*. The grandest plan seriously considered for the purpose of communicating with extraterrestrial civilizations, *Project Cyclops* would consist of 1,000 to 2,500 radio antennae, each 30 meters (100 feet) in diameter. This huge array, some 16 kilometers in extent, would be capable of detecting signals over a large portion of the radio spectrum, from planetary systems 1,000 or more parsecs away.

FIG. 32.13. A FIRST STEP FOR
EARTHLINGS? This is an artist's concept of a
permanent space platform, to which Space
Shuttles would make frequent visits. Perhaps such
a facility will be the next stage of the earth's rise
to galactic dominance.

a million stars could be scanned for artificial emissions over a broad wavelength region in a few years' time. By searching over a wide range of wavelengths, the scientists planning this effort hope to eliminate much of the guesswork about which is the most likely portion of the spectrum to be used by alien civilizations. This also enhances the chances of detecting accidental emissions, which, if they are anything like our own radio and television signals, may be spread over much of the spectrum. The *Project Cyclops* antennae could also be used to *send* powerful radio signals, allowing the earth to broadcast its own beacon for others to find, and providing for the possibility of two-way communications.

The principal problem in implementing *Cyclops,* or any other large-scale effort, is the enormous cost. The version of *Project Cyclops* just outlined is currently estimated to require some 10 billion dollars in expenditures. Such sums are rarely devoted to astronomy, and if they were, there would still be considerable debate over whether to spend it on this or on some other research programs that might appear to have greater certainty of success (Fig. 32.13). Presumably, alien astronomers in other solar systems would have to face the same questions, although we can hope that sufficiently advanced civilizations will have overcome the limitations of their planetary resources, and will be willing and able to tackle such a massive task.

Perspective

In this chapter we have introduced a bit of speculation, well-founded perhaps in the discussion of life on earth, but less so in the sections on the possibilities of life elsewhere. We are now prepared to understand and appreciate new advances in astronomy; having completed our study of the present knowledge of the universe, we will be able to put into perspective the major new discoveries that are sure to come as technology

and theory continue to improve. The *Space Telescope,* the large optical telescopes now being designed, the comparable advances in radio, infrared, and X-ray techniques; all of these will inevitably lead to novel and unforeseen breakthroughs in our view of the universe. The next few decades will be exciting times for astronomy, and we can anticipate their coming with a sense of excitement and wonder.

Summary

1. Although there were early beliefs that life started on earth spontaneously or by primitive spores from space, most scientists today accept the theory that life began through natural, evolutionary processes.

2. Amino acids, fundamental components of living organisms, are formed readily in experiments designed to simulate early conditions on the earth.

3. The steps that led to the development of the necessary forms of RNA and DNA are not yet fully understood, and probably occurred over a very long period of time.

4. Fossil evidence provides a record of the evolutionary steps leading from the first primitive life forms to modern life.

5. The conditions that prevailed on the early earth have probably been duplicated on other planets in the galaxy, though probably not on other planets in the solar system.

6. It is often assumed that only earthlike life could develop, because carbon is nearly unique in the complexity of its chemistry. However, there have been suggestions that at least one other element (silicon) may have the necessary properties.

7. Estimates of the number N of technological civilizations now in the galaxy can be made, with great uncertainty, based on what is known of the formation rate of sunlike stars, and what is guessed for the probability of such stars' having planets with the proper conditions, and the probability that life, leading to intelligence and technology, will develop on these planets. Estimates range from $N = 1$ to $N = 10^9$.

8. Some attempts have been made to search for radio signals at 21 centimeters and other wavelengths, with no success.

9. The chances for detecting or being detected by other civilizations depend strongly on whether or not deliberate attempts are made to send signals.

10. Future projects to search for and to send signals have been planned, and will be sufficiently powerful to have a high probability of success if a large number of civilizations exist, but they may not be funded and put into operation in the foreseeable future.

Review Questions

1. Suppose life did start on earth as a result of a chance arrival of microscopic spores from space. If these spores travel through space with velocities typical of interstellar clouds, then their speeds might be about 20 km/sec. Calculate how long it would take a spore to travel at this speed from the nearest star (1.4 parsecs away) to earth. How long would it take for a spore to cross the galactic disk, a distance of 30,000 parsecs?

2. Summarize the evolution of the earth's atmosphere, as outlined in chapter 7. Does this provide a possible clue to look for in determining whether life exists on a planet that might be discovered orbiting a distant star?

3. We have noted here and in chapter 12 that some of the conditions thought necessary for the formation of life exist in the atmosphere of Jupiter. Some key conditions are probably absent, however. What are they?

4. Based on your answer to question 3, do you think life is likely to have formed inside of interstellar clouds? (Recall that rather complex organic molecules have been observed in dark clouds.) Explain.

5. Why would a planet only slightly closer to the sun than the earth probably not be able to support life? (Hint: Recall what you learned in chapter 9 about the origin of the contrasting conditions on the earth and on Venus.)

6. Repeat the calculation of N, the number of technological civilizations in the galaxy, with all parameters the same as adopted in the text, except that the probability of life beginning on an earthlike planet is only 10^{-4}, rather than 1 (assume that the lifetime L is 10^{10} years).

7. What is N if all the values used in the text are adopted, except that L is set equal to 10,000 years?

8. Would you expect technological civilizations in the

galaxy to be concentrated along the spiral arms, or distributed more or less uniformly throughout the disk? Explain your answer.

9. Would you expect life to be as likely to arise on planets orbiting Population II stars as on those orbiting Population I stars? Explain.

10. Why do you think it makes such a big difference whether or not a deliberate attempt is made by a civilization to send radio signals; that is, explain why deliberate signals can be detected at much greater distances than accidental signals such as radio and television entertainment broadcasts.

Additional Readings

Abt, H. A. 1979. The companions of sun-like stars. *Scientific American* 236(4):96.

Ball, John A. 1980. Extraterrestrial intelligence: where is everybody? *American Scientist* 68:656.

Goldsmith, D., ed. 1980. *The quest for extraterrestrial life*. Mill Valley, Cal.: University Science Books.

O'Neill, G. K. 1974. The colonization of space. *Physics Today* 27(9):32.

Pollard, W. G. 1979. The prevalence of earth-like planets. *American Scientist* 67:653.

Rood, R. T., and Trefil, J. S. 1981. *Are we alone?* New York: Scribner.

Sagan, C., and Drake, F. 1975. The search for extraterrestrial intelligence. *Scientific American* 232(5):80.

Sagan, C., and Shklovskii, I. S. 1966. *Intelligent life in the universe*. San Francisco: Holden-Day.

Trying to Psyche Out Extraterrestrial Civilizations

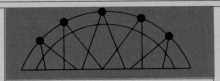

Frank D. Drake

Dr. Drake, a radio astronomer with research interests in solar-system and galactic astronomy, has served as a distinguished member of the faculty at Cornell University since 1964, following his tenure in senior posts at the U.S. National Radio Astronomy Observatory and at the Jet Propulsion Laboratory of the California Institute of Technology. At Cornell he has served as the director of the National Astronomy and Ionospheric Center, and holds the Goldwin Smith Professorship of Astronomy. Despite his impressive record in solar-system research and science administration, Dr. Drake is probably best known for his work on the question of extraterrestrial civilizations. He pioneered the modern methods of treating the subject, instigated many of the searches that have been attempted, and is a central figure in all organized studies of the question. His equation describing the number of civilizations in the galaxy as a function of various probabilities that can be independently assessed has set the stage for most serious discussion on the subject.

The chances of success and the cost and effort of a search for radio signals from other worlds depend on N, the number of detectable civilizations. You have read here of the thinking which leads to estimates of N, and as you have seen, more than one scenario for the development of civilizations

has been invoked to predict N. On the one hand there are rather reserved people who feel that we should only assume that what has happened on our earth and in our civilization is typical, and that vast projects in cosmic engineering are not likely. Such people conclude that there are indeed many civilizations like ours, populating planets like ours and perhaps a few space colonies near their planets. These civilizations can be found most easily and perhaps only by searching diligently with radio telescopes.

There is another school of thought which says that civilizations like ours will accede to a compulsion to sustain ever larger populations of their creatures. If so, they must eventually establish living room beyond their own planet, first in suitable places in their planetary system, and later,

in a heroic enterprise, on planets of other stars. Even at a slow rate of colonization of one star after another, the rate at which populations increase, if they are anything like us, will lead to the colonization of every star in the galaxy in a time of a hundred million years or so, or about 1 percent of the age of the galaxy. Clearly, if even one such civilization had arisen in the past, their colonists would be here by now and we would know of them. This might be called the "cancer" scenario: all it takes is one "malignant" civilization, and it quickly spreads to every star in the galaxy. And yet they have not come to us. If one accepts this picture as realistic, it implies that there are, for some reason, very few civilizations and that we may be alone or at least the most advanced.

Which of these basic scenarios is right? It is important to know, for one scenario says that a treasure of discoveries awaits us if we mount a respectable search for extraterrestrial intelligence. The other says we would be wasting our money and talents. There is nothing to find: just wait awhile and *we* will be the extraterrestrials. Thinking about it, you will see that the choice between these scenarios depends not, as we would very much wish, on the laws of physics or the arrangement of the universe. Rather, it depends on the actual priorities, needs, values, and even social systems and governments of advanced civilizations.

The price to be paid by a civilization to adopt various modes of life is enormous, and choices will not be made casually. Let's look at the interstellar colonization scenarios. Despite what we have been led to believe by science fiction, movies, and television, the colonization of the planets, and especially the stars, is an unbelievably costly undertaking, no matter how advanced you are. It is easy to compute that the cost in energy alone required to send a single colonist to a nearby star system in a time of 100 years, which already seems long, is enough energy to provide a good life to 200,000 people, if they are like us. Take into account the inefficiencies in rockets and fuel production, and the price goes up to the cost of supporting 2,000,000 people luxuriously throughout their lives. Send 100 colonists to start a colony, and the cost is equal to the cost of supporting roughly the whole population of the United States for all their lives. In other words, to launch a minimal colony at reasonable speed would require, in effect, the shutting down of America for 100 years. Would we do that?

Now you may say, "But they may have access to cheap energy sources." True, but energy will never be free; it must not only be generated, it must be stored and transmitted. There will always be an economic question as to how to get the most benefit from a given amount of energy. For a small fraction of the cost of sending a small colony to the stars, you could create glorious living space by enclosing the arctic in heated domes, or building floating cities on the oceans, or building a multitude of large colonies in space near the home planet.

What will they do? The decisions are not at all obvious. My own feeling is that they will choose to remain near home. They may embark on marvelous enterprises. They may build great cities in the sky. But they will be within sight of home, and not at distant alpha Centauri or their equivalent. A truly intelligent civilization will use its resources to the greatest benefit for the most creatures; and that is done by acknowledging the monstrous penalty in costs which would be paid if one tried to flaunt the laws of physics and relativity. Only do what can be done "slowly," that is, "slowly" compared to the speed of light!

Of course, others may think otherwise. Perhaps it is worth the cost to experience the adventure of starting anew on another star. Perhaps you decide to support only a small elite population, allotting enormous resources to each individual. The personal ration of materials and energy may be enough to provide for interstellar travel.

Who can say? And that is exactly my point. The laws of physics and the great distances which separate the stars have made it difficult for civilizations to choose what is good for them. It will depend on their goals and philosophies. It will be hard for *them* to decide, and surely impossible for *us* to decide what they may do. Until we learn the histories of many civilizations, we will be in no position to psyche out the extraterrestrials.

Thus the search for extraterrestrial intelligence for now must be an experimental one. Mind alone will not work to give us answers. Only the lengthy scanning of huge telescope mirrors and giant radio antennae can. We should not be disappointed; science has always worked best when we experimented with open minds. Now we should get on with the job.

• A P P E N D I X E S •

Appendix 1. Physical and Mathematical Constants

CONSTANT	SYMBOL	VALUE
Speed of light	c	2.9979250×10^{10} cm/sec
Gravitation constant	G	6.670×10^{-8} dyn cm^2/gm
Planck constant	h	6.62620×10^{-27} erg sec
Electron mass	m_e	9.10956×10^{-28} gm
Proton mass	m_p	1.672661×10^{-24} gm
Stefan-Boltzmann constant	σ	5.66596×10^{-5} erg/cm^{2}deg^{4}sec
Wien constant	W	0.289789
Boltzmann constant	k	1.38062×10^{-16} erg/deg
Astronomical unit	AU	1.495979×10^8 km
Parsec	pc	3.085678×10^{13} km = 3.261633 light years
Light year	ly	9.460530×10^{12} km
Solar mass	$M_\odot$	1.9891×10^{33} gm
Solar radius	$R_\odot$	6.9600×10^{10} cm
Solar luminosity	$L_\odot$	3.827×10^{33} erg/sec
Earth mass	$M_\oplus$	$5,9742 \times 10^{27}$ gm
Earth radius	$R_\oplus$	6378.140 km
Tropical year (equinox to equinox)		365.241219878 days
Sidereal year (with respect to stars)		365.256366 days = 3.155815×10^7 sec

Appendix 2. The Elements and Their Abundances

ELEMENT	SYMBOL	ATOMIC NUMBER	ATOMIC WEIGHT[1]	ABUNDANCE[2]
Hydrogen	H	1	1.0080	1.00
Helium	He	2	4.0026	0.063
Lithium	Li	3	6.941	1.55×10^{-9}
Beryllium	Be	4	9.0122	1.41×10^{-11}
Boron	B	5	10.811	2.00×10^{-10}
Carbon	C	6	12.0111	0.000372
Nitrogen	N	7	14.0067	0.000115
Oxygen	O	8	15.9994	0.000676
Fluorine	F	9	18.9984	3.63×10^{-8}
Neon	Ne	10	20.179	3.72×10^{-5}
Sodium	Na	11	22.9898	1.74×10^{-6}
Magnesium	Mg	12	24.305	3.47×10^{-5}
Aluminum	Al	13	26.9815	2.51×10^{-6}
Silicon	Si	14	28.086	3.55×10^{-5}
Phosphorus	P	15	30.9738	3.16×10^{-7}
Sulfur	S	16	32.06	1.62×10^{-5}
Chlorine	Cl	17	35.453	2×10^{-7}
Argon	Ar	18	39.948	4.47×10^{-6}
Potassium	K	19	39.102	1.12×10^{-7}
Calcium	Ca	20	40.08	2.14×10^{-6}
Scandium	Sc	21	44.956	1.17×10^{-9}
Titanium	Ti	22	47.90	5.50×10^{-8}
Vanadium	V	23	50.9414	1.26×10^{-8}
Chromium	Cr	24	51.996	5.01×10^{-7}
Manganese	Mn	25	54.9380	2.63×10^{-7}
Iron	Fe	26	55.847	2.51×10^{-5}
Cobalt	Co	27	58.9332	3.16×10^{-8}
Nickel	Ni	28	58.71	1.91×10^{-6}
Copper	Cu	29	63.546	2.82×10^{-8}
Zinc	Zn	30	65.37	2.63×10^{-8}
Gallium	Ga	31	69.72	6.92×10^{-10}
Germanium	Ge	32	72.59	2.09×10^{-9}
Arsenic	As	33	74.9216	2×10^{-10}
Selenium	Se	34	78.96	3.16×10^{-9}
Bromine	Br	35	79.904	6.03×10^{-10}
Krypton	Kr	36	83.80	1.6×10^{-9}
Rubidium	Rb	37	85.4678	4.27×10^{-10}
Strontium	Sr	38	87.62	6.61×10^{-10}
Yttrium	Y	39	88.9059	4.17×10^{-11}
Zirconium	Zr	40	91.22	2.63×10^{-10}

(continued on next page)

[1]The atomic weight of an element is its mass in *atomic mass units*. An atomic mass unit is defined as one-twelfth of the mass of the most common isotope of carbon, and has the value 1.660531×10^{-24} grams. In general, the atomic weight of an element is approximately equal to the total number of protons and neutrons in its nucleus.

[2]The abundances are given in terms of the number of atoms of each element compared with hydrogen, and are based on the composition of the sun. For very rare elements, particularly those toward the end of the list, the abundances can be quite uncertain, and the values given should not be considered exact.

Appendix 2. The Elements and Their Abundances
(Continued)

ELEMENT	SYMBOL	ATOMIC NUMBER	ATOMIC WEIGHT	ABUNDANCE
Niobium	Nb	41	92.906	2.0×10^{-10}
Molybdenum	Mo	42	95.94	7.94×10^{-11}
Technetium	Tc	43	98.906	—
Ruthenium	Ru	44	101.07	3.72×10^{-11}
Rhodium	Rh	45	102.905	3.55×10^{-11}
Palladium	Pd	46	106.4	3.72×10^{-11}
Silver	Ag	47	107.868	4.68×10^{-12}
Cadmium	Cd	48	112.40	9.33×10^{-11}
Indium	In	49	114.82	5.13×10^{-11}
Tin	Sn	50	118.69	5.13×10^{-11}
Antimony	Sb	51	121.75	5.62×10^{-12}
Tellurium	Te	52	127.60	1×10^{-10}
Iodine	I	53	126.9045	4.07×10^{-11}
Xenon	Xe	54	131.30	1×10^{-10}
Cesium	Cs	55	132.905	1.26×10^{-11}
Barium	Ba	56	137.34	6.31×10^{-11}
Lanthanum	La	57	138.906	6.46×10^{-11}
Cerium	Ce	58	140.12	4.37×10^{-11}
Praseodymium	Pr	59	140.908	4.27×10^{-11}
Neodymium	Nd	60	144.24	6.61×10^{-11}
Promethium	Pm	61	146	—
Samarium	Sm	62	150.4	4.57×10^{-11}
Europium	Eu	63	151.96	3.09×10^{-12}
Gadolinium	Gd	64	157.25	1.32×10^{-11}
Terbium	Tb	65	158.925	2.63×10^{-12}
Dysprosium	Dy	66	162.50	1.29×10^{-11}
Holmium	Ho	67	164.930	3.1×10^{-12}
Erbium	Er	68	167.26	5.75×10^{-12}
Thulium	Tm	69	168.934	2.69×10^{-12}
Ytterbium	Yb	70	170.04	6.46×10^{-12}
Lutetium	Lu	71	174.97	6.92×10^{-12}
Hafnium	Hf	72	178.49	6.3×10^{-12}
Tantalum	Ta	73	180.948	2×10^{-12}
Tungsten	W	74	183.85	3.72×10^{-10}
Rhenium	Re	75	186.2	1.8×10^{-12}
Osmium	Os	76	190.2	5.62×10^{-12}
Iridium	Ir	77	192.2	1.62×10^{-10}
Platinum	Pt	78	195.09	5.62×10^{-11}
Gold	Au	79	196.967	2.09×10^{-12}
Mercury	Hg	80	200.59	1×10^{-9}
Thallium	Tl	81	204.37	1.6×10^{-12}
Lead	Pb	82	207.19	7.41×10^{-11}
Bismuth	Bi	83	208.981	6.3×10^{-12}
Polonium	Po	84	210	—
Astatine	At	85	210	—

(continued on next page)

Appendix 2. The Elements and Their Abundances
(Continued)

ELEMENT	SYMBOL	ATOMIC NUMBER	ATOMIC WEIGHT	ABUNDANCE
Radon	Rn	86	222	—
Francium	Fr	87	223	—
Radium	Ra	88	226.025	—
Actinium	Ac	89	227	—
Thorium	Th	90	232.038	6.61×10^{-12}
Protactinium	Pa	91	230.040	—
Uranium	U	92	238.029	4.0×10^{-12}
Neptunium	Np	93	237.048	—
Plutonium	Pu	94	242	—
Americium	Am	95	242	—
Curium	Cm	96	245	—
Berkelium	Bk	97	248	—
Californium	Cf	98	252	—
Einsteinium	Es	99	253	—
Fermium	Fm	100	257	—
Mendelevium	Md	101	257	—
Nobelium	No	102	255	—
Lawrencium	Lr	103	256	—

Appendix 3. Temperature Scales

At the most basic level, temperature can be defined in terms of the motion of particles in a gas (or a solid or liquid as well). We have an intuitive idea of what heat is, and we all are familiar with at least one scale for measuring temperature.

The most commonly used scales are somewhat arbitrarily defined, with zero points not representing any truly fundamental physical basis. The popular *Fahrenheit* scale, for example, has water freezing at a temperature of 32°F and boiling at 212°F. On this scale, absolute zero, the lowest possible temperature (where all molecular motions cease), is −459°F.

The *centigrade* (or *Celsius*) scale is perhaps more well-founded, although it is based on the freezing and boiling points of water, rather than the more fundamental absolute zero. In this system, the freezing point is defined as 0°C, and 100°C is the boiling point. This scale has the advantage over the Fahrenheit scale that there are exactly 100° between the freezing and boiling points, rather than 180°, as in the Fahrenheit system. To convert from Fahrenheit to centigrade, we must subtract 32° first, and then multiply the remainder by 100/180, or 5/9. For example, 50°F is equal to 5/9 × (50 − 32) = 10°C. To convert from centigrade to Fahrenheit, first multiply by 9/5 and then add 32°. Thus, −10°C = 9/5 × (−10) + 32 = 14°F. On the centigrade scale, absolute zero occurs at −273°C.

The temperature scale preferred by scientists is a modification of the centigrade system. In this system, named after its founder, the British physicist Lord Kelvin, the same degree is used as in the centigrade scale; that is, one degree is equal to one one-hundredth of the difference between the freezing and boiling points of water. The zero point is different than on the centigrade scale, however; it is set equal to absolute zero. Hence on this scale water freezes at 273°K and boils at 373°K. Comfortable room temperature is around 300°K. In modern usage, the degree symbol (°) is dropped, and we speak simply of temperatures in units of Kelvins.

Appendix 4. Radiation Laws

In chapter 5 we described several laws that apply to continuous radiation from hot objects such as stars. These laws were discussed in general terms, and a few simple applications were explained. In this appendix, the same laws are given in more precise mathematical form, and their use in that form is illustrated.

Wien's Law

In general terms, Wien's law states that the wavelength of maximum emission from a glowing object is inversely proportional to its temperature. Mathematically, this may be written as:

$$\lambda_{max} \propto 1/T,$$

where λ_{max} is the wavelength of strongest emission, T is the surface temperature of the object, and $\propto$ is a special mathematical symbol meaning "is proportional to."

Experimentation can determine the *proportionality constant*, specifying the exact relationship between λ_{max} and T, and Wien did this, finding

$$\lambda_{max} = W/T,$$

where W has the value 0.29 if λ_{max} is measured in centimeters and T in degrees absolute. With this equation it is possible to calculate λ_{max}, given T, or vice versa. Thus if we measure the spectrum of a star and find that it emits most strongly at a wavelength of 2,000 Å $= 2 \times 10^{-5}$ cm, then we can solve for the temperature:

$$T = W/\lambda_{max} = 0.29/2 \times 10^{-5} = 14,500°K.$$

This is a relatively hot star, and would appear blue to the eye. Note that it was necessary to measure the spectrum in ultraviolet wavelengths in order to find λ_{max}.

When solving problems implementing Wien's law, we can always use the equation form, as we have just done. Often, however, it is more convenient to compare the properties of two objects by considering the ratio of the temperatures or of the wavelengths of maximum emission. This, in effect, is what we did in the text, when we compared two objects of different temperatures in order to determine how their values compared; for example, we said that if one object is twice as hot as another, its value of λ_{max} is half that of the other.

We can see how this ratio technique works by writing the equation for Wien's law separately for object 1 and object 2:

$$\lambda_{max_1} = W/T_2, \text{ and } \lambda_{max_2} = W/T_2.$$

Now we can divide one equation by the other:

$$\frac{\lambda_{max_1} = (W/T_1)}{\lambda_{max_2} = (W/T_2)}, \text{or}$$

$$\lambda_{max_1}/\lambda_{max_2} = T_2/T_1.$$

The numerical factor W has canceled out, and we are left with a simple expression relating the values of λ_{max} and T for the two objects. Now we see that if $T_1 = 2T_2$ (that is, object 1 is twice as hot as object 2), then

$$\lambda_{max_1}/\lambda_{max_2} = T_2/2T_2 = 1/2,$$

or λ_{max} for object 1 is one-half that for object 2. In this extremely simple example, it probably would have been easier to just work it out in our heads, but what if we have a case where one object is 3.368 times hotter than the other, for example?

A great deal can be learned from making comparisons in this way. Astronomers often use the sun as the standard for comparison, expressing various quantities in terms of the solar values. Comparisons are also useful when the numerical constants (such as Wien's constant in the previous examples) are not known. If the trick of comparing is kept in mind, it is often possible to quickly work out answers to astronomical questions, simply by carrying around in one's head a few numbers describing the sun.

The Stefan-Boltzmann Law

As discussed in the text, the energy emitted by a glowing object is proportional to T^4, where T is the surface temperature. This can be written mathematically as

$$E = \sigma T^4,$$

where σ stands for a proportionality constant that has the value 5.7×10^{-5}, if the centimeter-gram-second metric units are used.

If we now consider how much surface area a star has, then the total energy it emits, called its *luminosity* and usually denoted L, is

$$L = \text{surface area} \times E$$
$$= 4\pi R^2 \sigma T^4,$$

where R is the radius of the star. This equation is called the Stefan-Boltzmann law.

As in other cases, we can use the law directly in this form, or we may choose to compare properties of stars by writing the equation separately for two stars and then dividing. If we do this, we find

$$\frac{L_1}{L_2} = \frac{R_1^2 \, T_1^4}{R_2^2 \, T_2^4},$$

The constant factors 4π and σ have canceled out.

As an example of how to use this expression, suppose we determine that a particular star has twice the temperature of the sun, but only one-half the radius, and we wish to know how this star's luminosity compares with that of the sun. If we designate the star as object 1, and the sun as object 2, then

$$L_1 = \left(\frac{1}{2}\right)^2 2^4 = 1/4 \times 16 = 4.$$

This star has four times the luminosity of the sun.

The Stefan-Boltzmann law is particularly useful because it relates three of the most important properties of stars to each other.

The Planck Function

The radiation laws described here and in the text are actually specific forms of a much more general law, discovered by the great German physicist Max Planck. Wien's law, the Stefan-Boltzmann law, and some others not mentioned in the text bear a similar relation to Planck's law as the laws of planetary motion discovered by Kepler do to Newton's mechanics. Kepler's laws were first discovered by observation, but with Newton's laws of motion it is possible to derive Kepler's laws theoretically. In the same fashion, the radiation laws discussed so far were found experimentally, but can be derived mathematically from the much more general and powerful Planck's law.

Planck's law is usually referred to as the Planck function, a mathematical relationship between intensity and wavelength (or frequency) that describes the spectrum of any glowing object at a given temperature. The Planck function specifically applies only to objects that radiate solely because of their temperature, and do not reflect any light or have any spectral lines. The popular term for such objects is *blackbody,* and we often refer to radiation from such an object as *blackbody radiation* or, more commonly, as *thermal radiation,* the term used in the text. Stars are not perfect radiators, but to a close approximation can be treated as such; hence, in the text, we apply Wien's law and the Stefan-Boltzmann law to stars (and even planets in certain circumstances) without pointing out the fact that to do so is only approximately correct.

The form of the Planck function for the radiation intensity B as a function of wavelength is

$$B = 2hc^2\lambda^{-5}/(e^{hc/\lambda kT} - 1),$$

where h is the Planck constant, c is the speed of light, and k is the Boltzmann constant (the values of all three are tabulated in appendix 1), λ is the wavelength (in cm), and T is the temperature of the object (on the absolute scale). The symbol e represents the base of the natural logarithm, something not used elsewhere in the text; for the present purpose, this may be regarded simply as a mathematical constant with the value 2.718.

In terms of frequency ν rather than wavelength, the expression is

$$B = 2h\nu^3c^{-2}/(e^{h\nu/kT} - 1).$$

Either expression may be used to calculate the spectrum of continuous radiation from a glowing object at a specific temperature. In practice, the Planck function is used in a wide assortment of theoretical calculations that call for knowledge of the intensity of radiation so that its effects on physical conditions such as ionization can be assessed.

Appendix 5. Major Telescopes of the World*

OBSERVATORY	LOCATION	TELESCOPE
Special Astrophysical Observatory	Mount Pastukhov, USSR	6.0-m Bol'shoi Teleskop Azimutal'nyi
Palomar Observatory	Palomar Mountain, California	5.08-m George Ellery Hale Telescope
Smithsonian Observatory	Mount Hopkins, Arizona	4.5-m Multiple Mirror Telescope
[Royal Greenwich Observatory]	[Canary Islands]	[4.2-m William Herschel Telescope]
Cerro Tololo Inter-American Observatory	Cerro Tololo, Chile	4.0-m
Anglo-Australian Observatory	Siding Spring Mountain, Australia	3.9-m Anglo-Australian Telescope
Kitt Peak National Observatory	Kitt Peak, Arizona	3.8-m Nicholas U. Mayall Telescope
Royal Observatory Edinburgh	Mauna Kea, Hawaii	3.8-m United Kingdom Infrared Telescope
Canada-France-Hawaii Observatory	Mauna Kea, Hawaii	3.6-m Canada-France-Hawaii Telescope
European Southern Observatory	Cerro La Silla, Chile	3.57-m
Max Planck Institute (Heidelberg)	Calar Alto, Spain	3.5-m
Italian National Astronomical Observatory	[Italy; final site to be chosen]	[3.5-m]
[Iraqi National Observatory]	[Mount Korek]	[3.5-m]
Lick Observatory	Mount Hamilton, California	3.05-m C. Donald Shane Telescope
Mauna Kea Observatory	Mauna Kea, Hawaii	3.0-m NASA Infrared Telescope
McDonald Observatory	Mount Locke, Texas	2.7-m
Haute Provence Observatory	St. Michele, France	2.6-m
Crimean Astrophysical Observatory	Simferopol, USSR	2.6-m
Byurakan Astrophysical Observatory	Mount Aragatz, Armenian S.S.R.	2.6-m
Mount Wilson Las Campanas Observatory	Mt. Wilson, California	2.5-m Hooker Telescope
Mount Wilson Las Campanas Observatory	Las Campanas, Chile	2.5-m Irenée du Pont Telescope
Royal Greenwich Observatory	Canary Islands	2.5-m Isaac Newton Telescope
Wyoming Infrared Observatory	Mount Jelm, Wyoming	2.3-m Wyoming Infrared Telescope
Steward Observatory	Kitt Peak, Arizona	2.3-m
Mauna Kea Observatory	Mauna Kea, Hawaii	2.2-m
Max Planck Institute (Heidelberg)	Calar Alto, Spain	2.2-m
Max Planck Institute (Heidelberg)	LaSilla, Chile	2.2-m
[Mexican National Observatory]	[San Pedro Martir, Baja California]	[2.16-m]
La Plata Observatory	La Plata, Argentina	2.15-m
Kitt Peak National Observatory	Kitt Peak, Arizona	2.1-m
McDonald Observatory	Mount Locke, Texas	2.1-m Otto Struve Telescope
Karl Schwarzschild Observatory	Tautenburg, East Germany	2.0-m (largest Schmidt telescope)

*Telescopes described in brackets are planned or under construction, but not yet in operation.

Appendix 6. Planetary and Satellite Data

Orbital Data for the Planets

PLANET	SEMIMAJOR AXIS	SIDEREAL PERIOD	ORBITAL ECCENTRICITY[1]	INCLINATION OF ORBITAL PLANE	ROTATION PERIOD	TILT OF AXIS
Mercury	0.387	0.241	0.206	7°0′15″	58^{d}65	<28°
Venus	0.723	0.615	0.007	3°23′40″	243	3°
Earth	1.000	1.000	0.017	0°0′0″	23^{h}56^m	23°27′
Mars	1.523	1.881	0.093	1°51′0″	24^{h}37^m	23°59′
Jupiter	5.203	11.86	0.048	1°18′17″	9^{h}50^m	3°5′
Saturn	9.539	29.46	0.056	2°29′33″	10^{h}39^m	26°44′
Uranus	19.18	84.01	0.047	0°46′23″	16^{h}10^m	97°55′
Neptune	30.07	164.8	0.009	1°46′22″	18^{h}12^m	28°48′
Pluto	39.44	248.8	0.250	17°10′12″	6^{d}39	35°

[1]The eccentricity of an orbit is defined as the ratio of the distance between foci to the semimajor axis. In practice, it is related to the perihelion distance P and the semimajor axis a by $P = a(1-e)$, where e is the eccentricity; and to the aphelion distance A by $A = a(1+e)$.

Physical Data for the Planets

PLANET	MASS[1]	DIAMETER[1]	DENSITY (g/cc)	SURFACE GRAVITY (G)	ESCAPE VELOCITY (KM/SEC)	TEMPERATURE	ALBEDO
Mercury	0.0558	0.382	5.42	0.38	4.2	100–700 K	0.06
Venus	0.851	0.951	5.3	0.90	10.3	730	0.76
Earth	1.000	1.000	5.517	1.00	11.2	200–300	0.39
Mars	0.107	0.532	3.96	0.38	5.0	130–290	0.15
Jupiter	318.1	10.79	1.33	2.64	60	130	0.51
Saturn	95.2	8.91	0.68	1.13	36	95	0.50
Uranus	14.6	4.10	1.2	1.07	21	95	0.66
Neptune	17.2	3.82	1.7	1.08	24	50	0.62
Pluto	0.0019	0.16–.47	0.5–0.8	0.024–.034	0.9–1.0	40	0.25

[1]The masses and diameters are given in units of the earth's mass and diameter, which are 5.974×10^{27} grams and 12,734 kilometers, respectively.

Satellites

PLANET	SATELLITE	SEMIMAJOR AXIS (KM)	PERIOD	DIAMETER	MASS	DENSITY
Earth	Moon	3.84×10^5	27^{d}322	3476 km	7.35×10^{25}	3.34 gm/cm^3
Mars	Phobos	9.4×10^3	0.3189	28 × 22 × 20	1.93×10^{18}	2:
	Deimos	2.35×10^4	1.2624	15 × 12 × 11	3.85×10^{17}	2:
Jupiter	J–14	1.237×10^5	0.297	40		
	Amalthea	1.813×10^5	0.489	240		
	Io	4.126×10^5	1.769	3,640	8.89×10^{22}	3.53

(continued on next page)

Satellites (continued)

PLANET	SATELLITE	SEMIMAJOR AXIS (KM)	PERIOD	DIAMETER	MASS	DENSITY
	Europa	6.709×10^5	3.551	3,130	4.85×10^{22}	3.03
	Ganymede	1.070×10^6	7.155	5,280	1.49×10^{23}	1.93
	Callisto	1.880×10^6	16.689	4,840	1.07×10^{23}	1.70
	Leda	1.11×10^7	240	2–14		
	Himalia	1.15×10^7	250.6	170		
	Lysithia	1.17×10^7	260	6–32		
	Elara	1.17×10^7	260.1	80		
	Ananke	2.07×10^7	617	6–28		
	Carme	2.24×10^7	692	8–40		
	Pasiphae	2.33×10^7	735	8–46		
	Sinope	2.37×10^7	758	6–36		
Saturn	S–15	1.7014×10^5	0.602	40×20		
	S–14	1.3937×10^5	0.613	$140 \times 100 \times 80$		
	S–13	1.4172×10^5	0.629	$110 \times 90 \times 70$		
	S–10	1.5144×10^5	0.694	$140 \times 120 \times 100$		
	S–11	1.5150×10^5	0.695	$220 \times 200 \times 160$		
	Mimas	1.8552×10^5	0.942	392	4.5×10^{22}	1.44
	Enceladus	2.3807×10^5	1.370	500	8.4×10^{22}	1.16
	S–16	2.9467×10^5	1.888	$34 \times 28 \times 26$		
	S–17	2.9467×10^5	1.888	$34 \times 22 \times 22$		
	Tethys	2.9467×10^5	1.888	1,060	7.6×10^{23}	1.21
	Dione	3.7744×10^5	2.737	1,120	1.05×10^{24}	1.43
	S–12	3.7811×10^5	2.739	$36 \times 32 \times 30$		
	Rhea	5.2713×10^5	4.518	1,530	2.44×10^{24}	1.33
	Titan	1.222×10^6	15.945	5,150	1.35×10^{26}	1.88
	Hyperion	1.481×10^6	21.277	$410 \times 260 \times 220$		
	Iapetus	3.561×10^6	79.331	1,460	1.88×10^{24}	1.16
	Phoebe	1.295×10^7	550.45	220		
Uranus	Miranda	1.301×10^5	1.414	600		
	Ariel	1.918×10^5	2.5204	1,600		
	Umbriel	2.673×10^5	4.1442	1,100		
	Titania	4.387×10^5	8.7059	2,000		
	Oberon	5.866×10^5	13.463	1,800		
Neptune	Triton	6.536×10^5	5.877	7,200–8,000		
	Nereid	5.570×10^6	365.2	600		
Pluto	Charon	1.8×10^4	6.39			

Appendix 7. The Fifty Brightest Stars

	STAR	SPECTRAL TYPE	APPARENT MAGNITUDE	DISTANCE	POSITION (1980) RIGHT ASCENSION	DECLINATION
α Eri	Archernar	B3 V	0.51	36 pc.	01^{h}37^m.0	−57°20′
α UMi	Polaris	F8 Ib	1.99	208	02 12.5	+89 11
α Per	Mirfak	F5 Ib	1.80	175	03 22.9	+49 47
α Tau	Aldebaran	K5 III	0.86	21	04 34.8	+16 28
β Ori	Rigel	B8 Ia	0.14	276	05 13.6	−08 13
α Aur	Capella	G8 III	0.05	14	05 15.2	+45 59
γ Ori	Bellatrix	B2 III	1.64	144	05 24.0	+06 20
β Tau	Elnath	B7 III	1.65	92	05 25.0	+28 36
ε Ori	Alnilam	B0 Ia	1.70	490	05 35.2	−01 13
ζ Ori	Alnitak	O9.5 Ib	1.79	490	05 39.7	−01 57
α Ori	Betelgeuse	M2 Iab	0.41	159	05 54.0	+07 24
β Aur	Menkalinan	A2 V	1.86	27	05 58.0	+44 57
β CMa		B1 II–III	1.96	230	06 21.8	−17 56
α Car	Canopus	F0 Ib–II	−0.72	30	06 23.5	−52 41
γ Gem	Alhena	A0 IV	1.93	32	06 36.6	+16 25
α CMa	Sirius	A1 V	−1.47	2.7	06 44.2	−16 42
ε CMa	Adhara	B2 II	1.48	209	06 57.8	−28 57
δ CMa		F8 Ia	1.85	644	07 07.6	−26 22
α Gem	Castor	A1 V	1.97	14	07 33.3	+31 56
α CMi	Procyon	F5 IV–V	0.37	3.5	07 38.2	+05 17
β Gem	Pollux	K0 III	1.16	11	07 44.1	+28 05
γ Vel		WC8	1.83	160	08 08.9	−47 18
ε Car	Avior	K3 III?	1.90	104	08 22.1	−59 26
δ Vel		A2 V	1.95	23	08 44.2	−54 38
β Car	Miaplacidus	A1 III	1.67	26	09 13.0	−69 38
α Hya	Alphard	K4 III	1.98	29	09 26.6	−08 35
α Leo	Regulus	B7 V	1.36	26	10 07.3	+12 04
γ Leo		K0 III	1.99	28	10 18.8	+19 57

(continued on next page)

Appendix 7. The Fifty Brightest Stars
(Continued)

	STAR	SPECTRAL TYPE	APPARENT MAGNITUDE	DISTANCE	POSITION (1980) RIGHT ASCENSION	DECLINATION
α UMa	Dubhe	K0 III	1.81	32	11 02.5	+ 61 52
α Cru A	Acrux	B0.5 IV	1.39	114	12 25.4	− 62 59
α Cru B	Acrux	B1 V	1.86	114	12 25.4	− 62 59
γ Cru	Gacrux	M4 III	1.69	67	12 30.1	− 57 00
β Cru		B0.5 III	1.28	150	12 46.6	− 59 35
ε UMa	Alioth	A0p	1.79	21	12 53.2	+ 56 04
α Vir	Spica	B1 V	0.91	67	13 24.1	− 11 03
η UMa	Alkaid	B3 V	1.87	64	13 46.8	+ 49 25
β Cen	Hadar	B1 III	0.63	150	14 02.4	− 60 16
α Boo	Arcturus	K2 III	− 0.06	11	14 14.8	+ 19 17
α Cen A	Rigil Kentaurus	G2 V	0.01	1.3	14 38.4	− 60 46
α Cen B	Rigil Kentaurus	K4 V	1.40	1.3	14 38.4	− 60 46
α Sco	Antares	M1 Ib	0.92	160	16 28.2	− 26 23
α TrA	Atria	K2 Ib	1.93	25	16 46.5	− 68 60
λ Sco	Shaula	B1 V	1.60	95	17 32.3	− 37 05
θ Sco		F0 Ib	1.86	199	17 35.9	− 42 59
ε Sgr	Kaus Australis	B9.5 III	1.81	38	18 22.9	− 34 24
α Lyr	Vega	A0 V	0.04	8	18 36.2	+ 38 46
α Aql	Altair	A7 IV–V	0.77	5	19 49.8	+ 08 49
α Pav	Peacock	B2.5 V	1.95	95	20 24.1	− 56 48
α Cyg	Deneb	A2 Ia	1.26	491	20 40.7	+ 45 12
α Gru	Al Na'ir	B7 IV	1.76	20	22 06.9	− 47 04
α PsA	Fomalhaut	A3 V	1.15	7	22 56.5	− 29 44

Appendix 8. The Constellations

NAME	GENITIVE	ABBREVIATION	POSITION RIGHT ASCENSION	DECLINATION
Andromeda	Andromedae	And	01^h	+40°
Antlia	Antliae	Ant	10	−35
Apus	Apodis	Aps	16	−75
Aquarius	Aquarii	Aqr	23	−15
Aquila	Aquilae	Aql	20	+05
Ara	Arae	Ara	17	−55
Aries	Arietis	Ari	03	+20
Auriga	Aurigae	Aur	06	+40
Boötes	Bootis	Boo	15	+30
Caelum	Caeli	Cae	05	−40
Camelopardalis	Camelopardalis	Cam	06	−70
Cancer	Cancri	Cnc	09	+20
Canes Venatici	Canum Venaticorum	CVn	13	+40
Canis Major	Canis Majoris	CMa	07	−20
Canis Minor	Canis Minoris	CMi	08	+05
Capricornus	Capricorni	Cap	21	−20
Carina	Carinae	Car	09	−60
Cassiopeia	Cassiopeiae	Cas	01	+60
Centaurus	Centauri	Cen	13	−50
Cepheus	Cephei	Cep	22	+70
Cetus	Ceti	Cet	02	−10
Chamaeleon	Chamaeleonis	Cha	11	−80
Circinis	Circini	Cir	15	−60
Columba	Columbae	Col	06	−35
Coma Berenices	Comae Berenices	Com	13	+20
Corona Australis	Coronae Australis	CrA	19	−40
Corona Borealis	Coronae Borealis	CrB	16	+30
Corvus	Corvi	Crv	12	−20
Crater	Crateris	Crt	11	−15
Crux	Crucis	Cru	12	−60
Cygnus	Cygni	Cyg	21	+40
Delphinus	Delphini	Del	21	+10
Dorado	Doradus	Dor	05	−65
Draco	Draconis	Dra	17	+65
Equuleus	Equulei	Equ	21	+10
Eridanus	Eridani	Eri	03	−20
Fornax	Fornacis	For	03	−30
Gemini	Geminorum	Gem	07	+20
Grus	Gruis	Gru	22	−45
Hercules	Herculis	Her	17	+30
Horologium	Horologii	Hor	03	−60
Hydra	Hydrae	Hya	10	−20
Hydrus	Hydri	Hyi	02	−75
Indus	Indi	Ind	21	−55

(continued on next page)

Appendix 8. The Constellations
(Continued)

NAME	GENITIVE	ABBREVIATION	POSITION	
			RIGHT ASCENSION	DECLINATION
Lacerta	Lacertae	Lac	22	+ 45
Leo	Leonis	Leo	11	+ 15
Leo Minor	Leonis Minoris	LMi	10	+ 35
Lepus	Leporis	Lep	06	− 20
Libra	Librae	Lib	15	− 15
Lupus	Lupi	Lup	15	− 45
Lynx	Lincis	Lyn	08	+ 45
Lyra	Lyrae	Lyr	19	+ 40
Mensa	Mensae	Men	05	− 80
Microscopium	Microscopii	Mic	21	− 35
Monoceros	Monocerotis	Mon	07	− 05
Musca	Muscae	Mus	12	− 70
Norma	Normae	Nor	16	− 50
Octans	Octantis	Oct	22	− 85
Ophiuchus	Ophiuchi	Oph	17	00
Orion	Orionis	Ori	05	+ 05
Pavo	Pavonis	Pav	20	− 65
Pegasus	Pegasi	Peg	22	+ 20
Perseus	Persei	Per	03	+ 45
Phoenix	Phoenicis	Phe	01	− 50
Pictor	Pictoris	Pic	06	− 55
Pisces	Piscium	Psc	01	+ 15
Piscis Austrinus	Piscis Austrini	PsA	22	− 30
Puppis	Puppis	Pup	08	− 40
Pyxis	Pyxidis	Pyx	09	− 30
Reticulum	Reticuli	Ret	04	− 60
Sagitta	Sagittae	Sge	20	+ 10
Sagittarius	Sagittarii	Sgr	19	− 25
Scorpius	Scorpii	Sco	17	− 40
Sculptor	Sculptoris	Scl	00	− 30
Scutum	Scuti	Sct	19	− 10
Serpens	Serpentis	Ser	17	00
Sextans	Sextantis	Sex	10	00
Taurus	Tauri	Tau	04	+ 15
Telescopium	Telescopii	Tel	19	− 50
Triangulum	Trianguli	Tri	02	+ 30
Triangulum Australe	Trianguli Australi	TrA	16	− 65
Tucana	Tucanae	Tuc	00	− 65
Ursa Major	Ursae Majoris	UMa	11	+ 50
Ursa Minor	Ursae Minoris	UMi	15	+ 70
Vela	Velorum	Vel	09	− 50
Virgo	Virginis	Vir	13	00
Volans	Volantis	Vol	08	− 70
Vulpecula	Vulpeculae	Vul	20	+ 25

Appendix 9. Mathematical Treatment of Stellar Magnitudes

Logarithmic Representation

In the text, magnitudes were discussed in terms of the brightness ratios between stars of different magnitudes. We generally avoided talking about two stars that differ by a fraction of a magnitude, because in such cases it is not simple to calculate the brightness ratio corresponding to the magnitude difference. If star A is 0.5 magnitudes brighter than star B, for example, what is the brightness ratio? Or, if star C is a factor of 48.76 fainter than star D, what is the difference in magnitudes?

Astronomers use an exact mathematical relationship between magnitude differences and brightness ratios, written as

$$m_1 - m_2 = 2.5 \log (b_2/b_1),$$

where m_1 and m_2 are the magnitudes of two stars, and b_2/b_1 is the ratio of their brightnesses. Log (b_2/b_1) represents the logarithm of this ratio; a logarithm is the power to which 10 must be raised to give this ratio. Thus, if $b_2/b_1 = 100$, log $(b_2/b_1) = \log (10^2) = 2$, because 10 must be raised to the second power to give 100. The magnitude difference is 2.5 log (100) = 2.5 × 2 = 5. Similarly, if $b_2/b_1 = 0.001$, then log $(b_2/b_1) = \log(0.001) = \log (10^{-3}) = -3$, and in this case the magnitude difference is 2.5 × −3 = −7.5 (the minus sign indicates that star 1 is brighter than star 2 in this example).

The method works equally well in cases where the power of 10 is not a whole number, as in $b_D/b_C = 48.76$. Here log $(b_D/b_C) = \log (48.76) = 1.69$ (we can usually find this by consulting tables of logarithms or by using a scientific calculator). In this example, the magnitude difference is 2.5 log$(b_D/b_C) = 2.5 \log(48.76) = 2.5 × 1.69 = 4.23$, so star C is 4.23 magnitudes fainter than star D.

The equation can be used in other ways as well; solving for b_2/b_1 yields

$$b_2/b_1 = 10^{(m_1 - m_2)/2.5}$$
$$= 10^{0.4(m_1 - m_2)}$$

Thus, if we know that the magnitudes of two stars differ by $m_1 - m_2$, then we multiply this difference by 0.4 and raise 10 to the power $0.4(m_1 - m_2)$, again using a calculator or tables, to get the brightness ratio b_2/b_1. As a simple example, suppose $m_1 - m_2 = 5$; then $0.4 (m_1 - m_2) = 0.4 × 5 = 2$, and $10^{0.4(m_1 - m_2)} = 10^2 = 100$, as we knew it should. As a more complex example, consider the stars Betelgeuse (magnitude +0.41) and Deneb (magnitude +1.26). From the equation just cited, we see that Betelgeuse is $10^{0.4(1.26 - 0.41)} = 10^{0.34} = 2.19$ times brighter than Deneb.

Although in most cases it is possible to follow the discussions of magnitudes and brightness ratios in the text without using this exact mathematical technique, it will still be useful to be familiar with it.

The Distance Modulus

Whenever we know both the apparent and absolute magnitudes of a star, a comparison of the two will give its distance. In the text we learned how to make this calculation:

1. Convert the difference m-M between apparent and absolute magnitude into a brightness ratio; that is, a numerical factor indicating how much brighter or fainter the star would appear at 10 parsecs distance than its actual distance;

2. Using the inverse-square law, determine the change in distance required to produce this change in brightness;

3. Multiply this distance factor by 10 parsecs to find the distance to the star.

To do calculations mentally in this way can be laborious, especially in cases where the magnitude difference does not correspond neatly to a simple numerical factor, as it did in the examples given in the text. Hence astronomers use a mathematical equation expressing the relationship between distance and the distance modulus m-M. This equation, which works equally well for all cases, is

$$d = 10^{1 + .2(m-M-A_v)}$$

where d is the distance in parsecs to a star whose apparent magnitude is m and whose absolute magnitude is M.

In a simple example, where $m = 9$ and $M = -6$, we have

$$d = 10^{1 + .2 (15)}$$
$$= 10^{1 + 3}$$
$$= 10^4 \text{ parsecs}$$

Now let us try a more complex case. Suppose the star is an M2 main-sequence star, so that $M = 13$, as found from the H–R diagram. The apparent magnitude is $m = 16$. Our equation tells us that

$$d = 10^{1 + .2(16 - 13)}$$
$$= 10^{1 + .6}$$
$$= 10^{1.6}$$
$$= 39.8 \text{ parsecs.}$$

Of course, we must use a slide rule, calculator, or mathematical tables, but this method is still relatively straightforward compared with the steps outlined earlier, which involve mental calculations.

The Effect of Extinction on the Distance Modulus

When we discussed distance-determination techniques (in chapter 19), we ignored the effects of interstellar extinction. In any method that depends on the apparent brightness of a star, however, extinction can be important, particularly for very distant stars. Because the effect of extinction is to make a star appear fainter than it otherwise would, the tendency is to overestimate distances if no allowance is made for it.

Recall that in the spectroscopic-parallax technique, the distance modulus m-M is used to find the distance to a star from the equation given earlier:

$$d = 10^{1 + .2(m\text{-}M)},$$

where m is the apparent magnitude, M is the absolute magnitude, and d is the distance to the star in parsecs.

To correct this equation for extinction, we add a term A_v, which refers to the extinction (in magnitudes) in visual light. Thus, if the extinction toward a particular star makes that star appear 2 magnitudes fainter than it otherwise would, then A_v = 2. If we insert this into the equation, we find

$$d = 10^{1 + .2(m\text{-}M\text{-}A_v)}.$$

Let us consider a simple example, a star whose apparent magnitude is m = 12.4 and whose absolute magnitude is M

= 2.4. First, let us calculate its distance if extinction is ignored:

$$d = 10^{1 + .2(12.4 - 2.4)} = 10^3 = 1{,}000 \text{ parsecs.}$$

Suppose we determine that the extinction in the direction of this star amounts to 1 magnitude. Now the distance is

$$d = 10^{1 + .2(12.4 - 2.4 - 1)} = 10^{2.8} = 631 \text{ parsecs.}$$

One magnitude is a modest amount of extinction, yet by neglecting it, we overestimated the distance to this star by almost 60 percent. We can see from this example that extinction can have a drastic effect on distance estimates.

It is worthwhile to add a note about how the extinction A_v is determined. In chapter 25 we referred to measuring interstellar reddening by comparing the observed B-V color index with what it would be if the star suffered no extinction. In other words, it is possible to determine how much redder, in terms of the B-V color index, a star appears to be because of extinction. To carry that step further, astronomers define a *color excess* called $E(B\text{-}V)$, which is the difference between the observed and intrinsic values of B-V:

$$E(B\text{-}V) = (B\text{-}V)_{observed} - (B\text{-}V)_{intrinsic}.$$

Studies of the variation of interstellar extinction with wavelength show that the extinction at the visual wavelength is approximately equal to three times the color excess:

$$A_v = 3\, E(B\text{-}V).$$

Hence determination of excess reddening leads to an estimate of A_v, and this in turn can be used in determining the distance, as shown above.

Appendix 10. The Proton-Proton Chain and the CNO Cycle

In the text we did not explain the details of the reactions that convert hydrogen into helium in stellar cores, although they are quite simple. To do so here, we will use the notation of the nuclear physicist. This is basically a shorthand in symbols. For example, a helium nucleus, containing two protons and two neutrons, is designated $^4_2$He, the subscript indicating the *atomic number* (the number of protons) and the superscript the *atomic weight* (the total number of protons and neutrons). Similarly, a hydrogen nucleus is $^1_1$H, and deuterium, a form of hydrogen with an extra neutron in the nucleus, is $^2_1$H. A special symbol (ν) is used for the *neutrino,* a massless subatomic particle emitted in some reactions, and e^+ indicates a *positron,* which is equivalent to an electron, but with positive electrical charge. The symbol γ indicates a gamma ray, a very short-wavelength photon of light emitted in some reactions.

Using this system, we can now spell out the proton-proton chain:

$$^1_1\text{H} + {}^1_1\text{H} \rightarrow {}^2_1\text{H} + e^+ + \nu$$

$$^2_1\text{H} + {}^1_1\text{H} \rightarrow {}^3_2\text{He} + \gamma.$$

The $^3_2$He particle is a form of helium, but not the common type. Once we have this particle, it will combine with another:

$$^3_2\text{He} + {}^3_2\text{He} \rightarrow {}^4_2\text{He} + {}^1_1\text{H} + {}^1_1\text{H}.$$

We end up with a normal helium nucleus. A total of six hydrogen nuclei went into the reaction (remember, the first two steps had to occur twice, in order to produce two $^3_2$He particles for the final reaction), and there were two left at the end, so the result is the conversion of four hydrogen nuclei into one helium nucleus.

The CNO cycle, which dominates at higher temperatures, is more complex, involving not only carbon, but also nitrogen and oxygen. Each of these elements has more than one form, with differing numbers of neutrons. Some of these *isotopes* are unstable, and spontaneously emit positrons, decaying into other species in the process. Here is the CNO cycle:

$$^{12}_{6}\text{C} + {}^2_1\text{H} \rightarrow {}^{13}_{7}\text{N} + \gamma$$

$$^{13}_{7}\text{N} \rightarrow {}^{13}_{6}\text{C} + e^+ + \nu$$

$$^{13}_{6}\text{C} + {}^1_1\text{H} \rightarrow {}^{14}_{7}\text{N} + \gamma$$

$$^{14}_{7}\text{N} + {}^1_1\text{H} \rightarrow {}^{15}_{8}\text{O} + \gamma$$

$$^{15}_{8}\text{O} \rightarrow {}^{15}_{7}\text{N} + e^+ + \nu$$

$$^{15}_{7}\text{N} + {}^1_1\text{H} \rightarrow {}^{12}_{6}\text{C} + {}^4_2\text{He}.$$

Here we end up with a helium nucleus and a carbon nucleus, although the particles going into the reaction were four hydrogen nuclei and a carbon nucleus. Along the way three isotopes of nitrogen and one of oxygen were created and then converted into something else, leaving neither element at the end. As in the proton-proton chain, the net result is the conversion of four hydrogen nuclei into one helium nucleus.

Appendix 11. Detected Interstellar Molecules[1]

NUMBER OF ATOMS	SYMBOL	NAME	NUMBER OF ATOMS	SYMBOL	NAME
2	H_2	Molecular hydrogen		HOCN*	Cyanic acid*
	C_2	Diatomic carbon		HNCS	Isothiocyanic acid
	Ch	Methylidyne		C_3N	Cyanoethynyl
	CH^+	Methylidyne ion		H_2CS	Thioformaldehyde
	CN	Cyanogen	5	[CH_4]	[Methane]
	CO	Carbon monoxide		C_4H	Butadynyl
	CO^+	Carbon monoxide ion		HCO_2H	Formic acid
				CH_2CO	Ketene
	CS	Carbon monosulfide		HC_3N	Cyanoacetylene
	OH	Hydroxyl		NH_2CN	Cyanamide
	NO	Nitric oxide		CH_2NH	Methanimine
	NS	Nitrogen sulfide	6	CH_3OH	Methanol
	SiO	Silicon monoxide		CH_3CN	Methyl cyanide
	SiS	Silicon sulfide		CH_3SH	Methyl mercaptan
	SO	Sulfur monoxide		NH_2CHO	Formamide
3	C_2H	Ethynyl	7	CH_2CHCN	Vinyl cyanide
	HCN	Hydrogen cyanide		[CH_2CH_2O]	[Ethylene oxide]
	HNC	Hydrogen isocyanide		CH_3C_2H	Methylacetylene
	HCO	Formyl		CH_3CHO	Acetaldehyde
	HCO^+	Formyl ion		CH_3NH_2	Methylamine
	N_2H^+	Protonated nitrogen		HC_5N	Cyanodiacetylene
	HNO	Nitroxyl	8	$HCOOCH_3$	Methyl formate
	H_2O	Water	9	CH_3CH_3O	Dimethyl ether
	HCS^+	Thioformyl ion		CH_3CH_2CN	Ethyl cyanide
	H_2S	Hydrogen sulfide		CH_3CH_2OH	Ethanol
	OCS	Carbonyl sulfide		HC_7N	Cyano-hexa-tri-yne
	SO_2	Sulfur dioxide	10	[NH_2CH_2COOH]	[Glycene]
4	H_2CO	Formaldehyde	11	HC_9N	Cyano-octa-tetra-yne
	NH_3	Ammonia	13	$HC_{11}N$	Cyano-deca-penta-yne
	HNCO	Isocyanic acid			

[1]This list does not include isotopic variations; that is, molecules that are identical except that one or more atoms are in rare isotopic forms, such as deuterium in place of hydrogen, or ^{13}C instead of the much more common ^{14}C. Bracketed [] species are at present tentatively identified, but not yet confirmed.

*An alternative possible identification of this molecule is $HOCO^+$.

Appendix 12. Galaxies of the Local Group

GALAXY[1]	TYPE[2]	ABSOLUTE MAGNITUDE	POSITION (1950) RIGHT ASCENSION	POSITION (1950) DECLINATION
M31 (Andromeda)	Sb	−21.1	00^h40^m0	+41°00′
Milky Way	Sbc	−20.5	17 42.5	−28 59
M33 = NGC 598	Sc	−18.9	01 31.1	+30 24
Large Magellanic Cloud	Irr	−18.5	05 24	−69 50
IC 10	Irr	−17.6	00 17.6	+59 02
Small Magellanic Cloud	Irr	−16.8	00 51	−73 10
M32 = NGC 221	E2	−26.4	00 40.0	+40 36
NGC 205	E6	−16.4	00 37.6	+41 25
NGC 6822	Irr	−15.7	19 42.1	−14 53
NGC 185	Dwarf E	−15.2	00 36.1	+48 04
NGC 147	Dwarf E	−14.9	00 30.4	+48 14
IC 1613	Irr	−14.8	01 02.3	+01 51
WLM	Irr	−14.7	23 59.4	−15 44
Fornax	Dwarf sph	−13.6	02 37.5	−34 44
Leo A	Irr	−13.6	09 56.5	+30 59
IC 5152	Irr	−13.5	21 59.6	−51 32
Pegasus	Irr	−13.4	23 26.1	+14 28
Sculptor	Dwarf sph	−11.7	00 57.5	−33 58
And I	Dwarf sph	−11	00 42.8	+37 46
And II	Dwarf sph	−11	01 13.6	+33 11
And III	Dwarf sph	−11	00 32.7	+36 14
Aquarius	Irr	−11	20 44.1	−13 02
Leo I	Dwarf sph	−11	10 05.8	+12 33
Sagittarius	Irr	−10	19 27.1	−17 47
Leo II	Dwarf sph	−9.4	11 10.8	+22 26
Ursa Minor	Dwarf sph	−8.8	15 08.2	+67 18
Draco	Dwarf sph	−8.6	17 19.4	+57 58
Carina	Dwarf sph		06 40.4	−50 55
Pisces	Irr	−8.5	00 01.2	+21 37

[1]Galaxy names are derived from a variety of sources, including catalogues (designating galaxies as M, NGC, and IC, for example), and colloquial names bestowed by the discoverer. Many in this list are simply named after the constellation where they are found.

[2]The galaxy types listed here are described in the text, except for the *Dwarf sph* designation, which stands for *dwarf spheroidal* and refers to dwarf galaxies that do not fit easily into the category of dwarf ellipticals. Note that the absolute magnitudes of some of the dwarf spheroidals are comparable to those of the brightest individual stars in our galaxy.

• G L O S S A R Y •

ABSOLUTE MAGNITUDE: The magnitude a star would have if it were precisely ten parsecs away from the sun.

ABSOLUTE ZERO: The temperature at which all molecular or atomic motion stops, equal to $-273°C$ or $-459°F$.

ABSORPTION LINE: A wavelength at which light is absorbed, producing a dark feature in the spectrum.

ACCELERATION: Any change in the state of rest or motion of a body; either a change in speed or direction.

ACCRETION DISK: A rotating disk of gas surrounding a compact object (such as a neutron star or black hole), formed by material falling inward.

ALBEDO: The fraction of incident light that is reflected from a surface such as that of a planet.

ALPHA-CAPTURE REACTION: A nuclear fusion reaction in which an alpha particle merges with an atomic nucleus. A typical example is the formation of ^{16}O by the fusion of an alpha particle with ^{12}C.

ALPHA PARTICLE: A nucleus of ordinary helium, containing two protons and two neutrons.

AMINO ACID: A complex organic molecule of the type that forms proteins. Amino acids are fundamental constituents of all living matter.

ANDROMEDA GALAXY: The large spiral galaxy located some 700,000 parsecs from the sun; the most distant object visible to the unaided eye.

ÅNGSTROM (Å): The unit generally used in measuring wavelengths of visible and ultraviolet light; one Ångstrom is equal to 10^{-8} centimeters.

ANGULAR DIAMETER: The diameter of an object as seen in the sky, measured in units of angle.

ANGULAR MOMENTUM: A measure of the mass, radius, and rotational velocity of a rotating or orbiting body. In the simple case of an object in circular orbit, the angular momentum is equal to the mass of the object times its distance from the center of the orbit times its orbital speed.

ANNUAL MOTIONS: Motions in the sky caused by the earth's orbital motion about the sun. These include the seasonal variations of the sun's latitude, and the sun's motion through the zodiac.

ANNULAR ECLIPSE: A solar eclipse that occurs when the moon is near its greatest distance from the earth, so that its angular diameter is slightly smaller than that of the sun, and a ring, or annulus, of the sun's disk is visible surrounding the disk of the moon.

ANTICYCLONE: A rotating wind system around a high-pressure area. On the earth, an anticyclone rotates clockwise in the northern hemisphere, and counterclockwise in the southern hemisphere.

ANTIMATTER: Matter composed of the antiparticles of ordinary matter. For each subatomic particle, there is an antiparticle that is its opposite in such properties as electrical charge, but its equivalent in mass.

Matter and antimatter, if combined, annihilate each other, producing energy in the form of gamma rays according to the formula $E = mc^2$.

APHELION: The point in the orbit of a solar-system object where it is farthest from the sun.

APOLLO ASTEROID: An asteroid whose orbit brings it closer than 1 astronomical unit to the sun.

ASTEROID: Any of the thousands of small, irregular bodies orbiting the sun, primarily in the asteroid belt, which is located between the orbits of Mars and Jupiter.

ASTHENOSPHERE: The deep portions of the earth's mantle, below the zone (the lithosphere) where convection currents are thought to operate. The term is also applied to similar zones in the interiors of the moon and other planets.

ASTROLOGY: The ancient belief that earthly affairs and human lives are influenced by the positions of the sun, moon, and planets with respect to the zodiac.

ASTROMETRIC BINARY: A double star recognized as such because the visible star or stars undergo periodic motion that is detected by astrometric measurements.

ASTROMETRY: The science of accurately measuring stellar positions.

ASTRONOMICAL UNIT (AU): A unit of distance used in astronomy, equal to the average distance between the sun and the earth. One AU is equal to 1.4959787×10^8 kilometers.

ATOMIC NUMBER: The number of protons in the nucleus of an element. It is the atomic number that defines the identity of an element.

ATOMIC WEIGHT: The mass of an atomic nucleus in atomic mass units (one atomic mass unit is defined as the average mass of the protons and neutrons in a nucleus of ordinary carbon, ^{12}C). For most atoms, the atomic weight is approximately equal to the total number of protons and neutrons in the nucleus.

AUTUMNAL EQUINOX: The point where the sun crosses the celestial equator from north to south, around September 21. See also *equinox* and *vernal equinox*.

BARRED SPIRAL GALAXY: A spiral galaxy whose nucleus has linear extensions on opposing sides, giving it a barlike shape. The spiral arms usually appear to emanate from the ends of the bar.

BASALT: An igneous silicate rock, common in regions formed by lava flows on the earth, the moon, and probably the other terrestrial planets.

BETA DECAY: A spontaneous nuclear reaction in which a neutron decays into a proton and an electron, with a neutrino being emitted also. The term has been generalized to mean any spontaneous reaction in which an electron and a neutrino (or their antimatter equivalents) are emitted.

BIG BANG: A term referring to any theory of cosmogony in which the universe began at a single point and was very hot initially, and has been expanding from that state since.

BINARY STAR: A double-star system in which the two stars orbit a common center of mass.

BLACK HOLE: An object that has collapsed under its own gravitation to such a small radius that its gravitational force traps photons of light.

BODE'S LAW: Also known as the Titius-Bode relation, this law is a simple numerical sequence that approximately represents the relative distances of the inner seven planets from the sun. Four is added to each number in the sequence 0,3,6,12, . . . , and then each is divided by ten, resulting in the sequence 0.4, 0.7, 1.0, 1.6, . . . , which is approximately the sequence of planetary distances from the sun in astronomical units.

BOLIDE: An extremely bright meteor that explodes in the upper atmosphere.

BOLOMETRIC MAGNITUDE: A magnitude in which all wavelengths of light are included.

BRECCIA: Lunar rocks consisting of pebbles and soil fused together by meteorite impacts.

BURSTER: A sporadic source of intense X rays, probably consisting of a neutron star onto which new matter falls at irregular intervals.

CARBONACEOUS CHONDRITE: A meteorite containing chondrules, with a high abundance of carbon and other volatile elements. Carbonaceous chondrites, thought to be very old, have apparently been unaltered since the formation of the solar system.

CASSEGRAIN FOCUS: A focal arrangement for a reflecting telescope. A convex secondary mirror reflects the image through a hole in the primary mirror to a focus at the bottom of the telescope tube. This arrangement is commonly used in situations where relatively lightweight instruments for analyzing the light are attached directly to the telescope.

CATASTROPHIC THEORY: Any theory in which observed phenomena are attributed to sudden changes in conditions or to the intervention of an outside force or body.

CELESTIAL EQUATOR: The imaginary circle formed by the intersection of the earth's equatorial plane with the celestial sphere. The celestial equator is the reference line for north-south (that is, declination) measurements in the standard equatorial coordinate system.

CELESTIAL POLE: The point on the celestial sphere directly overhead at either of the earth's poles.

CELESTIAL SPHERE: The imaginary sphere formed by the sky. It is a convenient device for discussing and measuring the positions of astronomical objects.

CENTER OF MASS: In a binary star system, or in any system consisting of several objects, this is the point about which the mass is "balanced"; that is, the point which moves with a constant velocity through space, while the individual bodies in the system move about it.

CEPHEID VARIABLE: A pulsating variable star, of a class named after the prototype δ Cephei. Cepheid variables obey a period-luminosity relationship and are therefore useful as distance indicators. There are two classes of Cepheid variables, the so-called classical Cepheids, which belong to Population I, and the Population II Cepheids, also known as W Virginis stars.

CHONDRITE: A stony meteorite containing chondrules.

CHONDRULE: A spherical inclusion in certain meteorites, usually composed of silicates, and always of very great age.

CHROMOSPHERE: A thin layer of hot gas just outside the photosphere in the sun and other cool stars. The temperature in the chromosphere rises from about 4,000 K at its inner edge to 10,000 or 20,000 K at its outer boundary. The chromosphere is characterized by the strong red emission line of hydrogen.

CLOSED UNIVERSE: A possible state of the universe. In this state, the expansion of the universe will eventually be reversed; it is characterized by positive curvature, being finite in extent but having no boundaries.

CNO CYCLE: A nuclear-fusion-reaction sequence in which hydrogen nuclei are combined to form helium nuclei, and in which other nuclei, such as isotopes of carbon, oxygen, and nitrogen, appear as catalysts or by-products. The CNO cycle is dominant in the cores of stars on the upper main sequence.

COLOR INDEX: The difference B-V between the blue (B) and visual (V) magnitudes of a star. If B is less than V (that is, the star is brighter in blue than in visual light), then the star has a negative color index, and is a relatively hot star. If B is greater than V, the color index is positive, and the star is relatively cool.

COMA: The extended, glowing region that surrounds the nucleus of a comet.

COMET: An interplanetary body, composed of loosely bound rocky and icy material, that forms a glowing head and extended tail when it enters the inner solar system.

CONFIGURATION: The position of a planet or the moon relative to the sun-earth line.

CONFUSION LIMIT: A natural, fundamental limitation on the astronomer's ability to probe the universe, that cannot be overcome by technological improvement.

CONJUNCTION: The alignment of two celestial bodies on the sky. In connection with the planets, a conjunction is the alignment of a planet with the sun. An inferior conjunction occurs when an inferior planet is directly between the sun and the earth, and a superior conjunction occurs when any planet is directly behind the sun as seen from the earth.

CONSTELLATION: A prominent pattern of bright stars, historically associated with mythological figures. In modern usage, each constellation incorporates a precisely defined region of the sky.

CONTINENTAL DRIFT: The slow motion of the continental masses over the surface of the earth, caused by the motions of the earth's tectonic plates, which in turn are probably caused by convection in the underlying asthenosphere.

CONTINUOUS RADIATION: Electromagnetic radiation that is emitted in a smooth distribution with wavelength, without spectral features such as emission and absorption lines.

CONVECTION: The transport of energy by fluid motions occurring in gases, in liquids, or in semirigid material such as the earth's mantle. These motions are usually driven by the buoyancy of heated material, which tends to rise while cooler material descends.

CORONA: The very hot, extended outer atmosphere of the sun and other cool main-sequence stars. The high temperature in the corona $(1 - 2 \times 10^6$ K) is probably caused by the dissipation of mechanical energy from the convective zone just below the photosphere.

COSMIC BACKGROUND RADIATION: The primordial radiation field that fills the universe. It was created in the form of gamma rays at the time of the big bang, but has since cooled so that today its temperature is 3 K and its peak wavelength is near 1.1 millimeters (in the microwave portion of the spectrum). Also known as the 3-degree background radiation.

COSMIC RAY: A rapidly moving atomic nucleus from space. Some cosmic rays are produced in the sun, whereas others come from interstellar space and probably originate in supernova explosions.

COSMOLOGICAL CONSTANT: A term added to the field equations by Einstein in order to allow solutions in which the universe was static; that is, neither expanding nor contracting. Although the need for the term disappeared when it was discovered that the universe is expanding, the cosmological constant is retained in the field equations by modern cosmologists, but is usually assigned the value zero.

COSMOLOGICAL PRINCIPLE: The postulate, put forth by most cosmologists, that the universe is both homogeneous and isotropic; it is sometimes stated that the universe looks the same to all observers everywhere.

COSMOLOGICAL REDSHIFT: A Doppler shift toward longer wavelengths that is caused by a galaxy's motion of recession, which in turn is caused by the expansion of the universe.

COSMOGONY: The study of the origins of the universe.

COSMOLOGY: The study of the universe as a whole.

COUDÉ FOCUS: A focal arrangement for a reflecting telescope. The image is reflected by a series of mirrors to a remote, fixed location where a massive, immovable instrument can be used to analyze it.

CYCLONE: A rotating wind system, often associated with storms on the earth, about a low-pressure center. On the earth, cyclones rotate counterclockwise in the northern hemisphere, and clockwise in the southern hemisphere.

DECLINATION: The coordinate in the equatorial system that measures positions in the north-south direction, with the celestial equator as the reference line. Declinations are measured in units of degrees, minutes, and seconds of arc.

DEFERENT: The large circle centered on or near the earth on which the epicycle for a given planet moved, in the geocentric theory of the solar system developed by ancient Greek astronomers such as Hipparchus and Ptolemy.

DEGENERATE GAS: A gas in which either free electrons or free neutrons are as densely spaced as allowed by laws of quantum mechanics. Such a gas has extraordinarily high density, and its pressure is not dependent on temperature, as it is in an ordinary gas. Degenerate electron gas provides the pressure that supports white dwarfs against collapse, and degenerate neutron gas similarly supports neutron stars.

DEUTERIUM: An isotope of hydrogen, containing in its nucleus one proton and one neutron.

DIFFERENTIAL GRAVITATIONAL FORCE: A gravitational force acting on an extended object, such that the portions of the object closer to the source of gravitation feel a stronger force than the portions farther away. Such a force, also known as a tidal force, acts to deform or disrupt the object, and is responsible for many phenomena, ranging from synchronous rotation of moons or double stars to planetary ring systems to the disruption of galaxies in clusters.

DIFFERENTIATION: The sinking of relatively heavy elements into the core of a planet or other body. Differentiation can occur only in fluid bodies, so any planet that has undergone this process must once have been at least partially molten.

DISTANCE MODULUS: The difference $m\text{-}M$ between the apparent and absolute magnitudes for a given star. This difference, which must be corrected for the effects of interstellar extinction, is a direct measure of the distance to the star.

DIURNAL MOTION: Any motion related to the rotation of the earth.

Diurnal motions include the daily risings and settings of all celestial objects.

DOPPLER EFFECT: The shift in wavelength of light that is caused by relative motion between the source of light and the observer. The Doppler shift, $\Delta\lambda$, is defined as the difference between the observed and rest (laboratory) wavelengths for a given spectral line.

DWARF ELLIPTICAL GALAXY: A member of a class of small spheroidal galaxies, similar to standard elliptical galaxies except for their small size and low luminosity. Dwarf galaxies are probably the most common in the universe, but cannot be detected at distances beyond the Local Group of galaxies.

DWARF NOVA: A close binary-star system containing a white dwarf; material from the companion star falls onto the other at sporadic intervals, creating brief nuclear outbursts.

ECLIPSE: An occurrence in which one object is partially or totally blocked from view by another, or passes through the shadow of another.

ECLIPSING BINARY: A double-star system in which one or both stars are periodically eclipsed by the other as seen from earth. This situation can occur only when the orbital plane of the binary is viewed edge-on from the earth.

ECLIPTIC: The plane of the earth's orbit about the sun, which is approximately the plane of the solar system as a whole. The apparent path of the sun across the sky is the projection of the ecliptic onto the celestial sphere.

ELECTROMAGNETIC FORCE: The force created by the interaction of electric and magnetic fields. The electromagnetic force can be either attractive or repulsive, and is important in countless situations in astrophysics.

ELECTROMAGNETIC RADIATION: Waves consisting of alternating electric and magnetic fields. Depending on the wavelength, these waves may be known as gamma rays, X rays, ultraviolet radiation, visible light, infrared radiation, or radio radiation.

ELECTROMAGNETIC SPECTRUM: The entire array of electromagnetic radiation, arranged according to wavelength.

ELECTRON: A tiny, negatively charged particle that orbits the nucleus of an atom. The charge is equal and opposite to that of a proton in the nucleus, and in a normal atom the number of electrons and protons is equal, so that the overall electrical charge is zero. It is the electrons that emit and absorb electromagnetic radiation, by making transitions between fixed energy levels.

ELLIPSE: A geometrical shape such that the sum of the distances from any point on it to two fixed points called foci is constant. In any bound system where two objects orbit a common center of mass, their orbits are ellipses, with the center of mass at one focus.

ELLIPTICAL GALAXY: One of a class of galaxies characterized by smooth spheroidal forms, few young stars, and little interstellar matter.

EMISSION LINE: A wavelength at which radiation is emitted, creating a bright line in the spectrum.

EMISSION NEBULA: A cloud of interstellar gas that glows by the light of emission lines. The source of excitation that causes the gas to emit may be radiation from a nearby star, or heating by any of a variety of mechanisms.

ENDOTHERMIC REACTION: Any reaction, nuclear or chemical, that requires more energy to occur than is produced.

ENERGY: The ability to do work. Energy can be in either kinetic form, when it is a measure of the motion of an object, or potential form, when it is stored but capable of being released into kinetic form.

EPICYCLE: A small circle on which a planet revolves, which in turn orbits another, distant body. Astronomers in ancient times used epicycles in theories of the solar system in order to devise a cosmology with the earth at the center, but that also accurately accounted for the observed planetary motions.

EQUATORIAL COORDINATES: The astronomical coordinate system in which positions are measured with respect to the celestial equator (in the north-south direction) and with respect to a fixed direction (in the east-west direction). The coordinates used are declination (north-south, in units of angle) and right ascension (east-west, in units of time).

EQUINOX: Either of two points on the sky where the planes of the ecliptic and the earth's equator intersect. When the sun is at one of these two points, the lengths of night and day on the earth are equal. See also *autumnal equinox* and *vernal equinox*.

ERG: A unit of energy that is equal to the kinetic energy of an object with a mass of two grams moving at a speed of one centimeter per second. An erg is defined technically as the work required to raise a mass of one gram through a distance of one centimeter under a gravitational field equal to that at the earth's surface.

ESCAPE VELOCITY: The velocity required for an object to escape the gravitational field of a body such as a planet. In a more technical sense, the escape velocity is the velocity at which the kinetic energy of the object equals its gravitational potential energy; if the object moves any faster, its kinetic energy exceeds its potential energy, and the object can escape the gravitational field.

EVENT HORIZON: The "surface" of a black hole; the boundary of the region from within which no light can escape.

EVOLUTIONARY THEORY: Any theory in which observed phenomena are thought to have arisen as a result of natural processes, requiring no outside intervention or sudden changes.

EXCITATION: A state in which one or more electrons of an atom or ion are in energy levels above the lowest possible one.

FLUORESCENCE: The emission of light at a particular wavelength following excitation of the electron by absorption of light at another, shorter, wavelength.

FOCUS: (1) The point at which light collected by a telescope is brought together to form an image; (2) one of two fixed points that define an ellipse (see also *ellipse*).

FORCE: Any agent or phenomenon that produces acceleration of a mass.

FREQUENCY: The rate (in units of Hertz, or cycles per second) at which electromagnetic waves pass a fixed point. The frequency, usually designated ν, is related to the wavelength λ and the speed of light c by $\nu = c/\lambda$.

FUSION REACTION: A nuclear reaction in which atomic nuclei combine to form more massive nuclei.

GALACTIC CLUSTER: A loose cluster of stars located in the disk or spiral arms of the galaxy.

GAMMA RAY: A photon of electromagnetic radiation, whose wavelength is very short and whose energy is very high. Radiation whose wavelength is less than one Ångstrom is usually considered to be gamma-ray radiation.

GEGENSCHEIN: The diffuse glowing spot, seen on the ecliptic opposite the sun's direction, created by sunlight reflected off of interplanetary dust.

GLOBULAR CLUSTER: A large, spherical cluster of stars located in the halo of the galaxy. These clusters, containing up to several hundred thousand members, are thought to be among the oldest objects in the galaxy.

GRAM: A unit of mass equal to the quantity of mass contained in one cubic centimeter of water.

GRANULATION: The spotty appearance of the solar surface (that is, of the photosphere) caused by convection in the layers just below.

GRAVITATIONAL REDSHIFT: A shift toward long wavelengths caused by the effect of a gravitational field on photons of light. Photons escaping a gravitational field lose energy to the field, which results in the redshift.

GREATEST ELONGATION: The greatest angular distance from the sun that an inferior planet can reach, as seen from earth.

GREENHOUSE EFFECT: The trapping of heat near the surface of a planet by atmospheric molecules (such as carbon dioxide) that absorb infrared radiation emitted by the surface.

H II REGION: A volume of ionized gas surrounding a hot star (see also *emission nebula*).

H-R DIAGRAM: See *Hertzsprung-Russell diagram*.

HALF-LIFE: The time required for half of the nuclei of an unstable (that is, radioactive) isotope to decay.

HALO: (1) The extended outer portions of a galaxy such as the Milky Way. The halo is thought to contain a large fraction of the total mass of the galaxy, mostly in the form of dim stars and interstellar gas, (2) the extensive cloud of gas surrounding the head of a comet.

HELIUM FLASH: A rapid burst of nuclear reactions in the degenerate core of a moderate-mass star in the hydrogen shell-burning phase. The flash occurs when the core temperature reaches a sufficiently high temperature to trigger the triple-alpha reaction.

HERTZ (HZ):   A unit of frequency used in describing electromagnetic radiation; one hertz is equal to one cycle or wave per second.

HERTZSPRUNG-RUSSELL DIAGRAM: A diagram on which stars are represented according to their absolute magnitudes (on the vertical axis) and spectral types (on the horizontal axis). Because the physical properties of stars are interrelated, stars do not fall randomly on such a diagram, but instead lie in well-defined regions according to their state of evolution. Very similar diagrams can be constructed that show luminosity instead of absolute magnitude, and temperature or color index in place of spectral type.

HIGH-VELOCITY STAR: A star whose velocity relative to the solar system is large. As a rule, high-velocity stars are Population II objects following orbital paths that are highly inclined to the plane of the galactic disk.

HOMOGENEOUS: Having the quality of being uniform in properties throughout. In astronomy, this term is often applied to the universe as a whole, which is postulated to be homogeneous.

HORIZONTAL BRANCH: A sequence of stars in the H–R diagram of a globular cluster, extending horizontally across the diagram to the left from the red-giant region. These are probably stars undergoing helium burning in their cores, by the triple-alpha reaction.

HUBBLE CONSTANT: The numerical factor, usually denoted H, that describes the rate of expansion of the universe. It is the proportionality constant in the Hubble law $v = Hd$, which relates the speed of recession of a galaxy (v) to its distance (d). The present value of H is not well known; estimates range between 55 and 90 km/sec/Mpc.

HYDROSTATIC EQUILIBRIUM: The state of balance between gravitational and pressure forces that exists at all points inside any stable object such as a star or planet.

IGNEOUS ROCK: A rock that was formed by cooling and hardening from a molten state.

IMPACT CRATER: A crater formed on the surface of a terrestrial planet or a satellite by the impact of a meteoroid or planetesimal.

INERTIA: The tendency of an object to remain in its state of rest or uniform motion; this tendency is directly related to the mass of the object.

INFERIOR PLANET: A planet whose orbit lies closer to the sun than that of the earth; namely, Mercury and Venus.

INFRARED RADIATION: Electromagnetic radiation in the wavelength region just longer than that of visible light; that is, radiation whose wavelength lies roughly between 7,000 Å and 0.01 centimeter.

INTERFEROMETRY: The use of interference phenomena in electromagnetic waves to precisely measure positions or to achieve gains in resolution. Interferometry in radio astronomy entails the use of two or more antennae to overcome the normally very coarse resolution of a single radio telescope; in visible-light observations, the goal is to eliminate the distorting effects of the earth's atmosphere.

INTERSTELLAR CLOUD: A region of relatively high density in the interstellar medium. Interstellar clouds have densities ranging between 1 and 10^6 particles per cubic centimeter, and in aggregate, contain most of the mass in interstellar space.

INTERSTELLAR EXTINCTION: The obscuration of starlight by interstellar dust. Light is scattered off of dust grains, so that a distant star appears dimmer than it otherwise would. The scattering process is most effective at short (blue) wavelengths, so that stars seen through interstellar dust appear reddened and dimmed.

INVERSE-SQUARE LAW: Any law describing a force or other phenomenon that decreases in strength as the square of the distance from some central reference point. The term *inverse-square law* is often used by itself to mean the law stating that the intensity of light emitted by a source such as a star diminishes as the square of the distance from the source.

ION: Any subatomic particle with a nonzero electrical charge. In standard practice, the term *ion* is usually applied only to positively charged particles such as atoms missing one or more electrons.

IONIZATION: Any process by which an electron or electrons are freed from an atom or ion. Generally, ionization occurs in two ways: by the absorption of a photon with sufficient energy, or by collision with another particle.

IONOSPHERE: The zone of the earth's upper atmosphere, between 80- and 500-km altitude, where charged subatomic particles (chiefly protons and electrons) are trapped by the earth's magnetic field. See also *Van Allen belts.*

ISOTOPE: Any form of a given chemical element. Different isotopes of the same element have the same number of protons in their nuclei, but different numbers of neutrons.

ISOTROPIC: Having the property of appearing the same in all directions. In astronomy, this term is often postulated to apply to the universe as a whole.

KELVIN: A unit of temperature equal to one one-hundredth of the difference between the freezing and boiling points of water, and used in a scale whose zero point is absolute zero. A Kelvin is usually denoted by K.

KILOPARSEC (KPC):   A unit of distance equal to 1,000 parsecs.

KINETIC ENERGY: The energy of motion. The kinetic energy of a moving object is equal to one-half times its mass times the square of its velocity.

KIRKWOOD'S GAPS: Narrow gaps in the asteroid belt created by orbital resonance with Jupiter.

LIGHT-GATHERING POWER: The ability of a telescope to collect light from an astronomical source. The light-gathering power is directly related to the area of the primary mirror or lens.

LIMB DARKENING: The dark region around the edge of the visible disk of the sun or a planet, caused by a decrease in temperature with height in the atmosphere.

LIQUID METALLIC HYDROGEN: Hydrogen in a state of semirigidity that can exist only under conditions of extremely high pressure, as in the interiors of Jupiter and Saturn.

LITHOSPHERE: The layer in the earth, moon, and terrestrial planets that includes the crust and the outer part of the mantle.

LOCAL GROUP: The cluster of about thirty galaxies to which the Milky Way belongs.

LUMINOSITY: The total energy emitted by an object per second; that is, the power of the object. For stars the luminosity is usually measured in units of ergs per second.

LUMINOSITY CLASS: One of several classes to which a star can be assigned on the basis of certain luminosity indicators in its spectrum. The classes range from I for supergiants to V for main-sequence stars (also known as dwarfs).

LUNAR MONTH: The synodic period of the moon, equal to 27 days 7 hours 43 minutes 11.5 seconds.

L WAVE: A type of seismic wave that travels only over the surface of the earth.

MAGELLANIC CLOUDS: The two irregular galaxies that are the nearest neighbors of the Milky Way; they are visible to the unaided eye in the southern hemisphere.

MAGMA: Molten rock from the earth's interior.

MAGNETIC BRAKING: The slowing of the spin of a young star (such as the early sun) by magnetic forces exerted on the surrounding ionized gas.

MAGNETIC DYNAMO: A rotating internal zone inside the sun or a planet, thought to carry the electrical currents that create the solar or planetary magnetic field.

MAGNETOSPHERE: A region, surrounding a star or planet, that is permeated by the magnetic field of that body.

MAGNITUDE: A measure of the brightness of a star. It is based on a system established by Hipparchus, in which stars were ranked according to how bright they appeared to the unaided eye. In the modern system, a difference of five magnitudes corresponds exactly to a brightness ratio of 100, so that a star of a given magnitude has a brightness that is $100^{1/5} = 2.512$ times that of a star one magnitude fainter.

MAIN SEQUENCE: The strip in the H-R diagram (running from upper left to lower right) where most stars, those which are converting hydrogen to helium by nuclear reactions in their cores, are found.

MAIN-SEQUENCE FITTING: A distance-determination technique in which an H-R diagram for a cluster of stars is compared with a standard H-R diagram to establish the absolute-magnitude scale for the cluster H-R diagram.

MAIN-SEQUENCE TURN-OFF: In an H-R diagram for a cluster of stars, the point where the main sequence turns off toward the upper right. The main-sequence turn-off, showing which stars in the cluster have evolved to become red giants, is an indicator of the age of the cluster.

MANTLE: The semirigid outer portion of the earth's interior, extending from roughly the midway point nearly to the surface, and consisting of the mesosphere (the lower portion) and the asthenosphere.

MARE (pl. MARIA): Any of several extensive, smooth lowland areas on the surface of the moon or Mercury that were created by extensive lava flows early in the history of the solar system.

MASCON: A contraction of *mass concentration,* referring to any of several locations on the near side of the moon where regions of enhanced density are found.

MASS: The quantity of matter contained in a body or object, measurable only by virtue of its inertia or the force it exerts when subjected to a gravitational field.

MASS-TO-LIGHT RATIO: The mass of a galaxy, in units of solar masses, divided by its luminosity, in units of the sun's luminosity. The mass-to-light ratio is an indicator of the relative quantities of Population I and Population II stars in a galaxy.

MEAN SOLAR DAY: The average length of the solar day, as measured throughout the year; the mean solar day is precisely equal to twenty-four hours.

MECHANICS: The study of motions and gravitational interactions of objects.

MEGAPARSEC (MPC): A unit of distance equal to 10^6 parsecs.

MERIDIAN: The great circle on the celestial sphere that passes through both poles and directly overhead; that is, the north-south line directly overhead.

MESOSPHERE: (1) The layer of the earth's atmosphere between roughly 50- and 80-km altitude, where the temperature decreases with height; (2) the layer below the asthenosphere in the earth's mantle.

METAMORPHIC ROCK: A rock formed by heat and pressure in the earth's interior.

METEOR: A bright streak or flash of light created when a meteoroid enters the earth's atmosphere from space.

METEORITE: The remnant of a meteoroid that survives a fall through the earth's atmosphere and reaches the ground.

METEOROID: A small interplanetary body.

METEOR SHOWER: A period during which meteors are seen with high frequency, occurring when the earth passes through a swarm of meteoroids.

MICROMETEORITE: A microscopically small meteorite.

MICROWAVE BACKGROUND: See *cosmic background radiation* and *3-degree background radiation.*

MILKY WAY: Historically, the diffuse band of light stretching across the sky; our cross-sectional view of the disk of our galaxy. In modern usage, the term *Milky Way* refers to our galaxy as a whole.

NEUTRINO: A subatomic particle, without mass or electrical charge, that is emitted in certain nuclear reactions.

NEUTRON: A subatomic particle with no electrical charge and a mass nearly equal to that of the proton; neutrons and protons are the chief components of the atomic nucleus.

NEUTRON STAR: A very compact, dense stellar remnant whose interior consists entirely of neutrons, and which is supported against collapse by degenerate neutron gas pressure.

NEWTONIAN FOCUS: A focal arrangement for reflecting telescopes. A flat mirror is used to reflect the image through a hole in the side of the telescope tube.

NONTHERMAL RADIATION: Radiation not caused solely by the temperature of an object. The term is most often applied to sources of continuous radiation such as synchrotron radiation.

NOVA: A star that temporarily flares up in brightness, most likely as the result of nuclear reactions caused by the deposition of new nuclear fuel on the surface of a white dwarf in a binary system. See also *recurrent nova.*

NUCLEUS: The central, dense concentration in an atom, comet, or galaxy.

OB ASSOCIATION: A group of young stars whose luminosity is dominated by O and B stars.

OCCAM'S RAZOR: The principle that the simplest explanation of any natural phenomenon is most likely the correct one.

OORT CLOUD: The cloud of cometary bodies hypothesized to be orbiting the sun at a great distance, from which comets originate.

OPEN UNIVERSE: A possible state of the universe. In this state, the expansion of the universe will never stop; it is characterized by negative curvature, being infinite in extent and having no boundaries.

OPPOSITION: A planetary configuration in which a superior planet is positioned exactly in the opposite direction from the sun, as seen from earth.

OPTICAL BINARY: A pair of stars that happen to appear near each other on the sky, but which are not in orbit; not a true binary.

ORBITAL RESONANCE: A situation in which the periods of two orbiting bodies are simple multiples of each other, so that they are frequently aligned, and gravitational forces exerted by the outer body may move the inner body out of its original orbit. This is one mechanism thought responsible for creating gaps in the rings of Saturn and for creating Kirkwood's gaps in the asteroid belt.

ORGANIC MOLECULE: Any of a large class of carbon-bearing molecules that are found in living matter.

PALEOMAGNETISM: Vestigial traces or artifacts of ancient magnetic fields.

PARALLAX: Any apparent shift in position caused by an actual motion or displacement of the observer. See also *stellar parallax.*

PARSEC (PC): A unit of distance equal to the distance to a star whose stellar parallax is one second of arc. One parsec is equal to 206,265 AU, 3.03×10^{13} kilometers, or 3.26 light years.

PECULIAR VELOCITY: The deviation of a star's velocity from perfect circular motion about the galactic center.

PENUMBRA: (1) The light, outer part of a shadow, such as the portion of the earth's shadow where the moon is not totally obscured during a lunar eclipse; (2) the light, outer portion of a sunspot.

PERFECT COSMOLOGICAL PRINCIPLE: The postulate, adopted by advocates of the so-called steady-state theory of cosmology, stating that the universe is homogeneous and isotropic with respect to space and time. It is commonly stated that the universe looks the same to all observers everywhere, in all directions and at all times.

PERIHELION: The point in the orbit of any sun-orbiting body where it most closely approaches the sun.

PERMAFROST: A permanent layer of ice just below the surface of certain regions on the earth and probably on Mars.

PHOTOMETER: A device, usually with a photoelectric cell, for measuring the brightnesses of astronomical objects.

PHOTON: A particle of light having wave properties but also acting as a discrete unit.

PHOTOSPHERE: The visible surface layer of the sun and stars; the layer from which continuous radiation escapes and where absorption lines form.

PLANCK CONSTANT: The numerical factor h relating the frequency v of a photon to its energy E in the expression $E = hv$. The Planck constant has the value $h = 6.62620 \times 10^{-27}$ erg sec.

PLANCK FUNCTION (also known as the *Planck law*): The mathematical expression describing the continuous thermal spectrum of a glowing object. For a given temperature, the Planck function specifies the intensity of radiation as a function of either frequency or wavelength.

PLANETARY NEBULA: A cloud of glowing, ionized gas, usually taking the form of a hollow sphere or shell, ejected by a star in the late stages of its evolution.

PLANETESIMAL: A small (diameter up to several hundred kilometers) solar-system body of the type that first condensed from the solar nebula. Planetesimals are thought to have been the principal bodies that combined to form the planets.

PLATE TECTONICS: A general term referring to the motions of lithospheric plates over the surface of the earth or other terrestrial planets. See also *continental drift*.

POLARIZATION: The preferential alignment of the magnetic and electric fields in electromagnetic radiation.

POPULATION I: The class of stars with relatively high abundances of heavy elements. These stars are generally found in the disk and spiral arms of spiral galaxies, and are relatively young. The term *Population I* is also commonly applied to other components of galaxies associated with the star formation, such as the interstellar material.

POPULATION II: The class of stars with relatively low abundances of heavy elements. These stars are generally found in a spheroidal distribution about the galactic center and throughout the halo, and are relatively old.

POSITRON: A subatomic particle with the same mass as the electron, but with a positive electrical charge; the antiparticle of the electron.

POTENTIAL ENERGY: Energy that is stored, and which may be converted into kinetic energy under certain circumstances. In astronomy, the most common form of potential energy is gravitational potential energy.

PRECESSION: The slow shifting of star positions on the celestial sphere, caused by the 26,000-year periodic wobble of the earth's rotation axis.

PRIMARY MIRROR: The principal light-gathering mirror in a reflecting telescope.

PRIME FOCUS: A focal arrangement (in a reflecting telescope) in which the image is allowed to form inside the telescope structure at the focal point of the primary mirror, so that no secondary mirror is needed.

PROGRADE MOTION: Orbital or spin motion in the forward or "normal" direction; in the solar system, this is counterclockwise as viewed looking down from above the north pole.

PROPER MOTION: The motion of a star across the sky, usually measured in units of arcseconds per year.

PROTON-PROTON CHAIN: The sequence of nuclear reactions in which four hydrogen nuclei combine (through intermediate steps involving deuterium and ^{3}He) to form one helium nucleus. The proton-proton chain is responsible for energy production in the cores of stars on the lower main sequence.

PROTOSTAR: A star in the process of formation, specifically one that has entered the slow gravitational contraction phase.

PULSAR: A rapidly rotating neutron star that emits periodic pulses of electromagnetic radiation, probably by the emission of beams of radiation from the magnetic poles, which sweep across the sky as the star rotates.

P WAVE: A seismic wave that is a compressional, or density, wave. P waves can travel through both solid and liquid portions of the earth, and are the first to reach any location remote from an earthquake site.

QSO: See *quasistellar object*.

QUADRATURE: The configuration where a superior planet or the moon is 90 degrees away from the sun, as seen from the earth.

QUANTUM: The amount of energy associated with a photon, equal to hv, where h is the Planck constant, and v is the frequency. The quantum is the smallest amount of energy that can exist at a given frequency.

QUANTUM MECHANICS: The physics of atomic structure and the behavior of subatomic particles, based on the principle of the quantum.

QUASAR: See *quasistellar object*.

QUASISTELLAR OBJECT: Any of a class of extragalactic objects also known as quasars, characterized by emission lines with very large redshifts. The quasistellar objects are thought to lie at great distances, in which case they existed only at earlier times in the history of the universe; they may be cores of young galaxies.

RADIATION PRESSURE: Pressure created by light hitting a surface.

RADIATIVE TRANSPORT: The transport of energy, inside of a star or in other situations, by radiation.

RADIOACTIVE DATING: A technique for estimating the age of material such as rock, based on the known initial isotopic composition and the known rate of radioactive decay for unstable isotopes originally present.

RADIO GALAXY: Any of a class of galaxies whose luminosity is greatest in radio wavelengths. Radio galaxies are usually large elliptical galaxies, with synchrotron radiation emitted from one or more pairs of lobes located on opposite sides of the visible galaxy.

RAY: A bright streak of ejecta emanating from an impact crater, especially on the moon or on Mercury.

RECURRENT NOVA: A star known to flare up in nova outbursts more than once. A recurrent nova is thought to be a binary system containing a white dwarf and a mass-losing star, in which the white dwarf sporadically flares up when material falls onto it from the companion.

RED GIANT: A star that has completed its core hydrogen-burning stage, and has begun hydrogen shell-burning, causing its outer layers to become very extended and cool.

REFLECTING TELESCOPE: A telescope that brings light to a focus using mirrors.

REFLECTION NEBULA: An interstellar cloud containing dust that shines by light reflected from a nearby star.

REFRACTING TELESCOPE: A telescope that uses lenses to bring light to a focus.

REFRACTORY: The property of being able to exist in solid form under conditions of very high temperature. Refractory elements are characterized by a high temperature of vaporization; they are the first to condense into solid form when a gas cools, as in the solar nebula.

REGOLITH: The layer of debris on the surface of the moon created by the impact of meteorites; the lunar surface layer.

RESOLUTION: In an image, the ability to separate closely spaced features; that is, the clarity or fineness of the image. In a spectrum, the ability to separate features that are close together in wavelength.

REST WAVELENGTH: The wavelength of a spectral line as measured in a laboratory, when there is no relative motion between source and observer.

RETROGRADE MOTION: Orbital or spin motion in the opposite direction from prograde motion; in the solar system, retrograde motions are clockwise as seen from above the north pole.

RIGHT ASCENSION: The east-west coordinate in the equatorial coordinate system. The right ascension is measured in units of hours, minutes, and seconds to the east from a fixed direction in the sky, which itself is defined as the line of intersection of the ecliptic and the celestial equator.

RILLE: A type of winding, sinuous valley commonly found on the moon.

ROCHE LIMIT: The point near a massive body such as a planet or star, inside of which the tidal forces acting on an orbiting body exceed the gravitational force holding that body together. The location of the Roche limit depends on the size of the orbiting body.

RR LYRAE VARIABLE: A member of a class of pulsational variable stars named after the prototype star, RR Lyrae. These stars are blue-white giants with pulsational periods of less than one day, and are Population II objects found primarily in globular clusters.

SAROS CYCLE: An 18-year, 11-day repeating pattern of solar and lunar eclipses caused by a combination of the tilt of the lunar orbit with respect to the ecliptic and the precession of the plane of the moon's orbit.

SCATTERING: The random reflection of photons by particles such as atoms or ions in a gas, or dust particles in interstellar space.

SCHWARZSCHILD RADIUS: The radius within which an object has collapsed at the point when light can no longer escape the gravitational field, as the object becomes a black hole.

SECONDARY MIRROR: The second mirror in a reflecting telescope (after the primary mirror), usually either convex, to reflect the image out of a hole in the bottom of the telescope to the cassegrain focus or along the telescope mount axis to the coudé focus; or flat, to reflect the image out of the side of the telescope to the Newtonian focus.

SEDIMENTARY ROCK: A rock formed by the deposition and hardening of layers of sediment, usually either underwater or in an area subject to flooding.

SEEING: The blurring and distortion of point sources of light such as stars, caused by turbulent motions in the earth's atmosphere.

SEISMIC WAVE: A wave created in a planetary or satellite interior, usually caused by an earthquake.

SELECTION EFFECT: The tendency for a conclusion based on observations to be influenced by the method used to select the objects for observation. An example was the early belief that all quasars are radio sources, when the principal method used to discover quasars was to look for radio sources and then to find out whether they had other properties associated with quasars.

SEMIMAJOR AXIS: One-half of the major, or long axis of an ellipse.

SEYFERT GALAXY: Any of a class of spiral galaxies with unusually bright, blue-colored nuclei. First recognized by Carl Seyfert.

SHEAR WAVE: A wave that consists of transverse motions; that is, motions perpendicular to the direction of wave travel.

SIDEREAL DAY: The rotation period of the earth with respect to the stars (or as seen by a distant observer) equal to 23 hours 56 minutes 4.091 seconds.

SIDEREAL PERIOD: The orbital or rotation period of any object with respect to the fixed stars, or as seen by a distant observer.

SOLAR DAY: The synodic rotation period of the earth with respect to the sun; that is, the length of time from one local noon, when the sun is on the meridian, to the next local noon.

SOLAR FLARE: An explosive outburst of ionized gas from the sun, usually accompanied by X-ray emission and the injection of large quantities of charged particles into the solar wind.

SOLAR MOTION: The deviation of the sun's velocity from perfect circular motion about the center of the galaxy; that is, the sun's peculiar velocity.

SOLAR NEBULA: The primordial gas and dust cloud from which the sun and planets condensed.

SOLAR WIND: The stream of charged subatomic particles flowing steadily outward from the sun.

SOLSTICE: The occasion when the sun, as viewed from the earth, reaches its farthest northern point (the summer solstice) or its farthest southern point (the winter solstice).

SPECTROGRAM: A photograph of a spectrum.

SPECTROGRAPH: An instrument for recording the spectra of astronomical bodies or other sources of light.

SPECTROSCOPE: An instrument allowing an observer to view the spectrum of a source of light.

SPECTROSCOPIC BINARY: A binary system recognized as a binary because its spectral lines undergo periodic Doppler shifts as the orbital motions of the two stars cause them to move toward and away from the earth. If lines of only one star are seen, it is a *single-lined spectroscopic binary;* if lines of both stars are seen, it is a *double-lined spectroscopic binary.*

SPECTROSCOPIC PARALLAX: The technique of distance determination for stars in which the absolute magnitude is inferred from the H-R diagram and then compared with the observed apparent magnitude to yield the distance.

SPECTROSCOPY: The science of analyzing the spectra of stars or other sources of light.

SPECTRUM (pl. *spectra*): An arrangement of electromagnetic radiation according to wavelength.

SPECTRUM BINARY: A binary system recognized as a binary because its spectrum contains lines of two stars of different spectral types.

SPIN-ORBIT COUPLING: A simple relationship between the orbital and spin periods of a satellite or planet, caused by tidal forces that have slowed the rate of rotation of the orbiting body. Synchronous rotation is the simplest and most common form of spin-orbit coupling.

SPIRAL DENSITY WAVE: A spiral wave pattern in a rotating, thin disk, such as the rings of Saturn or the plane of a spiral galaxy like the Milky Way.

SPIRAL GALAXY: Any of a large class of galaxies exhibiting a disk with spiral arms.

STANDARD CANDLE: A general term for any astronomical object whose absolute magnitude can be inferred from its other observed characteristics, and which is therefore useful as a distance indicator.

STEADY-STATE THEORY: The theory of cosmology in which the universe is thought to have had no beginning and is postulated not to change with time.

STEFAN-BOLTZMANN LAW: The law of continuous radiation stating that for a spherical glowing object such as a star, the luminosity is proportional to the square of the radius and the fourth power of the temperature.

STEFAN'S LAW: An experimentally derived law of continuous radiation stating that the energy emitted by a glowing body per square centimeter of surface area is proportional to the fourth power of the absolute temperature.

STELLAR PARALLAX: The apparent annual shifting of position of a nearby star with respect to more distant background stars. The term *stellar parallax* is often assumed to mean the parallax angle, which is one-half of the total angular motion a star undergoes. See also *parallax* and *parsec*.

STELLAR WIND: Any stream of gas flowing outward from a star, including the very rapid winds from hot, luminous stars; the intermediate-velocity, rarefied winds from stars like the sun; and the slow, dense winds from cool supergiant stars.

STRATOSPHERE: The layer of the earth's atmosphere between 10 and 50 kilometers in altitude, where the temperature increases with height.

SUBDUCTION: The process in which one tectonic plate is submerged below another along a line where two plates collide. A subduction zone is usually characterized by a deep trench and an adjoining mountain range.

SUPERCLUSTER: A cluster of clusters of galaxies.

SUPERGIANT: A star in its late stages of evolution which is undergoing shell burning and is therefore very extended in size and very cool; an extremely large giant.

SUPERGRANULATION: The pattern of large cells seen in the sun's chromoshpere, when viewed in the light of the strong emission line of ionized hydrogen.

SUPERIOR PLANET: Any planet whose orbit lies beyond the earth's orbit around the sun.

SUPERNOVA: The explosive destruction of a massive star that occurs when all sources of nuclear fuel have been consumed, and the star collapses catastrophically.

S WAVE: A type of seismic wave that is a transverse, or shear, wave, and which can travel only through rigid materials.

SYNCHRONOUS ROTATION: A situation in which the rotational and orbital periods of an orbiting body are equal, so that the same side is always facing the companion object.

SYNCHROTRON RADIATION: Continuous radiation produced by rapidly moving electrons traveling along lines of magnetic force.

SYNODIC PERIOD: The orbital or rotational period of an object as seen by an observer on the earth. For the moon or a planet, the synodic period is the interval between repetitions of the same phase or configuration.

TECTONIC ACTIVITY: Geophysical processes involving motions of tectonic plates and associated volcanic and earthquake activity. See also *plate tectonics* and *continental drift*.

THERMAL RADIATION: Continuous radiation emitted by any object whose temperature is above absolute zero.

THERMAL SPECTRUM: The spectrum of continuous radiation from a body that glows because it has a temperature above absolute zero; a thermal spectrum is one that is described mathematically by the Planck function.

THERMOSPHERE: The layer of the earth's atmosphere above the mesosphere, extending upward from a height of 80 kilometers, where the temperature rises with altitude.

3-DEGREE BACKGROUND RADIATION: The radiation field filling the universe, left over from the big bang. See also *cosmic background radiation*.

TIDAL FORCE: A gravitational force that tends to stretch or distort an extended object. See *differential gravitational force*.

TOTAL ECLIPSE: Any eclipse in which the eclipsed body is totally blocked from view or totally immersed in shadow.

TRANSIT TELESCOPE: A telescope designed to point straight overhead and accurately measure the times at which stars cross the meridian.

TRANSVERSE WAVE: See *shear wave*.

TRIPLE-ALPHA REACTION: A nuclear fusion reaction in which three helium nuclei (or alpha particles) combine to form a carbon nucleus.

TROJAN ASTEROID: Any of several asteroids orbiting the sun at stable positions in the orbit of Jupiter, either 60 degrees ahead of the planet or 60 degrees behind it.

TROPOSPHERE: The lowest temperature zone in the earth's atmosphere, extending from the surface to a height of about 10 kilometers, in which the temperature decreases with altitude.

T TAURI STAR: A young star still associated with the interstellar material from which it formed, typically exhibiting brightness variations and a stellar wind.

24-HOUR ANISOTROPY: The daily fluctuation in the observed temperature of the cosmic background radiation caused by a combination of the earth's motion with respect to the background and its rotation.

21-CM LINE: The emission line of atomic hydrogen, whose wavelength is 21.11 centimeters, emitted when the spin of the electron with respect to that of the proton reverses itself. The 21-cm line is the most widely used and effective means of tracing the distribution of interstellar gas in the Milky Way and other galaxies.

ULTRAVIOLET: The portion of the electromagnetic spectrum between roughly 100 Ångstroms and 3,000 Ångstroms.

UMBRA: (1) The dark inner portion of a shadow, such as the part of the earth's shadow where the moon is in total eclipse during a lunar eclipse; (2) the dark central portion of a sunspot.

VAN ALLEN BELTS: Zones in the earth's magnetosphere where charged particles are confined by the earth's magnetic field. There are two main belts, one centered at an altitude of roughly 1.5 times the earth's radius, and the other between 4.5 and 6.0 times the earth's radius.

VELOCITY CURVE: A plot showing the orbital velocity of stars in a spiral galaxy versus distance from the galactic center.

VELOCITY DISPERSION: A measure of the average velocity of particles or stars in a group or cluster with random internal motions. In globular clusters and elliptical galaxies, the velocity dispersion can be used to infer the central mass.

VERNAL EQUINOX: The occasion when the sun crosses the celestial equator from south to north, usually occurring around March 21. See also *autumnal equinox* and *equinox*.

VISUAL BINARY: Any binary system in which both stars can be seen through a telescope or on photographs.

VOLATILE: The property of being easily vaporized. Volatile elements stay in gaseous form except at very low temperatures; they did not condense into solid form during the formation of the solar system.

WAVELENGTH: The distance between wavecrests in any type of wave.

WHITE DWARF: The compact remnant of a low-mass star, supported against further gravitational collapse by the pressure of the degenerate electron gas that fills its interior.

WIEN'S LAW: An experimentally discovered law applicable to thermal continuum radiation which states that the wavelength of maximum emission intensity is inversely proportional to the absolute temperature.

X RAY: A photon of electromagnetic radiation in the wavelength interval between about 1 Ångstrom and 100 Ångstroms.

ZEEMAN EFFECT: The broadening or splitting of spectral lines caused by the presence of a magnetic field in the gas where the lines are formed.

ZENITH: The point directly overhead.

ZERO-AGE MAIN SEQUENCE: The main sequence in the H-R diagram formed by stars that have just begun their hydrogen-burning lifetimes, and have not yet converted any significant fraction of their core mass into helium. The zero-age main sequence forms the lower left boundary of the broader band representing the general main sequence.

ZODIAC: A band circling the celestial sphere along the ecliptic, broad enough to encompass the paths of all the planets visible to the naked eye. In some usages, the sequence of constellations lying along the ecliptic.

ZODIACAL LIGHT: A diffuse band of light visible along the ecliptic near sunrise and sunset, created by sunlight scattered off of interplanetary dust.

• I N D E X •

absorption lines, see spectral lines
    interstellar, 390–391
abundance gradients, galactic, 403–404
acceleration, 50, 64–65
accretion disks, 361, 472
accretion theory of solar system formation,
    266
Adams, J. C., 219
Alfonsine tables, 36
α Canis Majoris, see Sirius
alpha-capture reactions, 344
α Orionis, see Betelgeuse
Amalthea, 190
American Astronomical Society, 6
Anaxagoras, 25, 28
Anaximander, 24, 28
Andromeda galaxy, 434–436, 448, Color
    Plate 21
Angstrom, 76
angular measures, 23, 24
angular momentum, 69
annual motions, 8–11
antarctic circle, see latitude zones
anticyclone, 115, 183–184
antimatter, 470–471
aphelion, 69
amino acids, 238, 495–497, 501
Apollo asteroids, 230
Apollo program, 129
Apollonius, 28, 29
Arab astronomy, 36
arctic circle, see latitude zones
Ariel, 217
Aristarchus, 28–29
Aristotle, 26, 28, 233
Arrhenius, S., 495
asteroids, 228–233, 271
asthenosphere, 117
Astrea, 229
astrology, 10
astrometry, 278–281
astronomer, 6
astronomical unit (A.U.), 47
astrophysicist, 6
atoms

structure of, 85
aurorae, 124
    on Venus, 147
Aztec astronomy, 33

Baade, W., 404, 421
Babylonian astronomy, 22–23
barred spiral galaxies, see galaxies, spiral
    barred
Bell, J., 360
Bellatrix, 280
belts, on Jupiter, 180, 183–184
Bessell, F. W., 279
β Orionis, see Rigel
Betelgeuse, 27, 280, Color Plate 15
big bang, see universe, evolution of
Bighorn medicine wheel, 35
binary stars, see stars, binary
binary X-ray sources, see stars, binary, X-
    ray
black dwarfs, 353
black holes, 349–350, 362–366, 382–383,
    413, 471–472, 482, 490–491
blink comparator, 224
BL Lac objects, 464–466
blueshift, see Doppler effect
Bode, J., 229
Bode's law, 229, 232
Bohr, N., 85
Bok, B. J., 413
bolide, 237
Brahe, T., 42–45, 233, 358
brecchia, 133
Buffon, G. L., 265
bursters, 362, 384
Bunsen, R., 83, 284
butterfly diagram, 258

calculus, 62–63
Callisto, 188–189, 275, Color Plate 10
Caloris Planitia, 175–177

Cannon, A. J., 284–287
Caracol tower, 33
carbonaceous chondrites, 238
cassegrain focus, 93–94
Cassini, G. D., 196
celestial equator, see equatorial
    coordinates
celestial sphere, 3–5, 7
Centaurus A, 458, Color Plate 23
center of mass, 68–69
Cepheid variables, 291, 374–375, 420,
    424–426
Ceres, 229
Cerro Tololo Inter-American Observatory,
    97, Color Plate 3
Chamberlain, T. C., 265
chaotic terrain, 160–161
Chaldean tables, 23
Charon, 223–225
Chichen Itza, 33
Chinese astronomy, 31–32
chondrites, 238
chondrules, 238
chromatic aberration, 92
chromosphere
    of the sun, 251, 252
    of stars, see stars, chromospheres of
cluster variables, see RR Lyrae stars
color index, 283, 388–389
color-magnitude diagrams, 298, 325
Comet Bennett, Color Plate 13
Comet Kohoutek, Color Plate 13
comets, 4, 18, 228, 233–237
    coma of, 235
    evolution of, 236–237
    nucleus of, 235
    tail of, 235, 236
configurations,
    of the moon, 12–14
    of the planets, 15–17
confusion limits, 394, 468
conjunction, see planets, motions of
constellations, 23, 27
continental drift, see tectonic activity
continuous radiation, 78–83

537

(continued from copyright page)

chart of the heavens, courtesy Field Museum of Natural History, photograph by Von Del Chamberlain. **Fig. 2.18.** E. E. Barnard Observatory photograph by G. Emerson. **Fig. 2.19.** J. A. Eddy. **Fig. 2.20.** E. C. Krupp.

Chapter 3

Fig. 3.1. City Museum, Torun, Poland, courtesy O. Gingerich. **Fig. 3.3.** The Bettmann Archive. **Fig. 3.4.** Historical Picture Services, Inc. **Fig. 3.5.** The Bettmann Archive. **Fig. 3.8.** O. Gingerich. **Fig. 3.11.** The Granger Collection. **Fig. 3.12.** Owen Gingerich. **Fig. 3.13.** The Granger Collection. **Fig. 3.14.** Yerkes Observatory. **Fig. 3.15.** NASA photograph. **Fig. 3.17.** O. Gingerich.

Chapter 4

Fig. 4.1. The Granger Collection. **Fig. 4.2.** The Granger Collection. **Fig. 4.6.** NASA photograph. **Fig. 4.8.** NASA photograph.

Chapter 5

Fig 5.2. The Granger Collection.

Chapter 6

Fig. 6.3. Yerkes Observatory. **Fig. 6.5.** The Bettmann Archive. **Fig. 6.6.** University of Wyoming. **Fig. 6.8.** University of Wyoming. **Fig. 6.9.** © Association of Universities for Research in Astronomy, Inc., The Kitt Peak National Observatory. **Fig. 6.10.** A. G. D. Phillip. **Fig. 6.11.** MMTO: University of Arizona and Smithsonian Institution. **Astronomical Insight (6.1)** © 1980 Anglo-Australian Telescope Board. **Figs. 6.13. and 6.14.** NASA photographs. **Fig. 6.15.** W. C. Cash. **Fig. 6.16.** Harvard-Smithsonian Center for Astrophysics. **Fig. 6.17.** Neils Bohr Library. **Fig. 6.18.** The National Radio Astronomy Observatory, operated by Associated Universities, Inc. under contract with the National Science Foundation. **Fig. 6.19.** The National Radio Astronomy Observatory, operated by Associated Universities, Inc. under contract with the National Science Foundation.. **Fig. 6.20.** University of California. **Fig. 6.21.** NASA photograph.

Chapter 7

Fig. 7.1. NASA photograph. **Fig. 7.4.** NASA photograph. **Fig. 7.5.** From F. K. Lutgens, and E. J. Tarbuck, 1979, *The Atmosphere: An Introduction to Meteorology.* Fig. 7.3., p. 150. Englewood Cliffs, Prentice-Hall. © Prentice-Hall. Reprinted with the permission of the publisher. **Fig. 7.12.** From F. S. Sawkins, C. G. Chase, D. C. Darby, and G. Rupp, 1978, *The Evolving Earth: A Text in Physical Geology,* Fig. 6.2, p. 163. New York; Macmillan. ©Macmillan. Reprinted with the permission of the publisher.

Chapter 8

Figs. 8.1. and 8.2. Mount Wilson and Las Campanas Observatories, Carnegie Institution of Washington. **Figs. 8.3 and 8.4.** NASA photographs. **Figs. 8.5, 8.6., and 8.7.** NASA photographs. **Figs. 8.8., 8.9., and 8.10.** NASA photographs. **Figs. 8.11. and 8.12.** NASA photographs. **Figs. 8.13., 8.14., and 8.15.** NASA photographs.

Chapter 9

Fig. 9.1. NASA photograph. **Figs. 9.3. and 9.4.** NASA photographs. **Fig. 9.6.** NASA photograph. **Fig. 9.8.** Laboratory for Atmospheric and Space Physics, University of Colorado, sponsored by NASA. **Fig. 9.9.** U. S. Geological Survey. **Fig. 9.10.** U. S. Geological Survey. **Figs. 9.11. and 9.12.** TASS from SOVFOTO.

Chapter 10

Fig. 10.1. Mount Wilson and Las Campanas Observatories, Carnegie Institution of Washington. **Fig. 10.2.** Historical Pictures Services, Inc. **Fig. 10.3.** Lowell Observatory photograph. **Fig. 10.4.** Mount Wilson and Las Campanas Observatories, Carnegie Institution of Washington. **Figs. 10.5. and 10.6.** NASA photographs. **Fig. 10.8.** NASA photograph. **Figs. 10.9., 10.10., and 10.11.** NASA photographs. **Figs. 10.12., 10.13., and 10.14.** NASA photographs. **Figs. 10.15., 10.16., and 10.17.** NASA photographs. **Fig. 10.18.** NASA photograph. **Figs. 10.19. and 10.20.** NASA photographs.

Chapter 11

Fig. 11.1. NASA photograph. **Fig. 11.5.** NASA photograph. **Fig. 11.7.** NASA photograph. **Figs. 11.8. and 11.10.** NASA photographs.

Chapter 12

Fig. 12.1. Mount Wilson and Las Campanas Observatories, Carnegie Institution of Washington. **Figs. 12.3. and 12.4.** NASA photographs. **Figs. 12.5. and 12.6.** NASA photographs. **Figs. 12.7. and 12.8.** NASA photographs. **Fig. 12.9.** NASA photographs. **Fig. 12.10.** NASA photograph. **Fig. 12.12.** NASA photograph. **Figs. 12.13., 12.14., and 12.15.** NASA photographs. **Figs. 12.16., 12.17., and 12.18.** NASA photographs. **Figs. 12. 19. and 12.20.** NASA photographs. **Figs. 12.22. and 12.23.** NASA photographs.

Chapter 13

Fig. 13.1. Mount Wilson and Las Campanas Observatories, Carnegie Institution of Washington. **Fig. 13.2.** Yerkes Observatory. **Fig. 13.3.** NASA photograph. **Figs. 13.4. and 13.5.** NASA photographs. **Fig. 13.6.** Data from Beatty, J. K., O'Leary, B., and Chaikin, A. eds. 1981, *The New Solar System* (Cambridge, England: Cambridge University Press). **Figs. 13.7. and 13.8.** NASA photographs. **Fig. 13.9.** NASA photograph. **Fig. 13.10.** NASA photograph. **Fig. 13.11.** Data from Beatty, J. K., O'Leary, B., and Chaikin, A. eds. 1981, *The New Solar System* (Cambridge, England: Cambridge University Press). **Fig. 13.12.** NASA photograph. **Figs. 13.13., 13.14., 13.15., and 13.16.** NASA photographs. **Figs. 13.17. and 13.18.** NASA photographs. **Fig. 13.19.** NASA photograph. **Fig. 13.20.** NASA photograph. **Fig. 13.21.** Laboratory for Atmospheric and Space Physics, University of Colorado, sponsored by NASA. **Figs. 13.22. and 13.23.** NASA photographs. **Fig. 13.24.** NASA photograph.

Chapter 14

Fig. 14.1. Project Stratoscope, Princeton University, supported by the National Science Foundation and NASA. **Fig. 14.3.** NASA photograph. **Figs. 14.4. and 14.5.** NASA photographs. **Fig. 14.6.** NASA photograph. **Fig. 14.8.** E. F. Guinan, C. C. Harris, and F. P. Maloney. **Fig. 14.9.** Lowell Observatory photograph. **Fig. 14.10.** Lowell Observatory photograph. **Fig. 14.11.** U. S. Naval Observatory photograph.

Chapter 15

Fig. 15.1. Eleanor F. Helin, Palomar Observatory. **Fig. 15.2.** NASA photograph. **Fig. 15.4.** The Granger Collection. **Fig. 15.6.** The Granger Collection. **Fig. 15.7.** Palomar Observatory, California Institute of Technology. **Fig. 15.9.** NASA photograph. **Fig. 15.10.** Fr. R. E. Royer. **Fig. 15.11.** New Mexico State University. **Fig. 15.12.** Yerkes Observatory. **Fig. 15.13.** Griffith Observatory, Ronald A. Oriti Collection. **Figs. 15.14. and 15.15.** Griffith Observatory, Ronald A. Oriti Collection. **Astronomical Insight (15.2)** Griffith Observatory. **Fig. 15.16.** Griffith Observatory, Ronald A. Oriti Collection. **Astronomical Insight (15.3)** E. E. Barnard Observatory photograph by G. Emerson. **Fig. 15.18.** Yerkes Observatory. **Fig. 15.19.** Yerkes Observatory. **Fig. 15.20.** Photo by S. Suyama. **Fig. 15.21.** NASA photograph.

Chapter 16

Fig. 16.1. Mount Wilson and Las Campanas Observatories, Carnegie Institution of Washington. **Fig. 16.5.** Project Stratoscope, Princeton University, sponsored by the National Science Foundation and NASA. **Fig. 16.6.** Mount Wilson and Las Campanas Observatories, Carnegie Institution of Washinton. **Fig. 16.7.** NASA photograph. **Fig. 16.8.** Sacramento Peak Observatory, operated by Associated Universities for Research in Astronomy, Inc. **Fig. 16.9.** High Altitude Observatory, National Center for Atmospheric Research, sponsored by the National Science Foundation. **Fig. 16.10.** NASA photograph. **Fig. 16.11.** High Altitude Observatory, National Center for Atmospheric Research, sponsored by the National Science Foundation. **Fig. 16.12.** NASA photograph. **Fig. 16.13.** High Altitude Observatory, National Center for Atmospheric Research, sponsored by the National Science Foundation. **Fig. 16.15.** Palomar Observatory, California Institute of Technology. **Fig. 16.16.** High Altitude Observatory, National Center for Atmospheric Research, sponsored by the National Science Foundation. **Fig. 16.17.** Prepared by Royal Greenwich Observatory, and reproduced with permission of the Science and Engineering Council. Courtesy J. A. Eddy. **Fig. 16.18.** NASA photograph. **Fig. 16.19.** J. A. Eddy.

Chapter 17

Fig. 17.7. From E. E. Barnard 1927, *A Photographic Atlas of Selected Regions of the Milky Way* (Carnegie Institution of Washington), photographed at the Mount Wilson Observatory.

Chapter 18

Fig. 18.1. Historical Pictures Services, Inc. **Fig. 18.8.** From Abt. H., Meinel, A., Morgan, W. W., and Tapscott, R. 1968, *An Atlas of Low-Dispersion Grating Stellar Spectra* (Tucson: Kitt Peak National Observatory). **Fig. 18.9.** Harvard College Observatory. **Fig. 18.10.** Harvard College Observatory. **Fig. 18.11.** Mount Wilson and Las Campanas Observatories, Carnegie Institution of Washington. **Astronomical Insight (18.2).** Harvard College Observatory.

Chapter 19

Fig. 19.3. Princeton University. **Fig. 19.4.** Estate of Henry Norris Russell. Reprinted with permission. **Fig. 19.12.** D. L. Lambert.

Chapter 20

Fig. 20.8. T. R. Ayres. **Fig. 20.12.** Data from C. J. Hansen.

Chapter 21

Fig. 21.1. From E. E. Barnard 1927, *A Photographic Atlas of Selected Regions of the Milky Way* (Carnegie Institution of Washington), photographed at the Mount Wilson Observatory. **Fig. 21.2.** E. E. Barnard Observatory photograph by G. Emerson. **Fig. 21.3.** Lick Observatory photograph. **Fig. 21.4.** Lick Observatory photograph. **Fig. 21.5.** Lick Observatory photograph. **Fig. 21.6.** P.J. Flower. **Fig. 21.8** Lick Observatory photograph. **Fig. 21.9.** Lick Observatory photograph. **Fig. 21.10.** Photograph by B. J. Bok, made at the prime focus of the 4-m reflector at the Cerro Tololo Inter-American Observatory. **Fig. 21.11.** Photograph by B. J. Bok, made at the prime focus of the 4-m reflector at the Cerro Tololo Inter-American Observatory. **Fig. 21.13.** Lick Observatory photograph, infrared data added by R. D. Gehrz, J. Hackwell, and G. Grasdalen. **Fig. 21.14.** Lick Observatory photograph.

Chapter 22

Fig. 22.3. Data from I. Iben. **Fig. 22.8.** U. S. Naval Observatory photograph. **Fig. 22.9.** Data from I. Iben. **Fig. 22.15.** Data from E.P.J. van den Heuvel, 1976, *IAU Symposium 73, Structure and Evolution of Close Binary Systems,* ed. by P. Eggleton, S. Mitton, and J. Whelan (Dodrecht, Riddel), p.35.

Chapter 23

Fig. 23.1. Palomar Observatory, California Institute of Technology. **Fig. 23.2.** G. A. Wegner. **Fig. 23.4** E. E. Barnard Observatory photograph by G. Emerson. **Fig. 23.5.** Palomar Observatory, California Institute of Technology. **Fig. 23.7.** Lick Observatory photograph. **Fig. 23.9.** Data from P. C. Joss. **Fig. 23.10.** Data from R. N. Manchester and J. H. Taylor 1977, *Pulsars* (San Francisco: W. H. Freeman). **Fig. 23.13.** Harvard-Smithsonian Center for Astrophysics. **Fig. 23.14.** Data from P. C. Joss. **Fig. 23.17.** Harvard-Smithsonian Center for Astrophysics.

Chapter 24

Fig. 24.1. Mount Wilson and Las Campanas Observatory, Carnegie Institution of Washington. **Fig. 24.3.** ©1980 Anglo-Australian Telescope Board. **Fig. 24.4.** Harvard College Observatory. **Fig. 24.7.** Harvard College Observatory. **Fig. 24.8.** Estate of H. Shapley. Reprinted with permission.. **Fig. 24.12.** E. E. Barnard Observatory photograph by G. Emerson. **Fig. 24.13.** The National Radio Astronomy Observatory, operated by Associated Universities, Inc. under contract with the National Science Foundation. **Fig. 24.15.** U. S. Naval Observatory photograph. **Fig. 24.17.** Harvard-Smithsonian Center for Astrophysics.

Chapter 25

Fig. 25.1. From E. E. Barnard 1927, *A Photographic Atlas of Selected Regions of the Milky Way* (Carnegie Institution of Washington), photographed at the Mount Wilson Observatory. **Fig. 25.9.** From E. E. Barnard 1927, *A Photographic Atlas of Selected Regions of the Milky Way* (Carnegie Institution of Washington),

photographed at the Mount Wilson Observatory. **Fig. 25.11.** Lick Observatory photograph. **Fig. 25.12.** From E. E. Barnard 1927, *A Photographic Atlas of Selected Regions of the Milky Way* (Carnegie Institution of Washington), photographed at the Yerkes Observatory. **Fig. 25.14.** E. E. Barnard Observatory photograph by G. Emerson. **Fig. 25.15.** C. Heiles. **Fig. 25.16.** NASA photograph. **Fig. 25.17.** *X-ray image:* Harvard-Smithsonian Center for Astrophysics; *Optical image;* K. Kamper and S. van den Bergh; *Radio image;* The National Radio Astronomy Observatory, operated by Associated Universities, Inc. under contract with the National Science Foundation. **Fig. 25.20.** Courtesy T. R. Gull and J. Heckathorn, photographed at the Kitt Peak National Observatory. **Fig. 25.21.** W. C. Cash.

Chapter 26

Fig. 26.1. © 1980 Anglo-Australian Telescope Board. **Fig. 26.6.** A. J. Kalnajs. **Fig. 26.8.** After T. M. Eneev, N. N. Kuzlov, and R. A. Sunyaev, 1973, *Astronomy and Astrophysics,* 22:41.

Chapter 27

Fig. 27.1. U. S. Naval Observatory photograph. **Fig. 27.2.** © 1980 Anglo-Australian Telescope Board. **Fig. 27.3.** U. S. Naval Observatory photograph. **Fig. 27.4.** Palomar Observatory, California Institute of Technology. **Fig. 27.7.** Palomar Observatory, California Institute of Technology. **Fig. 27.9.** Mount Wilson and Las Campanas Observatories, Carnegie Institution of Washington. **Fig. 27.10.** Mount Wilson and Las Campanas Observatories, Carnegie Institution of Washington. **Fig. 27.12.** Palomar Observatory, California Institute of Technology. **Fig. 27.13.** Palomar Observatory, California Institute of Technology. **Fig. 27.14.** U. S. Naval Observatory photograph, computer simulation supplied by A. Toomre and J. Toomre. **Fig. 27.15.** Palomar Observatory, California Institute of Technology. **Fig. 27.16.** The National Radio Astronomy Observatory, operated by Associated Universities, Inc. under contract with the National Science Foundation.

Chapter 28

Fig. 28.1. Palomar Observatory, California Institute of Technology. **Fig. 28.2.** Photograph by P. W. Hodge at the Boyden Observatory. **Fig. 28.3.** Photograph by P. W. Hodge at the Lick Observatory. **Fig. 28.4.** © Association of Universities for Research in Astronomy, Inc., the Cerro Tololo Inter-American Observatory. **Fig. 28.5.** Fr. R. E. Royer. **Fig. 28.6.** Mount Wilson and Las Campanas Observatories, Carnegie Institution of Washington. **Fig. 28.7.** H. Spinrad. **Fig. 28.8.** Palomar Observatory, California Institute of Technology. **Fig. 28.9.** Photograph from the Palomar Observatory, California Institute of Technology; computer simulation courtesy A. Toomre and J. Toomre. **Fig. 28.11.** The Kitt Peak National Observatory. **Fig. 28.13.** Harvard-Smithsonian Center for Astrophysics. **Fig. 28.14.** Illustration from *Sky and Telescope* magazine by Rob Hess, © 1982 Sky Publishing Corporation, by permission; also with permission of R. B. Tully.

Chapter 29

Fig. 29.1. Palomar Observatory, California Institute of Technology. **Fig. 29.4.** Estate of E. Hubble and M. Humanson. Reprinted with permission. **Fig. 29.5.** After J. Silk, 1980. *The Big Bang* (San Francisco: W. H. Freeman). **Fig. 29.10.** Bell Laboratories. **Fig. 29.11.** Data from D. P. Woody and P.L. Richards, 1981, *Astrophysical Journal,* 248:18. **Fig. 29.13.** NASA photograph.

Chapter 30

Fig. 30.1. © Association of Universities for Research in Astronomy, the Cerro Tololo Inter-American Observatory. **Fig. 30.2.** The National Radio Astronomy Observatory, operated by Associated Universities, Inc. under contract with the National Science Foundation. **Fig. 30.3.** After P. J. Hargrave and M. Ryle, 1974, *Monthly Notices of the Royal Astronomical Society,* 166:305. **Fig. 30.4.** JPL, photograph by H. Arp, processing by Jean L. Lorre, Image Processing Laboratory. **Fig. 30.5.** The National Radio Astronomy Observatory, operated by Associated Universities, Inc. under contract with the National Science Foundation. **Fig. 30.6.** JPL, photograph by H. Arp, processing by Jean J. Lorre, Image Processing Laboratory. **Fig. 30.7.** The National Radio Astronomy Observatory, operated by Associated Universities, Inc. under contract with the National Science Foundation. **Fig. 30.8.** Palomar Observatory, California Institute of Technology. **Fig. 30.9.** Palomar Observatory, California Institute of Technology.